RNA Regulation

Edited by
Robert A. Meyers

Related Titles

Meyers, R.A. (ed.)

Encyclopedia of Molecular Cell Biology and Molecular Medicine

Online ISBN: 9783527600908,
www.meyers-emcbmm.com

Meyers, R.A. (ed.)

Epigenetic Regulation and Epigenomics

2012
Print ISBN: 978-3-527-32682-2,
also available in digital formats

Meyers, R.A. (ed.)

Stem Cells

From Biology to Therapy

2013
Print ISBN: 978-3-527-32925-0,
also available in digital formats

Stamm, S., Smith, C., Lührmann, R. (eds.)

Alternative pre-mRNA Splicing

Theory and Protocols

2012
Print ISBN: 978-3-527-32606-8,
also available in digital formats

Meister, G.

RNA Biology

An Introduction

2011
Print ISBN: 978-3-527-32278-7

Hartmann, R.K., Bindereif, A., Schön, A., Westhof, E. (eds.)

Handbook of RNA Biochemistry

2nd Edition

2013
Print ISBN: 978-3-527-32764-5,
also available in digital formats

Kahl, G.

The Dictionary of Genomics, Transcriptomics and Proteomics

5th Edition

2014
Print ISBN: 978-3-527-32852-9,
also available in digital formats

Wu, J. (ed.)

Posttranscriptional Gene Regulation

RNA Processing in Eukaryotes

2013
Print ISBN: 978-3-527-32202-2,
also available in digital formats

RNA Regulation

Advances in Molecular Biology and Medicine

Edited by
Robert A. Meyers

Volume 1

Editor

Dr. Robert A. Meyers
Editor in Chief
RAMTECH LIMITED
122, Escalle Lane
Larkspur, CA 94939
United States

Cover

RNA, © Francis Repoila. Detail from figure 4 in chapter 3 "RNA-Mediated Control of Bacterial Gene Expression: Role of Regulatory Noncoding RNAs1".

■ **Limit of Liability/Disclaimer of Warranty:** While the publisher and author have used their best efforts in preparing this book, they make no representations or warranties with respect to the accuracy or completeness of the contents of this book and specifically disclaim any implied warranties of merchantability or fitness for a particular purpose. No warranty can be created or extended by sales representatives or written sales materials. The Advice and strategies contained herein may not be suitable for your situation. You should consult with a professional where appropriate. Neither the publisher nor authors shall be liable for any loss of profit or any other commercial damages, including but not limited to special, incidental, consequential, or other damages.

Library of Congress Card No.: applied for

British Library Cataloguing-in-Publication Data
A catalogue record for this book is available from the British Library.

Bibliographic information published by the Deutsche Nationalbibliothek
The Deutsche Nationalbibliothek lists this publication in the Deutsche Nationalbibliografie; detailed bibliographic data are available on the Internet at <http://dnb.d-nb.de>.

© 2014 Wiley-VCH Verlag GmbH & Co. KGaA, Boschstr. 12, 69469 Weinheim, Germany

Wiley-Blackwell is an imprint of John Wiley & Sons, formed by the merger of Wiley's global Scientific, Technical, and Medical business with Blackwell Publishing.

All rights reserved (including those of translation into other languages). No part of this book may be reproduced in any form – by photoprinting, microfilm, or any other means – nor transmitted or translated into a machine language without written permission from the publishers. Registered names, trademarks, etc. used in this book, even when not specifically marked as such, are not to be considered unprotected by law.

Print ISBN: 978-3-527-33156-7
ePDF ISBN: 978-3-527-66864-9
ePub ISBN: 978-3-527-66865-6
Mobi ISBN: 978-3-527-66866-3

Cover Design   Adam Design, Weinheim, Germany

Typesetting   Laserwords Private Limited, Chennai, India

Printing and Binding   Markono Print Media Pte Ltd, Singapore

Printed on acid-free paper.

Contents

Preface IX

Volume 1

Part I Molecular Cell Biology 1

1 **RNA Regulation in Apoptosis** 3
   Christopher von Roretz and Imed-Eddine Gallouzi

2 **Bacterial *trans*-Translation: From Functions to Applications** 47
   Emmanuel Giudice and Reynald Gillet

3 **RNA-mediated Control of Bacterial Gene Expression: Role of Regulatory non-Coding RNAs** 81
   Pierre Mandin, Alejandro Toledo-Arana, Aymeric Fouquier d'Hérouel and Francis Repoila

4 **Functions and Applications of RNA-Guided CRISPR-Cas Immune Systems** 117
   Rodolphe Barrangou and Philippe Horvath

5 **RNA Regulation in Myogenesis** 141
   Andrie Koutsoulidou, Nikolaos P. Mastroyiannopoulos and Leonidas A. Phylactou

6 **Regulation of Animal Gene Expression by Ingested Plant Small RNAs** 169
   Xi Chen, Lin Zhang and Chen-Yu Zhang

7 **RNA Interference in Animals** 185
   Mikiko C. Siomi

8 **To Translate or Degrade: Cytoplasmic mRNA Decision Mechanisms** 211
   Daniel Beisang and Paul R. Bohjanen

9	RNA Modification Yuri Motorin and Bruno Charpentier	237
10	Regulation of Gene Expression Anil Kumar, Sarika Garg and Neha Garg	285
11	RNA Silencing in Plants Charles W. Melnyk and C. Jake Harris	345
12	mRNA Stability Ashley T. Neff, Carol J. Wilusz and Jeffrey Wilusz	391
13	Visualization of RNA and RNA Interactions in Cells Natalia E. Broude	417
14	RNA as a Regulator of Chromatin Structure Yota Murakami	437
15	Intracellular RNA Localization and Localized Translation Florence Besse	471
16	tRNA Subcellular Dynamics Tohru Yoshihisa	513

Volume 2

Part II	Methods for Analysis and Manipulation of RNA-mediated Regulation	545
17	RNA Methodologies Robert E. Farrell, Jr.	547
18	Bacterial Vectors for RNAi Delivery to Cancer Cells Hermann Lage	585
19	RNAi Screening and Assays Marie Lundbæk and Pål Sætrom	605

Part III	Medical Applications	631
20	Translation Regulation by microRNAs in Acute Leukemia Christos K. Kontos, Diamantina Vasilatou, Sotirios G. Papageorgiou and Andreas Scorilas	633
21	Cancer Stem Cells Mei Zhang and Jeffrey M. Rosen	663
22	RNAi Synthetic Logic Circuits for Sensing, Information Processing, and Actuation Zhen Xie, Liliana Wroblewska and Ron Weiss	687

23	**Emerging Clinical Applications and Pharmacology of RNA** *Sailen Barik and Vira Bitko*	711
24	**RNAi to Treat Chronic Hepatitis C Infection** *Usman Ali Ashfaq, Saba Khaliq and Shah Jahan*	751
25	**RNAi Gene Therapy to Combat HIV-1 Infection** *Pierre Corbeau*	781
26	**Virus-Encoded microRNAs** *Lee Tuddenham and Sébastien Pfeffer*	807
	Index	849

Preface

Our compendium is written for university undergraduates, graduate students, faculty, and investigators at research institutes. There are 26 articles and a length of over 800 pages, and as such it is the largest in-depth, up-to-date treatment of RNA Regulation presently available.

This *RNA Regulation* compendium differs in content and quality from all others available in five ways: (i) the overall coverage was approved by our Board, which includes 11 Nobel Prize laureates; (ii) the selection of each article and author was validated by reviewers from major university research centers; (iii) each article was then reviewed by peers from other universities; (iv) a glossary of terms with definitions is provided at the beginning of each article; and (v) the articles average 30 printed pages – which provides several times the depth of other such compendia.

RNA has long been regarded as a molecule that can function either as a messenger (mRNA) and/or as part of the translational machinery (tRNA, rRNA), and in fact RNA dominates the functions that define the state of a given cell at any moment. Subsequently, it became clear that RNAs are versatile molecules that not only play key roles in many important biological processes such as splicing, editing, protein export and the degradation of proteins and other RNAs, but also – like enzymes – can act catalytically. RNAi serves as an umbrella term, encompassing several phenomena discovered in a diversity of eukaryotes and prokaryotes, all of which are mediated by short noncoding RNA species. RNAi has been shown to be important in regulating gene expression by several mechanisms, from transcriptional silencing to mRNA degradation and translational repression. It has been determined that the genomes of all eukaryotes are almost entirely transcribed, generating an enormous number of nonprotein-coding RNAs (ncRNAs). It is increasingly evident that many of these RNAs have regulatory functions exhibiting cell type-specific expression, localization to subcellular compartments, and association with human diseases. Advances in understanding the basic molecular cell biology of RNA regulation is leading towards medical applications in cancer, aging, neurological disease, cardiac disease and infectious diseases. In addition, we cover the utilization of RNA interference for sensing endogenous molecular signals, processing and actuation in mammalian cells, and medical applications aimed towards the creation of smart molecular devices for therapeutic purposes, as well as elucidating genetic regulation and cellular functions.

Our coverage includes RNA interference and the delivery of therapeutic RNA to cells, as well as the basis for some of the most recent RNA breakthroughs such as the bacterial origin for harnessing the CRISPR system to target the destruction of specific genes in human cells and thereby provide a method for RNA-guided next-generation genome editing. We also provide an introduction to the recently emphasized investigation into extracellular RNA (exRNA), and whether it can be harnessed for the diagnosis and treatment of human diseases.

The 26 chapters are arranged in three sections: **Molecular Cell Biology**, which includes RNA regulation chapters on bacteria, plants, fungi, and humans and RNA stability and modification as well as transport and regulation in development; **Methods for Analysis and Manipulation of RNA-Mediated Regulation**, which includes vectors for delivery to cells, pharmacokinetics of RNA and methods for the purification of chemically stable and biologically functional RNA; and **Medical Applications**, for example, in cancer, infectious diseases, neurodegenerative disorders and cardiac disease. In fact, noncoding RNAs – functional RNA molecules that do not encode proteins such as microRNAs (miRNAs) – are discussed in relation to regulation of the cancer stem cell properties of self-renewal and differentiation, and thus also provide a potentially new therapeutic approach.

Our team hopes that you, the reader, will benefit from our hard work, finding the content useful in your research and educational. We wish to thank our Managing Editor, Sarah Mellor, as well as our Executive Editor, Gregor Cicchetti, for both their advice and hard work during the course of this project.

Larkspur, CA, USA
January 2014

Robert A. Meyers
RAMTECH Limited

Part I
Molecular Cell Biology

1
RNA Regulation in Apoptosis

Christopher von Roretz and Imed-Eddine Gallouzi
McGill University, Department of Biochemistry and Rosalind and Morris Goodman Cancer Research Centre, McIntyre Medical Building Room 915B, 3655 Promenade Sir William Osler, Montreal, Quebec, H3G 1Y6 Canada

1	**Balancing Life and Death** 6	
2	**The Coordinated Process of Apoptosis** 6	
2.1	Caspase Cascade 7	
2.2	Extrinsic Activation of Caspases 7	
2.3	Intrinsic Activation of Caspases 9	
2.3.1	Mitochondrial Release of Cytochrome c 9	
2.3.2	Activation of the Apoptosome 9	
2.4	Caspase Substrates 10	
3	**Protein Regulators of Apoptosis** 11	
3.1	Regulators of Cytochrome c Release 11	
3.1.1	Bcl-2 Family Proteins 11	
3.1.2	BH3-Only Proteins 11	
3.2	Targeting Caspase Activation 12	
3.2.1	DISC Inhibition 13	
3.2.2	Regulation of Apoptosome Activity 13	
3.3	Caspase Inhibitor of Apoptosis Proteins 13	
3.3.1	IAPs 14	
3.3.2	Inhibitors of IAPs 14	
3.4	Post-Translational Regulation of Apoptotic Factors 14	
4	**Transcriptional Regulation of Apoptosis** 15	
4.1	p53 and p73 15	
4.2	E2F Family 16	
4.3	FOXO Family 16	
4.4	Additional Transcriptional Regulators of Apoptosis 16	

RNA Regulation: Advances in Molecular Biology and Medicine, First Edition. Edited by Robert A. Meyers.
© 2014 Wiley-VCH Verlag GmbH & Co. KGaA. Published 2014 by Wiley-VCH Verlag GmbH & Co. KGaA.

5	**Post-Transcriptional Regulation of Apoptosis**	**17**
5.1	Splicing	17
5.1.1	Splicing to Regulate Cytochrome c Release	18
5.1.2	Splicing-Mediated Regulation of Caspases	19
5.1.3	Nuclear Export of mRNA	21
5.2	Translation	21
5.2.1	*cis*-Elements and *trans*-Acting Factors that Regulate Translation	21
5.2.2	HuR as a Regulator of Translation	22
5.2.3	IRES-Mediated Translation and Apoptosis	23
5.2.4	microRNAs as Regulators of the Translation of Apoptotic Factors	24
5.2.5	miRNAs as Inhibitors of Apoptosis	24
5.2.6	Pro-Apoptotic Roles of miRNA	26
5.3	mRNA Turnover	27
5.3.1	The Destabilizing A/U Rich Element	27
5.3.2	ARE-Mediated Turnover of Apoptosis-Related mRNAs	28
5.3.3	Alternate Destabilizing Elements	29
5.3.4	Destabilizing miRNAs	29
6	**Discussion**	**30**
6.1	Rearranging Deckchairs on the Titanic?	30
6.2	All a Matter of Time	31
	Acknowledgments	**32**
	References	**32**

Keywords

Alternative splicing
Splicing of pre-mRNA to generate mature mRNA involves the removal of noncoding segments of messenger RNA known as *introns*. The factors involved in this process determine if certain splice sites are skipped or not, however, yielding different splice variants.

Bcl-2 family proteins
The B-cell lymphoma 2 (Bcl-2) family of proteins all contain one or more Bcl-2 homology (BH) domains. Members of the Bcl-2 family may be either pro- or anti-apoptotic, as they influence mitochondrial outer membrane permeabilization, leading to the release of cytochrome c into the cytoplasm.

Caspase
Cysteine-dependent, aspartic acid-specific proteases (caspases) are generally considered the effectors of apoptotic cell death, cleaving a variety of substrates in order to lead to the organized death of a cell.

Cytochrome c
Cytochrome c (cyt c) is a small polypeptide that plays a role in the electron transport chain as an electron carrier under normal conditions. When released from the mitochondria following mitochondrial outer membrane permeabilization, cytochrome c binding to Apaf-1 protein enables structural changes which ultimately permit the activation of caspase-9.

IRES-mediated translation
Internal ribosomal entry site (IRES)-mediated translation is the process whereby mRNA translation is initiated through the recruitment of translation factors, and consequently, ribosomes, to the mRNA independently of the m7G 5′ cap.

miRNA
microRNA (miRNA) is a single-stranded RNA of approximately 22 nt that can induce the inhibition of translation or decay of a target mRNA upon binding to a mRNA for which a specific miRNA maintains a certain degree of complementarity.

mRNA turnover
The turnover of messenger RNA is the stability of a particular mRNA, which can be modulated by various factors that promote either mRNA decay or stabilization.

Post-transcriptional regulation
Subsequent to the transcription of genetic material to generate messenger RNA, various mechanisms including splicing, localization, translation, and mRNA turnover, can influence the expression of the gene encoded by a target mRNA.

> The organized process of apoptotic cell death is tightly regulated, with many protein factors promoting or inhibiting the activity of its key players. A growing number of studies have shown that these numerous regulators of apoptosis are themselves regulated at the level of RNA. Through transcription, the levels of mRNA encoding these factors can be increased or decreased. Less studied is the post-transcriptional regulation of expression for these pro- and anti-apoptotic players. Alternative splicing, effects on translation, and the modulation of mRNA turnover can all influence protein levels of the broad cast of factors involved in apoptosis. While most of the studies delineating these mechanisms have not examined explicitly the

RNA regulation of these factors during apoptosis, a small collection of data does suggest that post-transcriptional regulation of apoptotic modulators occurs during the cell death process, thus hinting at a previously underappreciated role for RNA regulation in apoptosis.

1
Balancing Life and Death

All cellular processes depend on balance. Metabolism, motility, and growth all depend on juggling factors that promote or perturb each of these events. Perhaps the most significant of cellular processes, however, is that of cell death when the decision is made – either by the victim cell or by its environment – that life should cease and one, possibly final, balance is tilted either to seal the fate of this cell through death, or to allow it to survive despite assaulting conditions by activating survival mechanisms. At a cellular level, with numerous factors regulating each and every pathway, balance is a matter of life or death.

Several types of cell death exist [1], and while these were initially classified based on their degree of organization, during recent years such a broad-stroke classification has been challenged [1, 2]. While autophagic cell death and necrosis/necroptosis have gained attention more recently [1,3–7], the best known form of cell death is apoptosis. Derived from the Greek word for "falling off," as with the leaves from a tree [8], apoptosis has been the most well-studied class of cell death, and is typically considered the default method of ending the life of a cell. In essence, apoptosis is an organized cell death whereby specific processes take place to coordinate the destruction of a cell in a practical manner, to avoid harm to neighboring cells. In fact, given the individualized focus of this process, apoptosis is often referred to as *"cell suicide"* [3], even though it can be triggered both internally by the victim cell and by external stimuli from neighboring cells or the surrounding environment. Numerous pathways are activated during apoptosis (see below) which integrate many pro- and anti-apoptotic factors to weigh in on the decision to survive or to engage death in response to a stress stimulus. While the activity of these factors has often been studied at great lengths, over the past few years evidence has been mounting to suggest that the expression of apoptotic factors is also an important determinant of apoptotic cell death. Transcription to produce mRNAs encoding these factors has been studied to significant lengths. Conversely, there have been limited attempts at understanding the post-transcriptional mechanisms that influence the expression of apoptotic factors. Of these studies, only a fraction has specifically examined how these regulatory mechanisms may directly influence the progression of apoptosis. In this chapter, attention will be focused on outlining the link between RNA and apoptosis, and a summary provided of how transcription – and, more specifically, post-transcriptional events – may contribute to apoptotic cell death.

2
The Coordinated Process of Apoptosis

Apoptosis will not occur spontaneously, and generally begins with a trigger. The diversity of sources that can activate apoptosis is broad, and ranges from chemical activators, endocrine signaling from neighboring cells, or internal sensors of cell damage [9–13]. Typically, the activation of apoptosis is divided into two categories: (i) that which is triggered externally, known as *extrinsic-induced apoptosis*; and (ii) that which results from internal signals, termed *intrinsic-induced apoptosis* [14]. Each of these involves a different sequence of activated factors, although the end goal is the same – to enable the activation of a specific family of proteases known as *caspases* (Fig. 1).

2.1
Caspase Cascade

Caspases are cysteine-dependent, aspartic acid-specific proteases that cleave protein substrates after an aspartate residue [15]. Caspases exist as inactive zymogens in normal cellular conditions, known as *procaspases* [14, 16]. Structurally, their activation may be achieved by cleaving off a precursor region, and also by cleaving their catalytic domain to yield two regions that then interact. Additionally, caspase activity depends on their dimerization.

Upon the activation of one caspase, such as caspase-8 or -9, a sequential activation of other caspases can occur, commonly known as the *"caspase cascade"* [17]. These upstream caspases are referred to as *"initiator"* caspases, given that they are those initially activated in response to apoptotic stimuli, and include caspases-2, -8, -9, and -10 [16]. Each of these initiator caspases then in turn activates other caspases. For example, caspase-9 is best described as cleaving procaspases-3 and -7 to generate their active forms [18, 19]. These secondary caspases have a much broader panel of targets and, as such, are referred to as *"effector"* or *"executioner"* caspases [20]. Caspases-3, -6, and -7 are generally considered as the effector caspases in apoptosis [16]. While initiator caspases are activated by being correctly oriented following the binding to adaptor proteins (described below), effector caspases await cleavage from initiator caspases. The reason behind this is that the initiator caspases exist as monomers in the cytosol, and require platforms to allow dimerization to occur, whereas effector caspases perpetually exist as dimers but depend on cleavage to become activated [14, 21].

2.2
Extrinsic Activation of Caspases

In extrinsic-induced apoptosis (Fig. 1), an external signal is needed to act on cellular receptors at the surface of a target cell. These receptors, given their purpose, have been termed *"death receptors,"* and include many members of the tumor necrosis factor (TNF) receptor family, such as the Fas and TNF-α receptors [22, 23]. Each of these receptors is designed to bind and be stimulated by the binding of specific substrates. For example, when Fas ligand is released by T cells, it will bind to the Fas receptor on target cells. Similarly, other secreted ligands such as TNF-α and TNF-related apoptosis-inducing ligand (TRAIL) will bind to their respective receptors [22]. The result of such binding is mechanistically similar for each of these cases. The binding of "death ligands" to "death receptors" triggers the trimerization of receptor monomers [24, 25]; this trimeric structure is then taken up by the cell through endocytosis and, in doing so, permits the recruitment of intermediary proteins, such as Fas-associated protein with death domain (FADD) [26]. FADD and other, related proteins such as death

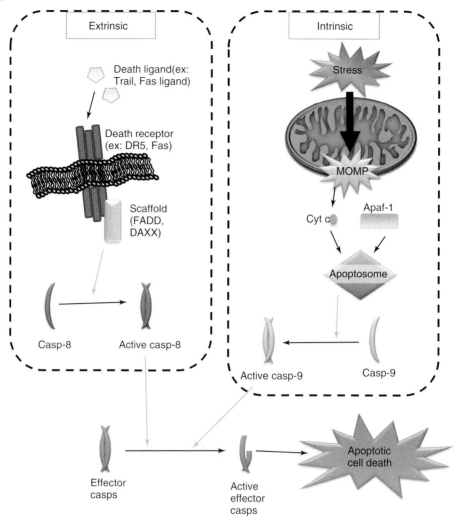

Fig. 1 Extrinsic and intrinsic pathways of apoptosis. In the extrinsic pathway of apoptotic activation, binding of a death ligand (e.g., Fas ligand, TRAIL) to a death receptor (DR5, Fas) triggers the recruitment of an adaptor protein (FADD, DAXX) and an initiator caspase such as procaspase-8 or -10. Activated initiator caspase can then activate effector caspases such as caspase-3 or -7. Intrinsic activation of apoptosis involves a stress signal which triggers the mitochondrial outer membrane permeabilization (MOMP), leading to the release of cytochrome c (cyt c) from the mitochondrion. With Apaf-1, dATP, and procaspase-9 the apoptosome is formed, which allows for the activation of caspase-9. This initiator caspase can also activate effector caspases.

domain-associated protein 6 (DAXX) and receptor-interacting protein-1 (RIP-1) are all capable of binding monomeric procaspases [27], in particular procaspase-8 and -10. This binding is achieved via either the death domain or the death effector domain (DED) of these adaptor proteins [28]. Collectively, the group of death ligand, receptor, DED-containing adaptor and pro-caspase form what is referred to as the Death-inducing signaling complex (DISC) (Fig. 1).

2.3 Intrinsic Activation of Caspases

While external signals can stimulate apoptosis through the mechanisms described above, caspases can also be activated through internal sensory mechanisms. DNA damage, for example, leads to the activation of kinases such as ataxia telangiectasia mutated (ATM) kinase, which can trigger either DNA repair or apoptosis through the phosphorylation of a variety of targets [29]. One crucial target of ATM is p53, a transcription factor that regulates the expression of numerous pro-apoptotic factors (see below). Under normal conditions, p53 is bound by Mdm2, which leads to the ubiquitination and degradation of p53 [30, 31]. The phosphorylation of p53 prevents Mdm2-binding, allowing this transcription factor to promote the apoptotic response [32]. ATM can also phosphorylate Mdm2 to further interfere with its association to p53 [33]. Several other pathways, responding to various stimuli, exist to promote apoptosis.

2.3.1 Mitochondrial Release of Cytochrome c

One of the key events in intrinsic-mediated apoptosis is the release of cytochrome c from the mitochondria into the cytoplasm (Fig. 1). Under normal conditions, cytochrome c plays an essential role in the electron transport chain, serving as a single electron carrier [34, 35]. Cytochrome c, a small protein of 104 amino acids, is normally located between the inner and outer membranes of the mitochondrion [36]. Specific channels that form in the outer membrane of the mitochondrion allow for cytochrome c to travel from the inter-membrane mitochondrial space to the cytoplasm; this is referred to as mitochondrial outer membrane permeabilization (MOMP) [37]. MOMP is often considered a key event in apoptosis activation, and is generally thought to be irreversible. Having reached the cytoplasm, cytochrome c plays a very different role as it is specifically capable of forming a very large complex, known as the *apoptosome* (Fig. 1), in view of its ability to advance apoptosis [14].

2.3.2 Activation of the Apoptosome

The core of the apoptosome is a heptameric structure comprised of seven monomers of a protein called apoptotic protease activating factor 1 (Apaf-1) [38]. Apaf-1 exists as an inactive monomer in the cytoplasm, but readily binds cytochrome c should it become available, for example following its release from the mitochondrion. When cytochrome c binds to Apaf-1 a conformational change occurs in this larger protein, allowing the association of several Apaf-1 monomers to form a seven-membered ring [39]. This structure is then capable of recruiting procaspase-9 monomers [14] that, like Apaf-1, circulate in the cytoplasm under normal conditions, without having any known activity. Once recruited to the Apaf-1 heptamer, procaspase-9 molecules are placed in close proximity, which allows for their dimerization, which is in fact sufficient for caspase-9 activation [40]. This entire process utilizes energy, however, and dATP is a key ingredient for caspase-9 activation and correct apoptosome formation. dATP must be recruited by Apaf-1 and converted to dADP in order for the apoptosome to assume its optimal structure capable of recruiting procaspase-9 [41].

The collective complex of cytochrome c, Apaf-1, procaspase-9, and dATP is termed the *apoptosome*, the kinetics and regulation of the formation of which have been thoroughly reviewed [14].

2.4
Caspase Substrates

Each caspase, whether an initiator or effector, has a set of targets. In some cases, the list of known targets for a caspase is very limited. Other than its two target procaspases (-3 and -7), limited evidence has demonstrated other apoptotic targets for caspase-9 [42]. Caspase-8, on the other hand, does not solely have procaspases as its targets; indeed, caspase-8 does cleave procaspase-3 to cause its activation, but also targets a pro-apoptotic protein known as *Bid* [43]. Under normal conditions, Bid is a 22 kDa protein that may be found in the cytosol [44]; however, upon cleavage by caspase-8, Bid is cleaved to generate truncated Bid (tBid). Numerous groups have shown that tBid is an important promoter for the release of cytochrome c from the mitochondria [43]. Hence, caspase-8 can in fact both directly activate an effector caspase (caspase-3) as well as activate the intrinsic apoptotic pathway, through cytochrome c release. Far more targets have been identified for effector caspases [20], which is consistent with their role as executioners of organized cell death. Ultimately, they are responsible for shutting down all functions of the target cell, and do so in an organized manner. Targets of caspases-3 and -7 include poly (ADP-ribose) polymerase (PARP), DNA fragmentation factor-45 (DFF-45, also known as inhibitor of caspase-activated deoxyribonuclease; ICAD), Mdm2, protein kinase Cδ (PKCδ), actin, lamin, and fodrin [45–55].

PARP is an important enzyme for the repair of DNA breaks [56]. It is believed that, by inactivating this enzyme, caspases halt any futile, energy-requiring attempts to repair DNA damage, and commit the cells to die [57]. ICAD is an inhibitor of the cytosolic endonuclease CAD that has been shown to fragment DNA, and by subsequently inactivating the ICAD, the caspases will promote DNA degradation [47–50]. As described above, Mdm2 is an inhibitor of the apoptotic transcription factor p53; hence, the cleavage of Mdm2 further enhances transcriptional events during cell death [51]. Finally, actin, fodrin, and lamin are all structural proteins, the cleavage of which during apoptosis is believed to be responsible for morphological changes within the cell [53–55].

In addition to mediating apoptosis by cleaving downstream targets, there is substantial support that caspases also act in feedback loops to further enhance the process. For example, caspases-3 and -7 are capable of cleaving the initiator caspases-8 and -9 [58].

Collectively, through self-perpetuating caspase cascades as well as the cleavage of specific key cellular players, caspases produce a state of organized death that makes the condensed cells recognizable, ideal targets for phagocytes or other neighboring cells [59]. Unlike many other physiological processes, apoptosis is distinguished for being irreversible, given the nature of its purpose, and this provides all the more reason for a tight control to exist over the process. Caspase activation, while an important component of the pathway, is just one piece of the puzzle. The regulation of apoptosis occurs at several levels, ranging from protein–protein interactions to controlling the expression of apoptotic players.

3
Protein Regulators of Apoptosis

The regulation of apoptosis involves numerous factors that directly or indirectly inhibit or activate caspases and other protein players of the process. Importantly, the expression of these factors must also be considered, as the upregulation or downregulation of any player that contributes to cell fate can tilt the balance between survival and apoptosis.

3.1
Regulators of Cytochrome c Release

One of the most significant levels at which apoptosis is regulated is through cytochrome c release. As noted above, both extrinsic and intrinsic apoptosis pathways trigger the release of cytochrome c through MOMP, which has been shown to then activate caspase-9 [60].

3.1.1 Bcl-2 Family Proteins

The actual release of cytochrome c from the mitochondrion requires the formation of a channel or pore, composed of Bax and Bak monomers [61]. Importantly, Bax and Bak can be inhibited in forming this channel if they bind to proteins such as B-cell lymphoma 2 (Bcl-2), Bcl-w, Mcl-1, or Bcl-x_L [62]. Structurally, the difference between the anti-apoptotic Bcl-2, Bcl-w, and Bcl-x_L proteins and the pro-apoptotic Bak and Bax is not large [63], as both contain four regions termed Bcl-2 homology (BH) domains that allow dimerization of the proteins and enable each of these factors to interact with one another. During the past decade, another class of players in the regulation of apoptosis has been identified, termed "BH3-only proteins" [44]. This naming is based on the fact that these proteins contain only the BH3 domain homologous to those found in Bax, Bcl-2, and other members of the Bcl-2 family of proteins.

3.1.2 BH3-Only Proteins

Several BH3-only proteins have been identified, including Bim, Bid, Bad, Bik, Noxa, and Puma [64–68]. These proteins have been shown as being important in the regulation of apoptosis, as their knockdown results in a resistance to apoptotic stimuli [69, 70]. The reason for this is that, with only their BH3 domains, Bad, Bim, Puma, and Noxa bind to anti-apoptotic Bcl-2 family members such as Bcl-2 and Bcl-x_L, and sequester them from inhibiting Bak and Bax [64–66]. Furthermore, evidence has shown that BH3-only proteins also promote MOMP by binding Bax to enhance its integration into the outer mitochondrial membrane [71] (Fig. 2). In fact, active caspase-8 promotes intrinsic apoptosis through cleavage of the BH-3 only protein Bid to generate tBid, which can then assist in promoting Bax oligomerization at the outer mitochondrial membrane [61, 67], thus allowing MOMP (Fig. 2). The importance of BH3-only proteins in the regulation of apoptosis has been further demonstrated during recent years as several chemical agents have been developed to mimic BH3-only proteins, and have been used successfully to induce apoptosis in cancer models, whether *in vitro*, *in vivo*, or even in human clinical trials [72–78].

These intricate regulations of MOMP closely influence the release of cytochrome c, and hence, caspase-9 activation. Importantly, other proteins can promote or inhibit caspase-activation above or below the release of cytochrome c.

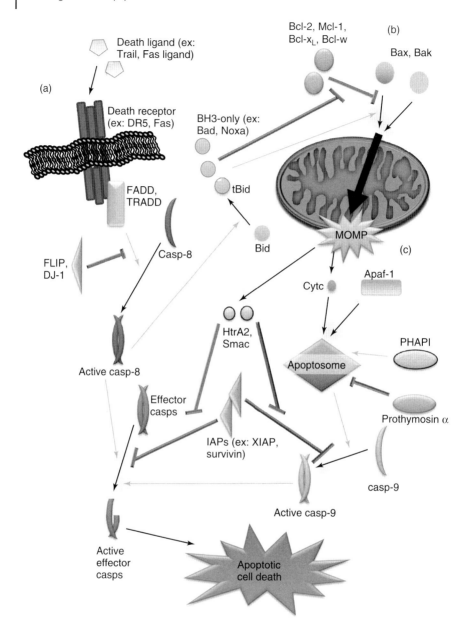

3.2
Targeting Caspase Activation

The activation machineries of initiator caspases are also the targets of regulation (Fig. 2). DISC can be inhibited by several factors [79, 80], including FLICE-like inhibitory protein (FLIP), a caspase-8 mimic that contains a nonfunctional catalytic domain [81, 82]. Likewise, the apoptosome is also subject to regulation by different pro- and anti-apoptotic

Fig. 2 The regulated pathways of apoptotic cell death. As noted in Fig. 1, apoptosis may be engaged either extrinsically (a) or intrinsically (b). The downstream cascades that result from external or internal apoptotic stimuli are complex and are subject to multiple levels of regulation. In particular, protein activators and inhibitors mediate both apoptotic pathways. (a) In the extrinsic pathway, FLIP and DJ-1 proteins can inhibit the activation of caspase-8 (casp-8) and caspase-10 following the binding of death ligands (such as TRAIL) to death receptors; (b) In intrinsic apoptosis, pro-apoptotic proteins Bax and Bak are inhibited by Bcl-2 family members such as Bcl-2, Mcl-1, and Bcl-xL. BH3-only members of the Bcl-2 family can block this inhibitory effect, or can activate Bax/Bak heterodimerization. Once the latter is achieved, mitochondrial outer membrane permeabilization (MOMP) occurs (c) and releases cytochrome c (cyt c), HtrA2/Omi, and Smac/DIABLO. Cyt c activates Apaf-1 to form the apoptosome complex, which leads to the activation of caspase-9. Inhibitors of apoptosis proteins (IAPs) such as XIAP and survivin can inhibit caspase activity, but are themselves negatively regulated by HtrA2/Omi and Smac/DIABLO. IAPs can also target effector caspases such as caspase-3 and -7. In this figure, the solid black arrows indicate movement or processing of a factor; faint arrows indicate the activation of a process; and thicker lines represent the inhibition of a process. In addition to the regulation shown here, virtually all of these players in the regulation and execution of apoptosis are further regulated at the transcriptional, post-transcriptional, or post-translational levels.

factors [83], both of which involve particular mechanisms and have undergone extensive investigation.

3.2.1 DISC Inhibition

FLIP, which is the best-characterized protein regulator of extrinsic-induced apoptosis, is capable of binding the DED region of adaptors such as FADD and tumor necrosis factor receptor type 1-associated death domain (TRADD) and, in doing so, prevents the recruitment of procaspases-8 and -10 [81, 82]. Another similar inhibitor is the DJ-1 protein, which has been found to bind DAXX as well as FADD (Fig. 2), also consequentially inhibiting the activation of caspase-8 [79, 80].

3.2.2 Regulation of Apoptosome Activity

Several other factors have also been identified as regulating apoptosis, at the level of the apoptosome (Fig. 2). One such protein is prothymosin α, which inhibits the apoptosome through a mechanism that involves binding to Apaf-1, but about which nothing more has been reported [83, 84]. Prothymosin α has also been linked to cell proliferation and cancer growth [85–89]. By comparison, a positive regulator of apoptosome activity that had no previous link to cell death (putative HLA-associated protein-I; PHAPI), has been identified recently. Previously known as *pp32*, PHAPI is an acidic-rich protein that has been reported to inhibit oncogene-induced tumorigenesis [90–92] and, intriguingly, is also implicated in apoptosis as an activator of the apoptosome [83]. Subsequently, PHAPI, in complex with Hsp70 and a cellular apoptosis susceptibility (CAS) protein, was found to play an important part in the exchange of dADP for dATP in apoptosome activation [93]. Kim *et al.* demonstrated that Hsp70, CAS, and PHAPI are all needed for optimal caspase-9 activation, and that the absence of CAS can produce a significant decrease in apoptotic cell death [93].

3.3 Caspase Inhibitor of Apoptosis Proteins

An entire series of apoptotic regulators exists downstream of caspase-activation, which focuses on the inhibition or promotion of caspase activity (Fig. 2). The most well-studied molecules that aim to inhibit caspase activity are known as inhibitors of apoptosis proteins (IAPs).

3.3.1 IAPs

Several members of the IAP family exist, including XIAP (X-linked inhibitor of apoptosis protein), cIAP1, cIAP2, and survivin [94, 95]. While each of these has unique targets and particular properties [96–99], the commonality between each is their ability to inhibit active caspases. In the case of XIAP, this is achieved through a direct interaction with caspases-9, -3, and -7 [98–102]. XIAP interacts at two sites with each of these caspases and, in doing so, blocks their catalytic sites (Fig. 2). The inhibitory mechanisms of other IAPs are less well understood; survivin has been suggested to form a complex with procaspase-9 that sequesters it from recruitment to Apaf-1 [103, 104]. It has also been suggested that IAPs may be responsible for the ubiquitination and degradation of caspases [105–107].

3.3.2 Inhibitors of IAPs

Just as BH3-only proteins exist to inhibit the inhibitors of MOMP, inhibitors of IAP proteins are also present in the cell, the roles of which are to sequester IAPs and allow normal caspase activity. A few such proteins have been described to exist in mammals during the past few years, including HtrA2/Omi and Smac/DIABLO (direct IAP-binding protein with low pI) [108–110]. These proteins were found to be released from the mitochondria during MOMP along with cytochrome c, and to promote apoptosis [13, 111], but in a distinct mechanism from the apoptosome-activating effect of cytochrome c. Both, HtrA2/Omi and Smac/DIABLO have in fact been shown to bind and block the inhibitory capacity of IAPs [13, 108, 112], thus enhancing caspase activity (Fig. 2).

3.4 Post-Translational Regulation of Apoptotic Factors

The many interactions between activators and inhibitors of apoptosis described above provide a complex control over the balance between life and death. Further regulation of this balance exists within the realm of post-translational modifications, to which many of these apoptotic factors may be subjected. The best studied of these is phosphorylation, which has been found to regulate a handful of Bcl-2 family proteins by either promoting or inhibiting their activity [113–119]. Caspase-activity is also influenced by phosphorylation [120]. Caspases-2, -3, -8, and -9 have all been documented as the substrates of kinases (this subject has been extensively reviewed by Kurokawa and Kornbluth [121]). Generally, the phosphorylation of caspases results in a decreased activity, such as when caspase-9 is phosphorylated by protein kinase B (PKB) [122]. Importantly, the phosphorylation of caspases allows an additional level of regulation and fine-tuning of the apoptotic response, since phosphatases such as protein phosphatase 2 (PP2A) can be used to promote apoptosis by alleviating inhibitory phosphates from caspases [121, 123, 124].

The ubiquitination of apoptotic factors has also been found to play a role in the regulation of this process. Ubiquitin is a 76-amino acid protein that is ligated onto the lysine residues of a target protein, and polyubiquitination of a protein typically signals for the degradation of this protein by the proteosome [125]. Of pertinence to apoptosis, several anti-apoptotic factors have been identified as ubiquitin ligases [105–107, 126], enzymes that are capable of attaching ubiquitin to targets, consequentially signaling for their degradation. XIAP, cIAP1, and cIAP2 are some examples of ubiquitin ligases, and are capable of signaling the degradation of caspases [105–107]. Mcl-1, FLIP, and numerous other apoptotic players are targeted by enzymes involved in the ubiquitin pathway [127–131].

The regulation of apoptosis through protein–protein interactions and modifications to prevent such interactions is a crucial facet of the tight control over the apoptotic process. One of the easiest ways to influence an imbalance in these various regulatory mechanisms is to modulate the expression of pro- and anti-apoptotic factors. Indeed, the regulation of apoptosis through gene expression is just as important a contributing factor to cell fate, if not more so, than are the interactions between activator and inhibitor proteins.

4
Transcriptional Regulation of Apoptosis

Regulating apoptosis by modulating gene expression is a question of regulating mRNA. From the moment it is generated until the time it is degraded, mRNA is the intermediate that encodes apoptotic factors. If provided with the correct conditions, mRNAs will lead to the production of pro- and anti-apoptotic players and, as such, should have a tremendous influence on the progression of this cell process. Intriguingly, only a small number of the studies characterizing the roles of mRNA in influencing apoptotic factors have examined specifically what occurs in response to lethal stress, to mimic apoptotic conditions.

4.1
p53 and p73

At the level of transcription to produce mRNAs, various studies have defined how transcription factors become active in response to stimuli, such as fatal triggers. While numerous apoptosis-related transcription factors have been identified, easily the most well known is p53 [132]. Although many stress signals have been shown to activate p53 function, the most common mechanism is for signals to activate kinases, such as ATM, ATR (ataxia telangiectasia and Rad3-related), and Chk2, which then phosphorylate p53 in order to reduce its association with the inhibitory Mdm2 [133].

The targets of p53 are diverse and linked to various effects on the cell. For example, p53 can halt the cell cycle by enhancing transcription of the cyclin-dependent kinase inhibitor p21$^{Waf1/Cip1}$, or by repressing the translation of cdc25c, which promotes mitosis [134–136]. The best-known involvement of p53, however, is through its regulation of apoptosis. Indeed, the ability of p53 to induce cell death is tied to its identification as a tumor suppressor and, given the breadth of targets that it regulates, it is not surprising that p53 is mutated or absent in more than half of all human cancers [137–140]. These targets include Bax, Bad, Bid, Fas, Apaf-1, Noxa, and Puma, the transcription of which are promoted by p53, and survivin and Bcl-2, for which transcription is inhibited [65, 132, 141–149]. p53 also promotes the expression of Mdm2, in order to exercise a tight regulation of its levels.

Importantly, a relative to p53 – named p73 – can also promote the expression of p21 and Mdm2 in response to stress [150]. In fact, p73 variants can also regulate apoptotic targets of p53, such as Bax and Noxa [151–153]. Similarly to p53, a complex regulation of p73 exists, and the existence of this alternate key transcriptional regulator of apoptotic factors indicates that cells possess various strategies to genetically respond to stress signals.

4.2
E2F Family

The E2F family of transcription factors is also strongly linked to stress response and the induction of apoptosis. E2F1, a transcriptional promoter, is the member of this family that has been best characterized for a role in influencing apoptotic targets, such as caspase-8 and Bid [154]. Growing evidence has linked the E2F1-mediated control of expression to include a number of other apoptotic targets that are also under the control of p53, such as Apaf-1, caspases, BH3-only proteins, survivin and Smac/DIABLO [155–164]. Importantly, E2F1 also promotes the transcription of p73, and can thus promote an even broader panel of apoptotic factors [165, 166].

4.3
FOXO Family

The forkhead transcription factor family (FKHR or FOXO) contains over 90 members that influence transcription in response to upstream Akt signaling [167, 168]. FOXO3a (also known as FKHR-L1) is a member of this family that has been linked to the regulation of apoptosis. One target of FOXO3a that has been identified is the BH3-only protein Bim [149]. The upregulation of Bim by FOXO3a has been shown to occur following a decrease in cytokine levels in T lymphocytes, possibly allowing the death of these cells when there is no longer an immune signal prompting their levels [169]. FOXO3a has also been shown to repress transcription of the caspase-8 inhibitor FLIP in response to a decrease in Akt signaling [170]. Akt is capable of phosphorylating FOXO3a, which leads to its cytoplasmic retention; hence, a decrease in Akt activation allows the dephosphorylated FOXO3a to translocate to the nucleus where it can exercise its transcriptional roles. The ability of FOXO3a to promote transcription of the death ligands TRAIL and Fas ligand in different cell systems [171–173], as well as of the BH3-only protein Puma, further emphasizes the involvement of this transcription factor in the regulation of apoptosis.

4.4
Additional Transcriptional Regulators of Apoptosis

One of the most widely studied stress response signaling pathways is that by which NF-κB is activated. Upon binding of TNF to the TNF receptor (TNFR), the proteins TRADD, RIP-1, and FADD associate with the receptor, and are capable of activating NF-κB. This survival transcription factor can then promote the expression of the anti-apoptotic factors cIAP1 and cIAP2 [174]. Expression of FLIP can also be regulated by NF-κB [175].

Other less-extensively studied transcription factors have also been described to influence the expression of apoptotic players. For example, in neurons the Brn-3a POU domain transcription factor can upregulate Bcl-2 in order to protect cells from neuronal apoptosis [176]. Bcl-6 has been shown to repress Bcl-x_L transcription following its own activation by AFX (ALL1 fused gene from chromosome X), which is another member of the FOXO family of transcription factors [177]. It is likely that many other factors also influence the expression of genes linked to apoptotic stress response. Furthermore, while the above-described networks influence the production of pro- and anti-apoptotic factors, investigations must still be conducted to determine the mechanisms that permit an altered transcription during the actual process of apoptotic cell death.

Importantly, in order to appreciate gene expression as a means of regulating cell death, transcription cannot be a sole focus of these investigations. The journey of a transcript (mRNA) following its production, up until translation is complete, includes several post-transcriptional regulatory steps at each of which expression modulation is possible.

5
Post-Transcriptional Regulation of Apoptosis

The production of factors involved in apoptosis is not only regulated by transcription. In fact, during the past few years many reports have described how post-transcriptional events can also influence the levels of pro- and anti-apoptotic factors [178–180]. The splicing, translation, and turnover of numerous mRNA factors have been investigated, and it has been revealed that these mechanisms may play important roles in regulating apoptotic cell death. Whilst, in general, these studies do not focus on post-transcriptional regulatory events during apoptosis, the insight that they provide on potential controls over cell death is still valuable.

An mRNA molecule is subject to several means of regulation before yielding the desired protein product (Fig. 3). Splicing of the message, potentially producing splice variants with differing function, can occur prior to nuclear export. The trafficking of mRNA to the cytoplasm, and then to an appropriate destination within the cytosol, will also influence gene expression. The recruitment of multiple ribosomes to a message results in the formation of multi-ribosome complexes known as *polysomes*, and the resulting translation of a message is subject to numerous cellular influences. Finally, the protein output from one mRNA copy depends on the stability of the transcript of interest, as mRNA turnover can also be enhanced or perturbed through the actions of a number of factors. Collectively, these mechanisms play a crucial part in regulating the expression and the resulting actions of factors involved in the apoptotic stress response.

5.1
Splicing

The splicing of mRNAs – and, specifically, alternative splicing – provides eukaryotes with an added level of regulation for gene expression and function. The processing of RNA transcripts following transcription involves the removal of introns in the pre-mRNA by the spliceosome, which is a complex of several small nuclear ribonucleoproteins (snRNPs) [180]. The spliceosome recognizes splice sites with the assistance of other RNA-binding proteins such as heteronuclear ribonucleoproteins (hnRNPs) and serine-arginine-rich proteins (SR proteins) [181, 182] (Fig. 3). Both SR proteins and hnRNPs can manipulate the location of splicing to yield different splice variants. While this would generally suggest that alternative splicing can result in different mRNAs encoding

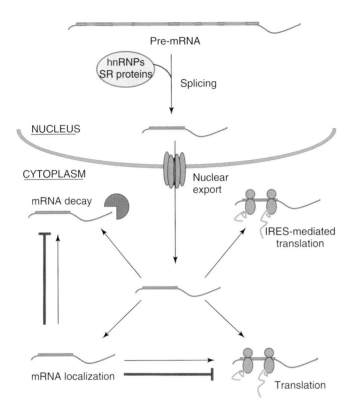

Fig. 3 Post-transcriptional regulation of gene expression. Gene expression may be regulated at several levels following transcription and addition of the 5′ methylguanylate cap and poly(A) tail. Within the nucleus, pre-mRNA undergoes splicing to remove introns, by use of hnRNPs and SR proteins. Mature mRNA is then exported into the cytoplasm. Once in the cytoplasmic compartment of the cell, transcripts can be localized to specific sites to enable localized translation (polarization), or repression of translation. Likewise, localization of an mRNA can promote or deter the decay of a message. The translation of an mRNA to produce protein product is also subject to regulation, and can be mediated through cap-dependent translation, or cap-independent translation, such as IRES-mediated translation. Finally, gene expression can also be regulated through the turnover of messages, as mRNA stability and decay can be influenced by a number of factors.

proteins with varying function, alternative splicing can also produce a "premature stop codon," which signals the decay of the target mRNA through a process known as nonsense-mediated decay (NMD) [180]. Hence, alternative splicing can influence not only the biological effects of a protein but also its global expression levels.

5.1.1 Splicing to Regulate Cytochrome c Release

Numerous factors involved in apoptosis have been identified as targets of alternative splicing [180], including death ligands, death receptors, caspases, members of the Bcl-2 family, and downstream caspase inhibitors and activators (much of this regulation has been summarized by Schwerk and Schulze-Osthoff [180]). Included in this list are the BH3-only proteins Bid and Bim. Indeed, it has been found that three splice variants of Bid are produced, each of which contains a different region of this pro-apoptotic factor [183]. In each of these three variants, the deletion or addition of domains elicits different effects. For example, Bid_S, which is a shorter version of Bid lacking its BH3 domain, inhibits the pro-apoptotic activity of tBid, while the other two extra long and extra short variants (Bid_{EL} and Bid_{ES}, respectively) promote apoptosis, although Bid_{ES} also appears to counter the effects of normal tBid [183]. While Bim has had numerous splice variants identified, three principal variants exist in the cells, as Bim_S, Bim_L, and Bim_{EL} [184–186]. Under normal conditions, Bim_{EL} is the most highly expressed, which correlates with the understanding that this isoform does not possess as strong a pro-apoptotic potential as do Bim_S and Bim_L [187]. During apoptosis, Bim_{EL} is actually degraded, while Bim_S and Bim_L are produced in larger quantities, thus realigning the purpose of Bim expression in these cells. These observations clearly suggest that alternative splicing could be active, and modulated, during the apoptotic response itself.

Perhaps the best-studied Bcl-2 family member with regards to splicing regulation is $Bcl-x_L$. The Bcl-x gene can produce two major variants, $Bcl-x_L$ and $Bcl-x_S$ [188]. As described earlier, $Bcl-x_L$ exercises an anti-apoptotic role as it can sequester Bax and Bak. By comparison, the $Bcl-x_S$ isoform lacks BH1 and BH2 domains and, as a consequence, can inhibit Bcl-2 and $Bcl-x_L$. The regulation controlling which Bcl-x isoform is produced depends on different cellular signals and cascades. For example, Fas activation can lead to a preferential production of $Bcl-x_S$ [189]. Differing signals will determine which SR proteins or hnRNPs are recruited to mediate the splicing of Bcl-x mRNA, and those selected for this event will govern which variant is generated. Recently, a mechanism was characterized to explain how DNA damage signals such a variation. Under normal conditions, signaling through PKC allows the SR protein SB1 to repress the 5′ splice site of $Bcl-x_S$ [190, 191], possibly by recruiting a repressor that has not yet been identified. In response to DNA damage, p53 signaling causes the removal of this repressor, permitting production of the apoptotic $Bcl-x_S$ isoform, and advancing cell death [191]. This is one of the first and only cases in which a mechanistic understanding of alternative splicing has been described in response to stress (Fig. 4).

Other regulators of MOMP have also been shown to be targets of regulation via alternative splicing. While two variants of Noxa were identified, the half-lives of their protein products supported the fact that they do not likely contribute to an effect in regulating apoptosis [192]. The pro-apoptotic Mcl-1 has had one alternative splice variant identified that contains only the third Bcl-2 homology domain (BH3) of Mcl-1 [193]. Not surprisingly, this variant promotes apoptosis similarly to BH3-only proteins [194].

5.1.2 Splicing-Mediated Regulation of Caspases

Caspases and downstream regulators of caspases are also subject to the regulation of gene expression through alternative splicing. Caspases-3 and -9 can be spliced to generate dominant negative isoforms that lack the catalytic sites of these enzymes [195–199], subsequently acting as anti-apoptotic factors. Regulation of the production of these splice variants has

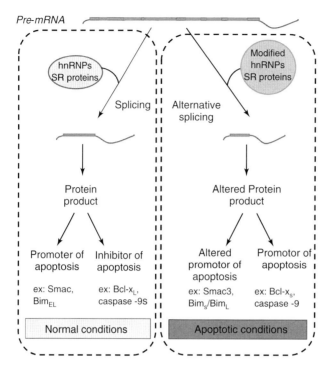

Fig. 4 Alternative splicing as a regulator of apoptosis. Several modulators of apoptosis are regulated at the post-transcriptional level of splicing. For certain of these factors, evidence suggests that alternative splicing produces protein variants that exercise altered effects on the progression of apoptosis. Under apoptotic conditions, alternate or modified splicing factors can produce variants that have opposing functions, or enhanced functions. References for the examples given may be found in the text.

been investigated, and in the case of caspase-9, has implicated hnRNP L as well as SR proteins such as SRp30a and SRSF1 [199–201]. Other caspases, such as -2 and -10, are also susceptible to alternative splicing, and as with what is seen for many of the proteins described above, opposite effects are mediated by these altered isoforms [180, 202, 203]. Consistently, certain alternate splicing isoforms of the caspase inhibitor survivin act in pro-apoptotic manners. In fact, four alternative splicing isoforms of survivin exist and, depending on the missing structural components, these can either inhibit the anti-apoptotic effect of full-length survivin or they can still succeed to inhibit cell death [204–208]. Importantly, mutations in p53, which regulates survivin expression, correlate with an increase in the production of anti-apoptotic survivin isoforms, without affecting the pro-apoptotic variants [209]. In the case of FLIP, the splice variant actually possesses a more potent effect than does the full-length protein [210], rather than producing an isoform with an opposing effect. Comparably, the Smac3 splice variant of Smac/DIABLO contributes to apoptosis by interfering with XIAP activity, similar to Smac/DIABLO, though possibly via a different mechanism [211].

The importance of splicing as a mechanism of post-transcriptional regulation that affects the apoptotic response has been emphasized by experiments in which an oligonucleotide successfully blocked the 5′ splice site of Bcl-x$_L$, resulting in an upregulation of the pro-apoptotic Bcl-x$_S$ [212, 213]. Recently, a similar approach has been used to induce death by enhancing expression of the pro-apoptotic isoform of Mcl-1 [194]. Clearly, this post-transcriptional process has important implications in regulating the factors that modulate apoptotic cell death. While some studies have described how stress signals can trigger alternative splicing to influence apoptosis [190, 191], the cellular mechanisms that enable differential regulation through this post-transcriptional event remain largely unexplored [180].

5.1.3 Nuclear Export of mRNA

Following transcription and the splicing of pre-mRNAs, a target message is covered with many RNA-binding proteins, several of which allow for the nuclear export of a message (Fig. 3). A handful of mRNA-export pathways exist [214–216], though each is under the control of regulatory mechanisms. Controlling the export of messages is another means of influencing gene expression, as mRNA translation occurs in the cytoplasmic compartment of the cell. To date however, there is no evidence supporting the involvement of the nuclear export of mRNAs encoding apoptotic players in the regulation of apoptosis. The conversion of RNA transcripts to polypeptides via translation is quite the opposite case, and many modulators of apoptosis may have their translation promoted or perturbed ,[178, 179]. This demonstrates that post-transcriptional mechanisms are conceivable targets for influencing apoptosis; hence, the possibility exists that in the future mRNA export will prove likewise to be targeted.

5.2
Translation

Considered the ultimate purpose of most mRNAs, translation converts genetic material into functional proteins that can execute countless cellular functions (Fig. 3). Whilst many of these functions in apoptosis have been described above, it is particularly important to appreciate *how* translation permits the production of these apoptotic factors, and how this imperative process is regulated. Similar to the case of alternative splicing, the expression of numerous apoptotic factors has been shown to undergo regulation at the level of translation, although the investigations made to date on this process during apoptosis are few in number [217–220].

5.2.1 *cis*-Elements and *trans*-Acting Factors that Regulate Translation

In order to specifically alter the translation of one message or one group of messages, regions in the 5′ or 3′ untranslated region (UTR) of the mRNA frequently encode *cis* elements to regulate this effect. This is certainly the case for the translation of Bcl-2, whereby the 5′ UTR of Bcl-2 mRNA has been shown to regulate its translation [221]; more recently, it was also shown that this involves the presence of a G-quadruplex element in the 5′ UTR of this message [222]. These authors noted that this element could reduce the translation of its message, potentially because of a need for helicases to enable an effective translation. It was not clear how the presence of this element factored into the regulation of Bcl-2 expression, however. It has also been shown that, under certain conditions, Bcl-x$_L$ and

Bcl-2 translation can be *increased*. Although the mechanism for this remains to be elucidated, Pardo *et al.* showed that *m*itogen-activated *e*xtracellularly regulated *k*inase *k*inase (MEK) can enhance the translation of both these targets, thus promoting cell survival [223].

p53 is also subject to translational regulation. The recruitment of p53 mRNA to heavy polysomes is inhibited by nucleolin and enhanced by the ribosomal protein L26 (RPL26) [217]. Importantly, in response to DNA damage, RPL26 relocalizes within the cell and binds to p53 mRNA, thus triggering an increase in p53 translation, while the inhibitory effect of nucleolin does not change, which suggests that this imbalance enables a greater p53 expression. The translation of Puma and Bad have also been shown to be enhanced in response to the activation of a kinase, MAP4K3, potentially through the activation of mTORC1, a promoter of translation [224]. How the translation of these messages is selectively enhanced compared to other messages is unclear, however. Similarly, the herbal compound Rocaglamide reduces FLIP translation by inhibiting the effect of the eukaryotic translation initiation factor (eIF) 4E [225]. However, the translation of other apoptosis-related factors was not decreased following treatment with this compound, indicating that more than one mechanism of action may be occurring.

The regulation of Mcl-1 via translation has been more mechanistically characterized. CUG triplet repeat RNA-binding protein 2 (CUGBP2) is an mRNA-binding protein which generally downregulates translation. Importantly, CUGBP2 was found to bind a U-rich sequence in the 3′ UTR of Mcl-1 mRNA, and inhibits the translation of this message [226]. As CUGBP2 expression is enhanced in response to ultraviolet (UV) and γ-irradiation [227], while simultaneously Mcl-1 protein expression decreases despite no major changes in mRNA steady-state levels, the inhibitory effect of CUGBP2 may be responsible for this effect [226].

5.2.2 HuR as a Regulator of Translation

Importantly, CUGBP2 is not the only mRNA-binding protein to target Mcl-1. Human antigen R (HuR), which can regulate the turnover and nuclear export of messages, also associates with Mcl-1 mRNA [179]. HuR is referred to as an A/U-rich element binding protein (AUBP), so-named for its association with mRNA elements rich in adenine and uracil (A/U-rich elements; AREs). While it remains to be seen how HuR and CUGBP2 may compete or collaborate in regulating the translation of Mcl-1 mRNA, one study has revealed that the activities of each of these two are linked [228]. Both, HuR and CUGBP2 can bind to the mRNA of cyclooxygenase-2 (COX-2), and while HuR promotes the translation of this message, CUGBP2 has an inhibitory effect, which even competitively inhibits the effect of HuR. This demonstrates that the interaction between mRNA-binding proteins in regulating gene expression at the post-transcriptional level is important.

Mcl-1 is but one of several apoptosis-related mRNAs with which HuR associates and regulates. In various studies, HuR has been found to associate with and regulate the translation of Prothymosin α, XIAP, cytochrome c, and p53 [218, 219, 229, 230]. Additionally, HuR can influence the mRNA stability of other factors, such as survivin, Bcl-2, and Hsp70, along with the deacetylase SIRT1 (NAD-dependent deacetylase sirtuin-1) which inhibits p53 and FOXO3a activity

[179, 231–233] (details of this stabilization are provided below). What is particularly important here regarding the studies on HuR as a regulator of translation, is that the effect it exercises varies from target to target. In response to UV light treatment, HuR enabled an increase in Prothymosin α recruitment to heavy polysomes and, consequentially, supported the translation of this message [218]. Intriguingly, this same UV stress also triggered an HuR-mediated promotion of the translation of p53 mRNA [219], which suggested that HuR can promote both pro- and anti-apoptotic factors. Data have also been acquired showing that HuR may influence translation by collaborating with other mRNA-binding proteins. In fact, the mRNA-binding protein hematopoietic zinc finger (Hzf) also associates with p53 mRNA, and cooperates with HuR to enhance the translation of this target [234]. In regulating cytochrome c translation, HuR competes with the RNA-binding protein TIA-1 (T-cell-restricted intracellular antigen-1) for association with this message [230]. Under normal conditions, HuR associates with the cytochrome c mRNA and promotes the translation of this message. In response to endoplasmic reticulum (ER) stress however, HuR loses association with this target, which may allow for a greater binding of TIA-1 to this message and a subsequent inhibition of TIA-1-mediated translation [230]. The role of HuR in promoting the expression of both pro- and anti-apoptotic players has sparked debate over whether HuR favors survival or death [235, 236]. This duality has been highlighted by observations that, during apoptosis, HuR is a substrate of caspases and that the cleavage products of HuR enhance apoptosis [236, 237]. To date, the characterization of this pro-apoptotic role for HuR has revealed an involvement of protein partners such as PHAPI [236, 238]. Currently, investigations are under way to determine whether the cleavage of HuR can also influence the mRNA-binding properties of this factor and, more importantly, if HuR can continue to influence the translation and stability of pro- and anti-apoptotic mRNAs.

5.2.3 IRES-Mediated Translation and Apoptosis

In the case of XIAP, HuR recruits this message to ribosomes directly in response to stress [229]. In this scenario, it is through binding an internal ribosomal entry site (IRES), and not an ARE, that HuR regulates its target. In fact, the possible role of IRESs in translation, particularly during apoptosis, has been extended; for example, it has been speculated that IRES-mediated translation may explain how certain mRNAs, such as cIAP1 and cIAP2 [239], are not only successfully translated in response to stress but also that their translation is *enhanced*. Following ER stress, PKR-like endoplasmic reticulum kinase (PERK) is activated and shuts off any general translation by phosphorylating the initiation factor eIF2α [240–242]. The presence of an IRES in cIAP1 suggests that IRES-mediated translation enables the selective enhancement of cIAP1 translation, despite this general reduction in cap-mediated translation [243].

Messages containing an IRES often have a complex 5′ UTR with secondary structure, and hence under normal conditions these elements most likely do not have any great effect on translation [244]. In specific instances where general translation is inhibited, however, IRES-mediated translation can become an important means of circumventing a global halt in translation

(Fig. 3). Considering that stress response and apoptotic signaling lead eventually to a cessation of general translation [245], it is not surprising that several mRNAs encoding factors linked to apoptotic cell death have been found to contain an IRES [229, 246, 247]. For example, Apaf-1 and Bcl-2 both contain an IRES. In response to stresses such as the chemotherapeutic etoposide or the inhibitor of oxidative phosphorylation arsenite, Bcl-2 translation increases due to the presence of this IRES [246]. In the case of Apaf-1, polypyrimidine tract binding protein (PTB) and upstream of N-ras protein (UNR), both of which are mRNA-binding proteins, are involved in enabling the translation of this pro-apoptotic factor in an IRES-dependent manner [247]. The role of these IRES *trans*-acting factors (ITAFs) likely involves the correct restructuring of their mRNA targets to allow for internal entry of the ribosome [248].

Interestingly, the process of apoptosis promotes IRES-mediated translation during the caspase cascade by activating ITAFs. Both, the initiation factor eIF4GI and death-associated protein 5 (DAP5) are subject to caspase-mediated cleavage [220]. In the case of eIF4G, this reduces general cap-dependent translation, yet cleavage of DAP5 *activates* this factor. The cleavage of DAP5 to yield DAP5/p86 removes an inhibitory region from this factor, and enhances IRES-mediated translation. This augments the translation of Apaf-1, XIAP, and c-myc, as well as DAP5 itself, as this factor contains an IRES and triggers its upregulation in a feedback loop [220].

While it remains to be seen if the mRNAs encoding other regulators and players in apoptosis contain IRESs, the regulation of translation during apoptosis does suggest the presence of intricate mechanisms to translate selective targets [245, 248]. The relatively recent discovery of IRES-mediated translation has helped to shed light on how stress response can alter the specific expression of mRNAs through translation (Fig. 5).

5.2.4 microRNAs as Regulators of the Translation of Apoptotic Factors

During recent years, another field that has been shown increasingly to influence the translation of many factors involved in apoptosis is that of interfering RNAs, such as microRNA (miRNA). miRNAs are produced following the transcription of specific noncoding regions of the genome to produce hairpin RNAs [249] which are processed through a defined series of pathways to produce short, single-stranded RNAs of approximately 22 nt in length. These miRNA products are each able to target one message (or more), based on their complementarity to a given mRNA sequence. On binding to their targets, the miRNAs may either induce translation inhibition or cause the decay of a bound mRNA [249]. Given the added level of complexity that miRNAs provide to the regulation of gene expression, many studies have recently been conducted to identify miRNAs that target specific mRNAs, and also to assess the nature of the targets of any given miRNA.

5.2.5 miRNAs as Inhibitors of Apoptosis

Numerous apoptotic players are targeted by miRNAs (Table 1) [250]; for example, the translation of Fas ligand and Fas receptor are inhibited by miR-21 and miR-23a, respectively [251–253]. The translation of FLIP mRNA is inhibited by miR-512-3p, and it was found that the DNA-damaging chemotherapeutic Taxol enhances expression of this miRNA to promote cell death through a reduction in FLIP protein [254]. miRNAs can also inhibit apoptosis in

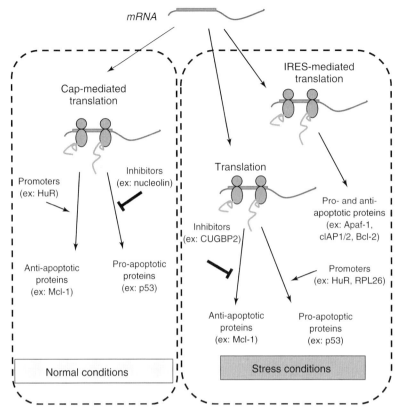

Fig. 5 Translation as a point of control over expression of apoptotic factors. Under normal conditions, the expression of anti-apoptotic factors is generally promoted, while that of pro-apoptotic factors is suppressed. This occurs at several levels, including for translation, whereby different factors influence the recruitment of mRNAs to translation machineries and the rate of translation. Under stress conditions, the translation of mRNAs can be altered to enable the expression of pro-apoptotic factors. Additionally, IRES-mediated translation occurs, which leads to cap-independent translation of both pro- and anti-apoptotic factors. Details and references for the examples given may be found in the text.

order to adequately regulate this death process. During morphogenesis of the eye, miR-24a has been shown to inhibit the expression of caspase-9 and Apaf-1, thus preventing cell death [255].

Given their potential to inhibit such a broad panel of apoptosis-related factors, it is not surprising that many miRNAs have been found to be upregulated in cancers. In therapy-resistant ovarian cancer, miR-125b expression is enhanced, which inhibits the translation of Bak [256]. Comparatively, in cervical carcinoma cells elevated levels of miR-886-5p inhibit Bax translation [257]. Puma is the target of suppression by miR-221 and miR-222 in glioblastomas, to inhibit apoptosis [258].

In the case of one important miRNA that is involved in regulating apoptosis, namely miR-21, several targets have been identified that include Bcl-2, p53, and Apaf-1 [259–261]. miR-21 overexpression

Tab. 1 microRNAs as post-transcriptional regulators of apoptotic factors miRNAs can influence gene expression by silencing the translation of target mRNAs, or by triggering the decay of certain messages. Numerous apoptotic factors are subject to regulation by miRNAs, as described in the text; these miRNAs are listed here based on whether they promote or inhibit apoptosis.

miRNA name	Target	Mechanism
Anti-apoptotic miRNAs		
miR-21	Bcl-2, Fas ligand, Apaf-1	Inhibit translation
miR-23a	Fas receptor	Inhibit translation
miR-24a	Caspase-9, Apaf-1	Inhibit translation
miR-125b	Bak	Inhibit translation
miR-221	Puma	Inhibit translation
miR-222	Puma	Inhibit translation
miR-512-3p	FLIP	Inhibit translation
miR-886-5p	Bax	Inhibit translation
let-7e	Caspase-3	Destabilize
Pro-apoptotic miRNAs		
let-7c	Bcl-x_L	Inhibit translation
let-7g	Bcl-x_L	Inhibit translation
miR-15a	Bcl-2	Inhibit translation
miR-16-1	Bcl-2	Inhibit translation
miR-29	Bcl-2, Mcl-1	Inhibit translation
miR-101	Mcl-1	Inhibit translation
miR-148a	Bcl-2	Inhibit translation
miR-193b	Mcl-1	Inhibit translation
miR-200bc/429 cluster	Bcl-2, XIAP	Inhibit translation
miR-203	Survivin	Inhibit translation
miR320a	Survivin	Inhibit translation
miR-494	Survivin	Inhibit translation
miR-497	Bcl-2	Inhibit translation
miR-122	Bcl-w	Destabilize
miR-542-3p	Survivin	Destabilize
miR-23a	XIAP	Destabilize

is detected in cancerous cells, and has been considered a marker for a poor prognosis in tongue squamous cell carcinomas [262, 263]. It has also been shown that the inhibition of miR-21 causes an increase in the expression of several apoptotic players, including several downstream targets of p53. This suggests that, by repressing the translation of p53, miR-21 may have a substantial inhibitory effect on apoptotic pathways [259]. In addition to p53, the targeting of specific factors such as Apaf-1 and Fas ligand by miR-21 proposes that this is a broad-scope regulator of apoptotic gene expression [251, 252, 259].

5.2.6 Pro-Apoptotic Roles of miRNA

miRNAs may also be used by a cell to enhance apoptosis [264]. In fact, several miRNAs have been shown to inhibit the translation of anti-apoptotic factors, including miR-29 and miR-148a [265,

266]. Not surprisingly, in cancers these miRNAs are in fact deregulated, thus causing an increase in the translation of anti-apoptotic factors such as Bcl-2 and Mcl-1 [265, 267, 268], as in the case of miR-29. miR-148a can also silence Bcl-2 expression, although in colorectal cancers it was found that the transcription factor MYB would bind and repress the production of miR-148a [266]. The expression levels of Bcl-2 and Mcl-1 are also regulated by several other miRNAs. For example, miR-193b and miR-101 repress Mcl-1 expression [269, 270], while miR-15a and miR-16-1 both inhibit Bcl-2 translation, with their downregulation being believed to protect chronic lymphocytic leukemia cells against apoptosis [271]. miR-497 and the miR-200bc/429 cluster also inhibit Bcl-2 expression [272, 273], with the latter also inhibiting the expression of XIAP. Beyond this, miR-7 was found to inhibit protein production and RNA levels for Bcl-2 mRNA, which suggests a possible role for miR-7 in mediating the destabilization of Bcl-2 message, and not simply inhibiting its translation [274]. Both, let-7c and let-7g are also implicated in death as they repress the expression of Bcl-x_L [275], as are miR-203, miR-494, and miR-320a by targeting survivin [276, 277].

The regulation of gene expression can involve RNA-binding proteins, IRES-mediated regulation, and miRNAs to modulate translation levels. Collectively, this ensemble creates a diverse series of regulators for this process, which clearly influences the expression of apoptotic players. Most importantly, as mRNAs can only be translated for as long as they are present, regulating the degradation of mRNA represents another means by which a cell can control protein expression (Fig. 3).

5.3 mRNA Turnover

Normal mRNA decay occurs by removing stabilizing structures from the message, such as the 5′ methyl guanosine cap or the 3′ poly(A) tail, after which exoribonucleases can access and degrade the message. Although the process of mRNA decay is well documented [278–280], certain messages are known to contain *cis* elements that influence the rate of decay, potentially conferring an increased or decreased stability on the mRNA in which they are contained [281, 282]. The influence on mRNA turnover of these *cis* elements depends on the *trans*-acting factors that associate with such elements, as both stabilizing and destabilizing mRNA-binding proteins exist. Controlling the stabilization and decay of specific messages encoding apoptotic players has proven to be an important mechanism for regulating the expression of these factors (Fig. 3), and the consequence of this on apoptosis has been described to varying degrees.

5.3.1 The Destabilizing A/U Rich Element

Among the elements that are known to influence mRNA decay rates, the most extensively studied is the ARE. Currently, several "classes" of AREs are known to exist [283]; these are based on the distribution of adenine and uracil nucleotides within the sequence, but they are generally located in the 3′ UTR of a message. AREs have been shown repeatedly to confer instability on mRNAs, which has often been demonstrated through their conjugation to an otherwise stable mRNA. For example, the initial studies aimed at characterizing an ARE involved linking the 3′ UTR of granulocyte-macrophage colony-stimulating factor (GM-CSF), which contains an ARE, to the stable mRNA of β-globin [284]. This resulted in a drastic reduction of the mRNA stability of the chimeric β-globin message, thus indicating the potency of these decay elements.

In seeking a greater understanding of how these elements confer instability, a number of AUBPs have been identified and described [282, 285]. The most frequently studied have included the destabilizing factors tristetraprolin (TTP) and KH-type splicing regulatory protein (KSRP), the stabilizer HuR, and ARE-binding factor 1 (AUF1), which has been described as both promoting and inhibiting mRNA stability, depending on the AUF1 splice variant [286, 287].

5.3.2 ARE-Mediated Turnover of Apoptosis-Related mRNAs

In terms of mRNAs encoding apoptotic factors, several studies have demonstrated the importance of mRNA turnover as a regulator of gene expression and the apoptotic response. Bim, Bcl-2, XIAP, survivin, and p53 are all subject to regulation by influences on the turnover of their mRNAs [231, 288–291]. Importantly, in each of these cases, AREs appear to play a role. In the case of Bim, heat shock protein cognate 70 (Hsc70) was found to bind the AREs of Bim mRNA, and to prevent its ARE-mediated decay [288]. In response to the release of cytokines such as interleukin (IL)-3 however, Hsc70 is sequestered to a separate protein complex, removing it from Bim mRNA and thereby enabling the decay of this message. The investigators speculated that this regulatory mechanism allowed for a cytokine-mediated survival through an inhibition of the anti-apoptotic activities of Bim [288].

The four AREs of Bcl-2 have been found to recruit more than one mRNA-binding protein. Before even determining which AUBPs may be responsible, the destabilizing AREs of Bcl-2 mRNA were found to enhance the decay of this message [289], and consequentially to reduce its expression, in response to PKC activation, as would be expected to occur during apoptosis. In response to UV radiation, the p45 isoform of the RNA-binding protein AUF1 (also known as *hnRNP D1*) localizes to the cytoplasm and associates with Bcl-2 mRNA via its AREs, and thus triggers a decrease in the half-life of this message [292]. Conversely, the mRNA-binding protein nucleolin was found to stabilize Bcl-2 mRNA under normal conditions, while in response to stress a phosphorylation of nucleolin may occur that could trigger its degradation, thereby removing its stabilizing role for Bcl-2 mRNA [293]. It has been proposed that these two AUBPs may work in opposition, in order to maintain a tight regulation on Bcl-2 expression (Fig. 6) [294]. Finally, HuR has also been found to associate with the AREs of Bcl-2 and to influence the mRNA levels of this message, although it is unclear whether HuR enhances the stability of this target, or inhibits it [295]. Furthermore, Bcl-2 protein appears to influence the fate of its respective mRNA, and may prevent HuR from stabilizing the Bcl-2 message [295]. The ambiguity surrounding these results implies that the coordinated activities of this collection of AUBPs must be considered.

HuR has also been identified as a regulator of other ARE-containing mRNAs that encode players in apoptosis. Both, XIAP and survivin are stabilized by HuR [231, 290], which suggests a role for HuR in promoting cell survival through an enhanced expression of the gene products of these mRNAs. Consideration must also be made for the fact that HuR was found to promote the stability of p53 mRNA

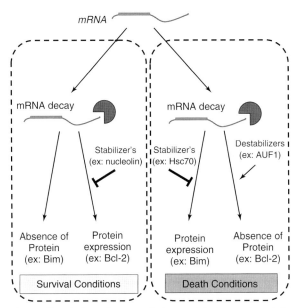

Fig. 6 Regulation of apoptosis by modulation of mRNA turnover. Certain protein regulators of apoptosis are regulated at the level of mRNA turnover. Certain messages contain destabilizing elements that lead to mRNA decay. If stabilizing proteins bind to a message, translation may occur to generate protein product. In the absence of a stabilizing protein, or upon binding of a destabilizing factor, the mRNA is subject to decay. This figure includes examples of these scenarios, with references included in the text.

[291], and that survivin is downregulated at the level of transcription by p53 [296]. Appropriately, the overexpression of HuR in fact decreases the steady-state levels of survivin mRNA, most likely through an enhancement of p53 expression [291]. In esophageal cancer cells lacking p53, however, the stabilizing effect of HuR on survivin mRNA enhances the expression of this potent anti-apoptotic factor, likely contributing to the apoptotic resistance of these cells [231].

5.3.3 Alternate Destabilizing Elements

AREs are not the only *cis* elements that have been shown to regulate the turnover of apoptotic mRNA. Specifically, Bcl-2 mRNA was found to be subject to destabilization by more than AREs [297, 298].

Located in the 3′ UTR of Bcl-2 mRNA is another *cis* element that is defined by repeats of CA nucleotides and also destabilizes Bcl-2 mRNA [297]; this latter effect was thought to result from the element's ability to recruit hnRNP L [298]. hnRNP L is a splicing factor that promotes Bcl-2 mRNA decay by assisting in the shortening of its poly(A) tail. This discovery highlighted the possibility that the regulation of apoptosis through influences on the turnover of mRNAs encoding its players may involve a number of factors and mechanisms which are, as yet, unidentified.

5.3.4 Destabilizing miRNAs

In the growing field of miRNA research, a variety of miRNAs have been implicated in regulating the turnover of messages encoding apoptotic factors. In addition to the above-described targets being subject to miRNA-mediated translation repression, several other miRNAs have been shown to influence the decay of mRNAs for apoptotic factors. For example, Bcl-w mRNA and protein levels are reduced when levels of miR-122 are elevated, which suggests that this miRNA inhibits Bcl-w expression at the level of mRNA stability, rather than at translation [299]. The same is true of the let-7e-mediated reduction of caspase-3 expression [300]. miR-542-3p may also be a regulator of mRNA stability, as it has been found to decrease the levels of survivin mRNA and protein [301], in similar fashion to the effect of miR-23a on XIAP mRNA and protein [302]. The role of miRNA in mediating a destabilization of messages is less well understood than their involvement in translation regulation; consequently, it is likely that other mechanisms of miRNA-mediated regulation of mRNA turnover will be discovered that impact on apoptotic cell death.

Collectively, several mechanisms of post-transcriptional regulation of gene expression have been found to regulate apoptotic factors. While this implies that apoptosis can likewise be influenced by these regulatory mechanisms, in many cases this has not been clearly established.

6
Discussion

6.1
Rearranging Deckchairs on the Titanic?

Apoptosis is an irreversible process that is engaged when a cell is committed to die, with activation of the apoptotic pathways, as described above, leading to an organized shutdown of cell function. It might be argued that, to modulate the expression of apoptotic factors during this collapse of the cell may not be relevant. Indeed, caspases lead to the specific activation of nucleases that degrade genomic DNA, in which case transcription may not be possible. Even mild stress can trigger an arrest in translation [281], and doing so would render any modulators of translation unnecessary. These concerns may explain why so few investigations have been undertaken to examine the regulation of expression of apoptotic factors during cell death.

Nonetheless, several studies have been initiated to address this concern, by suggesting that there is a relevance for post-transcriptional regulation as a modulator of apoptosis. As noted above, alterations of gene expression for apoptotic factors occur in response to different stresses. For example, treating cells with the apoptosis-inducing compound staurosporine triggers an alternative splicing of Bcl-x, to increase the relative amount of Bcl-x_S (pro-apoptotic) compared to Bcl-x_L (anti-apoptotic). A similar effect is seen when DNA damage is induced with oxaliplatin [190, 191]. When these studies were extended to investigate the nature of the endogenous mechanisms that could shift this balance, it was suggested that cells could indeed utilize post-transcriptional routes to influence cell death. Most importantly, in certain cancer cell lines staurosporine failed to trigger any enhancement in Bcl-x_S production [190], which may partly explain why such cells are resistant to chemotherapeutic agents.

The translation of p53 has also been found to be enhanced by apoptosis-inducing treatments. For example, Takagi *et al.* showed that 5 Gy of irradiation would produce an increase in p53 translation [217], while the treatment of cells with short-wavelength UV light (UVC) caused a similar effect [219]. Given the established role of p53 in promoting the transcription of numerous apoptotic factors, these data support the proposal that stresses can enhance the expression of apoptotic factors, at least initially. Intriguingly, UVC treatment also promoted recruitment of the mRNA encoding the anti-apoptotic factor Prothymosin α to polysomes, implying an enhancement in the translation of this factor [218]. This raises the possibility that a temporal progression may occur in response to stress whereby survival is supported initially; however, if a stress is too strong or becomes too potent (e.g., more prolonged UVC treatment), then apoptosis would be engaged. Data confirming that Bcl-2 translation is increased following stress treatments such as arsenite or etoposide also support this proposal [246]. Intriguingly, Bcl-2 translation under these conditions increases because of the presence of an IRES in Bcl-2 mRNA. Hence, Sherrill and coworkers proposed that this cap-independent translation would allow for the translation of Bcl-2 mRNA in spite of a stress signal, so that a cell would not immediately undergo apoptosis but rather would gauge its response [246]. Independently it was shown that, after treatment with ceramide, the stability of Bcl-2 mRNA was decreased [289]. These data led Sherrill et al. to suggest that these mechanisms of control over translation and turnover could collaborate to control Bcl-2 levels, depending on the duration and severity of stress.

The role of IRES-mediated translation in response to stress has received additional attention, as it is believed that this mechanism of translation will allow protein production even if general translation is halted [245]. The augmentation of protein levels of apoptotic factors during this process supports this model, while the fact that DAP5, an ITAF, becomes activated by caspase-mediated cleavage strengthens the notion [220].

6.2
All a Matter of Time

One noteworthy consideration that should be made in reviewing these data is the severity of the stresses employed. If a stress was sustainable and/or the cells were able to recover, then it was most likely insufficient to allow for conclusions to be drawn with regards to apoptosis. Rather, a more important consideration is the *timing* of these events; for example, a lethal stress such as a toxic compound or UV irradiation can induce death at different concentrations or exposure times. However, even if these stresses are lethal and lead to apoptotic cell death, studies should be focused on the progression of gene expression for apoptotic factors leading to the demise of a cell. Although translation may be halted during apoptosis, if such an arrest were to be delayed following the death stimulus and did not occur immediately, then there may be a sufficient opportunity to generate novel apoptotic protein factors, and thus, an equal opportunity to enhance or inhibit this process. Since mRNA decay is a perpetual process, the stabilization of a message – even during the early stages of apoptosis – would enable more mRNA molecules to be available for translation. Taken together, these possibilities suggest that the timing of gene expression modulation in apoptosis may be paramount for appreciating the roles of mRNA in apoptosis. Clearly, in order to proceed, investigations aimed at linking mRNA and apoptosis should pay equal attention to "when" as they should to "what."

Some evidence has suggested that mRNA regulation in apoptosis does occur, and can contribute to the apoptotic phenotype. However, substantially more data exist showing that apoptotic factors are subject to post-transcriptional regulatory mechanisms. It remains to be seen how many of these may occur in situations of lethal stress, and hence may contribute actually and/or practically to committing a cell to die. If they do contribute, then post-transcriptional mechanisms may provide additional points of check and balance to a system that is also tightly regulated through transcription, post-translational modifications, and various protein–protein interactions with activators and inhibitors. It is clear that apoptotic cell death makes use of many factors to determine cell fate, and that each of these factors is regulated in order to provide a tight control on this cellular process. Currently, numerous diseases have been identified that exploit such regulation to produce an abundance or deficiency of apoptosis, yielding tremendous physiological consequences [303–305]. Organized cell suicide is thought to exist in order to enable the controlled destruction of cells. However, given the substantial importance of such a role it is not surprising that an extensive network of mechanisms is employed to regulate the process, and that much attention has been focused on elucidating the protein regulators of apoptotic death. Nonetheless, a growing pool of evidence suggests that it is the mRNA precursors of these proteins that must be borne in mind when attempting to visualize the full picture of regulated death.

Acknowledgments

The authors are grateful for the thorough and critical review of the chapter conducted by Dr Sergio Di Marco. Imed-Eddine Gallouzi is the recipient of a Tier II Canada Research Chair.

References

1 Yuan, J., Kroemer, G. (2010) Alternative cell death mechanisms in development and beyond. *Genes Dev.*, **24**, 2592–2602.
2 Galluzzi, L., Vanden Berghe, T., Vanlangenakker, N., Buettner, S. *et al.* (2011) Programmed necrosis from molecules to health and disease. *Int. Rev. Cell Mol. Biol.*, **289**, 1–35.
3 Orrenius, S., Nicotera, P., Zhivotovsky, B. (2011) Cell death mechanisms and their implications in toxicology. *Toxicol. Sci.*, **119**, 3–19.
4 Vandenabeele, P., Galluzzi, L., Vanden Berghe, T., Kroemer, G. (2010) Molecular mechanisms of necroptosis: an ordered cellular explosion. *Nat. Rev. Mol. Cell Biol.*, **11**, 700–714.
5 Schweichel, J.U., Merker, H.J. (1973) The morphology of various types of cell death in prenatal tissues. *Teratology*, **7**, 253–266.
6 Levine, B., Yuan, J. (2005) Autophagy in cell death: an innocent convict? *J. Clin. Invest.*, **115**, 2679–2688.
7 Diaz, L.F., Chiong, M., Quest, A.F., Lavandero, S. *et al.* (2005) Mechanisms of cell death: molecular insights and therapeutic perspectives. *Cell Death Differ.*, **12**, 1449–1456.
8 Kerr, J.F., Wyllie, A.H., Currie, A.R. (1972) Apoptosis: a basic biological phenomenon with wide-ranging implications in tissue kinetics. *Br. J. Cancer*, **26**, 239–257.
9 Janssen, O., Qian, J., Linkermann, A., Kabelitz, D. (2003) CD95 ligand – death factor and costimulatory molecule? *Cell Death Differ.*, **10**, 1215–1225.
10 Ashkenazi, A. (2002) Targeting death and decoy receptors of the tumour-necrosis factor superfamily. *Nat. Rev. Cancer*, **2**, 420–430.
11 Gentile, M., Latonen, L., Laiho, M. (2003) Cell cycle arrest and apoptosis provoked by UV radiation-induced DNA damage are transcriptionally highly divergent responses. *Nucleic Acids Res.*, **31**, 4779–4790.
12 Stevenson, M.A., Pollock, S.S., Coleman, C.N., Calderwood, S.K. (1994) X-irradiation, phorbol esters, and H_2O_2

stimulate mitogen-activated protein kinase activity in NIH-3T3 cells through the formation of reactive oxygen intermediates. *Cancer Res.*, **54**, 12–15.

13 Ola, M.S., Nawaz, M., Ahsan, H. (2011) Role of Bcl-2 family proteins and caspases in the regulation of apoptosis. *Mol. Cell. Biochem.*, **351**, 41–58.

14 Riedl, S.J., Salvesen, G.S. (2007) The apoptosome: signalling platform of cell death. *Nat. Rev. Mol. Cell Biol.*, **8**, 405–413.

15 Alnemri, E.S., Livingston, D.J., Nicholson, D.W., Salvesen, G. et al. (1996) Human ICE/CED-3 protease nomenclature. *Cell*, **87**, 171.

16 Sakamaki, K., Satou, Y. (2009) Caspases: evolutionary aspects of their functions in vertebrates. *J. Fish Biol.*, **74**, 727–753.

17 Riedl, S.J., Shi, Y. (2004) Molecular mechanisms of caspase regulation during apoptosis. *Nat. Rev. Mol. Cell Biol.*, **5**, 897–907.

18 Pan, G., Humke, E.W., Dixit, V.M. (1998) Activation of caspases triggered by cytochrome c in vitro. *FEBS Lett.*, **426**, 151–154.

19 Li, P., Nijhawan, D., Budihardjo, I., Srinivasula, S.M. et al. (1997) Cytochrome c and dATP-dependent formation of Apaf-1/caspase-9 complex initiates an apoptotic protease cascade. *Cell*, **91**, 479–489.

20 Demon, D., Van Damme, P., Vanden Berghe, T., Vandekerckhove, J. et al. (2009) Caspase substrates: easily caught in deep waters? *Trends Biotechnol.*, **27**, 680–688.

21 Yi, C.H., Yuan, J. (2009) The Jekyll and Hyde functions of caspases. *Dev. Cell*, **16**, 21–34.

22 LeBlanc, H.N., Ashkenazi, A. (2003) Apo2L/TRAIL and its death and decoy receptors. *Cell Death Differ.*, **10**, 66–75.

23 Wiley, S.R., Schooley, K., Smolak, P.J., Din, W.S. et al. (1995) Identification and characterization of a new member of the TNF family that induces apoptosis. *Immunity*, **3**, 673–682.

24 Boldin, M.P., Goncharov, T.M., Goltsev, Y.V., Wallach, D. (1996) Involvement of MACH, a novel MORT1/FADD-interacting protease, in Fas/APO-1- and TNF receptor-induced cell death. *Cell*, **85**, 803–815.

25 Scaffidi, C., Kirchhoff, S., Krammer, P.H., Peter, M.E. (1999) Apoptosis signaling in lymphocytes. *Curr. Opin. Immunol.*, **11**, 277–285.

26 Medema, J.P., Scaffidi, C., Kischkel, F.C., Shevchenko, A. et al. (1997) FLICE is activated by association with the CD95 death-inducing signaling complex (DISC). *EMBO J.*, **16**, 2794–2804.

27 Baker, S.J., Reddy, E.P. (1998) Modulation of life and death by the TNF receptor superfamily. *Oncogene*, **17**, 3261–3270.

28 Ashkenazi, A., Dixit, V.M. (1998) Death receptors: signaling and modulation. *Science*, **281**, 1305–1308.

29 Shiloh, Y. (2001) ATM and ATR: networking cellular responses to DNA damage. *Curr. Opin. Genet. Dev.*, **11**, 71–77.

30 Kubbutat, M.H., Jones, S.N., Vousden, K.H. (1997) Regulation of p53 stability by Mdm2. *Nature*, **387**, 299–303.

31 Haupt, Y., Maya, R., Kazaz, A., Oren, M. (1997) Mdm2 promotes the rapid degradation of p53. *Nature*, **387**, 296–299.

32 Siliciano, J.D., Canman, C.E., Taya, Y., Sakaguchi, K. et al. (1997) DNA damage induces phosphorylation of the amino terminus of p53. *Genes Dev.*, **11**, 3471–3481.

33 Khosravi, R., Maya, R., Gottlieb, T., Oren, M. et al. (1999) Rapid ATM-dependent phosphorylation of MDM2 precedes p53 accumulation in response to DNA damage. *Proc. Natl Acad. Sci. USA*, **96**, 14973–14977.

34 Villani, G., Attardi, G. (1997) In vivo control of respiration by cytochrome c oxidase in wild-type and mitochondrial DNA mutation-carrying human cells. *Proc. Natl Acad. Sci. USA*, **94**, 1166–1171.

35 Villani, G., Greco, M., Papa, S., Attardi, G. (1998) Low reserve of cytochrome c oxidase capacity in vivo in the respiratory chain of a variety of human cell types. *J. Biol. Chem.*, **273**, 31829–31836.

36 Huttemann, M., Pecina, P., Rainbolt, M., Sanderson, T.H. et al. (2011) The multiple functions of cytochrome c and their regulation in life and death decisions of the mammalian cell: from respiration to apoptosis. *Mitochondrion*, **11**, 369–381.

37 Martinou, J.C., Youle, R.J. (2011) Mitochondria in apoptosis: Bcl-2 family members and mitochondrial dynamics. *Dev. Cell*, **21**, 92–101.

38. Zou, H., Henzel, W.J., Liu, X., Lutschg, A. et al. (1997) Apaf-1, a human protein homologous to *C. elegans* CED-4, participates in cytochrome c-dependent activation of caspase-3. *Cell*, **90**, 405–413.
39. Acehan, D., Jiang, X., Morgan, D.G., Heuser, J.E. et al. (2002) Three-dimensional structure of the apoptosome: implications for assembly, procaspase-9 binding, and activation. *Mol. Cell*, **9**, 423–432.
40. Boatright, K.M., Renatus, M., Scott, F.L., Sperandio, S. et al. (2003) A unified model for apical caspase activation. *Mol. Cell*, **11**, 529–541.
41. Riedl, S.J., Li, W., Chao, Y., Schwarzenbacher, R. et al. (2005) Structure of the apoptotic protease-activating factor 1 bound to ADP. *Nature*, **434**, 926–933.
42. Nakanishi, K., Maruyama, M., Shibata, T., Morishima, N. (2001) Identification of a caspase-9 substrate and detection of its cleavage in programmed cell death during mouse development. *J. Biol. Chem.*, **276**, 41237–41244.
43. Li, H., Zhu, H., Xu, C.J., Yuan, J. (1998) Cleavage of BID by caspase 8 mediates the mitochondrial damage in the Fas pathway of apoptosis. *Cell*, **94**, 491–501.
44. Wang, K., Yin, X.M., Chao, D.T., Milliman, C.L. et al. (1996) BID: a novel BH3 domain-only death agonist. *Genes Dev.*, **10**, 2859–2869.
45. Tewari, M., Quan, L.T., O'Rourke, K., Desnoyers, S. et al. (1995) Yama/CPP32 beta, a mammalian homolog of CED-3, is a CrmA-inhibitable protease that cleaves the death substrate poly(ADP-ribose) polymerase. *Cell*, **81**, 801–809.
46. Nicholson, D.W., Ali, A., Thornberry, N.A., Vaillancourt, J.P. et al. (1995) Identification and inhibition of the ICE/CED-3 protease necessary for mammalian apoptosis. *Nature*, **376**, 37–43.
47. Liu, X., Zou, H., Slaughter, C., Wang, X. (1997) DFF, a heterodimeric protein that functions downstream of caspase-3 to trigger DNA fragmentation during apoptosis. *Cell*, **89**, 175–184.
48. Liu, X., Li, P., Widlak, P., Zou, H., Luo, X. et al. (1998) The 40-kDa subunit of DNA fragmentation factor induces DNA fragmentation and chromatin condensation during apoptosis. *Proc. Natl Acad. Sci. USA*, **95**, 8461–8466.
49. Enari, M., Sakahira, H., Yokoyama, H., Okawa, K. et al. (1998) A caspase-activated DNase that degrades DNA during apoptosis, and its inhibitor ICAD. *Nature*, **391**, 43–50.
50. Sakahira, H., Enari, M., Nagata, S. (1998) Cleavage of CAD inhibitor in CAD activation and DNA degradation during apoptosis. *Nature*, **391**, 96–99.
51. Chen, L., Marechal, V., Moreau, J., Levine, A.J. et al. (1997) Proteolytic cleavage of the mdm2 oncoprotein during apoptosis. *J. Biol. Chem.*, **272**, 22966–22973.
52. Emoto, Y., Manome, Y., Meinhardt, G., Kisaki, H. et al. (1995) Proteolytic activation of protein kinase C delta by an ICE-like protease in apoptotic cells. *EMBO J.*, **14**, 6148–6156.
53. Mashima, T., Naito, M., Fujita, N., Noguchi, K. et al. (1995) Identification of actin as a substrate of ICE and an ICE-like protease and involvement of an ICE-like protease but not ICE in VP-16-induced U937 apoptosis. *Biochem. Biophys. Res. Commun.*, **217**, 1185–1192.
54. Rao, L., Perez, D., White, E. (1996) Lamin proteolysis facilitates nuclear events during apoptosis. *J. Cell Biol.*, **135**, 1441–1455.
55. Kothakota, S., Azuma, T., Reinhard, C., Klippel, A. et al. (1997) Caspase-3-generated fragment of gelsolin: effector of morphological change in apoptosis. *Science*, **278**, 294–298.
56. Satoh, M.S., Lindahl, T. (1992) Role of poly(ADP-ribose) formation in DNA repair. *Nature*, **356**, 356–358.
57. Lazebnik, Y.A., Kaufmann, S.H., Desnoyers, S., Poirier, G.G. et al. (1994) Cleavage of poly(ADP-ribose) polymerase by a proteinase with properties like ICE. *Nature*, **371**, 346–347.
58. Slee, E.A., Harte, M.T., Kluck, R.M., Wolf, B.B. et al. (1999) Ordering the cytochrome c-initiated caspase cascade: hierarchical activation of caspases-2, -3, -6, -7, -8, and -10 in a caspase-9-dependent manner. *J. Cell Biol.*, **144**, 281–292.
59. Erwig, L.P., Henson, P.M. (2008) Clearance of apoptotic cells by phagocytes. *Cell Death Differ.*, **15**, 243–250.
60. Kantari, C., Walczak, H. (2011) Caspase-8 and bid: caught in the act between death receptors and mitochondria. *Biochim. Biophys. Acta*, **1813**, 558–563.

61 Wei, M.C., Zong, W.X., Cheng, E.H., Lindsten, T. et al. (2001) Proapoptotic BAX and BAK: a requisite gateway to mitochondrial dysfunction and death. *Science*, **292**, 727–730.

62 Youle, R.J., Strasser, A. (2008) The BCL-2 protein family: opposing activities that mediate cell death. *Nat. Rev. Mol. Cell Biol.*, **9**, 47–59.

63 Kvansakul, M., Yang, H., Fairlie, W.D., Czabotar, P.E. et al. (2008) Vaccinia virus anti-apoptotic F1L is a novel Bcl-2-like domain-swapped dimer that binds a highly selective subset of BH3-containing death ligands. *Cell Death Differ.*, **15**, 1564–1571.

64 Yang, E., Zha, J., Jockel, J., Boise, L.H. et al. (1995) Bad, a heterodimeric partner for Bcl-XL and Bcl-2, displaces Bax and promotes cell death. *Cell*, **80**, 285–291.

65 Oda, E., Ohki, R., Murasawa, H., Nemoto, J. et al. (2000) Noxa, a BH3-only member of the Bcl-2 family and candidate mediator of p53-induced apoptosis. *Science*, **288**, 1053–1058.

66 Elangovan, B., Chinnadurai, G. (1997) Functional dissection of the pro-apoptotic protein Bik. Heterodimerization with anti-apoptosis proteins is insufficient for induction of cell death. *J. Biol. Chem.*, **272**, 24494–24498.

67 Eskes, R., Desagher, S., Antonsson, B., Martinou, J.C. (2000) Bid induces the oligomerization and insertion of Bax into the outer mitochondrial membrane. *Mol. Cell. Biol.*, **20**, 929–935.

68 Yu, J., Zhang, L., Hwang, P.M., Kinzler, K.W. et al. (2001) PUMA induces the rapid apoptosis of colorectal cancer cells. *Mol. Cell*, **7**, 673–682.

69 Zhang, L., Lopez, H., George, N.M., Liu, X. et al. (2011) Selective involvement of BH3-only proteins and differential targets of Noxa in diverse apoptotic pathways. *Cell Death Differ.*, **18**, 864–873.

70 Schmelzle, T., Mailleux, A.A., Overholtzer, M., Carroll, J.S. et al. (2007) Functional role and oncogene-regulated expression of the BH3-only factor Bmf in mammary epithelial anoikis and morphogenesis. *Proc. Natl Acad. Sci. USA*, **104**, 3787–3792.

71 Zong, W.X., Lindsten, T., Ross, A.J., MacGregor, G.R. et al. (2001) BH3-only proteins that bind pro-survival Bcl-2 family members fail to induce apoptosis in the absence of Bax and Bak. *Genes Dev.*, **15**, 1481–1486.

72 Holinger, E.P., Chittenden, T., Lutz, R.J. (1999) Bak BH3 peptides antagonize Bcl-xL function and induce apoptosis through cytochrome c-independent activation of caspases. *J. Biol. Chem.*, **274**, 13298–13304.

73 Tzung, S.P., Kim, K.M., Basanez, G., Giedt, C.D. et al. (2001) Antimycin A mimics a cell-death-inducing Bcl-2 homology domain 3. *Nat. Cell Biol.*, **3**, 183–191.

74 Tse, C., Shoemaker, A.R., Adickes, J., Anderson, M.G. et al. (2008) ABT-263: a potent and orally bioavailable Bcl-2 family inhibitor. *Cancer Res.*, **68**, 3421–3428.

75 Zhai, D., Jin, C., Satterthwait, A.C., Reed, J.C. (2006) Comparison of chemical inhibitors of antiapoptotic Bcl-2-family proteins. *Cell Death Differ.*, **13**, 1419–1421.

76 Nguyen, M., Marcellus, R.C., Roulston, A., Watson, M. et al. (2007) Small molecule obatoclax (GX15-070) antagonizes MCL-1 and overcomes MCL-1-mediated resistance to apoptosis. *Proc. Natl Acad. Sci. USA*, **104**, 19512–19517.

77 Vogler, M., Weber, K., Dinsdale, D., Schmitz, I. et al. (2009) Different forms of cell death induced by putative BCL2 inhibitors. *Cell Death Differ.*, **16**, 1030–1039.

78 Shamas-Din, A., Brahmbhatt, H., Leber, B., Andrews, D.W. (2011) BH3-only proteins: Orchestrators of apoptosis. *Biochim. Biophys. Acta*, **1813**, 508–520.

79 Junn, E., Taniguchi, H., Jeong, B.S., Zhao, X. et al. (2005) Interaction of DJ-1 with Daxx inhibits apoptosis signal-regulating kinase 1 activity and cell death. *Proc. Natl Acad. Sci. USA*, **102**, 9691–9696.

80 Fu, K., Ren, H., Wang, Y., Fei, E. et al. (2012) DJ-1 inhibits TRAIL-induced apoptosis by blocking pro-caspase-8 recruitment to FADD. *Oncogene*, **31**, 1311–1322.

81 Irmler, M., Thome, M., Hahne, M., Schneider, P. et al. (1997) Inhibition of death receptor signals by cellular FLIP. *Nature*, **388**, 190–195.

82 Thome, M., Schneider, P., Hofmann, K., Fickenscher, H. et al. (1997) Viral FLICE-inhibitory proteins (FLIPs) prevent apoptosis induced by death receptors. *Nature*, **386**, 517–521.

83 Jiang, X., Kim, H.E., Shu, H., Zhao, Y. et al. (2003) Distinctive roles of PHAP proteins and prothymosin-alpha in a death regulatory pathway. *Science*, **299**, 223–226.

84 Qi, X., Wang, L., Du, F. (2010) Novel small molecules relieve prothymosin alpha-mediated inhibition of apoptosome formation by blocking its interaction with Apaf-1. *Biochemistry*, **49**, 1923–1930.

85 Dominguez, F., Magdalena, C., Cancio, E., Roson, E. *et al.* (1993) Tissue concentrations of prothymosin alpha: a novel proliferation index of primary breast cancer. *Eur. J. Cancer*, **29A**, 893–897.

86 Tsitsiloni, O.E., Stiakakis, J., Koutselinis, A., Gogas, J. et al. (1993) Expression of alpha-thymosins in human tissues in normal and abnormal growth. *Proc. Natl Acad. Sci. USA*, **90**, 9504–9507.

87 Smith, M.R., al-Katib, A., Mohammad, R., Silverman, A. *et al.* (1993) Prothymosin alpha gene expression correlates with proliferation, not differentiation, of HL-60 cells. *Blood*, **82**, 1127–1132.

88 Szabo, P., Ehleiter, D., Whittington, E., Weksler, M.E. (1992) Prothymosin alpha expression occurs during G1 in proliferating B or T lymphocytes. *Biochem. Biophys. Res. Commun.*, **185**, 953–959.

89 Sburlati, A.R., Manrow, R.E., Berger, S.L. (1991) Prothymosin alpha antisense oligomers inhibit myeloma cell division. *Proc. Natl Acad. Sci. USA*, **88**, 253–257.

90 Chen, T.H., Brody, J.R., Romantsev, F.E., Yu, J.G. *et al.* (1996) Structure of pp32, an acidic nuclear protein which inhibits oncogene- induced formation of transformed foci. *Mol. Biol. Cell*, **7**, 2045–2056.

91 Brody, J.R., Kadkol, S.S., Mahmoud, M.A., Rebel, J.M. *et al.* (1999) Identification of sequences required for inhibition of oncogene- mediated transformation by pp32. *J. Biol. Chem.*, **274**, 20053–20055.

92 Bai, J., Brody, J.R., Kadkol, S.S., Pasternack, G.R. (2001) Tumor suppression and potentiation by manipulation of pp32 expression. *Oncogene*, **20**, 2153–2160.

93 Kim, H.E., Jiang, X., Du, F., Wang, X. (2008) PHAPI, CAS, and Hsp70 promote apoptosome formation by preventing Apaf-1 aggregation and enhancing nucleotide exchange on Apaf-1. *Mol. Cell*, **30**, 239–247.

94 Lopez, J., Meier, P. (2010) To fight or die – inhibitor of apoptosis proteins at the crossroad of innate immunity and death. *Curr. Opin. Cell Biol.*, **22**, 872–881.

95 Goyal, L. (2001) Cell death inhibition: keeping caspases in check. *Cell*, **104**, 805–808.

96 Duckett, C.S., Nava, V.E., Gedrich, R.W., Clem, R.J. *et al.* (1996) A conserved family of cellular genes related to the baculovirus iap gene and encoding apoptosis inhibitors. *EMBO J.*, **15**, 2685–2694.

97 Galban, S., Duckett, C.S. (2010) XIAP as a ubiquitin ligase in cellular signaling. *Cell Death Differ.*, **17**, 54–60.

98 Deveraux, Q.L., Reed, J.C. (1999) IAP family proteins – suppressors of apoptosis. *Genes Dev.*, **13**, 239–252.

99 Sun, C., Cai, M., Gunasekera, A.H., Meadows, R.P. *et al.* (1999) NMR structure and mutagenesis of the inhibitor-of-apoptosis protein XIAP. *Nature*, **401**, 818–822.

100 Huang, Y., Park, Y.C., Rich, R.L., Segal, D. *et al.* (2001) Structural basis of caspase inhibition by XIAP: differential roles of the linker versus the BIR domain. *Cell*, **104**, 781–790.

101 Riedl, S.J., Renatus, M., Schwarzenbacher, R., Zhou, Q. *et al.* (2001) Structural basis for the inhibition of caspase-3 by XIAP. *Cell*, **104**, 791–800.

102 Chai, J., Shiozaki, E., Srinivasula, S.M., Wu, Q. *et al.* (2001) Structural basis of caspase-7 inhibition by XIAP. *Cell*, **104**, 769–780.

103 Marusawa, H., Matsuzawa, S., Welsh, K., Zou, H. *et al.* (2003) HBXIP functions as a cofactor of survivin in apoptosis suppression. *EMBO J.*, **22**, 2729–2740.

104 Kelly, R.J., Lopez-Chavez, A., Citrin, D., Janik, J.E. *et al.* (2011) Impacting tumor cell-fate by targeting the inhibitor of apoptosis protein survivin. *Mol. Cancer*, **10**, 35.

105 Suzuki, Y., Nakabayashi, Y., Takahashi, R. (2001) Ubiquitin-protein ligase activity of X-linked inhibitor of apoptosis protein promotes proteasomal degradation of caspase-3 and enhances its anti-apoptotic effect in Fas-induced cell death. *Proc. Natl Acad. Sci. USA*, **98**, 8662–8667.

106 Schile, A.J., Garcia-Fernandez, M., Steller, H. (2008) Regulation of apoptosis

by XIAP ubiquitin-ligase activity. *Genes Dev.*, **22**, 2256–2266.

107. Choi, Y.E., Butterworth, M., Malladi, S., Duckett, C.S. et al. (2009) The E3 ubiquitin ligase cIAP1 binds and ubiquitinates caspase-3 and -7 via unique mechanisms at distinct steps in their processing. *J. Biol. Chem.*, **284**, 12772–12782.

108. Suzuki, Y., Imai, Y., Nakayama, H., Takahashi, K. et al. (2001) A serine protease, HtrA2, is released from the mitochondria and interacts with XIAP, inducing cell death. *Mol. Cell*, **8**, 613–621.

109. Martins, L.M., Iaccarino, I., Tenev, T., Gschmeissner, S. et al. (2002) The serine protease Omi/HtrA2 regulates apoptosis by binding XIAP through a reaper-like motif. *J. Biol. Chem.*, **277**, 439–444.

110. Green, D.R. (2000) Apoptotic pathways: paper wraps stone blunts scissors. *Cell*, **102**, 1–4.

111. Kominsky, D.J., Bickel, R.J., Tyler, K.L. (2002) Reovirus-induced apoptosis requires mitochondrial release of Smac/DIABLO and involves reduction of cellular inhibitor of apoptosis protein levels. *J. Virol.*, **76**, 11414–11424.

112. Huang, Y., Rich, R.L., Myszka, D.G., Wu, H. (2003) Requirement of both the second and third BIR domains for the relief of X-linked inhibitor of apoptosis protein (XIAP)-mediated caspase inhibition by Smac. *J. Biol. Chem.*, **278**, 49517–49522.

113. Datta, S.R., Katsov, A., Hu, L., Petros, A. et al. (2000) 14-3-3 proteins and survival kinases cooperate to inactivate BAD by BH3 domain phosphorylation. *Mol. Cell*, **6**, 41–51.

114. Harada, H., Becknell, B., Wilm, M., Mann, M. et al. (1999) Phosphorylation and inactivation of BAD by mitochondria-anchored protein kinase A. *Mol. Cell*, **3**, 413–422.

115. Kobayashi, S., Lee, S.H., Meng, X.W., Mott, J.L. et al. (2007) Serine 64 phosphorylation enhances the antiapoptotic function of Mcl-1. *J. Biol. Chem.*, **282**, 18407–18417.

116. Upreti, M., Galitovskaya, E.N., Chu, R., Tackett, A.J. et al. (2008) Identification of the major phosphorylation site in Bcl-xL induced by microtubule inhibitors and analysis of its functional significance. *J. Biol. Chem.*, **283**, 35517–35525.

117. Deng, X., Gao, F., Flagg, T., May, W.S. Jr (2004) Mono- and multisite phosphorylation enhances Bcl2's antiapoptotic function and inhibition of cell cycle entry functions. *Proc. Natl Acad. Sci. USA*, **101**, 153–158.

118. Kodama, Y., Taura, K., Miura, K., Schnabl, B. et al. (2009) Antiapoptotic effect of c-Jun N-terminal Kinase-1 through Mcl-1 stabilization in TNF-induced hepatocyte apoptosis. *Gastroenterology*, **136**, 1423–1434.

119. Maurer, U., Charvet, C., Wagman, A.S., Dejardin, E. et al. (2006) Glycogen synthase kinase-3 regulates mitochondrial outer membrane permeabilization and apoptosis by destabilization of MCL-1. *Mol. Cell*, **21**, 749–760.

120. Kitazumi, I., Tsukahara, M. (2011) Regulation of DNA fragmentation: the role of caspases and phosphorylation. *FEBS J*, **278**, 427–441.

121. Kurokawa, M., Kornbluth, S. (2009) Caspases and kinases in a death grip. *Cell*, **138**, 838–854.

122. Cardone, M.H., Roy, N., Stennicke, H.R., Salvesen, G.S. et al. (1998) Regulation of cell death protease caspase-9 by phosphorylation. *Science*, **282**, 1318–1321.

123. Alvarado-Kristensson, M., Andersson, T. (2005) Protein phosphatase 2A regulates apoptosis in neutrophils by dephosphorylating both p38 MAPK and its substrate caspase 3. *J. Biol. Chem.*, **280**, 6238–6244.

124. Alvarado-Kristensson, M., Melander, F., Leandersson, K., Ronnstrand, L. et al. (2004) p38-MAPK signals survival by phosphorylation of caspase-8 and caspase-3 in human neutrophils. *J. Exp. Med.*, **199**, 449–458.

125. Vucic, D., Dixit, V.M., Wertz, I.E. (2011) Ubiquitylation in apoptosis: a post-translational modification at the edge of life and death. *Nat. Rev. Mol. Cell Biol.*, **12**, 439–452.

126. Honda, R., Tanaka, H., Yasuda, H. (1997) Oncoprotein MDM2 is a ubiquitin ligase E3 for tumor suppressor p53. *FEBS Lett.*, **420**, 25–27.

127. Eckelman, B.P., Salvesen, G.S., Scott, F.L. (2006) Human inhibitor of apoptosis proteins: why XIAP is the black sheep of the family. *EMBO Rep.*, **7**, 988–994.

128. Silke, J., Kratina, T., Chu, D., Ekert, P.G. et al. (2005) Determination of cell survival by RING-mediated regulation of inhibitor

of apoptosis (IAP) protein abundance. *Proc. Natl Acad. Sci. USA*, **102**, 16182–16187.
129. Dogan, T., Harms, G.S., Hekman, M., Karreman, C. et al. (2008) X-linked and cellular IAPs modulate the stability of C-RAF kinase and cell motility. *Nat. Cell Biol.*, **10**, 1447–1455.
130. Chang, L., Kamata, H., Solinas, G., Luo, J.L. et al. (2006) The E3 ubiquitin ligase itch couples JNK activation to TNFalpha-induced cell death by inducing c-FLIP(L) turnover. *Cell*, **124**, 601–613.
131. Zhong, Q., Gao, W., Du, F., Wang, X. (2005) Mule/ARF-BP1, a BH3-only E3 ubiquitin ligase, catalyzes the polyubiquitination of Mcl-1 and regulates apoptosis. *Cell*, **121**, 1085–1095.
132. Zilfou, J.T., Lowe, S.W. (2009) Tumor suppressive functions of p53. *Cold Spring Harbor Perspect. Biol.*, **1**, a001883.
133. Appella, E., Anderson, C.W. (2001) Post-translational modifications and activation of p53 by genotoxic stresses. *Eur. J. Biochem.*, **268**, 2764–2772.
134. Brugarolas, J., Chandrasekaran, C., Gordon, J.I., Beach, D. et al. (1995) Radiation-induced cell cycle arrest compromised by p21 deficiency. *Nature*, **377**, 552–557.
135. Deng, C., Zhang, P., Harper, J.W., Elledge, S.J. et al. (1995) Mice lacking p21CIP1/WAF1 undergo normal development, but are defective in G1 checkpoint control. *Cell*, **82**, 675–684.
136. St Clair, S., Manfredi, J.J. (2006) The dual specificity phosphatase Cdc25C is a direct target for transcriptional repression by the tumor suppressor p53. *Cell Cycle*, **5**, 709–713.
137. Nigro, J.M., Baker, S.J., Preisinger, A.C., Jessup, J.M. et al. (1989) Mutations in the p53 gene occur in diverse human tumour types. *Nature*, **342**, 705–708.
138. Harris, C.C., Hollstein, M. (1993) Clinical implications of the p53 tumor-suppressor gene. *N. Engl. J. Med.*, **329**, 1318–1327.
139. Hollstein, M., Rice, K., Greenblatt, M.S., Soussi, T. et al. (1994) Database of p53 gene somatic mutations in human tumors and cell lines. *Nucleic Acids Res.*, **22**, 3551–3555.
140. Levine, A.J., Wu, M.C., Chang, A., Silver, A. et al. (1995) The spectrum of mutations at the p53 locus. Evidence for tissue-specific mutagenesis, selection of mutant alleles, and a "gain of function" phenotype. *Ann. N. Y. Acad. Sci.*, **768**, 111–128.
141. Wang, B., Xiao, Z., Ko, H.L., Ren, E.C. (2010) The p53 response element and transcriptional repression. *Cell Cycle*, **9**, 870–879.
142. Robles, A.I., Bemmels, N.A., Foraker, A.B., Harris, C.C. (2001) APAF-1 is a transcriptional target of p53 in DNA damage-induced apoptosis. *Cancer Res.*, **61**, 6660–6664.
143. Riley, T., Sontag, E., Chen, P., Levine, A. (2008) Transcriptional control of human p53-regulated genes. *Nat. Rev. Mol. Cell Biol.*, **9**, 402–412.
144. Thornborrow, E.C., Manfredi, J.J. (2001) The tumor suppressor protein p53 requires a cofactor to activate transcriptionally the human BAX promoter. *J. Biol. Chem.*, **276**, 15598–15608.
145. Thornborrow, E.C., Patel, S., Mastropietro, A.E., Schwartzfarb, E.M. et al. (2002) A conserved intronic response element mediates direct p53-dependent transcriptional activation of both the human and murine bax genes. *Oncogene*, **21**, 990–999.
146. Koutsodontis, G., Vasilaki, E., Chou, W.C., Papakosta, P. et al. (2005) Physical and functional interactions between members of the tumour suppressor p53 and the Sp families of transcription factors: importance for the regulation of genes involved in cell-cycle arrest and apoptosis. *Biochem. J.*, **389**, 443–455.
147. Flores, E.R., Tsai, K.Y., Crowley, D., Sengupta, S. et al. (2002) p63 and p73 are required for p53-dependent apoptosis in response to DNA damage. *Nature*, **416**, 560–564.
148. Jiang, P., Du, W., Heese, K., Wu, M. (2006) The Bad guy cooperates with good cop p53: Bad is transcriptionally up-regulated by p53 and forms a Bad/p53 complex at the mitochondria to induce apoptosis. *Mol. Cell. Biol.*, **26**, 9071–9082.
149. Mandal, M., Crusio, K.M., Meng, F., Liu, S. et al. (2008) Regulation of lymphocyte progenitor survival by the proapoptotic activities of Bim and Bid. *Proc. Natl Acad. Sci. USA*, **105**, 20840–20845.
150. Fang, L., Lee, S.W., Aaronson, S.A. (1999) Comparative analysis of p73 and p53

regulation and effector functions. *J. Cell Biol.*, **147**, 823–830.

151 Muller, M., Schilling, T., Sayan, A.E., Kairat, A. *et al.* (2005) TAp73/Delta Np73 influences apoptotic response, chemosensitivity and prognosis in hepatocellular carcinoma. *Cell Death Differ.*, **12**, 1564–1577.

152 Flinterman, M., Guelen, L., Ezzati-Nik, S., Killick, R. *et al.* (2005) E1A activates transcription of p73 and Noxa to induce apoptosis. *J. Biol. Chem.*, **280**, 5945–5959.

153 Zawacka-Pankau, J., Kostecka, A., Sznarkowska, A., Hedstrom, E. *et al.* (2010) p73 tumor suppressor protein: a close relative of p53 not only in structure but also in anti-cancer approach? *Cell Cycle*, **9**, 720–728.

154 Cao, Q., Xia, Y., Azadniv, M., Crispe, I.N. (2004) The E2F-1 transcription factor promotes caspase-8 and bid expression, and enhances Fas signaling in T cells. *J. Immunol.*, **173**, 1111–1117.

155 Bracken, A.P., Ciro, M., Cocito, A., Helin, K. (2004) E2F target genes: unraveling the biology. *Trends Biochem. Sci.*, **29**, 409–417.

156 Moroni, M.C., Hickman, E.S., Lazzerini Denchi, E., Caprara, G. *et al.* (2001) Apaf-1 is a transcriptional target for E2F and p53. *Nat. Cell Biol.*, **3**, 552–558.

157 Furukawa, Y., Nishimura, N., Satoh, M., Endo, H. *et al.* (2002) Apaf-1 is a mediator of E2F-1-induced apoptosis. *J. Biol. Chem.*, **277**, 39760–39768.

158 Nahle, Z., Polakoff, J., Davuluri, R.V., McCurrach, M.E. *et al.* (2002) Direct coupling of the cell cycle and cell death machinery by E2F. *Nat. Cell Biol.*, **4**, 859–864.

159 Hershko, T., Ginsberg, D. (2004) Up-regulation of Bcl-2 homology 3 (BH3)-only proteins by E2F1 mediates apoptosis. *J. Biol. Chem.*, **279**, 8627–8634.

160 Xie, W., Jiang, P., Miao, L., Zhao, Y. *et al.* (2006) Novel link between E2F1 and Smac/DIABLO: proapoptotic Smac/DIABLO is transcriptionally upregulated by E2F1. *Nucleic Acids Res.*, **34**, 2046–2055.

161 Jiang, Y., Saavedra, H.I., Holloway, M.P., Leone, G. *et al.* (2004) Aberrant regulation of survivin by the RB/E2F family of proteins. *J. Biol. Chem.*, **279**, 40511–40520.

162 Young, A.P., Nagarajan, R., Longmore, G.D. (2003) Mechanisms of transcriptional regulation by Rb-E2F segregate by biological pathway. *Oncogene*, **22**, 7209–7217.

163 Stanelle, J., Stiewe, T., Theseling, C.C., Peter, M. *et al.* (2002) Gene expression changes in response to E2F1 activation. *Nucleic Acids Res.*, **30**, 1859–1867.

164 Muller, H., Bracken, A.P., Vernell, R., Moroni, M.C. *et al.* (2001) E2Fs regulate the expression of genes involved in differentiation, development, proliferation, and apoptosis. *Genes Dev.*, **15**, 267–285.

165 Irwin, M., Marin, M.C., Phillips, A.C., Seelan, R.S. *et al.* (2000) Role for the p53 homologue p73 in E2F-1-induced apoptosis. *Nature*, **407**, 645–648.

166 Stiewe, T., Putzer, B.M. (2000) Role of the p53-homologue p73 in E2F1-induced apoptosis. *Nat. Genet.*, **26**, 464–469.

167 Kaufmann, E., Knochel, W. (1996) Five years on the wings of fork head. *Mech. Dev.*, **57**, 3–20.

168 Birkenkamp, K.U., Coffer, P.J. (2003) Regulation of cell survival and proliferation by the FOXO (Forkhead box, class O) subfamily of Forkhead transcription factors. *Biochem. Soc. Trans.*, **31**, 292–297.

169 Dijkers, P.F., Medema, R.H., Lammers, J.W., Koenderman, L. *et al.* (2000) Expression of the pro-apoptotic Bcl-2 family member Bim is regulated by the forkhead transcription factor FKHR-L1. *Curr. Biol.*, **10**, 1201–1204.

170 Skurk, C., Maatz, H., Kim, H.S., Yang, J. *et al.* (2004) The Akt-regulated forkhead transcription factor FOXO3a controls endothelial cell viability through modulation of the caspase-8 inhibitor FLIP. *J. Biol. Chem.*, **279**, 1513–1525.

171 Brunet, L.R., Beall, M., Dunne, D.W., Pearce, E.J. (1999) Nitric oxide and the Th2 response combine to prevent severe hepatic damage during *Schistosoma mansoni* infection. *J. Immunol.*, **163**, 4976–4984.

172 Modur, V., Nagarajan, R., Evers, B.M., Milbrandt, J. (2002) FOXO proteins regulate tumor necrosis factor-related apoptosis inducing ligand expression. Implications for PTEN mutation in prostate cancer. *J. Biol. Chem.*, **277**, 47928–47937.

173 You, H., Pellegrini, M., Tsuchihara, K., Yamamoto, K. *et al.* (2006) FOXO3a-dependent regulation of Puma in response to cytokine/growth factor withdrawal. *J. Exp. Med.*, **203**, 1657–1663.

174 Wang, C.Y., Mayo, M.W., Korneluk, R.G., Goeddel, D.V. et al. (1998) NF-kappaB antiapoptosis: induction of TRAF1 and TRAF2 and c-IAP1 and c-IAP2 to suppress caspase-8 activation. *Science*, **281**, 1680–1683.

175 Micheau, O., Lens, S., Gaide, O., Alevizopoulos, K. et al. (2001) NF-kappaB signals induce the expression of c-FLIP. *Mol. Cell. Biol.*, **21**, 5299–5305.

176 Smith, M.D., Ensor, E.A., Coffin, R.S., Boxer, L.M. et al. (1998) Bcl-2 transcription from the proximal P2 promoter is activated in neuronal cells by the Brn-3a POU family transcription factor. *J. Biol. Chem.*, **273**, 16715–16722.

177 Tang, T.T., Dowbenko, D., Jackson, A., Toney, L. et al. (2002) The forkhead transcription factor AFX activates apoptosis by induction of the BCL-6 transcriptional repressor. *J. Biol. Chem.*, **277**, 14255–14265.

178 Lima, R.T., Busacca, S., Almeida, G.M., Gaudino, G. et al. (2011) MicroRNA regulation of core apoptosis pathways in cancer. *Eur. J. Cancer*, **47**, 163–174.

179 Abdelmohsen, K., Lal, A., Kim, H.H., Gorospe, M. (2007) Posttranscriptional orchestration of an anti-apoptotic program by HuR. *Cell Cycle*, **6**, 1288–1292.

180 Schwerk, C., Schulze-Osthoff, K. (2005) Regulation of apoptosis by alternative pre-mRNA splicing. *Mol. Cell*, **19**, 1–13.

181 Manley, J.L., Tacke, R. (1996) SR proteins and splicing control. *Genes Dev.*, **10**, 1569–1579.

182 Graveley, B.R. (2000) Sorting out the complexity of SR protein functions. *RNA*, **6**, 1197–1211.

183 Renshaw, S.A., Dempsey, C.E., Barnes, F.A., Bagstaff, S.M. et al. (2004) Three novel Bid proteins generated by alternative splicing of the human Bid gene. *J. Biol. Chem.*, **279**, 2846–2855.

184 O'Connor, L., Strasser, A., O'Reilly, L.A., Hausmann, G. et al. (1998) Bim: a novel member of the Bcl-2 family that promotes apoptosis. *EMBO J.*, **17**, 384–395.

185 Miyashita, T., Shikama, Y., Tadokoro, K., Yamada, M. (2001) Molecular cloning and characterization of six novel isoforms of human Bim, a member of the proapoptotic Bcl-2 family. *FEBS Lett.*, **509**, 135–141.

186 Marani, M., Tenev, T., Hancock, D., Downward, J. et al. (2002) Identification of novel isoforms of the BH3 domain protein Bim which directly activate Bax to trigger apoptosis. *Mol. Cell. Biol.*, **22**, 3577–3589.

187 Mouhamad, S., Besnault, L., Auffredou, M.T., Leprince, C. et al. (2004) B cell receptor-mediated apoptosis of human lymphocytes is associated with a new regulatory pathway of Bim isoform expression. *J. Immunol.*, **172**, 2084–2091.

188 Boise, L.H., Gonzalez-Garcia, M., Postema, C.E., Ding, L. et al. (1993) bcl-x, a bcl-2-related gene that functions as a dominant regulator of apoptotic cell death. *Cell*, **74**, 597–608.

189 Chalfant, C.E., Ogretmen, B., Galadari, S., Kroesen, B.J. et al. (2001) FAS activation induces dephosphorylation of SR proteins; dependence on the de novo generation of ceramide and activation of protein phosphatase 1. *J. Biol. Chem.*, **276**, 44848–44855.

190 Revil, T., Toutant, J., Shkreta, L., Garneau, D. et al. (2007) Protein kinase C-dependent control of Bcl-x alternative splicing. *Mol. Cell. Biol.*, **27**, 8431–8441.

191 Shkreta, L., Michelle, L., Toutant, J., Tremblay, M.L. et al. (2011) The DNA damage response pathway regulates the alternative splicing of the apoptotic mediator Bcl-x. *J. Biol. Chem.*, **286**, 331–340.

192 Wang, Z., Sun, Y. (2008) Identification and characterization of two splicing variants of human Noxa. *Anticancer Res.*, **28**, 1667–1674.

193 Bingle, C.D., Craig, R.W., Swales, B.M., Singleton, V. et al. (2000) Exon skipping in Mcl-1 results in a bcl-2 homology domain 3 only gene product that promotes cell death. *J. Biol. Chem.*, **275**, 22136–22146.

194 Shieh, J.J., Liu, K.T., Huang, S.W., Chen, Y.J. et al. (2009) Modification of alternative splicing of Mcl-1 pre-mRNA using antisense morpholino oligonucleotides induces apoptosis in basal cell carcinoma cells. *J. Invest. Dermatol.*, **129**, 2497–2506.

195 Huang, Y., Shin, N.H., Sun, Y., Wang, K.K. (2001) Molecular cloning and characterization of a novel caspase-3 variant that attenuates apoptosis induced by proteasome inhibition. *Biochem. Biophys. Res. Commun.*, **283**, 762–769.

196 Vegran, F., Boidot, R., Oudin, C., Riedinger, J.M. et al. (2005) Implication of alternative splice transcripts of caspase-3

and survivin in chemoresistance. *Bull. Cancer*, **92**, 219–226.

197 Seol, D.W., Billiar, T.R. (1999) A caspase-9 variant missing the catalytic site is an endogenous inhibitor of apoptosis. *J. Biol. Chem.*, **274**, 2072–2076.

198 Srinivasula, S.M., Ahmad, M., Guo, Y., Zhan, Y. *et al.* (1999) Identification of an endogenous dominant-negative short isoform of caspase-9 that can regulate apoptosis. *Cancer Res.*, **59**, 999–1002.

199 Shultz, J.C., Goehe, R.W., Murudkar, C.S., Wijesinghe, D.S. *et al.* (2011) SRSF1 regulates the alternative splicing of caspase 9 via a novel intronic splicing enhancer affecting the chemotherapeutic sensitivity of non-small cell lung cancer cells. *Mol. Cancer Res.*, **9**, 889–900.

200 Shultz, J.C., Goehe, R.W., Wijesinghe, D.S., Murudkar, C. *et al.* (2010) Alternative splicing of caspase 9 is modulated by the phosphoinositide 3-kinase/Akt pathway via phosphorylation of SRp30a. *Cancer Res.*, **70**, 9185–9196.

201 Massiello, A., Chalfant, C.E. (2006) SRp30a (ASF/SF2) regulates the alternative splicing of caspase-9 pre-mRNA and is required for ceramide-responsiveness. *J. Lipid Res.*, **47**, 892–897.

202 Ng, P.W., Porter, A.G., Janicke, R.U. (1999) Molecular cloning and characterization of two novel pro-apoptotic isoforms of caspase-10. *J. Biol. Chem.*, **274**, 10301–10308.

203 Jiang, Z.H., Zhang, W.J., Rao, Y., Wu, J.Y. (1998) Regulation of Ich-1 pre-mRNA alternative splicing and apoptosis by mammalian splicing factors. *Proc. Natl Acad. Sci. USA*, **95**, 9155–9160.

204 Mahotka, C., Wenzel, M., Springer, E., Gabbert, H.E. *et al.* (1999) Survivin-deltaEx3 and survivin-2B: two novel splice variants of the apoptosis inhibitor survivin with different antiapoptotic properties. *Cancer Res.*, **59**, 6097–6102.

205 Zhu, N., Gu, L., Findley, H.W., Li, F. *et al.* (2004) An alternatively spliced survivin variant is positively regulated by p53 and sensitizes leukemia cells to chemotherapy. *Oncogene*, **23**, 7545–7551.

206 Badran, A., Yoshida, A., Ishikawa, K., Goi, T. *et al.* (2004) Identification of a novel splice variant of the human anti-apoptosis gene survivin. *Biochem. Biophys. Res. Commun.*, **314**, 902–907.

207 Caldas, H., Honsey, L.E., Altura, R.A. (2005) Survivin 2alpha: a novel Survivin splice variant expressed in human malignancies. *Mol. Cancer*, **4**, 11.

208 Vegran, F., Boidot, R., Bonnetain, F., Cadouot, M. *et al.* (2011) Apoptosis gene signature of Survivin and its splice variant expression in breast carcinoma. *Endocr. Relat. Cancer*, **18** (6), 783–792.

209 Vegran, F., Boidot, R., Oudin, C., Defrain, C. *et al.* (2007) Association of p53 gene alterations with the expression of antiapoptotic survivin splice variants in breast cancer. *Oncogene*, **26**, 290–297.

210 Bin, L., Li, X., Xu, L.G., Shu, H.B. (2002) The short splice form of Casper/c-FLIP is a major cellular inhibitor of TRAIL-induced apoptosis. *FEBS Lett.*, **510**, 37–40.

211 Fu, J., Jin, Y., Arend, L.J. (2003) Smac3, a novel Smac/DIABLO splicing variant, attenuates the stability and apoptosis-inhibiting activity of X-linked inhibitor of apoptosis protein. *J. Biol. Chem.*, **278**, 52660–52672.

212 Taylor, J.K., Zhang, Q.Q., Wyatt, J.R., Dean, N.M. (1999) Induction of endogenous Bcl-xS through the control of Bcl-x pre-mRNA splicing by antisense oligonucleotides. *Nat. Biotechnol.*, **17**, 1097–1100.

213 Leech, S.H., Olie, R.A., Gautschi, O., Simoes-Wust, A.P. *et al.* (2000) Induction of apoptosis in lung-cancer cells following bcl-xL anti-sense treatment. *Int. J. Cancer*, **86**, 570–576.

214 Herold, A., Klymenko, T., Izaurralde, E. (2001) NXF1/p15 heterodimers are essential for mRNA nuclear export in *Drosophila*. *RNA*, **7**, 1768–1780.

215 Braun, I.C., Herold, A., Rode, M., Conti, E. *et al.* (2001) Overexpression of TAP/p15 heterodimers bypasses nuclear retention and stimulates nuclear mRNA export. *J. Biol. Chem.*, **19**, 19.

216 Brennan, C.M., Gallouzi, I.E., Steitz, J.A. (2000) Protein ligands to HuR modulate its interaction with target mRNAs in vivo. *J. Cell Biol.*, **151**, 1–14.

217 Takagi, M., Absalon, M.J., McLure, K.G., Kastan, M.B. (2005) Regulation of p53 translation and induction after DNA damage by ribosomal protein L26 and nucleolin. *Cell*, **123**, 49–63.

218 Lal, A., Kawai, T., Yang, X., Mazan-Mamczarz, K. et al. (2005) Antiapoptotic function of RNA-binding protein HuR effected through prothymosin alpha. *EMBO J.*, **24**, 1852–1862.

219 Mazan-Mamczarz, K., Galban, S., De Silanes, I.L., Martindale, J.L. et al. (2003) RNA-binding protein HuR enhances p53 translation in response to ultraviolet light irradiation. *Proc. Natl Acad. Sci. USA*, **100**, 8354–8359.

220 Henis-Korenblit, S., Shani, G., Sines, T., Marash, L. et al. (2002) The caspase-cleaved DAP5 protein supports internal ribosome entry site-mediated translation of death proteins. *Proc. Natl Acad. Sci. USA*, **99**, 5400–5405.

221 Harigai, M., Miyashita, T., Hanada, M., Reed, J.C. (1996) A cis-acting element in the BCL-2 gene controls expression through translational mechanisms. *Oncogene*, **12**, 1369–1374.

222 Shahid, R., Bugaut, A., Balasubramanian, S. (2010) The BCL-2 5' untranslated region contains an RNA G-quadruplex-forming motif that modulates protein expression. *Biochemistry*, **49**, 8300–8306.

223 Pardo, O.E., Arcaro, A., Salerno, G., Raguz, S. et al. (2002) Fibroblast growth factor-2 induces translational regulation of Bcl-XL and Bcl-2 via a MEK-dependent pathway: correlation with resistance to etoposide-induced apoptosis. *J. Biol. Chem.*, **277**, 12040–12046.

224 Lam, D., Dickens, D., Reid, E.B., Loh, S.H. et al. (2009) MAP4K3 modulates cell death via the post-transcriptional regulation of BH3-only proteins. *Proc. Natl Acad. Sci. USA*, **106**, 11978–11983.

225 Bleumink, M., Kohler, R., Giaisi, M., Proksch, P. et al. (2011) Rocaglamide breaks TRAIL resistance in HTLV-1-associated adult T-cell leukemia/lymphoma by translational suppression of c-FLIP expression. *Cell Death Differ.*, **18**, 362–370.

226 Subramaniam, D., Natarajan, G., Ramalingam, S., Ramachandran, I. et al. (2008) Translation inhibition during cell cycle arrest and apoptosis: Mcl-1 is a novel target for RNA binding protein CUGBP2. *Am. J. Physiol. Gastrointest. Liver Physiol.*, **294**, G1025–G1032.

227 Mukhopadhyay, D., Houchen, C.W., Kennedy, S., Dieckgraefe, B.K. et al. (2003) Coupled mRNA stabilization and translational silencing of cyclooxygenase-2 by a novel RNA binding protein, CUGBP2. *Mol. Cell*, **11**, 113–126.

228 Sureban, S.M., Murmu, N., Rodriguez, P., May, R. et al. (2007) Functional antagonism between RNA binding proteins HuR and CUGBP2 determines the fate of COX-2 mRNA translation. *Gastroenterology*, **132**, 1055–1065.

229 Durie, D., Lewis, S.M., Liwak, U., Kisilewicz, M. et al. (2011) RNA-binding protein HuR mediates cytoprotection through stimulation of XIAP translation. *Oncogene*, **30**, 1460–1469.

230 Kawai, T., Lal, A., Yang, X., Galban, S. et al. (2006) Translational control of cytochrome c by RNA-binding proteins TIA-1 and HuR. *Mol. Cell. Biol.*, **26**, 3295–3307.

231 Donahue, J.M., Chang, E.T., Xiao, L., Wang, P.Y. et al. (2011) The RNA-binding protein HuR stabilizes survivin mRNA in human oesophageal epithelial cells. *Biochem. J.*, **437**, 89–96.

232 Gallouzi, I.E., Brennan, C.M., Steitz, J.A. (2001) Protein ligands mediate the CRM1-dependent export of HuR in response to heat shock. *RNA*, **7**, 1348–1361.

233 Abdelmohsen, K., Pullmann, R. Jr, Lal, A., Kim, H.H. et al. (2007) Phosphorylation of HuR by Chk2 regulates SIRT1 expression. *Mol. Cell*, **25**, 543–557.

234 Nakamura, H., Kawagishi, H., Watanabe, A., Sugimoto, K. et al. (2011) Cooperative role of the RNA-binding proteins Hzf and HuR in p53 activation. *Mol. Cell. Biol.*, **31**, 1997–2009.

235 Abdelmohsen, K., Gorospe, M. (2010) Post-transcriptional regulation of cancer traits by HuR. *Wiley Interdiscip. Rev. RNA*, **1**, 214–229.

236 Mazroui, R., Di Marco, S., Clair, E., von Roretz, C. et al. (2008) Caspase-mediated cleavage of HuR in the cytoplasm contributes to pp32/PHAP-I regulation of apoptosis. *J. Cell Biol.*, **180**, 113–127.

237 von Roretz, C., Gallouzi, I.E. (2010) Protein kinase RNA/FADD/caspase-8 pathway mediates the proapoptotic activity of the RNA-binding protein human antigen R (HuR). *J. Biol. Chem.*, **285**, 16806–16813.

238 von Roretz, C., Macri, A.M., Gallouzi, I.E. (2011) Transportin 2 regulates apoptosis

through the RNA-binding protein HuR. *J. Biol. Chem.*, **286**, 25983–25991.

239 Hamanaka, R.B., Bobrovnikova-Marjon, E., Ji, X., Liebhaber, S.A. *et al.* (2009) PERK-dependent regulation of IAP translation during ER stress. *Oncogene*, **28**, 910–920.

240 Harding, H.P., Zhang, Y., Ron, D. (1999) Protein translation and folding are coupled by an endoplasmic-reticulum-resident kinase. *Nature*, **397**, 271–274.

241 Harding, H.P., Novoa, I., Zhang, Y., Zeng, H. *et al.* (2000) Regulated translation initiation controls stress-induced gene expression in mammalian cells. *Mol. Cell*, **6**, 1099–1108.

242 Harding, H.P., Zhang, Y., Bertolotti, A., Zeng, H. *et al.* (2000) Perk is essential for translational regulation and cell survival during the unfolded protein response. *Mol. Cell*, **5**, 897–904.

243 Warnakulasuriyarachchi, D., Cerquozzi, S., Cheung, H.H., Holcik, M. (2004) Translational induction of the inhibitor of apoptosis protein HIAP2 during endoplasmic reticulum stress attenuates cell death and is mediated via an inducible internal ribosome entry site element. *J. Biol. Chem.*, **279**, 17148–17157.

244 Stoneley, M., Willis, A.E. (2004) Cellular internal ribosome entry segments: structures, trans-acting factors and regulation of gene expression. *Oncogene*, **23**, 3200–3207.

245 Holcik, M., Sonenberg, N. (2005) Translational control in stress and apoptosis. *Nat. Rev. Mol. Cell Biol.*, **6**, 318–327.

246 Sherrill, K.W., Byrd, M.P., Van Eden, M.E., Lloyd, R.E. (2004) BCL-2 translation is mediated via internal ribosome entry during cell stress. *J. Biol. Chem.*, **279**, 29066–29074.

247 Mitchell, S.A., Brown, E.C., Coldwell, M.J., Jackson, R.J. *et al.* (2001) Protein factor requirements of the Apaf-1 internal ribosome entry segment: roles of polypyrimidine tract binding protein and upstream of N-ras. *Mol. Cell. Biol.*, **21**, 3364–3374.

248 Spriggs, K.A., Bushell, M., Mitchell, S.A., Willis, A.E. (2005) Internal ribosome entry segment-mediated translation during apoptosis: the role of IRES-trans-acting factors. *Cell Death Differ.*, **12**, 585–591.

249 Filipowicz, W., Bhattacharyya, S.N., Sonenberg, N. (2008) Mechanisms of post-transcriptional regulation by microRNAs: are the answers in sight? *Nat. Rev. Genet.*, **9**, 102–114.

250 Garofalo, M., Condorelli, G.L., Croce, C.M., Condorelli, G. (2010) MicroRNAs as regulators of death receptors signaling. *Cell Death Differ.*, **17**, 200–208.

251 Sayed, D., He, M., Hong, C., Gao, S. *et al.* (2010) MicroRNA-21 is a downstream effector of AKT that mediates its antiapoptotic effects via suppression of Fas ligand. *J. Biol. Chem.*, **285**, 20281–20290.

252 Wang, K., Li, P.F. (2010) Foxo3a regulates apoptosis by negatively targeting miR-21. *J. Biol. Chem.*, **285**, 16958–16966.

253 Lin, H., Qian, J., Castillo, A.C., Long, B. *et al.* (2011) Effect of miR-23 on oxidant-induced injury in human retinal pigment epithelial cells. *Invest. Ophthalmol. Vis. Sci.*, **52**, 6308–6314.

254 Chen, G., Wang, Y., Zhou, M., Shi, H. *et al.* (2010) EphA1 receptor silencing by small interfering RNA has antiangiogenic and antitumor efficacy in hepatocellular carcinoma. *Oncol. Rep.*, **23**, 563–570.

255 Walker, J.C., Harland, R.M. (2009) microRNA-24a is required to repress apoptosis in the developing neural retina. *Genes Dev.*, **23**, 1046–1051.

256 Kong, F., Sun, C., Wang, Z., Han, L. *et al.* (2011) miR-125b confers resistance of ovarian cancer cells to cisplatin by targeting pro-apoptotic Bcl-2 antagonist killer 1. *J. Huazhong Univ. Sci. Technol. Med. Sci.*, **31**, 543–549.

257 Li, J.H., Xiao, X., Zhang, Y.N., Wang, Y.M. *et al.* (2011) MicroRNA miR-886-5p inhibits apoptosis by down-regulating Bax expression in human cervical carcinoma cells. *Gynecol. Oncol.*, **120**, 145–151.

258 Zhang, C.Z., Zhang, J.X., Zhang, A.L., Shi, Z.D. *et al.* (2010) MiR-221 and miR-222 target PUMA to induce cell survival in glioblastoma. *Mol. Cancer*, **9**, 229.

259 Papagiannakopoulos, T., Shapiro, A., Kosik, K.S. (2008) MicroRNA-21 targets a network of key tumor-suppressive pathways in glioblastoma cells. *Cancer Res.*, **68**, 8164–8172.

260 Wickramasinghe, N.S., Manavalan, T.T., Dougherty, S.M., Riggs, K.A. *et al.* (2009) Estradiol downregulates miR-21 expression

and increases miR-21 target gene expression in MCF-7 breast cancer cells. *Nucleic Acids Res.*, **37**, 2584–2595.

261 Dong, J., Zhao, Y.P., Zhou, L., Zhang, T.P. et al. (2011) Bcl-2 upregulation induced by miR-21 via a direct interaction is associated with apoptosis and chemoresistance in MIA PaCa-2 pancreatic cancer cells. *Arch. Med. Res.*, **42**, 8–14.

262 Li, J., Huang, H., Sun, L., Yang, M. et al. (2009) MiR-21 indicates poor prognosis in tongue squamous cell carcinomas as an apoptosis inhibitor. *Clin. Cancer Res.*, **15**, 3998–4008.

263 Li, S., Liang, Z., Xu, L., Zou, F. (2011) MicroRNA-21: a ubiquitously expressed pro-survival factor in cancer and other diseases. *Mol. Cell. Biochem.*, **360** (1-2), 147–158.

264 Saini, S., Yamamura, S., Majid, S., Shahryari, V. et al. (2011) MicroRNA-708 induces apoptosis and suppresses tumorigenicity in renal cancer cells. *Cancer Res.*, **71**, 6208–6219.

265 Xiong, Y., Fang, J.H., Yun, J.P., Yang, J. et al. (2010) Effects of microRNA-29 on apoptosis, tumorigenicity, and prognosis of hepatocellular carcinoma. *Hepatology*, **51**, 836–845.

266 Zhang, H., Li, Y., Huang, Q., Ren, X. et al. (2011) MiR-148a promotes apoptosis by targeting Bcl-2 in colorectal cancer. *Cell Death Differ.*, **18**, 1702–1710.

267 Mott, J.L., Kobayashi, S., Bronk, S.F., Gores, G.J. (2007) mir-29 regulates Mcl-1 protein expression and apoptosis. *Oncogene*, **26**, 6133–6140.

268 Zhang, Y.K., Wang, H., Leng, Y., Li, Z.L. et al. (2011) Overexpression of microRNA-29b induces apoptosis of multiple myeloma cells through down regulating Mcl-1. *Biochem. Biophys. Res. Commun.*, **414**, 233–239.

269 Chen, J., Zhang, X., Lentz, C., Abi-Daoud, M. et al. (2011) miR-193b regulates Mcl-1 in melanoma. *Am. J. Pathol.*, **179**, 2162–2168.

270 Su, H., Yang, J.R., Xu, T., Huang, J. et al. (2009) MicroRNA-101, down-regulated in hepatocellular carcinoma, promotes apoptosis and suppresses tumorigenicity. *Cancer Res.*, **69**, 1135–1142.

271 Cimmino, A., Calin, G.A., Fabbri, M., Iorio, M.V. et al. (2005) miR-15 and miR-16 induce apoptosis by targeting BCL2. *Proc. Natl Acad. Sci. USA*, **102**, 13944–13949.

272 Yadav, S., Pandey, A., Shukla, A., Talwelkar, S.S. et al. (2011) miR-497 and miR-302b regulate ethanol-induced neuronal cell death through BCL2 protein and cyclin D2. *J. Biol. Chem.*, **286**, 37347–37357.

273 Zhu, W., Xu, H., Zhu, D., Zhi, H. et al. (2011) miR-200bc/429 cluster modulates multidrug resistance of human cancer cell lines by targeting BCL2 and XIAP. *Cancer Chemother. Pharmacol.*, **69** (3), 723–731.

274 Xiong, S., Zheng, Y., Jiang, P., Liu, R. et al. (2011) MicroRNA-7 inhibits the growth of human non-small cell lung cancer A549 cells through targeting BCL-2. *Int. J. Biol. Sci.*, **7**, 805–814.

275 Shimizu, S., Takehara, T., Hikita, H., Kodama, T. et al. (2010) The let-7 family of microRNAs inhibits Bcl-xL expression and potentiates sorafenib-induced apoptosis in human hepatocellular carcinoma. *J. Hepatol.*, **52**, 698–704.

276 Saini, S., Majid, S., Yamamura, S., Tabatabai, L. et al. (2011) Regulatory role of mir-203 in prostate cancer progression and metastasis. *Clin. Cancer Res.*, **17**, 5287–5298.

277 Diakos, C., Zhong, S., Xiao, Y., Zhou, M. et al. (2010) TEL-AML1 regulation of survivin and apoptosis via miRNA-494 and miRNA-320a. *Blood*, **116**, 4885–4893.

278 Parker, R., Song, H. (2004) The enzymes and control of eukaryotic mRNA turnover. *Nat. Struct. Mol. Biol.*, **11**, 121–127.

279 Doma, M.K., Parker, R. (2007) RNA quality control in eukaryotes. *Cell*, **131**, 660–668.

280 Parker, R., Sheth, U. (2007) P bodies and the control of mRNA translation and degradation. *Mol. Cell*, **25**, 635–646.

281 von Roretz, C., Di Marco, S., Mazroui, R., Gallouzi, I.E. (2011) Turnover of AU-rich-containing mRNAs during stress: a matter of survival. *Wiley Interdiscip. Rev. RNA*, **2**, 336–347.

282 von Roretz, C., Gallouzi, I.E. (2008) Decoding ARE-mediated decay: is microRNA part of the equation? *J. Cell Biol.*, **181**, 189–194.

283 Chen, C.Y., Shyu, A.B. (1995) AU-rich elements: characterization and importance in mRNA degradation. *Trends Biochem. Sci.*, **20**, 465–470.

284 Shaw, G., Kamen, R. (1986) A conserved AU sequence from the 3' untranslated region of GM-CSF mRNA mediates selective mRNA degradation. *Cell*, **46**, 659–667.

285 von Roretz, C., Beauchamp, P., Di Marco, S., Gallouzi, I.E. (2011) HuR and myogenesis: being in the right place at the right time. *Biochim. Biophys. Acta*, **1813**, 1663–1667.

286 Sela-Brown, A., Silver, J., Brewer, G., Naveh-Many, T. (2000) Identification of AUF1 as a parathyroid hormone mRNA 3'-untranslated region-binding protein that determines parathyroid hormone mRNA stability. *J. Biol. Chem.*, **275**, 7424–7429.

287 Xu, N., Chen, C.Y., Shyu, A.B. (2001) Versatile role for hnRNP D isoforms in the differential regulation of cytoplasmic mRNA turnover. *Mol. Cell. Biol.*, **21**, 6960–6971.

288 Matsui, H., Asou, H., Inaba, T. (2007) Cytokines direct the regulation of Bim mRNA stability by heat-shock cognate protein 70. *Mol. Cell*, **25**, 99–112.

289 Schiavone, N., Rosini, P., Quattrone, A., Donnini, M. et al. (2000) A conserved AU-rich element in the 3' untranslated region of bcl-2 mRNA is endowed with a destabilizing function that is involved in bcl-2 down- regulation during apoptosis. *FASEB J.*, **14**, 174–184.

290 Zhang, X., Zou, T., Rao, J.N., Liu, L., Xiao, L. et al. (2009) Stabilization of XIAP mRNA through the RNA binding protein HuR regulated by cellular polyamines. *Nucleic Acids Res.*, **37**, 7623–7637.

291 Zou, T., Mazan-Mamczarz, K., Rao, J.N., Liu, L. et al. (2006) Polyamine depletion increases cytoplasmic levels of RNA-binding protein HuR leading to stabilization of nucleophosmin and p53 mRNAs. *J. Biol. Chem.*, **281**, 19387–19394.

292 Lapucci, A., Donnini, M., Papucci, L., Witort, E. et al. (2002) AUF1 Is a bcl-2 A + U-rich element-binding protein involved in bcl-2 mRNA destabilization during apoptosis. *J. Biol. Chem.*, **277**, 16139–16146.

293 Sengupta, S., Jang, B.C., Wu, M.T., Paik, J.H. et al. (2003) The RNA-binding protein HuR regulates the expression of cyclooxygenase-2. *J. Biol. Chem.*, **278**, 25227–25233.

294 Ishimaru, D., Zuraw, L., Ramalingam, S., Sengupta, T.K. et al. (2010) Mechanism of regulation of bcl-2 mRNA by nucleolin and A+U-rich element-binding factor 1 (AUF1). *J. Biol. Chem.*, **285**, 27182–27191.

295 Ghisolfi, L., Calastretti, A., Franzi, S., Canti, G. et al. (2009) B cell lymphoma (Bcl)-2 protein is the major determinant in bcl-2 adenine-uridine-rich element turnover overcoming HuR activity. *J. Biol. Chem.*, **284**, 20946–20955.

296 Hoffman, W.H., Biade, S., Zilfou, J.T., Chen, J. et al. (2002) Transcriptional repression of the anti-apoptotic survivin gene by wild type p53. *J. Biol. Chem.*, **277**, 3247–3257.

297 Lee, J.H., Jeon, M.H., Seo, Y.J., Lee, Y.J. et al. (2004) CA repeats in the 3'-untranslated region of bcl-2 mRNA mediate constitutive decay of bcl-2 mRNA. *J. Biol. Chem.*, **279**, 42758–42764.

298 Lee, D.H., Lim, M.H., Youn, D.Y., Jung, S.E. et al. (2009) hnRNP L binds to CA repeats in the 3'UTR of bcl-2 mRNA. *Biochem. Biophys. Res. Commun.*, **382**, 583–587.

299 Lin, C.J., Gong, H.Y., Tseng, H.C., Wang, W.L. et al. (2008) miR-122 targets an anti-apoptotic gene, Bcl-w, in human hepatocellular carcinoma cell lines. *Biochem. Biophys. Res. Commun.*, **375**, 315–320.

300 Peng, G., Yuan, Y., He, Q., Wu, W. et al. (2011) MicroRNA let-7e regulates the expression of caspase-3 during apoptosis of PC12 cells following anoxia/reoxygenation injury. *Brain Res. Bull.*, **86**, 272–276.

301 Yoon, S., Choi, Y.C., Lee, S., Jeong, Y. et al. (2010) Induction of growth arrest by miR-542-3p that targets survivin. *FEBS Lett.*, **584**, 4048–4052.

302 Siegel, C., Li, J., Liu, F., Benashski, S.E. et al. (2011) miR-23a regulation of X-linked inhibitor of apoptosis (XIAP) contributes to sex differences in the response to cerebral ischemia. *Proc. Natl Acad. Sci. USA*, **108**, 11662–11667.

303 Hanahan, D., Weinberg, R.A. (2000) The hallmarks of cancer. *Cell*, **100**, 57–70.

304 Roshal, M., Zhu, Y., Planelles, V. (2001) Apoptosis in AIDS. *Apoptosis*, **6**, 103–116.

305 Mattson, M.P. (2000) Apoptosis in neurodegenerative disorders. *Nat. Rev. Mol. Cell Biol.*, **1**, 120–129.

2
Bacterial *trans*-Translation: From Functions to Applications

Emmanuel Giudice[1] *and Reynald Gillet*[1,2]
[1]*Université de Rennes 1, CNRS UMR 6290 IGDR, "Translation and Folding Team", Campus de Beaulieu, 35042 Rennes cedex, France*
[2]*Institut Universitaire de France*

1	Introduction	49
2	Canonical Protein Synthesis in Bacteria	52
3	Situations Inducing Ribosome Stalling	52
3.1	Non-Stop mRNAs	52
3.2	No-Go mRNAs	53
3.2.1	Altered Reading of the mRNAs	54
3.2.2	Altered Reading of the Protein Undergoing Synthesis	54
3.2.3	Altered Conditions of Translation	55
4	Structure and Function of Isolated tmRNA	55
4.1	tmRNA Processing	55
4.2	tRNA-Like Domain	55
4.3	mRNA-Like Domain	55
4.4	Pseudoknots	57
4.5	Structural Variations of Two-Piece tmRNA	57
5	Structure and Function of SmpB	57
6	Other tmRNA Partners	58
6.1	Alanyl-tRNA Synthetase (AlaRS)	58
6.2	Elongation Factor EF-Tu	59
6.3	Ribosomal Protein S1	60
6.4	Ribonuclease R (RNase R)	61

RNA Regulation: Advances in Molecular Biology and Medicine, First Edition. Edited by Robert A. Meyers.
© 2014 Wiley-VCH Verlag GmbH & Co. KGaA. Published 2014 by Wiley-VCH Verlag GmbH & Co. KGaA.

7	***trans*-Translation: from Structures to Mechanistics** 61	
7.1	The Pre-Accommodation Step 61	
7.2	The Accommodation Step 63	
7.3	The Translocation Steps 64	
7.4	When tmRNA-SmpB Reaches the Edge of the Ribosome: What Next? 64	
8	**Other Physiological Roles Attributed to trans-Translation** 68	
8.1	Role of trans-Translation during the Cell Cycle 68	
8.2	Role of trans-Translation in Phage Growth 68	
9	**Alternative Pathways** 68	
10	**Medical Applications** 70	
10.1	*trans*-Translation as a Target for Protein Synthesis Inhibitors 70	
10.2	tmRNA–SmpB Mutants as Live Attenuated Vaccines 71	
10.3	Use of tmRNA as a Diagnostics Target 71	
11	**Concluding Remarks** 72	
	Acknowledgments 72	
	References 72	

Keywords

Ribosome
A large macromolecular complex composed of RNA and proteins that ensures protein synthesis (namely translation) in every cell of every organism. Each particle ("70S", in prokaryotes, "80S" in eukaryotes), is divided into two subunits: a large subunit that performs the links between the amino acids to form a polypeptide, and a small subunit that ensures decoding of the messenger RNA (mRNA).

Ribosome stalling
Ribosome stalling on messenger RNA (mRNA) transcripts is a serious issue for bacterial integrity. Stalling on "non-stop" mRNAs occurs when a ribosome reaches the end of a messenger RNA lacking an in-frame stop codon. Stalling on "no-go" mRNAs occurs on intact full-length mRNAs when ribosomes pause on internal sites, and is due to the reading of a problematic mRNA sequence or to the synthesis of a problematic protein motif.

SmpB
Small protein B is the main partner of tmRNA. It is indispensable to the process of *trans*-translation.

tmRNA

Transfer-messenger RNA (tmRNA) is a unique hybrid RNA molecule which has the properties of both a transfer RNA (tRNA) and an mRNA. It is five times bigger than a tRNA.

Trans-translation

A quality-control process performed in all bacteria by tmRNA-SmpB. It ensures the delivery of ribosomes stalled on mRNAs lacking stop codons, as well as destruction of the faulty proteins and mRNAs.

> Ribosome stalling is a serious issue for cell survival. In bacteria, the primary rescue system is *trans*-translation, performed by transfer-messenger RNA (tmRNA) and its protein partner small protein B (SmpB). This multitask process releases the stalled ribosomes while destroying the faulty protein and messenger RNA (mRNA). In this chapter, the cellular and molecular details of *trans*-translation are described, with special mention of technological developments and potential therapeutics suggested by the process.

1
Introduction

In every cell, several quality control pathways exist that ensure an accurate correspondence between the genetic information stored in DNA and the final proteins. In bacteria, when translation occurs on faulty messenger RNAs (mRNAs) lacking a stop codon, the ribosomes stall at the 3′-end of the mRNA and thus encode aberrant proteins. The primary rescue system that permits release of the ribosomes as well as elimination of the proteins and mRNAs, is *trans*-translation, mediated by transfer-messenger RNA (tmRNA) and small protein B (SmpB). The tmRNA-SmpB system is active in all eubacteria and in the eubacterial-like organelles of certain simple eukaryotes [1]. tmRNA is a remarkable hybrid molecule that contains both transfer RNA (tRNA) and mRNA activities. Like tRNA, tmRNA is loaded with an amino acid, although this is always an alanine. Thanks to SmpB, the tRNA-like domain (TLD) of tmRNA recognizes and enters the ribosomes that are stalled at the end of problematic mRNAs, and restarts translation. Protein synthesis then switches on the mRNA part of tmRNA. The truncated protein is extended by a short sequence of amino acids that tags it for destruction by cellular proteases. The incomplete mRNA is also kicked off to be destroyed by ribonucleases. Almost 20 years after the discovery of its tagging activity [2], tmRNA has finally come of age, and was designated "Molecule of the month" by the RCSB Protein Data Bank in January 2013 (http://www.rcsb.org). The aim of this chapter is to decipher the mechanistics of the process at the cellular and molecular levels, with particular attention being paid to the potential therapeutics and technological developments brought about by the process.

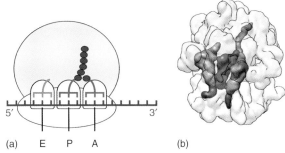

Fig. 1 Structure of the ribosome. (a) Schematic representation of a bacterial 70S ribosome with the 50S large subunit in light blue, the 30S small subunit in light yellow, the mRNA in dark blue, the nascent polypeptide and incoming amino acids in purple, and three A-, P-, and E-site tRNAs, in red, green, and orange, respectively; (b) Surface representation of the same ribosome, based on its X-ray structure [5]. For the mRNA, only the nucleotides inserted into the small subunit are represented. This and all other figures were made with VMD, UCSF Chimera, PowerPoint (Microsoft), and Photoshop (Adobe). For consistency, the same color codes and schematic representations are used consistently throughout this chapter.

Fig. 2 Overview of bacterial translation. In bacteria, ribosomes initiate translation on mRNA start sites characterized by a Shine–Dalgarno sequence positioned upstream to the AUG (or less commonly GUG or UUG) initiation codon. On coupling with the complementary sequence located at the 3′-end of the 16S rRNA (the anti-Shine–Dalgarno), the mRNA and small ribosomal subunit are bound and the start codon is correctly positioned in the P-site. A ternary complex made of the bacterial initiator fMet-tRNA^{fMet}, initiation factor 2, and GTP is then accurately positioned over the start codon with the help of two other initiation factors, IF1 and IF3. The large ribosomal subunit 50S binds to the complex, inducing GTP hydrolysis and release of the initiation factors. The elongation phase then starts and the ternary complex aa-tRNA^{aa}-EF-Tu · GTP binds to the newly formed 70S. If the mRNA codon and tRNA anti-codon match, EF-Tu hydrolyzes GTP. The aa-tRNA^{aa} accommodates in the A-site and EF-Tu · GDP is released. The nascent peptide is transferred to the A-site aminoacyl-tRNA, resulting in an elongation of the peptide by one amino acid. The ribosome oscillates between the classic and ratcheted conformation, due to the spontaneous rotation of the 30S subunit relative to the 50S. EFG · GTP binds to the ribosome GTPase center and GTP hydrolysis induces translocation of the tRNAs in the P- and E-sites, followed by the release of EFG · GDP. The process is repeated iteratively, causing elongation of the nascent peptide while the tRNAs move through the ribosomal A-, P-, and E-sites. Once the ribosome finally encounters a stop codon on the mRNA, a class I release factor (RF1 or RF2) binds the A-site and catalyzes the separation of the polypeptide chain from the P-site tRNA. The neosynthesized protein is released from the ribosome. A binary complex made from the class II release factor RF3, and GTP then facilitates the dissociation of class I factors and deacyl tRNAs from the ribosome. Finally, ribosome recycling factor (RRF) functions with EF-G to recycle the ribosome into subunits, allowing for a new round of translation. Abbreviations: mRNA, messenger RNA; SD, Shine–Dalgarno sequence; 30S, small ribosomal subunit; IF, initiation factor; fMet-tRNA, N-formylmethionine transfer RNA; GTP, guanosine-5′-triphosphate; 50S, large ribosomal subunit; aa-tRNA, aminoacyl transfer RNA; EF-G, elongation factor G; GDP, guanosine diphosphate; (P), peptidyl transfer; (R), ratchet; EF-Tu, elongation factor thermounstable; (T), translocation; CFM, cotranslational folding machinery; RF, release factor; RRF, ribosome recycling factor.

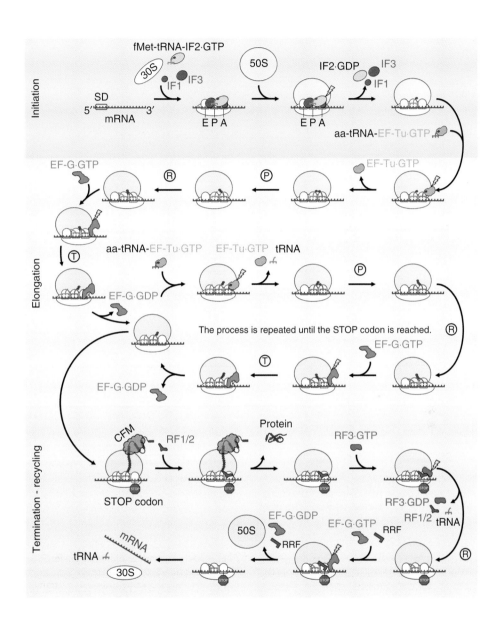

2
Canonical Protein Synthesis in Bacteria

The genetic code carried on DNA is made up of an alphabet of only four letters (A, T, G, C) representing nucleotides. For protein synthesis, the initial DNA's information is first transcribed into mRNAs, also made up of four nucleotides (A, U, G, C). The template section of the mRNA is then taken charge of by ribosomes, which translate each codon (groups of 3 nt) into one of 20 amino acids and build proteins. Prokaryotic ribosomes consist of two subunits which make up the whole of the 70S ribosome. The larger subunit (50S) is composed of the 5S and 23S ribosomal RNAs (rRNAs) and 34 proteins. The smaller subunit (30S) is made up of 16S rRNA and 21 proteins. The ribosome has three binding sites for tRNAs: the A-site, where the incoming aminoacyl-tRNA binds; the P-site, which holds the peptidyl-tRNA carrying the nascent polypeptide chain; and the E (exit) site, through which the deacylated P-site tRNA transits before being ejected (Fig. 1). Although eukaryotic ribosomes are larger they do not strongly differ [3], which argues in favor of a common protoribosomal ancestor [4].

Bacterial translation (Fig. 2) is a very dynamic process which can be divided into four sequential steps – initiation, elongation, termination, and recycling – and involves numerous protein and RNA partners that constantly bind and dissociate from the ribosome (see Ref. [6] for a complete review). Of course, any disruption of this fine-tuned process will jeopardize the viability of the bacteria, and this is why several lines of defense constantly operate to maintain translation accuracy. Among these, trans-translation is the major and most elegant pathway for overcoming ribosome stalling that results from the absence of a stop codon on the mRNA (for recent reviews, see Refs [1, 7–9]).

3
Situations Inducing Ribosome Stalling

Nonproductive translation complexes (NTCs), which are composed of the stalled ribosome, peptidyl-tRNA and mRNA, result from various translation events. Although bacteria occasionally use nascent polypeptide-induced translation arrest to regulate gene expression [10, 11], ribosome stalling is generally a serious issue for bacterial survival. Without release, most of the ribosomes would be sequestered into polysomes away from active translation and become inefficient within a single generation [9]; moreover, the incomplete polypeptides stuck on the ribosomal P-site might become harmful if released. Among the numerous situations triggering ribosome stalling, it is possible to distinguish two main pathways – the translation of non-stop or of no-go mRNAs. Various rescue systems, one of which is trans-translation, are required in order to deliver these stalled ribosomes (see also Sect. 9).

3.1
Non-Stop mRNAs

The most obvious of these situations is when ribosomes reach the 3'-end of "non-stop" mRNAs, where they stall due to an absence of the termination step (Fig. 3). This is generally caused by an altered mRNA that is devoid of a stop codon. The alteration can result from spontaneous mutations or from defects in

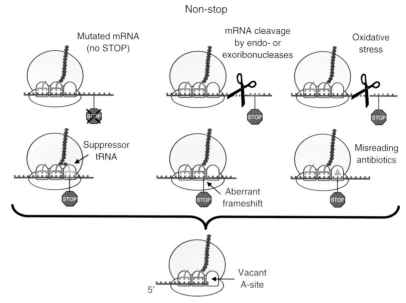

Fig. 3 Non-stop situations inducing ribosomal stalling. Several situations can lead to ribosomal stalling on a vacant A-site. The most obvious is the absence of a stop codon due to spontaneous mutations or errors of transcription. The mRNA can also be altered by the action of ribonucleases or by oxidative stress. Occasionally, stalling can also occur when a normal stop codon is erroneously translated due to the presence of non-sense suppressor tRNAs, aberrant frameshifts, or misreading drugs.

transcription because the RNA polymerase terminated transcription prematurely or did not transcribe the stop codon correctly [12]. Non-stop stalling can also be due to mRNA degradation caused by ribonucleases (endonucleases and/or processive $3' \rightarrow 5'$ exoribonucleases) or by base modifications such as oxidations, which are known to result in direct strand scission [13]. This degradation can occur either before the translation starts or during the advance of the ribosome on the mRNA. Occasionally, a situation similar to non-stop can be observed when a normal stop codon is translated (readthrough) due to the presence of nonsense suppressor tRNAs [14, 15], aberrant frameshifts [16], or misreading drugs [17]. The common result of all these situations is a ribosome stalled with 0–2 nt in the decoding (A-) site, preventing any recognition by tRNA anticodons or release factors (RFs). In bacteria, the first line of defense when ribosomes stall on non-stop mRNAs is the reaction of *trans*-translation. Alternative factors ArfA and ArfB are the standard alternate pathways of *trans*-translation at times when tmRNA-SmpB becomes scarce or overwhelmed, allowing for the very efficient (if less-sophisticated) release of ribosomes stalled at the 3′-end of non-stop mRNAs.

3.2 No-Go mRNAs

The other major problematic situation occurs on intact full-length mRNAs when ribosomes pause before the stop codon is read (Fig. 4). These "no-go" situations

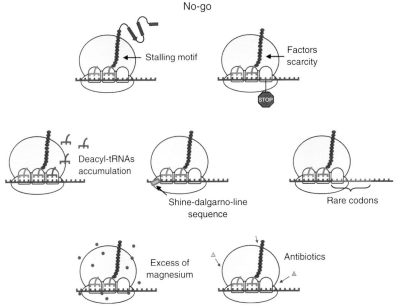

Fig. 4 No-go situations potentially inducing ribosomal stalling. No-go situations frequently occur due to the synthesis of stalling protein motifs; the scarcity of translation factors; binding of deacylated tRNAs during bacterial amino acid starvation; the presence of a Shine–Dalgarno-like sequence upstream from the pausing site; the presence of an excess of rare codons on the mRNA sequence; high magnesium concentrations; or antibiotics blocking peptidyl-transfer or translocation. Among the potential outcomes of stalling, endonucleolytic cleavage of the mRNA in the A-site by RelE or other factors can possibly convert the no-go situation into a non-stop one.

occur quite frequently in normally growing cells and can be due either to an altered reading of the mRNA, the synthesis of a problematic protein motif, or altered conditions of translation. As with the non-stop, these situations can also be rescued by *trans*-translation, but only if the mRNA is cleaved, leaving an empty A-site (for a more detailed review on this topic, see Ref. [12]).

3.2.1 Altered Reading of the mRNAs

The ribosome slows down on the mRNA due to: (i) a scarcity of translation-involved proteins, such as aminoacyl-tRNA synthetases or termination factors [18]; (ii) the binding of deacylated tRNAs to their corresponding codons, triggering the stringent response [19]; (iii) an excess of rare codons on the mRNA sequence [20–23]; or (iv) Shine–Dalgarno-(SD)-like sequences within coding sequences [24]. In the latter case, direct effects on *trans*-translation have to be demonstrated.

3.2.2 Altered Reading of the Protein Undergoing Synthesis

The ribosome cannot synthesize all of the protein sequences equally well, and some nascent peptides will strongly inhibit elongation due to: (i) the synthesis of polyproline sequences [25, 26]; (ii) cotranslational misfolding [12]; or (iii) the synthesis of specific motifs, such as the secretion monitor (SecM) "arrest sequence" (FxxxxWIxxxxGIRAGP in *Escherichia coli*) that regulates the expression of a downstream gene [27, 28] or TnaC, which is induced by high tryptophan concentrations through the binding of free tryptophan within the ribosome [29]. However, when the translational pausing is used to regulate gene expression, as it is for SecM or TnaC, the stalled ribosomes become refractory to *trans*-translation because the A-site is filled with charged tRNAs or RF [12, 30].

3.2.3 Altered Conditions of Translation

Finally, altered conditions of translation can also induce stalling. Among these should be noted: (i) the possible increased intracellular Mg^{2+} concentrations [31]; and/or (ii) antibiotics that block peptidyl-transfer or translocation [32].

4 Structure and Function of Isolated tmRNA

4.1 tmRNA Processing

The *ssrA* gene encoding tmRNA is present in all bacterial genomes [33]. Like the canonical tRNA gene, it encodes for a primary transcript that needs to be processed before giving birth to the fully mature molecule. In *E. coli*, the precursor is 457 nt long (Fig. 5), and is processed by the endonuclease RNase P at the 5′-terminus [34], whereas the 3′-terminus is first cleaved by endonucleases RNases III or E and then trimmed by exonucleases RNases T and/or PH [35]. This generates a 363 nt-long mature *E. coli* tmRNA that is aminoacylated by alanine at the conserved 3′-terminal CCA (cytidine-cytidine-adenosine) trinucleotide [36]. In most cases, tmRNA is composed of a long single strand of RNA that folds into several functional domains comprising a partial tRNA domain ("TLD," for tRNA-like domain), an internal open reading frame (ORF; "MLD," for mRNA like-domain), several RNA helices, and four pseudoknots (PKs) [37]. The total length of the tmRNA varies between about 260 and about 430 nt, depending on the cell species (Fig. 6).

4.2 tRNA-Like Domain

The TLD includes an acceptor stem formed by the interactions between the 5′- and 3′-ends of the mature tmRNA molecule. It terminates by a 3′-terminal CCA that is always aminoacylated by an alanine (see Sect. 6.1 and Refs [34, 38]). The tRNA-like structure includes a canonical T-stem–loop but a reduced D-loop devoid of a stem. It also contains the two modified nucleosides, 5-methyluridine and pseudouridine within the T loop, but not the dihydrouridine that usually is present in tRNAs [39, 40]. Specific interactions between the D- and T-loops are necessary for the binding and functioning of SmpB [41], while the T-stem and acceptor stem bind the elongation factor-thermounstable (EF-Tu).

4.3 mRNA-Like Domain

The MLD is a short internal ORF encoding a conserved tag added to the stalled peptide to promote its degradation. This tag is composed of between 8 and 35 residues, although 10 residues is the most common length (AANDENYALAA in *E. coli*,

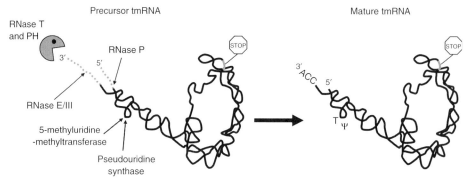

Fig. 5 tmRNA maturation. The *ssrA* gene encoding tmRNA is transcribed as a precursor. The 5'-end is processed by endoribonuclease P (RNase P), while the 3'-end is first cleaved by endoribonuclease III or E (RNase III/E) and then trimmed by exoribonucleases T or PH (RNase T/PH). As for canonical tRNAs, tRNA pseudouridine synthetase and (5-methyluridine)-methyltransferase modify two nucleosides within the T-loop of the tRNA-like domain. Once the maturation is complete, the aminoacyl tRNA synthetase (AlaRS) catalyzes the aminoacylation of tmRNA 3'-CCA with an alanine.

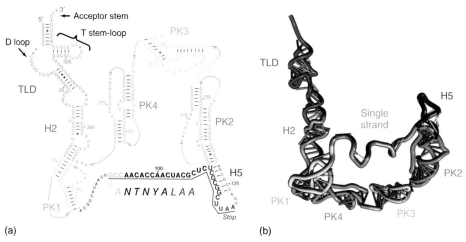

Fig. 6 tmRNA secondary and tertiary structures. (a) Secondary structure diagram of *T. thermophilus* tmRNA. Watson–Crick base pairs are connected by lines, and GU pairs are represented by dots. Domains are highlighted with colors: tRNA-like domain (TLD) in blue; helix 2 (H2) in red; pseudoknot 1 (PK1) in orange; helix 5 (H5) in brown; pseudoknot 2 (PK2) in green; pseudoknot 3 (PK3) in pink; and pseudoknot 4 (PK4) in cyan. The nucleotides within the internal ORF are underlined and shown in a larger font. The resume codon is highlighted in yellow, and the localization of the STOP codon is indicated. The corresponding amino acid sequence is also provided; (b) Three-dimensional molecular model of tmRNA based on homology modeling of each independent domain followed by flexible fitting into the cryo-EM density map of the accommodated step. EMDataBank entry: EMD-5188.

with the first A carried by tmRNA) [1]. Since tmRNA intervenes under conditions that are unfavorable to translation, the sequence encoded by the MLD is highly efficient and terminates by a stop-codon. After release, degradation of the tagged protein is performed by various enzymes. In the cytoplasm of *E. coli*, the main proteases taking over the tmRNA-tagged substrates are ClpXP, ClpAP, and FtsH [42], while in the periplasm it is Tsp, which is an energy-independent protease. Both, ClpXP and ClpAP consist of a tetradecamer of ClpP bound to a hexamer of ClpX or ClpA ATPase. The proteolytic active site of ClpP is composed of an internal chamber in which ClpX or ClpA translocate denatured tagged polypeptides after specific recognition and unfolding. Within the classical tagging sequence AANDENYALAA, ClpX binds the C-terminal residues "LAA," while ClpA binds the C-terminal residues ALA and makes additional contacts with the N-terminal residues "AA" [12]. FtsH is a hexameric protease anchored to the internal side of the cytoplasmic membrane which uses a proteolytic chamber to degrade tagged proteins [1]. The majority of degradation is performed by ClpXP, while FtsH only degrades a small subset of proteins present in the inner membrane.

4.4
Pseudoknots

tmRNA generally encompasses four PKs: PK1 is located upstream from the MLD, while PK2, PK3 and PK4 are downstream. PK1 is necessary for tmRNA tagging, but the replacement or interchange of any of the other PKs in *E. coli* does not seriously affect tmRNA functions [43]. Nonetheless, it was shown recently that the substitution of PK1 with a small and stable RNA hairpin still allows for tmRNA tagging *in vitro* or *in vivo* [44]. These data argue in favor of the participation of PK1 in stabilizing the region between the TLD and the MLD and preventing global misfolding of tmRNA, rather than having a direct role in ribosome binding [44, 45]. For the three other PKs the predominant role is the folding and maturation of tmRNA [46].

4.5
Structural Variations of Two-Piece tmRNA

In addition to the classic single-chain molecule, tmRNA also exists in alpha-proteobacteria, cyanobacteria, and some beta-proteobacteria lineages as a two-piece molecule. Two-piece tmRNA undergoes a maturation process similar to that observed for standard tmRNA, but as a result of gene circular permutation it is split into two molecules (Fig. 7) [33, 47]. It has a TLD and an MLD similar to those of standard one-piece tmRNAs, but with a break located in the loop containing the tag-reading frame and a dramatic reduction in PK numbers [48].

5
Structure and Function of SmpB

SmpB is a small basic protein (160 amino acids in *E. coli*) that is essential for *trans*-translation [49] (Fig. 8a). The *smpb* gene is present in all bacterial genomes with a high primary sequence conservation, and its deletion results in the same phenotypes as those observed in tmRNA-defective cells. All of the cellular roles of SmpB are linked in one way or another to *trans*-translation: it enhances tmRNA aminoacylation; prevents its degradation [50, 51]; and is required for tmRNA to bind the vacant A-sites of the stalled ribosomes [52]. The structure of

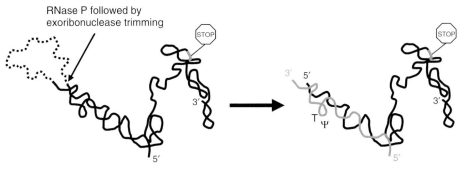

Fig. 7 Two-piece tmRNA maturation. Beside the classic single-chain molecule, tmRNA is also sometimes found as a two-piece molecule, due to independent gene circular permutation events. Similar processing at internal sites (illustrated as a dotted line) of permuted precursor tmRNA explains its physical splitting into two pieces (highlighted in black and gray). While the TLD is remarkably conserved, the MLD is drastically reduced and several pseudoknots are missing.

SmpB was first solved by nuclear magnetic resonance studies [53, 54]. The SmpB protein forms an oligonucleotide-binding (OB) fold made from three α-helices surrounding six antiparallel β-strands arranged in the typical closed β-barrel. This arrangement exposes two conserved RNA-binding domains on opposite sides of the protein. The protein also possesses a C-terminal tail (residues 131–160 in *E. coli*) that is predicted to form an α-helix but always appears unstructured in solution (Fig. 8b). This tail is essential for tmRNA tagging [55, 56], and was correctly predicted to bind to the 30S A-site [57–59] (see Sect. 7.1). As shown by further X-ray studies, SmpB binds with high specificity to the elbow region of the TLD, stabilizing the single-stranded D-loop in an extended conformation [60, 61]. This interaction leads to an increased elbow angle of approximately 120° (versus 90° in canonical tRNAs), in agreement with results from electrical birefringence studies [62]. This increased angle places SmpB where the anticodon and D stems of a tRNA should be, while the H2 helix of tmRNA (Fig. 8c) mimics a tRNA long-variable-arm.

6
Other tmRNA Partners

In addition to SmpB, tmRNA requires at least four other RNA-binding proteins to perform *trans*-translation: Alanyl tRNA synthetase (AlaRS); EF-Tu; S1; and ribonuclease R (RNase R) [64].

6.1
Alanyl-tRNA Synthetase (AlaRS)

tmRNA is always charged by AlaRS, a class II tRNA synthetase that catalyzes the esterification of alanine to $tRNA^{Ala}$. The specific recognition of $tRNA^{Ala}$ by AlaRS depends largely on the presence of the $G3 \cdot U70$ wobble base pair (found in the acceptor stem of all $tRNA^{Ala}$ isoacceptors) and of an adenosine at the discriminator position adjacent to the 3'-terminal CCA [65]. Strikingly, this $G \cdot U$ wobble base pair is also conserved in all *ssrA* sequences, making

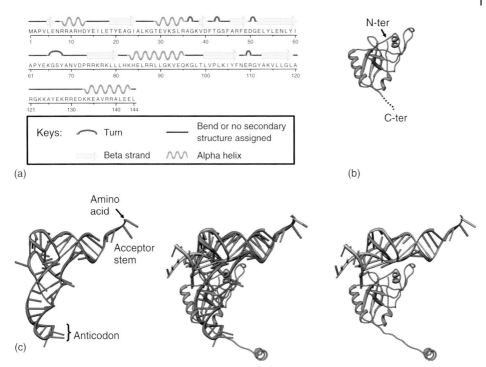

Fig. 8 SmpB, the handyman of tmRNA. (a) Secondary structure diagram of *T. thermophilus* SmpB, as observed in the X-ray atomic structure of a ribosome in complex with TLD–SmpB-EF-Tu [63]; (b) Atomic structure of *T. thermophilus* SmpB in solution. The N- and C-terminal ends are indicated. SmpB adopts an oligonucleotide-binding fold (OB fold) with a central β-barrel and three flanking α-helices. The C-terminal tail is unstructured in solution but folds into a fourth α-helix once SmpB is inserted into the ribosome; (c) Structural comparison between TLD-SmpB and tRNA in the pre-accommodated state. Left: tRNA in green; Right: TLD in red and SmpB in purple; Center: superposition of the two structures shows that the TLD resembles the upper part of a tRNA while SmpB replaces the tRNA anticodon stem–loop. Structural coordinates were taken from Protein Data Bank entries 4ABR for the TLD-SmpB and 2WRN for tRNA.

AlaRS the only amino-acid synthetase that takes charge of tmRNA *in vivo*. By swapping the upper half of the acceptor stem of tmRNA with tRNA^{His}, it was shown that the tmRNA^{His} mutant retains the ability to specifically tag stalled polypeptides, revealing that the first residue of the tag-peptide derives from the TLD of tmRNA [66]. The structural basis of the interaction between AlaRS and tmRNA is unknown. However, one of the SmpB loops might participate in the interaction with AlaRS, explaining why the protein promotes tmRNA alanylation [60].

6.2
Elongation Factor EF-Tu

EF-Tu is the most abundant protein in the bacterial cell, accounting for up to 5% of the total cellular protein. It brings the aminoacyl-transfer RNA (aa-tRNA) to the ribosome as part of a ternary complex, aa-tRNA-EF-Tu · GTP [67]. When the tRNA anticodon matches the mRNA codon, GTP hydrolysis by EF-Tu results in EF-Tu · GDP dissociation and aa-tRNA accommodation into the ribosomal A-site (see Fig. 2). As takes place in canonical translation, EF-Tu also interacts with tmRNA-SmpB to initiate *trans*-translation. The overall conformation of a *trans*-translating ribosome in a pre-accommodated stage closely resembles that of the equivalent complex of EF-Tu with acylated tRNA, including the conformation of the 3'-CCA end and the acceptor arm, as well as the T-arm portion that interacts with EF-Tu (see Sect. 7.1). Notably, EF-Tu shows no interaction with SmpB [63] More surprisingly, EF-Tu can also bind to charged or deacylated tmRNA in its GDP form [68, 69], and in that case it also interacts with regions outside the TLD. These unexpected interactions might serve to protect tmRNA from degradation.

6.3
Ribosomal Protein S1

S1, the product of the *rpsA* gene, is the longest and largest ribosomal protein. It is a key mRNA-binding protein in Gram-negative bacteria, weakly associating with the 30S small subunit and facilitating recognition of most mRNAs by ribosomes during the initiation step of translation [70, 71]. It consists of six homologous S1 domains, characteristic of the OB-fold family of RNA-binding proteins [72]. S1 binds to the ribosome via its N-terminal domain, while its elongated C-terminal RNA-binding domain protrudes into solution [73]. The involvement of S1 in *trans*-translation has been debated on numerous occasions. The high binding affinity of S1 to tmRNA (600-fold greater than that to tRNA) suggests that S1 plays an important role in *trans*-translation by forming complexes with free tmRNA and then by facilitating its binding to ribosomes [74]. Interactions between S1 and tmRNA trigger significant conformational changes throughout the RNA molecule (with the exception of the TLD), suggesting that the protein binds tmRNA through PK2 [75]. Accordingly, cryoelectron microscopy (cryo-EM) data showing that tmRNA would bind to the ribosome in a pre-accommodated step, with or without S1, represents an interesting clue to the putative role played by S1 during *trans*-translation [52, 76]. When the complex is prepared in the absence of S1, an extra density corresponding to the MLD of tmRNA is observed; this suggests a possible unwinding of that portion by S1 outside the ribosome, before *trans*-translation starts. In the pre-binding stage, S1 could therefore enter the ring of PKs (which has an inner diameter of about 80 Å), thereby converting tmRNA into its functional conformation. Subsequently, upon binding of the tmRNA to the stalled ribosome, S1 would be released and the ORF placed in the decoding site [52]. However, the physiological relevance of S1 during *trans*-translation remains elusive. Indeed, by using an *in vitro trans*-translation *Thermus thermophilus* system it was shown that S1 was not essential for the early steps of the process, even though it tightly binds to tmRNA [77]. Additional conflicting data derives from the absence of S1 within the low-G + C group of Gram-positive bacteria, though the idea cannot be ruled out that S1 homologous proteins might fulfill S1 roles in such bacteria. Recent data acquired from *Actinobacteria* group of Gram-positive bacteria have strongly confirmed that S1 is essential for *trans*-translation in these organisms.

For example, it has been shown that pyrazinamide, a first-line antituberculosis drug, functions by being transformed intracellularly into pyrazinoic acid (POA), a molecule that binds to S1 and consequently inhibits *trans*-translation [78]. Therefore, S1 is an attractive target for antibiotics inhibiting *trans*-translation (see Sect. 10).

6.4
Ribonuclease R (RNase R)

In *E. coli*, RNase R is a 92 kDa ubiquitous protein encoded by the *rnr* gene, and is involved in the degradation of different types of structured RNAs such as rRNA, small RNA, and mRNA [79, 80]. RNase R belongs to the RNase II superfamily of enzymes that degrade RNA hydrolytically in the 3′ → 5′ direction in a processive and sequence-independent manner. During *trans*-translation, problematic mRNA transcripts must be rapidly degraded in order to avoid a vicious cycle of being recruited over and over for new rounds of translation and subsequent *trans*-translation. RNase R has the unique capability to degrade defective mRNA in a *trans*-translation-dependent manner [81, 82]. It is recruited to ribosomes that are stalled on non-stop mRNAs thanks to its unique K-rich domain [81]. SmpB and tmRNA interact with RNase R both *in vitro* and *in vivo*, and the C-terminal region of RNase R is required for this interaction. More surprisingly, tmRNA and SmpB regulate RNase R. RNase R acetylation promotes the binding of tmRNA-SmpB to its C-terminal region, which in turn induces the instability of the enzyme [83]. In the absence of either tmRNA or SmpB, the half-life of RNase R and its level in exponential-phase cells are increased markedly [84]. In contrast, in *Streptococcus pneumoniae*, SmpB is regulated by RNase R which is, in turn, under the control of SmpB [85]. These proteins are therefore mutually dependent, confirming the tight crossregulation between *trans*-translation and RNase R.

7
trans-Translation: from Structures to Mechanistics

7.1
The Pre-Accommodation Step

The first glimpse into how tmRNA-SmpB interacts with the stalled ribosome was revealed by the cryo-EM-determined structure of the tmRNA-SmpB-EF-Tu·GDP pre-accommodation step [76]. These first images showed that the TLD is associated with EF-Tu and is brought to the ribosome in the same manner as an aminoacyl tRNA (Fig. 9). Originally, the SmpB molecule was placed in close contact with the TLD and the 50S subunit. However, the crystal structures of the TLD-SmpB in solution [60, 61] and results of chemical probing experiments [86] soon suggested that SmpB mimics the "codon–anticodon" pairing and D-stem–loop of a tRNA, while the TLD plays the role of its upper half. This positioning was confirmed in a second map of the pre-accommodated step which established the occurrence of two SmpBs in the complex: one binding to the 30S subunit close to the decoding site, and the other binding to the 50S subunit at the GTPase-associated center [57]. Previously, the number of SmpBs molecules involved in *trans*-translation has been quite controversial and has been the subject of numerous debates [7]. However, the model presented in this chapter now accounts

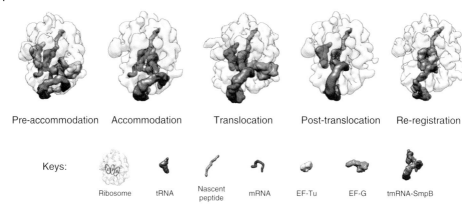

Fig. 9 Cryo-EM maps of the currently solved functional complexes. Left to right: Pre-accommodation: The alanyl-tmRNA–SmpB•EF-Tu · GTP quaternary complex enters the vacant A-site of a non-stop stalled ribosome. Accommodation: EF-Tu dissociates after GTP hydrolysis, allowing tmRNA–SmpB to occupy the A-site. Translocation: EF-G · GTP catalyzes the translocation of tmRNA–SmpB to the P-site; the ribosome is in a ratcheted state. Post-translocation: After dissociation of EF-G · GDP, the subunits return to normal positioning and the resume codon of tmRNA is correctly placed into the A-site. Re-registration: Translation switches on tmRNA internal ORF, and a new aminoacyl-tRNA binds to the resume codon. Structural coordinates were taken from the following EMDataBank entries: EMD-1312 for pre-accommodation; EMD-5188 for accommodation; EMD-5386 for translocation; EMD-5189 for post-translocation; and EMD-5234 for re-registration.

for a 1 : 1 molar ratio of SmpB and tmRNA, in accordance with the most recent crystallographic and cryo-EM studies of pre- and post-accommodated states (see below). The first cryo-EM maps also led to several interesting insights into how the huge tmRNA molecule would interact with the ribosome. The H2 helix mimics a tRNA variable arm and leans along the 30S subunit, pointing toward the beak and out of the ribosome. The remaining domains (PK1, the MLD, H5, PK2, PK3, and PK4) then wrap around the small unit beak, much like a lasso. Further details on how the tmRNA–SmpB complex identifies stalled ribosomes were recently revealed by the first crystal structure of ribosome complex with TLD–SmpB-EF-Tu [63]. The TLD–SmpB complex shows no major distortions after binding to the ribosome, acting in every way as a tRNA molecule (Fig. 10). The overall conformation of a *trans*-translating ribosome in a pre-accommodated stage closely resembles that of the equivalent complex of EF-Tu with amino-acylated tRNA. The similarities include the conformation of the 3′-CCA end, the acceptor arm, and the T-arm portion. SmpB interacts with the 30S subunit in order to bring the shoulder domain of the 30S subunit (containing the key residue G530) closer to its 3′ major domain (containing the decoding residues A1492 and A1493). In doing so, SmpB tricks the 30S subunit into adopting a "closed" conformation, in which the shoulder and head domains are rotated toward the subunit center, as if in the presence of a cognate codon–anticodon [87]. At the same time, the C-terminal tail of SmpB folds in an α-helix in the unoccupied mRNA pathway downstream of the

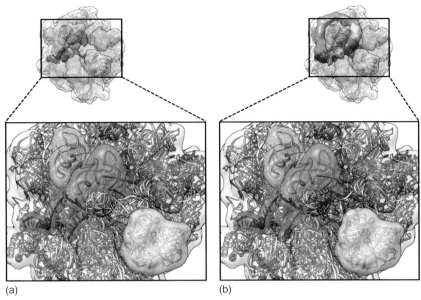

Fig. 10 Recognition of the ribosomal A-site ribosome by tmRNA–SmpB compared to a canonical tRNA. (a) Fitting of the X-ray atomic structure of the ribosomal complex with tRNA–EF-Tu·GDP [63, 88], and the cryo-EM map of the same complex [76, 89]. The inset shows a close-up view of the ribosomal A-site; (b) Fitting of the X-ray atomic structure of the ribosomal complex with TLD-SmpB–EF-Tu·GDP [63], and the cryo-EM map of the complete tmRNA–SmpB-EF-Tu·GDP complex [37]. The inset shows a close-up view of the ribosomal A-site. The C-terminal tail of SmpB folds into an α-helix inserted into the empty mRNA path. For clarity, the tmRNA envelope is not represented. Color codes used for the envelope and atomic structures: tmRNA, red; SmpB, purple; 30S small ribosomal subunit, gold; 50S large, light blue; A-site tRNA, yellow; P-site tRNA, green; E-site tRNA, orange; EF-Tu, pink; mRNA path, deep blue.

decoding center (Fig. 10b). In this way, the protein can undergo specific interactions with regions that are only accessible in the absence of mRNA, thus stabilizing SmpB and allowing for an accurate identification of the vacant A-site. The structure thus explains the functional relevance of the SmpB C-terminal tail to tmRNA tagging activity.

7.2
The Accommodation Step

Upon GTP hydrolysis, EF-Tu·GDP is released and the tmRNA–SmpB complex accommodates into the A-site (Fig. 9). The TLD contacts with the large ribosomal subunit resemble those of an accommodated canonical tRNA [90–93]. The D-loop interacts with helix H38, while the acceptor branch brings the CCA 3′-end into the peptidyl transfer center (PTC). The TLD swings into the A-site and SmpB follows, rotating by around 30° while still mimicking an anticodon stem–loop. As a consequence, helix H2 realigns toward the large subunit, where it interacts with protein L11. The ring of PKs remains wrapped around the beak of the small subunit and does not undergo large movements.

7.3
The Translocation Steps

After transpeptidation, the nascent peptide is elongated by one alanine, and the 30S subunit rotates spontaneously in an anticlockwise direction relative to the 50S. This ratchet-like motion brings tmRNA and tRNA into hybrid states of binding (respectively A/P and P/E). By using fusidic acid antibiotic, Spahn and coworkers trapped the 70S–tmRNA–EF-G complex in an intermediate state during the translocation reaction [94]. The structure (Fig. 9) revealed that the ribosome was once again in a state similar to that observed during canonical translation, but with a unique 12° tilt of the head. EF-G binds to SmpB in a very similar fashion as it does to tRNA. The tmRNA–SmpB is in a hybrid state with the TLD bound to the 50S P-site and SmpB still pointing toward the A-site. The opening of the inter-subunit B1a bridge (or "A-site finger") during the ratchet allows for the transit of H2 helix [92]. As a result, the A-site finger – mutations of which are known to alter tmRNA function [95] – interacts with PK1. Following the movement of the 30S head, the PK ring rotates and H5/PK2 comes in contact with proteins S2 and S3 at the surface of the 30S subunit. At the same time, the tmRNA internal ORF extends into the mRNA path.

After dissociation of EF-G·GDP, the subunits return to normal positioning and the resume codon of tmRNA is correctly placed into the A-site [92]; translation then switches on the tmRNA internal ORF, and a new aminoacyl-tRNA binds to the resume codon [91]. SmpB remains present in tmRNA–ribosome complexes at the different stages of tmRNA passage through the ribosome [96]. In these "post-translocation" and "re-registration" states (Fig. 9), the TLD and SmpB occupy the space of a regular tRNA in the P-site. The H2 helix is tightly inserted between the two ribosomal subunits, forming numerous contacts with them both. The conformation and orientation of the PK ring does not change much, with H5 and PK2 still lying at the surface of the 30S subunit. At this point, the internal ORF is only partially unfolded as the helix H5 is still present. However, the distance separating PK1 and H5 is increased, which suggests that the single strand connecting these two domains is fully extended during its insertion in the mRNA path. While this single strand part of the tmRNA is not directly observable in the cryo-EM structure, the ribosomal environment imposes sufficient restraint for it to be precisely modeled (Fig. 11). By using molecular dynamic flexible fitting, the groups of J. Frank and the present authors were able to show that the hydrophobic pocket at the bottom of the protein and the C-terminal tail of SmpB would interact with the upstream region of the tmRNA resume codon [91, 92]. These interactions were instrumental in placing the resume codon directly into the 30S decoding center, and explained the importance of the five residues upstream from the tmRNA resume codon [97–99], the SmpB C-terminal tail [100], and four specific residues on SmpB surface [101] in frame selection. A comparison of the accommodation and post-translocation electron density maps also confirmed that the truncated mRNA is released during the translocation of tmRNA to the P-site [92].

7.4
When tmRNA-SmpB Reaches the Edge of the Ribosome: What Next?

Although the remaining steps have yet to be observed, an accurate model for

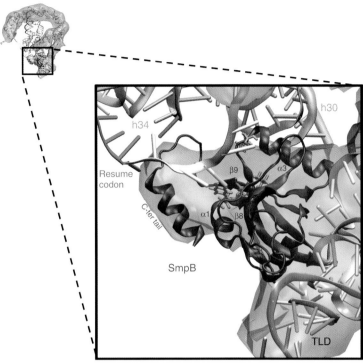

Fig. 11 mRNA swapping. Overview: Molecular model of tmRNA–SmpB in the cryo-EM density of the post-translocation complex. The inset shows a close-up view of SmpB frame selection. Modeling tmRNA–SmpB in the crowded environment of the ribosome indicates two different binding sites. While the narrow side of SmpB OB-fold connects to the elbow region of the TLD, a second binding site links the hydrophobic pocket at the bottom of the protein and the C-terminal tail of SmpB to the upstream region of the tmRNA resume codon. Color codes: Resume codon, green; helices H30 and H34 of the 16S rRNA, pink; tmRNA–SmpB envelope, gray; tmRNA, white; SmpB, purple. Locations of SmpB mutations known to affect frame choice [101] are highlighted in orange.

trans-translation can now be proposed by comparing it to canonical translation (Fig. 12). The formerly stalled peptide is transferred to the aminoacyl-tRNA recognizing the tmRNA resume codon. The 30S subunit then rotates spontaneously in an anticlockwise direction relative to the 50S, this ratchet-like motion bringing the tRNA and tmRNA to hybrid states of binding (A/P and P/E) and allowing for the opening of several bridges between the two subunits. The ribosome again oscillates spontaneously between the classic and ratcheted conformation until elongation factor G (EFG) · GTP binding. Upon GTP hydrolysis, the tRNA and tmRNA-SmpB are translocated into the P- and E-sites, respectively, whereupon EFG · GDP is released and a new cycle restarts. The process is repeated and the nascent peptide is elongated until the tmRNA stop codon is reached. Along the way, the TLD and H2 parts of tmRNA, as well as SmpB, are rapidly released. The messenger part of tmRNA is extended, leading to the destructuration of helix H5 [102, 103].

66 | *Bacterial trans-Translation: From Functions to Applications*

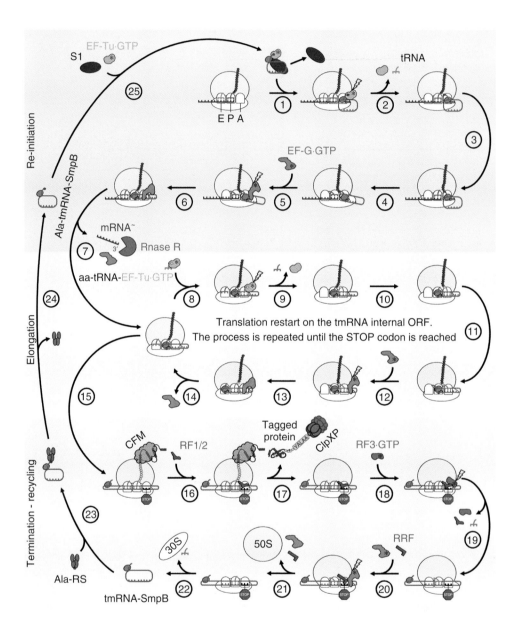

The tmRNA–SmpB complex remains in the vicinity of the ribosome because the internal ORF is in the mRNA path; however, once the stop codon is reached a class I release factor (RF) 1 or 2 binds to the A-site. The GGQ (tripeptide sequence Gly-Gly-Gln) motif of the RFs is positioned in the PTC and induces the nascent peptide hydrolysis from the P-site tRNA. The protein is released from the ribosome and immediately targeted by proteases because of the tagged C-terminal tail. RF3 in complex with GTP then binds the ribosome. Upon GTP hydrolysis, the deacyl-tRNA present in the E-site along with RF1/2 and RF3 are released. The small subunit rotates with respect to the large subunit and once again the ribosome adopts a ratcheted conformation. The ribosome recycling factors bind the ribosome

Fig. 12 Overview of *trans*-translation. Re-initiation: (1) pre-accommodation. Ala-tmRNA^{Ala}-SmpB–EF-Tu · GTP quaternary complex binds to a stalled ribosome. SmpB recognizes the vacant A-site. (2) SmpB simulates the codon–anticodon recognition and induces GTP hydrolysis. Ala-tmRNA^{Ala}-SmpB accommodates in the A-Site. EF-Tu · GDP and E-Site deacyl tRNA are released. (3) Peptidyl transfer: the nascent peptide is transferred from the P-Site tRNA to the Ala-tmRNA^{Ala}. The nascent peptide is elongated by one Ala. (4) Ratchet: the 30S subunit rotates spontaneously in an anticlockwise direction relative to the 50S. This ratchet-like motion brings TLD-SmpB and tRNA into hybrid states of binding (A/P and P/E respectively). (5) EF-G · GTP binds to the ribosome. (6) GTP hydrolysis. TLD-SmpB and tRNA are translocated to the P- and E-sites, respectively. tmRNA internal ORF is positioned in the A-site. (7) EF-G · GDP and non-stop mRNA release. Subsequent degradation of the non-stop mRNA by RNAse R. Elongation: translation restart on the tmRNA internal ORF. (8) aa-tRNA^{aa}-EF-Tu · GTP ternary complex binds to the ribosome. (9) The recognition of tmRNA internal ORF codon by the aminoacyl tRNA induces GTP hydrolysis. The aa-tRNA^{aa} is accommodated in the A-site. EF-Tu · GDP and E-site deacyl tRNA are released. (10) Peptidyl transfer. The nascent peptide is transferred to the incoming aa-tmRNA^{aa} and is elongated by one amino acid. (11) Ratchet. (12) EF-G · GTP binding. (13) GTP hydrolysis and translocation. (14) EF-G · GDP release. The process is repeated until the tmRNA STOP codon is reached. After the first cycle, TLD and SmpB are released from the E-site like deacyl tRNAs. Termination–recycling: (15) The tmRNA STOP codon is reached. (16) RF1 or RF2 recognize the STOP codon and bind to the A-site. (17) The class I release factor triggers the P-site tRNA deacylation. The neosynthetized peptide (if unfolded) or protein (if already folded by the CFM) that carries the tmRNA tag is released. A protease such as ClpXP recognizes the tag and degrades the potentially hazardous product. (18) Class II release factor binds to the ribosome. (19) GTP hydrolysis induces a ratchet-like movement and rapid dissociation of class I and II release factors and E-site deacyl-tRNA. (20) RRF and EF-G · GTP binding. (21) GTP hydrolysis. RRF acts as a wedge, inducing dissociation and recycling of the large ribosomal subunit. RRF and EF-G · GDP are also released. (22) Deacyl tmRNA–SmpB and tRNA dissociate from the small ribosomal subunit. The 30S can be used for a new round of translation. (23) Ala-RS charges the deacyl tmRNA–SmpB with a new alanine. (24) Ala-RS is released. (25) EF-Tu · GTP and S1 bind to Ala-tmRNA^{Ala}–SmpB, and the complex is ready to rescue another stalled ribosome. S1, small ribosomal subunit protein 1; EF-Tu, elongation factor thermounstable; GTP, guanosine-5′-triphosphate; t-RNA, deacyl transfer RNA; EF-G, elongation factor G; mRNA, non-stop mRNA. RNase R, exoribonuclease R; aa-tRNA, amino-acyl transfer RNA; CFM, cotranslational folding machinery; RF1/2, release factor 1 or 2; RF3, release factor 3; RRF, ribosome recycling factor; 50S, large ribosomal subunit; 30S, small ribosomal subunit; tmRNA–SmpB, deacyl transfer-messenger RNA and small protein B binary complex; AlaRS, alanyl-tRNA synthetase; ala-tmRNA–SmpB, alanyl transfer-messenger RNA and small protein B binary complex.

and recognize the tmRNA stop codon present in the A-site, and EF-G · GTP then binds to the ribosome GTPase center. Upon GTP hydrolysis the two subunits of the ratcheted ribosome revert to their original position. This motion is transmitted to the ribosome recycling factor (RRF) through specific contacts between EFG and the RRF tRNA-mimicry domains. However, in contrast to tRNA, RRF cannot translocate to the P-site, and so acts as a wedge and forces a dissociation of the large subunit; RRF and EFG · GDP are also released during this process. The deacyl-tRNA, deacyl-tmRNA–SmpB and ribosomal small subunits separate, and the tmRNA–SmpB can now be recycled. The AlaRS recognizes and charges the deacyl-tmRNA with an alanine. The aminoacylated tmRNA–SmpB complex is then taken charge of by EF-Tu · GTP and the protein S1, and a new round of trans-translation is begun.

8
Other Physiological Roles Attributed to trans-Translation

8.1
Role of trans-Translation during the Cell Cycle

Trans-translation plays an important role in cell-cycle control. In particular, tmRNA is required for correct timing of the $G_1 \to S$ phase transition in the bacterium Caulobacter crescentus, and its deletion leads to a delay in the initiation of DNA replication [104]. As a result, control of the cell cycle is tightly regulated by the timing of both synthesis and degradation of tmRNA: it increases in the late G_1 phase, peaks during the $G_1 \to S$ transition, and declines in the S phase [105]. This regulation is controlled by a fine-tuned balance between RNase R (that degrades tmRNA from its 3′-end) and SmpB (which specifically protects tmRNA from degradation) [106].

8.2
Role of trans-Translation in Phage Growth

Directed proteolysis may not be the only biological function of trans-translation [107]. tmRNA is required for the growth of certain phages, including the hybrid lambdaimmP22 phages in E. coli [108]. In that particular case, the charging of tmRNA with alanine, but not the degradation of tmRNA-tagged proteins, is essential for phage growth. Strikingly, these data led the authors to propose that the main biological role of trans-translation is not directed proteolysis but rather towards an increased translational efficiency through the removal of stalled ribosomes from the original mRNAs [107].

9
Alternative Pathways

Intriguingly, although required by some bacteria, trans-translation is not always necessary to maintain cell viability [1], which suggests the existence of alternative rescue systems. Indeed, alternative ribosome rescue factors A (ArfA, formerly YhdL) and B (ArfB, formerly YaeJ) have been recently discovered in many bacteria (including E. coli), revealing a far more complicated network than previously expected [109] (Fig. 13).

ArfA is a small protein that joins with RF2 to take over the rescue of stalled ribosomes in the absence of trans-translation [110]. The deletion of both genes that code for tmRNA and ArfA is lethal for prokaryotic cells. The positioning

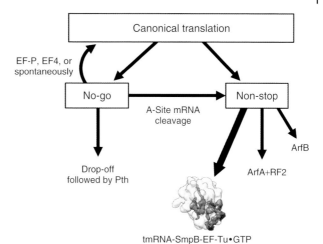

Fig. 13 Bacterial translation quality control machinery. During translation the ribosome can stall in two different ways: when slowed-down on an intact mRNA ("no-go" stalling), or in the absence of a termination codon at the 3′-end of the mRNA ("non-stop" stalling). No-go stalling can evolve in three different ways. First, ribosomes return to canonical translation spontaneously or through the action of factors such as EF-P or EF4. Second, the ribosomal subunits dissociate and peptidyl-tRNA hydrolase (Pth) separates the nascent peptide from the tRNA. Third, endonucleolytic cleavage of the mRNA in the A-site by RelE toxin or other factors converts the no-go situation into a non-stop one. Non-stop stalling is primarily rescued by *trans*-translation mediated by tmRNA–SmpB. When tmRNA–SmpB becomes scarce or overwhelmed, recycling is taken over by ArfA in cooperation with RF2, and/or to a lower extent by ArfB. Structural coordinates were taken from Protein Data Bank entry 3IYQ for tmRNA-SmpB.

of ArfA in the ribosome in the presence of RF2 is not yet known, but when joining the stalled ribosome the RF2 catalytic unit is expected to enter the heart of the PTC and trigger release of the stalled peptide. Interestingly, the mRNA that encodes ArfA is cleaved by ribonuclease III and subsequently downregulated by tmRNA–SmpB. Therefore, the ArfA protein is correctly translated from its truncated mRNA only when *trans*-translation is inefficient [111].

ArfB is similar in sequence and structure to a catalytic domain of class I RFs [112, 113]. As does SmpB [63], ArfB carries a C-terminal tail that binds to the empty mRNA entry channel of the small subunit, thus discriminating between the active and stalled ribosomes [114]. In contrast to ArfA, ArfB possesses its own catalytic domain and does not need the help of a class I RF to separate the nascent peptide from the P-site tRNA. When positioned optimally in the PTC, ArfB catalyzes peptidyl-tRNA hydrolysis and release of the stalled peptide. A double deletion of ArfB and tmRNA is not lethal.

Ribosomal stalling sometimes occurs on intact "no-go" mRNAs, before the stop codon is reached (see Sect. 3.2). If mRNAs are not cleaved, then four possible pathways may open up (Fig. 13, left). In the first pathway, stalled ribosomes return spontaneously to canonical translation if the cellular conditions improve, whereas in the second pathway the stalled ribosomes are remobilized. Elongation factor 4 (EF4, formerly LepA) stimulates protein synthesis under conditions of high Mg^{2+} that form no-go ribosomal states, which are due to an impairment of EF-G-dependent translocation [31]. EF4 reactivates protein synthesis by provoking a back-translocation that results in a pre-translocational state that is then correctly translocated by EF-G [31]. In the third pathway, the stalled ribosomes are recognized by an alleviating factor that facilitates peptidyl transfer. This situation was recently described for EF-P, a highly conserved and ubiquitous translation factor that enhances the translation of proline codon stretches [115, 116]. It should be noted that EF-P is active only on polyproline stretches and not on other amino acids encoded by rare codons. In the fourth pathway, stalled ribosomes are recycled by a premature drop-off of peptidyl-tRNAs and subsequent release of the polypeptide; this reaction is mediated by the enzyme peptidyl-tRNA hydrolase (Pth), which hydrolyzes the ester bond that links the peptide and tRNA [117].

However, no-go mRNAs can also become non-stop and trigger *trans*-translation or ArfA/B rescuing! A major role is then played by ribonucleases and bacterial toxins that cleave stalled mRNAs in the ribosomal A-site (Fig. 13, center) [19, 118–121].

During bacterial amino acid starvation, ribosomes stall on intact mRNAs because of the binding of uncharged tRNAs to their A-sites, triggering the stringent response. In response, stringent factor RelA binds to the ribosomes and triggers (p)ppGpp synthesis. This downregulates the transcription of elements involved in translation and upregulates the transcription of enzymes involved in amino acid synthesis [19]. This decrease in protein synthesis reduces the amounts of RelB that is part of the *relBE* toxin/antitoxin system. The RelE toxin is stable, whereas the RelB antitoxin is labile [122]. Therefore, the decrease in RelB concentration frees the stable RelE that in turn binds the stalled ribosome and specifically cleaves mRNA in the A-site. This leaves a vacant A-site and converts the complexes into non-stop mRNA stalled ribosomes that can be targeted by tmRNA-SmpB, ArfA, or ArfB. Other toxin/antitoxin systems, such as the *mazEF* module (*mazF* encodes a stable toxin, MazF, and *mazE* encodes a labile antitoxin, MazE) act in a same manner [119].

However, A-site cleavage during translational stalling also occurs without the release of toxins [16]. A-site cleavage must therefore be dependent upon other factors, such as RNase II, the major $3' \rightarrow 5'$ exoribonuclease responsible for mRNA degradation in *E. coli* [123]. Presumably, this enzyme degrades stalled mRNA, starting downstream from the ribosome and continuing to its leading edge, and thus facilitating subsequent cleavage of the A-site codon by an as-yet unidentified RNase. It has been suggested that the ribosome might cleave the mRNA itself, but this hypothesis is unproven [16, 18, 124, 125].

10
Medical Applications

10.1
trans-Translation as a Target for Protein Synthesis Inhibitors

Taking into account the absence of *trans*-translation in eukaryotes, tmRNA–

SmpB is a promising target for novel antibiotics, or for increasing the activity of currently used protein synthesis inhibitors. Clearly, when it is essential to the survival of pathogenic bacteria (e.g., *Neisseria gonorrhoeae*, *Helicobacter pylori*, or *Shigella flexneri*), *trans*-translation machinery is an excellent specific target for the development of novel antibiotics [126–128].

When nonlethal, the deletion of tmRNA or SmpB induces the appearance of an hypersensitive phenotype in several bacterial species, including the cyanobacterium *Synechocystis* sp. strain PCC6803, *E. coli* and *Salmonella typhimurium* [32, 129]. These hypersensitive mutants are not viable in the presence of low doses of protein synthesis inhibitors (e.g., chloramphenicol, lincomycin, spiramycin, tylosin, erythromycin, and spectinomycin) that do not otherwise significantly affect the growth of wild-type cells. This indicates that ribosomes stalled by certain protein synthesis inhibitors can be recycled by tmRNA, and that tmRNA inactivation could serve as a useful therapeutic target to increase the sensitivity of pathogenic bacteria to antibiotics [32, 129, 130].

More surprisingly, mutants deleted for tmRNA are also more sensitive than wild-type cells to antibiotics not targeting translation, such as inhibitors of cell-wall synthesis. This is most likely because these drugs induce an overall stress on the bacteria that is handled more efficiently when *trans*-translation is active [131].

Last, but not least, it was shown recently that pyrazinamide – a mainstay of combination therapy for tuberculosis [132] – inhibits *trans*-translation [78], making this process a potentially promising target for other drugs. In this particular case, the ribosomal protein S1 becomes the target of POA, the active form of the pyrazinamide prodrug. Interestingly, POA only inhibits *trans*-translation and not canonical translation, and this inhibition depends strictly on wild-type *M. tuberculosis* S1. This confirms that the inhibition of *trans*-translation in any fashion would serve as an excellent approach to developing novel antibacterial agents [133].

10.2
tmRNA–SmpB Mutants as Live Attenuated Vaccines

The impairment of *trans*-translation leads to a wide variety of phenotypes in different bacterial species (see Sect. 10.1). Notably, it can lead to the attenuation of virulence of several pathogens, including *Yersinia pestis* (the causative agent of plague) and *Francisella tularensis* (the etiological agent of the zoonotic disease, tularemia). Consistent with this, the use of attenuated mutants of *Y. pestis* or *F. tularensis* deleted for *SmpB–SsrA* genes has been shown recently to induce a strong antibody response in mice [134, 135]. The defined *SmpB–SsrA* mutation could thus be a very promising candidate for future live-cell vaccine development [136].

10.3
Use of tmRNA as a Diagnostics Target

rRNA molecules (particularly 16S rRNAs) are used routinely for the phylogenetic or *in situ* identification of bacteria. However, in some instances rRNA cannot be used as a target to distinguish between phylogenetically closely related bacteria [137], and tmRNA may represent an attractive alternative. Indeed, the high sequence conservation at both 5′- and 3′-ends of tmRNA genes, along with their small size, allows for the design of universal primers that could amplify the central part of *ssrA* genes [138]. Moreover, unlike 16S rRNA the tmRNA internal sequences display considerable divergence between species. A first phylogenetic analysis of *ssrA* genes from 24 strains of Gram-positive bacteria with low G + C content (namely *Bacillus, Lactococcus, Lactobacillus, Streptococcus, Staphylococcus, Leuconostoc, Listeria,* and *Enterococcus*) showed the tmRNAs to be more divergent than 16S RNA, making tmRNA a promising tool for species identification [138]. The efficacy of this method was confirmed on *E. coli* bacteria that were visualized specifically with fluorescence *in-situ* hybridization (FISH) using tmRNA-targeted probes [138].

Since then, the method has been employed for the detection of many species. For example, *Listeria monocytogenes* was identified in foods by combining culture enrichment and real-time polymerase chain reaction (PCR) [139]; a real-time PCR test (*RiboSEQ*) was used on vaginal swabs from pregnant women to identify group B streptococci [140]; a rapid identification of *Staphylococcus aureus* was made in raw milk [141]; and *Streptococcus pneumoniae* was found in total RNA templates [142].

Similarly, an assay integrating real-time quantitative PCR (Q-PCR) with a high-resolution melting curve analysis (HRMA) was recently developed and assessed for the rapid identification of six *Listeria* species in contaminated food samples [143]. The identification of *L. monocytogenes* from cultures, using chemiluminescent labeled probes to target tmRNA, is also a promising diagnostic area [137].

11
Concluding Remarks

Almost 20 years after the discovery of *trans*-translation, the structures and functions of tmRNA–SmpB are now rather well understood, and have given rise to a clear model of the processes that occur at the molecular level. Following the "golden age" of genetics, biochemical and structural studies, it is now time for therapeutic and technological developments to take over. In this respect, the recent discovery of *trans*-translation as a target for pyrazinamide – the current mainstay of tuberculosis treatment – has confirmed the great potential of *trans*-translation and surely will continue to whet the appetite for future investigations in that direction?

Acknowledgments

The authors wish to thank the members of the "Translation and Folding" laboratory for many stimulating discussions and criticisms, and Juliana Berland for her insightful comments on the manuscript. These studies were supported by grants from the "Agence Nationale pour la Recherche" (ANR-08JCJC-0027-01 and ANR-09-MIE) and from the "Institut Universitaire de France."

References

1 Moore, S.D., Sauer, R.T. (2007) The tmRNA system for translational surveillance and ribosome rescue. *Annu. Rev. Biochem.*, **76**, 101–124.

2 Keiler, K.C., Waller, P.R., Sauer, R.T. (1996) Role of a peptide tagging system in degradation of proteins synthesized from damaged messenger RNA. *Science*, **271**, 990–993.

3 Wilson, D.N., Doudna Cate, J.H. (2012). The structure and function of the eukaryotic

4 Belousoff, M.J., Davidovich, C., Zimmerman, E., Caspi, Y. et al. (2010) Ancient machinery embedded in the contemporary ribosome. *Biochem. Soc. Trans.*, **38**, 422–427.

5 Selmer, M., Dunham, C.M., Murphy, F.V.T., Weixlbaumer, A. et al. (2006) Structure of the 70S ribosome complexed with mRNA and tRNA. *Science*, **313**, 1935–1942.

6 Schmeing, T.M., Ramakrishnan, V. (2009) What recent ribosome structures have revealed about the mechanism of translation. *Nature*, **461**, 1234–1242.

7 Felden, B., Gillet, R. (2011) SmpB as the handyman of tmRNA during trans-translation. *RNA Biol.*, **8**, 440–449.

8 Hayes, C.S., Keiler, K.C. (2010) Beyond ribosome rescue: tmRNA and co-translational processes. *FEBS Lett.*, **584**, 413–419.

9 Healey, D., Miller, M., Woolstenhulme, C., Buskirk, A. (2011) The mechanism by which tmRNA rescues stalled ribosomes, in: Rodnina, M.V., Green, R., Wintermeyer, W. (Eds), *Ribosomes Structure, Function, and Dynamics*, Springer, Vienna, pp. 361–373.

10 Ito, K., Chiba, S., Pogliano, K. (2010) Divergent stalling sequences sense and control cellular physiology. *Biochem. Biophys. Res. Commun.*, **393**, 1–5.

11 Wilson, D.N., Beckmann, R. (2011) The ribosomal tunnel as a functional environment for nascent polypeptide folding and translational stalling. *Curr. Opin. Struct. Biol.*, **21**, 274–282.

12 Janssen, B.D., Hayes, C.S. (2012) The tmRNA ribosome-rescue system. *Adv. Protein Chem. Struct. Biol.*, **86**, 151–191.

13 Poulsen, H.E., Specht, E., Broedbaek, K., Henriksen et al. (2012) RNA modifications by oxidation: a novel disease mechanism? *Free Radic. Biol. Med.*, **52**, 1353–1361.

14 Collier, J., Binet, E., Bouloc, P. (2002) Competition between SsrA tagging and translational termination at weak stop codons in *Escherichia coli*. *Mol. Microbiol.*, **45**, 745–754.

15 Ueda, K., Yamamoto, Y., Ogawa, K., Abo, T. et al. (2002) Bacterial SsrA system plays a role in coping with unwanted translational readthrough caused by suppressor tRNAs. *Genes Cells*, **7**, 509–519.

16 Hayes, C.S., Sauer, R.T. (2003) Cleavage of the A site mRNA codon during ribosome pausing provides a mechanism for translational quality control. *Mol. Cell*, **12**, 903–911.

17 Abo, T., Ueda, K., Sunohara, T., Ogawa, K. et al. (2002) SsrA-mediated protein tagging in the presence of miscoding drugs and its physiological role in *Escherichia coli*. *Genes Cells*, **7**, 629–638.

18 Li, X., Yagi, M., Morita, T., Aiba, H. (2008) Cleavage of mRNAs and role of tmRNA system under amino acid starvation in *Escherichia coli*. *Mol. Microbiol.*, **68**, 462–473.

19 Wendrich, T.M., Blaha, G., Wilson, D.N., Marahiel, M.A. et al. (2002) Dissection of the mechanism for the stringent factor RelA. *Mol. Cell*, **10**, 779–788.

20 Elf, J., Nilsson, D., Tenson, T., Ehrenberg, M. (2003) Selective charging of tRNA isoacceptors explains patterns of codon usage. *Science*, **300**, 1718–1722.

21 Hayes, C.S., Bose, B., Sauer, R.T. (2002) Stop codons preceded by rare arginine codons are efficient determinants of SsrA tagging in *Escherichia coli*. *Proc. Natl Acad. Sci. USA*, **99**, 3440–3445.

22 Li, X., Hirano, R., Tagami, H., Aiba, H. (2006) Protein tagging at rare codons is caused by tmRNA action at the 3′ end of nonstop mRNA generated in response to ribosome stalling. *RNA*, **12**, 248–255.

23 Roche, E.D., Sauer, R.T. (1999) SsrA-mediated peptide tagging caused by rare codons and tRNA scarcity. *EMBO J.*, **18**, 4579–4589.

24 Li, G.W., Oh, E., Weissman, J.S. (2012) The anti-Shine-Dalgarno sequence drives translational pausing and codon choice in bacteria. *Nature*, **484**, 538–541.

25 Hayes, C.S., Bose, B., Sauer, R.T. (2002) Proline residues at the C terminus of nascent chains induce SsrA tagging during translation termination. *J. Biol. Chem.*, **277**, 33825–33832.

26 Woolstenhulme, C.J., Parajuli, S., Healey, D.W., Valverde, D.P. et al. (2013) Nascent peptides that block protein synthesis in bacteria. *Proc. Natl Acad. Sci. USA*, **110** (10), E878–E887.

27 Collier, J., Bohn, C., Bouloc, P. (2004) SsrA tagging of *Escherichia coli* SecM at its translation arrest sequence. *J. Biol. Chem.*, **279**, 54193–54201.

28 Sunohara, T., Jojima, K., Tagami, H., Inada, T. et al. (2004) Ribosome stalling during translation elongation induces cleavage of mRNA being translated in *Escherichia coli*. *J. Biol. Chem.*, **279**, 15368–15375.

29 Gong, F., Yanofsky, C. (2002) Instruction of translating ribosome by nascent peptide. *Science*, **297**, 1864–1867.

30 Garza-Sanchez, F., Janssen, B.D., Hayes, C.S. (2006) Prolyl-tRNA(Pro) in the A-site of SecM-arrested ribosomes inhibits the recruitment of transfer-messenger RNA. *J. Biol. Chem.*, **281**, 34258–34268.

31 Pech, M., Karim, Z., Yamamoto, H., Kitakawa, M. et al. (2011) Elongation factor 4 (EF4/LepA) accelerates protein synthesis at increased Mg^{2+} concentrations. *Proc. Natl Acad. Sci. USA*, **108**, 3199–3203.

32 Vioque, A., de la Cruz, J. (2003) Trans-translation and protein synthesis inhibitors. *FEMS Microbiol. Lett.*, **218**, 9–14.

33 Keiler, K.C., Shapiro, L., Williams, K.P. (2000) tmRNAs that encode proteolysis-inducing tags are found in all known bacterial genomes: a two-piece tmRNA functions in *Caulobacter*. *Proc. Natl Acad. Sci. USA*, **97**, 7778–7783.

34 Komine, Y., Kitabatake, M., Yokogawa, T., Nishikawa, K. et al. (1994) A tRNA-like structure is present in 10Sa RNA, a small stable RNA from *Escherichia coli*. *Proc. Natl Acad. Sci. USA*, **91**, 9223–9227.

35 Li, Z., Pandit, S., Deutscher, M.P. (1998) 3′ exoribonucleolytic trimming is a common feature of the maturation of small, stable RNAs in *Escherichia coli*. *Proc. Natl Acad. Sci. USA*, **95**, 2856–2861.

36 Lin-Chao, S., Wei, C.L., Lin, Y.T. (1999) RNase E is required for the maturation of ssrA RNA and normal ssrA RNA peptide-tagging activity. *Proc. Natl Acad. Sci. USA*, **96**, 12406–12411.

37 Felden, B., Himeno, H., Muto, A., McCutcheon, J.P. et al. (1997) Probing the structure of the *Escherichia coli* 10Sa RNA (tmRNA). *RNA*, **3**, 89–103.

38 Ushida, C., Himeno, H., Watanabe, T., Muto, A. (1994) tRNA-like structures in 10Sa RNAs of *Mycoplasma capricolum* and *Bacillus subtilis*. *Nucleic Acids Res.*, **22**, 3392–3396.

39 Felden, B., Hanawa, K., Atkins, J.F., Himeno, H. et al. (1998) Presence and location of modified nucleotides in *Escherichia coli* tmRNA: structural mimicry with tRNA acceptor branches. *EMBO J.*, **17**, 3188–3196.

40 Ranaei-Siadat, E., Fabret, C., Seijo, B., Dardel, F. et al. (2013) RNA-methyltransferase TrmA is a dual-specific enzyme responsible for C(5)-methylation of uridine in both tmRNA and tRNA. *RNA Biol.*, **10** (4), 572–578.

41 Barends, S., Bjork, K., Gultyaev, A.P. de Smit, M.H. et al. (2002) Functional evidence for D- and T-loop interactions in tmRNA. *FEBS Lett.*, **514**, 78–83.

42 Karzai, A.W., Roche, E.D., Sauer, R.T. (2000) The SsrA-SmpB system for protein tagging, directed degradation and ribosome rescue. *Nat. Struct. Biol.*, **7**, 449–455.

43 Nameki, N., Tadaki, T., Himeno, H., Muto, A. (2000) Three of four pseudoknots in tmRNA are interchangeable and are substitutable with single-stranded RNAs. *FEBS Lett.*, **470**, 345–349.

44 Tanner, D.R., Dewey, J.D., Miller, M.R., Buskirk, A.R. (2006) Genetic analysis of the structure and function of transfer messenger RNA pseudoknot 1. *J. Biol. Chem.*, **281**, 10561–10566.

45 Wower, I.K., Zwieb, C., Wower, J. (2009) *Escherichia coli* tmRNA lacking pseudoknot 1 tags truncated proteins in vivo and in vitro. *RNA*, **15**, 128–137.

46 Wower, I.K., Zwieb, C., Wower, J. (2004) Contributions of pseudoknots and protein SmpB to the structure and function of tmRNA in trans-translation. *J. Biol. Chem.*, **279**, 54202–54209.

47 Sharkady, S.M., Williams, K.P. (2004) A third lineage with two-piece tmRNA. *Nucleic Acids Res.*, **32**, 4531–4538.

48 Gaudin, C., Zhou, X., Williams, K.P., Felden, B. (2002) Two-piece tmRNA in cyanobacteria and its structural analysis. *Nucleic Acids Res.*, **30**, 2018–2024.

49 Karzai, A.W., Susskind, M.M., Sauer, R.T. (1999) SmpB, a unique RNA-binding protein essential for the peptide-tagging activity of SsrA (tmRNA). *EMBO J.*, **18**, 3793–3799.

50 Hanawa-Suetsugu, K., Takagi, M., Inokuchi, H., Himeno, H. et al. (2002) SmpB functions in various steps of trans-translation. *Nucleic Acids Res.*, **30**, 1620–1629.

51 Shimizu, Y., Ueda, T. (2006) SmpB triggers GTP hydrolysis of elongation factor Tu on ribosomes by compensating for the lack of codon-anticodon interaction during trans-translation initiation. *J. Biol. Chem.*, **281**, 15987–15996.

52 Gillet, R., Kaur, S., Li, W., Hallier, M. et al. (2007) Scaffolding as an organizing principle in *trans*-translation. The roles of small protein B and ribosomal protein S1. *J. Biol. Chem.*, **282**, 6356–6363.

53 Dong, G., Nowakowski, J., Hoffman, D.W. (2002) Structure of small protein B: the protein component of the tmRNA-SmpB system for ribosome rescue. *EMBO J.*, **21**, 1845–1854.

54 Someya, T., Nameki, N., Hosoi, H., Suzuki, S. et al. (2003) Solution structure of a tmRNA-binding protein, SmpB, from *Thermus thermophilus*. *FEBS Lett.*, **535**, 94–100.

55 Jacob, Y., Sharkady, S.M., Bhardwaj, K., Sanda, A. et al. (2005) Function of the SmpB tail in transfer-messenger RNA translation revealed by a nucleus-encoded form. *J. Biol. Chem.*, **280**, 5503–5509.

56 Sundermeier, T.R., Dulebohn, D.P., Cho, H.J., Karzai, A.W. (2005) A previously uncharacterized role for small protein B (SmpB) in transfer messenger RNA-mediated *trans*-translation. *Proc. Natl Acad. Sci. USA*, **102**, 2316–2321.

57 Kaur, S., Gillet, R., Li, W., Gursky, R. et al. (2006) Cryo-EM visualization of transfer messenger RNA with two SmpBs in a stalled ribosome. *Proc. Natl Acad. Sci. USA*, **103**, 16484–16489.

58 Kurita, D., Muto, A., Himeno, H. (2010) Role of the C-terminal tail of SmpB in the early stage of trans-translation. *RNA*, **16**, 980–990.

59 Nonin-Lecomte, S., Germain-Amiot, N., Gillet, R., Hallier, M. et al. (2009) Ribosome hijacking: a role for small protein B during trans-translation. *EMBO Rep.*, **10**, 160–165.

60 Bessho, Y., Shibata, R., Sekine, S., Murayama, K. et al. (2007) Structural basis for functional mimicry of long-variable-arm tRNA by transfer-messenger RNA. *Proc. Natl Acad. Sci. USA*, **104**, 8293–8298.

61 Gutmann, S., Haebel, P.W., Metzinger, L., Sutter, M. et al. (2003) Crystal structure of the transfer-RNA domain of transfer-messenger RNA in complex with SmpB. *Nature*, **424**, 699–703.

62 Stagg, S.M., Frazer-Abel, A.A., Hagerman, P.J., Harvey, S.C. (2001) Structural studies of the tRNA domain of tmRNA. *J. Mol. Biol.*, **309**, 727–735.

63 Neubauer, C., Gillet, R., Kelley, A.C., Ramakrishnan, V. (2012) Decoding in the absence of a codon by tmRNA and SmpB in the ribosome. *Science*, **335**, 1366–1369.

64 Saguy, M., Gillet, R., Metzinger, L., Felden, B. (2005) tmRNA and associated ligands: a puzzling relationship. *Biochimie*, **87**, 897–903.

65 Hou, Y.M., Schimmel, P. (1988) A simple structural feature is a major determinant of the identity of a transfer RNA. *Nature*, **333**, 140–145.

66 Nameki, N., Tadaki, T., Muto, A., Himeno, H. (1999) Amino acid acceptor identity switch of *Escherichia coli* tmRNA from alanine to histidine in vitro. *J. Mol. Biol.*, **289**, 1–7.

67 Kavaliauskas, D., Nissen, P., Knudsen, C.R. (2012) The busiest of all ribosomal assistants: elongation factor Tu. *Biochemistry*, **51**, 2642–2651.

68 Stepanov, V.G., Nyborg, J. (2003) tmRNA from *Thermus thermophilus*. Interaction with alanyl-tRNA synthetase and elongation factor Tu. *Eur. J. Biochem.*, **270**, 463–475.

69 Zvereva, M.I., Ivanov, P.V., Teraoka, Y., Topilina, N.I. et al. (2001) Complex of transfer-messenger RNA and elongation factor Tu. Unexpected modes of interaction. *J. Biol. Chem.*, **276**, 47702–47708.

70 Hajnsdorf, E., Boni, I.V. (2012) Multiple activities of RNA-binding proteins S1 and Hfq. *Biochimie*, **94**, 1544–1553.

71 Sorensen, M.A., Fricke, J., Pedersen, S. (1998) Ribosomal protein S1 is required for translation of most, if not all, natural mRNAs in *Escherichia coli* in vivo. *J. Mol. Biol.*, **280**, 561–569.

72 Bycroft, M., Hubbard, T.J., Proctor, M., Freund, S.M. et al. (1997) The solution structure of the S1 RNA binding domain: a member of an ancient nucleic acid-binding fold. *Cell*, **88**, 235–242.

73 Subramanian, A.R. (1983) Structure and functions of ribosomal protein S1. *Prog. Nucleic Acid Res. Mol. Biol.*, **28**, 101–142.

74 Wower, I.K., Zwieb, C.W., Guven, S.A., Wower, J. (2000) Binding and cross-linking of tmRNA to ribosomal protein S1, on and

off the *Escherichia coli* ribosome. *EMBO J.*, **19**, 6612–6621.
75 Bordeau, V., Felden, B. (2002) Ribosomal protein S1 induces a conformational change of tmRNA; more than one protein S1 per molecule of tmRNA. *Biochimie*, **84**, 723–729.
76 Valle, M., Gillet, R., Kaur, S., Henne, A. et al. (2003) Visualizing tmRNA entry into a stalled ribosome. *Science*, **300**, 127–130.
77 Takada, K., Takemoto, C., Kawazoe, M., Konno, T. et al. (2007) In vitro *trans*-translation of *Thermus thermophilus*: ribosomal protein S1 is not required for the early stage of trans-translation. *RNA*, **13**, 503–510.
78 Shi, W., Zhang, X., Jiang, X., Yuan, H. et al. (2011) Pyrazinamide inhibits trans-translation in *Mycobacterium tuberculosis*. *Science*, **333**, 1630–1632.
79 Cheng, Z.F., Deutscher, M.P. (2002) Purification and characterization of the *Escherichia coli* exoribonuclease RNase R. Comparison with RNase II. *J. Biol. Chem.*, **277**, 21624–21629.
80 Matos, R.G., Barbas, A., Gomez-Puertas, P., Arraiano, C.M. (2011) Swapping the domains of exoribonucleases RNase II and RNase R: conferring upon RNase II the ability to degrade ds RNA. *Proteins*, **79**, 1853–1867.
81 Ge, Z., Mehta, P., Richards, J., Karzai, A.W. (2010) Non-stop mRNA decay initiates at the ribosome. *Mol. Microbiol.*, **78**, 1159–1170.
82 Richards, J., Mehta, P., Karzai, A.W. (2006) RNase R degrades non-stop mRNAs selectively in an SmpB-tmRNA-dependent manner. *Mol. Microbiol.*, **62**, 1700–1712.
83 Liang, W., Deutscher, M.P. (2012) Transfer-messenger RNA-SmpB protein regulates ribonuclease R turnover by promoting binding of HslUV and Lon proteases. *J. Biol. Chem.*, **287**, 33472–33479.
84 Liang, W., Deutscher, M.P. (2010) A novel mechanism for ribonuclease regulation: transfer-messenger RNA (tmRNA) and its associated protein SmpB regulate the stability of RNase R. *J. Biol. Chem.*, **285**, 29054–29058.
85 Moreira, R.N., Domingues, S., Viegas, S.C., Amblar, M. et al. (2012) Synergies between RNA degradation and *trans*-translation in *Streptococcus pneumoniae*: cross regulation and co-transcription of RNasc R and SmpB. *BMC Microbiol.*, **12**, 268.
86 Kurita, D., Sasaki, R., Muto, A., Himeno, H. (2007) Interaction of SmpB with ribosome from directed hydroxyl radical probing. *Nucleic Acids Res.*, **35**, 7248–7255.
87 Ogle, J.M., Murphy, F.V., Tarry, M.J., Ramakrishnan, V. (2002) Selection of tRNA by the ribosome requires a transition from an open to a closed form. *Cell*, **111**, 721–732.
88 Schmeing, T.M., Voorhees, R.M., Kelley, A.C., Gao, Y.G. et al. (2009) The crystal structure of the ribosome bound to EF-Tu and aminoacyl-tRNA. *Science*, **326**, 688–694.
89 Schuette, J.C., Murphy, F.V.T., Kelley, A.C., Weir, J.R. et al. (2009) GTPase activation of elongation factor EF-Tu by the ribosome during decoding. *EMBO J.*, **28**, 755–765.
90 Cheng, K., Ivanova, N., Scheres, S.H., Pavlov, M.Y. et al. (2010) tmRNA.SmpB complex mimics native aminoacyl-tRNAs in the A site of stalled ribosomes. *J. Struct. Biol.*, **169**, 342–348.
91 Fu, J., Hashem, Y., Wower, I., Lei, J. et al. (2010) Visualizing the transfer-messenger RNA as the ribosome resumes translation. *EMBO J.*, **29**, 3819–3825.
92 Weis, F., Bron, P., Giudice, E., Rolland, J.P. et al. (2010) tmRNA-SmpB: a journey to the centre of the bacterial ribosome. *EMBO J.*, **29**, 3810–3818.
93 Weis, F., Bron, P., Rolland, J.P., Thomas, D. et al. (2010) Accommodation of tmRNA-SmpB into stalled ribosomes: a cryo-EM study. *RNA*, **16**, 299–306.
94 Ramrath, D.J., Yamamoto, H., Rother, K., Wittek, D. et al. (2012) The complex of tmRNA-SmpB and EF-G on translocating ribosomes. *Nature*, **485**, 526–529.
95 Crandall, J., Rodriguez-Lopez, M., Pfeiffer, M., Mortensen, B. et al. (2010) rRNA mutations that inhibit transfer-messenger RNA activity on stalled ribosomes. *J. Bacteriol.*, **192**, 553–559.
96 Shpanchenko, O.V., Zvereva, M.I., Ivanov, P.V., Bugaeva, E.Y. et al. (2005) Stepping transfer messenger RNA through the ribosome. *J. Biol. Chem.*, **280**, 18368–18374.
97 Lee, S., Ishii, M., Tadaki, T., Muto, A. et al. (2001) Determinants on tmRNA for initiating efficient and precise

trans-translation: some mutations upstream of the tag-encoding sequence of *Escherichia coli* tmRNA shift the initiation point of *trans*-translation in vitro. *RNA*, **7**, 999–1012.

98 Miller, M.R., Healey, D.W., Robison, S.G., Dewey, J.D. et al. (2008) The role of upstream sequences in selecting the reading frame on tmRNA. *BMC Biol.*, **6**, 29.

99 Williams, K.P., Martindale, K.A., Bartel, D.P. (1999) Resuming translation on tmRNA: a unique mode of determining a reading frame. *EMBO J.*, **18**, 5423–5433.

100 Miller, M.R., Liu, Z., Cazier, D.J., Gebhard, G.M. et al. (2011) The role of SmpB and the ribosomal decoding center in licensing tmRNA entry into stalled ribosomes. *RNA*, **17**, 1727–1736.

101 Watts, T., Cazier, D., Healey, D., Buskirk, A. (2009) SmpB contributes to reading frame selection in the translation of transfer-messenger RNA. *J. Mol. Biol.*, **391**, 275–281.

102 Bugaeva, E.Y., Surkov, S., Golovin, A.V., Ofverstedt, L.G. et al. (2009) Structural features of the tmRNA-ribosome interaction. *RNA*, **15**, 2312–2320.

103 Wower, I.K., Zwieb, C., Wower, J. (2005) Transfer-messenger RNA unfolds as it transits the ribosome. *RNA*, **11**, 668–673.

104 Keiler, K.C., Shapiro, L. (2003) TmRNA is required for correct timing of DNA replication in *Caulobacter crescentus*. *J. Bacteriol.*, **185**, 573–580.

105 Keiler, K.C., Shapiro, L. (2003) tmRNA in *Caulobacter crescentus* is cell cycle regulated by temporally controlled transcription and RNA degradation. *J. Bacteriol.*, **185**, 1825–1830.

106 Hong, S.J., Tran, Q.A., Keiler, K.C. (2005) Cell cycle-regulated degradation of tmRNA is controlled by RNase R and SmpB. *Mol. Microbiol.*, **57**, 565–575.

107 Withey, J.H., Friedman, D.I. (2003) A salvage pathway for protein structures: tmRNA and trans-translation. *Annu. Rev. Microbiol.*, **57**, 101–123.

108 Withey, J., Friedman, D. (1999) Analysis of the role of trans-translation in the requirement of tmRNA for lambdaimmP22 growth in *Escherichia coli*. *J. Bacteriol.*, **181**, 2148–2157.

109 Giudice, E., Gillet, R. (2013) The task-force that rescues stalled ribosomes in bacteria. *Trends Biochem. Sci.*, **38** (8), 403–411.

110 Chadani, Y., Ono, K., Ozawa, S., Takahashi, Y. et al. (2010) Ribosome rescue by *Escherichia coli* ArfA (YhdL) in the absence of *trans*-translation system. *Mol. Microbiol.*, **78**, 796–808.

111 Garza-Sanchez, F., Schaub, R.E., Janssen, B.D., Hayes, C.S. (2011) tmRNA regulates synthesis of the ArfA ribosome rescue factor. *Mol. Microbiol.*, **80**, 1204–1219.

112 Chadani, Y., Ono, K., Kutsukake, K., Abo, T. (2011) *Escherichia coli* YaeJ protein mediates a novel ribosome-rescue pathway distinct from SsrA- and ArfA-mediated pathways. *Mol. Microbiol.*, **80**, 772–785.

113 Handa, Y., Inaho, N., Nameki, N. (2011) YaeJ is a novel ribosome-associated protein in *Escherichia coli* that can hydrolyze peptidyl-tRNA on stalled ribosomes. *Nucleic Acids Res.*, **39**, 1739–1748.

114 Gagnon, M.G., Seetharaman, S.V., Bulkley, D., Steitz, T.A. (2012) Structural basis for the rescue of stalled ribosomes: structure of YaeJ bound to the ribosome. *Science*, **335**, 1370–1372.

115 Doerfel, L.K., Wohlgemuth, I., Kothe, C., Peske, F. et al. (2013) EF-P is essential for rapid synthesis of proteins containing consecutive proline residues. *Science*, **339**, 85–88.

116 Ude, S., Lassak, J., Starosta, A.L., Kraxenberger, T. et al. (2013) Translation elongation factor EF-P alleviates ribosome stalling at polyproline stretches. *Science*, **339**, 82–85.

117 Singh, N.S., Varshney, U. (2004) A physiological connection between tmRNA and peptidyl-tRNA hydrolase functions in *Escherichia coli*. *Nucleic Acids Res.*, **32**, 6028–6037.

118 Christensen, S.K., Gerdes, K. (2003) RelE toxins from bacteria and Archaea cleave mRNAs on translating ribosomes, which are rescued by tmRNA. *Mol. Microbiol.*, **48**, 1389–1400.

119 Christensen, S.K., Pedersen, K., Hansen, F.G., Gerdes, K. (2003) Toxin-antitoxin loci as stress-response-elements: ChpAK/MazF and ChpBK cleave translated RNAs and are counteracted by tmRNA. *J. Mol. Biol.*, **332**, 809–819.

120 Neubauer, C., Gao, Y.G., Andersen, K.R., Dunham, C.M. et al. (2009) The structural basis for mRNA recognition and cleavage by the ribosome-dependent endonuclease RelE. *Cell*, **139**, 1084–1095.

121 Pedersen, K., Zavialov, A.V., Pavlov, M.Y., Elf, J. et al. (2003) The bacterial toxin RelE displays codon-specific cleavage of mRNAs in the ribosomal A site. *Cell*, **112**, 131–140.

122 Wilson, D.N., Nierhaus, K.H. (2005) RelBE or not to be. *Nat. Struct. Mol. Biol.*, **12**, 282–284.

123 Garza-Sanchez, F., Shoji, S., Fredrick, K., Hayes, C.S. (2009) RNase II is important for A-site mRNA cleavage during ribosome pausing. *Mol. Microbiol.*, **73**, 882–897.

124 Garza-Sanchez, F., Gin, J.G., Hayes, C.S. (2008) Amino acid starvation and colicin D treatment induce A-site mRNA cleavage in *Escherichia coli*. *J. Mol. Biol.*, **378**, 505–519.

125 Sunohara, T., Jojima, K., Yamamoto, Y., Inada, T. et al. (2004) Nascent-peptide-mediated ribosome stalling at a stop codon induces mRNA cleavage resulting in nonstop mRNA that is recognized by tmRNA. *RNA*, **10**, 378–386.

126 Huang, C., Wolfgang, M.C., Withey, J., Koomey, M. et al. (2000) Charged tmRNA but not tmRNA-mediated proteolysis is essential for *Neisseria gonorrhoeae* viability. *EMBO J.*, **19**, 1098–1107.

127 Ramadoss, N.S., Zhou, X., Keiler, K.C. (2013) tmRNA is essential in *Shigella flexneri*. *PLoS ONE*, **8**, e57537.

128 Thibonnier, M., Thiberge, J.M., De Reuse, H. (2008) *Trans*-translation in *Helicobacter pylori*: essentiality of ribosome rescue and requirement of protein tagging for stress resistance and competence. *PLoS ONE*, **3**, e3810.

129 de la Cruz, J., Vioque, A. (2001) Increased sensitivity to protein synthesis inhibitors in cells lacking tmRNA. *RNA*, **7**, 1708–1716.

130 Andini, N., Nash, K.A. (2011) Expression of tmRNA in mycobacteria is increased by antimicrobial agents that target the ribosome. *FEMS Microbiol. Lett.*, **322**, 172–179.

131 Luidalepp, H., Hallier, M., Felden, B., Tenson, T. (2005) tmRNA decreases the bactericidal activity of aminoglycosides and the susceptibility to inhibitors of cell wall synthesis. *RNA Biol.*, **2**, 70–74.

132 Cole, S.T. (2011) Microbiology. Pyrazinamide – old TB drug finds new target. *Science*, **333**, 1583–1584.

133 Ramadoss, N.S., Alumasa, J.N., Cheng, L., Wang, Y. et al. (2013) Small molecule inhibitors of trans-translation have broad-spectrum antibiotic activity. *Proc. Natl Acad. Sci. USA*, **110**, 10282–10287.

134 Okan, N.A., Mena, P., Benach, J.L., Bliska, J.B. et al. (2010) The smpB-ssrA mutant of *Yersinia pestis* functions as a live attenuated vaccine to protect mice against pulmonary plague infection. *Infect. Immun.*, **78**, 1284–1293.

135 Svetlanov, A., Puri, N., Mena, P., Koller, A. et al. (2012) *Francisella tularensis* tmRNA system mutants are vulnerable to stress, avirulent in mice, and provide effective immune protection. *Mol. Microbiol.*, **85**, 122–141.

136 Bohez, L., Ducatelle, R., Pasmans, F., Haesebrouck, F. et al. (2007) Long-term colonisation-inhibition studies to protect broilers against colonisation with *Salmonella enteritidis*, using *Salmonella* Pathogenicity Island 1 and 2 mutants. *Vaccine*, **25**, 4235–4243.

137 Clancy, E., Glynn, B., Reddington, K., Smith, T. et al. (2012) Culture confirmation of *Listeria monocytogenes* using tmRNA as a diagnostics target. *J. Microbiol. Methods*, **88**, 427–429.

138 Schonhuber, W., Le Bourhis, G., Tremblay, J., Amann, R. et al. (2001) Utilization of tm-RNA sequences for bacterial identification. *BMC Microbiol.*, **1**, 20.

139 O'Grady, J., Ruttledge, M., Sedano-Balbas, S., Smith, T.J. et al. (2009) Rapid detection of *Listeria monocytogenes* in food using culture enrichment combined with real-time PCR. *Food Microbiol.*, **26**, 4–7.

140 Wernecke, M., Mullen, C., Sharma, V., Morrison, J. et al. (2009) Evaluation of a novel real-time PCR test based on the ssrA gene for the identification of group B streptococci in vaginal swabs. *BMC Infect. Dis.*, **9**, 148.

141 O'Grady, J., Lacey, K., Glynn, B., Smith, T.J. et al. (2009) tmRNA – a novel high-copy-number RNA diagnostic target – its application for *Staphylococcus aureus* detection using real-time NASBA. *FEMS Microbiol. Lett.*, **301**, 218–223.

142 Scheler, O., Glynn, B., Parkel, S., Palta, P. *et al.* (2009) Fluorescent labeling of NASBA amplified tmRNA molecules for microarray applications. *BMC Biotechnol.*, **9**, 45.

143 Jin, D., Luo, Y., Zhang, Z., Fang, W. *et al.* (2012) Rapid molecular identification of *Listeria* species by use of real-time PCR and high-resolution melting analysis. *FEMS Microbiol. Lett.*, **330**, 72–80.

3
RNA-mediated Control of Bacterial Gene Expression: Role of Regulatory non-Coding RNAs

Pierre Mandin[1], Alejandro Toledo-Arana[2], Aymeric Fouquier d'Hérouel[3], and Francis Repoila[4]

[1] Aix-Marseille Université, CNRS-LCB, UMR7283, 31 Chemin Joseph Aiguier, 13402 Marseille cedex 20, France
[2] CSIC – Universidad Pública de Navarra, Instituto de Agrobiotecnología, 31006 Pamplona, Spain
[3] Institute for Systems Biology, 401 Terry Avenue North, Seattle, WA, 98109, USA
[4] AgroParisTech, INRA, UMR1319 Micalis, Domaine de Vilvert, F-78350, Jouy-en-Josas, France

1	Regulatory RNAs in the RNA World	83
2	Discovery of Chromosomal Regulatory RNAs	84
3	**Intrinsic Regulatory RNA Elements: Leaders and Riboswitches**	**85**
3.1	Regulatory 5′UTRs Sensing Physico-Chemical Cues	85
3.1.1	RNA Thermometers to Adjust Global Physiology	85
3.1.2	RNA Thermometers to Trigger Specific Gene-Expression Programs	87
3.1.3	RNA Sensing pH	87
3.2	Regulatory 5′UTR Sensing Ions	89
3.3	Riboswitches: Regulatory 5′UTR Sensing Metabolites	91
3.4	T-Boxes: Regulatory 5′UTR Sensing tRNAs	93
3.5	Regulatory 5′UTR Sensing Other Biomolecules	94
4	**RNA Pairing to Other RNAs**	**95**
4.1	*cis*-Antisense RNAs	95
4.1.1	Mode of Action	95
4.1.2	Specific and Global Antisense-Dependent Regulations	97
4.2	*trans*-Acting Noncoding RNAs	98
4.2.1	Gene Expression Inhibition by ncRNAs	99

5) Pierre Mandin and Alejandro Toledo-Arana contributed equally to this work.

RNA Regulation: Advances in Molecular Biology and Medicine, First Edition. Edited by Robert A. Meyers.
© 2014 Wiley-VCH Verlag GmbH & Co. KGaA. Published 2014 by Wiley-VCH Verlag GmbH & Co. KGaA.

4.2.2	Activation of Gene Expression by ncRNAs 101
5	**RNA Changing Protein Activities 102**
5.1	Mimicking DNA Promoters 102
5.2	Mimicking Ribosome Binding Sequences 104
6	**Concluding Remarks 106**
	Acknowledgments 107
	References 107

Keywords

Regulatory noncoding RNA
A nonprotein-coding RNA molecule that controls biological processes by directly affecting the transcription, translation, or stability of another RNA, or by binding to proteins.

***trans*-Antisense ncRNA**
A regulatory RNA that acts on an RNA molecule (target) that is encoded at a different locus.

***cis*I-Antisense ncRNA**
A regulatory RNA that acts on the expression of the gene transcribed from the reverse complementary DNA strand.

Regulatory 5′ untranslated region
A noncoding RNA sequence located at the 5′ end of a protein coding sequence within a messenger RNA, and controlling its expression.

Open reading frame
A nucleic sequence that starts with a translation initiation codon and ends with a translation stop codon, which can potentially be translated by the ribosome into a specific amino acid chain.

Shine–Dalgarno sequence
An RNA sequence recognized and bound by the 16S ribosomal RNA. The sequence is located upstream of the translation initiation codon (4–12 nt) within the ribosome binding site.

Ribosome binding site
An RNA sequence recognized and bound by the 30S ribosomal subunit to initiate translation. It generally extends from about −30 nt to around +16 nt relative to the translation initiation codon.

RNA duplex or RNA hybrid
Perfect or partial double-stranded RNA segments provided by two different RNA molecules or a unique folded RNA.

Ribonucleoprotein complex
A structure formed by the association of an RNA molecule and a protein.

RNA–RNA interaction
The association of two RNA sequences when the nucleotide bases are complementary.

RNA–protein interaction
The association of a protein or a complex and an RNA molecule.

> Regulatory noncoding RNAs are found in 5′ untranslated regions of messengers RNAs, or as individual molecules transcribed from the complementary DNA strand to annotated genes (antisense RNAs) or from empty chromosomal regions (noncoding RNAs). They can regulate each step of gene expression by directly acting on transcription or translation processes via alternative folding, base-pairing to RNA targets, or binding to regulatory proteins.

1 Regulatory RNAs in the RNA World

The RNA world is constituted by a diversity of sequences, which encode dynamic structures and functions dedicated to carrying the genetic information and to controlling its expression. Among the many functions of ribonucleic acids, the RNA-dependent control of gene expression has been an intense field of research over the past years that has led to the identification of numerous types of regulatory RNA-based elements in the three kingdoms of life, and to the remarkable conclusion that RNA controls each step of gene expression, from transcription initiation to protein activity [1–4]. In bacteria, the first regulatory RNA elements were discovered in plasmids, phages, and transposons [5], and were for a long time considered as peculiar regulators of these extrachromosomal genetic elements. Over the past years, however, bacterial chromosomes have been shown to encode a plethora of potential regulatory RNAs, including major classes involved in physiological adaptations. Among these, only three major classes of regulatory RNA elements will be described in this chapter, namely regulatory 5′ untranslated RNA regions (r5′UTRs), antisense RNAs (asRNAs), and small non-coding RNAs (ncRNAs) [6–10]. Whilst the denominations used here may be confusing, as most of these regulators are nonprotein-coding RNAs, they may be distinguished according to their mode of action:

- RNAs acting intramolecularly, as they are embedded in the RNA sequences that they control (i.e., r5′UTRs).
- RNAs acting intermolecularly and transcribed from the reverse complementary strand to the RNA target they pair with (i.e., asRNAs).
- RNAs acting intermolecularly by binding to proteins or by pairing to other RNA targets transcribed at a different chromosomal locus (i.e., ncRNAs).

By examining selected examples, the physiological importance of the processes that these RNAs control and the strategies employed by them to exert their effects are illustrated in this chapter.

2 Discovery of Chromosomal Regulatory RNAs

Until the early 2000s, the discovery of regulatory RNAs encoded by bacterial chromosomes was fortuitous. In *Escherichia coli*, the first regulatory RNAs discovered as relatively abundant species [4.5S (Ffs), 6S (SsrS), tmRNA (SsrA), RNaseP, and Spot42] [11–14], by *in-vitro* transcription assays when studying neighboring genes (*OxyS*, *GcvB*, *CrpTic*), [15–17], due to their copurification with a studied protein (CsrB), [18], or by their effects in genetic screens using multicopy libraries (DicF, DsrA, MicF, RprA, RseX), [19–23]. Subsequently, functional studies performed in *E. coli* K12 [19, 21–27], coupled with the major discoveries accomplished in eukaryotes (for a review, see Ref. [28]), have highlighted the ubiquity and biological importance of regulatory RNAs. These conclusions have motivated the development and the combination of *in-silico* predictions and experimental approaches to "hunt" for chromosomally encoded regulatory RNAs, not only in *E. coli* K12 [29–34] but also in other species, including Gram-positive bacteria [35–37]. More recently, "tiling arrays" and "RNA sequencing" ("RNAseq") techniques have provided an unprecedented view of the bacterial transcriptome (i.e., the RNA set present in a cell at a specific time in a given environmental condition). This, in turn, has enabled the discovery of a plethora of unexpected RNA molecules, and the unveiling of a highly complex organization of genetic information. In addition, transcriptome studies have shown that bacterial chromosomes contain much more genetic information than originally anticipated, as indicated by the prominence of different transcripts. This may mirror the ubiquity of RNA-mediated regulation [38–42], in that:

- Annotated open reading frames (ORFs) can contain several promoters with the same orientation, giving rise to different RNA molecules with identical 3′ end sequences.
- DNA regions between annotated ORFs, which originally were thought to be devoid of genetic information, contain genes and express numerous putative regulatory RNAs (hundreds of ncRNAs per chromosome).
- Transcriptional antisense organizations are extremely frequent due to long overlapping UTRs at their 5′ or 3′

ends, short internal transcripts to annotated ORFs, or long transcripts overlapping entire operons. Up to 20–50% of annotated ORFs have an antisense organization, suggesting the important contribution of antisense-mediated control.

3
Intrinsic Regulatory RNA Elements: Leaders and Riboswitches

The majority of regulatory RNA elements known in bacteria are embedded within the 5′UTRs of mRNAs, and will be referred to here as ''r5′UTRs'' (where ''r'' indicates ''regulatory''), as suggested previously [43]. The r5′UTRs modulate the expression of their downstream protein coding sequence by affecting transcription elongation (also referred to as ''transcription attenuation''), translation efficacy, or RNA stability [44, 45]. The regulatory effects of r5′UTRs rely on two mutually exclusive conformational structures (''conformers''): one which imposes the effect; and another which is silent. The switch from one conformer to the other is mediated by a signal that is specifically detected or bound by the r5′UTR. This signal may be of a physico-chemical nature (e.g., temperature, pH, ionic strength), a metabolite (i.e., small organic compounds), or a transfer RNA (tRNA). Diverse classes of r5′UTRs can be distinguished according to the nature of the signal sensed, although a current ''semantic drift'' tends to nominate all r5UTRs as ''riboswitches'' (see Ref. [44]; this distinction will be maintained in this chapter). Generally, r5′UTRs responding to the same signal are common to mRNAs of different genes (e.g., riboswitches), and in many cases a functional relationship exists between proteins encoded by these mRNAs and the signal to which their 5′UTRs respond. For instance, the signal may be a metabolite and the mRNAs encode for transporters or enzymes belonging to metabolic pathway using this metabolite. Such functional interplays, based on direct interactions between signals and mRNAs, give rise to regulatory feedback loops and the orchestration of genetic programs. The fact that gene expression control via r5′UTRs is found in the three kingdoms of life suggests an ancestral mode of regulation that has evolved a sequence diversity allowing the orchestration of numerous fundamental biological functions in a robust manner; indeed, more than 4% of bacterial genes are believed to have a r5′UTR-controlled expression [44–50].

3.1
Regulatory 5′UTRs Sensing Physico-Chemical Cues

The folding dynamics of an RNA molecule relies on hydrogen bonds, and changes in the physico-chemical parameters capable of affecting such bonds (e.g., temperature, pH, osmolarity) should also impact on the RNA function.

3.1.1 RNA Thermometers to Adjust Global Physiology

An ability to sense and respond to temperature is a universal cellular process, and most forms of macromolecule (e.g., DNA, RNA, proteins, lipids) have evolved structural domains that are capable of sensing temperature variations [51–55]. The different structural elements identified within RNA thermometers all feature a common functional pattern [56]. RNA thermometers are r5′UTRs that affect translation due to alternative foldings that either occlude, or not, the ribosome binding site (RBS) and, more specifically, the

Shine–Dalgarno (SD) sequence and/or the translation initiation codon ("AUG"). At nonpermissive temperatures, the secondary structures trap the RBS into a double-stranded RNA helix that prevents its utilization by the ribosome; at permissive temperatures, the RBS is not involved in the secondary structure and translation can occur (Fig. 1). Among RNA thermometers, *rpoH* and *cspA* mRNAs respond to heat-shock and cold-shock, respectively, and are widespread across bacterial species as they ensure essential adaptations in response to temperature. These regulatory RNA systems have been studied in great detail through elegant experimental approaches in *E. coli* [57–61]. The *rpoH* mRNA encodes the sigma factor RpoH (σ^{32} or σ^H) that controls the transcription initiation of genes belonging to the heat-shock regulon. This regulon contains more than 100 genes, including those that are essentially involved in protein damage control and folding; this forces the cell to keep RpoH under sharp regulation to cope with vital demands. Following a heat-shock, RpoH levels are increased as a result of an increased stability of the protein and an enhanced translation of the *rpoH* mRNA. The thermocontrol operating on *rpoH* mRNA translation is due to a complex folding of the RNA region that extends from −19 to +247 nt relative to the AUG of *rpoH*. In vitro, this complex secondary structure is not used by the ribosome at low temperature (30 °C), but at high temperature (42 °C) the ribosome forms an efficient translation initiation complex, demonstrating that the RNA structure is sufficient to mediate the thermocontrol without any additional factor. In vivo, genetic analyses have established that a temperature upshift transiently melts the secondary structure of *rpoH* mRNA, rendering the SD sequence accessible to ribosomes and enabling translation [60, 61].

In *E. coli* and most bacteria, in addition to the heat-shock regulon, the cold-shock response enables adaptation to drastic temperature decreases. When the temperature suddenly falls by at least 10 °C (cold-shock), most bacteria stop growing for 3–5 h, and gene expression ceases transiently; exceptions to this are "cold-shock genes," the expression of which is increased. Among the latter genes, the *cspA*

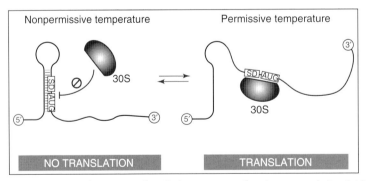

Fig. 1 Functional patterns of RNA thermometers. Left: At a nonpermissive temperature, the conformer traps the ribosome binding site in a double-stranded RNA that prevents the 30S ribosomal subunit (30S) from binding and forming a translation initiation complex; Right: At a permissive temperature, the conformer allows the formation of an efficient translation initiation complex.

gene encodes the CspA protein, a major coordinator of the cold-shock response and a multifunctional protein that acts as an RNA chaperone, a single-stranded DNA/RNA-binding protein, and a transcription antiterminator (see Refs [62, 63] and references therein). In response to cold-shock, an increased CspA synthesis results from controls acting at the transcriptional and post-transcriptional levels via an interplay of *trans*- and *cis*-regulatory factors [62, 63]. The post-transcriptional induction of CspA results from a stabilization of the *cspA* mRNA and an increased translation efficacy that involves the RNA thermometer element and translation initiation factor 1 IF1 and IF3 [58, 64]. The *cspA* mRNA folds into two major conformers with a structural transition at a threshold temperature of ∼20 °C. At 37 °C, the 5′UTR folds into a succession of bulges and helices encompassing the SD sequence, while the ∼80 nt downstream result in a conformer that is inefficient for translation. At low temperatures (i.e., below 20 °C), the RNA conformer exposes the SD sequence in a loop of a weak stem, while the AUG is involved in an imperfect stem flanked by single strands, which renders this region unstable and efficient for translation. In addition, the *cspA* conformer obtained at low temperature can melt and adopt a "high-temperature structure" if incubated at 37 °C, but the opposite switch is not possible (from 37 to 10 °C, for example), and a "low-temperature structure" can only be achieved by co-transcriptional folding [58]. This indicates that: (i) during cold-shock, CspA synthesis would occur only from mRNAs synthesized at low temperature, and not from pre-existing conformers folded at high temperatures; and (ii) the 37 °C conformer of *cspA* is thermodynamically more stable than its translatable counterpart that is formed at low temperatures [58].

3.1.2 RNA Thermometers to Trigger Specific Gene-Expression Programs

The expression of virulence genes in response to temperature changes is a common theme in many pathogens that infect warm-blooded animals: the host's body-temperature, which is generally higher than the outside temperature, serves as signal to launch the infection program [65]. The *prfA*- and *lcrF* mRNAs encode master transcriptional activators, PrfA and LcrF, at the top of the virulence gene expression cascade in Listeria monocytogenes and Yersiniae (i.e., Yersinia pestis, Y. pseudotuberculosis, and Y. enterocolitica), respectively [66–68]. Similarly to *rpoH*- and *cspA* mRNAs, the thermoregulation of *prfA*- and *lcrF* mRNAs relies on structural switches occurring between high (in-host) and low temperatures (out-of-host). In both Listeria sp. and Yersinia sp., *prfA*- and *lcrF* conformers are normally translated and PrfA and LcrF are produced at 37 °C. At nonpermissive temperatures, outside the body (i.e., <30 °C), the *prfA*- and *lcrF* mRNAs fold into a single stable RNA helix that traps the SD sequence and thereby abolishes translation [66, 68]. Recently, it was shown in animal models that the *lcrF* RNA thermometer ensures the correct expression of LcrF- and LcrF-dependent genes in terms of amounts and timing, thereby optimizing the infection efficiency of Y. pseudotuberculosis [66].

3.1.3 RNA Sensing pH

Bacteria such as E. coli maintain their cytoplasmic pH in a range of 7.4 to 7.8 while living in environments where the external pH values may range from 5.5 to 9. During pH-shifts, a variety of mechanisms are employed by bacteria to

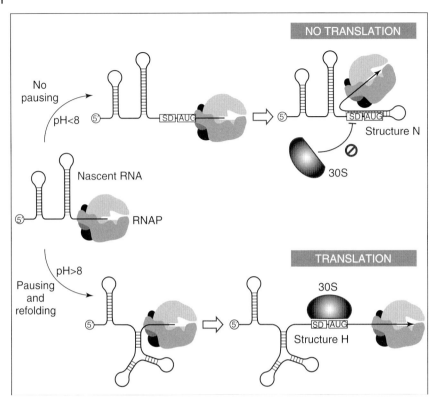

Fig. 2 Model of PRE-*alx* RNA response to the pH. Below pH 8 (neutral pH), transcription proceeds without pause and the nascent PRE-*alx* RNA folds into its most stable conformer where the SD sequence is trapped in an RNA helix that renders the mRNA inefficient for translation (structure N). The transcription elongation complex is represented by RNAP (RNA polymerase). Under high-pH conditions (i.e., pH > 8), pauses of the transcription elongation complex allow the nascent PRE-*alx* RNA to fold into a conformer, thus exposing the SD sequence to the ribosome that binds to it and forms an efficient translation initiation complex (structure H). The abbreviations are as shown in Fig. 1.

maintain homeostasis in the cytoplasm [69, 70], though to date only a single example of an ncRNA element responding to pH appears to have been described [71]. In *E. coli*, the *alx* ORF (originally named *ygjT*) encodes a putative transporter that is induced in response to an alkaline environment. The *alx* transcript is preceded by an r5′UTR named the "PRE" RNA element. In addition to the pH-dependent regulation of *alx* at the transcriptional and post-transcriptional levels, PRE modulates the translation efficiency of *alx* via an interplay between RNA folding dynamics and transcription elongation (Fig. 2). When cells are living under neutral conditions (pH <8), PRE-*alx* folds into a translationally inactive conformer that traps the RBS into a double-stranded RNA helix ("structure N"). In contrast, when cells are living under alkaline conditions (pH >8), the PRE-*alx* conformer folds into a structure that is efficient for translation ("structure H"), and the Alx protein is produced. Thermodynamically, structure N is more stable than H; moreover, when N has been synthesized and folded it cannot "switch" spontaneously into H, even in alkaline conditions. Indeed, the level of pH controls a pausing of the RNA polymerase (RNAP) elongation complex at two distinct sites during the synthesis of PRE RNA. At high pH, pauses occur that allow the nascent PRE-*alx* RNA to adopt a conformation compatible with translation, whereas at neutral pH there are no pauses and the rate of synthesis of PRE-*alx* RNA does not allow a correct folding to prevent the SD sequence from becoming a double-stranded structure (Fig. 2), [71]. Although the mechanism by which the RNAP elongation complex responds to pH is unknown, many regulatory processes mediated by r5′UTRs are known to be kinetically controlled and occur on the nascent RNA as they affect the ongoing transcription. Most likely, transcription pausing serves as a common process to mediate regulatory effects on RNA folding and/or RNA synthesis.

3.2
Regulatory 5′UTR Sensing Ions

Metal ions play important structural roles and act as cofactors for many biological molecules. Yet, although their role is vital, their presence in the cell is tightly controlled, as an excess may cause oxidative stress. Magnesium, in its cationic form Mg^{2+}, is one of the most prevalent ions in the cell (concentration 1–2 mM), and r5′UTRs sensing Mg^{2+} have been predicted to precede the three major classes of Mg^{2+} transporters known in bacteria [72]. Among these, the r5′UTRs of mRNAs *mgtE* in *Bacillus subtilis* and *mgtA* in *Salmonella enterica*, which encode an otherwise different class of transporters, have been shown to respond to Mg^{2+} when the ionic concentration is above a certain threshold [73, 74]. The results of crystallography experiments have indicated that approximately six ions would bind in a cooperative-dependent fashion to the regulatory RNA element of *mgtE* (termed "M Box"), thus rendering the system more sensitive to variations in Mg^{2+} than to a linear interaction mode [74].

In *S. enterica*, at high Mg^{2+} levels the long r5′UTR (264 nt) folds into a conformer, provoking a pause in the transcription elongation complex before it reaches the ORF MgtA. A Rho-dependent transcription termination process then takes place, after which the transcription complex dissociates. At low Mg^{2+} levels, the RNA adopts a conformation that allows the transcription to proceed into the *mgtA* encoding sequence; subsequently, the MgtA protein is produced and promotes uptake of Mg^{2+} [73, 75]. In addition to Mg^{2+}, *mgtA* expression responds to other signals, including high osmolarity and high intracellular proline concentrations (see Refs [76, 77] and references therein). Within the 264 nt preceding the ORF MgtA lies a proline-rich ORF of 18 amino acid residues, termed MgtL [77]. At high proline concentrations, the translation of MgtL favors transcription arrest into *mgtA* (Fig. 3a) [73, 75], but when the proline concentration falls during *mgtL* translation, those ribosomes stalled at proline codons allow a transcription elongation of *mgtA* to proceed. Hyperosmotic conditions have been shown to favor *mgtA* transcription in an *mgtL* proline-codon-dependent manner. Thus, low concentrations of proline and Mg^{2+}, as well as conditions of high osmolarity, act in synergistic fashion to ensure a rapid responsiveness to the system [77]. This, in turn, allows the r5′UTR of *mgtA* to integrate at least three different signals coordinating the transcription of *mgtA* accordingly.

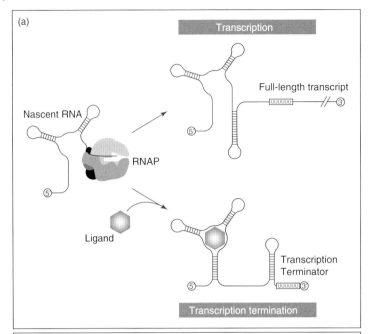

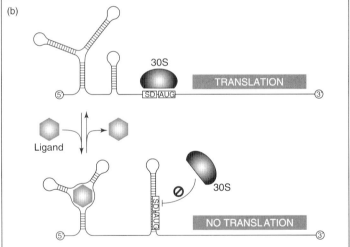

Fig. 3 Regulatory effects of r5'UTRs responding to ligands. The r5'UTR controls its downstream encoding sequence via either (a) transcriptional or (b) translational effects. Due to kinetic constraints, transcriptional effects must occur on the nascent RNA, whereas translational effects may occur on the full-length mRNA (see text for details). When the ligand binds to the RNA, the r5UTR folds into a structure that contains a transcription terminator (a), or that traps the SD sequence into a double-stranded RNA helix and prevents translation (b). In the examples provided, a negative effect is mediated by the ligand, but an opposite situation is possible; that is, binding of the ligand to the r5'UTR enables a folding that allows transcription or translation to proceed (see text for details). The ligand is represented by a gray diamond; abbreviations are as shown in Figs 1 and 2.

3.3 Riboswitches: Regulatory 5′UTR Sensing Metabolites

Riboswitches are ncRNA elements that are widespread among all domains of life, and are able to modulate the expression of their downstream protein-coding RNA sequences. The corresponding proteins are typically involved in metabolic pathways that consume, produce, or transport the metabolite molecules ("ligands") sensed by their respective riboswitches [78]. Typically, a riboswitch consists of two structurally linked RNA modules: (i) an "aptamer" that binds to the metabolite and is shared by riboswitches belonging to the same family; and (ii) an "expression platform" which features an alternate folding structure that responds to the interaction between the aptamer and its cognate ligand, thus switching "on" or "off" the expression of the downstream sequence [44, 46, 79]. In this way, riboswitches are able to monitor levels of specific key compounds of metabolism, and also to regulate downstream genes. With respect to the nature of the metabolite sensed, 24 different classes of riboswitches have been distinguished in bacteria to date [44]. Among the most-common riboswitches are those that sense purine nucleobases (e.g., guanine, prequeuosine, adenine), amino acids (e.g., lysine, glutamine, glycine), an aminosugar (e.g., glucosamine-6-phosphate, GlcN6P), the bacterial second messenger c-di-GMP, vitamins, and/or cofactors such as adenosylcobalamine (AdoCbl), thiamine pyrophosphate (TPP), flavin mononucleotide (FMN), and S-adenosylmethionine (SAM) (see Refs [44, 79] and references therein). In contrast to the expression platform, the aptamer domain is the most conserved part of the RNA element within the same class, within the same genome, and across species. Certain classes of riboswitches are unique in their ability to bind their corresponding metabolite(s), and are thus extremely well conserved even across domains, indicating strong physico-chemical and/or physiological constraints. For example, TPP is found in both archaea and plant species. In contrast, other metabolites can be recognized by distinct classes of aptamers; for instance, two different classes of riboswitches are employed to monitor the concentration of c-di-GMP, while no less than five are able to bind SAM.

The occurrence of each class of riboswitch within each genome varies across species [48, 78, 80], reflecting possible metabolic differences in the use of ubiquitous and vital molecules, and the involvement of other regulatory processes to monitor metabolite concentrations. For instance, the expression of *glmS*, a gene encoding the glucosamine synthase that produces GlcN6P (an essential compound for cell-wall synthesis), is under the control of a riboswitch, while GlcN6P serves as a cofactor to undertake self-cleavage (a unique example of a riboswitch behaving as a ribozyme) in the Gram-positive *Bacillus subtilis* and other related Firmicutes, while regulation is ensured via a *trans* ncRNA (GlmY) in the Gram-negative bacterium *E. coli* and the closely related Enterobacteriaceae [81–84]. For most cases described in bacteria, upon binding of the ligand to the aptamer, the riboswitch-mediated control operates either positively or negatively on the formation of: (i) a transcription terminator that stops the RNAP and thereby transcription elongation; or (ii) a secondary structure that sequesters the SD sequence of the mRNA and blocks translation initiation (Fig. 3b).

Although the significance of why riboswitch-mediated controls act on transcription or translation (or both) is not understood, an elegant series of experiments [85, 86] has demonstrated, nevertheless, that the mechanistic constraints and the timing imposed on the interaction step between the aptamer and the ligand, as well as the effects of this interaction on the conformational change of the expression platform, are exceedingly different. If the riboswitch-mediated regulation operates on transcription, then the aptamer–ligand interaction must occur on the nascent RNA, and within a period of time that is sufficiently short to allow transmission of the effect to the RNAP during elongation. Controls mediated by the FMN riboswitch *ribB* of *E. coli* and the M-box of *mgtA* in *S. enterica* (as described above) act via a Rho-dependent transcription termination that requires a pause of the transcription elongation complex that enables Rho to bind the nascent RNA, to translocate along the transcript to contact the RNAP, and to provoke its dissociation from DNA [75, 87, 88]. The importance of such a transcription pause was also described for the translationally competent PRE-*alx* RNA in response to alkali (see above), [71]. Taken together, the results of these studies have revealed that, in addition to the so-called "expression platforms," other RNA elements also control the progression rate of the transcription complex (Fig. 3a). In situations where effects operate on translation initiation by preventing ribosome binding, the aptamer–ligand interaction may occur at any time after formation of the aptamer, ranging from the very moment of its synthesis to the ongoing translation process on the already existing mRNA [85, 86].

A riboswitch-dependent control of gene expression may combine more sophisticated organizations of regulatory elements than a single RNA aptamer and an expression platform. For example:

- Consecutive riboswitches can be present within a unique 5′UTR, rendering the system more sensitive to changes in ligand amounts. For example, two glycine aptamers and a single expression platform precede the encoding sequence *gcvT* in *B. subtilis* and *Vibrio cholerae* [89, 90], while two independent TPP riboswitches are upstream of *tenA* in *Bacillus anthracis* [91].
- Two aptamers can sense the same ligand, and each platform will exert its effect at a distinct level: one on transcription and the other on translation. An example is the two SAM riboswitches preceding the *bhmT* encoding sequence of *Candidatus Pelagibacter ubique* [92].
- Riboswitches sensing and responding to different metabolites can be organized in tandem. Examples are the AdoCbl and SAM riboswitches in the 5′UTR of *metE* in *Bacillus clausii* [93].
- A riboswitch-mediated control may be associated with other regulatory elements that act at either transcriptional or post-transcriptional levels. An example is the expression control of the *ubiG-mccBA* operon in *Clostridium acetobutylicum*, which employs the association of a SAM riboswitch and a T-box, an RNA element binding a tRNA (see below) [94], or the utilization of ethanolamine by *Enterococcus faecalis*. The latter requires expression of the *eut* operon, which in turn relies on the interplay of an AdoCbl riboswitch and a transcriptional regulatory two-component system [95, 96].

3.4 T-Boxes: Regulatory 5′UTR Sensing tRNAs

T-boxes represent an abundant class of r5′UTRs that are known in many bacterial species but are most commonly found in Gram-positive bacteria, especially in Firmicutes [97, 98]. These regulatory elements interact with specific tRNAs and adopt an alternative conformation that depends on the loading state of the bound tRNA (i.e., charged or not with its corresponding aminoacyl residue). T-boxes regulate transcription elongation and, in some cases, the translation of downstream ORF(s). More than 90% of these regulatory RNA elements precede sequences encoding aminoacyl-tRNA synthetase (aaRS) and proteins, thus ensuring the biosynthesis and transport of amino acids [97, 99]. Each aaRS catalyzes the esterification reaction involving a specific amino acid and the acceptor sequence (NCCA) at the 3′ end of its cognate unloaded tRNA, to form the corresponding charged tRNA (aa-tRNA). The charged tRNAs are used by translating ribosomes to elongate nascent peptides. The specific interaction between a T-box and its corresponding tRNA involves a "specifier" sequence within the T-box and the anticodon of the tRNA (Fig. 4). Once this primary interaction has been established, the loading state of the tRNA dictates the following steps: (i) when the interaction is established by an aa-tRNA, a helix forms within the 3′ region of the T-box and stops the transcription elongation or occludes the RBS; and (ii) if the interaction involves an unloaded tRNA, then a specific interaction occurs between this previous helix and the free NCCA sequence at the 3′ end of the tRNA; this interaction destabilizes the terminator helix and the transcription or translation

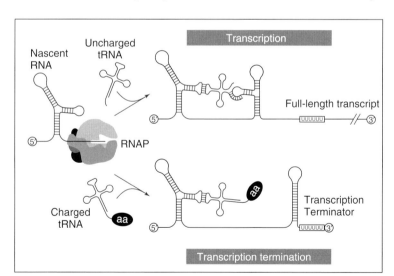

Fig. 4 T-box and tRNA-mediated regulation. The anticodon of the tRNA pairs to the "specifier" sequence of the T-box contained within the r5′UTR of the nascent RNA. When the paired tRNA is uncharged, the NCCA sequence of the tRNA 3′ end pairs to a sequence of the r5′UTR, and transcription elongation can proceed. When the tRNA paired to the specifier is charged (aa = amino acid), the NCCA sequence of the tRNA cannot pair to the r5′UTR and a transcription terminator is formed.

can proceed (Fig. 4). Thus, high levels of charged tRNAs, reflecting high amino acid levels, repress T-box-dependent gene expression. Conversely, high levels of unloaded tRNAs, reflecting low levels of cellular amino acid, will trigger the expression of genes required for the uptake and biosynthesis of amino acids, and the aa-tRNA synthesis [98–100].

By involving essential compounds of the translation machinery, T-boxes provide a regulatory feedback to the global translation phenomenon, and permanently assess the capacity of the cell to perform protein synthesis by gauging the [tRNA]:[aa-tRNA] ratio. This mode of regulation illustrates both the broad impact achieved by RNA-mediated regulation and the functional versatility of ncRNA molecules: (i) T-boxes affect transcription and/or translation; and (ii) these cis-regulatory ncRNA elements sense and respond to key ncRNAs of the translation machinery, namely tRNAs which ensure on the T-box a trans-acting regulatory role, as described for other ncRNAs (see below).

3.5
Regulatory 5′UTR Sensing Other Biomolecules

Based on sequence conservation and folding predictions, many classes of r5′UTRs have been described (http://rfam.sanger.ac.uk); indeed, approximately three new classes of riboswitches are currently reported each year [44]. Predictions of Rho-independent transcription terminators located within the 5′UTRs have shown a large occurrence in bacterial genomes (from ∼0.6 to 5% of predicted transcription units, depending on bacterial phyla). The location of such putative terminators suggests an effect on transcription elongation, as described above for demonstrated cis-regulatory RNA elements [48]. Although the majority of these putative r5′UTRs await demonstration of their regulatory effects and their capacity to specifically bind ligands, it is tempting to speculate that at least part of them would sense and respond to specific molecules. A few functional and evolutionary arguments strengthen this hypothesis. While intrinsic regulatory ncRNA elements certainly represent an ancient means of controlling genes, their conservation in the three domains of life highlights a highly efficient mode to regulate and provide feedback to metabolic pathways. Indeed, such an RNA-mediated control is reduced to the simplest form: the cis-regulatory RNA element and the ligand molecule, with no additional factor required. The conservation of such a system throughout evolution imposes low constraints, as only the effective recognition "RNA/ligand" and the RNA-dependent folding of the terminator must be retained. Furthermore, in vitro RNA evolution via "Systematic Evolution of Ligands by Exponential Enrichment" (SELEX) has amply demonstrated the extraordinary plasticity of the RNA molecule, in terms of sequence and structure, providing it with the ability to recognize efficiently any sort of molecule, and even to select RNA aptamers for various applications, including medical purposes [101–104]. The possibility of designing any desired aptamer under laboratory conditions by "replicating" the RNA molecule (vis SELEX) raises the scenario that evolution has most likely already employed that option, and strongly suggests that a large part of predicted r5′UTRs function by binding specific ligands and exerting regulatory effects.

4 RNA Pairing to Other RNAs

The majority of characterized regulatory RNAs act by base-pairing to different target RNA molecules. In this case, two classes have been distinguished:

- AsRNAs, transcribed from the opposite strand of their target RNA, regulator and target being perfectly complementary over, at least, a portion of their sequence (Fig. 5).
- *Trans*-acting RNAs (ncRNAs), encoded at different genomic loci than their RNA targets with which they form RNA duplexes containing imperfect matches. This latter class represents the most well-characterized bacterial regulatory RNAs [10, 105, 106].

4.1 *cis*-Antisense RNAs

One of the most intriguing features to be unveiled by transcriptomic studies has been the discovery of hundreds of novel asRNAs in bacterial chromosomes, ranging from a hundred to several thousand nucleotides in size, that may encompass entire operons on the opposite DNA strand (Fig. 5a).

4.1.1 Mode of Action

Transcriptional antisense organizations involve genes expressing various types of RNA: two mRNAs or two ncRNAs can be antisense to each other, while ncRNAs can be antisense to mRNAs (Fig. 5a). The biological significance of the abundance and the diversity of antisense organizations is largely unknown, however, and *a priori* there is no relationship between biological functions ensured by annotated genes and the fact that they have antisense genes [37, 39, 40, 42]. Nonetheless, the importance of an antisense-based regulation has been reported in several processes, including virulence, motility, conjugation, transposition, photosynthesis, phage development, and plasmid replication [5, 7, 39, 107–109]. Antisense-based regulation is a complex process that can affect multiple steps of gene expression (Fig. 5b), and requires perfect coordination as the antisense transcription units must be expressed simultaneously in the same cell. The effects of an antisense organization can be observed at different steps of gene expression. Opposite and overlapping transcriptional activities provoke mechanical constraints and decrease each other by a phenomenon termed "transcriptional interference." The latter results from:

- The "crash" of convergent transcription elongation complexes.
- The repression exerted by one transcription initiation complex formed on the formation of a second complex upstream and in the opposite direction ("promoter occlusion").
- The dissociation of a transcription initiation complex by collision with an elongating transcription complex coming from the opposite direction ("sitting-duck interference") [94, 110–112].

During transcription, the RNA duplex formed by an asRNA with its nascent RNA counterpart may provoke a premature transcription termination, thus preventing the synthesis of a functional transcript [113, 114]. When both complementary RNAs are synthesized, however, the RNA hybrid formed by the full-length asRNA and its RNA counterpart may: (i) affect

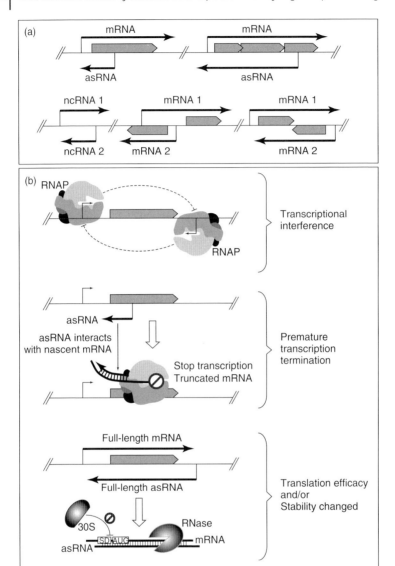

Fig. 5 Examples of transcriptional *cis*-antisense organizations and major antisense-mediated effects. (a) *Cis*-antisense organizations are due to genes transcribed from complementary DNA strands, the transcripts (thin arrows) of which are perfectly complementary at least to a portion of their sequences. The asRNAs can overlap mono- or polycistronic mRNAs (upper); two ncRNAs or two mRNAs can be antisense of each other (lower). The ORFs are represented by broad gray arrows; (b) Transcriptional organizations have three major consequences of gene expression if transcription units are expressed simultaneously in the same cell: (1) The transcriptional interference is due to the physical constraints exerted by each transcription complex on the other; (2) Premature transcription termination is caused by pairing of the nascent antisense RNAs; (3) Changes in the translation efficacy (30S ribosomal subunit) and/or the stability (RNase) of one transcript when it is paired, or not, to its antisense RNA.

the translation efficacy (e.g., the asRNA may pair to the RBS and prevent ribosome binding); or (ii) change the RNA stability (e.g., the RNA duplex may be a substrate for RNases [115–117]) or, in contrast, protect one RNA otherwise substrate for RNases [118].

4.1.2 Specific and Global Antisense-Dependent Regulations

The results of several studies have demonstrated the transcriptional control of asRNAs, enabling them to regulate the antisense transcription unit in response to specific cues. For example, in *S. enterica*, the presence of Mg^{2+} in the environment is sensed by the two-component system (TCS) PhoQ-PhoP, the histidine kinase sensor, and the cognate response regulator, respectively. At low Mg^{2+} concentrations, PhoP is phosphorylated and activates transcription of the *mgtCBR* operon, encoding an inner membrane protein (MgtC, which is required for intramacrophage survival, virulence in mice, and growth at low Mg^{2+}), an Mg^{2+} transporter (MgtB), and a proteolysis regulator of MgtC (MgtR), respectively. Within the intergenic region separating the ORFs MgtC and MgtB (but in the opposite direction), a 1200 nt-long asRNA (named AmgR) is expressed when Mg^{2+} resources are scarce, and this totally overlaps the *mgtC* encoding sequence [119]. Remarkably, PhoP also activates *amgR* transcription, though its binding affinity for the *amgR* promoter regions is lower than its affinity for the *mgtCBR* promoter, which suggests a sequential activation of the system. First, the *mgtCBR* transcription would be induced, after which *amgR* expression would be triggered to modulate *mgtCBR* expression according to the Mg^{2+} availability, which is reflected by the phosphorylated level of PhoP. AmgR was shown to decrease MgtC and MgtB protein levels by enhancing degradation of the *mgtC* portion of the *mgtCBR* operonic RNA [119]. Similarly, the TCS HP0165 (sensor kinase) and HP0166 (response regulator) regulate – in a pH-dependent manner – the bicistronic operon *ureA* and *ureB* that encode the urease subunits in *Helicobacter pylori*. The urease function is essential for this bacterium to achieve gastric colonization, as it enables the organism vitally to maintain a cytoplasmic and periplasmic neutral pH despite living in the extreme acidity of the stomach. The expression of *ureB* is turned off by an internal asRNA, *ureB*-sRNA (~290 nt), the transcription of which is increased at neutral pH via the HP0165–HP0166 system. In this case, the *ureB*-sRNA induction provokes a decrease in *ureAB* RNA levels (when urease activity is no longer necessary) [120]. A remarkable asRNA-dependent control has also been described in *Clostridium acetobutylicum*, in response to sulfur resources. Sulfur is an essential element at the crossroad of metabolic pathways (synthesis and degradation) that involve amino acids and nucleic bases, and the metabolite SAM. The *ubiG-mccB-mccA* operon is controlled by a T-box that responds to the amino acid cysteine (see above). At the end of the *ubiG-mccB-mccA* operon, and in the opposite direction, an asRNA originate and its transcription elongation depends on a SAM-mediated riboswitch mechanism (see above). When SAM levels are high (e.g., in the presence of methionine), the transcription terminator is formed and the synthesis of a full-length asRNA is blocked, thus allowing an entire transcription of the operonic transcript *ubiG-mccB-mccA*. In contrast, when SAM levels are low, the elongating complex of the asRNA is not halted, as this would cause transcription interference that would arrest transcription of the operon [94]. Although other antisense organizations have been described, in most cases the environmental cues to which they respond, the regulatory mechanisms that control their expression, and/or their significance, remain unknown [113, 117, 118, 121, 122].

Recently, a transcriptomic study performed in *Staphylococcus aureus*, including short (<50 nt) RNA molecules present in the cell, revealed that at least 75% of RNAs from annotated genes have an antisense counterpart. This widespread antisense transcription was shown to generate RNA duplexes substrate for RNase III, and to occur in diverse and evolutionarily distant Gram-positive bacteria. Although this phenomenon is not understood, it certainly corresponds to a novel layer of RNA-mediated regulation that may adjust mRNA levels on a genome-wide scale [40].

4.2
trans-Acting Noncoding RNAs

This category of RNAs (ncRNAs) is also commonly known as "small RNAs" (sRNAs), as they are relatively small in size (~50–500 nt). In this chapter, the decision was made to use the denomination "noncoding," which is believed more appropriate in this case as an RNA is not used by the translation machinery when exerting its regulatory effect on mRNA targets. Two classes of ncRNA have been characterized: (i) those that act on mRNAs; and (ii) those that bind to proteins (see below). In contrast to an asRNA, which pairs perfectly to RNA synthesized from the reverse complementary DNA strand, the ncRNA regulates gene expression by binding to mRNA(s) encoded in *trans* – that is, transcribed from a different locus in the genome. The ncRNA and its mRNA target form an RNA duplex via an imperfect base-pairing, resembling the microRNAs of eukaryotic organisms. The present understanding of ncRNA-mediated regulation derives mostly from studies with *E. coli* and *Salmonella* sp., where approximately 30 ncRNAs have been functionally characterized. Although several functional studies have been conducted in other bacterial species, target identification has not followed the pace of RNA discovery via tiling-arrays and RNAseq. Structural and sequence criteria have proved to be insufficient for distinguishing those ncRNAs that target mRNAs from those that bind proteins. It is to be expected, therefore, that among the numerous identified ncRNAs (e.g., more than 100 in *E. coli* alone), only a fraction will fit the definition of "*trans* ncRNAs.*"* It must be borne in mind, however, that the present general view on the mechanistic aspects of ncRNA-mediated regulation in bacteria may be biased, as the majority of present knowledge on ncRNA is based on observations in *E. coli* and *Salmonella* sp., which are two very close relatives in the family Enterobacteriaceae. For instance, Hfq – an RNA chaperone related to the Sm eukaryotic protein family – has been shown necessary for the action of most ncRNAs on their mRNA targets, notably by facilitating the formation of ncRNA/mRNA duplexes in enterobacteria [123, 124]. In addition, the stability of most ncRNAs is decreased in the *hfq* mutant, indicating that the protein protects the ncRNAs from the action of RNase. However, *hfq* is present in only about 50% of bacterial species and, even when present, it has been shown as dispensable for ncRNA/mRNA duplex formation in several cases, including *E. coli* [36, 125–128]. One interesting hypothesis here is that Hfq may not be required when ncRNAs form strong interactions with their mRNA targets [129]. Alternatively, the presence of functionally equivalent protein(s) remains to be identified.

4.2.1 Gene Expression Inhibition by ncRNAs

Most of the ncRNAs characterized exert their effect by inhibiting translation and/or promoting mRNA decay. As a general rule, these ncRNAs base-pair to their mRNA targets in a region overlapping or close to the RBS, and the RNA duplex formed prevents the ribosome access to the SD sequence and subsequent translation (Fig. 6a). Although the untranslated mRNA becomes unprotected against RNase attack, degradation of the mRNA target requires an active recruitment of RNases.

The RyhB-mediated regulation is possibly one of the most thoroughly characterized. RyhB was first identified in E. coli through genome-wide searches [29, 34], and was shown subsequently to regulate negatively more than 20 mRNA targets, including mRNAs encoding iron-using proteins. This RyhB-mediated control responds to iron (Fe^{2+}) starvation and is Hfq-dependent [130, 131]. Under growth conditions where Fe^{2+} is not limiting, *ryhB* transcription is directly repressed by Ferric uptake regulator (Fur). In a negative feedback loop, RyhB pairs to *fur* mRNA and represses its translation [132]. Thus, one of the major roles of RyhB is to limit the amount of iron-using proteins when iron is scarce, and to ensure its availability for essential iron-requiring proteins [133].

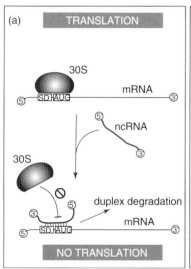
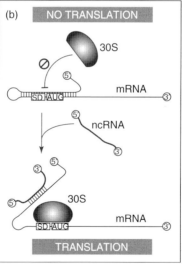

Fig. 6 Major effects mediated by *trans*-acting ncRNAs on mRNA targets; (a) translation repression and (b) translation activation. (a) In the absence of the ncRNA, the mRNA is competent for translation. The ncRNA targets a sequence of the mRNA (generally the RBS or a sequence in close vicinity) to form a duplex that no longer allows binding of the 30S ribosomal subunit to initiate translation. Generally, the ncRNA–mRNA duplex is degraded; (b) In the absence of the ncRNA, the 5′UTR of the target mRNA folds into a structure that occludes ribosomal access to the RBS ("translation off"). The ncRNA targets a portion of this folded 5′UTR region and pairs to it, releasing the RBS and rendering it competent for translation ("translation on"). The abbreviations are as shown in Figs 1 and 2.

RyhB was the first ncRNA shown to recruit the endoribonuclease E (RNaseE), associated with the degradosome, and to promote degradation of the mRNA target and the ncRNA itself [134]. Subsequently, RNaseE was shown to be involved in the degradation of RNA duplexes formed by a large number of the Hfq using ncRNAs that repress mRNA targets, at least in enterobacteria. Both, Hfq and RNaseE seem to share the same sequence/structure affinity for RNAs, namely A/U-rich sequences preceded by a stem loop. RNaseE and Hfq have each been shown to interact *in vivo*, and RNaseE is supposed to reach the RNA duplex via its recruitment by Hfq [135]. However, translation inhibition does not necessarily imply duplex degradation, and the coupling between both events is not fully understood [136, 137].

It is interesting to note that in *S. aureus*, the ncRNA RNAIII mediates translation repression in an Hfq-independent manner and provokes duplex degradation by RNase III for numerous of its targets [125, 128, 138]. RNase III is a double-strand-specific endoribonuclease; thus, in an Hfq-independent mechanism its affinity for RNA duplexes may help to bypass the requirement for an auxiliary protein to recruit the RNAse at the duplex.

Remarkably, ncRNA degradation due to pairing to its mRNA target can be exploited to control the levels and effects of the regulatory ncRNA. For instance, in the absence of a chito-oligosaccharide, the ncRNA ChiX (also termed *MicM*) is constitutively transcribed and induces degradation of the *chiP* mRNA that encodes a porin which allows uptake of the sugar. When chito-oligosaccharide is present in the medium, the chitobiose operon *chb* (*chbBCARFG*) is massively induced. The *chb* mRNA pairs with high affinity to the transcription terminator of ChiX and provokes degradation of the ncRNA. Thus, by mimicking a target for ChiX, the *chb* mRNA relieves the ChiX-dependent repression of *chiP* and enables consumption of the sugar [139, 140].

A surprisingly high number of mRNAs encoding outer-membrane proteins (OMPs) are targets of ncRNAs. Of these ncRNAs, MicF was the first to be described as a regulator of gene expression acting by base-pairing to the RBS and inhibiting the translation of *ompF* mRNA that encodes the most abundant OMP in *E. coli* [22]. Almost twenty years later, many other ncRNAs have been found to regulate OMPs, including MicC regulating OmpC [141], RseX regulating OmpC and OmpA [20], IpeX regulating OmpC and OmpF [142], the paralogous ncRNAs OmrA and OmrB that regulate at least four OMPs [143], and SdsR regulating OmpD [144]. Interestingly, the majority of these ncRNAs repress the expression of their mRNA targets by pairing to a region close to or overlapping the RBS. Among these, MicA and RybB ncRNAs were shown to regulate more than 30 mRNA targets in *E. coli*, including the major OMPs, namely OmpA, OmpC, OmpD, OmpF, and OmpX [145–150]. RybB and OmpA are both under the transcriptional control of the transcription sigma factor σ^E, which activates the expression of about 100 genes in response to the stress envelope. This regulatory network allows a dynamic control of the envelope homeostasis by endowing indirect repressor functions to a transcriptional activator [145]. The fact that numerous ncRNAs are devoted to the repression of OMPs remains unclear, but the ncRNA-dependent expression could allow a fine-tuning of the porin balance and a response to multiple signals in a tailored manner.

4.2.2 Activation of Gene Expression by ncRNAs

ncRNAs can also positively regulate gene expression by activating translation and/or enhancing the stability of their mRNA targets. However, only a few examples of this mode of action have been described, perhaps indicating that positive regulation is less common. In most cases, gene expression activation occurs by the displacement of an internal inhibitory element contained within the 5′UTR of the mRNA target. One predominant category of such a mode of action is termed an "anti-antisense" mechanism. In the absence of ncRNA, the 5′UTR folds in a structure that prevents the ribosome accessing the SD sequence and hence represses translation. When expressed, the ncRNA binds to the inhibitory structure of the 5′UTR, freeing the RBS and allowing translation (Fig. 6b). The stability of the mRNA can also be increased, though it is difficult to determine whether this is an effect due solely to the protection of the mRNA by the translating ribosomes, or to a specific RNase protection mechanism. The most thoroughly studied of these cases is that of the sigma factor RpoS in *E. coli*. RpoS is the major stress response sigma factor in *E. coli*, and is involved in the transcriptional control of a few hundreds of genes. The expression of RpoS is complex and regulated at many levels, and at least three ncRNAs activate the translation of *rpoS* mRNA [151]. Originally, the *rpoS* translation was shown to be impaired in an *hfq* mutant due to a stem-loop structure within the *rpoS* 5′UTR that occluded the SD sequence [152–154]. Subsequently, it was observed that DsrA (a previously identified ncRNA) had an activating effect on the expression of *rpoS* at a post-transcriptional level [23, 26]. Through an elegant mutational analysis, it was shown that DsrA could pair to the 5′UTR of the *rpoS* mRNA region that otherwise would form the inhibitory stem-loop structure [155]. Subsequently, two other ncRNAs – RprA and ArcZ – were shown to activate *rpoS* translation through the same mechanism as DsrA [21, 156]. Interestingly, all three ncRNAs bind to the same region in the *rpoS* 5′UTR, indicating that this is probably the most energetically favorable position to "open" the stem-loop [157]. The fact that *rpoS* translation is controlled by different ncRNAs provides different inputs to RpoS expression, as DsrA, RprA, and ArcZ are induced by distinct environmental signals of low temperature, membrane, and anaerobic/aerobic stresses, respectively [26, 156, 158, 159]. Although the three ncRNAs act through the same general mechanism, their binding properties to the *rpoS* mRNA differ significantly *in vitro*, indicating that they may impact differently on the activation of *rpoS* translation. In addition, these ncRNAs have other mRNA targets that they control negatively [160–162]; this suggests that, besides the RpoS-dependent response, they may also coordinate different regulatory pathways, providing to the cell a flexibility to adapt simultaneously to different cues. The ncRNA OxyS has been reported to inhibit *rpoS* translation when it is expressed at high levels [163]. Although the mechanism for this repression remains unclear, it may be due to an indirect effect of OxyS on the available levels of Hfq for the positive regulators, DsrA, RprA, and ArcZ [156, 164, 165]. Other examples of this anti-antisense activation mode have been reported, including the RyhB-dependent activation of the *shiA* mRNA, encoding a shikimate permease in *E. coli*, and the RNAIII-dependent activation of the *hla* mRNA, encoding the α-toxin in *S. aureus* [166, 167].

Activation by a ncRNA can also be achieved by stabilizing the mRNA target, whereby the ncRNA–mRNA duplex protects the mRNA from RNase action. For instance, in group A *Streptococcus*, the expression of secreted streptokinase (SKA, a virulence factor), depends on the ncRNA FasX that stabilizes the *ska* mRNA by pairing to the 5′UTR [168]. Unfortunately, the mechanism involved is not fully understood and additional examples must be identified to ascertain if this represents a common means of enhancing mRNA stability.

An interesting example of a ncRNA that increases translation and stabilizes its mRNA target is provided by the ncRNA VR-RNA, which acts on *colA* mRNA (encoding collagenase) in *Clostridium perfringens*. Typically, VR-RNA pairs to the *colA* 5′UTR and induces an endonucleolytic cleavage immediately downstream of the duplex. Such cleavage disrupts an otherwise inhibitory structure that sequesters the RBS and, in the absence of VR-RNA, *colA* mRNA is rapidly degraded [169].

5
RNA Changing Protein Activities

RNAs are major components of ribonucleoprotein (RNP) complexes in the cell, where they carry out vital functions involved in the expression of genetic information, including DNA replication [170, 171]. For instance, ribosome assembly operates under a strict timeline that is orchestrated by conformational changes of rRNAs, and interactions between ribosomal proteins and rRNAs. The translation process within the ribosome is catalyzed by rRNAs, and proteins provide an architectural environment where RNA actors (rRNAs, tRNAs, and mRNAs) gather and react [172]. Another example of the central role played by RNA in the functioning of a RNP is during the transcription termination process. When the termination process is Rho-independent, the formation of a stable stem-loop structure, followed by a run of "U"s, is a signal for the RNAP to stop polymerizing RNA and dissociate; when the process depends on Rho protein, RNA sequences and/or structures are responsible for the pausing of the elongation complex that allows Rho to translocate along the RNA and to interact with the RNAP to dissociate the complex [88, 173]. Also, in the *trans*-translation process, the ncRNA coined "tmRNA" (or "SsrA") acts as a tRNA to enter into the ribosome, and encodes a 10-amino acid unit as a mRNA. The translation of these 10 amino acids both releases ribosomes from truncated mRNAs and tags aberrant proteins by an addressing signal for proteolysis [174]. Besides these essential roles, very few examples of ncRNAs acting as regulators of protein activities, *per se*, have been reported. These ncRNAs control the activity of pleiotropic proteins acting on transcription or translation by directly binding nucleic acids (DNA or RNA). ncRNAs mimic these targeted nucleic acids and form RNPs, where they act as competitive inhibitors and have profound impacts on gene expression reprogramming.

5.1
Mimicking DNA Promoters

The amount of RNA produced by each transcription unit in the cell results from a strict and complex control. In *E. coli*, the 184 nt-long ncRNA 6S (or "SsrS") binds specifically to the RNAP holoenzyme that contain the housekeeping transcription sigma factor σ^{70} ($E\sigma^{70}$). 6S folds in a single stem-loop structure that mimics a DNA promoter and competes with σ^{70}-dependent promoters to bind Eσ^{70}. The complex formed by 6S and Eσ^{70} (6 S – Eσ^{70}) is more stable than the normal Eσ^{70} complex, and the 6 S – Eσ^{70} form of the holoenzyme no longer binds to σ^{70}-promoters. The biogenesis of 6S is tightly controlled via transcription and RNA processing, and causes an accumulation throughout growth that reaches elevated levels when the bacteria enter stationary phase, at which time 80–90% of Eσ^{70} are sequestered in a 6 S – Eσ^{70} complex [27, 175]. Thus, over the growth period, the increase in the 6 S – Eσ^{70} complex leads to a reduction in the pool of active Eσ^{70} and thus imposes an increasing competition between σ^{70}-dependent promoters to be utilized by Eσ^{70}: this process results in both direct and indirect transcription inhibition of 6S-dependent genes (i.e., "6S-sensitive" genes) in cells as they enter and remain in the stationary phase (Fig. 7), [176–178]. When growth resumes, however, the increase in intracytoplasmic nucleotide concentration triggers an initiation of transcription within the 6 S – Eσ^{70} complex. For this, Eσ^{70} utilizes 6S as template and produces a 14–20 nt-long RNA (pRNA), the synthesis of which induces a structural rearrangement that dissociates the Eσ^{70}–6S–pRNA complex; this enables the cell to recover a functional pool of Eσ^{70} that will be required to undertake exponential growth (Fig. 7). The 6S–pRNA complex released from the holoenzyme is subsequently degraded, although pRNA synthesis and/or the sRNA molecule itself may be functionally important during

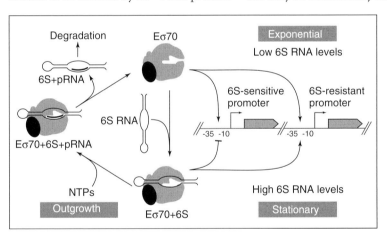

Fig. 7 6S-mediated regulation of the vegetative RNAP during growth. During the exponential growth phase, the vegetative holoenzyme Eσ^{70} transcribes all Eσ^{70}-dependent promoters. Over the growth period, 6S accumulates and binds tightly to Eσ^{70}. During the early stationary phase more than 80% of Eσ^{70} is sequestered into an Eσ^{70}–6S complex which is able to use 6S-resistant promoters but not 6S-sensitive promoters. Thus, 6S provokes a reprogramming of the main transcriptional activity by switching off certain promoters (6S-sensitive) and increasing competition between others promoters to be used by Eσ^{70} (6S-resistant). On outgrowth from the stationary phase, an increased level of nucleotide triphosphates (NTPs) enables Eσ^{70} to behave as an RNA-dependent RNA polymerase, and 6S is used as a template to synthesize pRNA. The complex Eσ^{70}–6S–pRNA then dissociates; the duplex 6S–pRNA is degraded while the holoenzyme Eσ^{70} reuses all of the Eσ^{70}-dependent promoters, allowing the cell to undertake exponential growth.

or after outgrowth, as a 6S mutant deficient for pRNA synthesis has a decreased viability [176, 179–181]. Although the sequence is poorly conserved, the secondary structure of the 6S RNA is extremely well preserved across the bacterial kingdom, highlighting structural constraints due to a functional conservation of the ncRNA. In certain species, including *B. subtilis*, *Legionella pneumophila*, and *Prochlorococcus marinus* MED4, two distinct 6S RNAs have been identified; this suggests that each RNA may ensure a different function by targeting different RNAP complexes, or by having different effects on the activity of the vegetative holoenzyme, as suggested for *B. subtilis* [42, 176, 182–185]. Recently, the results of several studies have underscored the important physiological functions ensured by the 6S regulatory system, for example, efficient outgrowth and viability under laboratory conditions in *E. coli*, and optimal intracellular development in *L. pneumophila* [176, 177, 186]. Additional studies are required, however, to provide a better understanding of this fascinating and complex system, which has a profound impact on gene expression.

5.2
Mimicking Ribosome Binding Sequences

The metabolism of carbon involves a balance between opposite fluxes in response to environmental conditions and resources, whereby *catabolism* first allows the degradation of organic molecules and the production of energy that, ultimately, is consumed by *anabolism*, enabling the synthesis of cellular components. Based on its crucial importance, the metabolism of carbon is intimately connected to other metabolic pathways and physiological traits, so as to impose drastic switches in gene expression and a tight coordination via global regulatory networks. Among these global and crucial regulatory networks, the Csr/Rsm system (carbon storage regulator/regulator of secondary metabolite) ensures connections between carbon metabolism and diverse traits in many species of γ-proteobacteria, such as motility, biofilm formation, virulence, and quorum sensing [187–191]. The regulatory circuitry is composed of at least three components connected in a feedback loop (Fig. 8): (i) a translation regulatory protein, the activity of which is directly inhibited by ncRNAs; (ii) a TCS that is controlled by the translational regulatory protein and regulates the transcription of ncRNAs; and (iii) ncRNAs themselves.

The regulatory protein includes one to four RNA-binding proteins of the CsrA/Rsm family (CsrA, RsmA, RsmI, and RsmE) that repress metabolic pathways or biological processes that otherwise would be induced upon entry into the stationary phase (e.g., glucogenesis and glycogen synthesis, biofilm development), while stimulating pathways expressed during the exponential phase (e.g., glycolysis, use of particular carbon sources, chemotaxis, motility, type III secretion system in pathogenic species). Although, at present, the details of the activation process are unclear, it may be due to the repression of a specific translation repressor or to an increased mRNA stability. In contrast, the repression process has been shown due to the binding as a homodimer of the CsrA/Rsm protein to one or several sites (GGA motifs) within, or in the vicinity of, the SD sequences of mRNAs, where they compete directly with the 30S ribosome subunit. There may be exceptions to this mechanism, however, as it was first described for *E. coli*, where CsrA binds to the encoding sequence of *sdiA* mRNA [192].

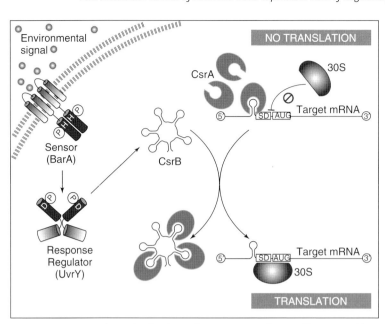

Fig. 8 Csr/Rsm and Crc feedback regulatory loops. For the Csr/Rsm system, in response to environmental cues sensed by the two-component systems BarA/UvrY, UvrY activates the transcription of ncRNA encoding genes "*csrB*-like." These ncRNAs bind to the repressor CsrA and counteract its translational regulatory effects. In addition to the Csr/Rsm systems, in Pseudomonads, the Crc system functions in a similar manner and involves unrelated regulators: CbrA/CbrB is the two-component system, CrcZ is the ncRNA, and Crc is the translation regulator.

In agreement with a pleiotropic effect of CsrA, more than 700 different mRNAs have been found to bind the protein [193].

The translation-repressing effect of CsrA/Rsm proteins is inhibited by CsrB-like ncRNAs that range in size from ~110 to 479 nt and feature repetitions of GGA sequences located in single-stranded regions of the RNA molecules [189, 194]. The number of these ncRNAs depends on the species; for example, a single copy is found in *Erwinia carotovora* (RsmB), two copies in *Pseudomonas aeruginosa* (RsmY and Z), *S. enterica*, and *E. coli* (CsrB and C), and three copies in *Vibrio cholerae* (CsrB, C, and D), *Acinetobacter baumannii* and *Pseudomonas fluorescens* (RsmX, Y, and Z), [189, 194]. The third element of the feedback regulatory loop is the TCS, which comprises the sensor kinase BarA and the cognate response regulator UvrY that stimulates transcription of the CsrB-like ncRNAs genes. The BarA/UvrY system is conserved within γ-proteobacteria, and is found under different designations, "GacS/GacA," "LetS/LetA," "VarS/VarA," or "RepA/RepB" [189]. How the BarA/UvrY TCS is activated is not fully understood, but it is known to respond to various stimuli, including weak acids (formates and acetates), quorum sensing molecules, and an imbalance of tricarboxylic acid cycle intermediates. Once triggered by BarA, UvrY activates the transcription of CsrB-like ncRNAs that bind and sequester CsrA/Rsm proteins, thus decreasing the CsrA/Rsm-mediated effect on regulated genes. The feedback of the system is ensured by a positive effect of CsrA/Rsm on the BarA/UvrY expression, which in turn increases the abundance of ncRNAs and antagonizes the CsrA/Rsm-mediated effects (Fig. 8).

How the CsrA/Rsm protein activates the TCS remains unclear. Additional inputs would allow the Csr/Rsm system to integrate other cues and to coordinate carbon fluxes. For instance, crosstalk certainly exists between UvrY and other sensor kinases or factors that would allow the system to respond to environmental cues not detected by BarA. As a component of the regulatory circuitry, the turnover of CsrB-like ncRNA(s) can be a crucial parameter. In *E. coli*, CsrB and CsrC are degraded in an RNase E-dependent process that is strictly controlled by the CsrD protein [195]. The way in which CsrD modulates CsrB and CsrC stability is also unknown, but it may be an important factor as the regulatory effects of the CsrA/Rsm system are intimately dependent on the amounts of its components. In addition to the CsrA/Rsm system, the most likely species-specific mechanisms contribute towards modulating exactly how such physiological constraints are satisfied [188, 194].

Among the Pseudomonadaceae, the carbon Crc (catabolite repression control) system coordinates the cells' carbon metabolism by adapting it to carbon sources that provide the most efficient growth. The Crc regulatory circuitry is very similar to the CsrA/Rsm system, but the components involved are different and nonhomologous. The RNA-binding protein, Crc, affects the expression of more than 130 genes, and has pleiotropic effects due to the control exerted on transcriptional regulators. Crc binds A-rich single-stranded sequences (AAxAAxAA, where x is A, C, or U) in the vicinity of the SD sequence, and prevents binding of the 30S ribosomal subunit to initiate translation [196, 197]. mRNAs regulated by Crc encode functions that allow antibiotic and stress resistance, biofilm development, and the utilization of "poor" carbon sources, including glucose transporters (contrary to most related γ-protobacteria, glucose is a non-preferred sugar for *Pseudomonas*) and other functions involved in the assimilation and transport of sugars and amino acids [198]. In *Pseudomonas putida*, two ncRNAs – CrcY and CrcZ, each of which contains six Crc-binding sites – control the level of free Crc by direct binding and sequestration. In *P. aeruginosa*, only CrcZ is present and no CrcY was found. Hence, the following functioning model has been proposed [199]. When bacteria are grown with a "poor" carbon source, levels of CrcY and CrcZ are high, free levels of Crc protein are low, and translation of mRNA targets occurs; conversely, when bacteria encounter a "good" carbon source, levels of CrcY and CrcZ drop, free Crc protein increases, and translation is repressed. The TCS CbrA/CbrB (sensor kinase and response regulator, respectively) regulates the CrcZ amount by activating the transcription initiation of the *crcZ* gene, but not *crcY* [199]. The regulatory aspects related to the synthesis and turnover of CrcY and CrcZ are not yet totally understood, however, and different stimuli may affect the activity of the Crc system [200, 201].

6
Concluding Remarks

In this chapter, the biological functions regulated by ncRNA molecules acting in *trans*, encoded in antisense organizations, and/or by those contained within 5′UTRs, have been described. There is indeed a tremendous diversity of processes where ncRNAs play essential roles by controlling cellular responses and coordinating them with other metabolic pathways, such as carbon fluxes and virulence or translation and amino acid synthesis [106, 202]. In addition to the functional importance of ncRNAs, impressive amounts of ncRNA elements have been shown present in bacteria, whether as individual transcription units or as cis-elements of protein-coding RNA sequences. For example, in *Mycobacterium tuberculosis* a quantitative study based on RNAseq indicated that, apart from ribosomal RNAs, two-thirds of RNA sequences are not expressed from the annotated genome sequences, which suggests that the amount of RNA that encodes proteins represents a very small fraction of the transcriptome (<10% of total RNA) [38]. Whether or not all of these ncRNA sequences are regulatory elements is unknown, but it is tempting to speculate that in some way they may act globally by modulating RNA turnover and the nucleotide triphosphate pool. Furthermore, the abundance and diversity of RNA molecules expressed from bacterial genomes are in sharp contrast to the limited number of ncRNAs, asRNAs, and r5′UTRs, where biological functions and targeted genes have been assigned. However, it is highly likely that, in future, novel regulatory RNA classes and new mechanisms will emerge from the RNA world.

Acknowledgments

P.M. is supported by the "Centre National de la Recherche Scientifique" (CNRS, France), and would like to thank F. Barras for his support. A.T.-A. is supported by a "Ramón y Cajal" contract from the "Consejo Superior de Investigaciones Cientificas" and Grants BFU2011-23222 and ERA-NET Pathogenomics PIM2010EPA-00606 from the Spanish Ministry of Economy and Competitiveness. A.F.dH. is supported by the Institute for Systems Biology (Seattle, WA, USA). F.R. is supported by the "Institut National de la Recherche Agronomique" (INRA, France). The authors thank Françoise Wessner and David Halpern for comments on the manuscript. Examples provided to illustrate this review have been selected from an abundant literature, and apologies are offered to colleagues whose work has not been cited.

References

1 Carninci, P. (2010) RNA dust: where are the genes? *DNA Res.*, **17**, 51–59.
2 Collins, L.J. (2011) The RNA infrastructure: an introduction to ncRNA networks. *Adv. Exp. Med. Biol.*, **722**, 1–19.
3 Gottesman, S. (2004) The small RNA regulators of *Escherichia coli*: roles and mechanisms. *Annu. Rev. Microbiol.*, **58**, 303–328.
4 Storz, G., Altuvia, S., Wassarman, K.M. (2005) An abundance of RNA regulators. *Annu. Rev. Biochem.*, **74**, 199–217.
5 Wagner, E.G., Flardh, K. (2002) Antisense RNAs everywhere? *Trends Genet.*, **18**, 223–226.
6 Felden, B., Vandenesch, F., Bouloc, P., Romby, P. (2011) The *Staphylococcus aureus* RNome and its commitment to virulence. *PLoS Pathog.*, **7**, e1002006.
7 Georg, J., Hess, W.R. (2011) cis-Antisense RNA, another level of gene regulation in bacteria. *Microbiol. Mol. Biol. Rev.*, **75**, 286–300.
8 Guell, M., Yus, E., Lluch-Senar, M., Serrano, L. (2011) Bacterial transcriptomics: what is beyond the RNA horiz-ome? *Nat. Rev. Microbiol.*, **9**, 658–669.
9 Romby, P., Charpentier, E. (2010) An overview of RNAs with regulatory functions

in gram-positive bacteria. *Cell. Mol. Life Sci.*, **67**, 217–237.

10 Waters, L.S., Storz, G. (2009) Regulatory RNAs in bacteria. *Cell*, **136**, 615–628.

11 Griffin, B.E. (1971) Separation of ^{32}P-labelled ribonucleic acid components. The use of polyethylenimine-cellulose (TLC) as a second dimension in separating oligoribonucleotides of '4.5 S' and 5 S from *E. coli*. *FEBS Lett.*, **15**, 165–168.

12 Hindley, J. (1967) Fractionation of ^{32}P-labelled ribonucleic acids on polyacrylamide gels and their characterization by fingerprinting. *J. Mol. Biol.*, **30**, 125–136.

13 Jain, S.K., Gurevitz, M., Apirion, D. (1982) A small RNA that complements mutants in the RNA processing enzyme ribonuclease P. *J. Mol. Biol.*, **162**, 515–533.

14 Ikemura, T., Dahlberg, J.E. (1973) Small ribonucleic acids of *Escherichia coli*. I. Characterization by polyacrylamide gel electrophoresis and fingerprint analysis. *J. Biol. Chem.*, **248**, 5024–5032.

15 Altuvia, S., Weinstein-Fischer, D., Zhang, A., Postow, L. et al. (1997) A small, stable RNA induced by oxidative stress: role as a pleiotropic regulator and antimutator. *Cell*, **90**, 43–53.

16 Okamoto, K., Freundlich, M. (1986) Mechanism for the autogenous control of the *crp* operon: transcriptional inhibition by a divergent RNA transcript. *Proc. Natl. Acad. Sci. U.S.A.*, **83**, 5000–5004.

17 Urbanowski, M.L., Stauffer, L.T., Stauffer, G.V. (2000) The *gcvB* gene encodes a small untranslated RNA involved in expression of the dipeptide and oligopeptide transport systems in *Escherichia coli*. *Mol. Microbiol.*, **37**, 856–868.

18 Liu, M.Y., Gui, G., Wei, B., Preston, J.F. III et al. (1997) The RNA molecule CsrB binds to the global regulatory protein CsrA and antagonizes its activity in *Escherichia coli*. *J. Biol. Chem.*, **272**, 17502–17510.

19 Bouche, F., Bouche, J.P. (1989) Genetic evidence that DicF, a second division inhibitor encoded by the *Escherichia coli dicB* operon, is probably RNA. *Mol. Microbiol.*, **3**, 991–994.

20 Douchin, V., Bohn, C., Bouloc, P. (2006) Down-regulation of porins by a small RNA bypasses the essentiality of the regulated intramembrane proteolysis protease RseP in *Escherichia coli*. *J. Biol. Chem.*, **281**, 12253–12259.

21 Majdalani, N., Chen, S., Murrow, J., St John, K. et al. (2001) Regulation of RpoS by a novel small RNA: the characterization of RprA. *Mol. Microbiol.*, **39**, 1382–1394.

22 Mizuno, T., Chou, M.Y., Inouye, M. (1984) A unique mechanism regulating gene expression: translational inhibition by a complementary RNA transcript (micRNA). *Proc. Natl Acad. Sci. USA*, **81**, 1966–1970.

23 Sledjeski, D., Gottesman, S. (1995) A small RNA acts as an antisilencer of the H-NS-silenced *rcsA* gene of *Escherichia coli*. *Proc. Natl Acad. Sci. USA*, **92**, 2003–2007.

24 Altuvia, S., Zhang, A., Argaman, L., Tiwari, A. et al. (1998) The *Escherichia coli* OxyS regulatory RNA represses *fhlA* translation by blocking ribosome binding. *EMBO J.*, **17**, 6069–6075.

25 Masse, E., Gottesman, S. (2002) A small RNA regulates the expression of genes involved in iron metabolism in Escherichia coli. *Proc. Natl Acad. Sci. USA*, **99**, 4620–4625.

26 Sledjeski, D.D., Gupta, A., Gottesman, S. (1996) The small RNA, DsrA, is essential for the low temperature expression of RpoS during exponential growth in *Escherichia coli*. *EMBO J.*, **15**, 3993–4000.

27 Wassarman, K.M., Storz, G. (2000) 6S RNA regulates *E. coli* RNA polymerase activity. *Cell*, **101**, 613–623.

28 Lai, E.C. (2003) microRNAs: runts of the genome assert themselves. *Curr. Biol.*, **13**, R925–R936.

29 Argaman, L., Hershberg, R., Vogel, J., Bejerano, G. et al. (2001) Novel small RNA-encoding genes in the intergenic regions of *Escherichia coli*. *Curr. Biol.*, **11**, 941–950.

30 Chen, S., Lesnik, E.A., Hall, T.A., Sampath, R. et al. (2002) A bioinformatics based approach to discover small RNA genes in the *Escherichia coli* genome. *Biosystems*, **65**, 157–177.

31 Rivas, E., Eddy, S.R. (2001) Noncoding RNA gene detection using comparative sequence analysis. *BMC Bioinformatics*, **2**, 8.

32 Tjaden, B., Saxena, R.M., Stolyar, S., Haynor, D.R. et al. (2002) Transcriptome analysis of *Escherichia coli* using high-density oligonucleotide probe arrays. *Nucleic Acids Res.*, **30**, 3732–3738.

33 Vogel, J., Bartels, V., Tang, T.H., Churakov, G. et al. (2003) RNomics in *Escherichia coli* detects new sRNA species and indicates parallel transcriptional output in bacteria. *Nucleic Acids Res.*, **31**, 6435–6443.

34 Wassarman, K.M., Repoila, F., Rosenow, C., Storz, G. et al. (2001) Identification of novel small RNAs using comparative genomics and microarrays. *Genes Dev.*, **15**, 1637–1651.

35 Pichon, C., Felden, B. (2005) Small RNA genes expressed from *Staphylococcus aureus* genomic and pathogenicity islands with specific expression among pathogenic strains. *Proc. Natl Acad. Sci. USA*, **102**, 14249–14254.

36 Mandin, P., Repoila, F., Vergassola, M., Geissmann, T. et al. (2007) Identification of new noncoding RNAs in *Listeria monocytogenes* and prediction of mRNA targets. *Nucleic Acids Res.*, **35**, 962–974.

37 Fouquier d'Heroeul, A., Wessner, F., Halpern, D., Ly-Vu, J. et al. (2011) A simple and efficient method to search for selected primary transcripts: non-coding and antisense RNAs in the human pathogen *Enterococcus faecalis*. *Nucleic Acids Res.*, **39**, e46.

38 Arnvig, K.B., Comas, I., Thomson, N.R., Houghton, J. et al. (2011) Sequence-based analysis uncovers an abundance of non-coding RNA in the total transcriptome of *Mycobacterium tuberculosis*. *PLoS Pathog.*, **7**, e1002342.

39 Toledo-Arana, A., Dussurget, O., Nikitas, G., Sesto, N. et al. (2009) The Listeria transcriptional landscape from saprophytism to virulence. *Nature*, **459**, 950–956.

40 Lasa, I., Toledo-Arana, A., Dobin, A., Villanueva, M. et al. (2011) Genome-wide antisense transcription drives mRNA processing in bacteria. *Proc. Natl Acad. Sci. USA*, **108**, 20172–20177.

41 Nicolas, P., Mader, U., Dervyn, E., Rochat, T. et al. (2012) Condition-dependent transcriptome reveals high-level regulatory architecture in *Bacillus subtilis*. *Science*, **335**, 1103–1106.

42 Sharma, C.M., Hoffmann, S., Darfeuille, F., Reignier, J. et al. (2010) The primary transcriptome of the major human pathogen *Helicobacter pylori*. *Nature*, **464**, 250–255.

43 Livny, J., Waldor, M.K. (2010) Mining regulatory 5′UTRs from cDNA deep sequencing datasets. *Nucleic Acids Res.*, **38**, 1504–1514.

44 Breaker, R.R. (2011) Prospects for riboswitch discovery and analysis. *Mol. Cell*, **43**, 867–879.

45 Merino, E., Yanofsky, C. (2005) Transcription attenuation: a highly conserved regulatory strategy used by bacteria. *Trends Genet.*, **21**, 260–264.

46 Henkin, T.M. (2008) Riboswitch RNAs: using RNA to sense cellular metabolism. *Genes Dev.*, **22**, 3383–3390.

47 Loh, E., Dussurget, O., Gripenland, J., Vaitkevicius, K. et al. (2009) A trans-acting riboswitch controls expression of the virulence regulator PrfA in *Listeria monocytogenes*. *Cell*, **139**, 770–779.

48 Naville, M., Gautheret, D. (2010) Premature terminator analysis sheds light on a hidden world of bacterial transcriptional attenuation. *Genome Biol.*, **11**, R97.

49 Vitreschak, A.G., Rodionov, D.A., Mironov, A.A., Gelfand, M.S. (2004) Riboswitches: the oldest mechanism for the regulation of gene expression? *Trends Genet.*, **20**, 44–50.

50 Winkler, W.C., Breaker, R.R. (2005) Regulation of bacterial gene expression by riboswitches. *Annu. Rev. Microbiol.*, **59**, 487–517.

51 Drlica, K., Perl-Rosenthal, N.R. (1999) DNA switches for thermal control of gene expression. *Trends Microbiol.*, **7**, 425–426.

52 Guisbert, E., Yura, T., Rhodius, V.A., Gross, C.A. (2008) Convergence of molecular, modeling, and systems approaches for an understanding of the *Escherichia coli* heat shock response. *Microbiol. Mol. Biol. Rev.*, **72**, 545–554.

53 Hurme, R., Rhen, M. (1998) Temperature sensing in bacterial gene regulation – what it all boils down to. *Mol. Microbiol.*, **30**, 1–6.

54 Klinkert, B., Narberhaus, F. (2009) Microbial thermosensors. *Cell. Mol. Life Sci.*, **66**, 2661–2676.

55 Mansilla, M.C., Cybulski, L.E., Albanesi, D., de Mendoza, D. (2004) Control of membrane lipid fluidity by molecular thermosensors. *J. Bacteriol.*, **186**, 6681–6688.

56 Kortmann, J., Narberhaus, F. (2012) Bacterial RNA thermometers: molecular zippers and switches. *Nat. Rev. Microbiol.*, **10**, 255–265.

57 Brandi, A., Pietroni, P., Gualerzi, C.O., Pon, C.L. (1996) Post-transcriptional regulation of CspA expression in *Escherichia coli*. *Mol. Microbiol.*, **19**, 231–240.

58 Giuliodori, A.M., Di Pietro, F., Marzi, S., Masquida, B. *et al.* (2010) The *cspA* mRNA is a thermosensor that modulates translation of the cold-shock protein CspA. *Mol. Cell*, **37**, 21–33.

59 Goldenberg, D., Azar, I., Oppenheim, A.B. (1996) Differential mRNA stability of the *cspA* gene in the cold-shock response of *Escherichia coli*. *Mol. Microbiol.*, **19**, 241–248.

60 Morita, M., Kanemori, M., Yanagi, H., Yura, T. (1999) Heat-induced synthesis of σ^{32} in *Escherichia coli*: structural and functional dissection of *rpoH* mRNA secondary structure. *J. Bacteriol.*, **181**, 401–410.

61 Morita, M.T., Tanaka, Y., Kodama, T.S., Kyogoku, Y. *et al.* (1999) Translational induction of heat shock transcription factor σ^{32}: evidence for a built-in RNA thermosensor. *Genes Dev.*, **13**, 655–665.

62 Horn, G., Hofweber, R., Kremer, W., Kalbitzer, H.R. (2007) Structure and function of bacterial cold shock proteins. *Cell. Mol. Life Sci.*, **64**, 1457–1470.

63 Phadtare, S., Severinov, K. (2010) RNA remodeling and gene regulation by cold shock proteins. *RNA Biol.*, **7**, 788–795.

64 Giuliodori, A.M., Brandi, A., Gualerzi, C.O., Pon, C.L. (2004) Preferential translation of cold-shock mRNAs during cold adaptation. *RNA*, **10**, 265–276.

65 DiRita, V.J., Mekalanos, J.J. (1989) Genetic regulation of bacterial virulence. *Annu. Rev. Genet.*, **23**, 455–482.

66 Bohme, K., Steinmann, R., Kortmann, J., Seekircher, S. *et al.* (2012) Concerted actions of a thermo-labile regulator and a unique intergenic RNA thermosensor control *Yersinia* virulence. *PLoS Pathog.*, **8**, e1002518.

67 Hoe, N.P., Goguen, J.D. (1993) Temperature sensing in *Yersinia pestis*: translation of the LcrF activator protein is thermally regulated. *J. Bacteriol.*, **175**, 7901–7909.

68 Johansson, J., Mandin, P., Renzoni, A., Chiaruttini, C. *et al.* (2002) An RNA thermosensor controls expression of virulence genes in *Listeria monocytogenes*. *Cell*, **110**, 551–561.

69 Padan, E., Bibi, E., Ito, M., Krulwich, T.A. (2005) Alkaline pH homeostasis in bacteria: new insights. *Biochim. Biophys. Acta*, **1717**, 67–88.

70 Slonczewski, J.L., Fujisawa, M., Dopson, M., Krulwich, T.A. (2009) Cytoplasmic pH measurement and homeostasis in bacteria and archaea. *Adv. Microb. Physiol.*, **55**, 1–79, 317.

71 Nechooshtan, G., Elgrably-Weiss, M., Sheaffer, A., Westhof, E. *et al.* (2009) A pH-responsive riboregulator. *Genes Dev.*, **23**, 2650–2662.

72 Ramesh, A., Winkler, W.C. (2010) Magnesium-sensing riboswitches in bacteria. *RNA Biol.*, **7**, 77–83.

73 Cromie, M.J., Shi, Y., Latifi, T., Groisman, E.A. (2006) An RNA sensor for intracellular Mg(2+). *Cell*, **125**, 71–84.

74 Dann, C.E., Wakeman, C.A., Sieling, C.L., Baker, S.C. III *et al.* (2007) Structure and mechanism of a metal-sensing regulatory RNA. *Cell*, **130**, 878–892.

75 Hollands, K., Proshkin, S., Sklyarova, S., Epshtein, V. *et al.* (2012) Riboswitch control of Rho-dependent transcription termination. *Proc. Natl Acad. Sci. USA*, **109**, 5376–5381.

76 Cromie, M.J., Groisman, E.A. (2010) Promoter and riboswitch control of the Mg^{2+} transporter MgtA from *Salmonella enterica*. *J. Bacteriol.*, **192**, 604–607.

77 Park, S.Y., Cromie, M.J., Lee, E.J., Groisman, E.A. (2010) A bacterial mRNA leader that employs different mechanisms to sense disparate intracellular signals. *Cell*, **142**, 737–748.

78 Barrick, J.E., Breaker, R.R. (2007) The power of riboswitches. *Sci. Am.*, **296**, 50–57.

79 Garst, A.D., Batey, R.T. (2009) A switch in time: detailing the life of a riboswitch. *Biochim. Biophys. Acta*, **1789**, 584–591.

80 Weinberg, Z., Wang, J.X., Bogue, J., Yang, J. *et al.* (2010) Comparative genomics reveals 104 candidate structured RNAs from bacteria, archaea, and their metagenomes. *Genome Biol.*, **11**, R31.

81 Barrick, J.E., Breaker, R.R. (2007) The distributions, mechanisms, and structures of metabolite-binding riboswitches. *Genome Biol.*, **8**, R239.

82 Gorke, B., Vogel, J. (2008) Noncoding RNA control of the making and breaking of sugars. *Genes Dev.*, **22**, 2914–2925.

83 Ferre-D'Amare, A.R. (2011) Use of a coenzyme by the *glmS* ribozyme-riboswitch suggests primordial expansion of RNA chemistry by small molecules. *Philos. Trans. R. Soc. Lond. B, Biol. Sci.*, **366**, 2942–2948.

84 Winkler, W.C., Nahvi, A., Roth, A., Collins, J.A. et al. (2004) Control of gene expression by a natural metabolite-responsive ribozyme. *Nature*, **428**, 281–286.

85 Lemay, J.F., Desnoyers, G., Blouin, S., Heppell, B. et al. (2011) Comparative study between transcriptionally- and translationally-acting adenine riboswitches reveals key differences in riboswitch regulatory mechanisms. *PLoS Genet.*, **7**, e1001278.

86 Wickiser, J.K., Winkler, W.C., Breaker, R.R., Crothers, D.M. (2005) The speed of RNA transcription and metabolite binding kinetics operate an FMN riboswitch. *Mol. Cell*, **18**, 49–60.

87 Perdue, S.A., Roberts, J.W. (2011) σ^{70}-dependent transcription pausing in *Escherichia coli*. *J. Mol. Biol.*, **412**, 782–792.

88 Peters, J.M., Vangeloff, A.D., Landick, R. (2011) Bacterial transcription terminators: the RNA 3′-end chronicles. *J. Mol. Biol.*, **412**, 793–813.

89 Huang, L., Serganov, A., Patel, D.J. (2010) Structural insights into ligand recognition by a sensing domain of the cooperative glycine riboswitch. *Mol. Cell*, **40**, 774–786.

90 Mandal, M., Lee, M., Barrick, J.E., Weinberg, Z. et al. (2004) A glycine-dependent riboswitch that uses cooperative binding to control gene expression. *Science*, **306**, 275–279.

91 Welz, R., Breaker, R.R. (2007) Ligand binding and gene control characteristics of tandem riboswitches in *Bacillus anthracis*. *RNA*, **13**, 573–582.

92 Poiata, E., Meyer, M.M., Ames, T.D., Breaker, R.R. (2009) A variant riboswitch aptamer class for S-adenosylmethionine common in marine bacteria. *RNA*, **15**, 2046–2056.

93 Sudarsan, N., Hammond, M.C., Block, K.F., Welz, R. et al. (2006) Tandem riboswitch architectures exhibit complex gene control functions. *Science*, **314**, 300–304.

94 Andre, G., Even, S., Putzer, H., Burguiere, P. et al. (2008) S-box and T-box riboswitches and antisense RNA control a sulfur metabolic operon of *Clostridium acetobutylicum*. *Nucleic Acids Res.*, **36**, 5955–5969.

95 Fox, K.A., Ramesh, A., Stearns, J.E., Bourgogne, A. et al. (2009) Multiple posttranscriptional regulatory mechanisms partner to control ethanolamine utilization in *Enterococcus faecalis*. *Proc. Natl Acad. Sci. USA*, **106**, 4435–4440.

96 Baker, K.A., Perego, M. (2011) Transcription antitermination by a phosphorylated response regulator and cobalamin-dependent termination at a B riboswitch contribute to ethanolamine utilization in *Enterococcus faecalis*. *J. Bacteriol.*, **193**, 2575–2586.

97 Gutierrez-Preciado, A., Henkin, T.M., Grundy, F.J., Yanofsky, C. et al. (2009) Biochemical features and functional implications of the RNA-based T-box regulatory mechanism. *Microbiol. Mol. Biol. Rev.*, **73**, 36–61.

98 Vitreschak, A.G., Mironov, A.A., Lyubetsky, V.A., Gelfand, M.S. (2008) Comparative genomic analysis of T-box regulatory systems in bacteria. *RNA*, **14**, 717–735.

99 Green, N.J., Grundy, F.J., Henkin, T.M. (2010) The T box mechanism: tRNA as a regulatory molecule. *FEBS Lett.*, **584**, 318–324.

100 Grundy, F.J., Winkler, W.C., Henkin, T.M. (2002) tRNA-mediated transcription antitermination in vitro: codon-anticodon pairing independent of the ribosome. *Proc. Natl Acad. Sci. USA*, **99**, 11121–11126.

101 Dua, P., Kim, S., Lee, D.K. (2011) Nucleic acid aptamers targeting cell-surface proteins. *Methods*, **54**, 215–225.

102 Topp, S., Reynoso, C.M., Seeliger, J.C., Goldlust, I.S. et al. (2010) Synthetic riboswitches that induce gene expression in diverse bacterial species. *Appl. Environ. Microbiol.*, **76**, 7881–7884.

103 Paige, J.S., Wu, K.Y., Jaffrey, S.R. (2011) RNA mimics of green fluorescent protein. *Science*, **333**, 642–646.

104 Weigand, J.E., Suess, B. (2009) Aptamers and riboswitches: perspectives in biotechnology. *Appl. Microbiol. Biotechnol.*, **85**, 229–236.

105 Gripenland, J., Netterling, S., Loh, E., Tiensuu, T. et al. (2010) RNAs: regulators of bacterial virulence. *Nat. Rev. Microbiol.*, **8**, 857–866.

106 Repoila, F., Darfeuille, F. (2009) Small regulatory non-coding RNAs in bacteria: physiology and mechanistic aspects. *Biol. Cell*, **101**, 117–131.

107 Brantl, S. (2007) Regulatory mechanisms employed by cis-encoded antisense RNAs. *Curr. Opin. Microbiol.*, **10**, 102–109.

108 Thomason, M.K., Storz, G. (2010) Bacterial antisense RNAs: how many are there, and what are they doing? *Annu. Rev. Genet.*, **44**, 167–188.

109 Weaver, K.E. (2007) Emerging plasmid-encoded antisense RNA regulated systems. *Curr. Opin. Microbiol.*, **10**, 110–116.

110 Callen, B.P., Shearwin, K.E., Egan, J.B. (2004) Transcriptional interference between convergent promoters caused by elongation over the promoter. *Mol. Cell*, **14**, 647–656.

111 Crampton, N., Bonass, W.A., Kirkham, J., Rivetti, C. et al. (2006) Collision events between RNA polymerases in convergent transcription studied by atomic force microscopy. *Nucleic Acids Res.*, **34**, 5416–5425.

112 Palmer, A.C., Ahlgren-Berg, A., Egan, J.B., Dodd, I.B. et al. (2009) Potent transcriptional interference by pausing of RNA polymerases over a downstream promoter. *Mol. Cell*, **34**, 545–555.

113 Giangrossi, M., Prosseda, G., Tran, C.N., Brandi, A. et al. (2010) A novel antisense RNA regulates at transcriptional level the virulence gene *icsA* of *Shigella flexneri*. *Nucleic Acids Res.*, **38**, 3362–3375.

114 Stork, M., Di Lorenzo, M., Welch, T.J., Crosa, J.H. (2007) Transcription termination within the iron transport-biosynthesis operon of *Vibrio anguillarum* requires an antisense RNA. *J. Bacteriol.*, **189**, 3479–3488.

115 Duhring, U., Axmann, I.M., Hess, W.R., Wilde, A. (2006) An internal antisense RNA regulates expression of the photosynthesis gene *isiA*. *Proc. Natl Acad. Sci. USA*, **103**, 7054–7058.

116 Krinke, L., Wulff, D.L. (1990) RNase III-dependent hydrolysis of lambda *cII-O* gene mRNA mediated by lambda OOP antisense RNA. *Genes Dev.*, **4**, 2223–2233.

117 Opdyke, J.A., Fozo, E.M., Hemm, M.R., Storz, G. (2011) RNase III participates in GadY-dependent cleavage of the *gadX-gadW* mRNA. *J. Mol. Biol.*, **406**, 29–43.

118 Stazic, D., Lindell, D., Steglich, C. (2011) Antisense RNA protects mRNA from RNase E degradation by RNA-RNA duplex formation during phage infection. *Nucleic Acids Res.*, **39**, 4890–4899.

119 Lee, E.J., Groisman, E.A. (2010) An antisense RNA that governs the expression kinetics of a multifunctional virulence gene. *Mol. Microbiol.*, **76**, 1020–1033.

120 Wen, Y., Feng, J., Scott, D.R., Marcus, E.A. et al. (2011) A cis-encoded antisense small RNA regulated by the HP0165-HP0166 two-component system controls expression of *ureB* in *Helicobacter pylori*. *J. Bacteriol.*, **193**, 40–51.

121 Frank, K.L., Barnes, A.M., Grindle, S.M., Manias, D.A. et al. (2012) Use of recombinase-based in vivo expression technology to characterize *Enterococcus faecalis* gene expression during infection identifies in vivo-expressed antisense RNAs and implicates the protease Eep in pathogenesis. *Infect Immun.*, **80**, 539–549.

122 Sevilla, E., Martin-Luna, B., Gonzalez, A., Gonzalo-Asensio, J.A. et al. (2011) Identification of three novel antisense RNAs in the fur locus from unicellular cyanobacteria. *Microbiology*, **157**, 3398–3404.

123 Sobrero, P., Valverde, C. (2012) The bacterial protein Hfq: much more than a mere RNA-binding factor. *Crit. Rev. Microbiol.* (e-pub ahead of print).

124 Vogel, J., Luisi, B.F. (2011) Hfq and its constellation of RNA. *Nat. Rev. Microbiol.*, **9**, 578–589.

125 Boisset, S., Geissmann, T., Huntzinger, E., Fechter, P. et al. (2007) *Staphylococcus aureus* RNAIII coordinately represses the synthesis of virulence factors and the transcription regulator Rot by an antisense mechanism. *Genes Dev.*, **21**, 1353–1366.

126 Darfeuille, F., Unoson, C., Vogel, J., Wagner, E.G. (2007) An antisense RNA inhibits translation by competing with standby ribosomes. *Mol. Cell*, **26**, 381–392.

127 Heidrich, N., Brantl, S. (2007) Antisense RNA-mediated transcriptional attenuation in plasmid pIP501: the simultaneous interaction between two complementary loop

128 Geisinger, E., Adhikari, R.P., Jin, R., Ross, H.F. et al. (2006) Inhibition of rot translation by RNAIII, a key feature of agr function. *Mol. Microbiol.*, **61**, 1038–1048.

129 Jousselin, A., Metzinger, L., Felden, B. (2009) On the facultative requirement of the bacterial RNA chaperone, Hfq. *Trends Microbiol.*, **17**, 399–405.

130 Salvail, H., Lanthier-Bourbonnais, P., Sobota, J.M., Caza, M. et al. (2010) A small RNA promotes siderophore production through transcriptional and metabolic remodeling. *Proc. Natl Acad. Sci. USA*, **107**, 15223–15228.

131 Benjamin, J.A., Desnoyers, G., Morissette, A., Salvail, H. et al. (2010) Dealing with oxidative stress and iron starvation in microorganisms: an overview. *Can. J. Physiol. Pharmacol.*, **88**, 264–272.

132 Vecerek, B., Moll, I., Blasi, U. (2007) Control of Fur synthesis by the non-coding RNA RyhB and iron-responsive decoding. *EMBO J.*, **26**, 965–975.

133 Jacques, J.F., Jang, S., Prevost, K., Desnoyers, G. et al. (2006) RyhB small RNA modulates the free intracellular iron pool and is essential for normal growth during iron limitation in *Escherichia coli*. *Mol. Microbiol.*, **62**, 1181–1190.

134 Masse, E., Majdalani, N., Gottesman, S. (2003) Regulatory roles for small RNAs in bacteria. *Curr. Opin. Microbiol.*, **6**, 120–124.

135 Morita, T., Maki, K., Aiba, H. (2005) RNase E-based ribonucleoprotein complexes: mechanical basis of mRNA destabilization mediated by bacterial noncoding RNAs. *Genes Dev.*, **19**, 2176–2186.

136 Morita, T., Mochizuki, Y., Aiba, H. (2006) Translational repression is sufficient for gene silencing by bacterial small noncoding RNAs in the absence of mRNA destruction. *Proc. Natl Acad. Sci. USA*, **103**, 4858–4863.

137 Prevost, K., Desnoyers, G., Jacques, J.F., Lavoie, F. et al. (2011) Small RNA-induced mRNA degradation achieved through both translation block and activated cleavage. *Genes Dev.*, **25**, 385–396.

138 Huntzinger, E., Boisset, S., Saveanu, C., Benito, Y. et al. (2005) *Staphylococcus aureus* RNAIII and the endoribonuclease III coordinately regulate *spa* gene expression. *EMBO J.*, **24**, 824–835.

139 Figueroa-Bossi, N., Valentini, M., Malleret, L., Bossi, L. (2009) Caught at its own game: regulatory small RNA inactivated by an inducible transcript mimicking its target. *Genes Dev.*, **23**, 2004–2015.

140 Overgaard, M., Johansen, J., Moller-Jensen, J., Valentin-Hansen, P. (2009) Switching off small RNA regulation with trap-mRNA. *Mol. Microbiol.*, **73**, 790–800.

141 Chen, S., Zhang, A., Blyn, L.B., Storz, G. (2004) MicC, a second small-RNA regulator of Omp protein expression in *Escherichia coli*. *J. Bacteriol.*, **186**, 6689–6697.

142 Castillo-Keller, M., Vuong, P., Misra, R. (2006) Novel mechanism of *Escherichia coli* porin regulation. *J. Bacteriol.*, **188**, 576–586.

143 Guillier, M., Gottesman, S. (2006) Remodelling of the *Escherichia coli* outer membrane by two small regulatory RNAs. *Mol. Microbiol.*, **59**, 231–247.

144 Frohlich, K.S., Papenfort, K., Berger, A.A., Vogel, J. (2011) A conserved RpoS-dependent small RNA controls the synthesis of major porin OmpD. *Nucleic Acids Res.*, **40**, 3623–3640.

145 Gogol, E.B., Rhodius, V.A., Papenfort, K., Vogel, J. et al. (2011) Small RNAs endow a transcriptional activator with essential repressor functions for single-tier control of a global stress regulon. *Proc. Natl Acad. Sci. USA*, **108**, 12875–12880.

146 Rasmussen, A.A., Eriksen, M., Gilany, K., Udesen, C. et al. (2005) Regulation of *ompA* mRNA stability: the role of a small regulatory RNA in growth phase-dependent control. *Mol. Microbiol.*, **58**, 1421–1429.

147 Udekwu, K.I., Darfeuille, F., Vogel, J., Reimegard, J. et al. (2005) Hfq-dependent regulation of OmpA synthesis is mediated by an antisense RNA. *Genes Dev.*, **19**, 2355–2366.

148 Johansen, J., Eriksen, M., Kallipolitis, B., Valentin-Hansen, P. (2008) Down-regulation of outer membrane proteins by noncoding RNAs: unraveling the cAMP-CRP- and σ^E-dependent CyaR-*ompX* regulatory case. *J. Mol. Biol.*, **383**, 1–9.

149 Johansen, J., Rasmussen, A.A., Overgaard, M., Valentin-Hansen, P. (2006) Conserved small non-coding RNAs that belong to the σ^E regulon: role in down-regulation of outer membrane proteins. *J. Mol. Biol.*, **364**, 1–8.

150 Thompson, K.M., Rhodius, V.A., Gottesman, S. (2007) σ^E regulates and is regulated by a small RNA in *Escherichia coli*. *J. Bacteriol.*, **189**, 4243–4256.

151 Battesti, A., Majdalani, N., Gottesman, S. (2011) The RpoS-mediated general stress response in *Escherichia coli*. *Annu. Rev. Microbiol.*, **65**, 189–213.

152 Brown, L., Elliott, T. (1996) Efficient translation of the RpoS sigma factor in *Salmonella typhimurium* requires host factor I, an RNA-binding protein encoded by the *hfq* gene. *J. Bacteriol.*, **178**, 3763–3770.

153 Brown, L., Elliott, T. (1997) Mutations that increase expression of the *rpoS* gene and decrease its dependence on *hfq* function in *Salmonella typhimurium*. *J. Bacteriol.*, **179**, 656–662.

154 Muffler, A., Fischer, D., Hengge-Aronis, R. (1996) The RNA-binding protein HF-I, known as a host factor for phage Qbeta RNA replication, is essential for *rpoS* translation in *Escherichia coli*. *Genes Dev.*, **10**, 1143–1151.

155 Majdalani, N., Cunning, C., Sledjeski, D., Elliott, T. *et al.* (1998) DsrA RNA regulates translation of RpoS message by an anti-antisense mechanism, independent of its action as an antisilencer of transcription. *Proc. Natl Acad. Sci. USA*, **95**, 12462–12467.

156 Mandin, P., Gottesman, S. (2010) Integrating anaerobic/aerobic sensing and the general stress response through the ArcZ small RNA. *EMBO J.*, **29**, 3094–3107.

157 Soper, T., Mandin, P., Majdalani, N., Gottesman, S. *et al.* (2010) Positive regulation by small RNAs and the role of Hfq. *Proc. Natl Acad. Sci. USA*, **107**, 9602–9607.

158 Repoila, F., Gottesman, S. (2001) Signal transduction cascade for regulation of RpoS: temperature regulation of DsrA. *J. Bacteriol.*, **183**, 4012–4023.

159 Majdalani, N., Hernandez, D., Gottesman, S. (2002) Regulation and mode of action of the second small RNA activator of RpoS translation, RprA. *Mol. Microbiol.*, **46**, 813–826.

160 Jorgensen, M.G., Nielsen, J.S., Boysen, A., Franch, T. *et al.* (2012) Small regulatory RNAs control the multi-cellular adhesive lifestyle of *Escherichia coli*. *Mol. Microbiol.*, **84**, 36–50.

161 Lease, R.A., Cusick, M.E., Belfort, M. (1998) Riboregulation in *Escherichia coli*: DsrA RNA acts by RNA: RNA interactions at multiple loci. *Proc. Natl Acad. Sci. USA*, **95**, 12456–12461.

162 Papenfort, K., Said, N., Welsink, T., Lucchini, S. *et al.* (2009) Specific and pleiotropic patterns of mRNA regulation by ArcZ, a conserved, Hfq-dependent small RNA. *Mol. Microbiol.*, **74**, 139–158.

163 Zhang, A., Altuvia, S., Tiwari, A., Argaman, L. *et al.* (1998) The OxyS regulatory RNA represses *rpoS* translation and binds the Hfq (HF-I) protein. *EMBO J.*, **17**, 6061–6068.

164 Moon, K., Gottesman, S. (2011) Competition among Hfq-binding small RNAs in *Escherichia coli*. *Mol. Microbiol.*, **82**, 1545–1562.

165 Hussein, R., Lim, H.N. (2011) Disruption of small RNA signaling caused by competition for Hfq. *Proc. Natl Acad. Sci. USA*, **108**, 1110–1115.

166 Morfeldt, E., Taylor, D., von Gabain, A., Arvidson, S. (1995) Activation of alpha-toxin translation in *Staphylococcus aureus* by the trans-encoded antisense RNA, RNAIII. *EMBO J.*, **14**, 4569–4577.

167 Prevost, K., Salvail, H., Desnoyers, G., Jacques, J.F. *et al.* (2007) The small RNA RyhB activates the translation of *shiA* mRNA encoding a permease of shikimate, a compound involved in siderophore synthesis. *Mol. Microbiol.*, **64**, 1260–1273.

168 Ramirez-Pena, E., Trevino, J., Liu, Z., Perez, N. *et al.* (2010) The group A *Streptococcus* small regulatory RNA FasX enhances streptokinase activity by increasing the stability of the *ska* mRNA transcript. *Mol. Microbiol.*, **78**, 1332–1347.

169 Obana, N., Shirahama, Y., Abe, K., Nakamura, K. (2010) Stabilization of *Clostridium perfringens* collagenase mRNA by VR-RNA-dependent cleavage in 5′ leader sequence. *Mol. Microbiol.*, **77**, 1416–1428.

170 Sharp, P.A. (2009) The centrality of RNA. *Cell*, **136**, 577–580.

171 Tomizawa, J., Selzer, G. (1979) Initiation of DNA synthesis in *Escherichia coli*. *Annu. Rev. Biochem.*, **48**, 999–1034.

172 Shajani, Z., Sykes, M.T., Williamson, J.R. (2011) Assembly of bacterial ribosomes. *Annu. Rev. Biochem.*, **80**, 501–526.

173 Santangelo, T.J., Artsimovitch, I. (2011) Termination and antitermination: RNA polymerase runs a stop sign. *Nat. Rev. Microbiol.*, **9**, 319–329.

174 Moore, S.D., Sauer, R.T. (2007) The tmRNA system for translational surveillance and ribosome rescue. *Annu. Rev. Biochem.*, **76**, 101–124.

175 Kim, K.S., Lee, Y. (2004) Regulation of 6S RNA biogenesis by switching utilization of both sigma factors and endoribonucleases. *Nucleic Acids Res.*, **32**, 6057–6068.

176 Cavanagh, A.T., Sperger, J.M., Wassarman, K.M. (2012) Regulation of 6S RNA by pRNA synthesis is required for efficient recovery from stationary phase in *E. coli* and *B. subtilis*. *Nucleic Acids Res.*, **40**, 2234–2246.

177 Neusser, T., Polen, T., Geissen, R., Wagner, R. (2010) Depletion of the non-coding regulatory 6S RNA in *E. coli* causes a surprising reduction in the expression of the translation machinery. *BMC Genomics*, **11**, 165.

178 Shephard, L., Dobson, N., Unrau, P.J. (2010) Binding and release of the 6S transcriptional control RNA. *RNA*, **16**, 885–892.

179 Beckmann, B.M., Hoch, P.G., Marz, M., Willkomm, D.K. et al. (2012) A pRNA-induced structural rearrangement triggers 6S-1 RNA release from RNA polymerase in *Bacillus subtilis*. *EMBO J.*, **31**, 1727–1738.

180 Gildehaus, N., Neusser, T., Wurm, R., Wagner, R. (2007) Studies on the function of the riboregulator 6S RNA from *E. coli*: RNA polymerase binding, inhibition of in vitro transcription and synthesis of RNA-directed de novo transcripts. *Nucleic Acids Res.*, **35**, 1885–1896.

181 Wassarman, K.M., Saecker, R.M. (2006) Synthesis-mediated release of a small RNA inhibitor of RNA polymerase. *Science*, **314**, 1601–1603.

182 Axmann, I.M., Holtzendorff, J., Voss, B., Kensche, P. et al. (2007) Two distinct types of 6S RNA in *Prochlorococcus*. *Gene*, **406**, 69–78.

183 Barrick, J.E., Sudarsan, N., Weinberg, Z., Ruzzo, W.L. et al. (2005) 6S RNA is a widespread regulator of eubacterial RNA polymerase that resembles an open promoter. *RNA*, **11**, 774–784.

184 Trotochaud, A.E., Wassarman, K.M. (2005) A highly conserved 6S RNA structure is required for regulation of transcription. *Nat. Struct. Mol. Biol.*, **12**, 313–319.

185 Weissenmayer, B.A., Prendergast, J.G., Lohan, A.J., Loftus, B.J. (2011) Sequencing illustrates the transcriptional response of *Legionella pneumophila* during infection and identifies seventy novel small non-coding RNAs. *PLoS ONE*, **6**, e17570.

186 Faucher, S.P., Friedlander, G., Livny, J., Margalit, H. et al. (2010) *Legionella pneumophila* 6S RNA optimizes intracellular multiplication. *Proc. Natl Acad. Sci. USA*, **107**, 7533–7538.

187 Babitzke, P., Baker, C.S., Romeo, T. (2009) Regulation of translation initiation by RNA binding proteins. *Annu. Rev. Microbiol.*, **63**, 27–44.

188 Heroven, A.K., Bohme, K., Dersch, P. (2012) The Csr/Rsm system of *Yersinia* and related pathogens: a post-transcriptional strategy for managing virulence. *RNA Biol.*, **9**.

189 Lapouge, K., Schubert, M., Allain, F.H., Haas, D. (2008) Gac/Rsm signal transduction pathway of gamma-proteobacteria: from RNA recognition to regulation of social behaviour. *Mol. Microbiol.*, **67**, 241–253.

190 Sonnleitner, E., Haas, D. (2011) Small RNAs as regulators of primary and secondary metabolism in *Pseudomonas* species. *Appl. Microbiol. Biotechnol.*, **91**, 63–79.

191 Timmermans, J., Van Melderen, L. (2010) Post-transcriptional global regulation by CsrA in bacteria. *Cell. Mol. Life Sci.*, **67**, 2897–2908.

192 Yakhnin, H., Baker, C.S., Berezin, I., Evangelista, M.A. et al. (2011) CsrA represses translation of sdiA, which encodes the N-acylhomoserine-L-lactone receptor of *Escherichia coli*, by binding exclusively within the coding region of sdiA mRNA. *J. Bacteriol.*, **193**, 6162–6170.

193 Edwards, A.N., Patterson-Fortin, L.M., Vakulskas, C.A., Mercante, J.W. et al. (2011) Circuitry linking the Csr and stringent

194 Babitzke, P., Romeo, T. (2007) CsrB sRNA family: sequestration of RNA-binding regulatory proteins. *Curr. Opin. Microbiol.*, **10**, 156–163.

195 Suzuki, K., Babitzke, P., Kushner, S.R., Romeo, T. (2006) Identification of a novel regulatory protein (CsrD) that targets the global regulatory RNAs CsrB and CsrC for degradation by RNase E. *Genes Dev.*, **20** (18), 2605–2617.

196 Moreno, R., Martinez-Gomariz, M., Yuste, L., Gil, C. *et al.* (2009) The *Pseudomonas putida* Crc global regulator controls the hierarchical assimilation of amino acids in a complete medium: evidence from proteomic and genomic analyses. *Proteomics*, **9**, 2910–2928.

197 Moreno, R., Marzi, S., Romby, P., Rojo, F. (2009) The Crc global regulator binds to an unpaired A-rich motif at the *Pseudomonas putida alkS* mRNA coding sequence and inhibits translation initiation. *Nucleic Acids Res.*, **37**, 7678–7690.

198 Rojo, F. (2010) Carbon catabolite repression in *Pseudomonas*: optimizing metabolic versatility and interactions with the environment. *FEMS Microbiol. Rev.*, **34**, 658–684.

199 Moreno, R., Fonseca, P., Rojo, F. (2011) Two small RNAs, CrcY and CrcZ, act in concert to sequester the Crc global regulator in *Pseudomonas putida*, modulating catabolite repression. *Mol. Microbiol.*, **83**, 24–40.

200 Fonseca, P., Moreno, R., Rojo, F. (2012) *Pseudomonas putida* growing at low temperature shows increased levels of CrcZ and CrcY sRNAs, leading to reduced Crc-dependent catabolite repression. *Environ. Microbiol.* (e-pub ahead of print).

201 Zhang, X.X., Liu, Y.H., Rainey, P.B. (2010) CbrAB-dependent regulation of *pcnB*, a poly(A) polymerase gene involved in polyadenylation of RNA in *Pseudomonas fluorescens*. *Environ. Microbiol.*, **12**, 1674–1683.

202 Toledo-Arana, A., Repoila, F., Cossart, P. (2007) Small noncoding RNAs controlling pathogenesis. *Curr. Opin. Microbiol.*, **10**, 182–188.

4
Functions and Applications of RNA-Guided CRISPR-Cas Immune Systems

Rodolphe Barrangou[1] and Philippe Horvath[2]
[1] *North Carolina State University, Department of Food, Bioprocessing and Nutrition Sciences, Raleigh, NC 27695, USA*
[2] *DuPont Nutrition and Health, BP10, Dangé-Saint-Romain, F-86220, France*

1	Introduction to CRISPR-Cas Immune Systems	119
1.1	Initial Discovery and Tipping Point	119
1.2	Elements of CRISPR-Cas Systems	122
1.3	Diversity of CRISPR-Cas Systems	123
2	Mechanism of Action of CRISPR-Mediated Interference	124
2.1	Adaptation	124
2.2	crRNA Biogenesis	125
2.3	Interference	126
3	Exploitation of the Versatile CRISPR-Cas Systems	127
3.1	Typing and Epidemiological Studies	127
3.2	Genome Integrity and crRNA-Mediated Interference	129
3.3	RNA-Guided Next-Generation Genome Editing	131
4	Summary and Outlook	132
	Acknowledgments	133
	References	133

RNA Regulation: Advances in Molecular Biology and Medicine, First Edition. Edited by Robert A. Meyers.
© 2014 Wiley-VCH Verlag GmbH & Co. KGaA. Published 2014 by Wiley-VCH Verlag GmbH & Co. KGaA.

Keywords

Cas
<u>C</u>RISPR-<u>a</u>ssociated genes, that encode a functionally diverse family of proteins involved in immune marker acquisition, small interfering RNA biogenesis, and target nucleic acid cleavage.

Cas9
A signature protein in Type II systems involved in small interfering RNA biogenesis and target DNA cleavage, which can be reprogrammed for sequence-specific, RNA-guided DNA cleavage.

Cascade
<u>C</u>RISPR-<u>As</u>sociated Complex for <u>A</u>ntiviral <u>DE</u>fense, which forms a ribonucleoprotein complex together with CRISPR RNA (crRNA) to guide interference and DNA cleavage in Type I systems.

CRISPR
<u>C</u>lustered <u>R</u>egularly <u>I</u>nterspaced <u>S</u>hort <u>P</u>alindromic <u>R</u>epeats, a peculiar family of DNA repeats that are the defining hallmark of CRISPR-Cas immune systems.

crRNA
Mature, small, noncoding <u>C</u>RISPR <u>RNA</u>, which comprise spacer sequences flanked by partial CRISPR repeats, that guide the Cas machinery for complementary target nucleic acid interference.

Interference
A sequence-specific process by which the crRNA:Cas ribonucleoprotein complex targets and cleaves complementary nucleic acid.

■ Clustered regularly interspaced short palindromic repeats (CRISPRs), together with CRISPR-associated sequences (*cas*) constitute the CRISPR-Cas adaptive immune system in bacteria and archaea. Adaptive immunity is built into CRISPR arrays through the uptake of small pieces of invasive nucleic acids, such as viruses and plasmids. Acquired immunity is subsequently mediated by small interfering RNAs transcribed from these loci, that guide specific cleavage of complementary sequences by nucleases. Studies have established that CRISPR loci and their RNA-guided interference machinery can be exploited for a broad array of applications. Adaptive immunity can be built against viruses, or to preclude the uptake of undesirable sequences. The inheritable and hypervariable nature of these loci can be used to track the phylogenetic path of an organism and reveal the evolutionary interplay between hosts and their viruses. Recently, a new customizable genome editing system was developed based on these versatile interfering RNAs to specifically guide nucleases for sequence cleavage.

1
Introduction to CRISPR-Cas Immune Systems

1.1
Initial Discovery and Tipping Point

Clustered Regularly Interspaced Short Palindromic Repeats (CRISPRs) constitute a relatively recently documented DNA repeats family. These idiosyncratic genetic loci were first discovered in *Escherichia coli* K12 in 1987, in an intergenic region adjacent to the *iap* gene [1]. This peculiar locus included a set of short (29 nt) identical direct repeats that were interspaced regularly by short (32 nt), diverse, and seemingly random sequences. These loci were subjected to relatively little attention until the early 2000s, when studies documented their widespread occurrence in bacterial and archaeal genomes, ultimately establishing the CRISPR acronym [2]. As the genomes of a diversity of bacteria and archaea were determined and publicly released, it became readily apparent that these loci were widespread in prokaryotes, and that these arrays of DNA repeats were often adjacent to a set of CRISPR-*a*ssociated *s*equences (*cas* genes) [2]. Ever since, several comprehensive studies of bacterial and archaeal genomes have established that CRISPR-Cas systems generally occur in almost half of bacterial genomes (about 45%) and the large majority of archaeal genomes (about 90%), as documented in the CRISPR database [3].

Although the origin and function of CRISPR loci remained mysterious for nearly 20 years after their initial discovery, three milestone studies conducted in 2005 (see Tab. 1, which lists a subjective set of 25 milestone CRISPR studies) concurrently showed that the apparently random spacer sequences that interspace CRISPR repeats show significant homology to virus and plasmid sequences [4–6]. This somewhat refocused attention away from the highly conserved CRISPR repeats toward the highly diverse CRISPR spacers, and almost immediately led to the thoughtful hypothesis that CRISPR loci and their associated Cas machinery may be an RNAi-mediated prokaryotic immune system, based on the spacer origin and *in silico* predictions on the putative functionalities of Cas proteins [7]. Shortly thereafter, in 2007, the first experimental evidence was reported that CRISPR-Cas systems provide adaptive immunity against viruses in the lactic acid bacterium *Streptococcus thermophilus* [8]. Specifically, it was shown that novel spacers were acquired from the invasive phages in a polarized manner, to confer immunity against viruses that carry complementary DNA sequences. It was also established that *cas* genes were necessary for both spacer acquisition (immunization) and to carry out spacer-encoded interference (immunity) [8]. In 2008, two subsequent milestones studies showed that CRISPR-based immunity is mediated by small interfering RNAs [9], and that CRISPR can also provide immunity against plasmids [10]. Specifically, it was

Tab. 1 Milestone CRISPR studies.

Authors	Article title	Journal	Year
Jansen et al.	Identification of genes that are associated with DNA repeats in prokaryotes	Mol. Microbiol.	2002
Mojica et al.	Intervening sequences of regularly spaced prokaryotic repeats derive from foreign genetic elements	J. Mol. Evol.	2005
Pourcel et al.	CRISPR elements in Yersinia pestis acquire new repeats by preferential uptake of bacteriophage DNA, and provide additional tools for evolutionary studies	Microbiology	2005
Bolotin et al.	Clustered regularly interspaced short palindrome repeats (CRISPRs) have spacers of extrachromosomal origin	Microbiology	2005
Haft et al.	A guild of 45 CRISPR-associated (Cas) protein families and multiple CRISPR/Cas subtypes exist in prokaryotic genomes	PLoS Comput. Biol.	2005
Makarova et al.	A putative RNA-interference-based immune system in prokaryotes: computational analysis of the predicted enzymatic machinery, functional analogies with eukaryotic RNAi, and hypothetical mechanisms of action	Biol. Direct	2006
Barrangou et al.	CRISPR provides acquired resistance against viruses in prokaryotes	Science	2007
Deveau et al.	Phage response to CRISPR-encoded resistance in Streptococcus thermophiles	J. Bacteriol.	2008
Andersson et al.	Virus population dynamics and acquired virus resistance in natural microbial communities	Science	2008
Brouns et al.	Small CRISPR RNAs guide antiviral defense in prokaryotes	Science	2008
Marraffini et al.	CRISPR interference limits horizontal gene transfer in staphylococci by targeting DNA	Science	2008
Mojica et al.	Short motif sequences determine the targets of the prokaryotic CRISPR defence system	Microbiology	2009
Hale et al.	RNA-guided RNA cleavage by a CRISPR RNA–Cas protein complex	Cell	2009
Marraffini et al.	Self versus non-self discrimination during CRISPR RNA-directed immunity	Nature	2010
Haurwitz et al.	Sequence- and structure-specific RNA processing by a CRISPR endonuclease	Science	2010
Garneau et al.	The CRISPR/Cas bacterial immune system cleaves bacteriophage and plasmid DNA	Nature	2010

Tab. 1 (continued)

Authors	Article title	Journal	Year
Deltcheva et al.	CRISPR RNA maturation by trans-encoded small RNA and host factor RNase III	Nature	2011
Makarova et al.	Evolution and classification of the CRISPR-Cas systems	Nat. Rev. Microbiol	2011
Semenova et al.	Interference by clustered regularly interspaced short palindromic repeat (CRISPR) RNA is governed by a seed sequence	Proc. Natl Acad. Sci. USA	2011
Wiedenheft et al.	RNA-guided complex from a bacterial immune system enhances target recognition through seed sequence interactions	Proc. Natl Acad. Sci. USA	2011
Sapranauskas et al.	The Streptococcus thermophilus CRISPR/Cas system provides immunity in Escherichia coli	Nucleic Acids Res.	2011
Datsenko et al.	Molecular memory of prior infections activates the CRISPR/Cas adaptive bacterial immunity system	Nat. Commun.	2012
Jinek et al.	A programmable dual-RNA-guided DNA endonuclease in adaptive bacterial immunity	Science	2012
Mali et al.	RNA-guided human genome engineering via Cas9	Science	2013
Jiang et al.	RNA-guided editing of bacterial genomes using CRISPR-Cas systems	Nat. Biotechnol.	2013

shown in E. coli that small noncoding interfering CRISPR RNAs (crRNAs) guide the CRISPR-associated complex for antiviral defense (Cascade) [9]. A study in Staphylococcus epidermidis showed that CRISPR can interfere with plasmid transfer, implicating DNA as the target [10]. In the meantime, the acceleration of genome sequencing efforts and the implication of CRISPR loci in host–virus evolutionary dynamics in natural systems using metagenomics [11, 12] documented the widespread occurrence and importance of CRISPR loci. In the span of three years, the ability of CRISPR-Cas systems to provide adaptive RNA-guided immunity was established, setting the stage for a new era of studies on this exciting topic and fascinating system, focusing on the molecular basis of interference, the various roles of CRISPR-Cas systems in bacteria and archaea, and the establishment of model systems. A number of extensive, insightful, and focused manuscripts have reviewed the rapidly evolving CRISPR literature [13–25]. Here, an overview is provided of the core elements of CRISPR-Cas systems, and their diversity and mechanisms of action are highlighted.

1.2 Elements of CRISPR-Cas Systems

There are two major sets of sequences that constitute the CRISPR-Cas systems: the repeat-spacer array, and the *cas* genes (Fig. 1). The basic defining feature of a CRISPR locus is the CRISPR repeat sequence. This quintessential repeat is typically short (28–40 nt in most cases), though it can vary between 23 and 55 nt. Within a CRISPR locus, these sequences are very highly conserved in terms of sequence and length. Often, the sequence is partially palindromic, with the predicted ability to form secondary hairpin structures [26]. In some cases, the end of the last repeat of the CRISPR array is slightly different with one to ten single nucleotide polymorphisms (SNPs).

The CRISPR repeats are physically separated by sequences called *"spacers,"* that are highly conserved in terms of length, but highly diverse in terms of sequence and origin. Indeed, these sequences derive from foreign genetic elements – where they are called *proto-spacers* – such as plasmids and viruses, that invade the host over the course of time. Proto-spacer sequences are "sampled" from invading nucleic acids that carry CRISPR-targeting short sequence motifs called "proto-spacer

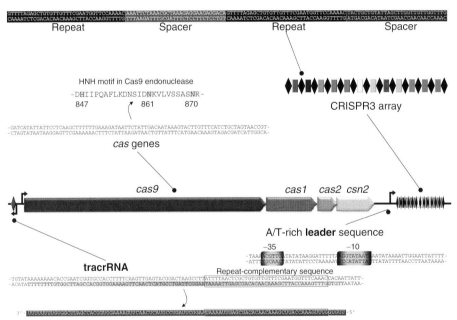

Fig. 1 Overview of the *Streptococcus thermophilus* CRISPR3-Cas system. The core elements of the *S. thermophilus* DGCC7710 CRISPR3 locus (accession number HQ712120) are outlined and detailed, including *cas* genes (middle, left) and the repeat-spacer array (middle right). *cas* genes include the signature *cas9* encoding an endonuclease, the universal *cas1* and *cas2* genes, and the Type II-A *csn2* gene. CRISPR repeats are shown as black diamonds, and CRISPR spacers are represented using colored boxes. Some sequences for the repeat-spacer array are provided (top, right). The HNH motif for the Cas9 endonuclease is highlighted (top left). The tracrRNA sequence is shown, including a portion partially complementary to the CRISPR repeat (bottom left). Putative -35 and -10 promoter boxes within the A/T-rich leader region are highlighted (bottom, right).

adjacent motifs" (PAMs) [27–30]. These sequences are typically 2–5 nt recognition motifs critical for CRISPR-mediated interference. Proto-spacers are likely spliced from invasive nucleic acids and are specifically integrated at the "leader" end of the repeat-spacer array, in a polarized manner. The leader is generally an A/T-rich sequence located upstream of the first CRISPR repeat, which typically includes promoter elements and binding sites for transcriptional regulators, and has been implicated in promoting transcription of the repeat-spacer array [31–33].

In the large majority of cases, CRISPR loci are relatively short, with an average of about 30 repeats, representing approximately 1.6 kb. In some extreme cases, however, they can be quite large and carry several hundreds of repeats. For organisms that carry CRISPR loci, it is typical that one or two CRISPR-Cas systems occur, though there are some exceptions, especially in archaea and extremophiles, for which genomes can carry almost two dozen CRISPR loci [3].

While CRISPR repeats are relatively conserved in terms of length and sometimes even sequence, *cas* genes are extremely diverse in terms of occurrence, number, and sequence [34, 35]. Indeed, Cas make up a highly polymorphic protein family which carries out a wide array of molecular functions [13–25, 34, 35]. Cas proteins overall bear a broad range of biochemical motifs that drive interaction with nucleic acids, such as RNA recognition motifs (RRMs), DNA-binding motifs (HTH, HNH), helicases, and nuclease motifs, which all reflect their functional roles in nucleic acid binding and cleavage through endonuclease and exonuclease activities.

1.3
Diversity of CRISPR-Cas Systems

Although CRISPR-Cas systems are phylogenetically widespread in bacteria and nearly ubiquitous in archaea, there are extensive differences in their occurrence and content [13–25]. Several attempts at categorizing, comparing, and contrasting *cas* genes [34–36] and the proteins they encode have led to the establishment of a (nearly unanimous) CRISPR-Cas system nomenclature, primarily based on *cas* sequences, phylogeny, and functional roles in the various stages of CRISPR-mediated immunity [35]. This nomenclature has established three main types and ten subtypes (subject to expansion as new subtypes are identified) of CRISPR-Cas systems [35, 36], based on universal and signature genes. Several well-characterized model systems have provided a basis for molecular and

Tab. 2 CRISPR-Cas model systems.

Type	Signature	Subtype	Model host
I	Cas3	I-A	*Sulfolobus solfataricus*
		I-E	*Escherichia coli*
		I-F	*Pseudomonas aeruginosa*
II	Cas9	II-A	*Streptococcus thermophiles*
		II-A	*Streptococcus pyogenes*
III	Cas10	III-A	*Staphylococcus epidermidis*
		III-B	*Pyrococcus furiosus*

genetic studies across these three types (Tab. 2).

There are two universally occurring *cas* genes, *cas1* and *cas2* (implicated in novel spacer acquisition [37–40]), that are shared across the three CRISPR-Cas types. In addition, each CRISPR-Cas type has been associated with a unique signature gene which selectively occurs in a given type, namely *cas3* for Type I, *cas9* for Type II, and *cas10* for Type III systems [35].

The occurrence and distribution of various CRISPR-Cas systems vary widely, with Type I systems occurring most frequently in bacteria, Type II systems thus far exclusively identified in bacteria, and Type III systems occurring most frequently in archaea [19, 35]. The absence of any correlation between phylogeny and CRISPR occurrence, and CRISPR presence inconsistencies observed within genera and species, are explained by a combination of propensity for decay and horizontal gene transfer [36, 41].

2
Mechanism of Action of CRISPR-Mediated Interference

The results of early studies have established that CRISPR-Cas systems function in three basic stages (Fig. 2):

- *Adaptation*, where proto-spacers derived from invasive elements are integrated as new spacers into the repeat-spacer array for immunization [8].
- *crRNA biogenesis*, where the repeat-spacer array is transcribed and processed into small interfering crRNAs [9].
- *Interference*, where crRNAs guide sequence-specific Cas nucleases towards homologous nucleic acids for cleavage [42].

While these general steps are shared across all three main CRISPR-Cas system types, the proteins involved in these phases and the mechanistic details that drive these processes can vary significantly, especially with the involvement of distinct signature genes and the proteins that they encode.

2.1
Adaptation

Generally, *adaptation* is the immunization step during which a novel spacer sequence is integrated in a polarized manner at the leader end of the repeat-spacer array. The spacer is directly derived from a proto-spacer sequence in the invasive nucleic acid, which is associated with a sampling PAM motif (see Fig. 2).

Relatively few molecular details have been uncovered thus far with regards to the specific roles of Cas proteins in proto-spacer sampling, novel repeat synthesis, and integration of the new repeat-spacer unit at the leader end of the array, though it is generally believed that the two universal *cas1* and *cas2* genes are implicated in spacer acquisition [37–40, 43]. Specifically, the DNA endonuclease activity of Cas1 may be involved in foreign DNA processing [37]. Nevertheless, it was shown recently that the first CRISPR repeat is the template for the novel repeat synthesis [40]. It was also established that the wild-type spacer content may influence subsequent spacer acquisition by a phenomenon called *"priming"* [44, 45], whereby partial homology between existing CRISPR spacers and invasive nucleic acid may recruit Cas proteins and direct novel spacer acquisition on the same strand, and/or in the immediate vicinity,

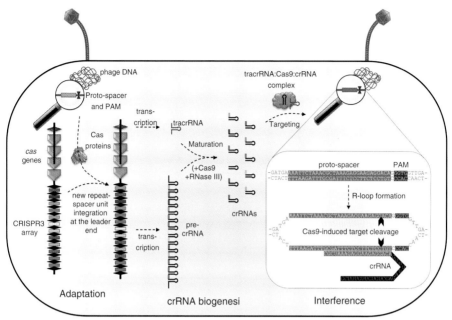

Fig. 2 CRISPR-immunity mechanism of action. *Adaptation* (left): a new repeat-spacer unit is integrated at the leader end of the CRISPR locus, following phage exposure, whereby the new spacer is derived from a phage region (proto-spacer) associated with a PAM. *crRNA biogenesis* (center): the repeat-spacer array is first transcribed into a long pre-crRNA, which is then processed by the tracrRNA:Cas9:RNAse III machinery to generate mature crRNAs. *Interference* (right): the tracrRNA:crRNA:Cas9 ribonucleoprotein complex targets homologous DNA, generates an R-loop, and provides dsDNA cleavage within the proto-spacer, exactly 3 nt away from the 3′ extremity of the spacer, which is adjacent to the PAM.

by targeting a neighboring PAM. Two models were proposed for new spacer acquisition in the Types I–E systems, namely naïve acquisition and priming acquisition, which requires Cas1, Cas2, Cascade-crRNA, and Cas3 [43]. This is consistent with a report revealing a strong bias in spacer sampling in a phage-host model system [46].

2.2 crRNA Biogenesis

Generally, *crRNA biogenesis* is the small interfering RNA manufacturing step during which the repeat-spacer array is transcribed into a long primary precursor CRISPR RNA (pre-crRNA), and subsequently processed into mature, small interfering crRNAs [9, 31–33].

In Type I systems, the CRISPR-associated complex for antiviral defense (Cascade) has been implicated in the processing of pre-crRNA into crRNA [9]. Specifically, within the Cascade complex, the endoribonuclease Cas6 is involved in pre-crRNA cleavage within the CRISPR repeat sequence, to generate a mature crRNA [47]. These processing events are typically precise, and cleavage occurs at the base of the stem–loop structure generated by the palindrome within the CRISPR repeat sequence [39, 48]. This generates a crRNA consisting of a full spacer flanked by a short (8 nt) 5′-handle and a (17–21 nt) 3′-stem–loop, both derived from the CRISPR repeat. Extensive structural and biochemical studies have documented the composition and structure of the Cascade seahorse-like structure, which encompasses a Cas7 hexameric backbone, a Cse1 tail, and a Cas6e head [49–53].

In Type II systems, the *cas*9 signature gene encodes a large protein involved in crRNA biogenesis [54–57]. This process necessitates a trans-encoded CRISPR RNA (tracrRNA; see Fig. 2) [54], which is partially complementary to a section of the CRISPR repeat sequence (over approximately 25 nt), and drives pre-crRNA cleavage within the CRISPR repeat, and subsequently a secondary cleavage within the spacer, involving the dicing housekeeping endoribonuclease RNAse III [54]. This three-component tracrRNA:Cas9:RNAse III system is responsible for the biogenesis of mature crRNAs.

In Type III systems, the pre-crRNA undergoes two sequential maturation steps: cleavage within the CRISPR repeat by Cas6 to generate 67 nt crRNAs followed by 3′-end trimming by Cascade, in a ruler-mechanism manner [20, 58–61], to generate 39–45 nt mature crRNAs.

Because phage lytic cycles can be extremely aggressive and rapid, it is obvious that functional CRISPR-Cas systems need to readily mount a phage immune response upon infection. Studies substantiating the constitutive expression of *cas* genes and CRISPR repeat-spacer arrays [62], together with the large amounts of crRNA and Cas proteins found in cells growing under standard laboratory conditions, suggest that CRISPR-Cas immune systems are permanently patrolling the host cytoplasm [54, 63]. This is consistent with the ability of CRISPR-Cas systems to generate large amounts of crRNA [64]. Nevertheless, several studies in *E. coli* have established that Type I–E systems can be subjected to extensive transcriptional regulation (including repression and activation) by H-NS, LeuO, and cAMP-CRP [32, 33, 65].

2.3
Interference

In general, *interference* is the immune step during which mature crRNAs specifically guide Cas nucleases toward complementary nucleic acids for cleavage, resulting in destruction of the invasive genetic element.

For Type I systems, the Cascade:crRNA complex specifically directs the Cas machinery toward homologous target DNA. Mechanistically, this ribonucleoprotein complex triggers R-loop dsDNA cleavage. Interference relies on PAM recognition, enthalpically coupled with 7–8 nt *seed* sequence crRNA:target DNA pairing which drives R-loop formation [66, 67], and strand separation by the Cascade:crRNA:Cas3 effector complex. Subsequently, target DNA degradation is performed by the signature 3′–5′ nuclease Cas3, which carries helicase (DExH and HelC domains) and ssDNA cleavage (HD nuclease) activities [35, 49, 50, 67–75].

For Type II systems, target dsDNA cleavage is mediated by the HNH and RuvC domains of Cas9, each nicking a strand of the target DNA in the immediate vicinity of the PAM [42, 56, 57]. The molecular processes driving DNA cleavage are identical for phage, plasmid, and synthetic DNA, and can vary between Type II systems based on their individual components. Diversity within Type II systems is illustrated by PAM sequence variation, which include NNAGAAW, NNGGNG, and NGG for the

CRISPR1 system of *S. thermophilus*, the CRISPR3 system of *S. thermophilus*, and the CRISPR1 system of *Streptococcus pyogenes*, respectively.

For Type III systems, the Cas10 signature nuclease is implicated in target nucleic acid interference. In Type III-A systems [10, 60], the target is DNA and interference relies on the Csm protein complex, whereas for Type III-B systems [58, 59, 61], the target is RNA and interference relies on the Cmr [Cas RAMP (Repeat-Associated Mysterious Proteins) module] protein complex [76]. Cognate nucleic acid targeting by Csm and Cmr complexes, and Cas6 target cleavage [47], are both homologous to Cascade-mediated targeting in Type I systems. Notably, Type III systems may not rely on PAM sequences for target interference, as opposed to Type I and Type II systems.

3
Exploitation of the Versatile CRISPR-Cas Systems

The overall polymorphism of CRISPR-Cas systems illustrates their functional diversity, the many roles they play in bacteria and archaea, and the broad application ranges over which they can be exploited.

3.1
Typing and Epidemiological Studies

Originally, notwithstanding the very limited understanding of the biological role(s) of CRISPR-Cas systems, their genetic hypervariable nature was used for the genotyping of bacteria, based on the number of repeats found in CRISPR loci, and the spacer sequences, and this was initially referred to as *spacer oligotyping* (spoligotyping) [77]. Early focus on pathogenic species including *Mycobacterium tuberculosis* and *Yersinia pestis* [6, 78–80] quickly substantiated their potential as molecular typing targets. Since then, their hypervariable nature in epidemiologically relevant *E. coli* [81–83] and *Salmonella enterica* [84–86] have further strengthened their ability to segregate nearly identical strains, and has provided critical insights into the phylogenetic relationships between isolates in a wide diversity of genera and species [28, 87–89].

Because new spacers are acquired in a strictly polarized manner, determination of the CRISPR spacer content provides a sequential record of an organism's exposure to invasive elements from their environment, and documents the historic origin and path of a given strain, including shared ancestry between (even unknown) isolates (Fig. 3). The distribution of CRISPR loci in numerous bacteria of medical and clinical importance will be instrumental for their broad use in diagnostic kit development, and field outbreak investigations. Their actual epidemiological value for clinical studies and food safety monitoring, however, will eventually depend heavily on their activity as an immune system, and broad distribution across a wide range of pathogenic genera and species of interest. Also, the long-term success of CRISPR-based typing schemes will depend on the accompanying sequencing throughput and fidelity, as well as the convenience and interpretation value of the accompanying assembly and visualization tools that are limited by the current bioinformatic solutions, especially for complex and/or large datasets.

CRISPR-based genotyping has also proven quite useful in the tracking and genetic tagging of industrially relevant organisms, whereby the origin

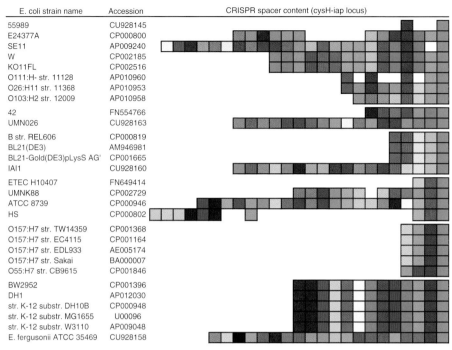

Fig. 3 *Escherichia coli* CRISPR-based typing. The spacer content of the CRISPR locus adjacent to the *E. coli iap* gene is shown (repeats are not shown), with unique color combinations representing different CRISPR spacer sequences. Shared spacers at the ancestral end of the locus (right) indicate common origin. Sets of shared ancestral spacers provide a basis for phylogenetic clustering. Crossed squares reflect internal spacer deletions.

of a proprietary strain can be readily determined and easily documented. With the widespread occurrence of CRISPR loci in many genomes of lactic acid bacteria used in food processing [21, 90], CRISPR-based typing has potential for industrial strain genotyping, for strains used both as starter cultures (notably streptococci and lactobacilli) and probiotics (notably lactobacilli and bifidobacteria) [90–95]. Noteworthy, because a particular sequential addition of spacers in CRISPR loci is statistically highly unlikely to occur twice independently (depending on the occurrence of PAMs in invasive genetic elements), specific CRISPR arrays can be used for genetic tagging and the tracking of proprietary industrial strains [21]. Although this is highly useful in organisms with active systems, it is important to note that CRISPR loci can also evolve through internal spacer deletions, and may be subjected to decay [96–98].

Beyond individual strain tracking and typing, CRISPR loci have tremendous potential for the determination of diversity and evolution in mixed populations in natural habitats and samples. Indeed, at an early stage in the metagenomic investigations of microbial diversity in natural samples, it became clear that CRISPR locus reconstruction can provide unique insights into the composition, diversity, and evolution of microbial populations [11, 12]. Because CRISPR-Cas systems are critical in providing immunity against invasive elements such as viruses, it was established that they can reveal the coevolutionary dynamics of host populations and their viral counterparts. This documented arms race actually is a driving force of host and virus genome evolution, and has since been confirmed by numerous studies, covering various natural niches such as acid mine drainage, microbial mats, hot springs, hydrothermal vents, and animal-associated microbiomes [99–106]. Further theoretical investigations, mathematical modeling [97, 107–109], and studies of model systems in laboratory environments have confirmed that CRISPR-Cas systems can play critical roles in the evolutionary dynamics of hosts and viral populations [46, 110]. Actually, in systems where CRISPR immunity plays a key role in phage resistance, it was shown that adaptive CRISPR-encoded immunity can rapidly impact and direct phage genome evolution trajectory. Indeed, escape viral populations specifically mutate sampled proto-spacers and PAM sequences [by SNPs and INDELs (INsertions and/or DELetions)] to circumvent CRISPR spacer acquisition [19, 27, 110, 111]. CRISPR-inactivating genes were also recently discovered in the genomes of phages that can escape CRISPR-immunity [112]. Because of the diversity and rate at which novel spacers can be acquired, CRISPR-immunity can impact phage genome evolution on the generational scale, and rapidly drive viral genome evolution to a dead-end in closed systems where multiple CRISPR loci are concurrently active. Also, even with limited knowledge or information about viral sequences, it is possible to investigate viral populations indirectly through the spacer content of their hosts [113, 114].

As the present still limited understanding of natural microbial diversity and population composition extends, it is anticipated that CRISPR analyses will allow the true biodiversity of natural samples to be probed, and for host–virus dynamics to be investigated, even in complex biosystems.

3.2
Genome Integrity and crRNA-Mediated Interference

Because the functional role of CRISPR-Cas systems was originally established in the dairy bacterium *S. thermophilus* [8], which is widely used as a starter culture in the manufacture of yogurt and certain cheeses, it was exploited relatively quickly as a means to iteratively build up resistance against bacteriophages that were widely problematic in industrial dairy product manufacture [21] (Fig. 4). This has been implemented in a wide array of *S. thermophilus* strains used globally in yogurt manufacture, and the exploitation of CRISPR-based immunity has led to an extension of the lifespan of highly valuable dairy cultures that convey important industrial manufacturing properties (acidification rates, product consistency) as well as organoleptic properties (texture, flavor) [21]. Surprisingly, although a broad range of phage resistance systems have been exploited for decades by starter culture manufacturers, phage have remained a problem in the dairy industry, and CRISPR immunity has expanded the arsenal of industrial scientists that battle industrial phages on a permanent basis [8, 21, 27, 28, 42, 115, 116].

Mechanistically, cultures are composed of naturally generated bacteriophage-insensitive mutants (BIMs) that are screened for new CRISPR spacer acquisition. Several rounds of phage challenges are used to build immunity over time

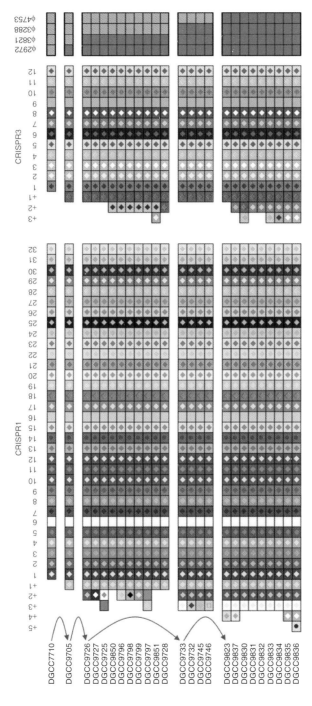

Fig. 4 Iterative build up of CRISPR immunity against bacteriophage. Upon exposure of *S. thermophilus* strain DGCC7710 to virulent phages, bacteriophage insensitive mutants (BIMs) are generated, that acquire novel spacers (boxes with unique color combinations represent unique spacer sequences; CRISPR repeats are omitted, only spacers are represented) in two active CRISPR loci, namely CRISPR1 (left) and CRISPR3 (right). Following exposure to φ2972, a variant which has acquired one novel spacer in each active locus is then exposed to a second phage, φ3821. Several rounds of CRISPR BIM selection and subsequent phage exposure eventually yield an immunized variant, DGCC9836, which has acquired five new spacers in CRISPR1 and two new spacers in CRISPR3, that provide resistance against the range of phages that were used in iterative challenges (right, resistant = R, green; sensitive = S, red).

through iterative spacer additions (see Fig. 4). Spacers that target widespread and/or highly conserved phage sequences can be selected for, leading to increases in both phage resistance spectrum and depth. This "CRISPerization" process has the noteworthy advantage of generating *isogenic* variants that maintain their original phenotypes, as opposed to more "classical" mutagenesis and/or screening methods that occasionally generate variants with altered and/or defective phenotypes.

Beyond dairy cultures, the presence of CRISPR-Cas systems in several bacteria relevant to the food, feed, biotechnology, and medical industries opens avenues for industrial strain lifespan extension, protecting the genome integrity of highly valuable and/or unique strains, by natural and/or genetic engineering approaches. Indeed, while it is readily possible to select natural CRISPR variants in organisms that carry active systems [42, 116, 117], several groups have shown that it is relatively easy to artificially engineer functional spacers that target undesirable genetic elements, and have also established that whole functional systems can be transferred to organisms devoid of CRISPR systems, or that do not carry functional CRISPR loci. Today, a point in time has been reached at which artificial spacers can be designed *in silico*, and readily artificially manufactured through DNA synthesis and integrated into active systems, that can be transferred across distant phylogenetic boundaries [55]. With studies documenting the ability of CRISPR-Cas systems to target plasmids and interfere with plasmid transfer [10, 42, 44, 45, 55–57], and to naturally acquire antibiotic resistance genes-targeting spacers that preclude the subsequent uptake of plasmids carrying homologous sequences [42], this can obviously be leveraged to maintain microbial genome integrity over time, and to preclude the uptake and dissemination of undesirable genetic elements such as antibiotic resistance cassettes, mobile genetic elements [118, 119], pathogenic traits (such as toxins and virulence) [120], and even lysogenic prophages [121, 122]. Though active CRISPR-Cas systems can readily limit the spread of antibiotic resistance markers, it is important to bear in mind that inactive or dormant CRISPR loci may not impact plasmid dissemination [123]. With the established negative correlation between the occurrence of CRISPR-Cas systems and pathogenicity of enterococci [124] and streptococci [125], there is potential to vaccinate strains of interest to generate more robust and safer variants that could potentially be used to outcompete pathogenic organisms. Similar applications can be envisioned in archaea, where CRISPR-Cas systems also have the ability to readily interfere with viruses and plasmids [19, 76, 126].

Notwithstanding the focus on phage resistance and plasmid uptake control, CRISPR-Cas systems also have documented roles in regulation of intracellular RNA levels and gene regulation [127–129], biofilm formation [130, 131], and resistance to stress [132]. It is also possible that additional and/or elusive roles are yet to be uncovered for these versatile systems.

3.3 RNA-Guided Next-Generation Genome Editing

The latest set of milestone CRISPR studies has established a new range of CRISPR-derived molecular machines that provide RNA-guided programmable cleavage. This new genome editing tool is based on leveraging and adapting the Cas and crRNA machinery of Type II systems that is responsible for target DNA cleavage [42, 54] (see Fig. 2).

Specifically, following studies which showed that crRNA-mediated interference would result in a specific cleavage of the target phage or plasmid DNA [42, 116], and with a mechanistic understanding that in Type II systems the tracrRNA:crRNA:Cas9 complex would guide this cleavage [54, 57], recent studies have shown that Cas9 can be reprogrammed using customized crRNAs to generate DNA nicks or dsDNA cleavage [56, 57]. A molecular *tour de force* reported by Jinek *et al.* [56] showed that Cas9 can actually be reprogrammed using a *single chimeric RNA* molecule to direct sequence-specific DNA cleavage. This flexible chimeric single guide RNA (sgRNA), which includes portions of crRNA and tracrRNA, can guide Cas9 towards any sequence of interest associated with the correct PAM. Early in 2013, it was shown that this versatile system, in combination with the error-prone nonhomologous end-joining machinery, can be implemented in complex organisms – including human cells – for genome editing [133, 134].

Subsequent studies conducted by both Cong *et al.* [133] and Mali *et al.* [134] showed simultaneously that the *S. pyogenes* Type II CRISPR-Cas system from Jinek *et al.* [56] can be engineered to specifically generate sequence-specific mutations in various human and mouse cell lines, including pluripotent stem cells. Further, multiplexing validation was provided, thus opening the door for concurrent mutations and the editing of distinct genomic loci. Similar findings were reported immediately thereafter, in human cells [135], zebrafish [136], and bacteria [137].

Similar on-going studies are implementing this next-generation genome editing system in plants and yeast. Noteworthy, this flexible and reprogrammable system can be exploited for genome editing, genome stacking, genome shuffling, and genome engineering in, potentially, any organism of interest [138]. In many ways, RNA-guided endonucleases (RGENs) combine the flexibility of RNAi with the documented efficiency and fidelity of "classical" restriction enzymes. Altogether, this paves the way for the next generation of biotechnological applications, as an alternative to zinc finger nucleases (ZFNs), transcription activator-like effector nucleases (TALENs), and homing endonucleases, with advantages that include flexibility, implementation timeline, specificity, accuracy, and targeting frequency. The milestone studies performed by Cong *et al.* and Mali *et al.*, as well as immediately subsequent studies illustrating the power of RNA-guided Cas9 nucleases, have opened up new avenues for human and mammalian genome engineering for medical and biotechnological purposes. Likewise, implementing this system in plants for crop protection and seed enhancement will lead to the next breakthrough in agricultural sustainability and next-generation seeds.

4
Summary and Outlook

CRISPR-Cas systems constitute a fascinating adaptive prokaryotic immune system which provides RNA-mediated interference against viruses and plasmids through sequence-specific cleavage. Notwithstanding the diversity of CRISPR-Cas systems that are widespread in bacteria and almost ubiquitous in archaea, CRISPR-Cas systems generally operate in three stages: (i) *adaptation* by the acquisition of new spacers derived from invasive elements that build up DNA-encoded inheritable immunity; (ii) *crRNA biogenesis* by transcription of the repeat-spacer array, followed by processing of the transcript into mature, small interfering RNAs; and (iii) *interference* by sequence-specific nucleases guided by the crRNAs. Several studies have shown that these versatile systems and their various components can be used in a diversity of applications, including: building immunity against viruses or plasmid DNA uptake; exploiting their hypervariable nature for precise genotyping; and revealing the arms race between hosts and their viruses.

Despite the relatively short history of CRISPR research, several distinct examples already exist of concurrent sets of milestone studies from different teams at the same time, reflecting convergent scientific foci and like-minded (perhaps talkative) individuals that assemble the CRISPR puzzle at the same pace. These concurrent sets of studies include: the 2005 reports that CRISPR spacers derive from foreign genetic elements [4–6]; the 2011 reports that a seed sequence drives R-loop formation [66, 67]; the 2012 reports that Type II systems can be reprogrammed for DNA cleavage [56, 57]; and the 2013 reports that RNA-guided Cas endonucleases can be reprogrammed for genome editing [133–137]. Noteworthy, this novel versatile RNA-guided nuclease family opens up new avenues for next-generation genome editing, and will likely increase the visibility of CRISPR-Cas systems. The rapidly evolving CRISPR literature doubtless reflects the amazing overall potential of these fantastic systems [139], hinting at a bright future.

Acknowledgments

The authors would like to acknowledge their DuPont colleagues and many academic collaborators with whom they have had the privilege and pleasure to explore CRISPR-Cas systems over the past decade, notably Moineau *et al.*, Banfield *et al.*, Siksnys *et al.*, Roberts *et al.*, Dudley *et al.*, Levin *et al.*, Bhaya *et al.*, Koonin *et al.*, Terns *et al.*, and van der Oost *et al.*

References

1 Ishino, Y., Shinagawa, H., Makino, K., Amemura, M. (1987) Nucleotide sequence of the *iap* gene, responsible for alkaline phosphatase isozyme conversion in *Escherichia coli*, and identification of the gene product. *J. Bacteriol.*, **169**, 5429–5433.

2 Jansen, R., van Embden, J.D., Gaastra, W., Schouls, L.M. (2002) Identification of genes that are associated with DNA repeats in prokaryotes. *Mol. Microbiol.*, **43**, 1565–1575.

3 Grissa, I., Vergnaud, G., Pourcel, C. (2007) The CRISPRdb database and tools to display CRISPRs and to generate dictionaries of spacers and repeats. *BMC Bioinform.*, **8**, 172.

4 Bolotin, A., Quinquis, B., Sorokin, A., Ehrlich, S.D. (2005) Clustered regularly interspaced short palindrome repeats (CRISPRs) have spacers of extrachromosomal origin. *Microbiology*, **151**, 2551–2561.

5 Mojica, F.J., Díez-Villaseñor, C., García-Martínez, J., Soria, E. (2005) Intervening sequences of regularly spaced prokaryotic repeats derive from foreign genetic elements. *J. Mol. Evol.*, **60**, 174–182.

6 Pourcel, C., Salvignol, G., Vergnaud, G. (2005) CRISPR elements in *Yersinia pestis* acquire new repeats by preferential uptake of bacteriophage DNA, and provide additional tools for evolutionary studies. *Microbiology*, **151**, 653–663.

7 Makarova, K.S., Grishin, N.V., Shabalina, S.A., Wolf, Y.I. (2006) A putative RNA-interference-based immune system in prokaryotes: computational analysis of the

predicted enzymatic machinery, functional analogies with eukaryotic RNAi, and hypothetical mechanisms of action. *Biol. Direct.*, **1**, 7.

8 Barrangou, R., Fremaux, C., Deveau, H., Richards, M. (2007) CRISPR provides acquired resistance against viruses in prokaryotes. *Science*, **315**, 1709–1712.

9 Brouns, S.J., Jore, M.M., Lundgren, M., Westra, E.R. (2008) Small CRISPR RNAs guide antiviral defense in prokaryotes. *Science*, **321**, 960–964.

10 Marraffini, L.A., Sontheimer, E.J. (2008) CRISPR interference limits horizontal gene transfer in staphylococci by targeting DNA. *Science*, **322**, 1843–1845.

11 Andersson, A.F., Banfield, J.F. (2008) Virus population dynamics and acquired virus resistance in natural microbial communities. *Science*, **320**, 1047–1050.

12 Tyson, G.W., Banfield, J.F. (2008) Rapidly evolving CRISPRs implicated in acquired resistance of microorganisms to viruses. *Environ. Microbiol.*, **10**, 200–207.

13 Horvath, P., Barrangou, R. (2010) CRISPR/Cas, the immune system of bacteria and archaea. *Science*, **327**, 167–170.

14 Karginov, F.V., Hannon, G.J. (2010) The CRISPR system: small RNA-guided defense in bacteria and archaea. *Mol. Cell*, **37**, 7–19.

15 Labrie, S.J., Samson, J.E., Moineau, S. (2010) Bacteriophage resistance mechanisms. *Nat. Rev. Microbiol.*, **8**, 317–327.

16 Marraffini, L.A., Sontheimer, E.J. (2010) CRISPR interference: RNA-directed adaptive immunity in bacteria and archaea. *Nat. Rev. Genet.*, **11**, 181–190.

17 Al-Attar, S., Westra, E.R., van der Oost, J., Brouns, S.J. (2011) Clustered regularly interspaced short palindromic repeats (CRISPRs): the hallmark of an ingenious antiviral defense mechanism in prokaryotes. *Biol. Chem.*, **392**, 277–289.

18 Bhaya, D., Davison, M., Barrangou, R. (2011) CRISPR-Cas systems in bacteria and archaea: versatile small RNAs for adaptive defense and regulation. *Annu. Rev. Genet.*, **45**, 273–297.

19 Garrett, R.A., Shah, S.A., Vestergaard, G., Deng, L. (2011) CRISPR-based immune systems of the Sulfolobales: complexity and diversity. *Biochem. Soc. Trans.*, **39**, 51–57.

20 Terns, M.P., Terns, R.M. (2011) CRISPR-based adaptive immune systems. *Curr. Opin. Microbiol.*, **14**, 321–327.

21 Barrangou, R., Horvath, P. (2012) CRISPR: new horizons in phage resistance and strain identification. *Annu. Rev. Food Sci. Technol.*, **3**, 143–162.

22 Jore, M.M., Brouns, S.J., van der Oost, J. (2012) RNA in defense: CRISPRs protect prokaryotes against mobile genetic elements. *Cold Spring Harbor Perspect. Biol.*, **4**, a003657.

23 Richter, C., Chang, J.T., Fineran, P.C. (2012) Function and regulation of clustered regularly interspaced short palindromic repeats (CRISPR)/CRISPR associated (Cas) systems. *Viruses*, **4**, 2291–2311.

24 Westra, E.R., Swarts, D.C., Staals, R.H., Jore, M.M. (2012) The CRISPRs, they are a-changin': how prokaryotes generate adaptive immunity. *Annu. Rev. Genet.*, **46**, 311–339.

25 Wiedenheft, B., Sternberg, S.H., Doudna, J.A. (2012) RNA-guided genetic silencing systems in bacteria and archaea. *Nature*, **482**, 331–338.

26 Kunin, V., Sorek, R., Hugenholtz, P. (2007) Evolutionary conservation of sequence and secondary structures in CRISPR repeats. *Genome Biol.*, **8**, R61.

27 Deveau, H., Barrangou, R., Garneau, J.E., Labonté, J. (2008) Phage response to CRISPR-encoded resistance in *Streptococcus thermophilus*. *J. Bacteriol.*, **190**, 1390–1400.

28 Horvath, P., Romero, D.A., Coûté-Monvoisin, A.C., Richards, M. (2008) Diversity, activity, and evolution of CRISPR loci in *Streptococcus thermophilus*. *J. Bacteriol.*, **190**, 1401–1412.

29 Mojica, F.J., Díez-Villaseñor, C., García-Martínez, J., Almendros, C. (2009) Short motif sequences determine the targets of the prokaryotic CRISPR defence system. *Microbiology*, **155**, 733–740.

30 Almendros, C., Guzmán, N.M., Díez-Villaseñor, C., García-Martínez, J. (2012) Target motifs affecting natural immunity by a constitutive CRISPR-Cas system in *Escherichia coli*. *PLoS ONE*, **7**, e50797.

31 Lillestol, R.K., Shah, S.A., Brugger, K., Redder, P. (2009) CRISPR families of the crenarchaeal genus *Sulfolobus*: bidirectional

transcription and dynamic properties. *Mol. Microbiol.*, **72**, 259–272.

32 Pougach, K., Semenova, E., Bogdanova, E., Datsenko, K.A. (2010) Transcription, processing and function of CRISPR cassettes in *Escherichia coli*. *Mol. Microbiol.*, **77**, 1367–1379.

33 Pul, U., Wurm, R., Arslan, Z., Geissen, R. (2010) Identification and characterization of *E. coli* CRISPR RNA processing in *Pectobacterium atrosepticum*. *RNA Biol.*, **8**, 517–528.

34 Haft, D.H., Selengut, J., Mongodin, E.F., Nelson, K.E. (2005) A guild of 45 CRISPR-associated (Cas) protein families and multiple CRISPR/Cas subtypes exist in prokaryotic genomes. *PLoS Comput. Biol.*, **1**, e60.

35 Makarova, K.S., Haft, D.H., Barrangou, R., Brouns, S.J. (2011) Evolution and classification of the CRISPR-Cas systems. *Nat. Rev. Microbiol.*, **9**, 467–477.

36 Makarova, K.S., Aravind, L., Wolf, Y.I., Koonin, E.V. (2011) Unification of Cas protein families and a simple scenario for the origin and evolution of CRISPR-Cas systems. *Biol. Direct*, **6**, 38.

37 Wiedenheft, B., Zhou, K., Jinek, M., Coyle, S.M. (2009) Structural basis for DNase activity of a conserved protein implicated in CRISPR-mediated genome defense. *Structure*, **17**, 904–912.

38 Babu, M., Beloglazova, N., Flick, R., Graham, C. (2011) A dual function of the CRISPR-Cas system in bacterial antivirus immunity and DNA repair. *Mol. Microbiol.*, **79**, 484–502.

39 Nam, K.H., Ding, F., Haitjema, C., Huang, Q. (2012) Double-stranded endonuclease activity in *B. halodurans* clustered regularly interspaced short palindromic repeats (CRISPR)-associated Cas 2 protein. *J. Biol. Chem.*, **287**, 35943–35952.

40 Yosef, I., Goren, M.G., Qimron, U. (2012) Proteins and DNA elements essential for the CRISPR adaptation process in *Escherichia coli*. *Nucleic Acids Res.*, **40**, 5569–5576.

41 Godde, J.S., Bickerton, A. (2006) The repetitive DNA elements called CRISPRs and their associated genes: evidence of horizontal transfer among prokaryotes. *J. Mol. Evol.*, **62**, 718–729.

42 Garneau, J.E., Dupuis, M.È., Villion, M., Romero, D.A. (2010) The CRISPR/Cas bacterial immune system cleaves bacteriophage and plasmid DNA. *Nature*, **468**, 67–71.

43 Fineran, P.C., Charpentier, E. (2012) Memory of viral infections by CRISPR-Cas adaptive immune systems: acquisition of new information. *Virology*, **434**, 202–209.

44 Datsenko, K.A., Pougach, K., Tikhonov, A., Wanner, B.L. (2012) Molecular memory of prior infections activates the CRISPR/Cas adaptive bacterial immunity system. *Nat. Commun.*, **3**, 945.

45 Swarts, D.C., Mosterd, C., van Passel, M.W., Brouns, S.J. (2012) CRISPR interference directs strand specific spacer acquisition. *PLoS ONE*, **7**, e35888.

46 Paez-Espino, D., Morovic, W., Sun, C.L., Thomas, B.C. (2013) Strong bias in the bacterial CRISPR elements that confer immunity to phage. *Nat. Commun.*, **4**, 1430.

47 Li, M., Liu, H., Han, J., Liu, J. (2013) Characterization of CRISPR RNA biogenesis and Cas6 cleavage-mediated inhibition of a provirus in the haloarchaeon *Haloferax mediterranei*. *J. Bacteriol.*, **195**, 867–875.

48 Haurwitz, R.E., Jinek, M., Wiedenheft, B., Shou, K. (2010) Sequence- and structure-specific RNA processing by a CRISPR endonuclease. *Science*, **329**, 1355–1358.

49 Jore, M.M., Lundgren, M., van Duijn, E., Bultema, J.B. (2011) Structural basis for CRISPR RNA-guided DNA recognition by Cascade. *Nat. Struct. Mol. Biol.*, **18**, 529–536.

50 Lintner, N.G., Kerou, M., Brumfield, S.K., Graham, S. (2011) Structural and functional characterization of an archaeal clustered regularly interspaced short palindromic repeat (CRISPR)-associated complex for antiviral defense (CASCADE). *J. Biol. Chem.*, **286**, 21643–21656.

51 Wiedenheft, B., Lander, G.C., Zhou, K., Jore, M.M. (2011) Structures of the RNA-guided surveillance complex from a bacterial immune system. *Nature*, **477**, 486–489.

52 Sashital, D.G., Wiedenheft, B., Doudna, J.A. (2012) Mechanism of foreign DNA selection in a bacterial adaptive immune system. *Mol. Cell*, **46**, 606–615.

53. Richter, C., Gristwood, T., Clulow, J.S., Fineran, P.C. (2012) In vivo protein interactions and complex formation in the *Pectobacterium atrosepticum* subtype I-F CRISPR/Cas system. *PLoS ONE*, **7**, e49549.
54. Deltcheva, E., Chylinski, K., Sharma, C.M., Gonzales, K. (2011) CRISPR RNA maturation by *trans*-encoded small RNA and host factor RNase III. *Nature*, **471**, 2470–2480.
55. Sapranauskas, R., Gasiunas, G., Fremaux, C., Barrangou, R. (2011) The *Streptococcus thermophilus* CRISPR/Cas system provides immunity in *Escherichia coli*. *Nucleic Acids Res.*, **39**, 9275–9282.
56. Jinek, M., Chylinski, K., Fonfara, I., Hauer, M. (2012) A programmable dual-RNA-guided DNA endonuclease in adaptive bacterial immunity. *Science*, **337**, 816–821.
57. Gasiunas, G., Barrangou, R., Horvath, P., Siksnys, V. (2012) Cas9-crRNA ribonucleoprotein complex mediates specific DNA cleavage for adaptive immunity in bacteria. *Proc. Natl Acad. Sci. USA*, **109**, E2579–E2586.
58. Carte, J., Pfister, N.T., Compton, M.M., Terns, R.M. (2008) Binding and cleavage of CRISPR RNA by Cas6. *RNA*, **16**, 2181–2188.
59. Hale, C.R., Zhao, P., Olson, S., Duff, M.O. (2009) RNA-guided RNA cleavage by a CRISPR RNA-Cas protein complex. *Cell*, **139**, 945–956.
60. Hatoum-Aslan, A., Maniv, I., Marraffini, L.A. (2011) Mature clustered, regularly interspaced, short palindromic repeats RNA (crRNA) length is measured by a ruler mechanism anchored at the precursor processing site. *Proc. Natl Acad. Sci. USA*, **108**, 21218–21222.
61. Hale, C.R., Majumdar, S., Elmore, J., Pfister, N. (2012) Essential features and rational design of CRISPR RNAs that function with the Cas RAMP module complex to cleave RNAs. *Mol. Cell*, **45**, 292–302.
62. Bernick, D.L., Cox, C.L., Dennis, P.P., Lowe, T.M. (2012) Comparative genomic and transcriptional analyses of CRISPR systems across the genus *Pyrobaculum*. *Front. Microbiol.*, **3**, 251.
63. Young, J.C., Dill, B.D., Pan, C., Hettich, R.L. (2012) Phage-induced expression of CRISPR-associated proteins is revealed by shotgun proteomics in *Streptococcus thermophilus*. *PLoS ONE*, **7**, e38077.
64. Djordjevic, M., Djordjevic, M., Severinov, K. (2012) CRISPR transcript processing: a mechanism for generating a large number of small interfering RNAs. *Biol. Direct*, **7**, 24.
65. Westra, E.R., Pul, U., Heidrich, N., Jore, M.M. (2010) H-NS-mediated repression of CRISPR-based immunity in *Escherichia coli* K12 can be relieved by the transcription activator LeuO. *Mol. Microbiol.*, **77**, 1380–1393.
66. Semenova, E., Jore, M.M., Datsenko, K.A., Semenova, A. (2011) Interference by clustered regularly interspaced short palindromic repeat (CRISPR) RNA is governed by a seed sequence. *Proc. Natl Acad. Sci. USA*, **108**, 10098–10103.
67. Wiedenheft, B., van Duijn, E., Bultema, J.B., Waghmare, S.P. (2011) RNA-guided complex from a bacterial immune system enhances target recognition through seed sequence interactions. *Proc. Natl Acad. Sci. USA*, **108**, 10092–10097.
68. Beloglazova, N., Petit, P., Flick, R., Brown, G. (2011) Structure and activity of the Cas3 HD nuclease MJ0384, an effector enzyme of the CRISPR interference. *EMBO J.*, **30**, 4616–4627.
69. Howard, J.A., Delmas, S., Ivančić-Baće, I., Bolt, E.L. (2011) Helicase dissociation and annealing of RNA-DNA hybrids by *Escherichia coli* Cas3 protein. *Biochem. J.*, **439**, 85–95.
70. Mulepati, S., Bailey, S. (2011) Structural and biochemical analysis of nuclease domain of clustered regularly interspaced short palindromic repeat (CRISPR-associated protein 3 (Cas3). *J. Biol. Chem.*, **286**, 31896–31903.
71. Sinkunas, T., Gasiunas, G., Femaux, C., Barrangou, R. (2011) Cas3 is a single-stranded DNA nuclease and ATP-dependent helicases in the CRISPR/Cas immune system. *EMBO J.*, **30**, 1335–1342.
72. Sternberg, S.H., Haurwitz, R.E., Doudna, J.A. (2012) Mechanism of substrate selection by a highly specific CRISPR endoribonuclease. *RNA*, **18**, 661–672.
73. Westra, E.R., van Erp, P.B.G., Kunne, T., Wong, S.P. (2012) CRISPR immunity relies on the consecutive binding and degradation

of negatively supercoiled invader DNA by Cascade and Cas3. *Mol. Cell.*, **46**, 595–605.

74. Sinkunas, T., Gasiunas, G., Waghmare, A.P., Dickman, M.J. (2013) In vitro reconstitution of Cascade-mediated CRISPR immunity in *Streptococcus thermophilus*. *EMBO J.*, **32**, 385–394.

75. Cady, K.C., O'Toole, G.A. (2011) Non-identity-mediated CRISPR-bacteriophage interaction mediated via the Csy and Cas3 proteins. *J. Bacteriol.*, **193**, 3433–3445.

76. Deng, L., Garrett, R.A., Shah, S.A., Peng, X. (2013) A novel interference mechanism by a type IIIB CRISPR-Cmr module in *Sulfolobus*. *Mol. Microbiol.*, **87**, 1088–1099.

77. Groenen, P.M., Bunschoten, A.E., van Soolingen, D., van Embden, J.D. (1993) Nature of DNA polymorphism in the direct repeat cluster of *Mycobacterium tuberculosis*; application for strain differentiation by a novel typing method. *Mol. Microbiol.*, **10**, 1057–1065.

78. Vergnaud, G., Li, Y., Gorgé, O., Cui, Y. (2007) Analysis of the three *Yersinia pestis* CRISPR loci provides new tools for phylogenetic studies and possibly for the investigation of ancient DNA. *Adv. Exp. Med. Biol.*, **603**, 327–338.

79. Cui, Y., Li, Y., Gorgé, O., Platonov, M.E. (2008) Insight into microevolution of *Yersinia pestis* by clustered regularly interspaced short palindromic repeats. *PLoS ONE*, **3**, e2652.

80. Zhang, J., Abadia, E., Refregier, G., Tafaj, S. (2010) *Mycobacterium tuberculosis* complex CRISPR genotyping: improving efficiency, throughput and discriminative power of 'spoligotyping' with new spacers and a microbead-based hybridization assay. *J. Med. Microbiol.*, **59**, 285–294.

81. Díez-Villaseñor, C., Almendros, C., García-Martínez, J., Mojica, F.J. (2010) Diversity of CRISPR loci in *Escherichia coli*. *Microbiology*, **156**, 1351–1361.

82. Delannoy, S., Beutin, L., Burgos, Y., Fach, P. (2012) Specific detection of enteroaggregative hemorrhagic *Escherichia coli* O104:H4 strains using the CRISPR locus as target for a diagnostic real-time PCR. *J. Clin. Microbiol.*, **50**, 3485–3492.

83. Delannoy, S., Beutin, L., Fach, P. (2012) Use of clustered regularly interspaced short palindromic repeat sequence polymorphisms for specific detection of enterohemorrhagic *Escherichia coli* strains of serotypes O26:H11, O45:H2, O103:H2, O111:H8, O121:H19, O145:H28, and O157:H7 by real-time PCR. *J. Clin. Microbiol.*, **50**, 4035–4040.

84. Liu, F., Barrangou, R., Gerner-Smidt, P., Ribot, E.M. (2011) Novel virulence gene and clustered regularly interspaced short palindromic repeat (CRISPR) multilocus sequence typing scheme for subtyping of the major serovars of *Salmonella enterica* subsp. *enterica*. *Appl. Environ. Microbiol.*, **77**, 1946–1956.

85. Liu, F., Kariyawasam, S., Jayarao, B.M., Barrangou, R. (2011) Subtyping *Salmonella enterica* serovar *enteritidis* isolates from different sources by using sequence typing based on virulence genes and clustered regularly interspaced short palindromic repeats (CRISPRs). *Appl. Environ. Microbiol.*, **77**, 4520–4526.

86. Fabre, L., Zhang, J., Guigon, G., Le Hello, S. (2012) CRISPR typing and subtyping for improved laboratory surveillance of *Salmonella* infections. *PLoS ONE*, **7**, e36995.

87. Mokrousov, I., Limeschenko, E., Vyazovaya, A., Narvskaya, O. (2007) *Corynebacterium diphtheriae* spoligotyping based on combined use of two CRISPR loci. *Biotechnol. J.*, **2**, 901–906.

88. Rezzonico, F., Smits, T.H., Duffy, B. (2011) Diversity, evolution, and functionality of clustered regularly interspaced short palindromic repeat (CRISPR) regions in the fire blight pathogen *Erwinia amylovora*. *Appl. Environ. Microbiol.*, **77**, 3819–3829.

89. McGhee, G.C., Sundin, G.W. (2012) *Erwinia amylovora* CRISPR elements provide new tools for evaluating strain diversity and for microbial source tracking. *PLoS ONE*, **7**, e41706.

90. Horvath, P., Coûté-Monvoisin, A.C., Romero, D.A., Boyaval, P. (2009) Comparative analysis of CRISPR loci in lactic acid bacteria genomes. *Int. J. Food Microbiol.*, **131**, 62–70.

91. Barrangou, R., Briczinski, E.P., Traeger, L.L., Loquasto, J.R. (2009) Comparison of the complete genome sequences of *Bifidobacterium animalis* subsp. *lactis*

DSM 10140 and Bl-04. *J. Bacteriol.*, **191**, 4144–4151.

92 Ventura, M., Turroni, F., Lima-Mendez, G., Foroni, E. (2009) Comparative analyses of prophage-like elements present in bifidobacterial genomes. *Appl. Environ. Microbiol.*, **75**, 6929–6936.

93 Loquasto, J.R., Barrangou, R., Dudley, E.G., Roberts, R.F. (2011) Short communication: the complete genome sequence of *Bifidobacterium animalis* subspecies *animalis* ATCC 25527(T) and comparative analysis of growth in milk with *B. animalis* subspecies *lactis* DSM 10140(T). *J. Dairy Sci.*, **94**, 5864–5870.

94 Broadbent, J.R., Neeno-Eckwall, E.C., Stahl, B., Tandee, K. (2012) Analysis of the *Lactobacillus casei* supragenome and its influence in species evolution and lifestyle adaptation. *BMC Genomics*, **13**, 533.

95 Stahl, B., Barrangou, R. (2012) Complete genome sequences of probiotic strains *Bifidobacterium animalis* subsp. *lactis* B420 and Bi-07. *J. Bacteriol.*, **194**, 4131–4132.

96 Gudbergsdottir, S., Deng, L., Chen, Z., Jensen, J.V. (2011) Dynamic properties of the *Sulfolobus* CRISPR/Cas and CRISPR/Cmr systems when challenged with vector-borne viral and plasmid genes and protospacers. *Mol. Microbiol.*, **79**, 35–49.

97 Weinberger, A.D., Sun, C.L., Pluciński, M.M., Denef, V.J. (2012) Persisting viral sequences shape microbial CRISPR-based immunity. *PLoS Comput. Biol.*, **8**, e1002475.

98 Stern, A., Keren, L., Wurtzel, O., Amitai, G. (2010) Self-targeting by CRISPR: gene regulation or autoimmunity? *Trends Genet.*, **26**, 335–340.

99 Heidelberg, J.F., Nelson, W.C., Schoenfeld, T., Bhaya, D. (2009) Germ warfare in a microbial mat community: CRISPRs provide insights into the co-evolution of host and viral genomes. *PLoS ONE*, **4**, e4169.

100 Snyder, J.C., Bateson, M.M., Lavin, M., Young, M.J. (2010) Use of cellular CRISPR (clusters of regularly interspaced short palindromic repeats) spacer-based microarrays for detection of viruses in environmental samples. *Appl. Environ. Microbiol.*, **76**, 7251–7258.

101 Anderson, R.E., Brazelton, W.J., Baross, J.A. (2011) Using CRISPRs as a metagenomic tool to identify microbial hosts of a diffuse flow hydrothermal vent viral assemblage. *FEMS Microbiol. Ecol.*, **77**, 120–133.

102 Pride, D.T., Sun, C.L., Salzman, J., Rao, N. (2011) Analysis of streptococcal CRISPRs from human saliva reveals substantial sequence diversity within and between subjects over time. *Genome Res.*, **21**, 126–136.

103 Berg Miller, M.E., Yeoman, C.J., Chia, N., Tringe, S.G. (2012) Phage-bacteria relationships and CRISPR elements revealed by a metagenomic survey of the rumen microbiome. *Environ. Microbiol.*, **14**, 207–227.

104 Delaney, N.F., Balenger, S., Bonneaud, C., Marx, C.J. (2012) Ultrafast evolution and loss of CRISPRs following a host shift in a novel wildlife pathogen, *Mycoplasma gallisepticum*. *PLoS Genet.*, **8**, e1002511.

105 Rho, M., Wu, Y.W., Tang, H., Doak, T.G. (2012) Diverse CRISPRs evolving in human microbiomes. *PLoS Genet.*, **8**, e1002441.

106 Stern, A., Mick, E., Tirosh, I., Sagy, O. (2012) CRISPR targeting reveals a reservoir of common phages associated with the human gut microbiome. *Genome Res.*, **22**, 1985–1994.

107 Levin, B.R. (2010) Nasty viruses, costly plasmids, population dynamics, and the conditions for establishing and maintaining CRISPR-mediated adaptive immunity in bacteria. *PLOS Genet.*, **6**, e1001171.

108 Childs, L.M., Held, N.L., Young, M.J., Whitaker, R.J. (2012) Multiscale model of CRISPR-induced coevolutionary dynamics: diversification at the interface of Lamarck and Darwin. *Evolution*, **66**, 2015–2029.

109 Levin, B.R., Moineau, S., Bushman, M., Barrangou, R. (2013) The population and evolutionary dynamics of phage and bacteria with CRISPR-mediated immunity. *PLoS Genet.*, **9**, e1003312.

110 Sun, C.L., Barrangou, R., Thomas, B.C., Horvath, P. (2012) Phage mutations in response to CRISPR diversification in a bacterial population. *Environ. Microbiol.*, **15**, 463–470.

111 Semenova, E., Nagornykh, M., Pyatnitskiy, M., Artamonova, I.I. (2009) Analysis of CRISPR system function in plant pathogen *Xanthomonas oryzae*. *FEMS Microbiol. Lett.*, **296**, 110–116.

112 Bondy-Denomy, J., Pawluk, A., Maxwell, K.L., Davidson, A.R. (2013) Bacteriophage genes that inactivate the CRISPR/Cas bacterial immune system. *Nature*, **493**, 429–432.

113 Held, N.L., Whitaker, R.J. (2009) Viral biogeography revealed by signatures in *Sulfolobus islandicus* genomes. *Environ. Microbiol.*, **11**, 457–466.

114 Pride, D.T., Salzman, J., Relman, D.A. (2012) Comparisons of clustered regularly interspaced short palindromic repeats and viromes in human saliva reveal bacterial adaptations to salivary viruses. *Environ. Microbiol.*, **14**, 2564–2576.

115 Mills, S., Griffin, C., Coffey, A., Meijer, W.C. (2010) CRISPR analysis of bacteriophage-insensitive mutants (BIMs) of industrial *Streptococcus thermophilus* – implications for starter design. *J. Appl. Microbiol.*, **108**, 945–955.

116 Magadán, A.H., Dupuis, M.È., Villion, M., Moineau, S. (2012) Cleavage of phage DNA by the *Streptococcus thermophilus* CRISPR3-cas system. *PLoS ONE*, **7**, e40913.

117 van der Ploeg, J.R. (2009) Analysis of CRISPR in *Streptococcus mutans* suggests frequent occurrence of acquired immunity against infection by M102-like bacteriophages. *Microbiology*, **155**, 1966–1976.

118 Brüggemann, H., Lomholt, H.B., Tettelin, H., Kilian, M. (2012) CRISPR/cas loci of type II *Propionibacterium acnes* confer immunity against acquisition of mobile elements present in type I *P. acnes*. *PLoS ONE*, **7**, e34171.

119 Lopez-Sanchez, M.J., Sauvage, E., Da Cunha, V., Clermont, D. (2012) The highly dynamic CRISPR1 system of *Streptococcus agalactiae* controls the diversity of its mobilome. *Mol. Microbiol.*, **85**, 1057–1071.

120 Louwen, R., Horst-Kreft, D., de Boer, A.G., van der Graaf, L. (2012) A novel link between *Campylobacter jejuni* bacteriophage defence, virulence and Guillain-Barré syndrome. *Eur. J. Clin. Microbiol. Infect. Dis.*, **32**, 207–226.

121 Edgar, R., Qimron, U. (2010) The *Escherichia coli* CRISPR system protects from lambda lysogenization, lysogens, and prophage induction. *J. Bacteriol.*, **192**, 6291–6294.

122 Nozawa, T., Furukawa, N., Aikawa, C., Watanabe, T. (2011) CRISPR inhibition of prophage acquisition in *Streptococcus pyogenes*. *PLoS ONE*, **6**, e19543.

123 Touchon, M., Charpentier, S., Pognard, D., Picard, B. (2012) Antibiotic resistance plasmids spread among natural isolates of *Escherichia coli* in spite of CRISPR elements. *Microbiology*, **158**, 2997–3004.

124 Palmer, K.L., Gilmore, M.S. (2010) Multidrug-resistant enterococci lack CRISPR-cas. *MBio*, **1**, e00227–e00210.

125 Bikard, D., Hatoum-Aslan, A., Mucida, D., Marraffini, L.A. (2012) CRISPR interference can prevent natural transformation and virulence acquisition during *in vivo* bacterial infection. *Cell Host Microbe*, **12**, 177–186.

126 Manica, A., Zebec, Z., Teichmann, D., Schleper, C. (2011) In vivo activity of CRISPR-mediated virus defence in a hyperthermophilic archaeon. *Mol. Microbiol.*, **80**, 481–491.

127 Aklujkar, M., Lovley, D.R. (2010) Interference with histidyl-tRNA synthetase by a CRISPR spacer sequence as a factor in the evolution of *Pelobacter carbinolicus*. *BMC Evol. Biol.*, **10**, 230.

128 Jorth, P., Whiteley, M. (2012) An evolutionary link between natural transformation and CRISPR adaptive immunity. *MBio*, **3**, e00309–e00312.

129 Qi, L., Haurwitz, R.E., Shao, W., Doudna, J.A. (2012) RNA processing enables predictable programming of gene expression. *Nat. Biotechnol.*, **30**, 1002–1006.

130 Zegans, M.E., Wagner, J.C., Cady, K.C., Murphy, D.M. (2009) Interaction between bacteriophage DMS3 and host CRISPR region inhibits group behaviors of *Pseudomonas aeruginosa*. *J. Bacteriol.*, **191**, 210–219.

131 Palmer, K.L., Whiteley, M. (2011) DMS3-42: the secret to CRISPR-dependent biofilm inhibition in *Pseudomonas aeruginosa*. *J. Bacteriol.*, **193**, 3431–3432.

132 Perez-Rodriguez, R., Haitjema, C., Huang, Q., Nam, K.H. (2011) Envelope stress is a trigger of CRISPR RNA-mediated DNA silencing in *Escherichia coli*. *Mol. Microbiol.*, **79**, 584–599.

133 Cong, L., Ran, F.A., Cox, D., Lin, S. (2013) Multiplex genome engineering using CRISPR/Cas systems. *Science*, **339**, 819–823.

134 Mali, P., Yang, L., Esvelt, K.M., Aach, J. (2013) RNA-guided human genome engineering via Cas9. *Science*, **339**, 823–826.

135 Cho, S.W., Kim, S., Kim, J.M., Kim, J.S. (2013) Targeted genome engineering in human cells with the Cas9 RNA-guided endonuclease. *Nat. Biotechnol.*, **31**, 230–232.

136 Hwang, W.Y., Fu, Y., Reyon, D., Maeder, M.L. (2013) Efficient genome editing in zebrafish using a CRISPR-Cas system. *Nat. Biotechnol.*, **31**, 227–229.

137 Jiang, W., Bikard, D., Cox, D., Zhang, F. (2013) RNA-guided editing of bacterial genomes using CRISPR-Cas systems. *Nat. Biotechnol.*, **31**, 233–239.

138 Barrangou, R. (2012) RNA-mediated programmable DNA cleavage. *Nat. Biotechnol.*, **30**, 836–838.

139 Barrangou, R., van der Oost, J. (Eds) (2013) *CRISPR-Cas Systems: RNA-mediated Adaptive Immunity in Bacteria and Archaea*, Springer, ISBN: 978-3-642-34656-9.

5
RNA Regulation in Myogenesis

Andrie Koutsoulidou, Nikolaos P. Mastroyiannopoulos, and Leonidas A. Phylactou
The Cyprus Institute of Neurology & Genetics, Department of Molecular Genetics, Function & Therapy, P.O. Box 2346, 1683 Nicosia, Cyprus

1	**Myogenic Program** 142	
1.1	The Core Transcriptional Program of Skeletal Muscle Cell Differentiation 143	
1.2	Regulation of Skeletal Muscle Cell Differentiation 143	
2	**RNA Regulation in Myogenesis** 145	
2.1	mRNA Turnover Regulation 145	
2.1.1	Regulation of mRNA Turnover during Myogenesis 147	
2.2	Alternative Splicing 150	
2.2.1	Alternative Splicing during Myogenesis 151	
2.3	Regulation of mRNA Translation 153	
2.3.1	RNA-Binding Proteins Regulate mRNA Translation during Myogenesis 153	
2.4	Noncoding RNAs 154	
2.4.1	MicroRNAs (miRNAs) as Post-Transcriptional Regulators 154	
2.4.2	Long Noncoding RNA 159	
3	**Conclusions** 159	
	List of Abbreviations 160	
	References 161	

RNA Regulation: Advances in Molecular Biology and Medicine, First Edition. Edited by Robert A. Meyers.
© 2014 Wiley-VCH Verlag GmbH & Co. KGaA. Published 2014 by Wiley-VCH Verlag GmbH & Co. KGaA.

Keywords

Myogenesis
The process by which proliferating myoblasts differentiate into mature myotubes

mRNA turnover
The half-life of the mRNAs

Alternative splicing
The process by which the exons of the RNA are rejoined in different combinations

Noncoding RNA
An RNA molecule that is not translated into a protein

MicroRNA
A small noncoding regulatory RNA molecule that is endogenously expressed in eukaryotes

■ Skeletal muscle cell differentiation, also known as *myogenesis*, is a complex process that directs proliferating myoblasts to differentiate into mature myotubes which are eventually organized into myofibers. Myogenesis involves a series of regulatory events that occur at both transcriptional and post-transcriptional levels. For the proper control of myogenesis, a significant impact has been ascribed to the regulation of mRNAs that are specifically expressed in muscle tissue, as well as mRNAs that are essential for the formation and function of muscle. Regulation of the mRNAs' half-lives, alternative splicing and translation are the most important examples that occur during myogenesis. Regulatory noncoding RNAs, such as microRNAs, have been also determined to play a crucial role during the process of skeletal muscle cell differentiation. The precise and tight regulation of myogenesis at the mRNA level is essential for the correct function of muscle, and changes in the regulation of mRNAs frequently lead to diseases.

1
Myogenic Program

Skeletal muscle cell differentiation, known as *myogenesis*, is a powerful and tightly regulated process that directs myoblasts, the proliferating muscle cells, to differentiate into mature myotubes. During the course of myogenesis, the mononucleated myoblasts are stimulated to initiate the expression of a unique set of myogenic differentiation-specific genes, permanently withdraw from the cell cycle, and fuse together to form multinucleated myotubes which, ultimately, are organized into myofibers [1] (Fig. 1).

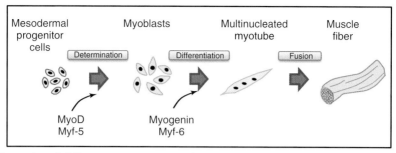

Fig. 1 Skeletal muscle cell differentiation. The myogenic determination myogenic regulatory factors (MRFs), MyoD, and Myf-5, are required for the commitment of the proliferating mesodermal progenitor cells to the myogenic lineage. The committed cells (myoblasts) proliferate and differentiate into multinucleated myotubes under the action of the differentiation MRFs, myogenin, and Myf-6. Multinucleated myotubes eventually fuse into muscle fibers.

1.1
The Core Transcriptional Program of Skeletal Muscle Cell Differentiation

Skeletal myogenesis is initiated and regulated mainly by the coordinate expression of a family of skeletal muscle-specific transcription factors, the myogenic regulatory factors (MRFs), which are essential for both the determination and maintenance of skeletal muscle. The first member of the MRFs to be identified was the *MyoD* gene [2]. MyoD was isolated due to its ability to convert non-muscle cells to stable myoblasts at high frequency [2–4]. Subsequently, three more MRFs – Myf-5, myogenin, and Myf-6 (or MRF4 or Herculin) – were identified and cloned [5–10]. In all cases, an overexpression of the MRFs converted a number of non-muscle cell lines, such as fibroblasts, to myoblasts, implying their crucial role in myogenic lineage determination and differentiation [11]. Each of the four MRFs was found to heterodimerize with E proteins, a family of ubiquitously and broadly expressed transcription factors, both *in vitro* and *in vivo*, and also to bind to the consensus hexanucleotide sequence 5′-CANNTG-3′ (where N corresponds to any nucleotide), termed *E-box*, which are present in the promoters of many skeletal muscle-specific genes and thereby initiate and promote myogenesis [12]. Furthermore, in some cases, MRFs were found to promote the myogenic program by forming homodimers and heterodimers with other myogenic factors, such as the myocyte enhancer factor (MEF), and the Id family of transcription factors.

1.2
Regulation of Skeletal Muscle Cell Differentiation

During muscle cell differentiation, myoblasts – the mononucleated proliferating muscle cells – are stimulated to permanently withdraw from the cell cycle and fuse together to form multinucleated myotubes, which in turn are ultimately organized into myofibers. Several studies have uncovered a firm relationship between the factors that regulate the progression and exit of the cell cycle and the expression of the myogenic transcription factors [13].

The cell cycle is a tightly regulated process that takes place in living cells, leading to their division and proliferation.

Its progression is controlled through different cyclins and cyclin-dependent kinases (cdks) which are expressed constitutively throughout the cell cycle and form active complexes, thus regulating the progress of the cell cycle [13]. Cyclin D1 preferentially activates cdk4, and is expressed during the transition from the G_1 interphase to the S phase, where DNA synthesis starts [13]. The expression of cdk4 and most cyclins, including Cyclin Dl, Cyclin A, Cyclin B2, and Cyclin C, was found to be reduced to undetectable levels in differentiated myotubes [14, 15]. Overexpression studies showed that Cyclin D1 inhibits activation of the transcription of muscle-specific genes, including *MyoD*, thus promoting cell cycle progression and an inhibition of muscle cell differentiation [14]. Interestingly, Cyclin D3 expression was shown to be induced during muscle cell differentiation [14, 15]. During the proliferation of myoblasts, the transcriptional activity of the MRFs is suppressed. Consistent with the notion that the ectopic expression of Cyclin D1 represses MyoD from activating muscle-specific genes, overexpression of the myogenic transcription factors inhibits cell cycle progression, suggesting that the myogenic transcription factors and the cell cycle regulatory proteins regulate one another's activities and functions [13].

During the withdrawal of myoblasts from the cell cycle, the expression of cyclin/cdk inhibitors and retinoblastoma protein (pRb) are upregulated. An upregulation of cdk inhibitors is necessary for terminal myogenic differentiation, and they have been reported to play an important role primarily in cell cycle withdrawal, but also in resistance to apoptosis and in the stability of MyoD [16, 17]. A well-established cyclin/cdk inhibitor involved during muscle cell differentiation is p21, which was determined to be markedly upregulated at the mRNA and protein levels, as well as at its activity levels during skeletal muscle cell terminal differentiation [18, 19]. Furthermore, p21 was found to coimmunoprecipitate and form complexes with cdk2 and cdk4 in myotubes [18, 20]. Overexpression of the p21 inhibitor induced muscle-specific gene expression in cells maintained in high concentrations of serum [21]. The induction of p21 was ascribed to MyoD transcription factor. Specifically, the expression of MyoD was shown to activate p21 expression during differentiation of murine muscle cells and in non-myogenic cell lines [18, 22]. A hallmark of p21 regulation during skeletal myogenesis is the maintenance of its expression and activity during restimulation with mitogen-rich growth media, which is consistent with the irreversible nature of cell cycle exit that accompanies the terminal skeletal muscle cell differentiation [18, 19, 22]. In addition to p21, other cyclin/cdk inhibitors that possibly function to initiate and/or maintain the post-mitotic state of muscle cells have been determined. For example, p18 is a cdk inhibitor found to be highly expressed in skeletal muscle cells, and its expression was determined to be markedly increased during *in vitro* muscle cell differentiation [23]. These observations suggested that p18 may play an important role during initiation and/or maintenance of the permanent cell cycle arrest associated with the terminal skeletal muscle cell differentiation [23].

The *Rb* gene is a tumor-suppressor gene that is known to inhibit cell cycle progression and maintain the post-mitotic state of the cells [24]. Rb activity is regulated by cdks. In particular, in cycling cells cdks phosphorylate Rb and inhibit its cell cycle inhibitory activity by blocking the interaction of Rb with other cellular and viral proteins that promote cell proliferation [25]. These include transcription factors of the E2F family, which are required to activate many of the genes necessary for cell proliferation [25]. The decision of cell division is accompanied by the dephosphorylation of Rb protein, which in turn binds and sequesters E2F transcription factors, thus inhibiting cell cycle progression [13]. Electrophoretic mobility shift assays have shown that E2F–pRB complexes are present in fully differentiated myotubes in the mouse muscle cell line C2C12 and in isolated skeletal muscle [26]. Moreover, *in vivo* studies in mice have shown that Rb protects myotubes from apoptosis [27].

The permanent exit of the proliferating myoblasts from the cell cycle is accompanied by a cascade of events in order to form the muscle fibers. The four MRFs are the master regulatory factors during the process of myogenesis. *In vivo* analyses using gene targeting experiments have provided much insight into the role and functions of MyoD, Myf-5, myogenin, and Myf-6 during myogenesis [1]. The generation of mice with null mutations in the four MRFs, whether individually and in combinations, have provided evidence of a hierarchical relationship among the four MRFs during the process of skeletal muscle formation [1]. In general, MRFs can be divided into two subgroups, depending on their function and time of expression. The early expressed primary MRFs – MyoD and Myf-5 – are responsible for the establishment of muscle identity, whereas the later-expressed secondary MRFs – myogenin and Myf-6 – play a crucial role as differentiation factors [1] (Fig. 1). Additionally, gene targeting experiments have provided evidence for the existence of potential redundancy among the MRFs, since some MRFs can substitute for one another's functions [1, 28].

2
RNA Regulation in Myogenesis

The regulation of gene expression is crucial to achieve precise developmental- and tissue-specific control of all the cellular processes. Gene expression is regulated at various transcriptionally and post-transcriptionally levels, including gene transcription, RNA splicing, RNA polyadenylation and capping, RNA trafficking, RNA stability, RNA translation, protein processing, and protein stability. Genes are transcribed in the nucleus into long heterogeneous nuclear RNAs, known as *pre-mRNAs*, which serve as precursors for smaller 5′ capped and 3′ polyadenylated mRNAs that are then exported to the cytoplasm [29]. In recent years, microarray analysis and high-throughput sequencing have revealed that an extensive complicated regulation occurs during post-transcriptional pre-mRNA processing, thereby providing new insights into the understanding of molecular cell biology and disease [29]. The myogenic program has been determined to be tightly regulated at the mRNA level (Fig. 2).

2.1
mRNA Turnover Regulation

Among the post-transcriptional regulation events, the control of mRNA turnover and stability is emerging as a crucial paradigm of gene regulation. It is an essential control mechanism during cell division and cellular differentiation. Through the regulation of the mRNA half-life, the overall gene expression is controlled and, as a result, the levels of protein expression in the cell

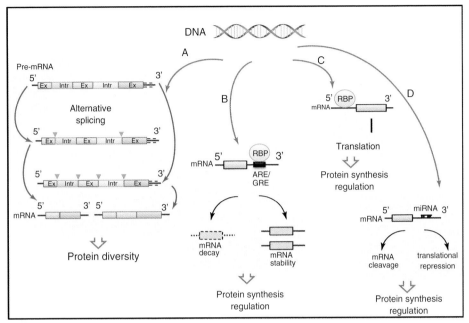

Fig. 2 RNA regulation in myogenesis. The mRNAs of many myogenic factors are regulated during myogenic program through several mechanisms. Route A: Pre-mRNAs are alternative spliced to produce different mRNAs, thus promoting protein diversity; Route B: The turnover of mRNAs is regulated through either decreasing or increasing their half-life, thus leading to mRNA decay or increasing mRNA stability, respectively. Specific RNA-binding proteins (RBPs) bind to AREs and/or GREs present in the 3′UTR regions of the target mRNAs, thus affecting the mRNAs' half-lives and consequently protein synthesis; Route C: RBPs also influence the mRNA translation during muscle cell differentiation by binding to 5′ regions of the target mRNAs. As a result, synthesis of the proteins is regulated; Route D: MiRNAs bind to the 3′UTRs of their target mRNAs and cause either mRNA cleavage or repression of protein translation, thereby inhibiting protein synthesis.

are also regulated. Aberrant regulation of mRNA turnover control contributes to disorders such as malignancy, inflammation, and immunopathology [30]. The mRNA stability is regulated by specific sequences, referred to as *cis-acting elements*, located in the 3′ untranslated region (3′UTR) of RNA, together with specialized proteins, known as *trans-acting factors*. *Trans*-acting factors bind to *cis*-acting elements and thereby control gene expression. The most important *trans*-acting factors that regulate the mRNA turnover are the RNA-binding proteins (RBPs). Best characterized among the *cis*-acting elements that influence mRNA stability include AU-rich elements (AREs) and GU-rich elements (GREs) that are usually found in the 3′UTR of labile mRNAs and play an important role in the regulation of 3′ processing. Some ARE-binding proteins, such as tristetraprolin, butyrate response factor 1 (BRF1) and KH-domain Splicing Regulatory Protein (KSRP), promote the mRNA decay, whereas others – including hu antigen R (HuR) – promote mRNA stabilization. However, there are ARE-binding proteins (such as AUF1) that promote either mRNA decay or stabilization, depending on the cellular context or the expression of individual isoform [31].

2.1.1 Regulation of mRNA Turnover during Myogenesis

The tight regulation of mRNA turnover is an important regulation in all cell types. In skeletal muscle, several mRNAs which encode for critical myogenic molecules or signaling factors that are involved in myogenic program have been determined to be regulated through either decreasing or increasing their half-lives, thus leading to mRNA decay or increasing mRNA stability, respectively. Aberrant muscle mRNA turnover regulation has been correlated to many diseases such as sarcopenia and oculopharyngeal muscular dystrophy [32, 33].

mRNA Decay during Muscle Differentiation
The process of myogenic differentiation includes the activation of a series of signal transduction pathways. A signaling pathway that plays an essential role in muscle cell differentiation involves p38 mitogen-activated protein kinase (MAPK). The p38 MAPK signaling pathway functions in many cellular processes including cell growth, cell cycle arrest, cell differentiation, and apoptosis [34]. Many of the downstream targets of p38 signaling are lineage-specific or play important roles during development [34]. Several studies have established a key role for p38 MAPK pathway in skeletal muscle differentiation. Specifically, p38 activation was found to induce the upregulation of myogenic factors and accelerate myotube formation. p38 was found to enhance the transcriptional activities of MEF2A and MEF2C by direct phosphorylation, in muscle cells *in vivo* [35, 36]. Moreover, p38 was shown to phosphorylate the E47, a member of the E proteins family which is a partner of MyoD, and thereby inducing the formation of MyoD–E47 heterodimers. These heterodimers bind to the E-boxes located on muscle-specific promoters and activate the transcription of muscle-specific genes [37]. Additionally, p38 MAPK pathway was found to regulate muscle-specific promoters by targeting chromatin-remodeling enzymes [38]. Briata and coworkers have shown that the activation of p38 causes the stabilization of labile myogenic transcripts in C2C12 myoblasts through the phosphorylation of the KSRP RBP [39]. C2C12 cells are a well-established mouse model system for skeletal muscle differentiation *in vitro*. In growth medium containing high-serum, C2C12 cells proliferate as mononuclear myotubes and, upon induction of differentiation in a low-serum differentiation medium, they fuse to form multinuclear myotubes. KSRP is an ARE-binding protein which interacts with the exosome and the poly(A) ribonuclease (PARN) deadenylase, thus promoting the rapid decay of several ARE-containing mRNAs both *in vitro* and *in vivo* [40]. In proliferating C2C12 myoblasts, KSRP was found to bind to the AREs present on MyoD, myogenin and p21 mRNAs, thus recruiting the decay machinery and promoting rapid mRNA decay [39, 41]. Upon muscle cell differentiation, p38 was shown to phosphorylate KSRP and impair its interaction with ARE-containing mRNAs. As a result, its mRNA destabilization function was diminished [39].

In addition to the AREs, many of the labile mRNAs expressed in muscle also contain GREs in their 3′UTRs [42]. The mRNAs that are enriched for GREs in myoblasts were found to be associated with CUG-binding protein 1 (CUGBP1) [42]. CUGBP1 is an RBP which has been extensively studied for its role as a regulator of alternative splicing. CUGBP1, however, was also found to influence mRNA turnover through interactions with GREs located in the 3′UTRs of its target mRNAs [43]. Specifically, CUGBP1 was determined to bind to the GREs of its mRNA substrates and to recruit deadenylases, such as PARN, thus promoting rapid poly(A) shortening and mRNA decay [44]. The muscle mRNAs associated with CUGBP1 were determined to have short lives and to encode factors involved in cell cycle regulation, the regulation of transcription, the establishment and maintenance of chromatin architecture, RNA processing, protein localization, signaling, and apoptosis [44]. Stable knockdown of CUGBP1 in C2C12 cells showed a significant stabilization of four unstable transcripts, *Ppp1r15b*, *Rnd3*, *Smad7*, and *Myod1*, with the latter being dramatically stable during myogenic differentiation through an association with HuR RBP [44, 45]. Smad7 was found to play an important role during myogenesis initiation through a Smad7–MyOD positive feedback loop and abrogation of the myostatin-mediated repression of muscle cell differentiation [46]. Remarkably, a subset of CUGBP1-associated mRNAs was found to be bound by HuR and/or Pumilio homolog 1 RBP in other cell types, which implied the existence of coordinated or competitive binding of RBPs during mRNA turnover regulation [44]. The transcripts bound by both CUGBP1 and HuR encode for factors involved in cell cycle regulation, post-transcriptional regulation and mRNA processing, whereas the transcripts bound by CUGBP1 and Pumilio homolog 1 encode for factors linked with cell proliferation [44].

mRNA Stability during Muscle Differentiation MyoD, myogenin, and the cyclin-dependent inhibitor p21 expression occurs concurrently in differentiating muscle cells and regenerating adult muscle through transcriptional mechanisms and post-transcriptional mRNA stabilization by specific RBPs [45, 47, 48]. MyoD, myogenin, and p21 mRNAs contain AREs in their 3′UTRs. The RBP HuR was shown to interact with the AREs located on the MyoD, myogenin, and p21 mRNAs, thus regulating their stability which is necessary for the process of muscle cell differentiation [45]. HuR is the more ubiquitously expressed member of the Hu-family of proteins, which are the homologs of "*Drosophila*" embryonic lethal abnormal visual protein, and characterized as an mRNA-stabilizing RBP [49–51]. The importance of HuR during the process of muscle cell differentiation was further supported by its subcellular localization throughout the process. Specifically, in actively proliferating C2C12 myoblasts, HuR was determined to be almost exclusively localized to the nucleus whereas, upon induction of differentiation, HuR was dramatically accumulated in the cytoplasm where it remained during the process of differentiation [45]. It finally returned to the nucleus upon the completion of muscle cell differentiation [45]. Similarly, *in vivo* experiments showed that cytoplasmic HuR increases during skeletal muscle regeneration and declines when regeneration is completed [45]. These observations coincide with the expression of the MyoD, myogenin, and p21 proteins. Overexpression studies have revealed that C2C12 cells expressing elevated levels of HuR had higher steady-levels of MyoD, myogenin, and p21 mRNAs compared to control cells. Furthermore, HuR overexpression resulted in myotube formation earlier in the process of differentiation [45].

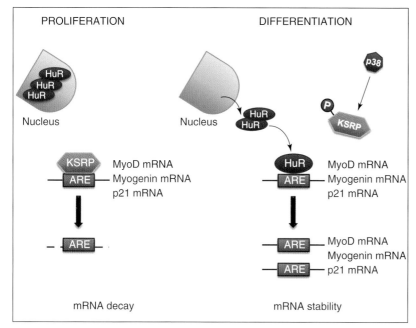

Fig. 3 Interplay between HuR and KSRP RNA-binding proteins. During myoblast proliferation, HuR RNA-binding protein is localized to the nucleus, whereas KSRP RNA-binding protein binds to AREs of MyoD, myogenin and p21 mRNAs and promotes rapid mRNA decay. Upon differentiation, p38 phosphorylates KSRP and diminishes its mRNA destabilization function. HuR, however, accumulates in the cytoplasm, binds to MyoD, myogenin and p21 mRNAs and increases their stability.

Considering that, during muscle cell proliferation, KSRP was found to promote a rapid mRNA decay of MyoD, myogenin and p21, and upon differentiation KSRP was shown to become phosphorylated and its mRNA destabilization function to diminish, it can be concluded that an interplay exists between HuR and KSRP in the regulation of MyoD, myogenin, and p21 mRNA decay which regulates the transition of the proliferating myoblasts to differentiation [39, 41] (Fig. 3).

MyoD and p21 mRNAs were also found to interact specifically with a second RBP known as *NF90* [48]. The latter is predominantly localized in the nucleus of diverse cells and tissues, including skeletal muscle [48]. *NF90*-deficient mice displayed a perinatal lethality associated with muscle weakness, thereby revealing a role of NF90 in myogenic differentiation that may possibly be explained by its role in MyoD and p21 mRNA stability [48].

Myogenin mRNA stability was also found to be regulated by the RNA-binding motif protein 24 (Rbm24) [52]. Rbm24 is an RBP found to be preferentially expressed in cardiomyocytes derived from embryonic stem cells, as well as in adult cardiac and skeletal muscle tissues [53, 54]. Knockdown and overexpression studies have shown that Rbm24 plays a critical role in myogenesis by promoting muscle cell differentiation *in vitro* [54]. Jin and coworkers used RBP immunoprecipitation followed by microarray analysis to show that myogenin is one of the potential targets of Rbm24 during C2C12 muscle cell differentiation [52]. Subsequent *in vitro* experiments proved that Rbm24 increases myogenin mRNA stability through direct binding to the 3′UTR of the myogenin mRNA and thereby controls, at least in part, the course of myogenic differentiation [52]. A paralog of Rbm24 – Rbm38, which is also known as *RNPC1* – was found to promote muscle cell differentiation through the regulation of cell cycle arrest [54]. Specifically, RBP immunoprecipitation assays showed that Rbm38 binds directly to the p21 transcript and increases its stability, thereby inducing cell cycle arrest and consequently promoting myogenic differentiation *in vitro* [54, 55].

The alternative spliced transcript of acetylcholinesterase (AChE$_T$) was found to be predominantly expressed in skeletal muscle where it forms non-amphiphilic tetramers that interact with collagen tail subunit required for the neuromuscular junctions [56]. Although the levels of AChE transcripts were determined to be dramatically elevated *in vitro* during myogenic differentiation, this cannot be attributed only to an increased transcriptional activity [57, 58]. Other post-transcriptional mechanisms have been determined to regulate the elevated levels of the AChE transcripts in differentiating muscle cells [58]. Specifically, AChE mRNAs were found to be stabilized by the direct interactions of HuR with AREs located on the 3′UTR of the AChE mRNA in differentiated muscle cells [58]. Overexpression and *AChE* gene silencing experiments verified that post-transcriptional regulation of AChE by HuR RBP plays important role in the later stages of muscle cell differentiation [58].

2.2
Alternative Splicing

In higher eukaryotes, most genes that encode proteins contain long noncoding or intervening sequences, known as *introns*, that are transcribed in the pre-mRNA and removed in a fundamental process called *splicing*; the latter process involves the appropriate intramolecular joining of the protein-coding sequences, the exons [59]. Splicing has revealed the enormous complexity and fundamental importance of this process in the regulation of eukaryotic gene expression [59]. It is a precise two-step transesterification reaction that is catalyzed by the spliceosome, a large multicomponent and highly dynamic complex [59–61]. The spliceosome complex consists of five small nuclear ribonucleoprotein particles (snRNPs), each of which contains a uridine-rich small nuclear RNA and several proteins [60]. Additionally, it contains 100–200 non-snRNP protein components [60]. Various components of the spliceosome recognize the *cis*-acting sequences that direct the process of splicing [60].

A single pre-mRNA can be potentially spliced in multiple different patterns to produce two or more distinct mRNA variants, a process known as *alternative splicing*. This powerful process contributes significantly to the complexity of the proteome and the diversity of cell- and tissue-specific protein expression profiles, and also results in an expanded transcriptome [62, 63]. Deep sequencing-based expression analyses in human tissues have shown that approximately 95% of multi-exon genes in humans are subjected to alternative splicing [64]. The process of alternative splicing is primarily regulated by a variety of RBPs that bind pre-mRNAs near variably used splice sites and modulate the efficiency of their recognition by the spliceosome, as a means of regulating the process [65]. A number of *cis*-elements are involved in the regulation of alternative splicing [60]. Indeed, alterations in the *cis*- or *trans*-regulation of alternative splicing can cause multiple pathologies as a result of general or specific aberrant pre-mRNA processing, thus underscoring the fundamental importance of this regulatory process [66–69]. In addition, cooperative or antagonistic action between the regulatory factors modulates the alternative splicing events.

2.2.1 Alternative Splicing during Myogenesis

Alternative splicing is enriched in skeletal muscle with a number of genes expressed in muscle to undergo the process. By using splicing sensitive microarrays, Bland and colleagues identified about 100 alternative spliced regions that are subjected to robust alternative splicing events during C2C12 differentiation [70]. Most of the strongly regulated splicing transitions tested in C2C12 were found to be dependent upon and integrated into the myogenic differentiation program [70]. Furthermore, alternative splicing events were also found to be induced by cell cycle withdrawal, independent of the myogenic program, and implying that separate alternative splicing networks are regulated by distinct mechanisms during muscle differentiation [70]. Computational analysis allowed the identification of multiple sequence motifs that are significantly enriched and conserved within the intronic regions surrounding the regulated alternative regions. Many of the alternative splicing events that occur during myogenic program showed different position- and time-dependent effects for several known splicing regulators. A number of RBPs have been shown to be involved in the regulation of alternative splicing occurring in muscle. The CELF (CUG-BP and ETR-3-like factor) family, polypyrimidine tract binding protein (PTB or PTBp1), muscleblind-like (MBNL) family, heterogeneous nuclear ribonucleoprotein H and Forkhead box (FOX) family are examples of the RBPs determined to regulate muscle alternative splicing events [71, 72]. Some of the alternative splicing factors undergo changes in nuclear abundance, implying that they play important roles in regulating myogenic alternative splicing events [70].

In vertebrates, three MBNL homologs have been identified, namely MBNL1, MBNL2 and MBNL3 [73]. MBNL1 and MBNL3 were each found to act antagonistically during terminal myogenic differentiation. Specifically, MBNL1 expression was found to increase significantly during myogenic differentiation and act as a pro-myogenic factor, whereas MBNL3 was found to decrease during myogenic differentiation and function as an inhibitor of terminal muscle differentiation [74–77]. *In vitro* experiments have shown that MBNL3 inhibits the transcription of various muscle-specific genes, including MEF2A, and disrupts MyoD-dependent gene transcription in differentiating C2C12 myoblasts [77]. MEF2 is an essential family of transcription factors that function with MyoD to regulate muscle-specific gene transcription and promote myogenic differentiation. In vertebrates, four MEF2 proteins have been identified – MEF2A, MEF2B, MEF2C, and MEF2D – each of which is encoded by a unique gene [78–81]. All of the MEF2 transcription factors contain a common DNA-binding and dimerization domain, but MEF2 transcripts are alternatively spliced within their transactivation domain and produce splicing isoforms [77, 82]. Specifically, the *MEF2* gene contains a short and highly conserved β-exon between exons 6 and 7 which is alternatively spliced into the mature message [82]. During muscle cell differentiation, each *mef2* gene isotype was shown to be subjected to alternative splicing to include the β-exon and produce Mef2 isoforms that are expressed predominantly in striated muscle and in the brain [82]. Lee and coworkers showed that MBNL3 binds directly to *Mef2D* intron 7 sequences and inhibits β-exon splicing during muscle differentiation in C2C12 myoblasts [82].

Chloride channel 1 (CLCN1) is a voltage-gated CLCN enriched in muscle which plays important role in stabilizing the resting potential of muscle membrane [83]. Alternative splicing events regulate the expression and function of CLCN1 in skeletal muscle [83]. The inclusion of exon 7A of the *CLCN1* gene results in reduced CLCN1 function [83]. Aberrant alternative splicing of the *CLCN1* gene leads to disorders such as myotonic dystrophy type 1 (DM1). Specifically, in DM1 patients the abnormal inclusion of alternative spliced exons 6B and/or 7A and the retention of intron 2 of *CLCN1* gene have been reported. All three MBNL proteins, MBNL1, MBNL2, and MBNL3, were shown to inhibit the inclusion of exon 7A [83]. Specifically, MBNL1 was shown to directly interact with the 5′ end of exon 7A which contains the UGCU(U/G)Y (where Y corresponds to C or U) motif that has been reported as a consensus binding sequence for MBNL1 [83]. In contrast, CELF3, CELF4, CELF5, and CELF6 RBPs, were determined to promote the inclusion of exon 7A, thus reducing the function of CLCN1 [83]. Among these CELF proteins, CELF4 is widely expressed in many tissues including muscle [71]. The inclusion of exon 7A of the *CLCN1* gene by CELF proteins might possibly play a role in keeping the expression of *CLCN1* gene at low levels in tissues such as brain, through a splicing-mediated regulation of gene expression [83]. CUGBP1 and ETR-3 were not detected to affect the inclusion of exon 7A of *CLCN1* gene [83]. Further analysis of the regulatory mechanism of exon 7A splicing revealed that the antagonistic regulation by MBNL1 and CELF4 is mediated by distinct regions of the *CLCN1* gene [83].

Tropomyosin is a protein that associates with actin in muscle fibers and controls the binding of myosin, thus regulating muscle contraction. The alternative splicing of α-tropomyosin results in nine isoforms, two of which – the skeletal and smooth muscle isoforms – differ by the selective inclusion of exon 2 and terminal exon 9 [84]. Rbm4 was found to be highly expressed in skeletal muscle and to promote the inclusion of the skeletal muscle-specific α-tropomyosin exons [85]. Furthermore, Rbm4 was shown to compete with a common PTB, binding to intronic CU-rich elements adjacent to the regulated exons and thereby antagonizing the suppressive activity of splicing factor PTB in selection of the skeletal muscle-specific α-tropomyosin exon, providing a hint that Rbm4 may modulate muscle cell type-specific splicing of other transcripts during the myogenic program [85, 86]. β-tropomyosin contains two exons – exon 6A and exon 6B – that are spliced in a mutually exclusive manner, depending on the tissue [87]. mRNAs from non-muscle cells, smooth muscle and myoblasts retain exon 6A, whereas mRNAs from striated muscle cells and myotubes retain exon 6B [87]. Interestingly, the inclusion of exon 6A or 6B was found to be dependent on the myogenic differentiation program. Specifically, during the course of the myogenic program, the splicing pattern from exon 6A inclusion was shown to switch towards exon 6B inclusion [87]. A complex array of regulatory elements located upstream and downstream of the muscle-specific exon 6B was shown to exclude its splicing in non-muscle cells [88–91]. PTB is an RBP that favors binding to UUCU elements which often exist within the polypyrimidine tracts of introns [86]. PTB was found to bind upstream of exon 6B and to mediate its splicing repression from β-tropomyosin pre-mRNA in HeLa cells, both *in vitro* and *ex-vivo* [92]. During myogenic differentiation, however, the CELF proteins, CUGBP1 and ETR3, were found to act on exon 6B and activate its inclusion [87]. The regulation of exon 6B inclusion during myogenic differentiation, however, was ascribed not only to CELF proteins but also to PTB. Specifically, CUGBP1 and PTB were determined to act antagonistically for the inclusion of exon 6B during myogenic differentiation, possibly due to changes in protein levels and/or the transition of PTB isoforms during myogenic program [87].

Fox-1 is a tissue-specific RBP that binds specifically to a pentanucleotide UGCAUG sequence *in vitro* [93]. In zebrafish, *fox-1* is expressed during muscle development, whereas its mouse homolog gene is expressed in brain, heart, and skeletal muscle [93]. Jin and colleagues showed that Fox-1 induces a muscle-specific splicing of human mitochondrial ATP synthase γ-subunit (F1γ) pre-mRNA through binding to the GCAUG sequence [93]. Human F1 mRNA was found to be alternatively spliced in a tissue-specific manner [93]; specifically, exon 9 was found to be excluded from the splicing product in a muscle-specific manner [93]. Fox-1 was determined to induce muscle-specific exon skipping of mammalian F1 transcript, irrespective of the type of cell [93].

2.3
Regulation of mRNA Translation

The translation of mRNA into protein represents the final step in the gene expression pathway which can be regulated by a number of mechanisms, usually at the level of initiation of the process [94]. RBPs have been determined to bind to transcribed mRNAs and to regulate the initiation of translation by a variety of mechanisms, either by activating or repressing translation [95, 96].

2.3.1 RNA-Binding Proteins Regulate mRNA Translation during Myogenesis

The RBPs, which act as *trans*-acting factors in mRNA regulation mechanisms, have been also determined to influence mRNA translation during muscle cell differentiation. For instance, CUGBP1 was found to increase the translation of p21 and Mef2a transcripts, which are both critical factors during the course of myogenic differentiation [95, 96]. Specifically, Timchenko and coworkers showed that CUGBP1 binds to GCN repeats present in the 5′ region of p21 mRNA during muscle cell differentiation, and thereby increases translation of the p21 cell cycle inhibitor in both a cell-free translation system and in cultured cells [95]. The effect of CUGBP1 on p21 mRNA translation was also determined to be altered in DM1 muscle cells due to a failure of the cells to accumulate CUGBP1 in the cytoplasm [95]. Consequently, CUGBP1 could not bind to p21 mRNA, and DM1 muscle cells failed to increase p21 protein levels during differentiation, probably contributing to the muscle cell differentiation impairment observed in DM1 [95]. Further studies performed on transgenic mice overexpressing CUGBP1 showed that, in addition to the p21 mRNA, CUGBP1 also induces the translation of Mef2a mRNA in skeletal muscle [96]. Specifically, CUGBP1 was found to bind GCA_9 located on Mef2a mRNA and to induce its translation both *in vitro* and *in vivo* [95, 96]. Knockdown and *in-situ* experiments have shown that CUGBP1 is a critical factor required for the increase of MEF2A and p21 protein levels during the differentiation course of C2C12 myoblasts, thus promoting myogenesis [96]. Therefore, the abnormal overexpression of CUGBP1 observed in DM1 could possibly contribute to the development of muscle deficiency through an induction of MEF2A and p21 translation which, in turn, inhibit myogenesis [96].

2.4
Noncoding RNAs

One of the greatest surprises of the high-throughput transcriptome analyses conducted during the past years has been the discovery that the mammalian genome is pervasively transcribed into many different complex families of RNA. Indeed, it is becoming largely accepted that the noncoding portion of the genome, rather than its coding counterpart, is likely to account for the greater complexity of higher eukaryotes. Many new functions have been assigned to noncoding RNAs, both in the nucleus and the cytoplasm.

2.4.1 MicroRNAs (miRNAs) as Post-Transcriptional Regulators

A novel class of small endogenous noncoding regulatory RNA molecules is that of microRNAs (miRNAs), which regulate the expression of genes post-transcriptionally. These have been found in a wide range of eukaryotes, and to date a wide range of regulatory functions has been attributed to miRNAs such as the regulation of numerous biological processes, including development, cell fate determination, cell proliferation and differentiation, signal transduction, apoptosis, and organ development [97]. MiRNAs precisely regulate the timing of a cellular event, either alone or in cooperation with other miRNAs. Moreover, a single miRNA can regulate several distinct mRNAs, thus functioning as an efficient molecular switch [98]. A number of miRNAs are expressed in a tissue- or developmental stage-specific manner [99]. Among many other tissues, miRNAs were determined to be expressed in skeletal muscle, where they were found to play a critical role during myogenesis [100–103].

Biogenesis and Mechanisms of Function of miRNAs miRNA is initially transcribed in the cell nucleus by RNA polymerase II as a long primary miRNA (pri-miRNA). The pri-miRNA transcript is subsequently cleaved in the nucleus by Drosha (an RNAse III enzyme) and its partner DGCR8/Pasha, to produce a hairpin structure of approximately 60–100 nt in length, termed the precursor miRNA (pre-miRNA). Pre-miRNA is then exported into the cytoplasm via Exportin-5 in a Ran-GTP-dependent manner [104]. In the cytoplasm, pre-miRNA is transferred into the RNA-induced silencing complex (RISC) for further processing to produce the mature miRNA. The cutting enzyme of the RISC is Dicer, which further cleaves the pre-miRNA to generate a short double-stranded miRNA product of about 22 nt length, consisting of the mature miRNA "guide" strand and the "passenger" miRNA strand [105]. Within the RISC, Dicer cooperates with other proteins, including the members of the Argonaute (AGO) family, of which only Ago2 has the capacity to cleave the target mRNA, the human immunodeficiency virus-1 transactivation response RBP, which is required for the Ago2 recruitment to the RISC, PW182, a P-body protein and possibly other proteins [106–112]. Following cleavage of the pre-miRNA, the miRNA:miRNA* duplex is unwound and the "guide" strand is incorporated onto AGO protein to form the programmed miRNA-induced silencing complex (miRISC) [113]. Driven by partial complementarity between the miRNA and its target mRNA binding site, two to eight nucleotides of the mature miRNA, which comprises the "seed" region of the miRNA, bind to their complementary sequences present in the 3′UTR of their target mRNAs. This results in protein repression, either by an inhibition of protein translation or by cleavage of the target mRNA [105, 113–115]. Protein translation or mRNA degradation through the action of miRISC depends on the level of sequence complementarity, or on whether the specific AGO protein is of a cleaving or noncleaving nature [116] (Fig. 4).

miRNAs in Skeletal Muscle Differentiation

An ever-growing number of miRNAs has been found to be expressed in muscle tissue during muscle cell differentiation *in vitro*, in embryonic muscles in the developing myotome, and in adult muscle during regeneration [117–119]. The greatest amount of information regarding the role of miRNAs during myogenesis has been obtained from studies of four muscle-specific miRNAs, miR-1, miR-133a, miR-133b, and miR-206, which were all determined to be induced during muscle cell differentiation [101–103]. However, an increasing number of miRNAs continue to be discovered in muscle and to play important roles in either promoting or inhibiting myogenesis. For instance, miR-181 was found to be expressed in embryonic stem cells and in C2C12 cells during muscle differentiation [120]. MiR-181 was determined to be required for skeletal muscle cell terminal differentiation, during which it was also found to regulate the expression of homeobox protein Hox-A11, a protein which was involved in skeletal muscle development and was also shown to repress MyoD expression [120, 121]. MiR-221 and miR-222 were found to be highly expressed in proliferating myoblasts and to be downregulated in differentiated myotubes in various muscle cells, thus playing a role in the progression of muscle cell differentiation and the establishment of the differentiated phenotype [122]. Another miRNA that was shown to be induced during muscle differentiation and to promote myogenesis was miR-24 [123]. Transforming growth factor-β was found to inhibit myogenesis via an inhibition of miR-24 at the transcriptional level [123]. MiR-214 expression was also determined to be upregulated during the differentiation of C2C12 and mouse primary cells, while its overexpression accelerated muscle gene expression and promoted myogenic differentiation [124]. MiR-378 was shown to be upregulated during C2C12 differentiation, and its expression levels to be regulated by MyoD [125]. The induction of miR-378 also promoted muscle cell differentiation by repressing MyoR, an antagonist of MyoD [125, 126]. In contrast, miR-155 expression was determined to decrease during C2C12 differentiation,

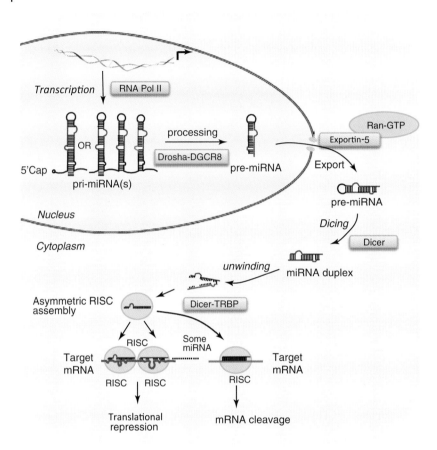

Fig. 4 MiRNA biogenesis and function. MiRNAs are transcribed in the nucleus by RNA pol II as pri-miRNAs, which are processed by Drosha-DGCR8 complex into pre-miRNA. Pre-miRNA is exported to the cytoplasm, via Exportin-5 in a Ran-GTP-dependent manner, where it is diced by Dicer to generate a miRNA duplex. Following processing, the mature miRNA duplex is preferentially incorporated into the RISC where it binds its target mRNA causing either repression of translation or degradation of the target mRNA.

and its overexpression was shown to inhibit muscle differentiation by repressing MEF2A expression [127].

The Involvement of Muscle-Specific miRNAs, miR-1, miR-133a, miR-133b, and miR-206, in Skeletal Muscle Differentiation
Among the miRNAs found to be expressed in muscle, the most extensively studied have been the so-called *myomiRs*, all of which are members of the miR-1/miR-206 and miR-133a/133b families. In humans, these miRNAs originate from three distinct loci, and are organized under bicistronic clusters on each chromosome. Specifically, there are three bicistronic clusters: miR-1-1/133a-2 clustered on chromosome 20; miR-1-2/133a-1 clustered on chromosome 18; and miR-206/133b clustered on chromosome 6 [128]. MiR-1-1 and miR-1-2 differ in their pre-miRNA sequences but have

identical mature sequences. The mature miR-1 nucleotide sequence differs from that of miR-206 by 4 nucleotides [103]. MiR-133a-1 and miR-133a-2 also have the same mature sequences, and differ from miR-133b by a single nucleotide at the 3'end of miR-133b [129]. Both, miR-1 and miR-133a are highly expressed in skeletal and cardiac muscles, whereas miR-206 and miR-133b are expressed specifically in skeletal muscle [101, 103]. All members of the miR-1/miR-206 and miR-133 families were determined to be induced during skeletal muscle differentiation, implying a critical role in the regulation of the process [101, 103, 130]. Interestingly, although all of these miRNAs are expressed during muscle differentiation, they appeared to have opposing effects during the process. Specifically, cell culture experiments showed that miR-1 and miR-206 promote muscle cell differentiation, whereas miR-133 drives myoblast proliferation [101, 103]. Additionally, the *in vitro* adenoviral overexpression of miR-1 and miR-206, both individually and in combination, as well as myosin heavy chain (MyHC) immunostaining experiments in satellite cells, showed that each miRNA restricts satellite cells at the proliferation stage and thus positively regulates their differentiation [130]. Moreover, miR-1 and miR-206 knockdown experiments have verified the important regulatory role of these miRNAs in satellite cell proliferation *in vivo* [130]. The mechanisms of these opposite effects were ascribed to the different miRNAs that are targeted by miR-1, miR-133a, miR-133b, and miR-206.

There is strong evidence that *miR-1, miR-133a, miR-133b*, and *miR-206* gene expression is directly regulated by the MRFs and key myogenic factors in muscle development [131]. For this reason, they can be considered good candidates for regulation of the myogenic program. Chromatin immunoprecipitation (ChIP)-on-ChIP analyses have shown that MyoD and myogenin transcription factors bind to the upstream regions of these myogenic miRNAs in differentiated C2C12 cells, and possibly mediate their upregulation detected in differentiated myotubes [132, 133]. Gain-of-function and double *in-situ* hybridization experiments performed in chicken embryos have demonstrated that an ectopic expression of the four MRFs individually induces the expression of miR-1 and miR-206 in the neural tube of chicken embryos *in vivo* [134]. Additionally to the MRFs, other transcription factors have been implicated during the expression of the four miRNAs; these include MEF2, which cooperates with MyoD, and the FOX O3 transcription factor [135, 136].

Overexpression and downregulation experiments, in addition to luciferase assays, revealed that myomiRs target the mRNAs of a series of proteins. MiR-1 was shown to bind to the 3'UTR of the histone deacetylase 4 (HDAC4) and to suppress its protein expression during growth and differentiation conditions [101, 137, 138]. HDAC4 was found to inhibit muscle differentiation and the expression of skeletal muscle genes, mainly by repressing the transcriptional activity of MEF2 [139]. HDAC4 downregulation by miR-1 overexpression was shown to contribute to the induction of *in vitro* muscle cell differentiation [101]. MiR-206 was determined to directly downregulate the largest catalytic subunit of DNA polymerase α (Pola1) at both mRNA and protein levels [103]. Pola1 is a replicative polymerase that is believed to be very important for cell proliferation and quiescence [103]. Hence, the downregulation of Pola1 during skeletal muscle differentiation by miR-206 may inhibit cell proliferation and contribute to cell cycle quiescence upon the process [103]. Furthermore, Rosenberg and colleagues have shown that miR-206 targets *follistatin-like 1* and *utrophin* 3′UTR regions, thereby downregulating their expression [133]. Follistatin-like 1 is a secreted glycoprotein that is involved in the regulation of function of endothelial cells, and also in the growth of blood vessels in muscle [140]. Utrophin is a ubiquitous protein shown to be expressed at the neuromuscular junctions of adult skeletal muscle, and which shares functional redundancy with dystrophin, a protein responsible for the development of Duchenne muscular dystrophy (DMD) [141]. Both, miR-206 and miR-1 were found to inhibit the expression of connexin 43 gap junction protein during muscle cell differentiation without, affecting its mRNA levels [117]. Connexin 43 was shown to be present predominantly prior to skeletal muscle cell fusion, and was rapidly downregulated following the induction of differentiation [117, 142, 143]. MiR-1 and miR-206 have also been shown to directly repress Pax7 protein expression post-transcriptionally in satellite cells, resulting in the inhibition of skeletal muscle progenitor cells [130]. Additionally, Hirai and colleagues showed that the activation of miR-1 and miR-206 decreased the levels of Pax3 protein by binding to the 3′UTR of Pax3 mRNA [144]. Pax3 downregulation by miR-1 and miR-206 was found to control somite myogenesis by stabilizing the commitment and subsequent differentiation of myoblasts, both *in vitro* and *in vivo*, during embryonic development [145]. Chen and coworkers showed that miR-133 binds and represses the expression of serum response factor (SRF) protein expression, without affecting mRNA levels, during growth and differentiation conditions [101]. SRF has been identified as an important factor during muscle proliferation and differentiation *in vitro* and *in vivo* [146, 147]. Furthermore, miR-133 was shown to repress, post-transcriptionally, the expression of insulin-like growth factor 1 (IGF-1) receptor by directly binding to its 3′UTR during C2C12 myogenic differentiation, thus negatively regulating the phosphatidylinositol 3-kinase (PI3K)/Akt pathway [136, 148]. The IGF signaling pathway was shown to potentially promote myogenesis in cultured muscle cells, and is required for normal skeletal muscle development during mouse embryogenesis [149, 150]. The binding of IGF to its corresponding receptor, the latter becomes phosphorylated and induces several intracellular signaling pathways, such as the PI3K/Akt pathway, which was shown to mediate the stimulatory effect of IGFs on muscle differentiation [149, 151, 152]. Therefore, the effect of miR-133 during muscle differentiation may also be mediated through signaling pathways, including the IGF signaling pathway. Although target prediction software has revealed an extensive list for these four myomiRs, the targets have to date been verified experimentally only in skeletal muscle tissue. Nonetheless, it would be of great interest to determine any other targets and to determine their precise mechanism of action during the development of muscle.

The complexity of miRNA pathways is further supported by the feedback loops in which miRNAs participate to control cellular proliferation and differentiation. For instance, as mentioned above, SRF is a direct target of miR-133; however, SRF was found to directly activate transcription of the *miR-1* gene in cardiogenesis [153]. An overexpression of IGF-1 – the ligand of the direct miR-133 target IGF-1 receptor – in differentiating C2C12 cells caused an accelerated expression of miR-133 and a concomitant increase in cell density and differentiated myotubes, most likely through myogenin induction [148]. Furthermore, a feedback loop between miR-1 expression and the IGF-1 signal transduction pathway was determined, since miR-1 was found to regulate the expression of IGF-1 receptor, whereas miR-1 transcription is regulated by the FOX O3 transcription factor, which acts downstream of IGF-1 signaling pathway [136, 148]. These findings imply a strong balanced regulation among miRNA expression and their function.

2.4.2 Long Noncoding RNA

Although many studies have contributed towards uncovering the function of many small noncoding RNAs such as miRNAs, very little is known to date regarding the long noncoding RNA counterpart of the transcriptome. Many thousands of long antisense, intergenic and intronic noncoding RNAs are transcribed from mammalian genomes [154]. Specific cell types, tissues, and developmental conditions showed specific patterns of long noncoding RNA expression [154–156].

Long Noncoding RNA in Myogenesis By conducting a detailed analysis of the genomic region of miR-206/133b, Cesana and coworkers discovered the existence of a muscle-specific long noncoding RNA which was termed *linc-MD1* [157]. Subsequent RT-PCR analysis showed that linc-MD1 was localized in the cytoplasm and was polyadenylated [157]. Further RT-PCR and *in situ* analyses showed that, under proliferating conditions, linc-MD1 was not expressed but rather was induced upon myoblast differentiation, specifically three days after differentiation induction [157]. The results of RNA interference and overexpression experiments showed that linc-MD1 would promote the process of muscle differentiation [157]. Two miRNAs expressed in muscle tissue, miR-133 and miR-135, were found to bind to linc-MD1 and to downregulate its expression [157]. Furthermore, linc-MD1, by binding miR-133 and miR-135, was determined to act as a competitive endogenous RNA for the targets of these miRNAs, including MEF2C [157]. Further experiments performed in human primary myoblasts showed that linc-MD1 RNA is also expressed in human muscle cells, and that it modulates miR-133 and miR-135 targets, contributing to the timing control of myoblast differentiation [157]. The importance of linc-MD1 in human muscle cell differentiation was further supported by experiments performed in DMD myoblasts, which showed a reduced and delayed muscle cell differentiation [157]. In particular, linc-MD1 levels were found to be strongly reduced in DMD cells, while its overexpression restored the expression of myogenin and MyHC (a marker of late muscle differentiation) towards control levels [157].

3
Conclusions

Myogenesis is a tightly and precisely regulated process which occurs during muscle development, growth, and regeneration. A number of molecular mechanisms have been determined to regulate skeletal muscle cell differentiation. Interestingly, several studies have unraveled crucial mechanisms by which myogenesis is regulated at the mRNA level. Regulations of the turnover of several mRNAs which encode for important myogenic molecules or signaling factors that are involved in myogenesis, such as MyoD, myogenin, and p21, have been determined. Specific RBPs, including KSRP, HuR, and CUGBP1, been found to either decrease or increase the mRNA stability in muscle. Alternative splicing is enriched in skeletal muscle with a number of genes expressed in muscle, such as Mef2 and CLCN1, to undergo the process. Many of the alternative splicing events that occur during the myogenic program showed different position- and time-dependent effects for several known splicing regulators. RBPs, such as the CELF and MBNL families, have been shown to regulate muscle alternative splicing events in muscle. Moreover, the translation of transcripts necessary for muscle cell differentiation was shown to be regulated by RBPs. Over the past years, RNA molecules that are not translated into proteins, known as *noncoding RNAs*, have been shown in higher eukaryotes to function as regulators of the myogenic program. Specifically, an ever-growing number of miRNAs has been found to be expressed in muscle tissue, of which miR-1, miR-133a, miR-133b and miR-206 are muscle-specific. Although many regulatory functions have been attributed to miRNAs during myogenesis, many other functions remain to be determined. More recently, a muscle-specific long noncoding RNA was also shown to have an important role in muscle cell differentiation, thus adding greater complexity to the regulation of the process.

In conclusion, RNA regulation involves many crucial pathways for the correct development and function of muscle. An aberrant regulation of the mRNAs in muscle can give rise to a range of muscle-related disorders, thus signifying the importance of RNA regulation in myogenesis. Clearly, unraveling the mechanisms by which mRNAs in muscle are regulated will provide a greater insight for decrypting many unknown aspects of the myogenic program, and help to provide an understanding of the pathogenesis of muscle-related disorders.

List of Abbreviations

MRF	Myogenic regulatory factor
MEF	Myocyte enhancer factor
Cdk	Cyclin-dependent kinase
Rb	Retinoblastoma
3'UTR	3' untranslated region
ARE	AU-rich element
GRE	GU-rich element
KSRP	KH-domain splicing regulatory protein
HuR	Hu antigen R
MAPK	Mitogen-activated protein kinase
PARN	Poly (A) ribonuclease
CUGBP1	CUG-binding protein 1
Rbm	RNA-binding motif protein
AChE	Acetylcholinesterase
snRNPs	Small nuclear ribonucleoprotein particles
PTB or PTBp1	Polypyrimidine tract binding protein
MBNL	Muscleblind-like
FOX	Forkhead box
DM1	Myotonic dystrophy type 1
miRNA	MicroRNA
pri-miRNA	Primary miRNA
pre-miRNA	Precursor miRNA
RISC	RNA-induced silencing complex
AGO	Argonaute

miRISC	MiRNA-induced silencing complex
MyHC	Myosin heavy chain
HDAC4	Histonedeacetylase 4
Polα1	DNA polymerase α
DMD	Duchenne muscular dystrophy
SRF	Serum response factor
IGF-1	Insulin-like growth factor 1
PI3K	Phosphatidylinositol 3-kinase

References

1 Sabourin, L.A., Rudnicki, M.A. (2000) The molecular regulation of myogenesis. *Clin. Genet.*, **57** (1), 16–25.

2 Davis, R.L., Weintraub, H., Lassar, A.B. (1987) Expression of a single transfected cDNA converts fibroblasts to myoblasts. *Cell*, **51** (6), 987–1000.

3 Blau, H.M., Chiu, C.P., Webster, C. (1983) Cytoplasmic activation of human nuclear genes in stable heterocaryons. *Cell*, **32** (4), 1171–1180.

4 Wright, W.E. (1984) Induction of muscle genes in neural cells. *J. Cell Biol.*, **98** (2), 427–435.

5 Braun, T., Buschhausen-Denker, G., Bober, E., Tannich, E. et al. (1989) A novel human muscle factor related to but distinct from MyoD1 induces myogenic conversion in 10T1/2 fibroblasts. *EMBO J.*, **8** (3), 701–709.

6 Edmondson, D.G., Olson, E.N. (1989) A gene with homology to the myc similarity region of MyoD1 is expressed during myogenesis and is sufficient to activate the muscle differentiation program. *Genes Dev.*, **3** (5), 628–640.

7 Wright, W.E., Sassoon, D.A., Lin, V.K. (1989) Myogenin, a factor regulating myogenesis, has a domain homologous to MyoD. *Cell*, **56** (4), 607–617.

8 Rhodes, S.J., Konieczny, S.F. (1989) Identification of MRF4: a new member of the muscle regulatory factor gene family. *Genes Dev.*, **3** (12B), 2050–2061.

9 Braun, T., Bober, E., Winter, B., Rosenthal, N. et al. (1990) Myf-6, a new member of the human gene family of myogenic determination factors: evidence for a gene cluster on chromosome 12. *EMBO J.*, **9** (3), 821–831.

10 Miner, J.H., Wold, B. (1990) Herculin, a fourth member of the MyoD family of myogenic regulatory genes. *Proc. Natl Acad. Sci. USA*, **87** (3), 1089–1093.

11 Perry, R.L., Rudnick, M.A. (2000) Molecular mechanisms regulating myogenic determination and differentiation. *Front. Biosci.*, **5**, D750–D767.

12 Lassar, A.B., Davis, R.L., Wright, W.E., Kadesch, T. et al. (1991) Functional activity of myogenic HLH proteins requires hetero-oligomerization with E12/E47-like proteins in vivo. *Cell*, **66** (2), 305–315.

13 Molkentin, J.D., Olson, E.N. (1996) Defining the regulatory networks for muscle development. *Curr. Opin. Genet. Dev.*, **6** (4), 445–453.

14 Rao, S.S., Chu, C., Kohtz, D.S. (1994) Ectopic expression of cyclin D1 prevents activation of gene transcription by myogenic basic helix-loop-helix regulators. *Mol. Cell. Biol.*, **14** (8), 5259–5267.

15 Rao, S.S., Kohtz, D.S. (1995) Positive and negative regulation of D-type cyclin expression in skeletal myoblasts by basic fibroblast growth factor and transforming growth factor beta. A role for cyclin D1 in control of myoblast differentiation. *J. Biol. Chem.*, **270** (8), 4093–4100.

16 Reynaud, E.G., Pelpel, K., Guillier, M., Leibovitch, M.P. et al. (1999) p57(Kip2) stabilizes the MyoD protein by inhibiting cyclin E-Cdk2 kinase activity in growing myoblasts. *Mol. Cell. Biol.*, **19** (11), 7621–7629.

17 Wang, J., Walsh, K. (1996) Resistance to apoptosis conferred by Cdk inhibitors during myocyte differentiation. *Science*, **273** (5273), 359–361.

18 Guo, K., Wang, J., Andres, V., Smith, R.C. et al. (1995) MyoD-induced expression of p21 inhibits cyclin-dependent kinase activity upon myocyte terminal differentiation. *Mol. Cell. Biol.*, **15** (7), 3823–3829.

19 Andres, V., Walsh, K. (1996) Myogenin expression, cell cycle withdrawal, and phenotypic differentiation are temporally separable events that precede cell fusion upon myogenesis. *J. Cell Biol.*, **132** (4), 657–666.

20 Wang, J., Walsh, K. (1996) Inhibition of retinoblastoma protein phosphorylation by myogenesis-induced changes in the subunit

composition of the cyclin-dependent kinase 4 complex. *Cell Growth Differ.*, **7** (11), 1471–1478.
21. Skapek, S.X., Rhee, J., Spicer, D.B., Lassar, A.B. (1995) Inhibition of myogenic differentiation in proliferating myoblasts by cyclin D1-dependent kinase. *Science*, **267** (5200), 1022–1024.
22. Halevy, O., Novitch, B.G., Spicer, D.B., Skapek, S.X. *et al.* (1995) Correlation of terminal cell cycle arrest of skeletal muscle with induction of p21 by MyoD. *Science*, **267** (5200), 1018–1021.
23. Franklin, D.S., Xiong, Y. (1996) Induction of p18INK4c and its predominant association with CDK4 and CDK6 during myogenic differentiation. *Mol. Biol. Cell*, **7** (10), 1587–1599.
24. Gu, W., Schneider, J.W., Condorelli, G., Kaushal, S. *et al.* (1993) Interaction of myogenic factors and the retinoblastoma protein mediates muscle cell commitment and differentiation. *Cell*, **72** (3), 309–324.
25. Weinberg, R.A. (1995) The retinoblastoma protein and cell cycle control. *Cell*, **81** (3), 323–330.
26. Corbeil, H.B., Whyte, P., Branton, P.E. (1995) Characterization of transcription factor E2F complexes during muscle and neuronal differentiation. *Oncogene*, **11** (5), 909–920.
27. Zacksenhaus, E., Jiang, Z., Chung, D., Marth, J.D. *et al.* (1996) pRb controls proliferation, differentiation, and death of skeletal muscle cells and other lineages during embryogenesis. *Genes Dev.*, **10** (23), 3051–3064.
28. Weintraub, H. (1993) The MyoD family and myogenesis: redundancy, networks, and thresholds. *Cell*, **75** (7), 1241–1244.
29. Licatalosi, D.D., Darnell, R.B. (2010) RNA processing and its regulation: global insights into biological networks. *Nat. Rev. Genet.*, **11** (1), 75–87.
30. Bevilacqua, A., Ceriani, M.C., Capaccioli, S., Nicolin, A. (2003) Post-transcriptional regulation of gene expression by degradation of messenger RNAs. *J. Cell. Physiol.*, **195** (3), 356–372.
31. Raineri, I., Wegmueller, D., Gross, B., Certa, U. *et al.* (2004) Roles of AUF1 isoforms, HuR and BRF1 in ARE-dependent mRNA turnover studied by RNA interference. *Nucleic Acids Res.*, **32** (4), 1279–1288.
32. Abu-Baker, A., Rouleau, G.A. (2007) Oculopharyngeal muscular dystrophy: recent advances in the understanding of the molecular pathogenic mechanisms and treatment strategies. *Biochim. Biophys. Acta*, **1772** (2), 173–185.
33. Ma, J.F., Hall, D.T., Gallouzi, I.E. (2012) The impact of mRNA turnover and translation on age-related muscle loss. *Ageing Res. Rev.*, **11**, 432–441.
34. Keren, A., Tamir, Y., Bengal, E. (2006) The p38 MAPK signaling pathway: a major regulator of skeletal muscle development. *Mol. Cell. Endocrinol.*, **252** (1–2), 224–230.
35. Zetser, A., Gredinger, E., Bengal, E. (1999) p38 mitogen-activated protein kinase pathway promotes skeletal muscle differentiation. Participation of the Mef2c transcription factor. *J. Biol. Chem.*, **274** (8), 5193–5200.
36. Zhao, M., New, L., Kravchenko, V.V., Kato, Y. *et al.* (1999) Regulation of the MEF2 family of transcription factors by p38. *Mol. Cell. Biol.*, **19** (1), 21–30.
37. Lluis, F., Ballestar, E., Suelves, M., Esteller, M. *et al.* (2005) E47 phosphorylation by p38 MAPK promotes MyoD/E47 association and muscle-specific gene transcription. *EMBO J.*, **24** (5), 974–984.
38. Simone, C., Forcales, S.V., Hill, D.A., Imbalzano, A.N. *et al.* (2004) p38 pathway targets SWI-SNF chromatin-remodeling complex to muscle-specific loci. *Nat. Genet.*, **36** (7), 738–743.
39. Briata, P., Forcales, S.V., Ponassi, M., Corte, G. *et al.* (2005) p38-dependent phosphorylation of the mRNA decay-promoting factor KSRP controls the stability of select myogenic transcripts. *Mol. Cell*, **20** (6), 891–903.
40. Gherzi, R., Lee, K.Y., Briata, P., Wegmuller, D. *et al.* (2004) A KH domain RNA binding protein, KSRP, promotes ARE-directed mRNA turnover by recruiting the degradation machinery. *Mol. Cell*, **14** (5), 571–583.
41. Apponi, L.H., Corbett, A.H., Pavlath, G.K. (2011) RNA-binding proteins and gene regulation in myogenesis. *Trends Pharmacol. Sci.*, **32** (11), 652–658.
42. Lee, J.E., Lee, J.Y., Wilusz, J., Tian, B. *et al.* (2010) Systematic analysis of cis-elements in unstable mRNAs demonstrates that

CUGBP1 is a key regulator of mRNA decay in muscle cells. *PLoS ONE*, **5** (6), e11201.

43 Vlasova, I.A., Tahoe, N.M., Fan, D., Larsson, O. *et al.* (2008) Conserved GU-rich elements mediate mRNA decay by binding to CUG-binding protein 1. *Mol. Cell*, **29** (2), 263–270.

44 Moraes, K.C., Wilusz, C.J., Wilusz, J. (2006) CUG-BP binds to RNA substrates and recruits PARN deadenylase. *RNA*, **12** (6), 1084–1091.

45 Figueroa, A., Cuadrado, A., Fan, J., Atasoy, U. *et al.* (2003) Role of HuR in skeletal myogenesis through coordinate regulation of muscle differentiation genes. *Mol. Cell. Biol.*, **23** (14), 4991–5004.

46 Kollias, H.D., Perry, R.L., Miyake, T., Aziz, A. *et al.* (2006) Smad7 promotes and enhances skeletal muscle differentiation. *Mol. Cell. Biol.*, **26** (16), 6248–6260.

47 Shi, L., Zhao, G., Qiu, D., Godfrey, W.R. *et al.* (2005) NF90 regulates cell cycle exit and terminal myogenic differentiation by direct binding to the 3'-untranslated region of MyoD and p21WAF1/CIP1 mRNAs. *J. Biol. Chem.*, **280** (19), 18981–18989.

48 Puri, P.L., Sartorelli, V. (2000) Regulation of muscle regulatory factors by DNA-binding, interacting proteins, and post-transcriptional modifications. *J. Cell. Physiol.*, **185** (2), 155–173.

49 Deschenes-Furry, J., Angus, L.M., Belanger, G., Mwanjewe, J. *et al.* (2005) Role of ELAV-like RNA-binding proteins HuD and HuR in the post-transcriptional regulation of acetylcholinesterase in neurons and skeletal muscle cells. *Chem. Biol. Interact.*, **157-158**, 43–49.

50 Good, P.J. (1995) A conserved family of elav-like genes in vertebrates. *Proc. Natl Acad. Sci. USA*, **92** (10), 4557–4561.

51 Darnell, R.B. (1996) Onconeural antigens and the paraneoplastic neurologic disorders: at the intersection of cancer, immunity, and the brain. *Proc. Natl Acad. Sci. USA*, **93** (10), 4529–4536.

52 Jin, D., Hidaka, K., Shirai, M., Morisaki, T. (2010) RNA-binding motif protein 24 regulates myogenin expression and promotes myogenic differentiation. *Genes Cells*, **15** (11), 1158–1167.

53 Terami, H., Hidaka, K., Shirai, M., Narumiya, H. *et al.* (2007) Efficient capture of cardiogenesis-associated genes expressed in ES cells. *Biochem. Biophys. Res. Commun.*, **355** (1), 47–53.

54 Miyamoto, S., Hidaka, K., Jin, D., Morisaki, T. (2009) RNA-binding proteins Rbm38 and Rbm24 regulate myogenic differentiation via p21-dependent and -independent regulatory pathways. *Genes Cells*, **14** (11), 1241–1252.

55 Shu, L., Yan, W., Chen, X. (2006) RNPC1, an RNA-binding protein and a target of the p53 family, is required for maintaining the stability of the basal and stress-induced p21 transcript. *Genes Dev.*, **20** (21), 2961–2972.

56 Tsim, K.W., Choi, R.C., Xie, H.Q., Zhu, J.T. *et al.* (2008) Transcriptional control of different subunits of AChE in muscles: signals triggered by the motor nerve-derived factors. *Chem. Biol. Interact.*, **175** (1-3), 58–63.

57 Angus, L.M., Chan, R.Y., Jasmin, B.J. (2001) Role of intronic E- and N-box motifs in the transcriptional induction of the acetylcholinesterase gene during myogenic differentiation. *J. Biol. Chem.*, **276** (20), 17603–17609.

58 Deschenes-Furry, J., Belanger, G., Mwanjewe, J., Lunde, J.A. *et al.* (2005) The RNA-binding protein HuR binds to acetylcholinesterase transcripts and regulates their expression in differentiating skeletal muscle cells. *J. Biol. Chem.*, **280** (27), 25361–25368.

59 Montes, M., Becerra, S., Sanchez-Alvarez, M., Sune, C. (2012) Functional coupling of transcription and splicing. *Gene*, **501** (2), 104–117.

60 McManus, C.J., Graveley, B.R. (2011) RNA structure and the mechanisms of alternative splicing. *Curr. Opin. Genet. Dev.*, **21** (4), 373–379.

61 Wahl, M.C., Will, C.L., Luhrmann, R. (2009) The spliceosome: design principles of a dynamic RNP machine. *Cell*, **136** (4), 701–718.

62 Maniatis, T., Tasic, B. (2002) Alternative pre-mRNA splicing and proteome expansion in metazoans. *Nature*, **418** (6894), 236–243.

63 Sanchez, S.E., Petrillo, E., Kornblihtt, A.R., Yanovsky, M.J. (2011) Alternative splicing at the right time. *RNA Biol.*, **8** (6), 954–959.

64 Pan, Q., Shai, O., Lee, L.J., Frey, B.J. *et al.* (2008) Deep surveying of alternative splicing complexity in the human transcriptome by

65 Kalsotra, A., Cooper, T.A. (2011) Functional consequences of developmentally regulated alternative splicing. *Nat. Rev. Genet.*, **12** (10), 715–729.

66 Caceres, J.F., Kornblihtt, A.R. (2002) Alternative splicing: multiple control mechanisms and involvement in human disease. *Trends Genet*, **18** (4), 186–193.

67 Garcia-Blanco, M.A., Baraniak, A.P., Lasda, E.L. (2004) Alternative splicing in disease and therapy. *Nat. Biotechnol.*, **22** (5), 535–546.

68 Licatalosi, D.D., Darnell, R.B. (2006) Splicing regulation in neurologic disease. *Neuron*, **52** (1), 93–101.

69 Wang, G.S., Cooper, T.A. (2007) Splicing in disease: disruption of the splicing code and the decoding machinery. *Nat. Rev. Genet.*, **8** (10), 749–761.

70 Bland, C.S., Wang, E.T., Vu, A., David, M.P. et al. (2010) Global regulation of alternative splicing during myogenic differentiation. *Nucleic Acids Res.*, **38** (21), 7651–7664.

71 Ladd, A.N., Charlet, N., Cooper, T.A. (2001) The CELF family of RNA binding proteins is implicated in cell-specific and developmentally regulated alternative splicing. *Mol. Cell. Biol.*, **21** (4), 1285–1296.

72 Kuroyanagi, H. (2009) Fox-1 family of RNA-binding proteins. *Cell. Mol. Life Sci.*, **66** (24), 3895–3907.

73 Kanadia, R.N., Urbinati, C.R., Crusselle, V.J., Luo, D. et al. (2003) Developmental expression of mouse muscleblind genes Mbnl1, Mbnl2 and Mbnl3. *Gene Expr. Patterns*, **3** (4), 459–462.

74 Squillace, R.M., Chenault, D.M., Wang, E.H. (2002) Inhibition of muscle differentiation by the novel muscleblind-related protein CHCR. *Dev. Biol.*, **250** (1), 218–230.

75 Fernandez-Costa, J.M., Llamusi, M.B., Garcia-Lopez, A., Artero, R. (2011) Alternative splicing regulation by Muscleblind proteins: from development to disease. *Biol. Rev. Camb. Philos. Soc.*, **86** (4), 947–958.

76 Miller, J.W., Urbinati, C.R., Teng-Umnuay, P., Stenberg, M.G. et al. (2000) Recruitment of human muscleblind proteins to (CUG)(n) expansions associated with myotonic dystrophy. *EMBO J.*, **19** (17), 4439–4448.

77 Lee, K.S., Smith, K., Amieux, P.S., Wang, E.H. (2008) MBNL3/CHCR prevents myogenic differentiation by inhibiting MyoD-dependent gene transcription. *Differentiation*, **76** (3), 299–309.

78 Breitbart, R.E., Liang, C.S., Smoot, L.B., Laheru, D.A. et al. (1993) A fourth human MEF2 transcription factor, hMEF2D, is an early marker of the myogenic lineage. *Development*, **118** (4), 1095–1106.

79 Martin, J.F., Schwarz, J.J., Olson, E.N. (1993) Myocyte enhancer factor (MEF) 2C: a tissue-restricted member of the MEF-2 family of transcription factors. *Proc. Natl Acad. Sci. USA*, **90** (11), 5282–5286.

80 McDermott, J.C., Cardoso, M.C., Yu, Y.T., Andres, V. et al. (1993) hMEF2C gene encodes skeletal muscle- and brain-specific transcription factors. *Mol. Cell. Biol.*, **13** (4), 2564–2577.

81 Martin, J.F., Miano, J.M., Hustad, C.M., Copeland, N.G. et al. (1994) A Mef2 gene that generates a muscle-specific isoform via alternative mRNA splicing. *Mol. Cell. Biol.*, **14** (3), 1647–1656.

82 Zhu, B., Ramachandran, B., Gulick, T. (2005) Alternative pre-mRNA splicing governs expression of a conserved acidic transactivation domain in myocyte enhancer factor 2 factors of striated muscle and brain. *J. Biol. Chem.*, **280** (31), 28749–28760.

83 Kino, Y., Washizu, C., Oma, Y., Onishi, H. et al. (2009) MBNL and CELF proteins regulate alternative splicing of the skeletal muscle chloride channel CLCN1. *Nucleic Acids Res.*, **37** (19), 6477–6490.

84 Perry, S.V. (2001) Vertebrate tropomyosin: distribution, properties and function. *J. Muscle Res. Cell Motil.*, **22** (1), 5–49.

85 Lin, J.C., Tarn, W.Y. (2005) Exon selection in alpha-tropomyosin mRNA is regulated by the antagonistic action of RBM4 and PTB. *Mol. Cell. Biol.*, **25** (22), 10111–10121.

86 Lin, J.C., Tarn, W.Y. (2011) RBM4 down-regulates PTB and antagonizes its activity in muscle cell-specific alternative splicing. *J. Cell Biol.*, **193** (3), 509–520.

87 Sureau, A., Sauliere, J., Expert-Bezancon, A., Marie, J. (2011) CELF and PTB proteins modulate the inclusion of the beta-tropomyosin exon 6B during myogenic differentiation. *Exp. Cell. Res.*, **317** (1), 94–106.

88 Gallego, M.E., Balvay, L., Brody, E. (1992) cis-acting sequences involved in exon selection in the chicken beta-tropomyosin gene. *Mol. Cell. Biol.*, **12** (12), 5415–5425.

89 Goux-Pelletan, M., Libri, D., d'Aubenton-Carafa, Y., Fiszman, M. et al. (1990) In vitro splicing of mutually exclusive exons from the chicken beta-tropomyosin gene: role of the branch point location and very long pyrimidine stretch. *EMBO J.*, **9** (1), 241–249.

90 Clouet d'Orval, B., d'Aubenton Carafa, Y., Sirand-Pugnet, P., Gallego, M. et al. (1991) RNA secondary structure repression of a muscle-specific exon in HeLa cell nuclear extracts. *Science*, **252** (5014), 1823–1828.

91 Libri, D., Balvay, L., Fiszman, M.Y. (1992) In vivo splicing of the beta tropomyosin pre-mRNA: a role for branch point and donor site competition. *Mol. Cell. Biol.*, **12** (7), 3204–3215.

92 Sauliere, J., Sureau, A., Expert-Bezancon, A., Marie, J. (2006) The polypyrimidine tract binding protein (PTB) represses splicing of exon 6B from the beta-tropomyosin pre-mRNA by directly interfering with the binding of the U2AF65 subunit. *Mol. Cell. Biol.*, **26** (23), 8755–8769.

93 Jin, Y., Suzuki, H., Maegawa, S., Endo, H. et al. (2003) A vertebrate RNA-binding protein Fox-1 regulates tissue-specific splicing via the pentanucleotide GCAUG. *EMBO J.*, **22** (4), 905–912.

94 Gebauer, F., Hentze, M.W. (2004) Molecular mechanisms of translational control. *Nat. Rev. Mol. Cell Biol.*, **5** (10), 827–835.

95 Timchenko, N.A., Iakova, P., Cai, Z.J., Smith, J.R. et al. (2001) Molecular basis for impaired muscle differentiation in myotonic dystrophy. *Mol. Cell. Biol.*, **21** (20), 6927–6938.

96 Timchenko, N.A., Patel, R., Iakova, P., Cai, Z.J. et al. (2004) Overexpression of CUG triplet repeat-binding protein, CUGBP1, in mice inhibits myogenesis. *J. Biol. Chem.*, **279** (13), 13129–13139.

97 Huang, Y., Shen, X.J., Zou, Q., Wang, S.P. et al. (2011) Biological functions of microRNAs: a review. *J. Physiol. Biochem.*, **67** (1), 129–139.

98 Sayed, D., Abdellatif, M. (2011) MicroRNAs in development and disease. *Physiol. Rev.*, **91** (3), 827–887.

99 Krol, J., Loedige, I., Filipowicz, W. (2010) The widespread regulation of microRNA biogenesis, function and decay. *Nat. Rev. Genet.*, **11** (9), 597–610.

100 Ge, Y., Chen, J. (2011) MicroRNAs in skeletal myogenesis. *Cell Cycle*, **10** (3), 441–448.

101 Chen, J.F., Mandel, E.M., Thomson, J.M., Wu, Q. et al. (2006) The role of microRNA-1 and microRNA-133 in skeletal muscle proliferation and differentiation. *Nat. Genet.*, **38** (2), 228–233.

102 Koutsoulidou, A., Mastroyiannopoulos, N.P., Furling, D., Uney, J.B. et al. (2011) Expression of miR-1, miR-133a, miR-133b and miR-206 increases during development of human skeletal muscle. *BMC Dev. Biol.*, **11**, 34.

103 Kim, H.K., Lee, Y.S., Sivaprasad, U., Malhotra, A. et al. (2006) Muscle-specific microRNA miR-206 promotes muscle differentiation. *J. Cell Biol.*, **174** (5), 677–687.

104 He, L., Hannon, G.J. (2004) MicroRNAs: small RNAs with a big role in gene regulation. *Nat. Rev. Genet.*, **5** (7), 522–531.

105 Davis-Dusenbery, B.N., Hata, A. (2010) Mechanisms of control of microRNA biogenesis. *J. Biochem.*, **148** (4), 381–392.

106 Chendrimada, T.P., Gregory, R.I., Kumaraswamy, E., Norman, J. et al. (2005) TRBP recruits the Dicer complex to Ago2 for microRNA processing and gene silencing. *Nature*, **436** (7051), 740–744.

107 Haase, A.D., Jaskiewicz, L., Zhang, H., Laine, S. et al. (2005) TRBP, a regulator of cellular PKR and HIV-1 virus expression, interacts with Dicer and functions in RNA silencing. *EMBO Rep.*, **6** (10), 961–967.

108 Liu, J., Rivas, F.V., Wohlschlegel, J., Yates, J.R. III et al. (2005) A role for the P-body component GW182 in microRNA function. *Nat. Cell Biol.*, **7** (12), 1261–1266.

109 Maniataki, E., Mourelatos, Z. (2005) A human, ATP-independent, RISC assembly machine fueled by pre-miRNA. *Genes Dev.*, **19** (24), 2979–2990.

110 MacRae, I.J., Ma, E., Zhou, M., Robinson, C.V. et al. (2008) In vitro reconstitution of the human RISC-loading complex. *Proc. Natl Acad. Sci. USA*, **105** (2), 512–517.

111 Meister, G., Landthaler, M., Patkaniowska, A., Dorsett, Y. et al. (2004) Human Argonaute2 mediates RNA

cleavage targeted by miRNAs and siRNAs. *Mol. Cell*, **15** (2), 185–197.

112 O'Carroll, D., Mecklenbrauker, I., Das, P.P., Santana, A. et al. (2007) A Slicer-independent role for Argonaute 2 in hematopoiesis and the microRNA pathway. *Genes Dev.*, **21** (16), 1999–2004.

113 Shruti, K., Shrey, K., Vibha, R. (2011) Micro RNAs: tiny sequences with enormous potential. *Biochem. Biophys. Res. Commun.*, **407** (3), 445–449.

114 Bartel, D.P. (2009) MicroRNAs: target recognition and regulatory functions. *Cell*, **136** (2), 215–233.

115 Starega-Roslan, J., Koscianska, E., Kozlowski, P., Krzyzosiak, W.J. (2011) The role of the precursor structure in the biogenesis of microRNA. *Cell. Mol. Life Sci.*, **68** (17), 2859–2871.

116 Ladomery, M.R., Maddocks, D.G., Wilson, I.D. (2011) MicroRNAs: their discovery, biogenesis, function and potential use as biomarkers in non-invasive prenatal diagnostics. *Int. J. Mol. Epidemiol. Genet.*, **2** (3), 253–260.

117 Anderson, C., Catoe, H., Werner, R. (2006) MIR-206 regulates connexin43 expression during skeletal muscle development. *Nucleic Acids Res.*, **34** (20), 5863–5871.

118 Sweetman, D., Rathjen, T., Jefferson, M., Wheeler, G. et al. (2006) FGF-4 signaling is involved in mir-206 expression in developing somites of chicken embryos. *Dev. Dyn.*, **235** (8), 2185–2191.

119 Yuasa, K., Hagiwara, Y., Ando, M., Nakamura, A. et al. (2008) MicroRNA-206 is highly expressed in newly formed muscle fibers: implications regarding potential for muscle regeneration and maturation in muscular dystrophy. *Cell Struct. Funct.*, **33** (2), 163–169.

120 Naguibneva, I., Ameyar-Zazoua, M., Polesskaya, A., Ait-Si-Ali, S. et al. (2006) The microRNA miR-181 targets the homeobox protein Hox-A11 during mammalian myoblast differentiation. *Nat. Cell Biol.*, **8** (3), 278–284.

121 Yamamoto, M., Kuroiwa, A. (2003) Hoxa-11 and Hoxa-13 are involved in repression of MyoD during limb muscle development. *Dev. Growth Differ.*, **45** (5–6), 485–498.

122 Cardinali, B., Castellani, L., Fasanaro, P., Basso, A. et al. (2009) Microrna-221 and microrna-222 modulate differentiation and maturation of skeletal muscle cells. *PLoS ONE*, **4** (10), e7607.

123 Sun, Q., Zhang, Y., Yang, G., Chen, X. et al. (2008) Transforming growth factor-beta-regulated miR-24 promotes skeletal muscle differentiation. *Nucleic Acids Res.*, **36** (8), 2690–2699.

124 Juan, A.H., Kumar, R.M., Marx, J.G., Young, R.A. et al. (2009) Mir-214-dependent regulation of the polycomb protein Ezh2 in skeletal muscle and embryonic stem cells. *Mol. Cell*, **36** (1), 61–74.

125 Gagan, J., Dey, B.K., Layer, R., Yan, Z. et al. (2011) MicroRNA-378 targets the myogenic repressor MyoR during myoblast differentiation. *J. Biol. Chem.*, **286** (22), 19431–19438.

126 Lu, J., Webb, R., Richardson, J.A., Olson, E.N. (1999) MyoR: a muscle-restricted basic helix-loop-helix transcription factor that antagonizes the actions of MyoD. *Proc. Natl Acad. Sci. USA*, **96** (2), 552–557.

127 Seok, H.Y., Tatsuguchi, M., Callis, T.E., He, A. et al. (2011) miR-155 inhibits expression of the MEF2A protein to repress skeletal muscle differentiation. *J. Biol. Chem.*, **286** (41), 35339–35346.

128 Chen, J.F., Callis, T.E., Wang, D.Z. (2009) microRNAs and muscle disorders. *J. Cell Sci.*, **122** (Pt 1), 13–20.

129 Townley-Tilson, W.H., Callis, T.E., Wang, D. (2010) MicroRNAs 1, 133, and 206: critical factors of skeletal and cardiac muscle development, function, and disease. *Int. J. Biochem. Cell Biol.*, **42** (8), 1252–1255.

130 Chen, J.F., Tao, Y., Li, J., Deng, Z. et al. (2010) microRNA-1 and microRNA-206 regulate skeletal muscle satellite cell proliferation and differentiation by repressing Pax7. *J. Cell Biol.*, **190** (5), 867–879.

131 Mok, G.F., Sweetman, D. (2011) Many routes to the same destination: lessons from skeletal muscle development. *Reproduction*, **141** (3), 301–312.

132 Rao, P.K., Kumar, R.M., Farkhondeh, M., Baskerville, S. et al. (2006) Myogenic factors that regulate expression of muscle-specific microRNAs. *Proc. Natl Acad. Sci. USA*, **103** (23), 8721–8726.

133 Rosenberg, M.I., Georges, S.A., Asawachaicharn, A., Analau, E. et al. (2006) MyoD inhibits Fstl1 and Utrn expression

by inducing transcription of miR-206. *J. Cell Biol.*, **175** (1), 77–85.
134 Sweetman, D., Goljanek, K., Rathjen, T., Oustanina, S. *et al.* (2008) Specific requirements of MRFs for the expression of muscle specific microRNAs, miR-1, miR-206 and miR-133. *Dev. Biol.*, **321** (2), 491–499.
135 Liu, N., Williams, A.H., Kim, Y., McAnally, J. *et al.* (2007) An intragenic MEF2-dependent enhancer directs muscle-specific expression of microRNAs 1 and 133. *Proc. Natl Acad. Sci. USA*, **104** (52), 20844–20849.
136 Elia, L., Contu, R., Quintavalle, M., Varrone, F. *et al.* (2009) Reciprocal regulation of microRNA-1 and insulin-like growth factor-1 signal transduction cascade in cardiac and skeletal muscle in physiological and pathological conditions. *Circulation*, **120** (23), 2377–2385.
137 Lu, J., McKinsey, T.A., Zhang, C.L., Olson, E.N. (2000) Regulation of skeletal myogenesis by association of the MEF2 transcription factor with class II histone deacetylases. *Mol. Cell*, **6** (2), 233–244.
138 McKinsey, T.A., Zhang, C.L., Lu, J., Olson, E.N. (2000) Signal-dependent nuclear export of a histone deacetylase regulates muscle differentiation. *Nature*, **408** (6808), 106–111.
139 Miska, E.A., Karlsson, C., Langley, E., Nielsen, S.J. *et al.* (1999) HDAC4 deacetylase associates with and represses the MEF2 transcription factor. *EMBO J.*, **18** (18), 5099–5107.
140 Ouchi, N., Oshima, Y., Ohashi, K., Higuchi, A. *et al.* (2008) Follistatin-like 1, a secreted muscle protein, promotes endothelial cell function and revascularization in ischemic tissue through a nitric-oxide synthase-dependent mechanism. *J. Biol. Chem.*, **283** (47), 32802–32811.
141 Weir, A.P., Morgan, J.E., Davies, K.E. (2004) A-utrophin up-regulation in mdx skeletal muscle is independent of regeneration. *Neuromuscular Disord.*, **14** (1), 19–23.
142 Proulx, A., Merrifield, P.A., Naus, C.C. (1997) Blocking gap junctional intercellular communication in myoblasts inhibits myogenin and MRF4 expression. *Dev. Genet.*, **20** (2), 133–144.
143 Balogh, S., Naus, C.C., Merrifield, P.A. (1993) Expression of gap junctions in cultured rat L6 cells during myogenesis. *Dev. Biol.*, **155** (2), 351–360.
144 Hirai, H., Verma, M., Watanabe, S., Tastad, C. *et al.* (2010) MyoD regulates apoptosis of myoblasts through microRNA-mediated down-regulation of Pax3. *J. Cell Biol.*, **191** (2), 347–365.
145 Goljanek-Whysall, K., Sweetman, D., Abu-Elmagd, M., Chapnik, E. *et al.* (2011) MicroRNA regulation of the paired-box transcription factor Pax3 confers robustness to developmental timing of myogenesis. *Proc. Natl Acad. Sci. USA*, **108** (29), 11936–11941.
146 Soulez, M., Rouviere, C.G., Chafey, P., Hentzen, D. *et al.* (1996) Growth and differentiation of C2 myogenic cells are dependent on serum response factor. *Mol. Cell. Biol.*, **16** (11), 6065–6074.
147 Li, S., Czubryt, M.P., McAnally, J., Bassel-Duby, R. *et al.* (2005) Requirement for serum response factor for skeletal muscle growth and maturation revealed by tissue-specific gene deletion in mice. *Proc. Natl Acad. Sci. USA*, **102** (4), 1082–1087.
148 Huang, M.B., Xu, H., Xie, S.J., Zhou, H. *et al.* (2011) Insulin-like growth factor-1 receptor is regulated by microRNA-133 during skeletal myogenesis. *PLoS ONE*, **6** (12), e29173.
149 Rommel, C., Bodine, S.C., Clarke, B.A., Rossman, R. *et al.* (2001) Mediation of IGF-1-induced skeletal myotube hypertrophy by PI(3)K/Akt/mTOR and PI(3)K/Akt/GSK3 pathways. *Nat. Cell Biol.*, **3** (11), 1009–1013.
150 Powell-Braxton, L., Hollingshead, P., Warburton, C., Dowd, M. *et al.* (1993) IGF-I is required for normal embryonic growth in mice. *Genes Dev.*, **7** (12B), 2609–2617.
151 Coolican, S.A., Samuel, D.S., Ewton, D.Z., McWade, F.J. *et al.* (1997) The mitogenic and myogenic actions of insulin-like growth factors utilize distinct signaling pathways. *J. Biol. Chem.*, **272** (10), 6653–6662.
152 Jiang, B.H., Zheng, J.Z., Vogt, P.K. (1998) An essential role of phosphatidylinositol 3-kinase in myogenic differentiation. *Proc. Natl Acad. Sci. USA*, **95** (24), 14179–14183.
153 Zhao, Y., Samal, E., Srivastava, D. (2005) Serum response factor regulates a muscle-specific microRNA that targets

Hand2 during cardiogenesis. *Nature*, **436** (7048), 214–220.
154 Mattick, J.S. (2011) The central role of RNA in human development and cognition. *FEBS Lett.*, **585** (11), 1600–1616.
155 Amaral, P.P., Mattick, J.S. (2008) Noncoding RNA in development. *Mamm. Genome*, **19** (7-8), 454–492.
156 Qureshi, I.A., Mattick, J.S., Mehler, M.F. (2010) Long non-coding RNAs in nervous system function and disease. *Brain Res.*, **1338**, 20–35.
157 Cesana, M., Cacchiarelli, D., Legnini, I., Santini, T. *et al.* (2011) A long noncoding RNA controls muscle differentiation by functioning as a competing endogenous RNA. *Cell*, **147** (2), 358–369.

6
Regulation of Animal Gene Expression by Ingested Plant Small RNAs

Xi Chen, Lin Zhang, and Chen-Yu Zhang
Nanjing University, Jiangsu Engineering Research Center for microRNA Biology and Biotechnology, State Key Laboratory of Pharmaceutical Biotechnology, School of Life Sciences, 22 Hankou Road, Nanjing 210093, China

1	**Introducing the Small RNAs** 171	
2	**Intracellular Regulation of Gene Expression by Small RNAs** 171	
2.1	Intracellular Regulation of Gene Expression by miRNAs 171	
2.2	Intracellular Regulation of Gene Expression by siRNAs 173	
3	**Intercellular Regulation of Gene Expression by Small RNAs** 173	
3.1	Intercellular Regulation of Gene Expression by miRNAs 173	
3.2	Intercellular Regulation of Gene Expression by siRNAs 174	
4	**Regulation of Animal Gene Expression by Ingested Plant miRNAs** 175	
4.1	Existence of Plant miRNAs in Animals 175	
4.2	Absorption of Exogenous Plant miRNAs by Animal Gastrointestinal Tract 176	
4.3	Regulation of Animal Gene Expression by Plant miRNAs 176	
4.4	Transport of Plant miRNAs by Microvesicles 177	
4.5	Physiological Effects of Plant miRNAs in Mammalian Systems 177	
5	**Regulation of Animal Gene Expression by Ingested siRNAs** 178	
6	**Issues Raised by Small RNA-Mediated Cross-Kingdom Regulation** 179	
6.1	Nutrients 179	
6.2	Medicine 179	
6.3	Evolution 180	
7	**Summary** 180	
	References 180	

RNA Regulation: Advances in Molecular Biology and Medicine, First Edition. Edited by Robert A. Meyers.
© 2014 Wiley-VCH Verlag GmbH & Co. KGaA. Published 2014 by Wiley-VCH Verlag GmbH & Co. KGaA.

Keywords

small RNA
Small RNAs are genome-transcribed or exogenous noncoding RNAs, usually 20–30 nt long. At least three classes of small RNAs have been identified, including siRNA, microRNA, and piRNA.

microRNA
microRNAs are a class of endogenous noncoding RNAs consisting of 19–24 nt that regulate the post-transcriptional silencing of protein-coding genes.

siRNA
siRNA is a class of double-stranded RNA molecules, usually 20–25 nt in length. siRNA plays many roles, but its most notable is in the RNA interference pathway, where it interferes with the expression of specific genes with complementary nucleotide sequence.

RNA interference
RNA interference refers to the inhibition of gene expression by small double-stranded RNA molecules.

Argonaute protein
Argonaute proteins are the catalytic components of the RNA-induced silencing complex, the protein complex responsible for the gene silencing phenomenon known as RNA interference. Argonaute proteins bind different classes of small noncoding RNAs, including microRNAs, siRNAs, and piRNAs.

Microvesicle
Microvesicles (sometimes called *exosomes*, *shedding vesicles*, or *microparticles*) are small membranous vesicles ranging from 30 to 1000 nm in size, that are shed from almost all cell types. Microvesicles play a role in intercellular communication, and can transport mRNAs, miRNAs, and proteins between cells.

Small RNAs, including microRNAs (miRNAs) and small interfering RNAs (siRNAs), have been found to be important regulators of gene expression. The function of small RNAs has been observed in many life processes, including proliferation, differentiation, and apoptosis. Surprisingly, recent studies have shown that plant small RNAs (especially those from food) can be delivered to recipient animal cells, where these exogenous RNAs function similarly to endogenous small RNAs, simultaneously regulating multiple target genes and biological processes. These findings indicate that small RNAs from one species may influence the physiology of other, distantly related, species. In this chapter, an overview is provided of the present scientific understanding of animal gene regulation by ingested plant small RNAs. Several examples of the unexpected – and often unexplained – complexity typical of the interactions between plants and animals through small RNAs are highlighted.

1
Introducing the Small RNAs

During the past decade, small RNAs have been recognized as crucial regulators of gene expression and genome function that play roles in almost every aspect of biology [1, 2]. The defining features of small RNAs are their short length (20–30 nt) and their association with members of the Argonaute (AGO) family of proteins. AGO proteins guide small RNAs to their regulatory targets, typically resulting in a reduced expression of target genes. In animals, three main classes of small RNAs are recognized, including: microRNAs (miRNAs); small interfering RNAs (siRNAs); and Piwi-interacting RNAs (piRNAs). In plants, only miRNAs and siRNAs have been identified. The classes of small RNAs differ with regard to their biogenesis, their mode of target regulation, and their participation in various biological functions. Despite their differences, these distinct small RNA pathways are interconnected; these pathways compete and collaborate as they regulate genes and protect the genome from external and internal threats.

2
Intracellular Regulation of Gene Expression by Small RNAs

Small RNAs have been recognized as important players in diverse aspects of biology. The most extensively studied among them are miRNAs that regulate target genes post-transcriptionally in animals and plants [3–5]. Another class of small RNAs that are closely related to the miRNAs are siRNAs; these are small RNAs that serve as key intermediates in the RNA interference (RNAi) pathway [6].

2.1
Intracellular Regulation of Gene Expression by miRNAs

miRNAs are a class of endogenous small noncoding RNAs, typically 22 nt in length, that act as post-transcriptional regulators of thousands of genes by binding to the 3' untranslated region (UTR) of specific messenger RNAs (mRNAs). Lee *et al.* discovered the first miRNA, *lin-4*, in a study of developmental timing in a nematode (*Caenorhabditis elegans*) system in 1993 [7]. These authors found that the *lin-4* gene did not encode a protein; instead, it encoded a small RNA molecule with sequence complementarity to the 3' UTR of another gene, *lin-14*. This complementarity was required for the correct regulation of lin-14 protein expression, but regulation of expression occurred without any noticeable changes in the levels of its mRNA [7, 8]. Seven years later, a second miRNA, known as *let-7*, was found to be implicated in the process of transition from late larval to adult cell fates [9]. It was later noted that the let-7 miRNA was conserved in many species [10]; this discovery prompted further exploration of this novel class of small RNAs and their crucial effects on eukaryotic gene regulation. Years later, an increasing number of miRNAs have been identified through both computational and experimental approaches in plants, animals, humans, and viruses. Over 20 000 mature miRNAs from 168 species are listed in the most recent version of the miRBase database (version 18.0), and approximately 1500 miRNAs have been identified in humans. Bioinformatic predictions have estimated that approximately 30% of gene products may be regulated by miRNAs. Several crucial steps in miRNA biogenesis and the molecules involved in these processes have now been well characterized; most of the miRNAs characterized thus far appear to be involved in cell proliferation, cell differentiation, apoptosis, and transformation [3, 11]. However, the functions of the vast majority of miRNAs remain to be elucidated.

In the canonical biogenesis pathway for animal miRNAs, genes encoding miRNAs are first transcribed by RNA polymerase II [12]. The hairpins of the primary transcripts (pri-miRNAs) are then recognized by a nuclear protein known as DiGeorge Syndrome Critical Region 8 (DGCR8) [13]. DGCR8 then associates with the Drosha enzyme to form an RNase III type endonuclease that cleaves the 5' and 3' ends of the pri-miRNAs to create hairpins approximately 70 nt long, referred to as *pre-miRNAs* [14]. Next, the pre-miRNAs are rapidly shuttled from the nucleus to the cytoplasm via the Exportin5/RAN–GTPase pathway [15]. Once in the cytoplasm, Dicer enzyme and its cofactors [TRBP (HIV-1 transactivating response RNA-binding protein) and PACT (protein activator of the interferon-induced protein kinase)] catalyze a second cleavage near the terminal loop of the pre-miRNAs; this cleavage event releases imperfect miRNA:miRNA* duplexes that are approximately 22 nt long [16–18]. The final steps in miRNA biogenesis are the assembly of mature miRNAs (guide strands) into a multi-subunit protein complex called the RNA-induced silencing complex (RISC) [19]; the miRNA* (passenger strand) is released and subsequently degraded [20]. Although the composition of RISC is not fully defined, it is clear that AGO proteins are crucial to its function [21]. The human genome contains eight AGO proteins, but only AGO2 has RNA cleavage activity and plays a key role in miRNA-mediated silencing [22]. For plant miRNAs, both cleavage events occur inside the nucleus and are catalyzed by the Dicer homolog Dicer-like 1 (DCL1) [23, 24]; this process is distinct from that in animal miRNAs, which are cleaved by two different enzymes inside and outside of the nucleus. Before the plant miRNA:miRNA* duplex is exported from the nucleus by the protein Hasty (HST, a homolog of Exportin5) [25, 26], the duplex 3' overhang is methylated by an RNA methyltransferase named Hua Enhancer1 (HEN1) [27, 28].

The miRNA-guided RISC complex can regulate gene expression through two different mechanisms. First, gene expression can be regulated through endonucleolytic cleavage of targeted mRNAs, a mechanism mainly observed in plants that requires a perfect complementarity between miRNAs and mRNAs. Second, regulation can occur as a result of imperfect base-pairing between the miRNA seed region (nucleotides 2–8 of the 5' end of miRNA) and the mRNA target site, a mechanism that is common in animals. This imperfect interaction can result in the inhibition of protein synthesis (most likely mediated by the GW182 protein) or mRNA degradation [29]. This degradation process involves the recruitment of deadenylase complexes and the removal of the mRNA 5' cap. Uncapped and deadenylated mRNAs are rapidly degraded by cellular 5' to 3' exoribonucleases [30, 31].

2.2 Intracellular Regulation of Gene Expression by siRNAs

siRNAs are a class of double-stranded RNA (dsRNA) molecules that are usually 21 nt in length. These RNA molecules can interfere with the translation of proteins by binding to specific mRNA sequences and promoting mRNA degradation. In doing so, they prevent the production of specific proteins based on the nucleotide sequences of their corresponding mRNA. The process is known as *RNAi*, but it may also be referred to as *siRNA silencing* or *siRNA knockdown*.

Each siRNA strand has a 5' phosphate group and a 3' hydroxyl group with two overhanging nucleotides at the 3' end. siRNA is produced from long dsRNA or hairpin looped RNA that are cleaved by Dicer after entering a cell. The siRNA is then incorporated into the RISC, which then "seeks out" an appropriate target mRNA, after which the siRNA unwinds. It is believed that the antisense strand directs degradation of the complementary strand of mRNA through a mechanism that employs a combination of endo- and exonuclease enzymes.

Although, generally, siRNAs are derived from an exogenously administered dsRNA, endogenous siRNAs (endo-siRNAs) that are naturally generated within cells have recently been discovered. In mammals, endogenous siRNAs are most abundant in germ cells, but in invertebrates and plants the siRNAs are more widespread. The latter provide protection against viruses, facilitate the degradation of overproduced mRNA, enable the removal of abortive mRNA transcripts, and prevent the disruption of genomic DNA by suppressing transposons.

3 Intercellular Regulation of Gene Expression by Small RNAs

In addition to their regulatory role within cells, small RNAs are involved in intercellular regulation in both plants and animals. The finding that viral-induced siRNAs and transgenes can move within plants is a key piece of evidence in support of this regulatory role, and accumulating scientific evidence suggests that small RNAs in plant can move between cells and have regulatory functions [32–34].

3.1 Intercellular Regulation of Gene Expression by miRNAs

Several recent studies have shown that mammalian miRNAs are not restricted to the intracellular environment. Indeed, mammalian miRNAs can be actively secreted into the extracellular environment through the release of small membranous vesicles called *microvesicles*, or by packaging with RNA-binding proteins (e.g., high-density lipoproteins). Extracellular miRNAs may function as secreted signaling molecules that influence recipient cell phenotypes [35, 36]. Whilst this type of intercellular regulation of gene expression by secreted miRNAs has been characterized extensively in mammalian cells [35, 36], neither the mechanisms responsible for the transport of these miRNAs, nor their mode of action as signaling molecules, is presently understood.

Secreted miRNAs have important roles in stem cell function, hematopoiesis, and immune regulation [37–41]. Several lines of evidence have suggested that two types of membrane-covered microvesicles released by various types of cells, exosomes and shedding vesicles, are important compartments involved in the transport of miRNAs, helping them to function efficiently in mammalian cells. For example, recent results have shown that monocytes can package miR-150 into exosomes and secrete these miRNAs into the blood. This exosome-transported miR-150 can be absorbed by endothelial cells, and exogenous miR-150 can specifically decrease the expression of its target gene c-Myb in endothelial cells, eventually leading to an enhanced endothelial cell migration [40]. Furthermore, it is believed that exosome and shedding vesicle packaging can allow miRNAs to avoid RNase digestion. Other results have also suggested that the miRNA packaging process is a selective rather than random event [40]. Recently, it was demonstrated that miRNAs contained in exosomes are released through ceramide-dependent secretory machinery [42]. In addition to exosomes and shedding vesicles, AGO2 [43] and high-density lipoproteins [44] have recently been proposed to act as carriers of miRNAs. These proteins may provide alternative, non-vesicular means of transporting miRNAs between cells.

3.2
Intercellular Regulation of Gene Expression by siRNAs

In plants and some animals, including *C. elegans*, siRNA has the ability to spread from cell to cell – a phenomenon referred to as *systemic RNAi* [45]. This mechanism can mediate passive cellular uptake and cell-to-cell distribution of siRNA across tissues and cell boundaries, leading to sequence-specific mRNA silencing in distant cells.

In plants, small RNAs move between cells through plasmodesmata and move over long distances through the phloem vascular tissue [46]. The mobility of endogenous small RNA (including miRNA) may serve as signals affecting plant cell functions, such as guiding the patterning of leaves and roots [32]. Small RNAs, such as dsRNA, can be transported by extracellular vesicles, and single-stranded RNA (ssRNA) can bind to phloem small RNA-binding protein 1 (PSRP1). For long-range movement, silencing signals can even initiate signal amplification through RNA-dependent RNA polymerases (PDRs) [47, 48].

In *C. elegans*, silencing triggered by injected, ingested, or locally expressed dsRNA can spread throughout the organism, enabling it to silence the targeted gene in all non-neuronal cells, including the germline, thus transmitting silencing to the next generation [49–52]. Systemic RNA interference defective protein 1 (SID-1) is a transmembrane protein that enables systemic RNAi in the nematode *C. elegans* [53], and it functions as a pore or channel that transports siRNA into and out of cells. SID-2 is another small transmembrane protein that appears to act as a receptor for the uptake of siRNA from the environment. The characterization of SID-2 activity in a variety of nematodes indicates that *C. elegans* SID-2 is required for environmental RNAi [54]. SID channels are present in a wide variety of invertebrate and vertebrate animals, and have some specificity for small RNA molecules. Although the siRNA spreading phenomenon is apparent in nematodes, systemic RNAi does not appear to be a common feature in mammals. However, recent reports have shown that the human SID-1 homolog, SIDT1, enhances siRNA uptake in human systems, resulting in increased siRNA-mediated gene-silencing efficacy [55–57]. Notably, Elhassan et al. have shown that SIDT1 facilitates rapid, contact-dependent, bidirectional siRNA, and miRNA transfer between human cells, resulting in target-specific non-cell-autonomous RNAi [58]. Furthermore, it was shown that the SIDT1-mediated intercellular transfer of miR-21, a famous oncogenic miRNA, contributes to the acquisition of drug resistance [58]. This observation suggests the possibility that SID channels may play a fundamental role in complex intercellular communication, most likely via exogenous small RNAs.

4
Regulation of Animal Gene Expression by Ingested Plant miRNAs

Previously, there was no evidence that regulation via different kingdom-derived molecules could be achieved. However, it was reported recently for the first time that food-derived exogenous plant miRNAs could pass through the mouse gastrointestinal tract and enter the circulation and various organs, especially the liver. Within the liver, the plant-derived miRNAs regulate target gene expression and physiological conditions across kingdoms [59]. This is an entirely new concept related to small RNA regulation and many other biological aspects. This type of cross-kingdom regulation between plants and animals may provide new insights into nutrition, evolution, medicine, and other disciplines.

4.1
Existence of Plant miRNAs in Animals

By using deep sequencing and a quantitative RT-PCR assay, numerous plant miRNAs were initially detected – including MIR156a and MIR168a – in the sera and tissues of humans and herbivorous animals [59]. Subsequently, oxidized deep sequencing was utilized to assess whether the identified plant miRNAs were 2'-O-methyl modified on their terminal nucleotide. In oxidized deep sequencing, animal miRNAs with free 2' and 3' hydroxyls were oxidized on their terminal nucleotide, preventing these miRNAs from ligating to the adapter and being sequenced; in contrast, plant miRNAs bearing 2'-O-methylated 3' ends were resistant to the oxidizing agent sodium periodate [28, 60]. It was shown subsequently that the putative plant miRNAs would be successfully sequenced after oxidation [59], which in turn suggested that they contained 2'-O-methylated 3' ends and were, therefore, genuine plant miRNAs.

It was assumed that the plant miRNAs found in animal systems were functional, and therefore the expression level of these plant miRNAs is an important issue. It was calculated that some plant miRNAs, such as MIR168a, are present in mammalian cells and tissues at concentrations similar to those of other endogenous miRNAs (e.g., miR-25) [59]. Hence, it is believed that exogenous plant miRNAs can reach a functional concentration in animals.

4.2
Absorption of Exogenous Plant miRNAs by Animal Gastrointestinal Tract

Because plant miRNAs are completely exogenous to animals, plant miRNAs detected in the sera and organs of animals should only be derived from the animal's food uptake. To test this hypothesis, mice were first fed with either a chow diet or rice, after which the levels of plant miRNAs were measured in the animals' sera and tissues [59]. In agreement with the observation that plant miRNA MIR168a levels were higher in rice than in chow [61], MIR168a levels were significantly higher in the serum and liver of rice-fed mice compared to chow-fed mice [59]. To exclude any interference by other materials, mice were then fed via gavage with either total RNA extracted from rice, or with synthetic MIR168a; subsequently, the levels of MIR168a in the animals' serum and liver were found also to be elevated after feeding [59]. Taken together, these results suggest that the plant miRNAs had accumulated in the sera and tissues of mice as a result of food intake, and that exogenous mature plant miRNAs in food could indeed be absorbed via the gastrointestinal tract.

The mechanism underlying the absorption of exogenous plant miRNAs is insufficiently understood, however. Plant-derived miRNAs must face many challenges as they are taken up by the body via food sources before reaching their target organs. For example, once inside the mammalian gastrointestinal tract, exogenous miRNAs face a number of extreme factors, including RNase, phagocytosis, and a low-pH environment. These unfavorable conditions require the miRNAs to adopt stable structures in order to protect themselves from degradation prior to reaching the recipient cells. Plant miRNAs are methylated on the 2'-hydroxyl group of the 3'-terminal nucleotide, which inhibits 3'-end uridylation and any subsequent 3'-to-5' exonuclease digestion. Consistent with this observation, it was shown that mature plant miRNAs have slower degradation rates compared to synthetic unmethylated miRNAs, which in turn suggests that methylation contributes to the stability of plant miRNAs *in vivo*. Furthermore, the mechanism by which intestinal epithelial cells absorb exogenous plant miRNAs is unclear. It has been hypothesized that SID-1/2 may be the transporter responsible for the uptake of exogenous plant miRNAs by these cells. To date, SID-1/2 proteins and their orthologs have been shown to mediate the uptake of many types of small RNAs, including double-stranded siRNAs, single-stranded pre-miRNAs, and single-stranded mature miRNAs [55–58]. Clearly, further investigations are required to determine whether SID-1/2 proteins are capable of transporting natural mature plant miRNAs, while mice with mutant SID proteins represent an ideal model for studying the dietary delivery of plant-derived miRNAs.

4.3
Regulation of Animal Gene Expression by Plant miRNAs

The low-density lipoprotein receptor adapter protein 1 (LDLRAP1) was also found to be a target of MIR168a in mammals [59]. Typically, plant MIR168a binds to the LDLRAP1 binding site located in exon 4, thereby inhibiting LDLRAP1 protein expression [59]. Although it is unknown precisely how plant-derived miRNAs regulate their target mRNAs in mammalian cells, mammalian miRNAs generally execute their function by an imperfect base-pairing to complementary sites in the 3' UTR of their target genes, either blocking translation or triggering degradation of the target mRNAs. To date, few reports have been made indicating that mammalian miRNA can bind to target gene exons [62–64]. Conversely, plant miRNAs tend to induce mRNA cleavage through perfect or near-perfect complementarity to their target sequences, which are often in mRNA-coding regions. The fact that MIR168a reduced LDLRAP1 protein levels in liver cells, but had no effect on the mRNA levels, indicates that it behaves – at least in some ways – in similar fashion to a mammalian miRNA. Yet, the fact that MIR168a shows near-perfect complementarity to its target sequence in the coding region of LDLRAP1 suggests that, in some respects, it behaves like a plant miRNA. Clearly, it would be of great interest to analyze the regulatory mechanisms of other plant miRNAs in mammalian cells.

4.4
Transport of Plant miRNAs by Microvesicles

Another important issue regarding plant miRNA functions in animal systems is how plant miRNAs are shuttled from the original cells to the target organs, such as the liver in mammals. It was shown previously that mammalian cells could selectively package miRNAs into microvesicles in response to different stimuli, and then secrete the microvesicles into the circulation [40]. Cell-derived microvesicles can efficiently deliver miRNAs to the recipient cells, where these exogenous miRNAs regulate the expression of target genes and the biological functions of recipient cells. Microvesicles may also deliver components of the RISC, such as AGO2, to ensure that the packaged miRNAs are active [40]. It has been hypothesized that the intestinal epithelial cells are able to take up plant miRNAs and package them into microvesicles; the plant miRNA-containing microvesicles could then be released into the circulatory system and delivered to recipient organs and cells. Consistent with this hypothesis, more than half of the MIR168a detected in animal serum was found inside microvesicles [59]. It was also shown that, although the microvesicles contained less MIR168a, they caused an increased repression of LDLRAP1 when compared to "free" MIR168a [59]. Furthermore, MIR168a was found to be associated with AGO2, both in microvesicles from human intestinal epithelial cells and in target liver cells [59]. Based on these observations, two conclusions can be drawn:

- AGO2-associated exogenous plant MIR168a represents an enriched fraction of active and functional miRNAs because it has a high affinity for AGO2, although the underlying mechanism of AGO2 assembly remains unknown.
- Plant miRNAs can utilize mammalian AGO2 to execute their functions, which are similar to mammalian miRNA functions.

4.5
Physiological Effects of Plant miRNAs in Mammalian Systems

It has been widely reported that the downregulation of LDLRAP1 increases plasma LDL levels [65, 66]. In the present authors' studies, a direct reduction of LDLRAP1 in mouse liver by RNAi led to a significantly elevated plasma level of low-density lipoprotein (LDL) [59], thus confirming that LDLRAP1 is a candidate gene that may be responsible for plasma LDL removal. Interestingly, an elevated level of MIR168a and a decreased level of LDLRAP1 in mouse liver were detected at 6 h after rice feeding [59]; this indicated that exogenous plant MIR168a from food intake can quickly alter the levels of LDLRAP1 in the mouse liver. The continuous downregulation of LDLRAP1 levels in the liver by MIR168a resulted in an elevation of plasma LDL-cholesterol levels [59], indicating that plant-derived MIR168a is physiologically relevant. The reduction in LDLRAP1 protein levels due to the rice diet, and the elevation of plasma LDL-cholesterol levels, could be largely reversed by MIR168a antisense oligoribonucleotides [59], thus confirming that the rice-mediated effect is specifically caused by MIR168a targeting LDLRAP1 in the liver. This conclusion was supported by the finding that a chow diet, when supplemented with mature MIR168a, caused a significant enhancement in the levels of mouse liver MIR168a and plasma LDL-cholesterol, but decreased the levels of mouse liver LDLRAP1 [59].

5
Regulation of Animal Gene Expression by Ingested siRNAs

Animal gene expression can also be regulated by siRNAs constructed in transgenic bacteria or plants. For example, feeding *C. elegans* with bacteria that express siRNA can lead to a systemic depletion of targeted mRNAs [67]. This finding indicates that bacteria and nematodes from different kingdoms can communicate with one another via a systemic RNAi mechanism. Furthermore, recent findings have suggested that siRNAs produced by transgenic plants may have an effect on insects. For example, in 2007 Mao *et al.* identified a cytochrome P450 gene, *CYP6AE14*, from the cotton bollworm *Helicoverpa armigera*; this gene allowed the herbivore to tolerate what would otherwise be inhibitory concentrations of the cotton metabolite gossypol [68]. It was shown that, when the worm larvae were fed plant material that expressed siRNA specific to CYP6AE14, the levels of this protein were decreased in the midgut and larval growth was retarded. These results confirmed that siRNA can be transmitted from plants to insects, which in turn suggests that plant material expressing siRNA could be used to protect crops from insect damage. Baum *et al.* also showed that transgenic corn plants engineered to express siRNAs against the V-ATPase of the western corn rootworm (*Diabrotica virgifera virgifera* Leconte) were more resistant to damage caused by rootworm feeding [69]. Cumulatively, these results suggest that it may be possible to exploit the RNAi pathway to control insect pests by expressing siRNA in transgenic plants and crops, although optimization studies must clearly be conducted before RNAi-mediated insect control will be considered efficient. Important factors that might influence the success of RNAi in specific organisms, and with respect to specific targets, include: siRNA concentration; nucleotide sequences; siRNA fragment length; persistence of silencing; and the life stage of the targeted insect during feeding. Nonetheless, several questions will also need to be addressed, including: How can any advantage be taken of the specificity of RNAi to develop crops that are resistant to pest insects but are not harmful to humans? And: What can be done to slow or eliminate the development of insect resistance to these methods? Clearly, whilst many challenges remain to be overcome in this new and exciting field, the findings to date have confirmed that organisms from different kingdoms in Nature can indeed communicate one with another through exogenous siRNAs.

6
Issues Raised by Small RNA-Mediated Cross-Kingdom Regulation

With the discovery that exogenous miRNAs or siRNAs derived from plants can regulate gene expression in animal cells, many issues (some related to our daily lives) are raised.

6.1
Nutrients

Because plant-derived small RNAs from food can be absorbed into animal systems and thereby regulate animal gene expression and affect animal physiology, these small RNAs may be equally as important as the six types of nutrient in food, namely water, proteins, free fatty acids, carbohydrates, vitamins, and mineral elements. Moreover, these small molecules may represent a previously unknown but essential bioactive compound in food, and clearly the function and power of plant-derived small RNAs in food should be re-evaluated. In this respect, certain questions remain to be answered:

- What is the scope and extent of food-derived small RNAs and their regulation of mammalian genes?
- What are the mammalian targets (other than LDLRAP1) of these small RNAs?
- What are the potential effects of these food-derived small RNAs on the metabolism and health of humans and animals?

Initially, due to the complexity of the physiology involved, a panel of organs and potential targets would need to be further examined in animal models. It would also be interesting to determine the contribution of individual small RNAs in regulating mammalian genes; an example of this would be to use engineered mice expressing individual plant-derived small RNAs.

6.2
Medicine

The results of these studies also imply that signaling and regulatory information is acquired through the foods that are consumed, and this may impact on the design of genetically engineered plants in the future, or lead to the identification of active ingredients in plants that have therapeutic effects. Indeed, engineered plants expressing siRNAs that specifically silence animal genes could be used to produce plant-derived small RNAs for gene therapy. More importantly, the fact that plant-derived small RNAs could affect human cellular behavior may partly explain the efficacy of Chinese herbal medicines that have been used to cure various diseases for thousands of years. In fact, some research groups have focused on isolating active and functional components from herbs. Beyond the components traditionally thought to be active in herbs, it is possible that small RNAs in the herbs may be preprogrammed to direct certain cells to initiate a healing process. Further studies are clearly required, however, to determine whether some such herbs contain unique small RNA signatures that are responsible for health-promoting properties.

6.3
Evolution

It is known that small RNAs can be transmitted from one species to another, facilitating cross-talk, communication and signal interference in different species, and even in cross-kingdom fashion. Whilst this novel mechanism may have important effects on the evolution of many extant species, it may also shed light on the present understanding of cross-kingdom (e.g., plant-animal) interactions and co-evolution. Surely, if plant miRNAs can function in animal cells, then it may be possible for animal miRNAs to act in plant cells? Within such an environment, plant roots might absorb animal miRNAs from animal waste or from animal-derived ingredients in organic fertilizers, and these miRNAs could then alter plant gene expression. Such pioneering studies may greatly enhance the present understanding of molecular signaling between species.

7
Summary

Although small RNAs were once thought to be unstable molecules, they have since been found to be transferred horizontally across species and kingdoms, and these findings suggest a novel role for these molecules in inter-species/inter-kingdom communication and coevolution. However, as this is an entirely new research area within the field of gene regulation, many problems must first be resolved with regards to the mechanisms and roles of exogenous small RNAs in both normal physiological and pathological contexts before proposals can be made of the implications of exogenous small RNA-mediated gene regulation.

References

1 Ghildiyal, M., Zamore, P.D. (2009) Small silencing RNAs: an expanding universe. *Nat. Rev. Genet.*, **10** (2), 94–108.
2 Moazed, D. (2009) Small RNAs in transcriptional gene silencing and genome defence. *Nature*, **457** (7228), 413–420.
3 Bartel, D.P. (2004) MicroRNAs: genomics, biogenesis, mechanism, and function. *Cell*, **116** (2), 281–297.
4 Du, T., Zamore, P.D. (2005) microPrimer: the biogenesis and function of microRNA. *Development*, **132** (21), 4645–4652.
5 Lee, Y.S., Dutta, A. (2009) MicroRNAs in cancer. *Annu. Rev. Pathol.*, **4**, 199–227.
6 Elbashir, S.M., Harborth, J., Lendeckel, W., Yalcin, A. *et al.* (2001) Duplexes of 21-nucleotide RNAs mediate RNA interference in cultured mammalian cells. *Nature*, **411** (6836), 494–498.
7 Lee, R.C., Feinbaum, R.L., Ambros, V. (1993) The *C. elegans* heterochronic gene lin-4 encodes small RNAs with antisense complementarity to lin-14. *Cell*, **75** (5), 843–854.
8 Wightman, B., Ha, I., Ruvkun, G. (1993) Posttranscriptional regulation of the heterochronic gene lin-14 by lin-4 mediates temporal pattern formation in *C. elegans*. *Cell*, **75** (5), 855–862.
9 Reinhart, B.J., Slack, F.J., Basson, M. and Pasquinelli, A.E. *et al.* (2000) The 21-nucleotide let-7 RNA regulates developmental timing in *Caenorhabditis elegans*. *Nature*, **403** (6772), 901–906.
10 Pasquinelli, A.E., Reinhart, B.J., Slack, F. Martindale, M.Q. *et al.* (2000) Conservation of the sequence and temporal expression of let-7 heterochronic regulatory RNA. *Nature*, **408** (6808), 86–89.
11 He, L., Hannon, G.J. (2004) MicroRNAs: small RNAs with a big role in gene regulation. *Nat. Rev. Genet.*, **5** (7), 522–531.
12 Lee, Y., Kim, M., Han, J., Yeom, K.H. *et al.* (2004) MicroRNA genes are transcribed by RNA polymerase II. *EMBO J.*, **23** (20), 4051–4060.

13 Gregory, R.I., Yan, K.P., Amuthan, G. Chendrimada, T. et al. (2004) The microprocessor complex mediates the genesis of microRNAs. *Nature*, **432** (7014), 235–240.

14 Lee, Y., Ahn, C., Han, J. Choi, H. et al. (2003) The nuclear RNase III Drosha initiates microRNA processing. *Nature*, **425** (6956), 415–419.

15 Lund, E., Guttinger, S., Calado, A. Dahlberg, J.E. et al. (2004) Nuclear export of microRNA precursors. *Science*, **303** (5654), 95–98.

16 Hutvagner, G., McLachlan, J., Pasquinelli, A.E. Balint, E. et al. (2001) A cellular function for the RNA-interference enzyme Dicer in the maturation of the let-7 small temporal RNA. *Science*, **293** (5531), 834–838.

17 Chendrimada, T.P., Gregory, R.I., Kumaraswamy, E., Norman, J. et al. (2005) TRBP recruits the Dicer complex to Ago2 for microRNA processing and gene silencing. *Nature*, **436** (7051), 740–744.

18 Lee, Y., Hur, I., Park, S.Y. Kim, Y.K. et al. (2006) The role of PACT in the RNA silencing pathway. *EMBO J.*, **25** (3), 522–532.

19 Schwarz, D.S., Hutvagner, G., Du, T. Xu, Z. et al. (2003) Asymmetry in the assembly of the RNAi enzyme complex. *Cell*, **115** (2), 199–208.

20 Khvorova, A., Reynolds, A., Jayasena, S.D. (2003) Functional siRNAs and miRNAs exhibit strand bias. *Cell*, **115** (2), 209–216.

21 Liu, J.D., Carmell, M.A., Rivas, F.V. Marsden, C.G. et al. (2004) Argonaute2 is the catalytic engine of mammalian RNAi. *Science*, **305** (5689), 1437–1441.

22 Hutvagner, G., Simard, M.J. (2008) Argonaute proteins: key players in RNA silencing. *Nat. Rev. Mol. Cell Biol.*, **9** (1), 22–32.

23 Reinhart, B.J., Weinstein, E.G., Rhoades, M.W. Bartel, B. et al. (2002) MicroRNAs in plants. *Genes. Dev.*, **16** (13), 1616–1626.

24 Tang, G., Reinhart, B.J., Bartel, D.P., Zamore, P.D. (2003) A biochemical framework for RNA silencing in plants. *Genes. Dev.*, **17** (1), 49–63.

25 Bollman, K.M., Aukerman, M.J., Park, M.Y. Hunter, C. et al. (2003) HASTY, the *Arabidopsis* ortholog of exportin 5/MSN5, regulates phase change and morphogenesis. *Development*, **130** (8), 1493–1504.

26 Park, M.Y., Wu, G., Gonzalez-Sulser, A., Vaucheret, H. et al. (2005) Nuclear processing and export of microRNAs in Arabidopsis. *Proc. Natl Acad. Sci. USA*, **102** (10), 3691–3696.

27 Park, W., Li, J., Song, R., Messing, J. et al. (2002) CARPEL F.A.C.T.O.R.Y. a Dicer homolog, and HEN1, a novel protein, act in microRNA metabolism in *Arabidopsis thaliana*. *Curr. Biol.*, **12** (17), 1484–1495.

28 Yu, B., Yang, Z., Li, J., Minakhina, S. et al. (2005) Methylation as a crucial step in plant microRNA biogenesis. *Science*, **307** (5711), 932–935.

29 Eulalio, A., Tritschler, F., Izaurralde, E. (2009) The GW182 protein family in animal cells: New insights into domains required for miRNA-mediated gene silencing. *RNA*, **15** (8), 1433–1442.

30 Wakiyama, M., Yokoyama, S. (2010) MicroRNA-mediated mRNA deadenylation and repression of protein synthesis in a mammalian cell-free system. *Prog. Mol. Subcell. Biol.*, **50**, 85–97.

31 Wu, E., Thivierge, C., Flamand, M. Mathonnet, G. et al. (2010) Pervasive and cooperative deadenylation of 3′UTRs by embryonic microRNA families. *Mol. Cell*, **40** (4), 558–570.

32 Carlsbecker, A., Lee, J.Y., Roberts, C.J. Dettmer, J. et al. (2010) Cell signalling by microRNA165/6 directs gene dose-dependent root cell fate. *Nature*, **465** (7296), 316–321.

33 Dunoyer, P., Schott, G., Himber, C. Meyer, D. et al. (2010) Small RNA duplexes function as mobile silencing signals between plant cells. *Science*, **328** (5980), 912–916.

34 Molnar, A., Melnyk, C.W., Bassett, A. Hardcastle, T.J. et al. (2010) Small silencing RNAs in plants are mobile and direct epigenetic modification in recipient cells. *Science*, **328** (5980), 872–875.

35 Chen, X., Liang, H., Zhang, J. Zen, K. et al. (2012) Secreted microRNAs: a new form of intercellular communication. *Trends Cell Biol.*, **22** (3), 125–132.

36 Chitwood, D.H., Timmermans, M.C. (2010) Small RNAs are on the move. *Nature*, **467** (7314), 415–419.

37 Hunter, M.P., Ismail, N., Zhang, X. Aguda, B.D. et al. (2008) Detection of microRNA expression in human peripheral blood microvesicles. *PLoS ONE*, **3** (11), e3694.

38 Yuan, A., Farber, E.L., Rapoport, A.L. Tejada, D. et al. (2009) Transfer of microRNAs by embryonic stem cell microvesicles. *PLoS ONE*, **4** (3), e4722.

39 Collino, F., Deregibus, M.C., Bruno, S. Sterpone, L. et al. (2010) Microvesicles derived from adult human bone marrow and tissue specific mesenchymal stem cells shuttle selected pattern of miRNAs. *PLoS ONE*, **5** (7), e11803.

40 Zhang, Y., Liu, D., Chen, X. Li, J. et al. (2010) Secreted monocytic miR-150 enhances targeted endothelial cell migration. *Mol. Cell*, **39** (1), 133–144.

41 Mittelbrunn, M., Gutierrez-Vazquez, C., Villarroya-Beltri, C. Gonzalez, S. et al. (2011) Unidirectional transfer of microRNA-loaded exosomes from T cells to antigen-presenting cells. *Nat. Commun.*, **2**, 282.

42 Kosaka, N., Iguchi, H., Yoshioka, Y. Takeshita, F. et al. (2010) Secretory mechanisms and intercellular transfer of microRNAs in living cells. *J. Biol. Chem.*, **285** (23), 17442–17452.

43 Turchinovich, A., Weiz, L., Langheinz, A., Burwinkel, B. (2011) Characterization of extracellular circulating microRNA. *Nucleic Acids Res.*, **39** (16), 7223–7233.

44 Vickers, K.C., Palmisano, B.T., Shoucri, B.M. Shamburek, R.D. et al. (2011) MicroRNAs are transported in plasma and delivered to recipient cells by high-density lipoproteins. *Nat. Cell Biol.*, **13** (4), 423–433.

45 Cohen, H.C., Xiong, M.P. (2011) Non-cell-autonomous RNA interference in mammalian cells: Implications for in vivo cell-based RNAi delivery. *J. RNAi Gene Silencing*, **7**, 456–463.

46 Brosnan, C.A., Voinnet, O. (2011) Cell-to-cell and long-distance siRNA movement in plants: mechanisms and biological implications. *Curr. Opin. Plant Biol.*, **14** (5), 580–587.

47 Sijen, T., Fleenor, J., Simmer, F. Thijssen, K.L. et al. (2001) On the role of RNA amplification in dsRNA-triggered gene silencing. *Cell*, **107** (4), 465–476.

48 Vaistij, F.E., Jones, L., Baulcombe, D.C. (2002) Spreading of RNA targeting and DNA methylation in RNA silencing requires transcription of the target gene and a putative RNA-dependent RNA polymerase. *Plant Cell*, **14** (4), 857–867.

49 Fire, A., Xu, S., Montgomery, M.K. Kostas, S.A. et al. (1998) Potent and specific genetic interference by double-stranded RNA in *Caenorhabditis elegans*. *Nature*, **391** (6669), 806–811.

50 Winston, W.M., Molodowitch, C., Hunter, C.P. (2002) Systemic RNAi in C. elegans requires the putative transmembrane protein SID-1. *Science*, **295** (5564), 2456–2459.

51 Whangbo, J.S., Hunter, C.P. (2008) Environmental RNA interference. *Trends Genet.*, **24** (6), 297–305.

52 Jose, A.M., Garcia, G.A., Hunter, C.P. (2011) Two classes of silencing RNAs move between *Caenorhabditis elegans* tissues. *Nat. Struct. Mol. Biol.*, **18** (11), 1184–1188.

53 Feinberg, E.H., Hunter, C.P. (2003) Transport of dsRNA into cells by the transmembrane protein SID-1. *Science*, **301** (5639), 1545–1547.

54 Winston, W.M., Sutherlin, M., Wright, A.J. Feinberg, E.H. et al. (2007) *Caenorhabditis elegans* SID-2 is required for environmental RNA interference. *Proc. Natl Acad. Sci. USA*, **104** (25), 10565–10570.

55 Duxbury, M.S., Ashley, S.W., Whang, E.E. (2005) RNA interference: a mammalian SID-1 homologue enhances siRNA uptake and gene silencing efficacy in human cells. *Biochem. Biophys. Res. Commun.*, **331** (2), 459–463.

56 Tsang, S.Y., Moore, J.C., Huizen, R.V. Chan, C.W. et al. (2007) Ectopic expression of systemic RNA interference defective protein in embryonic stem cells. *Biochem. Biophys. Res. Commun.*, **357** (2), 480–486.

57 Wolfrum, C., Shi, S., Jayaprakash, K.N. Jayaraman, M. et al. (2007) Mechanisms and optimization of in vivo delivery of lipophilic siRNAs. *Nat. Biotechnol.*, **25** (10), 1149–1157.

58 Elhassan, M.O., Christie, J., Duxbury, M.S. (2012) Homo sapiens systemic RNA interference-defective-1 transmembrane family member 1 (SIDT1) protein mediates contact-dependent small RNA transfer and microRNA-21-driven chemoresistance. *J. Biol. Chem.*, **287** (8), 5267–5277.

59 Zhang, L., Hou, D., Chen, X. Li, D. et al. (2012) Exogenous plant MIR168a specifically targets mammalian LDLRAP1: evidence of cross-kingdom regulation by microRNA. *Cell Res.*, **22** (1), 107–126.

60 Ohara, T., Sakaguchi, Y., Suzuki, T. Ueda, H. et al. (2007) The 3' termini of mouse Piwi-interacting RNAs are 2'-O-methylated. *Nat. Struct. Mol. Biol.*, **14** (4), 349–350.

61 Gazzani, S., Li, M., Maistri, S. Scarponi, E. et al. (2009) Evolution of MIR168 paralogs in Brassicaceae. *BMC Evol. Biol.*, **9**, 62.

62 Tay, Y., Zhang, J., Thomson, A.M. Lim, B. et al. (2008) MicroRNAs to Nanog, Oct4 and Sox2 coding regions modulate embryonic stem cell differentiation. *Nature*, **455** (7216), 1124–1128.

63 Huang, S., Wu, S., Ding, J. Lin, J. et al. (2010) MicroRNA-181a modulates gene expression of zinc finger family members by directly targeting their coding regions. *Nucleic Acids Res.*, **38** (20), 7211–7218.

64 Qin, W., Shi, Y., Zhao, B. Yao, C. et al. (2010) miR-24 regulates apoptosis by targeting the open reading frame (ORF) region of FAF1 in cancer cells. *PLoS ONE*, **5** (2), e9429.

65 Zuliani, G., Arca, M., Signore, A. Bader, G. et al. (1999) Characterization of a new form of inherited hypercholesterolemia: familial recessive hypercholesterolemia. *Arterioscler. Thromb. Vasc. Biol.*, **19** (3), 802–809.

66 Wilund, K.R., Yi, M., Campagna, F. Arca, M. et al. (2002) Molecular mechanisms of autosomal recessive hypercholesterolemia. *Hum. Mol. Genet.*, **11** (24), 3019–3030.

67 Timmons, L., Fire, A. (1998) Specific interference by ingested dsRNA. *Nature*, **395** (6705), 854.

68 Mao, Y.B., Cai, W.J., Wang, J.W. Hong, G.J. et al. (2007) Silencing a cotton bollworm P450 monooxygenase gene by plant-mediated RNAi impairs larval tolerance of gossypol. *Nat. Biotechnol.*, **25** (11), 1307–1313.

69 Baum, J.A., Chu, C.R., Rupar, M. Brown, G.R. et al. (2004) Binary toxins from *Bacillus thuringiensis* active against the western corn rootworm, *Diabrotica virgifera virgifera* LeConte. *Appl. Environ. Microbiol.*, **70** (8), 4889–4898.

7
RNA Interference in Animals*

Mikiko C. Siomi
Graduate School of Science, University of Tokyo,
2-11-16 Yayoi, Bunkyo-ku, Tokyo 113-0032, Japan

1	Introduction	187
2	**Small Interfering RNA (siRNA)-Mediated Gene Silencing**	**188**
2.1	Exogenous siRNA (exo-siRNA) Biogenesis	188
2.2	Endogenous siRNA (endo-siRNA) Biogenesis	191
2.3	Other Important Features of siRNA-Mediated Gene Silencing	193
3	**MicroRNA (miRNA)-Mediated Gene Silencing**	**194**
3.1	miRNA Biogenesis	194
3.2	miRNA Functions	196
4	**PIWI-Interacting RNA (piRNA)-Mediated Gene Silencing**	**199**
4.1	Primary piRNA Biogenesis and Function	199
4.2	Secondary piRNA Biogenesis and Function	201
5	Conclusions	203
	References	203

Keywords

Argonaute
Argonaute proteins are the core catalytic components of RNA interference (RNAi), and form the RNA-induced silencing complex (RISC) by interacting with a small RNA. Each animal species has one or more homologs, comprising the Argonaute family of proteins. The size of each Argonaute member varies, but is on average ∼100 kDa. Argonaute consists of four domains: the amino-terminal domain; the PAZ domain; the

---

*This is an updated version of a chapter previously published in: Meyers, R.A. (Ed.) Epigenetic Regulation and Epigenomics, 2013, ISBN 978-3-527-32682-2.

RNA Regulation: Advances in Molecular Biology and Medicine, First Edition. Edited by Robert A. Meyers.
© 2014 Wiley-VCH Verlag GmbH & Co. KGaA. Published 2014 by Wiley-VCH Verlag GmbH & Co. KGaA.

Mid domain; and the PIWI domain. The Mid and PAZ domains are the motifs that bind with the 5′ and 3′ ends of small RNA, respectively. The PIWI domain folds similar to RNase H, and serves as the catalytic domain of the protein.

Dicer

Dicer is a protein that processes small RNAs from their precursors in RNAi. Dicer contains two RNase III domains, with which Dicer processes long dsRNAs (i.e., small RNA precursors) into small RNA duplexes 20–23 nt long, by dicing the substrates from their termini. Dicer also contains a double-stranded RNA (dsRNA)-binding domain, which may contribute to recognizing the characteristic 2 nt overhang at the 3′ end of the substrate. In general, animals have one Dicer protein. However, *Drosophila* is an exception and has two Dicers – Dicer1 and Dicer2 – which process microRNA (miRNA) and small interfering RNA (siRNA), respectively, from their own precursors. In cells, Dicer is associated with a dsRNA-binding domain-containing protein, which acts as a stabilizer and/or enhancer of Dicer activity.

RISC

The RNA-induced silencing complex (RISC) is the core machinery and the catalytic engine of RNAi. The RISC interacts physically with target mRNAs, and silences them either by cleavage or inhibiting their translation. The RISC may function in transcriptional silencing. The core of the RISC is composed of one Argonaute protein and a single-stranded small RNA, 20–30 nt long. The RISC can contain other protein factors, such as GW182, when necessary.

> Each of the cells of which a living body is composed (with a few exceptions, such as spermatozoa, eggs, and anucleated and multinucleated cells) contains an identical set of genes. Yet, their expression is not identical; rather, it is regulated by multiple, complex mechanisms in a spatiotemporal and cell-specific manner. RNA interference (RNAi) is one of several gene-expression regulatory mechanisms that are necessary to generate cellular complexity and to orchestrate cellular events. RNAi is triggered by small RNAs, ranging from 20 to 30 nt in length, which show a high level of complementarity to their target genes. However, small RNA is unable to induce RNAi alone, and to accomplish the task it must form the core of the RNA–protein complex termed the RNA-induced silencing complex (RISC), specifically with Argonaute protein. Argonaute is the main effector protein in RNAi, whereby Argonaute is guided to its targets by bound small RNA. The fact that both Argonaute and small RNAs are highly conserved across species implies that RNAi is a fundamental, biological process that is required by living creatures. During recent years, the anticipation of RNAi as a new form of disease therapy has been steadily growing, based mainly on its high specificity and efficacy with regards to recognizing target genes and disrupting their expression.

1
Introduction

Gene expression in the individual cells of multicellular organisms is elaborately controlled by various mechanisms to initiate and maintain cellular differentiation, and this results in distinct gene expression profiles in each cell. One regulatory mechanism that is present in almost all eukaryotic organisms is that of RNAi (also known as *RNA silencing*), which provides a highly specific inhibition of gene expression through the complementary recognition of RNA targets by small RNAs [1–5]. The core of the RNA-induced silencing complex (RISC), the key machinery in RNAi, consists of a member of the Argonaute protein family and a small RNA [1–5].

Argonaute proteins are composed of four domains: the amino (N)-terminal domain; the PAZ (Piwi Argonaut and Zwille) domain; the Mid (Middle) domain; and the PIWI (P-element induced wimpy testis) domain [6–8]. The Mid and PAZ domains bind the 5′ and 3′ ends of small RNAs, respectively [8–10], while the PIWI domain folds in a similar fashion to RNase H and serves as the catalytic center of the protein [8, 11, 12]. Alone, Argonaute proteins are nonfunctional because they cannot target RNAs (though it has been reported very recently that, under certain conditions, Argonaute may be able to target genes without small RNA partners [13]; see Sect. 3.2 for details); however, upon association with small RNA, Argonaute proteins enable the inhibition of target gene expression, either: (i) by cleaving the mRNA using the PIWI domain that has RNAseH-like endonuclease or Slicer activity, thereby inducing translational inhibition; or (ii) by changing RNA stability or the chromatin structure via interactions with other proteins through the PIWI domain [1–5]. Eukaryotic organisms mostly possess more than one Argonaute protein [14]. A representative of unicellular eukaryotes, *Schizosaccharomyces pombe*, has only one Argonaute, whereas *Drosophila melanogaster* has five Argonautes. Likewise, eight and 27 Argonaute members are found in humans and nematodes, respectively. The functions of the individual Argonaute members are often nonredundant, which suggests their biological significance and independency in function.

Small RNAs in animals can be classified as small interfering RNAs (siRNAs), microRNAs (miRNAs), and PIWI-interacting RNAs (piRNAs) [1–3]. Each RNA shows unique characteristics with regard to size, origin, and tissue specificity. For example, the siRNAs are 20–23 nt long and can be either exogenous siRNAs (exo-siRNAs) or endogenous siRNAs (endo-siRNAs), depending on their precursor source. The exo-siRNAs are produced from double-stranded RNAs (dsRNAs) that are transcribed from transgenes or introduced into cells. The siRNAs contributing to antiviral mechanisms [15–18], and especially those identified in *Drosophila*, are considered exo-siRNAs, because they are derived from viral genomes. In contrast, endo-siRNAs arise from natural dsRNAs transcribed mostly from intergenic elements in the genome. The miRNAs are endogenous 21–23 nt-long RNAs, and are derived from their own genes in the genome. Hundreds of miRNAs are expressed in animals, mostly in a cell type-specific manner, and the number differs among species. piRNAs are endogenous 24–32 nt RNAs that, like endo-siRNAs, are derived from intergenic repetitive elements. In contrast to miRNAs and siRNAs, the piRNAs do not have their own genes; rather, they are expressed predominantly in the germline, whereas siRNAs and miRNAs are ubiquitously expressed. An overview of RNAi in animals, with attention focused mainly on small RNA biogenesis and function, will be provided in this chapter. Details will also be provided of the proteins that function in RNAi, such as Argonaute and Dicer.

2
Small Interfering RNA (siRNA)-Mediated Gene Silencing

In 1998, using *Caenorhabditis elegans* as a model system, Andrew Fire and Craig Mello demonstrated that dsRNA introduced into the bodies of worms, whether by injection or ingestion, were able to induce highly specific gene silencing [19]. This effect was termed RNA interference (RNAi), and the silencing efficacy of these dsRNAs was shown to be greater than that of single-stranded RNAs (ssRNAs). The perplexity then was how dsRNAs, the strands of which were already paired one with another, could recognize a third nucleic acid strand (i.e., the target gene). Today, the answer to this problem is well known, in that the dsRNAs serve as precursors of ss-siRNAs, which guide the Argonaute protein, the core factor of RNAi, by base-pairing with the target mRNAs [20–22]. During the early 1990s, a similar transgene-driven, sequence-specific gene silencing had been observed in plants and fungi; this process was referred to as *cosuppression* in plants, and as *quelling* in fungi, but the mechanisms involved were unclear [23, 24]. Subsequently, cosuppression and quelling were found to be mechanistically equivalent to RNAi, and the latter process was realized as being evolutionarily conserved among species. Today, the fundamental molecular mechanisms of siRNA production (or biogenesis) are well known, and involve Dicer [1–5, 25]. After processing, mature ss-siRNA is loaded onto a specific Argonaute member, Argonaute 2 (Ago2) in both *Drosophila* and humans, and onto ALG-1, ALG-2 or three other members in nematodes. Upon such association an active RISC is formed [26–28], within which Ago2 interacts with target RNAs through RNA–RNA base pairings, and cleaves them via its Slicer activity [20, 26, 29], such that the target genes are effectively downregulated.

In cells, Ago2 does not always wait for exo-siRNAs; rather, recent studies have shown that, in *Drosophila*, Ago2 also associates with siRNAs of cellular origins, which are termed *endo-siRNAs* [30–33]. The origin of endo-siRNAs includes retrotransposons and other intergenic repetitive elements, such as those in the pericentromeric and subtelomeric regions. In mice, endo-siRNA origins also include pseudogenes [34, 35]. This has revealed a new function for pseudogenes, namely to give rise to small RNAs and to downregulate their parental genes.

2.1
Exogenous siRNA (exo-siRNA) Biogenesis

Exo-siRNAs are processed by Dicer from long dsRNA precursors that consist of two RNA strands which are completely complementary to each other [25, 36]. Dicer binds the end of dsRNAs and cleaves them, approximately 21 nt from their end, using two RNaseIII domains [2, 37, 38] (Fig. 1). This "dicing" reaction produces ca. 21 nt siRNA duplexes that are in "phase," and all products possess a phosphate group and a 2 nt overhang at the 5′ and 3′ ends, respectively; these are typical signatures of RNaseIII processing. Because the length of siRNAs is defined

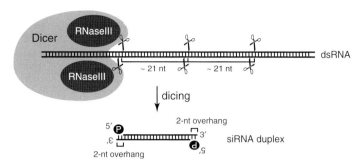

Fig. 1 siRNA processing by Dicer. Dicer contains two RNase III domains, with which Dicer processes long dsRNAs into siRNA duplexes (ca. 21 nt) by dicing the substrates from their termini. The resultant siRNA duplexes contain a 2-nt overhang at the 3' end, and a phosphate group at their 5' end.

Tab. 1 Key proteins in RNAi in mammals, flies, and nematodes.

Protein	Mammal	Fly[a]	Nematode[b]
RNase III	Dicer	Dicer1 Dicer2	DCR-1
	Drosha	Drosha	DRSH-1
Argonaute (AGO)	Ago1 Ago2 Ago3 Ago4 Ago5[c] (mouse)	Ago1 Ago2	ALG-1 ALG-2 Three others
Argonaute (PIWI)	MILI (HILI) MIWI (HIWI) MIWI2 (HIWI2) HIWI3 (human)	Ago3 Piwi Aub	ERGO-1 PRG-1 PRG-2
Cofactors of RNaseIII[d]	DGCR8 TRBP PACT	Pasha R2D2 Loqs	PASH-1 RDE-4

[a] Drosophila melanogaster.
[b] Caenorhabditis elegans.
[c] Possibly a pseudogene.
[d] dsRNA-binding domain-containing proteins.
PRG, PIWI-related genes; RDE, RNAi-defective.

by this dicing reaction, however, Dicer is considered a "molecular ruler."

Although many animals, including nematodes, mice, and humans, express only one Dicer (Table 1), *Drosophila* is known to contain two Dicers, namely Dicer1 and Dicer2 [39]. Although both Dicer1 and Dicer2 are ubiquitous, they are functionally nonredundant; Dicer2 is required for exo- and endo-siRNA biogenesis, while

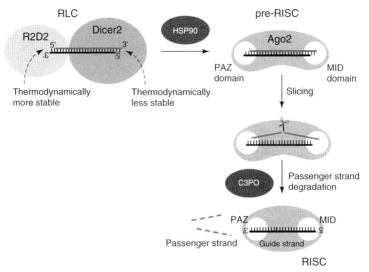

Fig. 2 RISC formation in *Drosophila*. The siRNA duplex in RLC is transferred to Argonaute protein, forming pre-RISC. Argonaute then cleaves one strand of the duplex (passenger strand) with its Slicer activity. The cleaved fragments are further processed by C3PO and disappear. Finally, an active RISC is formed.

Dicer1 is dispensable for siRNA biogenesis but is necessary for miRNA biogenesis (see Sect. 3.1).

Dicer associates with a protein containing dsRNA-binding domains (Table 1); in mammals either the TAR RNA-binding protein (TRBP; also known as TARBP2) or protein kinase interferon-inducible double-stranded RNA-dependent activator (PRKRA; also known as *PACT*) acts as the Dicer partner [2, 40, 41]. Functional differences between TRBP/TARBP2 and PRKRA/PACT remain elusive. In nematodes and flies, however, RDE-4 and R2D2 correspond to the counterparts of TARBP2/TRBP and PRKRA/PACT, respectively [42, 43]. The "dicing" activity of Dicer does not require its partner *in vitro*. In flies and humans, however, a lack of R2D2 and TARBP2/TRBP causes Dicer2 and Dicer, respectively, to be destabilized *in vivo*. In this case, the dsRNA-binding domain-containing proteins act, at least in part, as the Dicer stabilizer.

siRNA must be single-stranded in order to function in RNAi. Prior to the siRNA "unwinding" process, the siRNA duplex associates with Argonaute. In *Drosophila*, the Dicer2–R2D2 heterodimer functions at this step. R2D2 favors binding to one particular end of the siRNA duplex, which is thermodynamically more stable than the other end [44, 45]. Dicer2 may bind to the other end of the duplex (the end that is thermodynamically less stable than the one selected by R2D2), although detailed structural evidence is still awaited for certainty [46] (Fig. 2). This thermodynamic inequality, which is led at least by R2D2, contributes to the selection of one particular strand (the guide strand) over the other (the passenger strand) to function in RNAi. The Dicer2–R2D2 heterodimer is referred to as the RISC-loading complex (RLC) [29, 46]. The RLC may contain proteins other than Dicer2 and R2D2, but these remain (as yet) undetermined. In mammals, it remains unclear whether the

Dicer-TRBP/TARBP2 (or Dicer-PRKRA/PACT) complex functions according to thermodynamic inequality.

Argonaute in a form associated with an siRNA duplex is termed the *pre-RISC*. During pre-RISC formation, Dicer2 in the RLC may be replaced with Argonaute, which would concomitantly displace R2D2 from the duplex. This hierarchic organization is thought to greatly contribute to the selection of one siRNA strand over the other (Fig. 2), although pre-RISC formation does not appear to require the RLC in either flies or mammals. The heat shock protein ATPase HSP90, which is a chaperone for Argonaute, interacts and stimulates Argonaute to accommodate binding of the siRNA duplex, apparently by changing (most likely stretching) the conformation of Argonaute (Fig. 2), although structural evidence for this has not yet been provided [47, 48]. The inhibition of HSP90 function causes a failure of pre-RISC formation; thus, unlike the RLC, HSP90 is required for pre-RISC formation.

Argonaute cleaves one strand of the duplex in the same way as it cleaves the target RNA pairing with siRNA in RNAi (Fig. 2). Indeed, without this reaction the siRNA may not be efficiently made single-stranded, as Argonaute uses one particular strand of the duplex as a guide siRNA, the 5′ end of which is bound to the MID domain (Fig. 2). The other strand (the passenger strand) is cleaved by Argonaute–Slicer [49–51], after which the cleaved elements are further processed by the nuclease complex C3PO (also known as *Trax-Translin*). This seems to accelerate (but may not be necessary for) the siRNA unwinding process [52] (Fig. 2). Concomitantly, an active RISC is formed to silence the target genes.

Argonaute members that form the RISC with siRNA (siRISC) are ubiquitously expressed and, therefore, at least in theory the siRNA-mediated RNAi will be available in each cell. The ubiquitous Argonaute proteins are considered members of the "AGO" subgroup of the Argonaute family [6] (Table 1), whereas Argonaute proteins that are predominantly expressed in the germline are generically referred to as "*PIWI*" proteins [6] (see Sect. 44). In mammals, among the four AGO proteins Ago1, Ago2, Ago3, and Ago4, only Ago2 exhibits Slicer activity [20, 26, 53]. The other members contain the amino acid residues within the PIWI domain that are required for Slicer activity (the D-D-H triad), but do not show the activity [20, 26, 53]. Accordingly, it has been postulated that Ago2 – but not Ago1, 3 and 4 – may be able to form an active RISC with siRNA, unless other mechanisms are present that bypass Ago2 to help the duplex become single-stranded. Indeed, in mammalian cells all four AGO proteins were found to be loaded with guide siRNAs [20, 26], which suggests that Ago2-Slicer bypass mechanisms may act in mammalian cells. All four mammalian AGO proteins are also able to form a RISC with miRNA, and function in the miRNA-mediated RNAi pathway (see Sect. 3). Typically, *Drosophila* has two AGO members, namely Ago1 and Ago2. Although both have Slicer activity, Ago2 functions in siRNA-mediated RNAi, while Ago1 functions in miRNA-mediated RNAi [27, 50] (see Sect. 3).

2.2 Endogenous siRNA (endo-siRNA) Biogenesis

The precursors of endo-siRNAs mainly originate from the intergenic elements in the genome, including retrotransposons and their remnants [30–35]. Their precursors include long dsRNAs that are composed of sense and antisense, or converged, RNAs which have been transcribed from the same (i.e., cis) or different (i.e., trans) loci and very often contain mismatches across their entire length (Fig. 3a). In mice, endo-siRNAs also arise from transcripts of pseudogenes (in antisense), duplexed with their parental, functional, protein-coding genes (in sense) [34, 35]. Hairpin-type molecules with a long stem may serve as endo-siRNA precursors (Fig. 3a). However, as the stem becomes shortened during the processing it may be recognized as a miRNA precursor and processed by Dicer1 [54]. In this case, it appears that the length of the stem serves as the determinant of endo-siRNA or miRNA precursors.

In mammals, the molecular mechanisms of endo-siRNA biogenesis remain largely unknown, except that it at least requires Dicer, and that mature endo-siRNAs associate with Ago2, as do exo-siRNAs [34, 35]. In *Drosophila*, endo-siRNA biogenesis requires Dicer2 and Loquacious (Loqs; also known as *R3D1*), another dsRNA-binding domain protein [54, 55]. The *Loqs* gene gives rise to four Loqs isoforms, Loqs-PA to Loqs-PD, and the shortest isoform Loqs-PD is the one involved specifically in endo-siRNA production [54–58] (Fig. 3b). Loqs-PD has a unique C-terminal sequence, which serves

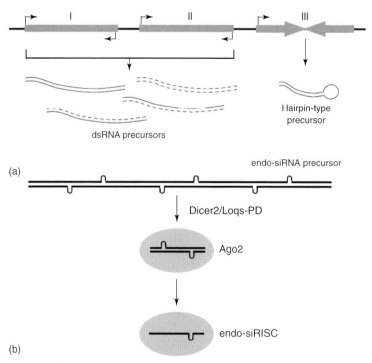

Fig. 3 Endo-siRNA biogenesis. (a) The transcripts from endo-siRNA loci form either long dsRNAs or hairpin-shaped endo-siRNA precursors; (b) In *Drosophila*, endo-siRNAs are processed from the precursors by the Dicer2/Loqs-PD complex. Endo-siRNA duplex is loaded onto Argonaute and then becomes single-stranded, forming an active endo-siRISC.

as the Dicer2-binding motif, and it is through this association that Loqs-PD enhances the dicing step. R2D2 may function also in endo-siRNA loading onto Ago2 [54], but the requirement for HSP90/ATP and Ago2–Slicer in endo-siRNA biogenesis remains elusive. In summary, both exo- and endo-siRNA biogenesis are similar in the sense that both require Dicer2 and R2D2, and that both products are loaded onto Ago2. However, each has unique points; for example, Loqs-PD is required only in endo-siRNA biogenesis.

2.3
Other Important Features of siRNA-Mediated Gene Silencing

Immediately after the discovery of RNAi in 1998, it was assumed that it would be improbable to induce RNAi in mammalian cells with long dsRNAs, because mammals are sensitive to dsRNA, which induces cell death by triggering an antiviral response [59]. However, Tom Tuschl and coworkers later showed that a minimum-length dsRNA (namely, an siRNA duplex) enabled RNAi even in mammalian cells [60]. This was possible simply because the siRNAs are short enough to escape the cellular defense system against alien dsRNAs. Subsequently, RNAi has been used widely as a useful tool, for example, to determine gene function in basic research. This technique is highly regarded because it replaces the expensive, time-consuming "gene knockout" strategies, although the expected effect by RNAi is "gene knockdown." The exo-siRNAs are designed with care to accomplish effective silencing, and therefore the target is obvious. Although off-target effects remain a concern, various methodologies are now available to minimize their obstructive effects. RNAi can also be used as a mechanism of defense against viral infection in *Drosophila* and nematodes [15–18]. Here, Dicer in the host cells processes the dsRNAs produced during viral replication, which consequently target viral genes. Some viruses produce proteins to interfere with the host cell defense mechanisms, and may successfully escape regulation by RNAi. Endo-siRNAs arising from transposable elements target their parental genes [30–35], although occasionally endo-siRNAs may arise from non-transposable intergenic elements, while the targets largely remain unclear.

Slicer-dependent target cleavage occurs across from the 10th and 11th nucleotides from the 5′ end of the guide siRNA [61] (Fig. 4). After cleavage, a phosphate group remains on the 5′ end of the downstream (3′ end) product. 2′-*O*-methyl modification on the ribose at the cleavage site of the target blocks Slicer activity [61]; a mismatch at the site will also block Slicer activity [61]. Consequently, the avoidance of mismatches and potentially interfering modifications at the slicing site is crucial for efficient RNAi and RISC formation. The crystal structure of archael Argonaute with a small DNA (Argonaute with a small RNA has not yet been crystallized) suggested how the RISC would determine the cleavage site and cleave the target RNAs [7, 62].

In *Drosophila*, both endo-siRNAs and exo-siRNAs are 2′-*O*-methylated at their 3′ end [63]. This particular type of small RNA modification was first observed in plants, where miRNAs are the recipients of a mono-methyl group [64]. The factor responsible for the miRNA modification is HEN1 [64]. In flies, the HEN1 homolog DmHEN1/Pimet accounts for the siRNA modification [63]. Whilst DmHEN1/Pimet is not essential in flies [65], the modification that it performs may contribute

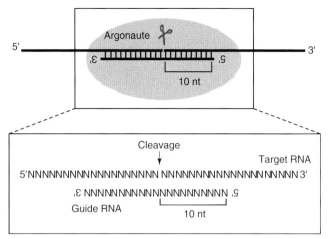

Fig. 4 Slicer-dependent target RNA cleavage. Slicer-dependent target RNA cleavage occurs across from the 10th and 11th nucleotides from the 5' end of the guide siRNA.

substantially to stabilizing small RNAs in vivo [63]. Typically, the fly miRNAs are not recipients of 2'-O-methylation by DmHEN1/Pimet, in contrast to the plant miRNAs [63, 65].

3 MicroRNA (miRNA)-Mediated Gene Silencing

miRNAs target transcripts of protein-coding genes in animals, and induce mainly instability and/or translational inhibition [66–69]. miRNA biogenesis resembles siRNA biogenesis, as it also requires Dicer cleavage before loading onto the RISC [1–5]. miRNA precursors are ssRNAs that fold into a hairpin-type structure, and must first be processed in the nucleus by another RNaseIII domain-containing nuclease, Drosha [70, 71]. The resultant secondary miRNA precursors are processed further by Dicer in the cytoplasm, yielding miRNA/miRNA* (miRNA-star) duplexes [25, 39, 72, 73]. Later, the mature miRNAs are loaded onto *Drosophila* Ago1 and all mammalian AGO proteins, Ago1, Ago2, Ago3, and Ago4, while *Drosophila* Ago2 has the ability to bind some miRNA* strands (20,26,51,54, see Sect. 3.1).

3.1 miRNA Biogenesis

miRNAs are encoded by their own genes [1–5], and miRNA gene sets overlap only partially among species, mainly because the gene numbers differ among species. For example, humans have approximately 1000 miRNA genes, whereas flies have fewer than 200. Some miRNA genes overlap with exons or introns of protein-coding or non-coding RNA genes, and the miRNA genes are mostly transcribed by RNA polymerase II [74]. The primary transcripts are abbreviated pri-miRNAs, the size of which is identical to that of the miRNA genes. The pri-miRNAs are first processed in the nucleus by Drosha [70, 71] (Fig. 5) which, like Dicer, associates with a dsRNA-binding domain-containing

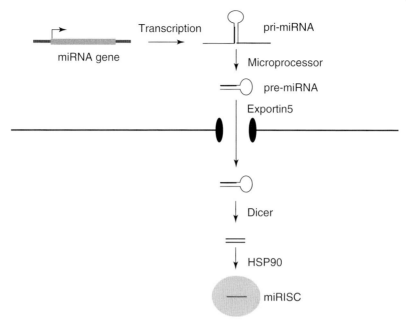

Fig. 5 miRNA biogenesis. miRNAs are produced from pri-miRNAs by two sequential steps by microprocessor and Dicer.

protein, the DiGeoge syndrome critical region gene 8 (DGCR8) in mammals. In flies, Pasha is the counterpart of DGCR8, and the Drosha–DGCR8 (or Pasha) heterodimer is referred to alternatively as the *microprocessor* [71, 75, 76]. Biochemical analyses have suggested that DGCR8 binds to the open terminus of the stem of the pri-miRNA, and that Drosha then cleaves approximately 11 nt from the open terminus [71]. The resultant molecule contains a 2 nt overhang, as does the siRNA duplex processed by Dicer.

Upon this reaction, the pri-miRNAs become approximately 70–80 nt hairpin-shaped pre-miRNAs, and are exported to the cytoplasm by Exportin5 and RanGTP [77] (Fig. 5). Exportin1 (also known as *Embargoed* in flies) may also contribute to pre-miRNA export [78]. In the cytoplasm, RanGTP is hydrolyzed to RanGDP, an event which releases pre-miRNAs from the export factors. In the cytoplasm, Dicer processes the pre-miRNAs (Fig. 5), and mammalian Dicer associates with TRBP/TARBP2 or PRKRA/PACT, as in siRNA processing. In flies, Dicer1 – but not Dicer2 – is responsible for producing miRNA duplexes from the precursors [39]. At this step, Dicer1 associates with Loqs-PB, the longest isoform of Loqs (or Loqs-PA, the second longest), which in turn enhances the dicing activity and substrate specificity of Dicer1 [56–58]. Dicer then binds the 2 nt overhang of the pre-miRNA hairpin structure and cleaves off the stem by dicing at a position about 22 nt from the terminus, thus generating the miRNA duplex [25, 39, 72, 73]. Consequently, Dicer is also considered a molecular ruler in miRNA processing.

HSP90 is required also for miRNA duplex loading onto Argonaute (Ago1 to Ago4 in mammals, and Ago1 in flies)

[48]. It has been suggested that HSP90 may stretch Argonaute so that it can accommodate the duplex. However, unlike the siRNA duplexes the miRNA duplexes contain mostly mismatches such that, after loading, the Argonaute protein is unable to cleave the passenger, or miRNA* strand. A likely model is that the miRNA* strand would be "flicked out" from Argonaute when Argonaute relieves the tension to return to its original conformation (it should be noted that some miRNA* is also loaded onto Argonaute proteins, as miRNA, to function in RNA silencing). Finally, the Argonaute–miRNA complex, or miRISC, becomes ready for silencing. Whether the RLC (see Sect. 2.1) -type complex is formed in the miRNA pathway remains obscure, but interestingly some miRNA* (passenger) strands tend to be loaded onto Ago2 in *Drosophila* [79–81], though the sorting mechanism involved remains undetermined.

Multiple noncanonical pathways that occur in a manner independent of Drosha or Dicer are now known [82]. The pre-miRNAs referred to as *"mirtrons"* (pre-miRNAs corresponding to introns) represent the Drosha-independent pathway [83, 84]. Because mirtrons correspond to one particular intron of a particular protein-coding gene, they are excised from their precursors by a conventional splicing step in the nucleus. The resultant molecules have a lariat structure and are subsequently debranched and refolded into a pre-miRNA-like structure; they then become a substrate for Dicer after being exported to the cytoplasm. miRNAs arising from small nucleolar RNAs (snoRNAs) also compose a subset of Drosha-independent miRNAs [85].

The maturation of miR-451 in mice and fish proceeds through Dicer-independent processing [86, 87]. Pri-miR-451 is conventionally processed by Drosha in the nucleus and exported to the cytoplasm; however, the stem of pre-miR-451 is too short to serve as a Dicer substrate. Alternatively, Ago2 captures the open terminal of the hairpin structure and cleaves at the middle of the 3′ stem, thus generating an approximately 30 nt ssRNA, while the 5′ end associates with Ago2. The miR-451 intermediate is then trimmed down to the regular miRNA size by unknown factors, such that the miR-451-RISC is formed.

Many proteins have been identified as regulatory factors of miRNA processing, including the RNA helicases DDX5 (also known as *P68*) and DDX17 (also known as *P72*), the RNA-binding proteins LIN28 and KHSRP, the splicing factors hnRNPA1 (heterogeneous nuclear ribonucleoprotein A1) and SRSF1 (also known as *SF2* or ASF; alternative splicing factor), the SMAD (signal-transducing adaptor protein) transcription factors, and the RNA-editing enzyme ADAR (adenosine deaminase acting on RNA) [88–95]. These factors associate either directly or indirectly with the microprocessor, or pri-/pre-miRNA, and either positively or negatively regulate the processing of a subset of, or specific, miRNAs.

3.2
miRNA Functions

The miRISC interacts with target RNAs through RNA–RNA base-pairing between the miRNA and its target. Seven nucleotides in the miRNA, spanning from the second to the eighth nucleotides from the 5′ end, greatly contribute to target recognition: this is termed the *"seed"* sequence (Fig. 6a) [66]. Other nucleotides may also contribute (albeit weakly) to

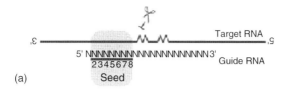

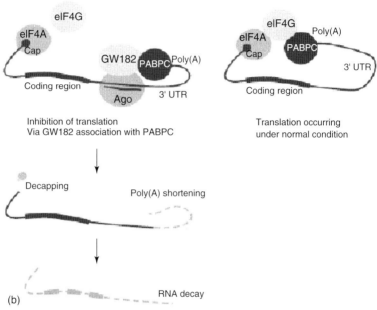

Fig. 6 miRNA target recognition and function. (a) Seven nucleotides in the miRNA, spanning from the second to the eighth nucleotides from the 5' end, termed the *seed sequence*, greatly contributes to target recognition; (b) During RNAi, PABPC is associated with miRISC and fails to interact with eIF4G, such that the translation of miRNA targets is inhibited. The miRNA targets then undergo a process of mRNA decay.

target recognition; however, because the seed sequence is so short the individual miRNAs would, in theory, be able to target multiple genes. The 3' untranslated region (UTR) of the target mRNAs, and in some cases also their coding region, has (potentially) one or more miRNA seed matches [66, 96] and, accordingly, it has been inferred that the majority of cellular genes are regulated by miRNAs [66, 97]. Indeed, approximately 60% of genes in humans are thought to be under miRNA regulation [66], although the real targets of individual miRNAs remain undetermined. miRNA target genes can be computationally predicted by algorithms, such as TargetScan and MicroCosm [66].

In animals, the RNA–RNA base-pairing between miRNAs and their targets is imperfect, with mismatches often occurring across the entire region (except for the seed sequence), and especially at the center of the duplex where the Argonaute–Slicer cleavage takes place (Fig. 6a). As a result, the miRNA-targeted RNAs show resistance to cleavage [98]. But how, then, can Argonaute induce silencing? Under these circumstances, Argonaute recruits protein(s) (by interacting with them) that have the ability to interfere with translation, and in so doing induces destabilization of the targets. The best-understood example is that of GW182 in *Drosophila*, and its three mammalian homologs TNRC6A, TNRC6B, and TNRC6C [68, 69, 99].

GW182 contains multiple characteristic domains, including a glycine-tryptophan (GW)-rich domain, a ubiquitin-associated (UBA) domain, a glutamine-rich (Q-rich) domain, a poly(A)-binding protein-binding motif 2 (PAM2), and an RNA-binding domain (RBD). GW182 interacts with Argonaute through the GW domain, and also with poly(A)-binding protein C (PABPC) mainly through the PAM2 domain. During translation, PABPC interacts with the translational factor eIF4G to accelerate the reaction (Fig. 6b). However, during RNAi, PABPC associates with miRISC and fails to interact with eIF4G, because GW182 occupies the site where eIF4G binds (Fig. 6b). Consequently, the translation of miRNAs targeted by the miRISC is inhibited. GW182 then recruits factors for deadenylation, namely the CCR4–CAF1–NOT complex, to initiate mRNA destabilization (Fig. 6b). At this point the poly(A) tails are trimmed and an efficient translation is unlikely to occur. The DCP1–DCP2 complex then removes the cap structure of the target mRNAs, which accelerates degradation of the target mRNAs by XRN1 nuclease. Both, the miRISC and GW182, together with the target RNA, are colocalized in processing bodies (P-bodies), which are cytoplasmic foci for mRNA decay. P-body localization is not necessary, however, for miRNA-mediated RNAi, as this is merely a downstream event of RNAi.

Very recently, Sharp and colleagues suggested that mouse Ago2 in embryonic stem (ES) cells might be able to recognize mRNA targets that lack homology to miRNAs [13]. The targets were found to contain guanine (G)-rich sequences; a proposed model was that G-rich sequences could give the Ago2–miRNA complexes a higher affinity for the targets, if they were to occur close to miRNA-binding sites. It remains unclear, however, if Ago2 is able to perform this by itself or if it requires an interacting factor such as FMRP, that has a preference to bind G-rich RNA sequences.

How, then, is the miRISC turnover regulated in cells? One way would be to replace the miRNA in the RISC, and in *C. elegans* the nuclease Xrn2 may be involved in this step as the stimulator of degrading miRNAs in the RISC [100]. Once the miRNA is degraded, Argonaute can associate with a new miRNA and thus silence a new set of targets. miRNA turnover may also be influenced by the degree of complementarity between the small RNAs and their targets; in *Drosophila*, when miRNAs are paired with perfectly complementary transcripts, they are tailed with additional nucleotides and destabilized [101].

Another mechanism of miRNA turnover would be to destabilize the whole RISC. In mouse ES cells, an E3 ubiquitin ligase (Lin41) induces RISC decay by ubiquitinating the Argonaute protein within the RISC [102]. The *Lin41* gene is one of the Let-7 miRNA targets, and Lin41 may contribute to a negative feedback mechanism by contributing to the decay of Argonaute protein. Two members of the TRIM-NHL family, which are regulators of cell proliferation and development (NHL-2 in *C. elegans* and TRIM32 in mice) are also known to play roles in regulating miRNA functions. Both, NHL-2 and TRIM32 contain a RING domain that confers ubiquitin ligase activity, as well as others such as the C-terminal NHL repeats; thus, NHL-2 and TRIM32 may regulate miRNA functions in similar fashion to Lin41, although the precise mechanism awaits further investigation [103].

miRNAs regulate the expression of genes that are involved in various cellular events, including differentiation and apoptosis, all of which are crucial for living creatures. Recent evidence has shown that miRNAs play a strong role in affecting cancer and metabolic disorders, and many cancerous cells express specific miRNAs aberrantly [104]. Conversely, changes found in the expression profiles of certain miRNAs can be used as biomarkers for diagnosis.

4
PIWI-Interacting RNA (piRNA)-Mediated Gene Silencing

In animal germlines, piRNAs guide PIWI proteins (Table 1) – but not AGO proteins – to their RNA targets by associating with them [3, 105–107]. One conserved function of the PIWI–piRNA complex or piRISC across species is to maintain the integrity of the germline genome from invasive transposable elements [3, 105–107]. The first "hint" of the existence of piRNAs appeared through studies on the silencing of *Stellate* protein-coding gene repeats in the *Drosophila* male germline [108]. An abundance of endogenous 23- to 29-nt RNAs was later revealed by a comprehensive small RNA profiling study in the *Drosophila* male germline and in embryos [109]. These small RNAs were mapped to genomic repetitive elements, including transposable elements and their remnants at the telomeric/centromeric regions, and associated specifically with PIWI proteins. PIWI–piRNA association was later confirmed in other animal species [110–114]. Intergenic regions to which piRNAs map are known as *piRNA clusters*, and these can extend over more than 150 kb. Mutations in PIWI genes and piRNA clusters cause a derepression of transposons in the germline, and consequently both are key factors for transposon repression [115–119]. In this regard, piRNAs are similar to endo-siRNAs, but their size and Argonaute partners differ one from another and so can be easily discriminated by these signatures. Similar to the fly siRNAs, the piRNAs in mice and flies are 2′-O-methylated at their 3′ end by HEN1 and DmHEN1/Pimet, respectively [65, 120, 121].

4.1
Primary piRNA Biogenesis and Function

piRNAs can be mapped to both genomic strands, which suggests a bidirectional transcription of the piRNA clusters [110, 111, 118]. However, some clusters – such as *flamenco* (*flam*) in *Drosophila* and pachytene piRNA clusters in mice – produce piRNAs exclusively from one genomic strand [111, 118]. Thus, piRNA precursors are likely single-stranded. As with the siRNAs, the piRNAs show a strong bias for uridine at position 1 (1U bias) [110, 111, 118]. However, unlike the siRNAs the piRNAs do not show any phasing pattern within a cluster sequence, and often overlap with each other. These piRNA signatures exclude the possibility of Dicer involvement in the pathway. Indeed, piRNAs accumulate in *dicer* mutant flies much as they do in wild-type flies.

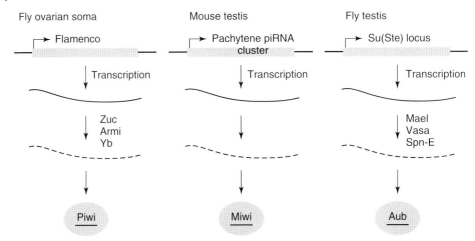

Fig. 7 Primary piRNA processing. *Flam*-piRNAs and *Su(Ste)*-piRNAs in *Drosophila*, and pachytene piRNAs in mice, are considered as primary piRNAs. The requirement of factors in the primary processing may vary in each case. In mice, the primary piRNA factors remain largely unknown.

Primary piRNAs are first produced from precursors (i.e., primary transcripts of piRNA clusters) through the primary piRNA processing pathway [3, 105–107]. Many primary piRNAs are subsequently amplified by the amplification loop or ping-pong pathway (see Sect. 4.2), and are considered secondary piRNAs (see Sect. 4.2).

piRNAs derived from *Drosophila flam* and mouse pachytene piRNAs are representative of primary piRNAs. *flam* contains remnants of transposons, mainly *gypsy*, *Idefix*, and *ZAM*, and is expressed specifically in somatic follicle cells of the ovaries [122]. *flam*-derived piRNAs are mostly antisense to transcripts of active transposons, and thus function as "antisense oligos" to mark the transposons for silencing. Ovarian follicle cells express Piwi but are devoid of Aubergine (Aub) and AGO3; thus, only Piwi is able to form the piRISC with *flam*-piRNAs (Fig. 7) [123]. Mouse pachytene piRNAs are derived from piRNA clusters during the pachytene stage of meiosis, and are associated with a PIWI member, Miwi, to form the piRISC (Fig. 7) [105]. The clusters that bear pachytene piRNAs rarely contain transposon sequences; thus, the function of pachytene piRNAs is not to target transposons and, indeed, mutations in *Miwi* do not lead to transposon derepression. Rather, the "real" targets of pachytene piRNAs remain unknown.

In fly ovarian follicle cells, primary piRNAs are also generated from mRNAs of protein-coding genes, mostly from the 3′ UTR [123, 124]. These piRNAs are generically termed *genic piRNAs*, and a subset of protein-coding genes serves as the source of genic piRNA production. It is conceivable to speculate that the machinery can discriminate the piRNA precursor genes from others, although the rules that govern such selection are currently unknown. The function of genic piRNAs also remains largely unknown; it is possible that genic piRNAs may function in regulating other coding genes, but their targets remain, as yet, unknown.

Somatic primary piRNA accumulation in flies requires Piwi as the piRNA binding partner and stabilizer, but not as the processing factor [123–126]. The putative RNA helicase Armitage (Armi), the endoribonuclease Zucchini (Zuc), and the Tudor- and helicase domain-containing protein Yb, are each required for somatic primary piRNA production (Fig. 7) [125, 126, 127, 128]. In *Drosophila* testes, piRNAs derived from *Suppressor of Stellate* (*Su(Ste)*) repeats that function in silencing *Stellate* are classified as primary piRNAs [129]. Although *Su(Ste)*-piRNAs and *tj*-derived (*tj*-) piRNAs are both primary, the factors they require are different: Vasa, Maelstrom (Mael), and Spindle-E (Spn-E) are necessary for accumulating *Su(Ste)*-piRNAs, but not *tj*-piRNAs (Fig. 7). The functions of Vasa, Mael, and Spn-E in *Su(Ste)*-piRNA production remain undetermined. Other factors responsible for primary piRNA production in *Drosophila* and other species also remain unknown.

4.2
Secondary piRNA Biogenesis and Function

A fraction of the cells residing in the gonads, such as germline cells of the *Drosophila* ovary and pre-meiotic spermatogonia in mice, have machinery to amplify primary piRNAs, referred to as the *ping-pong pathway* or the *amplification loop* (Fig. 8) [3, 105–107]. The resultant piRNAs are classified as secondary piRNAs. Secondary piRNAs do not exist in somatic follicle cells of the *Drosophila* ovary, and are possibly also not present in meiotic cells of mouse spermatocytes either, because of the unavailability of the ping-pong pathway.

Drosophila germline cells (but not somatic cells) in the ovary express all three PIWI proteins, and primary piRNAs produced in the cells are first loaded onto Piwi and Aub [116, 118, 130–133]. These piRNAs show a 1U bias, and are mostly antisense to transposon mRNAs. In contrast, Ago3 (the third member of the PIWI

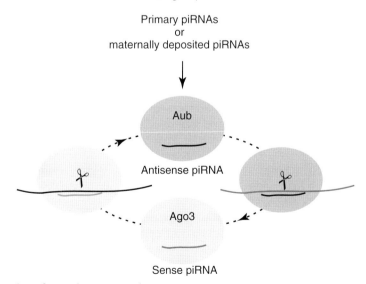

Fig. 8 The amplification loop for producing secondary piRNAs in *Drosophila*. Primary piRNAs and maternally deposited piRNAs are the triggers of the secondary piRNA production.

family)-associated piRNAs shows a strong bias for adenosine at the 10th nucleotide from their 5′ end (10A bias), and are predominantly in the sense orientation [118, 131]. Both, Ago3- and Aub-associated piRNAs are able to pair through their first 10 nt. The PIWI members exhibit Slicer activity, which cleaves target RNAs between their 10th and 11th nucleotides relative to the guide of small RNAs. These observations led to the "ping-pong" model, in which PIWI proteins reciprocally cleave their targets, constantly amplifying piRNAs by an ongoing cycle; the cleavage of complementary transcripts, guided by Aub-bound 1U primary piRNA, then leads to the generation of the 5′ end of secondary piRNAs that have a 10A bias, and are loaded onto Ago3. The enzymatic activity that generates the 3′ end of the secondary piRNAs is unknown. Ago3, loaded with secondary piRNAs, is then able to cleave complementary transcripts, which results in the generation of nascent piRNAs that correspond exactly to the original primary piRNA sequences; these are then loaded onto Aub. In the PIWI proteins, arginine (R) residues in particular are symmetrically dimethylated post-translationally by PRMT5 methyltransferase [134–136]. Through this particular modification, the PIWI proteins become associated with Tudor domain-containing proteins (Tud proteins), such as Tudor in flies and TDRD1 in mice, which control piRNA stability and PIWI subcellular localization in the germline [134–142]. The ping-pong pathway is also available in *Drosophila* testes and in the germline of other animals such as mice, rat, fish, and frogs [110–114, 134]. In *Drosophila*, the factors required for the amplification loop overlap only partially between the female and male: while female flies require *Spindle-E*, *Maelstrom*, and *Krimper*, male flies require *Vasa*, *Spindle-E*, and *Zucchini* [129, 143]. In mice, the mouse *Vasa homolog* (*Vmh*) gene is required for the amplification loop [144].

In mice, Mili and Miwi2 have roles in the ping-pong cycle and transposon repression, with Mili preferentially associating with primary piRNAs and Miwi2 with secondary piRNAs [145]. However, the orientation of piRNAs in mice is reversed compared to that in flies, with the majority of mouse primary piRNAs being sense and secondary piRNAs being antisense relative to transposons. In *Mili*-deficient mice that fail to generate primary piRNAs, Miwi2 remains unloaded [145]. A single PIWI might have dual roles in the amplification loop by accepting both primary and secondary piRNAs, as was shown for Mili expressed in postnatal spermatogonia without its Miwi2 partner [145].

Initiation of the amplification loop requires primary piRNAs that are mainly supplied by the primary processing pathway; however, this is not the only source, as demonstrated in studies of hybrid dysgenesis in *Drosophila* [105, 146]. This phenomenon is observed in crosses between two different fly stocks, which result in sterile progeny with severe gonadal dystrophy associated with an activation of specific transposons such as P- and I-elements. Hybrid dysgenesis is observed only when the males of one stock are crossed to females of the other stock, but no abnormalities are observed for the reciprocal cross. As the genotypes of the progeny of both crosses are identical, then epigenetic factors must be responsible for the phenotype. The transposon activation is caused by a failure in the maternal deposition of piRNAs, most likely in the form of piRISCs with PIWI which target specific, paternally deposited transposons. The maternal deposition of piRNAs into the egg provides the embryo with an initial piRNA pool to initiate the amplification cycle and build up resistance to transposons.

5
Conclusions

Today, major attempts to develop new RNAi-based drugs for therapeutic and/or diagnostic purposes is ongoing on an international scale. Such an approach is a direct reflection of the power that RNAi shows in gene silencing, and the convenience with which it can be induced by the administration of small RNAs with sequences complementary to the target genes. No other factors are required, because human cells already possess the machineries necessary for RNAi. Moreover, the triggering RNA molecules can be very natural – everybody has RNAs in their bodies! Therefore, by relying on RNA this form of therapy should cause little or no harm, unless the molecules have been extensively modified before their internal usage. Never the less, such therapy must be advanced with great care, so as to minimize any adverse side effects while maintaining, and hopefully maximizing, the efficacy. In this situation, a detailed knowledge of the molecular mechanisms of RNAi is indispensable. For example, it was shown recently that the administration of an excess of small RNA could kill otherwise healthy mice. In this case, the side effects were caused by small RNAs interfering with the miRNA-mediated pathway, because in mice the siRNA and miRNA pathways intersect. Although, currently, RNAi is still seen as mysterious, it is imperative that these investigations are continued towards clarifying the mechanisms involved. One final point here is to apologize to any colleagues whose relevant primary reports have not been cited due to space constraints.

References

1 Ghildiyal, M., Zamore, P.D. (2009) Small silencing RNAs: an expanding universe. *Nat. Rev. Genet.*, **10**, 94–108.
2 Kim, N.V., Han, J., Siomi, M.C. (2009) Biogenesis of small RNAs in animals. *Nat. Rev. Mol. Cell Biol.*, **10**, 126–139.
3 Malone, C.D., Hannon, G.J. (2009) Small RNAs as guardians of the genome. *Cell*, **136**, 656–668.
4 Siomi, H., Siomi, M.C. (2009) On the road to reading the RNAi code. *Nature*, **457**, 396–404.
5 Czech, B., Hannon, G.J. (2011) Small RNA sorting: matchmaking for Argonautes. *Nat. Rev. Genet.*, **12**, 19–31.
6 Hutvagner, G., Simard, M.J. (2008) Argonaute proteins: key players in RNA silencing. *Nat. Rev. Mol. Cell Biol.*, **9**, 22–32.
7 Wang, Y., Sheng, G., Juranek, S., Tuschl, T., Patel, D.J. (2008) Structure of the guide-strand-containing argonaute silencing complex. *Nature*, **456**, 209–213.
8 Parker, J.S. (2010) How to slice: snapshots of Argonaute in action. *Silence*, **1**, 3.
9 Lingel, A., Simon, B., Izaurralde, E., Sattler, M. (2003) Structure and nucleic-acid binding of the *Drosophila* Argonaute 2 PAZ domain. *Nature*, **426**, 465–469.
10 Frank, F., Sonenberg, N., Nagar, B. (2010) Structure basis for 5′-nucleotide base-specific recognition of guide RNA by human AGO2. *Nature*, **465**, 818–822.
11 Song, J.J., Smith, S.K., Hannon, G.J., Joshua-Tor, L. (2004) Crystal structure of Argonaute and its implications for RISC slicer activity. *Science*, **305**, 1434–1437.
12 Wang, Y., Juranek, S., Li, H., Sheng, G., Wardle, G.S., Tuschl, T., Patel, D.J. (2009) Nucleation, propagation and cleavage of target RNAs in Ago silencing complexes. *Nature*, **461**, 754–761.
13 Leung, A.K.L., Young, A.G., Bhutkar, A., Zheng, G.X., Bosson, A.D., Nielsen, C.B., Sharp, P.A. (2011) Genome-wide identification of Ago2 binding sites from mouse embryonic stem cells with and without mature microRNAs. *Nat. Struct. Mol. Biol.*, **18**, 237–245.
14 Carmell, M.A., Xuan, Z., Zhang, M.Q., Hannon, G.J. (2002) The Argonaute family:

tentacles that reach into RNAi, development; control, stem cell maintenance, and tumorigenesis. *Genes Dev.*, **16**, 2733–2742.

15 Galiana-Arnoux, D., Dostert, C., Schneemann, A., Hoffmann, J.A., Imler, J.L. (2006) Essential function *in vivo* for Dicer-2 in host defense against RNA viruses in *Drosophila*. *Nat. Immunol.*, **7**, 590–597.

16 van Rij, R.P., Saleh, M.C., Berry, B., Foo, C., Houk, A., Antoniewski, C., Andino, R. (2006) The RNA silencing endonuclease Argonaute 2 mediates specific antiviral immunity in *Drosophila melanogaster*. *Genes Dev.*, **20**, 2985–2995.

17 Wang, X.H., Aliyari, R., Li, W.X., Li, H.W., Kim K., Carthew, R., Atkinson, P., Ding, S.W. (2006) RNA interference directs innate immunity against viruses in adult *Drosophila*. *Science*, **312**, 452–454.

18 Ding, S.W., Voinnet, O. (2007) Antiviral immunity directed by small RNAs. *Cell*, **130**, 413–426.

19 Fire, A., Xu, S., Montgomery, M.K., Kostas, S.A., Driver, S.E., Mello, C.C. (1998) Potent and specific genetic interference by double-stranded RNA in *Caenorhabditis elegans*. *Nature*, **391**, 806–811.

20 Liu, J., Carmell, M.A., Rivas, F.V., Marsden, C.G., Thomson, J.M., Song, J.J., Hammond, S.M., Joshua-Tor, L., Hannon, G.J. (2004) Argonaute2 is the catalytic engine of mammalian RNAi. *Science*, **305**, 1437–1441.

21 Meister, G., Tuschl, T. (2004) Mechanisms of gene silencing by double-stranded RNA. *Nature*, **431**, 343–349.

22 Tomari, Y., Zamore, P.D. (2005) Machines for RNAi. *Genes Dev.*, **19**, 517–529.

23 Jorgensen, R.A. (1995) Cosuppression, flower color patterns, and metastable gene expression states. *Science*, **268**, 686–691.

24 Romano, N., Macino, G. (1992) Quelling: transient inactivation of gene expression in *Neurospora crassa* by transformation with homologous sequences. *Mol. Microbiol.*, **6**, 3343–3353.

25 Bernstein, E., Caudy, A.A., Hammond, S.M., Hannon, G.J. (2001) Role for a bidentate ribonuclease in the initiation step of RNA interference. *Nature*, **409**, 363–366.

26 Meister, G., Landthaler, M., Patkaniowska, A., Dorsett, Y., Teng, G., Tuschl, T. (2004) Human Argonaute2 mediates RNA cleavage targeted by miRNAs and siRNAs. *Mol. Cells*, **15**, 185–197.

27 Okamura, K., Ishizuka, A., Siomi, H., Siomi, M.C. (2004) Distinct roles for Argonaute proteins in small RNA-directed RNA cleavage pathways. *Genes Dev.*, **18**, 1655–1666.

28 Yigit, E., Batista, P.J., Bei, Y., Pang, K.M., Chen, C.C., Tolia, N.H., Joshua-Tor, L., Mitani, S., Simard M.J., Mello, C.C. (2006) Analysis of the *C. elegans* Argonaute family reveals that distinct Argonautes act sequentially during RNAi. *Cell*, **127**, 747–757.

29 Pham, J.W., Pellino, J.L., Lee, Y.S., Carthew, R.W., Sontheimer, E.J. (2004) A Dicer-2-dependent 80s complex cleaves targeted mRNAs during RNAi in *Drosophila*. *Cell*, **117**, 83–94.

30 Czech, C.D., Malone, C.D., Zhou, R., Stark, A., Shlingeheyde, C., Dus, M., Perrimon, N., Kellis, M., Wohlschlegel, J.A., Sachidanandam, R., Hannon, G.J., Brennecke, J. (2008) An endogenous small interfering RNA pathway in *Drosophila*. *Nature*, **453**, 798–802.

31 Ghildiyal, M., Seitz, H., Horwich, M.D., Li, C., Du, T., Lee, S., Xu, J., Kittler, E.L., Zapp, M.L., Weng, Z., Zamore, P.D. (2008) Endogenous siRNAs derived from transposons and mRNAs in *Drosophila* somatic cells. *Science*, **320**, 1077–1081.

32 Kawamura, Y., Saito, K., Kin, T., Ono, Y., Asai, K., Sunohara, T., Okada, T.N., Siomi, M.C., Siomi, H. (2008) *Drosophila* endogenous small RNAs bind to Argonaute2 in somatic cells. *Nature*, **453**, 793–797.

33 Okamura, K., Chung, W.J., Ruby, J.G., Guo, H., Bartel, D.P., Lai, E.C. (2008) The *Drosophila* hairpin RNA pathway generates endogenous short interfering RNAs. *Nature*, **453**, 803–806.

34 Watanabe, T., Totoki, Y., Toyoda, A., Kaneda, M., Kuramochi-Miyagawa, S., Obata, Y., Chiba, H., Kohara, Y., Kono, T., Nakano, T., Surani, M.A., Sakaki, Y., Sasaki, H. (2008) Endogenous siRNAs from naturally formed dsRNAs regulate transcripts in mouse oocytes. *Nature*, **453**, 539–543.

35 Tam, O.H., Aravin, A.A., Stein, P., Girard, A., Murchison, E.P., Cheloufi, S., Hodges, E., Anger, M., Sachidanandam, R., Schultz,

R.M., Hannon, G.J. (2008) Pseudogene-derived small interfering RNAs regulate gene expression in mouse oocytes. *Nature*, **453**, 534–538.

36 Hammond, S.M., Bernstein, E., Beach, D., Hannon, G.J. (2000) An RNA-directed nuclease mediates post-transcriptional gene silencing in *Drosophila* cells. *Nature*, **404**, 293–296.

37 Zhang, H., Kolb, F.A., Jaskiewicz, L., Westhof, E., Filipowicz, W. (2004) Single processing center models for human Dicer and bacterial RNase III. *Cell*, **118**, 57–68.

38 MacRae, I.J., Zhou, K., Li, F., Repic, A., Brooks, A.N., Cande, W.Z., Adams, P.D., Doudna, J.A. (2006) Structural basis for double-stranded RNA processing by Dicer. *Science*, **311**, 195–198.

39 Lee, Y.S., Nakahara, K., Pham, J.W., Kim, K., He, Z., Sontheimer, E.J., Carthew, R.W. (2004) Distinct roles for *Drosophila* Dicer-1 and Dicer-2 in the siRNA/miRNA silencing pathways. *Cell*, **117**, 69–81.

40 Gregory, R.I., Chendrimada, T.P., Cooch, N., Shiekhattar, R. (2005) Human RISC couples microRNA biogenesis an post transcriptional gene silencing. *Cell*, **123**, 631–640.

41 Lee, Y., Hur, I., Park, S.Y., Kim, Y.K., Suh, M.R., Kim, V.N. (2006) The role of PACT in the RNA silencing pathway. *EMBO J.*, **25**, 522–532.

42 Tabara, H., Sarkissian, M., Kelly, W.G., Fleenor, J., Grishok, A., Timmons, L., Fire, A., Mello, C.C. (2002) The *rde-4* gene, RNA interference, and transposon silencing in *C. elegans*. *Cell*, **99**, 123–132.

43 Liu, Q., Rand, T.A., Kalidas, S., Du, F., Kim, H.E., Smith, D.P., Wang, X. (2003) R2D2, a bridge between the initiation and effector steps of the *Drosophila* RNAi pathway. *Science*, **301**, 1921–1925.

44 Schwarz, D.S., Hutvàgner, G., Du, T., Xu, Z., Aronin, N., Zamore, P.D. (2003) Asymmetry in the assembly of the RNAi enzyme complex. *Cell*, **115**, 199–208.

45 Khvorova, A., Reynolds, A., Jayasena, S.D. (2003) Functional siRNAs and miRNAs exhibit strand bias. *Cell*, **115**, 209–216.

46 Tomari, Y., Matranga, C., Haley, B., Martinez, N., Zamore, P.D. (2004) A protein sensor for siRNA asymmetry. *Science*, **306**, 1377–1380.

47 Miyoshi, T., Takeuchi, A., Siomi, H., Siomi, M.C. (2010) A direct role for Hsp90 in pre-RISC formation in *Drosophila*. *Nat. Struct. Mol. Biol.*, **17**, 1024–1026.

48 Iwasaki, S., Kobayashi, M., Yoda, M., Sakaguchi, Y., Katsuma, S., Suzuki, T., Tomari, Y. (2010) Hsc70/Hsp90 chaperone machinery mediates ATP-dependent RISC loading of small RNA duplexes. *Mol. Cells*, **39**, 292–299.

49 Matranga, C., Tomari, Y., Shin, C., Bartel, D.P., Zamore, P.D. (2005) Passenger-strand cleavage facilitates assembly of siRNA into Ago2-containing RNAi enzyme complexes. *Cell*, **123**, 607–620.

50 Miyoshi, K., Tsukumo, H., Nagami, T., Siomi, H., Siomi, M.C. (2005) Slicer function of *Drosophila* Argonautes and its involvement in RISC formation. *Genes Dev.*, **19**, 2837–2848.

51 Rand, T.A., Petersen, S., Du, F., Wang, X. (2005) Argonaute2 cleaves the anti-guide strand of siRNA during RISC activation. *Cell*, **123**, 621–629.

52 Liu, Y., Ye, X., Jiang, F., Liang, C., Chen, D., Peng, J., Kinch, L.N., Grishin, N.V., Liu, Q. (2009) C3PO, an endoribonuclease that promotes RNAi by facilitating RISC activation. *Science*, **325**, 750–753.

53 Azuma-Mukai, A., Oguri, H., Mituyama, T., Qian, Z.R., Asai, K., Siomi, H., Siomi, M.C. (2007) Characterization of endogenous human Argonautes and their miRNA partners in RNA silencing. *Proc. Natl Acad. Sci. USA*, **105**, 7964–7969.

54 Miyoshi, K., Miyoshi, T., Hartig, J.V., Siomi, H., Siomi, M.C. (2010) Functional molecular mechanisms that funnel RNA precursors into endogenous small-interfering RNA and micro RNA biogenesis pathways in *Drosophila*. *RNA*, **16**, 506–515.

55 Hartig, J.V., Esslinger, S., Bottcher, R., Saito, K., Forstemann, K. (2009) Endo-siRNAs depend on a new isoform of loquacious and target artificially introduced, high-copy sequences. *EMBO J.*, **28**, 2932–2944.

56 Forstemann, K., Tomari, Y., Du, T., Vagin, V.V., Denli, A.M., Bratu, D.P., Klattenhoff, C., Theurkauf, W.E., Zamore, P.D. (2005) Normal microRNA maturation and germ-line stem cell maintenance requires Loquacious, a double-stranded RNA-binding domain protein. *PLoS Biol.*, **3**, e236.

57 Saito, K., Ishizuka, A., Siomi, H., Siomi, M.C. (2005) Processing of pre-microRNAs by the Dicer-1-Loquacious complex in Drosophila cells. *PLoS Biol.*, **3**, e235.

58 Jiang, F., Ye, X., Liu, X., Fincher, L., McKearin, D., Liu, Q. (2005) Dicer-1 and R3D1-L catalyze microRNA maturation in Drosophila. *Genes Dev.*, **19**, 1674–1679.

59 Stark, G.R., Kerr, I.M., Williams, B.R., Silverman, R.H., Schreiber, R.D. (1998) How cells respond to interferons. *Annu. Rev. Biochem.*, **67**, 227–264.

60 Elbashir, S.M., Harborth, J., Lendeckel, W., Yalcin, A., Weber, K., Tuschl, T. (2001) Duplexes of 21-nucleotide RNAs mediate RNA interference in cultured mammalian cells. *Nature*, **411**, 494–498.

61 Elbashir, S.M., Lendeckel, W., Tuschl, T. (2001) RNA interference is mediated by 21- and 22-nucleotide RNAs. *Genes Dev.*, **15** (1), 88–200.

62 Wang, Y., Juranek, S., Li, H., Sheng, G., Tuschl, T., Patel, D.J. (2008) Structure of an argonaute silencing complex with a seed-containing guide DNA and target RNA duplex. *Nature*, **456**, 921–926.

63 Horwich, M.D., Li, C., Matranga, C., Vagin, V., Farley, G., Wang, P., Zamore, P.D. (2007) The Drosophila RNA methyltransferase, DmHen1, modified germline piRNAs and single-stranded siRNAs in RISC. *Curr. Biol.*, **17**, 1265–1272.

64 Li, J., Yang, Z., Yu, B., Liu, J., Chen, X. (2005) Methylation protects miRNAs and siRNAs from a 3′-end uridylation activity in Arabidopsis. *Curr. Biol.*, **15**, 1501–1507.

65 Saito, K., Sakaguchi, Y., Suzuki, T., Suzuki, T., Siomi, H., Siomi, M.C. (2007) Pimet, the Drosophila homolog of HEN1, mediates 2′-O-methylation of Piwi-interacting RNAs at their 3′ ends. *Genes Dev.*, **21**, 1603–1608.

66 Bartel, D.P. (2009) MicroRNAs: Target recognition and regulatory functions. *Cell*, **136**, 215–233.

67 Guo, H., Ingolia, N.T., Weissman, J.S., Bartel, D.P. (2010) Mammalian microRNAs predominantly act to decrease target mRNA levels. *Nature*, **466**, 835–840.

68 Huntzinger, E., Izaurralde, E. (2010) Gene silencing by microRNAs: contributions of translational expression and mRNA decay. *Nat. Rev. Genet.*, **12**, 99–110.

69 Krol, J., Loedige, I., Filipowicz, W. (2010) The widespread regulation of microRNA biogenesis, function and decay. *Nat. Rev. Genet.*, **11**, 597–610.

70 Zheng, Y., Yi, R., Cullen, B. (2005) Recognition and cleavage of primary microRNA precursors by the nuclease processing enzyme Drosha. *EMBO J.*, **24**, 138–148.

71 Han, J., Lee, Y., Yeom, K.H., Nam, J.W., Heo, I., Rhee, J.K., Sohn, S.Y., Cho, Y., Zhang, B.T., Kim, V.N. (2006) Molecular basis for the recognition of primary microRNAs by the Drosha–DGCR8 complex. *Cell*, **125**, 887–901.

72 Hutvagner, G., McLachlan, J., Pasquinelli, A.E., Balint, E., Tuschl, T., Zamore, P.D. (2001) A cellular function for the RNA-interference enzyme Dicer in the maturation of the let-7 small temporal RNA. *Science*, **293**, 834–838.

73 Knight, S.W., Bass, B.L. (2001) A role for the RNaseIII enzyme DCR-1 in RNA interference and germ line development in Caenorhabditis elegans. *Science*, **293**, 2269–2271.

74 Lee, Y., Kim, M., Han, J., Yeom, K.H., Lee, S., Baek, S.H., Kim, V.N. (2004) MicroRNA genes are transcribed by RNA polymerase II. *EMBO J.*, **23**, 4051–4060.

75 Gregory, R.I., Yan, K.P., Amuthan, G., Chendrimada, T., Doratotaj, B., Cooch, N., Shiekhattar, R. (2004) The Microprocessor complex mediates the genesis of microRNAs. *Nature*, **432**, 235–240.

76 Denli, A.M., Tops, B.B., Plasterk, R.H., Ketting, R.F., Hannon, G.J. (2004) Processing of primary microRNAs by the Microprocessor complex. *Nature*, **432**, 231–235.

77 Lund, E., Guttinger, S., Calado, A., Darlberg, J.E., Kutay, U. (2004) Nuclear export of microRNA precursors. *Science*, **303**, 95–98.

78 Bussing, I., Yang, J.S., Lai, E.C., Grosshans, H. (2010) The nuclear export receptor XPO-1 supports primary miRNA processing in C. elegans and Drosophila. *EMBO J.*, **29**, 1830–1839.

79 Okamura, K., Liu, N., Lai, E.C. (2009) Distinct mechanisms for microRNA strand selection by Drosophila Argonautes. *Mol. Cells*, **36**, 431–444.

80 Czech, B., Zhou, R., Erlich, Y., Brennecke, J., Binari, R., Villalta, C., Gordon, A., Perrimon, N., Hannon, G.J. (2009) Hierarchical rules for Argonaute loading in Drosophila. *Mol. Cells*, **36**, 445–456.

81 Seitz, H., Ghildiyal, M., Zamore, P.D. (2008) Argonaute loading improves the 5′ precision of both MicroRNAs and their miRNA* strands in flies. *Curr. Biol.*, **18**, 147–151.

82 Siomi, H., Siomi, M.C. (2010) Post transcriptional regulation of microRNA biogenesis in animals. *Mol. Cells*, **38**, 323–332.

83 Okamura, K., Hagen, J.W., Duan, H., Tyler, D.M., Lai, E.C. (2007) The mirtron pathway generates microRNA-class regulatory RNAs in *Drosophila*. *Cell*, **130**, 89–100.

84 Ruby, J.G., Jan, C.H., Bartel, D.P. (2007) Intronic microRNA precursors that bypass Drosha processing. *Nature*, **448**, 83–86.

85 Ender, C., Krek, A., Friedlander, M.R., Beitzinger, M., Weinmann, L., Chen, W., Pfeffer, S., Rajewsky, N., Meister, G. (2008) A human snoRNA with microRNA-like functions. *Mol. Cells*, **32**, 519–528.

86 Cheloufi, S., Santos, C.O.D., Chong, M.M.W., Hannon, G.J. (2010) A dicer-independent miRNA biogenesis pathway that requires Ago catalysis. *Nature*, **465**, 584–589.

87 Cifuentes, D., Xue, H., Taylor, D.W., Patnode, H., Mishima, Y., Cheloufi, S., Ma, E., Mane, S., Hannon, G.J., Lawson, N.D., Wolfe, S.A., Giraldez, A.J. (2010) A novel miRNA processing pathway independent of Dicer requires Argonautes catalytic activity. *Science*, **328**, 1694–1698.

88 Guil, S., Caceres, J.F. (2007) The multifunctional RNA-binding protein hnRNP A1 is required for processing of miR-18a. *Nat. Struct. Mol. Biol.*, **14**, 591–596.

89 Davis, B.N., Hilyard, A.C., Lagna, G., Hata, A. (2008) SMAD proteins control DROSHA-mediated microRNA maturation. *Nature*, **454**, 56–61.

90 Viswanathan, S.R., Daley, G.Q., Gregory, R.I. (2008) Selective blockade of microRNA processing by Lin28. *Science*, **320**, 97–100.

91 Trabucchi, M., Briata, P., Garcia-Mayoral, M., Haase, A.D., Fillipowicz, W., Ramos, A., Gherzi, R., Rosenfeld, M.G. (2009) The RNA-binding protein KSRP promotes the biogenesis of a subset of microRNAs. *Nature*, **459**, 1010–1014.

92 Yang, W., Chendrimada, T.P., Wang, Q., Higuchi, M., Seeburg, P.H., Shiekhattar, R., Nishikura, K. (2006) Modulation of microRNA processing and expression through RNA editing by ADAR deaminases. *Nat. Struct. Mol. Biol.*, **13**, 13–21.

93 Fukuda, T., Yamagata, K., Fujiyama, S., Matsumoto, T., Koshida, I., Yoshimura, K., Mihara, M., Naitou, M., Endoh, H., Nakamura, T., Akimoto, C., Yamamoto, Y., Katagiri, T., Foulds, C., Takezawa, S., Kitagawa, H., Takeyama, K., O'Malley, B.W., Kato, S. (2007) DEAD-box RNA helicase subunits of the Drosha complex are required for processing of rRNA and a subset of microRNAs. *Nat. Cell Biol.*, **9**, 604–6611.

94 Heo, I., Joo, C., Kim, Y.K., Ha, M., Yoon, M.J., Cho, J., Yeom, K.H., Han, J., Kim, V.N. (2009) TUT4 in concert with Lin28 suppresses microRNA biogenesis through pre-microRNA uridylation. *Cell*, **138**, 696–708.

95 Suzuki, H.I., Yamagata, K., Sugimoto, K., Iwamoto, T., Kato, S., Miyazono, K. (2009) Modulation of microRNA processing by p53. *Nature*, **460**, 529–533.

96 Ameres, S.L., Martinez, J., Schroeder, R. (2007) Molecular basis for target RNA recognition and cleavage by human RISC. *Cell*, **130**, 101–112.

97 Hafner, M., Landthaler, M., Burger, L., Khorshid, M., Hausser, J., Berninger, P., Rothballer, A., Ascano, M. Jr, Jungkamp, A.C., Munschauer, M., Ulrich, A., Wardle, G.S., Dewell, S., Zavolan, M., Tuschl, T. (2010) Transcriptome-wide identification of RNA-binding protein and microRNA target sites by PAR-CLIP. *Cell*, **141**, 129–141.

98 Hutvagner, G., Zamore, P.D. (2002) A microRNA in a multiple-turnover RNAi enzyme complex. *Science*, **297**, 2056–2060.

99 Eulalio, A., Tritschler, F., Izaurralde, E. (2009) The GW182 protein family in animal cells: new insights into domains required for miRNA-mediated gene silencing. *RNA*, **15**, 1433–1442.

100 Chatterjee, S., Grosshans, H. (2009) Active turnover modulates mature microRNA activity in *Caenorhabditis* elegans. *Nature*, **461**, 546–549.

101 Ameres, S.L., Horwich, M.D., Hung, J.H., Xu, J., Ghildiyal, M., Weng, Z., Zamore, P.D. (2010) Target RNA-directed trimming and tailing of small silencing RNAs. *Science*, **328**, 1534–1539.

102 Rybak, A., Fuchs, H., Hadian, K., Smirnova, L., Wulczyn, E.A., Michel, G.,

Nitsch, R., Krappmann, D., Wulczyn, F.G. (2009) The let-7 target gene mouse lin-41 is a stem cell specific E3 ubiquitin ligase for the miRNA pathway protein Ago2. *Nat. Cell Biol.*, **11**, 1411–1420.

103 Hammell, C.M., Lubin, I., Boag, P.R., Blackwell, T.K., Ambros, V. (2009) nhl-2 Modulates microRNA activity in *Caenorhabditis elegans. Cell*, **136**, 926–938.

104 Schwamborn, J.C., Berezikov, E., Knoblich, J.A. (2009) The TRIM-NHL protein TRIM32 activates microRNAs and prevents self-renewal in mouse neural progenitors. *Cell*, **136**, 913–925.

105 Aravin, A.A., Hannon, G.J., Brennecke, J. (2007) The Piwi-piRNA pathway provides an adaptive defense in the transposon arms race. *Science*, **318**, 761–764.

106 Thomson, T., Lin, H. (2009) The biogenesis and function of PIWI proteins and piRNAs: progress and prospect. *Annu. Rev. Cell Dev. Biol.*, **25**, 355–376.

107 Siomi, M.C., Sato, K., Pezic, C., Aravin, A.A. (2011) PIWI-interacting small RNAs: the vanguard of genome defence. *Nat. Rev. Mol. Cell Biol.*, **12**, 246–258.

108 Aravin, A.A., Naumova, N.M., Tulin, A.V., Vagin, V.V., Rozovsky, Y.M., Gvozdev, V.A. (2001) Double-stranded RNA-mediated silencing of genomic tandem repeats and transposable elements in the *D. melanogaster* germline. *Curr. Biol.*, **11**, 1017–1027.

109 Aravin, A.A., Lagos-Quintana, M., Yalcin, A., Zavolan, M., Marks, D., Snyder, B., Gaasterland, T., Meyer, J., Tuschl, T. (2003) The small RNA profile during *Drosophila melanogaster* development. *Dev. Cell*, **5**, 337–350.

110 Aravin, A., Gaidatzis, D., Pfeffer, S., Lagos-Quintana, M., Landgraf, P., Iovino, N., Morris, P., Brownstein, M.J., Kuramochi-Miyagawa, S., Nakano, T., Chien, M., Russo, J.J., Ju, J., Sheridan, R., Sander, C., Zavolan, M., Tuschl, T. (2006) A novel class of small RNAs bind to MILI protein in mouse testes. *Nature*, **442**, 203–207.

111 Girard, A., Sachidanandam, R., Hannon, G.J., Carmell, M.A. (2006) A germline-specific class of small RNAs binds mammalian Piwi proteins. *Nature*, **442**, 199–202.

112 Grivna, S.T., Beyret, E., Wang, Z., Lin, H. (2006) A novel class of small RNAs in the mouse spermatogenic cells. *Genes Dev.*, **20**, 1709–1714.

113 Houwing, S., Berezikov, E., Ketting, R.F. (2008) Zili is required for germ cell differentiation and meiosis in zebrafish. *EMBO J.*, **27**, 2702–2711.

114 Lau, N.C., Seto, A.G., Kim, J., Kuramochi-Miyagawa, S., Nakano, T., Bartel, D.P., Kingston, R.E. (2006) Characterization of the piRNA complex from rat testes. *Science*, **313**, 363–367.

115 Cox, D.N., Chao, A., Baker, J., Chang, L., Qiao, D., Lin, H. (1998) A novel class of evolutionarily conserved genes defined by piwi are essential for stem cell self-renewal. *Genes Dev.*, **12**, 3715–3727.

116 Li, C., Vagin, V.V., Lee, S., Xu, J., Ma, S., Xi, H., Seitz, H., Horwich, M.D., Syrzycka, M., Honda, B.M., Kittler, E.L., Zapp, M.L., Klattenhoff, C., Schulz, N., Theurkauf, W.E., Weng, Z., Zamore, P.D. (2009) Collapse of germline piRNAs in the absence of Argonaute3 reveals somatic piRNAs in flies. *Cell*, **137**, 509–521.

117 Lin, H., Spradling, A.C. (1997) A novel group of pumilio mutations affects the asymmetric division of germline stem cells in the *Drosophila* ovary. *Development*, **124**, 2463–2476.

118 Brennecke, J., Aravin, A.A., Stark, A., Dus, M., Kellis, M., Sachidanandam, R., Hannon, G.J. (2007) Discrete small RNA-generating loci as master regulators of transposon activity in *Drosophila. Cell*, **128**, 1089–1103.

119 Schmidt, A., Palumbo, G., Bozzetti, M.P., Tritto, P., Pimpinelli, S., Schafer, U. (1999) Genetic and molecular characterization of sting, a gene involved in crystal formation and meiotic drive in the male germ line of *Drosophila melanogaster. Genetics*, **151**, 749–760.

120 Vagin, V.V., Sigova, A., Li, C., Seitz, H., Gvozdev, V., Zamore, P.D. (2006) A distinct small RNA pathway silences selfish genetic elements in the germline. *Science*, **313**, 320–324.

121 Kirino, Y., Mourelatos, Z. (2007) Mouse Piwi-interacting RNAs are 2′-O-methylated at their 3′ termini. *Nat. Struct. Mol. Biol.*, **14**, 347–348.

122 Prud'homme, N., Gans, M., Masson, M., Terzian, C., Bucheton, A. (1995) Flamenco, a gene controlling the gypsy retrovirus of *Drosophila melanogaster*. *Genetics*, **139**, 697–711.

123 Saito, K., Inagaki, S., Mituyama, T., Kawamura, Y., Ono, Y., Sakota, E., Kotani, H., Asai, K., Siomi, H., Siomi, M.C. (2009) A regulatory circuit for *piwi* by the large Maf gene *traffic jam* in *Drosophila*. *Nature*, **461**, 1296–1299.

124 Haase, A.D., Fenoglio, S., Muerdter, F., Guzzardo, P.M., Czech, B., Pappin, D.J., Chen, C., Gordon, A., Hannon, G.J. (2010) Probing the initiation and effector phases of the somatic piRNA pathway in *Drosophila*. *Genes Dev.*, **24**, 2499–2504.

125 Olivieri, D., Sykora, M.M., Sachidanandam, R., Mechtler, K., Brennecke, J. (2010) An *in vivo* RNAi assay identifies major genetic and cellular requirements for primary piRNA biogenesis in *Drosophila*. *EMBO J.*, **29**, 3301–3317.

126 Saito, K., Ishizu, H., Komai, H., Kotani, H., Kawamura, Y., Nishida, K.M., Siomi, H., Siomi, M.C. (2010) Roles for the Yb body components Armitage and Yb in primary piRNA biogenesis in *Drosophila*. *Genes Dev.*, **24**, 2493–2498.

127 Ipsaro, J.J., Haase, A., Knott, S.R., Joshua-Tor, L., Hannon, G.J., (2012) The structural biochemistry of Zucchini implicates it as a nuclease in piRNA biogenesis. *Nature*, **491**, 279–283.

128 Nishimasu, H., Ishizu, H., Saito, K., Fukuhara, S., Kamatani, M.K., Bonnefond, L., Matsumoto, N., Nishizawa, T., Nakanaga, K., Aoki, J., Ishitani, R., Siomi, H., Siomi, M.C., and Nureki, O. (2012) Structure and function of Zucchini endonuclease in piRNA biogenesis. *Nature*, **491**, 284–289.

129 Nagao, A., Mituyama, T., Huang, H., Chen, D., Siomi, M.C., Siomi, H. (2010) Biogenesis pathways of piRNAs loaded onto AGO3 in the *Drosophila* testis. *RNA*, **16**, 2503–2515.

130 Saito, K., Nishida, K.M., Mori, T., Kawamura, Y., Miyoshi, K., Nagami, T., Siomi, H., Siomi, M.C. (2006) Specific association of Piwi with rasiRNAs derived from retrotransposon and heterochromatic regions in the *Drosophila* genome. *Genes Dev.*, **20**, 2214–2222.

131 Nishida, K.M., Saito, K., Mori, T., Kawamura, Y., Nagami-Okada, T., Inagaki, S., Siomi, H., Siomi, M.C. (2007) Gene silencing mechanisms mediated by Aubergine piRNA complexes in *Drosophila* male gonad. *RNA*, **13**, 1911–1922.

132 Gunawardane, L.S., Saito, K., Nishida, K.M., Miyoshi, K., Kawamura, Y., Nagami, T., Siomi, H., Siomi, M.C. (2007) A slicer-mediated mechanism for repeat-associated siRNA 5′ end formation in *Drosophila*. *Science*, **315**, 1587–1590.

133 Malone, C.D., Brennecke, J., Dus, M., Stark, A., McCombie, W.R., Sachidanandam, R., Hannon, G.J. (2009) Specialized piRNA pathways act in germline and somatic tissues of the *Drosophila* ovary. *Cell*, **137**, 522–535.

134 Kirino, Y., Vourekas, A., Kim, N., de Lima Alves, F., Rappsilber, J., Klein, P.S., Jongens, T.A., Mourelatos, Z. (2009) Arginine methylation of Piwi proteins catalysed by dPRMT5 is required for Ago3 and Aub stability. *Nat. Cell Biol.*, **11**, 652–658.

135 Nishida, K.M., Okada, T.N., Kawamura, T., Mituyama, T., Kawamura, Y., Inagaki, S., Huang, H., Chen, D., Kodama, T., Siomi, H., Siomi, M.C. (2009) Functional involvement of Tudor and dPRMT5 in the piRNA processing pathway in *Drosophila* germlines. *EMBO J.*, **28**, 3820–3831.

136 Vagin, V.V., Wohlschlegel, J., Qu, J., Jonsson, Z., Huang, X., Chuma, S., Girard, A., Sachidanandam, R., Hannon, G.J., Aravin, A.A. (2009) Proteomic analysis of murine Piwi proteins reveals a role for arginine methylation in specifying interaction with Tudor family members. *Genes Dev.*, **23**, 1749–1762.

137 Chen, C., Jin, J., James, D.A., Adams-Cioaba, M.A., Park, J.G., Guo, Y., Tenaglia, E., Xu, C., Gish, G., Min, J., Pawson, T. (2009) Mouse Piwi interactome identifies binding mechanism of Tdrkh Tudor domain to arginine methylated Miwi. *Proc. Natl Acad. Sci. USA*, **106**, 20336–20341.

138 Reuter, M., Chuma, S., Tanaka, T., Franz, T., Stark, A., Pillai, R.S. (2009) Loss of the Mili-interacting Tudor domain-containing protein-1 activates transposons and alters the Mili-associated small RNA profile. *Nat. Struct. Mol. Biol.*, **16**, 639–646.

139 Shoji, M., Tanaka, T., Hosokawa, M., Reuter, M., Stark, A., Kato, Y., Kondoh, G., Okawa, K., Chujo, T., Suzuki, T., Hata, K., Martin, S.L., Noce, T., Kuramochi-Miyagawa, S., Nakano, T., Sasaki, H., Pillai, R.S., Nakatsuji, N., Chuma, S. (2009) The TDRD9-MIWI2 complex is essential for piRNA-mediated retrotransposon silencing in the mouse male germline. *Dev. Cell*, **17**, 775–787.

140 Vasileva, A., Tiedau, D., Firooznia, A., Müller-Reichert, T., Jessberger, R. (2009) Tdrd6 is required for spermiogenesis, chromatoid body architecture, and regulation of miRNA expression. *Curr. Biol.*, **19**, 630–639.

141 Wang, J., Saxe, J.P., Tanaka, T., Chuma, S., Lin, H. (2009) Mili interacts with tudor domain-containing protein 1 in regulating spermatogenesis. *Curr. Biol.*, **19**, 640–644.

142 Siomi, M.C., Mannen, T., Siomi, H. (2010) How does the royal family of tudor rule the PIWI-interacting RNA pathway? *Genes Dev.*, **24**, 636–646.

143 Nagao, A., Sato, K., Nishida, K.M., Siomi, H., Siomi, M.C. (2011) Gender-specific hierarchy in nuage localization of PIWI-interacting RNA factors in *Drosophila*. *Front. Gene.*, **2**, 55.

144 Kuramochi-Miyagawa, S., Watanabe, T., Gotoh, K., Takamatsu, K., Chuma, S., Kojima-Kita, K., Shiromoto, Y., Asada, N., Toyoda, A., Fujiyama, A., Totoki, Y., Shibata, T., Kimura, T., Nakatsuji, N., Noce, T., Sasaki, H., Nakano, T. (2010) MVH in piRNA processing and gene silencing of retrotransposons. *Genes Dev.*, **24**, 887–892.

145 Aravin, A.A., Sachidanandam, R., Bourchis, D., Schaefer, C., Pezic, D., Toth, K.F., Bestor, T., Hannon, G.J. (2008) A piRNA pathway primed by individual transposons is linked to *de novo* DNA methylation in mice. *Mol. Cells*, **31**, 785–799.

146 Brennecke, J., Malone, C.D., Aravin, A.A., Sachidanandam, R., Stark, A., Hannon, G.J. (2008) An epigenetic role for maternally inherited piRNAs in transposon silencing. *Science*, **322**, 1387–1392.

8
To Translate or Degrade: Cytoplasmic mRNA Decision Mechanisms

Daniel Beisang[1,3] *and Paul R. Bohjanen*[1,2,3]
[1]*University of Minnesota, Department of Microbiology, Minneapolis, MN 55455, USA*
[2]*University of Minnesota, Department of Medicine, Minneapolis, MN 55455, USA*
[3]*University of Minnesota, Center for Infectious Diseases and Microbiology Translational Research, Minneapolis, MN 55455, USA*

1	Introduction	213
2	Nonsense-Mediated Decay: The Removal of Defective mRNA	213
3	Regulation of 5′ Cap-Dependent Translation	215
4	**Regulation of mRNA Degradation**	**217**
4.1	ARE-Mediated mRNA Decay	218
4.2	GRE-Mediated mRNA Decay	221
4.3	Links between mRNA Decay and Translation	223
5	**Cellular Compartments That Determine mRNA Fate**	**223**
5.1	Stress Granules	224
5.2	P-Bodies	224
6	Micro-RNA-Mediated Regulation	225
7	Conclusions and Future Research	227
	References	228

RNA Regulation: Advances in Molecular Biology and Medicine, First Edition. Edited by Robert A. Meyers.
© 2014 Wiley-VCH Verlag GmbH & Co. KGaA. Published 2014 by Wiley-VCH Verlag GmbH & Co. KGaA.

Keywords

Nonsense-mediated decay
The selective cytoplasmic degradation of mRNA molecules that contain premature stop codons.

Translation
The process by which amino acids are assembled into polypeptides based on the nucleotide sequence encoded in mRNA.

mRNA decay
The process by which mRNA molecules are degraded.

AU-rich element (ARE)
An AU-rich sequence found in the 3'-untranslated region of certain mRNA molecules that regulates post-transcriptional processes such as intracellular localization, translation or mRNA decay.

GU-rich element (GRE)
A GU-rich sequence found in the 3' untranslated regions of certain mRNA molecules that regulates post-transcriptional processes such as translation or mRNA decay.

microRNA (miRNA)
A short ~21 nucleotide RNA molecule that hybridizes with mRNA molecules and influences the decay and/or translational efficiency of the target mRNA.

- The precise control of genetic information is crucial for an organism to survive and function. Post-transcriptional control at the levels of mRNA decay and translation has evolved as a crucial mechanism for ensuring accurate gene expression in eukaryotic organisms. These levels of regulation include the nonsense-mediated mRNA decay quality control pathway, cap-dependent translation, ARE- and GRE-mediated mRNA decay, and microRNA-mediated regulation of translation and mRNA decay. The importance of these mechanisms is highlighted by their dysregulation in a number of diseases, including neuromuscular disorders and cancer. In this chapter, attention is focused on the molecular machinery behind the decision to translate or degrade an mRNA molecule, with emphasis on how these mechanisms may be involved in human disease.

1
Introduction

In eukaryotes, genetic information encoded in chromosomal DNA is confined to the membrane-bound nucleus, but this information is transcribed into mRNA and exported to the cytoplasm, where genetic information, expressed as mRNA, is translated into proteins. The physical separation of transcription in the nucleus from translation in the cytoplasm has necessitated the evolution of complex regulatory mechanisms to control the proper usage of information encoded in mRNA to control protein expression [1]. Protein expression is regulated depending on the needs of the cell through post-transcriptional mechanisms, including the storage, degradation, or translation of mRNA. These mechanisms allow the cell to precisely and flexibly control the efficiency at which a given mRNA is translated, as well as to regulate the lifespan of an mRNA molecule in the cytoplasm [2]. The importance of the decision to translate or degrade mRNA is highlighted by the observation that many human diseases including viral infections [3–5], cancer [6–8], and neuromuscular disorders [9, 10] occur through disrupted post-transcriptional regulation. In recent years, the field of post-transcriptional regulation has made great strides in providing an understanding of the molecular biology of complex regulatory mechanisms that regulate the decision to translate or degrade mRNA, and is now beginning to leverage this knowledge to develop new therapies for human diseases. In this chapter the important pathways affecting the decision to translate or degrade an mRNA molecule are described, with particular attention being focused on nonsense-mediated decay (NMD), translation initiation, mRNA turnover, and microRNA (miRNA)-mediated regulation. In addition, a review is provided of the current knowledge of the molecules and mechanisms that regulate post-transcriptional gene expression.

2
Nonsense-Mediated Decay: The Removal of Defective mRNA

In eukaryotic cells, NMD is a mechanism by which erroneous mRNA that contains premature termination codons (PTCs) is rapidly eliminated; hence, NMD will function as a quality control mechanism to prevent the translation of truncated proteins. The first identification of NMD was made through investigations of mutations in patients with β-thalassemia that resulted in PTCs [11] which caused rapid mRNA degradation. The evolutionary driver behind NMD may be the great diversity of the eukaryotic proteome that arises through alternative splicing. Typically, more than 90% of human pre-mRNAs undergo splicing [12], and splicing events can be accompanied by erroneous or incomplete intron removal, resulting in cytoplasmic mRNAs that contain PTCs [13]. Consequently, NMD most likely evolved as a backup control mechanism to afford the cell greater flexibility in its alternative splicing program by providing a mechanism to remove erroneous transcripts. Interestingly, the cell has since adapted this quality control mechanism to regulate the expression of some correctly processed transcripts, including many selenoprotein mRNAs [14]. NMD occurs in all eukaryotes examined to date and is crucial to mammalian development, with mouse embryos that are genetically deficient in a critical NMD factor being found as nonviable [15].

Characteristically, NMD forms part of a quality control mechanism that occurs as mRNA is processed and prepared for translation. Upon export from the nucleus, mature mRNA molecules are bound by a host of proteins that interact with each other and circularize the molecule such that the 5′ and 3′ ends interact [16]. In this case, nuclear poly(A) binding protein nuclear 1 (PABPN1) binds to the poly(A) tail that is present on the vast majority of mRNAs and aids in shuttling of the transcript out of the nucleus [17]. In addition, the cap-binding complex – which is a heterodimer consisting of the CBP80 and CBP20 proteins – binds to the m^7GpppN cap structure at the 5′ end of mature mRNA [18]. The mRNA and its associated proteins (mRNA ribonucleoprotein; mRNP) then recruit ribosomes and undergo the first round of translation termed the *"pioneer round"* [19]. This pioneer round of translation represents one of the first cytoplasmic regulatory steps in a transcript's lifecycle. During the pioneer round, transcripts are interrogated to ensure that a transcript does not contain premature stop codons, which are often the result of erroneous nuclear splicing events [13]. In the nucleus, a complex of proteins, known as the exon–junction complex (EJC), remain on transcripts following splicing events [20] and serve as markers for the cytoplasmic machinery to determine the location of exon–exon junctions. During the pioneer round of translation, transcripts are scanned in a 5′ → 3′ manner and the EJCs interact with the NMD complex [21]. Typically, transcripts contain exon junctions in their coding region and not their 3′-untranslated region (UTR); hence, if the NMD machinery recognizes a stop codon greater than ∼50–55 nt upstream of an EJC, then that transcript is recognized as containing a PTC and is targeted for NMD [22].

An alternative mechanism proposed to explain the targeting of transcripts with PTCs is termed the *faux 3′-UTR model*, or distance from the poly(A) tail model. This mechanism has been studied extensively in yeast cells, and new evidence suggests that similar mechanisms may also be active in mammalian cells [23]. In this model, it is proposed that the cell recognizes transcripts with an abnormally large distance between the termination codon and the poly(A) tail, and targets these transcripts for NMD [24]. It is thought that the long distance between the PTC and the poly(A) tail prevents the physiologic interaction between poly(A) binding protein (PABP) and the ribosome [24]. This allows for NMD factors to interact with the ribosome and promote the decay of the aberrant transcript. Support for this model derives from studies showing that the tethering of PABP near a PTC prevents NMD of the substrate [25]. Additionally, ribosomes terminating at PTCs have different mRNA footprints than those terminating at normal termination codons, further supporting this model [25].

Transcripts that contain the cap-binding complex (predominantly newly transcribed transcripts) are recognized and scanned by the NMD machinery, which has at its core the RNA helicase and RNA-dependent ATPase protein UPF1 [26]. When UPF1 scans the transcript and finds a PTC, UPF1 is phosphorylated by an associated phosphatidylinositol 3-kinase-related kinase SMG1 [27]. This phosphorylation event is associated with formation of the "SURF" complex of proteins at the PTC, including SMG1, UPF1, and eRF1–eRF3 [28]. Subsequently, the phosphorylated UPF1 recruits the NMD factors SMG5, SMG6, and SMG7 to both recruit the PP2A phosphatase to dephosphorylate UPF1 and to promote degradation of the PTC-harboring transcript in an endonucleolytic, $5' \rightarrow 3'$ or $3' \rightarrow 5'$ direction [29–34]. In this manner, transcripts containing PTCs are eliminated, preventing translation of truncated proteins that may be harmful to the cell.

3 Regulation of 5′ Cap-Dependent Translation

Regulation of the efficiency of translation is an important level for the post-transcriptional control of gene expression (this process is depicted in Fig. 1). When a newly synthesized transcript is exported from the nucleus and undergoes the pioneer round of translation without detection of a PTC, then the mRNP undergoes significant remodeling to allow for the transcript to enter into the polysomes and undergo active translation. One of the changes to the mRNP is that the PABPN1 bound to the poly(A) tail initially upon export from the nucleus is exchanged for the mostly cytoplasmic poly(A) binding protein cellular 1 (PABPC1) [17]. Additionally, the cap-binding complex, composed initially of CBP20 and CBP80, is exchanged for eIF4E, which itself binds to the m^7G cap structure [17]. eIF4E in turn interacts with the scaffolding protein eIF4G, which serves to circularize the mRNP through simultaneous interaction with PABPC1 [35]. Also included in this new cap-binding complex is eIF4A, which has RNA helicase activity and is thought to help unwind the mRNA [36]. With this exchange of factors in the mRNP, the transcript is then poised to undergo translation.

Translation consists of four stages, including: (i) translation initiation; (ii) translation elongation; (iii) translation termination; and (iv) ribosome recycling [16]. It is thought that the majority of translational regulation occurs at the step of translation initiation [37]. Translation is initiated by the loading of a multi-protein complex called the *preinitiation complex* (also known as the 43S complex) onto the 5′ end of a transcript. The 43S complex consists of the 40S ribosomal subunit, eIF1, eIF1A, eIF3, and eIF5, methionine-loaded tRNA, and GTP-loaded eIF2 [38], and is recruited to a transcript through interactions between eIF3 of the 43S complex and eIF4G of the cap-binding complex [39]. Once bound, the 43S complex then scans the transcript $5' \rightarrow 3'$ until it encounters the (AUG) start codon, upon which the 60S ribosomal subunit is recruited and translational elongation begins [40]. eIF4G circularizes transcripts through simultaneous interaction with eIF4E and PABP, and this conformation is important for translational initiation [35]. eIF4G binds to eIF4E through a short domain which is conserved in a family of proteins called 4E-binding proteins (4EBPs) [41]. The 4EBPs function as negative regulators of translation by competing with eIF4G for binding to eIF4E [41]. The balance between 4EBPs and eIF4G binding to eIF4E, and thus the balance between translational inhibition and translational initiation, is largely affected by the phosphorylation status of the 4EBPs. Typically, 4EBPs are known to be phosphorylated via the PI3K/AKT/MTOR pathway, which is physiologically active in the setting of mitogens, growth factors, and some hormones [42, 43]. Phosphorylation of the 4EBPs decreases their ability to bind to eIF4E, and thus promotes the translation of many eIF4E-dependent proteins [44, 45].

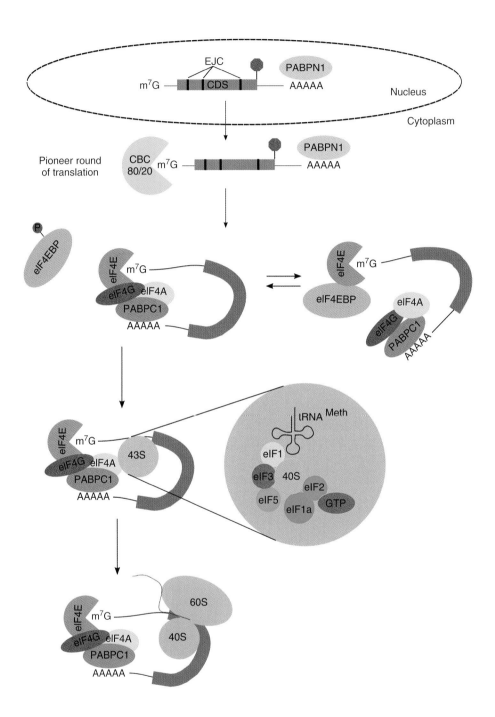

Several lines of evidence suggest that the upregulation of eIF4E-mediated translation promotes the development of malignancy. eIF4E is the rate-limiting initiation factor for translation, and the accurate regulation of its abundance is crucial to the health of a cell [46]. It has been shown that the overexpression of eIF4E in cell lines causes malignant transformation, as evidenced by their enhanced proliferation, altered cellular morphology, and growth in soft agar [47, 48]. Additionally it has been shown that eIF4E itself is overexpressed in a number of human cancers, including those of the colon, breast, bladder, and head and neck [49–52]. It has since been shown that eIF4E preferentially promotes the translation of proteins involved in malignancy-associated pathways such as cell cycle progression, angiogenesis, cell survival, and invasion [46]. These results have led to the classification of eIF4E as a bona fide oncogene.

The dysregulated expression of 4EBPs appears to also play a role in cancer by interacting with eIF4E. For example, 4EBPs act as tumor suppressors by reversing the malignant transformation of cells transformed by eIF4E overexpression [53–55]. Additionally, 4EBP levels have been shown to inversely correlate with susceptibility to oncogenesis, supporting their role as tumor suppressors [56]. The PI3K/AKT/MTOR pathway is aberrantly activated in a number of malignancies, and this leads to the aberrant phosphorylation of the 4EBPs in these tumors [43], resulting in a decreased ability to suppress the oncogenic activity of eIF4E.

The targeting of eIF4E pathways may be fruitful for the development of novel chemotherapeutics to treat cancer. Potential novel drug targets in this complicated pathway include inhibiting the eIF4E–eIF4G interaction [57] or inhibiting phosphorylation of the 4EBPs [58]. One approach with exceptional promise is to design synthetic analogs of the m^7G cap structure found at the 5′ end of mRNA [59]. Synthetic cap analogs could, in theory, compete with mature cellular mRNAs for binding to eIF4E, and thus reduce the pool of functional cytoplasmic eIF4E available to promote translation [60]. While these approaches are exciting for their broad therapeutic potential, several hurdles must still be overcome, including the establishment of an appropriate therapeutic window as all cells are dependent on these pathways for their survival.

4
Regulation of mRNA Degradation

In addition to tightly regulating the translational efficiency of mRNAs, cells have evolved intricate mechanisms to control

Fig. 1 Control of eIF4E-mediated translation. Upon export from the nucleus, cap-binding complex-bound mRNA undergoes the pioneer round of translation to ensure that a stop codon does not occur upstream of an exon–junction complex (EJC). If the transcript passes this quality control step, it undergoes extensive remodeling of its protein constituents, which now include eIF4E, eIF4G, eIF4A, and PABPC1. At this step, translation can be inhibited by increased functional eIF4E-BPs, which compete with eIF4G for binding to eIF4E and thus disrupt the circularized conformation of the mRNP. eIF4EBP binding to eIF4E is itself regulated by the phosphorylation status of the eIF4EBPs. The circularized mRNP is then able to recruit the 43S preinitiation complex which will scan the transcript 5′ → 3′ until it arrives at the start codon. Upon recognition of the start codon, the 60S ribosomal subunit is recruited to the transcript, and translational elongation occurs.

the cytoplasmic abundance of mRNAs as a means to control post-transcriptional gene expression. In particular, rapid mRNA degradation is a mechanism for turning off gene expression, by removing the pool of mRNA available for translation. The steady-state abundance of an mRNA species is a function of both its rate of production (transcription) and its rate of degradation. Classically, it was assumed that most of the regulation of mRNA abundance occurs at the level of transcription and that, upon export from the nucleus, a transcript's lifespan was stochastically determined by the availability of the RNA degradation machinery. A large body of evidence now exists to show that this assumption is incorrect; in fact, it is now understood that mRNA degradation is highly selective and tightly regulated in a manner that depends on the cellular and environmental context [61].

The importance of mRNA decay for the control of genetic information is illustrated by its role in the process of T lymphocyte activation [62, 63]. Within minutes to hours of receiving the appropriate activation signals, T lymphocytes display dramatic changes in their transcriptome [64] which allow the cell to undergo a cellular program characterized by a proliferative burst and altered apoptotic responses [65]. It has been shown that approximately 50% of the changes in T lymphocyte gene expression during activation are a result of changes in transcript half-life [66]. These changes have been quantified on a genome-wide scale, and individual transcript half-lives can change dramatically following activation [67].

The components which regulate a transcript's half-life include cis-elements and the trans-acting factors which recognize those elements [2]. cis-Elements are sequences harbored within an mRNA molecule that are recognized by trans factors. Decay-regulatory cis-elements typically occur in the 3′-UTR of transcripts and function by recruiting trans factors such as RNA-binding proteins and miRNAs [2, 68]. Whether a cis-element promotes transcript stability or instability is determined by the biochemical activity of the trans factor which binds to that element [2]. The best-characterized cis-elements that regulate mRNA degradation are the AU-rich element (ARE) and GU-rich element (GRE), as described in the following sections.

4.1
ARE-Mediated mRNA Decay

AREs were first discovered through a bioinformatic approach as an evolutionarily conserved sequence element in the 3′-UTR of the tumor necrosis factor α (TNFα) transcript and other cytokine transcripts [69]. A 51 nt AU-rich sequence from the granulocyte-macrophage colony stimulating factor transcript was shown to function as a mediator of mRNA degradation by destabilizing an otherwise stable β-globin transcript when it was inserted into its 3′-UTR [70]. Subsequently, numerous cytokine [TNF-α, interleukin (IL)-2, and interferon (IFN)-γ] and proto-oncogene (c-*fos*, v-*myc*) transcripts were identified as containing AREs [71]. Initially, the ARE was defined as a sequence containing overlapping AUUUA pentamers, but this definition has since been broadened to include several variations on this sequence. AREs are now organized into three classes depending on their sequence content, and biologic effect on harboring transcripts [72]. In this system, class I AREs contain one to three copies of scattered AUUUA motifs with a nearby U-rich region, and caused synchronous deadenylation. Class II AREs are composed of at least two overlapping copies of the UUAUUUA(U/A)(U/A) nonamer in a U-rich region, and caused asynchronous deadenylation. Finally, class III AREs were said to contain a U-rich region and other "poorly defined features," and caused synchronous deadenylation [72]. This classification system has proved to be useful in understanding the observed variability in the behavior of ARE-containing transcripts.

Given the heterogeneity in the structure of AREs, it is not surprising that there are multiple RNA-binding proteins capable of binding to ARE-containing transcripts. AU-rich element-binding proteins (AREBPs) include over 20 proteins from various families, with approximately 12 well-characterized members [73]. AREBPs can promote both transcript stability and transcript decay. The best-studied AREBPs that stabilize mRNA include the ubiquitously expressed HuR (ELAVL1, HuA) and the predominantly neuronal HuD (ELAVL4) [73–75]. These proteins are both members of the evolutionarily conserved family of Embryonic Lethal Abnormal Vision Like proteins which are composed of three RNA recognition motifs (RRMs) with the two N-terminal RRMs separated from the C-terminal domain by the HuR nucleocytoplasmic shuttling (HNS) domain involved in determining the protein's subcellular location [76]. These proteins bind to U-rich sequences (class III AREs) and promote transcript stability [76]. It is thought that these proteins promote stability through a competitive binding model where they compete for ARE-binding with other AREBPs that promote decay [77].

There are several examples of AREBPs that promote transcript decay, including tristetraprolin (TTP), BRF1/2, AUF1, and KSRP [78]. The best-studied member of this group of proteins is TTP, which binds preferentially to AREs with overlapping AUUUA multimers (class II AREs) and promotes a rapid decay of the harboring transcript through the recruitment of proteins containing RNAse activity [79, 80]. The importance of TTP in ARE-mediated mRNA decay was shown in mice that had been made genetically deficient in TTP. These mice developed a syndrome of cachexia, arthritis, and autoimmunity shortly after birth [81]. This constellation of symptoms was shown to have been a result of the aberrant stabilization of TNF-α mRNA and an upregulation of TNF-α protein expression due to a lack of TTP-mediated mRNA decay [81].

Studies investigating the mechanism of action of TTP have provided a model for mRNA degradation (as summarized in Fig. 2). Almost all mature, eukaryotic mRNAs contain a poly(A) tail at their $3'$ end [82]. The first step in mRNA decay is the shortening of this poly(A) tail (deadenylation) by the deadenylase enzymes, including the CCR4–NOT complex or the poly(A) ribonuclease (PARN) enzyme [83, 84]. TTP recruits PARN [85] and the NOT1 component of the CCR4–NOT complex [86]. Following a shortening of the poly(A) tail, the transcript can either enter a $3' \rightarrow 5'$ or $5' \rightarrow 3'$ decay pathway. In the $3' \rightarrow 5'$ pathway, the deadenylated transcript can then be targeted by the exosome, a large multiprotein complex with RNAse activity [80, 87]. In the $5' \rightarrow 3'$ decay pathway the deadenylated transcript (which now contains a short poly(A) tail) is bound by the Lsm1–7 complex, which in turn recruits the DCP1 and DCP2 enzymes [88]. DCP1/2 removes the m^7G cap, allowing for the transcript to be degraded

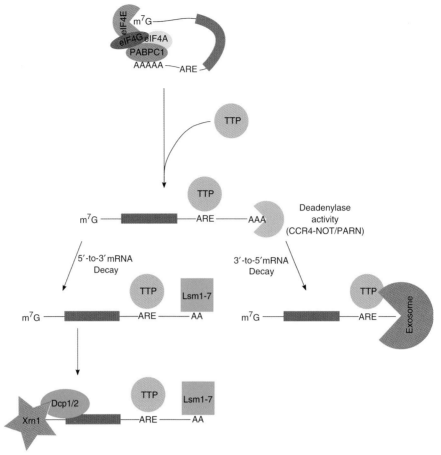

Fig. 2 Tristetraprolin (TTP)-mediated mRNA decay. TTP recognizes and binds to the ARE within the 3′-UTR of a number of cytokine and proto-oncogene transcripts. Upon binding, TTP is able to recruit cytoplasmic deadenylases including the CCR4–NOT complex and PARN. Removal of most of the poly(A) tail triggers the transcript for decay in either the 5′ → 3′ or 3′ → 5′ direction. In 5′ → 3′ decay, TTP recruits the Lsm1–7 complex which binds to the remaining, short, poly(A) tail. The Lsm1–7 complex in turn recruits the Dcp1/2 enzymes which remove the 5′-cap from transcripts, leaving them susceptible to Xrn1-mediated 5′ → 3′ mRNA decay. In the 3′ → 5′ decay pathway, TTP is able to recruit components of the exosome, which in turn degrades the target transcript.

in a 5′ → 3′ direction by the exoribonuclease 1 (XRN1) enzyme [89, 90]. TTP has been shown to recruit both XRN1 and the exosome subunit PM/SCL-100, which suggests that it is able to promote both 3′ → 5′ and 5′ → 3′ degradation [91].

It is interesting to note that, in addition to regulating mRNA turnover, AREBPs are also able to regulate the translational state of an mRNA. One example of this is the translational regulation by HuR (ELAVL1) of certain ARE-containing transcripts [92]. HuR has been shown to increase the translational efficiency of transcripts encoding p53 [92], cytochrome c [93], XIAP [94], and BCL2 [95], while decreasing the translational efficiency of those coding for p27 [96], MYC [97], and WNT5α [98]. While the biology behind this dichotomous behavior is poorly understood, the mechanism of HuR-mediated control of MYC has been well studied. In this example, HuR bound to the MYC 3'-UTR and recruited the RNA-induced silencing complex (RISC) loaded with the miRNA Let-7 [97]. As will be discussed in detail below, RISC is able to compete for binding to the 5'-cap structure with eIF4E, and thus inhibit translation [99]. However, whether this interaction with the miRNA pathway is a universal mechanism for HuR, or AREBP translational regulation, remains to be resolved.

4.2
GRE-Mediated mRNA Decay

The bioinformatic analysis of a genome-wide database of mRNA half-lives in resting and activated T lymphocytes [67] demonstrated that, while many short-lived transcripts contained AREs, many more did not [67]. This observation led research groups to carry out a search for novel conserved elements in the 3'-UTR of short-lived transcripts in T lymphocytes [100]. The search identified a GRE that was highly enriched in short-lived transcripts [100]. This 11 nt motif of the form UGUUUGUUUGU was shown to be bound by a 54 kDa protein in the cytoplasm of T lymphocytes [100]; the latter protein was subsequently identified as CUGBP and Etr3-Like Factor 1 (CELF1, CUGBP1) and shown to be responsible for mediating the decay of transcripts containing the GRE [100].

The binding specificity and function of CELF1 was clarified through subsequent experiments to identify the binding targets of CELF1 on a genome-wide level [101]. These experiments utilized a technique known as *RNA-immunoprecipitation*, which involved the immunopurification of an RNA-binding protein and subsequent analysis of the copurified RNA via high-throughput methodologies such as microarray or deep sequencing. These experiments were carried out to identify CELF1 target transcripts in human cervical carcinoma (HeLa) cells [101], mouse myoblast (C2C12) cells [102], and in resting and activated primary human T lymphocytes [103]. All of these experiments verified that CELF1 bound to the original GRE, and also identified a novel GU-repeat sequence that functioned to cause decay in a manner identical to the original GRE. Through these findings, the GRE was redefined to be the sequence UGU[G/U]UGU[G/U]UGU [101]. These experiments also shed insight into the function of CELF1-mediated regulation. An analysis of the biological functions of CELF1-bound transcripts in each of the different cell types revealed that CELF1 target transcripts are involved in cell cycle regulation, in the regulation of apoptosis, and in RNA post-transcriptional regulation [103].

The identification of CELF1 target transcripts in primary human T lymphocytes led to insight into the regulation of CELF1's activity. In resting T lymphocytes, approximately 1300 CELF1 target transcripts were identified, whereas only 150 CELF1 target transcripts were identified in activated T lymphocytes [103]. It was found that, upon T lymphocyte activation, CELF1 phosphorylation resulted in a reduced affinity of CELF1 for the GRE, the relative stabilization of target transcripts, and their accumulation in the cytoplasm [103]. Given that CELF1 target transcripts are involved in cell cycle progression and apoptosis, it is tempting to speculate that the inactivation of CELF1 through phosphorylation after T lymphocyte activation allows for the coordinated upregulation of an "activation" network of transcripts, causing the cell to acquire an activated phenotype. One intriguing line of investigation is whether the physiologic regulation of CELF1 allowing a transient proliferative burst following T lymphocyte activation is dysregulated in the setting of malignancy and promotes constitutive proliferation.

In addition to its role as a regulator of mRNA decay, CELF1 is involved in the regulation of both alternative splicing and translational control for a number of transcripts. Much of what is currently known about CELF1-mediated alternative splicing derives from the neuromuscular disease myotonic dystrophy type 1 (DM1). The pathogenesis of DM1 involves a PKC-mediated phosphorylation event on CELF1, causing increased CELF1 protein stability and abundance [104]. An increased functional CELF1 in the nucleus leads to the mis-splicing of a number of transcripts, including chloride channel 1 (CLCN1) [105], insulin receptor (INSR) [106], and troponin T (TNNT2) [107]. Interestingly, the mis-spliced genes correlate with the clinical symptoms observed in this disease, including muscle loss, insulin resistance, and neurologic impairment [108]. It has since been shown that CELF1 mediates a dramatic shift in the splicing patterns of cardiac transcripts at the fetal-to-adult developmental transition [109]. This finding suggests that CELF1 is involved in alternative splicing in both a physiologic and disease setting. It will be interesting to determine if CELF1 has a role in regulating alternative splicing programs in other tissues and diseases.

CELF1 is also involved in the translational regulation of a number of transcripts, and this regulation appears to involve multiple mechanisms. The best-studied example of CELF1-mediated translational regulation involves the promotion of context-specific isoform usage of the transcription factor CCAAT/enhancer-binding protein (CEBPβ) [110–112]. It has been shown in a rat model of partial hepatectomy that the phosphorylation of CELF1 promoted its association with the translational regulator eIF2. This complex subsequently promoted the selective translation of the liver-enriched inhibitory protein (LIP) isoform of CEBPβ [113]. CELF1 is also able to control the translation of the histone deacetylase complex 1 (HDAC1) mRNA. In a rat model of aging, the phosphorylation of CELF1 by cdk4 was increased with age. In turn, CELF1 phosphorylation increased the abundance of CELF1–eIF2 complexes, and led to CELF1–eIF2 binding to the 5′-UTR of HDAC1 mRNA and a resultant increase in HDAC1 protein in aging liver [114]. This same rat aging model was exploited to show that CELF1 phosphorylation promotes its increased interaction with a GC-rich sequence in the 5′-UTR of p21 mRNA [115]. Finally, CELF1 has been found to regulate the translation of the serine hydroxymethyltransferase (SHMT) [116] and cyclin-dependent kinase inhibitor p27 (Kip1) [117] transcripts. The regulation of SHMT and Kip1 was shown to be a result of CELF1 binding to an internal ribosomal entry site (IRES) in the 3′-UTR of these transcripts [116]. It is interesting to note that genome-wide studies to investigate the binding specificity of CELF1 have not found enrichment of any of the above-mentioned translational control-related binding motifs [101, 103]. This suggests that CELF1-mediated translational regulation may occur only in the context of specific transcripts under specific cellular conditions, and may represent the exception rather than the rule for CELF1 function.

4.3
Links between mRNA Decay and Translation

It has been proposed that there are codependencies between the regulation of mRNA decay and translation [118]. This concept is perhaps best exemplified by the previously discussed observation that HuR is able to promote the stability as well as the translational efficiency of its target transcripts [119]. Another example of this coregulation is in miRNA-mediated control, where the targeting of a transcript by a miRNA can promote either mRNA decay, or translational repression [120]. An example of a biochemical mechanism which elegantly ties together mRNA decay and translation derives from studies investigating the protein Pat1 in *Saccharomyces cerevisiae* [121]. Pat1 is a multifunctional protein which has been shown to be involved in promoting both translation repression during glucose deprivation as well as the decapping of mRNA and subsequent $5' \rightarrow 3'$ decay [121]. It has been shown that the dual function of this protein comes from separate functional domains of the Pat1 protein [122]. Through deletion and mutational studies in yeast, it was found that the C-terminal end of the protein is involved in promoting translational suppression. The middle portion of the protein was shown to promote mRNA decapping, most likely through recruitment of the Lsm1–7 complex. Through these separate domains, the Pat1 protein is able to influence both mRNA decay and translation, and thus serve as a common hub in the regulation of these separate biochemical pathways [122]. Overall, it seems that a theme is emerging where the regulation of mRNA decay can influence translation, and vice-versa, making it important in some cases to study these post-transcriptional regulatory steps in the context of one another, as opposed to in isolation.

5
Cellular Compartments That Determine mRNA Fate

During the course of studying RNA-binding proteins, and how they participate in post-transcriptional regulation, it was observed that some RNA-binding proteins could be induced to concentrate into two discrete cytoplasmic foci known as stress granules (SGs) and processing bodies (P-bodies) [123]. These evolutionarily conserved structures can be induced to form through a variety of cellular stresses that include radiation [124], ultraviolet light [125], and viral infection [126]. Both, SGs and P-bodies play important roles in shaping the cells' response to stress and the decision to either translate or degrade mRNA. The SGs function in the storage of translationally halted transcripts following cellular stress, while P-bodies participate in mRNA decay pathways. The constituents of these granules are very dynamic, and there is extensive movement of mRNA between SGs, P-bodies and polysomes [123], while the decision to store, degrade or translate mRNA is regulated by the interplay of these compartments.

5.1 Stress Granules

Cellular stress that triggers a global reduction in translation induces the formation of SGs, and cytoplasmic accumulations of RNA and protein that serve to store mRNA in a translationally silent state [127]. A variety of environmental triggers have been shown to trigger SG formation, including heat shock [128], hypoxia [129], arsenic exposure [130], knockdown of translational initiation factors [131], puromycin-induced ribosome-RNA dissociation [132], and others [127]. These triggers all involve the inhibition of translation at an early initiation phase, whereas blocking translation initiation at the later stage of 60S ribosomal subunit recruitment fails to induce SG formation [133].

The first proteins to be implicated in SG formation were TIA-1, TIAR, and PABP, which were found to aggregate in response to heat shock [134]. Over 25 proteins have since been observed to associate with SGs, and clues to the function of the SGs derive from the identity of many of these protein constituents [127]. Many SG factors, including eIF4A, eIF4G, and 40S ribosomal subunits, are involved in the process of translational initiation, supporting the hypothesis that SGs are involved in the halting of translation [127]. Interestingly, however, it has been observed that the localization of a transcript into a SG is not necessary for that transcript's translational inhibition [135, 136], and it is thought that the main determinant of a transcript's translational status is the group of proteins associated with it [127]. It is currently thought that the main function of SGs may be to serve as an area of increased concentration of these factors [127]. The increase in local concentrations of RNA-binding proteins and their mRNA targets would likely increase the rate of assembly of specific ribonucleoproteins (mRNPs), allowing for the efficient docking of these factors to stalled transcripts in response to stress. This would allow the cell to more efficiently alter its translational pattern and more quickly respond to a cellular stress, increasing the likelihood of the cell surviving the challenge.

Once a transcript is loaded into a SG, it can undergo one of two fates. The SGs are reversible structures, and a transcript which is stalled in translation in a SG can re-enter the translating pool of mRNA and be loaded into polysomes [123]. Alternatively, mRNA in the SGs can join a second, and compositionally distinct, cytoplasmic granule called P-bodies [123]. This exchange of mRNA cargo between SGs and P-bodies is thought to involve direct contact between SGs and P-bodies, and the orchestration of this cytoplasmic movement may be enacted through movement along the cytoskeletal microtubules [137, 138]. While the regulation of the decision to enter polysomes or P-bodies after exit from the SG is poorly understood, much progress has been made in the understanding of the components and function of P-bodies.

5.2 P-Bodies

P-bodies are cytoplasmic foci which contain aggregates of translationally repressed mRNA and RNA-binding proteins involved in translational suppression and mRNA decay [139]. The mRNA components of P-bodies seem to be translationally suppressed, as evidenced by a reduction in P-body size when ribosomes are tied to mRNA with the chemical cyclohexamide, and an increase in size upon stress-induced translational inhibition [140, 141]. The protein components of P-bodies share some similarities with SG components, but are especially enriched in proteins such as DCP1, DCP2, LSM1–7, CCR4–NOT complex, and XRN1 which are involved in 5′ → 3′ mRNA decay [139]. Components of the exosome, which participates in 3′ → 5′ decay, are notably absent from P-bodies, which suggests that the cell spatially separates these two decay pathways [142].

In addition to the 5′ → 3′ mRNA decay machinery, P-bodies contain protein constituents involved in NMD, miRNA, and ARE-mediated regulation [139]. As discussed above, premature stop codons trigger transcripts for NMD, and one of the proteins involved in this process is UPF1. In the process of recruiting decay factors to the aberrant transcript, UPF1 promotes a recruitment of the faulty transcript to P-bodies in a manner that may involve a direct interaction with the DCP enzymes [143]. Interestingly, UPF1 and the other NMD factors seem to be associated with P-bodies very transiently, which suggests that the NMD factors may serve as a delivery vehicle to drop the faulty transcripts off to P-bodies for degradation [30, 144]. In addition to NMD components, factors involved in miRNA regulation have been shown to associate with P-bodies, such as all four Ago (Argonaute) proteins [145, 146] and GW182 [147]. This suggests that some miRNA-mediated regulation may occur in P-bodies. One well-described model for this mode of regulation involves the control of the mRNA coding for CAT-1 which colocalizes with miR-122 in liver cells [148]. The miR-122-mediated translational suppression of CAT-1 is relieved under conditions of stress, upon which CAT-1 mRNA exits the P-body and enters the cytoplasmic polysomes. The exit of CAT-1 mRNA from the P-body was shown to depend on the ARE-binding protein HuR recognizing an ARE in the 3′-UTR of CAT-1 [148]. The results of this study provided a dramatic example of the crosstalk between miRNA and ARE-mediated regulation, and it is intriguing to speculate that P-bodies may provide a common physical hub for these two regulatory pathways to interact. In fact, many ARE-binding proteins – including TTP, BRF1, and BRF2 – have been shown to associate with P-bodies along with their bound, ARE-containing transcripts [149]. It is possible that these ARE-binding proteins also interact with the miRNA machinery through a P-body-mediated mechanism.

6
Micro-RNA-Mediated Regulation

Micro RNAs – one of the landmark discoveries in the field of biology over the past two decades – are short (∼21 nt) RNA molecules that are important regulators of gene expression in species ranging from plants to humans [150]. They were first discovered in 1993 by a group studying regulation of the expression of the *Caenorhabditis elegans* protein lin-14 [151], when the protein's abundance was found to correlate negatively with the expression of a 61 nt precursor RNA molecule from the *lin-4* gene. The precursor was subsequently processed into a 22 nt cytoplasmic RNA molecule with partial complementarity to several sites in the lin-14 3′-UTR that were crucial for this mode of regulation [151]. In 2000, miRNA-mediated control was recognized to be a broader phenomenon when the miRNA let-7 was discovered in *C. elegans*, and found to be conserved in both sequence and its developmental expression program in several species up through vertebrates [152, 153]. Since then, the field of miRNA has exploded such that over 20 000 unique miRNA sequences have now been identified in 168 species, with almost 2000 unique human miRNAs being recorded [154, 155]. These miRNAs are able to exert both translational and mRNA decay effects on the transcripts that they target, and comprise a complicated regulatory web that is only just beginning to be untangled.

miRNAs are processed from precursor molecules in a coordinated and evolutionarily conserved manner (see Fig. 3). First, the miRNA is transcribed in the nucleus by RNA polymerase II as precursor molecules termed *pri-miRNA* [156]; the latter then form hairpin structures which are recognized and cleaved at their base by the RNAse enzyme Drosha [157]. The Drosha-processed pri-miRNA is approximately 70 nt long, is referred to as a *pre-miRNA* and is exported to the cytoplasm where it is processed a second time by the cytoplasmic RNAse enzyme, Dicer [158, 159]. Dicer processes the pre-miRNA into an approximately 20 bp miRNA/miRNA* duplex molecule. By employing a currently poorly understood decision process, one strand of the miRNA/miRNA* duplex is chosen to be incorporated into the miRNA-induced silencing complex (miRISC) [160]. The RISC is a multiprotein complex, the key components of which include the miRNA itself, Ago proteins that interact directly with the miRNA [161], and a glycine-tryptophan protein of 182 kDa (GW182) which interacts with the Ago proteins and mediates some of the effects of miRNA recognition [162]. It should be noted that the miRNA only becomes functional following its incorporation into the RISC complex.

The rules regulating miRNA recognition of their target transcripts are still incompletely understood. However, it is known that complementarity between the seed region of the miRNA (which is composed of nucleotides 2–8) and sites generally within the 3′-UTR of transcripts is particularly important for determining whether a transcript will be targeted by a particular miRNA [163]. Upon recognition of its target transcript, miRNA-mediated regulation can promote either transcript degradation or translational suppression. In the case of miRNA-mediated transcript decay, the miRISC is able to induce deadenylation and endonucleic cleavage [164]. In the former case, deadenylation is mediated by GW182 recruiting the deadenylase enzymes CCR4 and CAF1 [162, 165, 166]. Additionally, when miRISCs containing Ago2 recognize mRNA harboring near-perfect complementary seed regions, Ago2 cleaves these mRNAs endonucleolytically and the resulting mRNA fragments are degraded in the $5' \rightarrow 3'$ and $3' \rightarrow 5'$ directions [167]. Alternatively, miRNA can promote a translational repression of their target transcripts. In this situation, it has been shown that the Ago proteins in the miRISC are able to bind to the m^7G-cap structure of the target mRNA through a highly conserved motif which has a similarity to the cap binding site in the eIF4E protein [99]. By competing with eIF4E for binding to the mRNA cap, the miRISC can reduce the rate of eIF4E-mediated translational initiation, and thus reduce the amount of protein produced in a transcript-specific manner. The cap-binding site on Ago proteins is critical for miRNA-induced translational inhibition, as mutations in this site on Ago2 prevent the protein's cap-binding activity and negate its ability to inhibit translation [99].

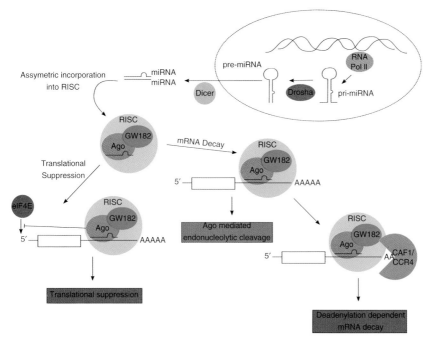

Fig. 3 miRNA-mediated regulation of translation and mRNA decay. Pri-miRNA molecules are transcribed from the genome by RNA polymerase II (RNA pol II). These hairpin structures are processed in the nucleus by the RNAse enzyme Drosha into pre-miRNA. Following export into the cytoplasm, Dicer cleaves the pre-miRNA into an ∼20 bp miRNA/miRNA* heteroduplex molecule. Through an unclear mechanism, one strand of the heteroduplex (miRNA) is incorporated into the RISC complex which contains the miRNA, the Argonaute protein (Ago), and the GW182 protein. The RISC complex then targets transcripts through complementarity between the miRNA and sequences found typically in the 3′-UTR of transcripts. Upon binding to mRNA, the RISC complex can subsequently promote translational suppression or transcript degradation. RISC promotes translational silencing by the Ago protein competitively binding to the 5′-cap structure, inhibiting eIF4E binding and subsequent translational activation. RISC can also promote mRNA decay through two separate mechanisms. Ago possesses endonuclease activity, and can cleave the transcript, triggering decay of the resultant cleavage products. Alternatively, GW182 is able to recruit deadenylases (CAF1 and CCR4) to the transcript, which in turn triggers deadenylation-dependent mRNA decay.

7 Conclusions and Future Research

The ability of a cell to quickly and accurately control the expression of its genetic material is crucial in order for the cell to respond to its environment. The control of transcript degradation and translational efficiency are two ways in which cells are able to control the flow of their genetic information. The individual regulatory pathways outlined above provide a complex and coherent regulatory scheme for ensuring the fidelity of the gene expression program. This scheme begins upon the export of a transcript to the cytoplasm, where it is scanned for the presence of mutations and splicing errors that might have introduced PTCs. These faulty transcripts are targeted for NMD so as to prevent their translation into truncated – and potentially toxic – proteins. If a transcript passes the NMD regulatory hurdle, it may then enter the actively translating pool of transcripts. This transition involves the exchange of a number of proteins to form the eIF4F complex and circularize the transcript. The efficiency with which the transcript is translated can be influenced by a number of regulatory mechanisms, including miRNA control. Additionally, under conditions of cellular stress, transcripts can exit the translating pool of transcripts and be reversibly shuttled to SGs for storage. A transcript will ultimately end its lifecycle by being degraded, either cytoplasmically or in P-bodies, and the importance of this stage in the lifecycle is underscored by the large number of regulatory pathways converging on this decision, including ARE/GRE-mediated decay and miRNAs.

While much progress has been made recently in understanding the biochemical pathways underlying the regulatory mechanisms outlined above, there are several lines of investigation which will likely lead to a better understanding of the rules of post-transcriptional decay. One such area currently under investigation is the interaction between ARE-mediated decay and miRNA-mediated regulation. Several examples have been reported of the coregulation of transcripts by ARE-binding proteins and miRNAs [168–171]. While these examples are important first steps in understanding how these pathways intersect, a more global understanding of the process will most likely derive from studies aimed at identifying – both computationally and experimentally – the binding targets of individual miRNAs and the physical relationship between miRNA and AREBP-binding sites. Another important area of these investigations is to understand how the dysregulation of mRNA decay – particularly GRE-mediated decay – contributes to malignancy. The network of transcripts targeted by CELF1 is highly enriched for cell cycle and apoptosis regulatory proteins [101, 103]. Thus, CELF1 represents an important node in the control network of these cellular processes [103], and may serve as a potential target for the development of novel chemotherapeutics. With the advent of novel techniques and technologies for the unraveling of genome-wide datasets, a number of advances are likely to be made in the field of post-transcriptional regulation in the coming decades. Indeed, as the field transitions from one of discovery to one of application, these advances will most likely have implications not only for normal physiology but also for the pathophysiology of diseases, ranging from neuromuscular disorders to cancer.

References

1 Keene, J.D., Tenenbaum, S.A. (2002) Eukaryotic mRNPs may represent posttranscriptional operons. *Mol. Cell*, **9**, 1161–1167.

2 Keene, J.D. (2007) RNA regulons: coordination of post-transcriptional events. *Nat. Rev. Genet.*, **8**, 533–543.

3 Sandri-Goldin, R.M., Mendoza, G.E. (1992) A herpesvirus regulatory protein appears to act post-transcriptionally by affecting mRNA processing. *Genes Dev.*, **6**, 848–863.

4 Noah, D.L., Twu, K.Y., Krug, R.M. (2003) Cellular antiviral responses against influenza A virus are countered at the posttranscriptional level by the viral NS1A protein via its binding to a cellular protein required for the 3′ end processing of cellular pre-mRNAS. *Virology*, **307**, 386–395.

5 Dolken, L., Perot, J., Cognat, V., Alioua, A. et al. (2007) Mouse cytomegalovirus microRNAs dominate the cellular small RNA profile during lytic infection and show features of posttranscriptional regulation. *J. Virol.*, **81**, 13771–13782.

6 Dixon, D.A., Tolley, N.D., King, P.H., Nabors, L.B. et al. (2001) Altered expression of the mRNA stability factor HuR promotes cyclooxygenase-2 expression in colon cancer cells. *J. Clin. Invest.*, **108**, 1657–1665.

7 Heinonen, M., Bono, P., Narko, K., Chang, S.H. et al. (2005) Cytoplasmic HuR expression is a prognostic factor in invasive ductal breast carcinoma. *Cancer Res.*, **65**, 2157–2161.

8 Vlasova, I.A., McNabb, J., Raghavan, A., Reilly, C. et al. (2005) Coordinate stabilization of growth-regulatory transcripts in T cell malignancies. *Genomics*, **86**, 159–171.

9 Jenal, M., Elkon, R., Loayza-Puch, F., van Haaften, G. et al. (2012) The poly(a)-binding protein nuclear 1 suppresses alternative cleavage and polyadenylation sites. *Cell*, **149**, 538–553.

10 Bhakar, A.L., Dolen, G., Bear, M.F. (2012) The pathophysiology of fragile X (and what it teaches us about synapses). *Annu. Rev. Neurosci.*, **35**, 417–443.

11 Chang, J.C., Kan, Y.W. (1979) beta 0 thalassemia, a nonsense mutation in man. *Proc. Natl Acad. Sci. USA*, **76**, 2886–2889.

12 Wang, E.T., Sandberg, R., Luo, S., Khrebtukova, I. et al. (2008) Alternative isoform regulation in human tissue transcriptomes. *Nature*, **456**, 470–476.

13 Lewis, B.P., Green, R.E., Brenner, S.E. (2003) Evidence for the widespread coupling of alternative splicing and nonsense-mediated mRNA decay in humans. *Proc. Natl Acad. Sci. USA*, **100**, 189–192.

14 Sun, X., Li, X., Moriarty, P.M., Henics, T. et al. (2001) Nonsense-mediated decay of mRNA for the selenoprotein phospholipid hydroperoxide glutathione peroxidase is detectable in cultured cells but masked or inhibited in rat tissues. *Mol. Biol. Cell*, **12**, 1009–1017.

15 Medghalchi, S.M., Frischmeyer, P.A., Mendell, J.T., Kelly, A.G. et al. (2001) Rent1, a trans-effector of nonsense-mediated mRNA decay, is essential for mammalian embryonic viability. *Hum. Mol. Genet.*, **10**, 99–105.

16 Maquat, L.E., Tarn, W.Y., Isken, O. (2010) The pioneer round of translation: features and functions. *Cell*, **142**, 368–374.

17 Sato, H., Maquat, L.E. (2009) Remodeling of the pioneer translation initiation complex involves translation and the karyopherin importin beta. *Genes Dev.*, **23**, 2537–2550.

18 Izaurralde, E., Lewis, J., McGuigan, C., Jankowska, M. et al. (1994) A nuclear cap binding protein complex involved in pre-mRNA splicing. *Cell*, **78**, 657–668.

19 Ishigaki, Y., Li, X., Serin, G., Maquat, L.E. (2001) Evidence for a pioneer round of mRNA translation: mRNAs subject to nonsense-mediated decay in mammalian cells are bound by CBP80 and CBP20. *Cell*, **106**, 607–617.

20 Le Hir, H., Izaurralde, E., Maquat, L.E., Moore, M.J. (2000) The spliceosome deposits multiple proteins 20-24 nucleotides upstream of mRNA exon-exon junctions. *EMBO J.*, **19**, 6860–6869.

21 Le Hir, H., Gatfield, D., Izaurralde, E., Moore, M.J. (2001) The exon-exon junction complex provides a binding platform for factors involved in mRNA export and nonsense-mediated mRNA decay. *EMBO J.*, **20**, 4987–4997.

22 Lykke-Andersen, J., Shu, M.D., Steitz, J.A. (2000) Human Upf proteins target an mRNA for nonsense-mediated decay when bound downstream of a termination codon. *Cell*, **103**, 1121–1131.

23 Hogg, J.R., Goff, S.P. (2010) Upf1 senses 3′UTR length to potentiate mRNA decay. *Cell*, **143**, 379–389.

24 Brogna, S., Wen, J. (2009) Nonsense-mediated mRNA decay (NMD) mechanisms. *Nat. Struct. Mol. Biol.*, **16**, 107–113.

25 Amrani, N., Ganesan, R., Kervestin, S., Mangus, D.A. et al. (2004) A faux 3′-UTR promotes aberrant termination and triggers nonsense-mediated mRNA decay. *Nature*, **432**, 112–118.

26 Chakrabarti, S., Jayachandran, U., Bonneau, F., Fiorini, F. et al. (2011) Molecular mechanisms for the RNA-dependent ATPase activity of Upf1 and its regulation by Upf2. *Mol. Cell*, **41**, 693–703.

27 Yamashita, A., Ohnishi, T., Kashima, I., Taya, Y. et al. (2001) Human SMG-1, a novel phosphatidylinositol 3-kinase-related protein kinase, associates with components of the mRNA surveillance complex and is involved in the regulation of nonsense-mediated mRNA decay. *Genes Dev.*, **15**, 2215–2228.

28 Kashima, I., Yamashita, A., Izumi, N., Kataoka, N. et al. (2006) Binding of a novel SMG-1-Upf1-eRF1-eRF3 complex (SURF) to the exon junction complex triggers Upf1 phosphorylation and nonsense-mediated mRNA decay. *Genes Dev.*, **20**, 355–367.

29 Ohnishi, T., Yamashita, A., Kashima, I., Schell, T. et al. (2003) Phosphorylation of hUPF1 induces formation of mRNA surveillance complexes containing hSMG-5 and hSMG-7. *Mol. Cell*, **12**, 1187–1200.

30 Unterholzner, L., Izaurralde, E. (2004) SMG7 acts as a molecular link between mRNA surveillance and mRNA decay. *Mol. Cell*, **16**, 587–596.

31 Nicholson, P., Yepiskoposyan, H., Metze, S., Zamudio Orozco, R. et al. (2010) Nonsense-mediated mRNA decay in human cells: mechanistic insights, functions beyond quality control and the double-life of NMD factors. *Cell. Mol. Life Sci.*, **67**, 677–700.

32 Okada-Katsuhata, Y., Yamashita, A., Kutsuzawa, K., Izumi, N. et al. (2012) N- and C-terminal Upf1 phosphorylations create binding platforms for SMG-6 and SMG-5:SMG-7 during NMD. *Nucleic Acids Res.*, **40**, 1251–1266.

33 Eberle, A.B., Lykke-Andersen, S., Muhlemann, O., Jensen, T.H. (2009) SMG6 promotes endonucleolytic cleavage of nonsense mRNA in human cells. *Nat. Struct. Mol. Biol.*, **16**, 49–55.

34 Chiu, S.Y., Serin, G., Ohara, O., Maquat, L.E. (2003) Characterization of human Smg5/7a: a protein with similarities to *Caenorhabditis elegans* SMG5 and SMG7 that functions in the dephosphorylation of Upf1. *RNA*, **9**, 77–87.

35 Wells, S.E., Hillner, P.E., Vale, R.D., Sachs, A.B. (1998) Circularization of mRNA by eukaryotic translation initiation factors. *Mol. Cell*, **2**, 135–140.

36 Svitkin, Y.V., Pause, A., Haghighat, A., Pyronnet, S. et al. (2001) The requirement for eukaryotic initiation factor 4A (eIF4A) in translation is in direct proportion to the degree of mRNA 5′ secondary structure. *RNA*, **7**, 382–394.

37 Sonenberg, N., Hinnebusch, A.G. (2009) Regulation of translation initiation in eukaryotes: mechanisms and biological targets. *Cell*, **136**, 731–745.

38 Gebauer, F., Hentze, M.W. (2004) Molecular mechanisms of translational control. *Nat. Rev. Mol. Cell Biol.*, **5**, 827–835.

39 Lamphear, B.J., Kirchweger, R., Skern, T., Rhoads, R.E. (1995) Mapping of functional domains in eukaryotic protein synthesis initiation factor 4G (eIF4G) with picornaviral proteases. Implications for cap-dependent and cap-independent translational initiation. *J. Biol. Chem.*, **270**, 21975–21983.

40 Pestova, T.V., Kolupaeva, V.G., Lomakin, I.B., Pilipenko, E.V. et al. (2001) Molecular mechanisms of translation initiation in eukaryotes. *Proc. Natl Acad. Sci. USA*, **98**, 7029–7036.

41 Marcotrigiano, J., Gingras, A.C., Sonenberg, N., Burley, S.K. (1999) Cap-dependent translation initiation in eukaryotes is regulated by a molecular mimic of eIF4G. *Mol. Cell*, **3**, 707–716.

42 Gingras, A.C., Gygi, S.P., Raught, B., Polakiewicz, R.D. et al. (1999) Regulation of 4E-BP1 phosphorylation: a novel two-step mechanism. *Genes Dev.*, **13**, 1422–1437.

43 Mamane, Y., Petroulakis, E., LeBacquer, O., Sonenberg, N. (2006) mTOR, translation initiation and cancer. *Oncogene*, **25**, 6416–6422.

44 Lin, T.A., Kong, X., Haystead, T.A., Pause, A. et al. (1994) PHAS-I as a link between mitogen-activated protein kinase and translation initiation. *Science*, **266**, 653–656.

45 Pause, A., Belsham, G.J., Gingras, A.C., Donze, O. et al. (1994) Insulin-dependent stimulation of protein synthesis by phosphorylation of a regulator of 5′-cap function. *Nature*, **371**, 762–767.

46 Mamane, Y., Petroulakis, E., Rong, L., Yoshida, K. et al. (2004) eIF4E – from translation to transformation. *Oncogene*, **23**, 3172–3179.

47 Lazaris-Karatzas, A., Montine, K.S., Sonenberg, N. (1990) Malignant transformation by a eukaryotic initiation factor subunit that binds to mRNA 5′ cap. *Nature*, **345**, 544–547.

48 Smith, M.R., Jaramillo, M., Tuazon, P.T., Traugh, J.A. et al. (1991) Modulation of the mitogenic activity of eukaryotic translation initiation factor-4E by protein kinase C. *New Biol.*, **3**, 601–607.

49 Kerekatte, V., Smiley, K., Hu, B., Smith, A. et al. (1995) The proto-oncogene/

translation factor eIF4E: a survey of its expression in breast carcinomas. *Int. J. Cancer*, **64**, 27–31.

50 Berkel, H.J., Turbat-Herrera, E.A., Shi, R., de Benedetti, A. (2001) Expression of the translation initiation factor eIF4E in the polyp-cancer sequence in the colon. *Cancer Epidemiol. Biomarkers Prev.*, **10**, 663–666.

51 Crew, J.P., Fuggle, S., Bicknell, R., Cranston, D.W. et al. (2000) Eukaryotic initiation factor-4E in superficial and muscle invasive bladder cancer and its correlation with vascular endothelial growth factor expression and tumour progression. *Br. J. Cancer*, **82**, 161–166.

52 Culjkovic, B., Borden, K.L. (2009) Understanding and targeting the eukaryotic translation initiation factor eIF4E in head and neck cancer. *J. Oncol.*, **2009**, 981679.

53 Rousseau, D., Gingras, A.C., Pause, A., Sonenberg, N. (1996) The eIF4E-binding proteins 1 and 2 are negative regulators of cell growth. *Oncogene*, **13**, 2415–2420.

54 Polunovsky, V.A., Gingras, A.C., Sonenberg, N., Peterson, M. et al. (2000) Translational control of the antiapoptotic function of Ras. *J. Biol. Chem.*, **275**, 24776–24780.

55 Li, S., Sonenberg, N., Gingras, A.C., Peterson, M. et al. (2002) Translational control of cell fate: availability of phosphorylation sites on translational repressor 4E-BP1 governs its proapoptotic potency. *Mol. Cell. Biol.*, **22**, 2853–2861.

56 Kim, Y.Y., Von Weymarn, L., Larsson, O., Fan, D. et al. (2009) Eukaryotic initiation factor 4E binding protein family of proteins: sentinels at a translational control checkpoint in lung tumor defense. *Cancer Res.*, **69**, 8455–8462.

57 Moerke, N.J., Aktas, H., Chen, H., Cantel, S. et al. (2007) Small-molecule inhibition of the interaction between the translation initiation factors eIF4E and eIF4G. *Cell*, **128**, 257–267.

58 Baumann, P., Mandl-Weber, S., Oduncu, F., Schmidmaier, R. (2009) The novel orally bioavailable inhibitor of phosphoinositol-3-kinase and mammalian target of rapamycin, NVP-BEZ235, inhibits growth and proliferation in multiple myeloma. *Exp. Cell Res.*, **315**, 485–497.

59 Kowalska, J., Lewdorowicz, M., Zuberek, J., Grudzien-Nogalska, E. et al. (2008) Synthesis and characterization of mRNA cap analogs containing phosphorothioate substitutions that bind tightly to eIF4E and are resistant to the decapping pyrophosphatase DcpS. *RNA*, **14**, 1119–1131.

60 Malina, A., Cencic, R., Pelletier, J. (2011) Targeting translation dependence in cancer. *Oncotarget*, **2**, 76–88.

61 Wilusz, C.J., Wormington, M., Peltz, S.W. (2001) The cap-to-tail guide to mRNA turnover. *Nat. Rev. Mol. Cell Biol.*, **2**, 237–246.

62 Chen, C.Y., Gherzi, R., Andersen, J.S., Gaietta, G. et al. (2000) Nucleolin and YB-1 are required for JNK-mediated interleukin-2 mRNA stabilization during T-cell activation. *Genes Dev.*, **14**, 1236–1248.

63 Anderson, P. (2008) Post-transcriptional control of cytokine production. *Nat. Immunol.*, **9**, 353–359.

64 Raghavan, A., Dhalla, M., Bakheet, T., Ogilvie, R.L. et al. (2004) Patterns of coordinate down-regulation of ARE-containing transcripts following immune cell activation. *Genomics*, **84**, 1002–1013.

65 Zhang, J., Cado, D., Chen, A., Kabra, N.H. et al. (1998) Fas-mediated apoptosis and activation-induced T-cell proliferation are defective in mice lacking FADD/Mort1. *Nature*, **392**, 296–300.

66 Cheadle, C., Fan, J., Cho-Chung, Y.S., Werner, T. et al. (2005) Control of gene expression during T cell activation: alternate regulation of mRNA transcription and mRNA stability. *BMC Genomics*, **6**, 75.

67 Raghavan, A., Ogilvie, R.L., Reilly, C., Abelson, M.L. et al. (2002) Genome-wide analysis of mRNA decay in resting and activated primary human T lymphocytes. *Nucleic Acids Res.*, **30**, 5529–5538.

68 Shyu, A.B., Wilkinson, M.F., van Hoof, A. (2008) Messenger RNA regulation: to translate or to degrade. *EMBO J.*, **27**, 471–481.

69 Caput, D., Beutler, B., Hartog, K., Thayer, R. et al. (1986) Identification of a common nucleotide sequence in the 3′-untranslated region of mRNA molecules specifying inflammatory mediators. *Proc. Natl Acad. Sci. USA*, **83**, 1670–1674.

70 Shaw, G., Kamen, R. (1986) A conserved AU sequence from the 3′ untranslated region of GM-CSF mRNA mediates selective mRNA degradation. *Cell*, **46**, 659–667.

71 Gillis, P., Malter, J.S. (1991) The adenosine-uridine binding factor recognizes the AU-rich elements of cytokine, lymphokine, and oncogene mRNAs. *J. Biol. Chem.*, **266**, 3172–3177.

72 Chen, C.Y., Shyu, A.B. (1995) AU-rich elements: characterization and importance in mRNA degradation. *Trends Biochem. Sci.*, **20**, 465–470.

73 Baou, M., Norton, J.D., Murphy, J.J. (2011) AU-rich RNA binding proteins in hematopoiesis and leukemogenesis. *Blood*, **118**, 5732–5740.

74 Deschenes-Furry, J., Perrone-Bizzozero, N., Jasmin, B.J. (2006) The RNA-binding protein HuD: a regulator of neuronal differentiation, maintenance and plasticity. *BioEssays*, **28**, 822–833.

75 Brennan, C.M., Steitz, J.A. (2001) HuR and mRNA stability. *Cell. Mol. Life Sci.*, **58**, 266–277.

76 Soller, M., Li, M., Haussmann, I.U. (2010) Determinants of ELAV gene-specific regulation. *Biochem. Soc. Trans.*, **38**, 1122–1124.

77 Raghavan, A., Robison, R.L., McNabb, J., Miller, C.R. et al. (2001) HuA and tristetraprolin are induced following T cell activation and display distinct but overlapping RNA binding specificities. *J. Biol. Chem.*, **276**, 47958–47965.

78 Schoenberg, D.R., Maquat, L.E. (2012) Regulation of cytoplasmic mRNA decay. *Nat. Rev. Genet.*, **13**, 246–259.

79 Lai, W.S., Carballo, E., Strum, J.R., Kennington, E.A. et al. (1999) Evidence that tristetraprolin binds to AU-rich elements and promotes the deadenylation and destabilization of tumor necrosis factor alpha mRNA. *Mol. Cell. Biol.*, **19**, 4311–4323.

80 Chen, C.Y., Gherzi, R., Ong, S.E., Chan, E.L. et al. (2001) AU binding proteins recruit the exosome to degrade ARE-containing mRNAs. *Cell*, **107**, 451–464.

81 Carballo, E., Lai, W.S., Blackshear, P.J. (1998) Feedback inhibition of macrophage tumor necrosis factor-alpha production by tristetraprolin. *Science*, **281**, 1001–1005.

82 Edmonds, M. (2002) A history of poly A sequences: from formation to factors to function. *Prog. Nucleic Acid Res. Mol. Biol.*, **71**, 285–389.

83 Shyu, A.B., Belasco, J.G., Greenberg, M.E. (1991) Two distinct destabilizing elements in the c-fos message trigger deadenylation as a first step in rapid mRNA decay. *Genes Dev.*, **5**, 221–231.

84 Garneau, N.L., Wilusz, J., Wilusz, C.J. (2007) The highways and byways of mRNA decay. *Nat. Rev. Mol. Cell Biol.*, **8**, 113–126.

85 Lai, W.S., Kennington, E.A., Blackshear, P.J. (2003) Tristetraprolin and its family members can promote the cell-free deadenylation of AU-rich element-containing mRNAs by poly(A) ribonuclease. *Mol. Cell. Biol.*, **23**, 3798–3812.

86 Sandler, H., Kreth, J., Timmers, H.T., Stoecklin, G. (2011) Not1 mediates recruitment of the deadenylase Caf1 to mRNAs targeted for degradation by tristetraprolin. *Nucleic Acids Res.*, **39**, 4373–4386.

87 Lykke-Andersen, S., Tomecki, R., Jensen, T.H., Dziembowski, A. (2011) The eukaryotic RNA exosome: same scaffold but variable catalytic subunits. *RNA Biol.*, **8**, 61–66.

88 Ingelfinger, D., Arndt-Jovin, D.J., Luhrmann, R., Achsel, T. (2002) The human LSm1-7 proteins colocalize with the mRNA-degrading enzymes Dcp1/2 and Xrn1 in distinct cytoplasmic foci. *RNA*, **8**, 1489–1501.

89 Franks, T.M., Lykke-Andersen, J. (2008) The control of mRNA decapping and P-body formation. *Mol. Cell*, **32**, 605–615.

90 Stoecklin, G., Mayo, T., Anderson, P. (2006) ARE-mRNA degradation requires the 5′-3′ decay pathway. *EMBO Rep.*, **7**, 72–77.

91 Hau, H.H., Walsh, R.J., Ogilvie, R.L., Williams, D.A. et al. (2007) Tristetraprolin recruits functional mRNA decay complexes to ARE sequences. *J. Cell. Biochem.*, **100**, 1477–1492.

92 Mazan-Mamczarz, K., Galban, S., Lopez de Silanes, I., Martindale, J.L. et al. (2003) RNA-binding protein HuR enhances p53 translation in response to ultraviolet light irradiation. *Proc. Natl Acad. Sci. USA*, **100**, 8354–8359.

93 Kawai, T., Lal, A., Yang, X., Galban, S. et al. (2006) Translational control of cytochrome c by RNA-binding proteins TIA-1 and HuR. *Mol. Cell. Biol.*, **26**, 3295–3307.

94 Durie, D., Lewis, S.M., Liwak, U., Kisilewicz, M. et al. (2011) RNA-binding protein HuR mediates cytoprotection through stimulation of XIAP translation. *Oncogene*, **30**, 1460–1469.

95 Ishimaru, D., Ramalingam, S., Sengupta, T.K., Bandyopadhyay, S. et al. (2009) Regulation of Bcl-2 expression by HuR in HL60 leukemia cells and A431 carcinoma cells. *Mol. Cancer Res.*, **7**, 1354–1366.

96 Kullmann, M., Gopfert, U., Siewe, B., Hengst, L. (2002) ELAV/Hu proteins inhibit p27 translation via an IRES element in the p27 5′UTR. *Genes Dev.*, **16**, 3087–3099.

97 Kim, H.H., Kuwano, Y., Srikantan, S., Lee, E.K. et al. (2009) HuR recruits let-7/RISC to repress c-Myc expression. *Genes Dev.*, **23**, 1743–1748.

98 Leandersson, K., Riesbeck, K., Andersson, T. (2006) Wnt-5a mRNA translation is suppressed by the Elav-like protein HuR in human breast epithelial cells. *Nucleic Acids Res.*, **34**, 3988–3999.

99 Kiriakidou, M., Tan, G.S., Lamprinaki, S., De Planell-Saguer, M. et al. (2007) An mRNA m7G cap binding-like motif within human Ago2 represses translation. *Cell*, **129**, 1141–1151.

100 Vlasova, I.A., Tahoe, N.M., Fan, D., Larsson, O. et al. (2008) Conserved GU-rich elements mediate mRNA decay by binding to CUG-binding protein 1. *Mol. Cell*, **29**, 263–270.

101 Rattenbacher, B., Beisang, D., Wiesner, D.L., Jeschke, J.C. et al. (2010) Analysis of CUGBP1 targets identifies GU-repeat sequences that mediate rapid mRNA decay. *Mol. Cell. Biol.*, **30**, 3970–3980.

102 Lee, J.E., Lee, J.Y., Wilusz, J., Tian, B. et al. (2010) Systematic analysis of cis-elements in unstable mRNAs demonstrates that CUGBP1 is a key regulator of mRNA decay in muscle cells. *PLoS ONE*, **5**, e11201.

103 Beisang, D., Rattenbacher, B., Vlasova-St Louis, I.A., Bohjanen, P.R. (2012) Regulation of CUG-binding protein 1 (CUGBP1) binding to target transcripts upon T cell activation. *J. Biol. Chem.*, **287**, 950–960.

104 Kuyumcu-Martinez, N.M., Wang, G.S., Cooper, T.A. (2007) Increased steady-state levels of CUGBP1 in myotonic dystrophy 1 are due to PKC-mediated hyperphosphorylation. *Mol. Cell*, **28**, 68–78.

105 Mankodi, A., Takahashi, M.P., Jiang, H., Beck, C.L. et al. (2002) Expanded CUG repeats trigger aberrant splicing of ClC-1 chloride channel pre-mRNA and hyperexcitability of skeletal muscle in myotonic dystrophy. *Mol. Cell*, **10**, 35–44.

106 Savkur, R.S., Philips, A.V., Cooper, T.A. (2001) Aberrant regulation of insulin receptor alternative splicing is associated with insulin resistance in myotonic dystrophy. *Nat. Genet.*, **29**, 40–47.

107 Philips, A.V., Timchenko, L.T., Cooper, T.A. (1998) Disruption of splicing regulated by a CUG-binding protein in myotonic dystrophy. *Science*, **280**, 737–741.

108 Klein, A.F., Gasnier, E., Furling, D. (2011) Gain of RNA function in pathological cases: Focus on myotonic dystrophy. *Biochimie*, **93**, 2006–2012.

109 Kalsotra, A., Xiao, X., Ward, A.J., Castle, J.C. et al. (2008) A postnatal switch of CELF and MBNL proteins reprograms alternative splicing in the developing heart. *Proc. Natl Acad. Sci. USA*, **105**, 20333–20338.

110 Timchenko, N.A., Welm, A.L., Lu, X., Timchenko, L.T. (1999) CUG repeat binding protein (CUGBP1) interacts with the 5′ region of C/EBPbeta mRNA and regulates translation of C/EBPbeta isoforms. *Nucleic Acids Res.*, **27**, 4517–4525.

111 Bae, E.J., Kim, S.G. (2005) Enhanced CCAAT/enhancer-binding protein beta-liver-enriched inhibitory protein production by Oltipraz, which accompanies CUG repeat-binding protein-1 (CUGBP1) RNA-binding protein activation, leads to inhibition of preadipocyte differentiation. *Mol. Pharmacol.*, **68**, 660–669.

112 Karagiannides, I., Thomou, T., Tchkonia, T., Pirtskhalava, T. et al. (2006) Increased CUG triplet repeat-binding protein-1 predisposes to impaired adipogenesis with aging. *J. Biol. Chem.*, **281**, 23025–23033.

113 Timchenko, N.A., Wang, G.L., Timchenko, L.T. (2005) RNA CUG-binding protein 1 increases translation of 20-kDa isoform of CCAAT/enhancer-binding protein beta by interacting with the alpha and beta subunits of eukaryotic initiation translation factor 2. *J. Biol. Chem.*, **280**, 20549–20557.

114 Wang, G.L., Salisbury, E., Shi, X., Timchenko, L. et al. (2008) HDAC1 cooperates with C/EBPalpha in the

115 Timchenko, N.A., Iakova, P., Cai, Z.J., Smith, J.R. et al. (2001) Molecular basis for impaired muscle differentiation in myotonic dystrophy. *Mol. Cell. Biol.*, **21**, 6927–6938.

116 Fox, J.T., Stover, P.J. (2009) Mechanism of the internal ribosome entry site-mediated translation of serine hydroxymethyltransferase 1. *J. Biol. Chem.*, **284**, 31085–31096.

117 Zheng, Y., Miskimins, W.K. (2011) CUG-binding protein represses translation of p27Kip1 mRNA through its internal ribosomal entry site. *RNA Biol.*, **8**, 365–371.

118 Jacobson, A., Peltz, S.W. (1996) Interrelationships of the pathways of mRNA decay and translation in eukaryotic cells. *Annu. Rev. Biochem.*, **65**, 693–739.

119 Srikantan, S., Gorospe, M. (2012) HuR function in disease. *Front. Biosci.*, **17**, 189–205.

120 Fabian, M.R., Sonenberg, N. (2012) The mechanics of miRNA-mediated gene silencing: a look under the hood of miRISC. *Nat. Struct. Mol. Biol.*, **19**, 586–593.

121 Marnef, A., Standart, N. (2010) Pat1 proteins: a life in translation, translation repression and mRNA decay. *Biochem. Soc. Trans.*, **38**, 1602–1607.

122 Pilkington, G.R., Parker, R. (2008) Pat1 contains distinct functional domains that promote P-body assembly and activation of decapping. *Mol. Cell. Biol.*, **28**, 1298–1312.

123 Balagopal, V., Parker, R. (2009) Polysomes, P bodies and stress granules: states and fates of eukaryotic mRNAs. *Curr. Opin. Cell Biol.*, **21**, 403–408.

124 Moeller, B.J., Cao, Y., Li, C.Y., Dewhirst, M.W. (2004) Radiation activates HIF-1 to regulate vascular radiosensitivity in tumors: role of reoxygenation, free radicals, and stress granules. *Cancer Cell*, **5**, 429–441.

125 Ivanov, P.A., Nadezhdina, E.S. (2006) Stress granules: RNP-containing cytoplasmic bodies springing up under stress. The structure and mechanism of organization. *Mol. Biol.*, **40**, 937–944.

126 Emara, M.M., Brinton, M.A. (2007) Interaction of TIA-1/TIAR with West Nile and dengue virus products in infected cells interferes with stress granule formation and processing body assembly. *Proc. Natl Acad. Sci. USA*, **104**, 9041–9046.

127 Buchan, J.R., Parker, R. (2009) Eukaryotic stress granules: the ins and outs of translation. *Mol. Cell*, **36**, 932–941.

128 Nover, L., Scharf, K.D., Neumann, D. (1989) Cytoplasmic heat shock granules are formed from precursor particles and are associated with a specific set of mRNAs. *Mol. Cell. Biol.*, **9**, 1298–1308.

129 Gardner, L.B., Corn, P.G. (2008) Hypoxic regulation of mRNA expression. *Cell Cycle*, **7**, 1916–1924.

130 Fujimura, K., Kano, F., Murata, M. (2008) Identification of PCBP2, a facilitator of IRES-mediated translation, as a novel constituent of stress granules and processing bodies. *RNA*, **14**, 425–431.

131 Mokas, S., Mills, J.R., Garreau, C., Fournier, M.J. et al. (2009) Uncoupling stress granule assembly and translation initiation inhibition. *Mol. Biol. Cell*, **20**, 2673–2683.

132 Kedersha, N., Cho, M.R., Li, W., Yacono, P.W. et al. (2000) Dynamic shuttling of TIA-1 accompanies the recruitment of mRNA to mammalian stress granules. *J. Cell Biol.*, **151**, 1257–1268.

133 Kedersha, N., Chen, S., Gilks, N., Li, W. et al. (2002) Evidence that ternary complex (eIF2-GTP-tRNA(i)(Met))-deficient preinitiation complexes are core constituents of mammalian stress granules. *Mol. Biol. Cell*, **13**, 195–210.

134 Kedersha, N.L., Gupta, M., Li, W., Miller, I. et al. (1999) RNA-binding proteins TIA-1 and TIAR link the phosphorylation of eIF-2 alpha to the assembly of mammalian stress granules. *J. Cell Biol.*, **147**, 1431–1442.

135 Kwon, S., Zhang, Y., Matthias, P. (2007) The deacetylase HDAC6 is a novel critical component of stress granules involved in the stress response. *Genes Dev.*, **21**, 3381–3394.

136 Buchan, J.R., Muhlrad, D., Parker, R. (2008) P bodies promote stress granule assembly in *Saccharomyces cerevisiae*. *J. Cell Biol.*, **183**, 441–455.

137 Aizer, A., Brody, Y., Ler, L.W., Sonenberg, N. et al. (2008) The dynamics of mammalian P body transport, assembly, and disassembly in vivo. *Mol. Biol. Cell*, **19**, 4154–4166.

138 Aizer, A., Shav-Tal, Y. (2008) Intracellular trafficking and dynamics of P bodies. *Prion*, **2**, 131–134.
139 Parker, R., Sheth, U. (2007) P bodies and the control of mRNA translation and degradation. *Mol. Cell*, **25**, 635–646.
140 Sheth, U., Parker, R. (2003) Decapping and decay of messenger RNA occur in cytoplasmic processing bodies. *Science*, **300**, 805–808.
141 Coller, J., Parker, R. (2005) General translational repression by activators of mRNA decapping. *Cell*, **122**, 875–886.
142 Stoecklin, G., Anderson, P. (2007) In a tight spot: ARE-mRNAs at processing bodies. *Genes Dev.*, **21**, 627–631.
143 Lykke-Andersen, J. (2002) Identification of a human decapping complex associated with hUpf proteins in nonsense-mediated decay. *Mol. Cell. Biol.*, **22**, 8114–8121.
144 Sheth, U., Parker, R. (2006) Targeting of aberrant mRNAs to cytoplasmic processing bodies. *Cell*, **125**, 1095–1109.
145 Liu, J., Valencia-Sanchez, M.A., Hannon, G.J., Parker, R. (2005) MicroRNA-dependent localization of targeted mRNAs to mammalian P-bodies. *Nat. Cell Biol.*, **7**, 719–723.
146 Sen, G.L., Blau, H.M. (2005) Argonaute 2/RISC resides in sites of mammalian mRNA decay known as cytoplasmic bodies. *Nat. Cell Biol.*, **7**, 633–636.
147 Eystathioy, T., Jakymiw, A., Chan, E.K., Seraphin, B. *et al.* (2003) The GW182 protein colocalizes with mRNA degradation associated proteins hDcp1 and hLSm4 in cytoplasmic GW bodies. *RNA*, **9**, 1171–1173.
148 Bhattacharyya, S.N., Habermacher, R., Martine, U., Closs, E.I. *et al.* (2006) Relief of microRNA-mediated translational repression in human cells subjected to stress. *Cell*, **125**, 1111–1124.
149 Kedersha, N., Stoecklin, G., Ayodele, M., Yacono, P. *et al.* (2005) Stress granules and processing bodies are dynamically linked sites of mRNP remodeling. *J. Cell Biol.*, **169**, 871–884.
150 Carthew, R.W., Sontheimer, E.J. (2009) Origins and mechanisms of miRNAs and siRNAs. *Cell*, **136**, 642–655.
151 Lee, R.C., Feinbaum, R.L., Ambros, V. (1993) The C. elegans heterochronic gene lin-4 encodes small RNAs with antisense complementarity to lin-14. *Cell*, **75**, 843–854.
152 Reinhart, B.J., Slack, F.J., Basson, M., Pasquinelli, A.E. *et al.* (2000) The 21-nucleotide let-7 RNA regulates developmental timing in *Caenorhabditis elegans*. *Nature*, **403**, 901–906.
153 Pasquinelli, A.E., Reinhart, B.J., Slack, F., Martindale, M.Q. *et al.* (2000) Conservation of the sequence and temporal expression of let-7 heterochronic regulatory RNA. *Nature*, **408**, 86–89.
154 Griffiths-Jones, S., Saini, H.K., van Dongen, S., Enright, A.J. (2008) miRBase: tools for microRNA genomics. *Nucleic Acids Res.*, **36**, D154–D158.
155 Griffiths-Jones, S., Grocock, R.J., van Dongen, S., Bateman, A. *et al.* (2006) miRBase: microRNA sequences, targets and gene nomenclature. *Nucleic Acids Res.*, **34**, D140–D144.
156 Lee, Y., Kim, M., Han, J., Yeom, K.H. *et al.* (2004) MicroRNA genes are transcribed by RNA polymerase II. *EMBO J.*, **23**, 4051–4060.
157 Lee, Y., Ahn, C., Han, J., Choi, H. *et al.* (2003) The nuclear RNase III Drosha initiates microRNA processing. *Nature*, **425**, 415–419.
158 Hutvagner, G., McLachlan, J., Pasquinelli, A.E., Balint, E. *et al.* (2001) A cellular function for the RNA-interference enzyme Dicer in the maturation of the let-7 small temporal RNA. *Science*, **293**, 834–838.
159 Bernstein, E., Caudy, A.A., Hammond, S.M., Hannon, G.J. (2001) Role for a bidentate ribonuclease in the initiation step of RNA interference. *Nature*, **409**, 363–366.
160 Schwarz, D.S., Hutvagner, G., Du, T., Xu, Z. *et al.* (2003) Asymmetry in the assembly of the RNAi enzyme complex. *Cell*, **115**, 199–208.
161 Okamura, K., Ishizuka, A., Siomi, H., Siomi, M.C. (2004) Distinct roles for Argonaute proteins in small RNA-directed RNA cleavage pathways. *Genes Dev.*, **18**, 1655–1666.
162 Eulalio, A., Huntzinger, E., Izaurralde, E. (2008) GW182 interaction with Argonaute is essential for miRNA-mediated translational repression and mRNA decay. *Nat. Struct. Mol. Biol.*, **15**, 346–353.

163 Lewis, B.P., Burge, C.B., Bartel, D.P. (2005) Conserved seed pairing, often flanked by adenosines, indicates that thousands of human genes are microRNA targets. *Cell*, **120**, 15–20.

164 Djuranovic, S., Nahvi, A., Green, R. (2012) miRNA-mediated gene silencing by translational repression followed by mRNA deadenylation and decay. *Science*, **336**, 237–240.

165 Behm-Ansmant, I., Rehwinkel, J., Doerks, T., Stark, A. *et al.* (2006) mRNA degradation by miRNAs and GW182 requires both CCR4:NOT deadenylase and DCP1:DCP2 decapping complexes. *Genes Dev.*, **20**, 1885–1898.

166 Fabian, M.R., Mathonnet, G., Sundermeier, T., Mathys, H. *et al.* (2009) Mammalian miRNA RISC recruits CAF1 and PABP to affect PABP-dependent deadenylation. *Mol. Cell*, **35**, 868–880.

167 Karginov, F.V., Cheloufi, S., Chong, M.M., Stark, A. *et al.* (2010) Diverse endonucleolytic cleavage sites in the mammalian transcriptome depend upon microRNAs, Drosha, and additional nucleases. *Mol. Cell*, **38**, 781–788.

168 Jing, Q., Huang, S., Guth, S., Zarubin, T. *et al.* (2005) Involvement of microRNA in AU-rich element-mediated mRNA instability. *Cell*, **120**, 623–634.

169 Trabucchi, M., Briata, P., Garcia-Mayoral, M., Haase, A.D. *et al.* (2009) The RNA-binding protein KSRP promotes the biogenesis of a subset of microRNAs. *Nature*, **459**, 1010–1014.

170 Asirvatham, A.J., Gregorie, C.J., Hu, Z., Magner, W.J. *et al.* (2008) MicroRNA targets in immune genes and the Dicer/Argonaute and ARE machinery components. *Mol. Immunol.*, **45**, 1995–2006.

171 Ma, F., Liu, X., Li, D., Wang, P. *et al.* (2010) MicroRNA-466l upregulates IL-10 expression in TLR-triggered macrophages by antagonizing RNA-binding protein tristetraprolin-mediated IL-10 mRNA degradation. *J. Immunol.*, **184**, 6053–6059.

9
RNA Modification

Yuri Motorin and Bruno Charpentier
Unité Mixte de Recherche 7365 Université de Lorraine-Centre National de la Recherche Scientifique, Laboratoire IMoPA, Ingénierie Moléculaire et Physiopathologie Articulaire, 9, avenue de la forêt de Haye, BP 184, 54505 Vandoeuvre-les-Nancy, France

1	**Origin and Occurrence of Modified Nucleotides in RNA**	**239**
1.1	The Discovery of Modified Nucleotides in RNA: A Brief Overview	242
1.2	The Post-Transcriptional Nature of RNA Modifications	243
1.3	Enzymes and Cofactors Implicated in RNA Modification: A Brief Overview	245
2	**Chemical Nature and Properties of Modified Nucleotides in RNA**	**246**
2.1	Chemical Reactivity	246
2.2	Base-Pairing Properties and Specific Conformation and RNA Folding	246
3	**Enzymes and Enzymatic Mechanisms of RNA Modification**	**249**
3.1	Two Concepts for Target Nucleotide Definition	249
3.2	Protein-Based RNA-Modifying Enzymes	250
3.2.1	RNA-Methyltransferases	250
3.2.2	RNA-Pseudouridine (ψ)-Synthases	250
3.2.3	Other Common Groups	252
3.3	RNA-Guided RNA-Modifying Complexes	253
3.3.1	C/D snoRNA-Guided RNA Methylation	253
3.3.2	H/ACA snoRNA-Guided Pseudouridylation	255
3.4	Catalytic Mechanisms of RNA Modification	256
3.4.1	Methylation	256
3.4.2	U $\rightarrow$ ψ Conversion: Pseudouridylation	258
4	**Functions of RNA Modifications**	**260**
4.1	Overall Stabilization of RNA Folding	260

RNA Regulation: Advances in Molecular Biology and Medicine, First Edition. Edited by Robert A. Meyers.
© 2014 Wiley-VCH Verlag GmbH & Co. KGaA. Published 2014 by Wiley-VCH Verlag GmbH & Co. KGaA.

4.2	Specific Effects of Chemical Reactivity or Conformation 261
4.3	RNA Thermostability 262
4.4	Functions of Modified Nucleotides in mRNA Decoding 263
4.5	rRNA Modifications and Antibiotic Resistance 266
4.6	Positive and Negative Determinants for RNA–Protein Recognition 266
4.7	Other Effects Outside of Translation 268
4.8	Modulation of Immune Response Using RNA-Modified Nucleotides 268
5	**Static and Regulated Character of RNA Modification** 269
5.1	Known Examples of Regulated RNA Modifications 269
5.2	Regulation of mRNA Translation by Modified Residues 269
6	**Molecular Pathologies Linked to RNA-Modification Defects** 271
6.1	Mitochondrial Diseases 272
6.1.1	Taurine Modification in Mitochondrial tRNA 272
6.1.2	Pseudouridine Modification of tRNAs by hPUS1 273
6.1.3	Mitochondrial 12S rRNA Mutation in Inherited Deafness 274
6.2	Link between Type II Diabetes and tRNA Modification 274
6.3	Dyskeratosis Congenita 274

References 275

Keywords

RNA modification
A post-transcriptional enzymatic event creating chemically different nucleotides in cellular RNAs.

RNA methylation
The enzymatic incorporation of CH_3- (Me-) groups at different positions of the base or 2′-OH of ribose in RNA.

SAM (AdoMet)
S-adenosyl-L-methionine, the almost universal Me-donor in biological methylation reactions.

RNA:MTase
RNA:methyltransferase, the enzyme that catalyzes the incorporation of Me-groups into the RNA chain.

RNA:ψ-synthase
RNA:pseudouridine synthase, the enzyme that catalyzes the conversion of uridine to pseudouridine in RNA.

snRNA
Small nuclear RNA, a component of the spliceosome in the eukaryotic cell.

snoRNA
Small nucleolar RNA; RNA of small size, of which two classes exist, snoRNA (C/D box and H/ACA box).

UsnRNA
U-rich small nuclear RNA.

Post-transcriptional RNA modification is a universally conserved and highly complex metabolic process in the living cell. To date, over 110 chemically different modifications have been identified and characterized in cellular RNA species, and all of these are formed by a highly specific enzymatic machinery which recognizes and modifies RNA targets. Cellular RNAs are modified by pure protein stand-alone enzymes, as well as by small nucleolar ribonucleoprotein (snoRNP) complexes containing small nucleolar RNAs (snoRNAs) guides. Modified nucleotides present in RNA play an important role in stabilizing the two- and three-dimensional structures of these molecules, as well as in the fine-tuning of numerous interactions between RNA itself and RNA-binding protein partners. Modifications present in the tRNA anticodon loop are crucially important for correct mRNA decoding during protein synthesis on the ribosome. Recent progress in the field has highlighted the regulatory character of certain modifications in RNA. This emerging concept of "RNA-epigenetics" can supply an additional level to the regulation of gene expression; indeed, the regulation (and deregulation) of RNA modification machineries may form the basis of some important human pathologies.

1 Origin and Occurrence of Modified Nucleotides in RNA

In addition to the canonical A, C, G, and U residues that are incorporated during RNA synthesis by RNA polymerases, the great majority of known RNA species contains an important number of so-called "modified nucleotides." The latter generally bear additional chemical groups at different base positions, and also at the 2′-OH moiety of ribose. All modified nucleotides in RNA appear as a result of a post-transcriptional modification process that forms an integral part of the complex RNA maturation pathways (see Fig. 1). With rare exceptions, almost any RNA species – including noncoding and coding RNAs – from all three kingdoms of life contain modified RNA nucleotides

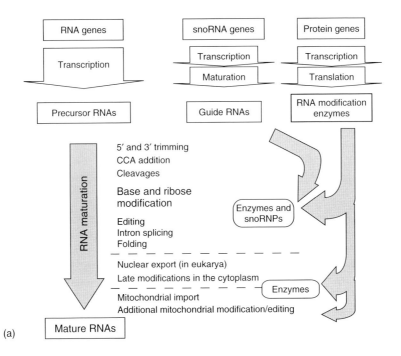

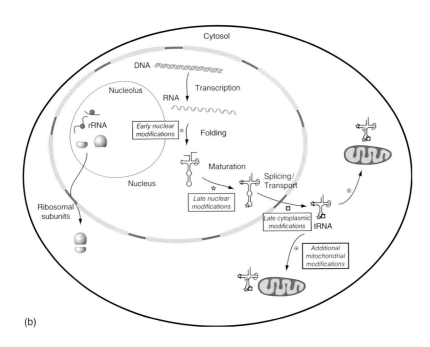

(a)

(b)

that are important for their structure and function. To date, more than 110 such modified residues have been discovered [1–5] (Fig. 2).

Amongst the various RNA species, eukaryotic tRNAs are most heavily modified, especially at the anticodon region, where a great diversity of modifications that are important for mRNA decoding during translation can be identified. Overall, in certain eukaryotic tRNA species, up to 25% of the nucleotides are modified.

Ribosomal RNAs from all organisms contain numerous modified nucleotides, which are mostly pseudouridines and 2′-O-methylations in Eukarya and Archaea (up to 90% of all modified residues), though base methylations have also been reported. As with tRNAs, the number of modified nucleotides in rRNAs increases from about 35 in a bacterium such as *Escherichia coli* to about 200 in humans and other higher eukaryotes. These differences are in fact related to the enzymatic mechanism of rRNA modification, which employs pure protein enzymes in bacteria, whereas eukaryotic rRNAs are mostly modified by small nucleolar ribonucleoprotein (snoRNP) complexes (see below).

Although small nuclear RNAs (UsnRNAs) and small nucleolar RNAs (snoRNAs) also contain a number of reported modifications, their diversity is significantly less [6], with only 11 different modified nucleotides having been identified in these species (for details, see: The RNA Modification Database; http://mods.rna.albany.edu/Introduction/Phylogenetic-Distribution).

In contrast, mRNAs and small regulatory RNAs have been found to bear modified nucleotides, though the proportion of modified residues present is generally quite low (<1%) [7, 8].

A significant proportion of the genes involved in RNA metabolism – notably those encoding hundreds of different enzymes catalyzing the modification of nucleotides of RNA – are present in the genomes of organisms in the three domains of life [9, 10]. Hence, many RNA-modifying enzymes belong to the minimal set of biomolecules necessary for cell life [11]. The conservation of these enzymes during the evolution of some modifications – in terms of both their identity and position in the RNAs – reinforces the principle of an essential role for these changes in the biology of the cell [12].

Fig. 1 (a) General organization of RNA maturation process in the cell. RNA genes are transcribed by appropriate RNA polymerases to give the precursor RNA molecules. In parallel, the sequences of snoRNA guides are transcribed as independent genes or as a part of the introns, and the corresponding transcripts undergo maturation. The RNA-modifying enzyme-encoding genes are transcribed and the corresponding mRNAs are translated to proteins. The resulting protein enzymes are either used as stand-alone activities, or incorporated into snoRNPs together with the snoRNA guides; (b) Cellular organization of tRNA and rRNA maturation and modification in the eukaryotic cell. Some events of maturation and modification may be common to other RNAs, such as snRNAs, snoRNAs and mRNAs. Some tRNA modifications appear rather early during pre-tRNA maturation, but the precursor is progressively modified in the nucleus. Some late modifications appear only in the cytoplasm, and are frequently related to intron removal from the pre-tRNA. A small fraction of the mitochondrial tRNA species is imported from the cytoplasm and may be further modified there. The intrinsic mitochondrial tRNAs are not shown.

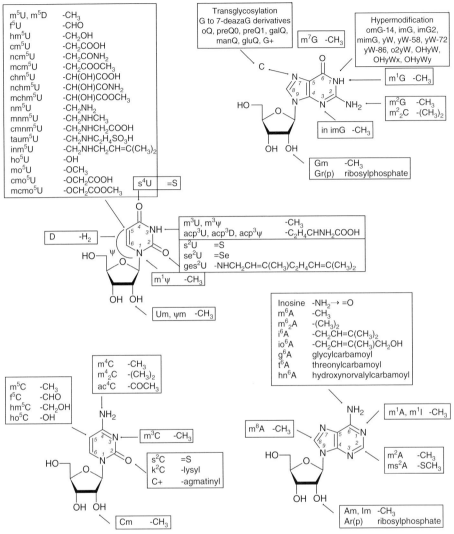

Fig. 2 An overview of the positions and chemistry of RNA modifications. Common abbreviations are used to designate the modified residue (the nature of the modifying group, position, identity of the parental nucleotide). The chemical structures of the modifying groups are indicated.

1.1
The Discovery of Modified Nucleotides in RNA: A Brief Overview

The first modified nucleotide to be identified (and designated at the time as a "minor" or "odd" nucleotide) was pseudouridine (5-(β-D-ribofuranosyl)uracil). It was discovered during the early 1950s and considered at one point to be a fifth RNA nucleotide (in addition to A, C, G, and U) due to its relative abundance in the total RNA fraction. Some time later, however, during the 1960s and 1970s many other modified RNA nucleotides were isolated and characterized [13, 14]. Subsequently, numerous tRNA species from various organisms were purified and sequenced on the RNA level. This laborious and time-consuming approach finally led to the establishment of the first RNA-modification database, but this was restricted only to tRNAs [15].

In parallel, abundant RNA species such as rRNAs and snRNAs were purified and analyzed, and this resulted in the identification of positions and the chemical nature of the presently modified nucleotides [16–18]. Studies of the identification of RNA modifications were essentially performed before the 1990s, which was an era of massive DNA sequencing and genetic engineering. Only a few additional RNA species have been analyzed more recently, the most important drawback being the amount of highly purified RNA species required. Methods for the identification and characterization of modified nucleotides in RNA have been extensively reviewed [19–21], and so will not be detailed at this point.

Data relating to the molecular structures and the presence of modified nucleotides in various RNA species are now maintained in the RNA Modification DataBase (http://mods.rna.albany.edu/) [7, 22].

More recently, bioinformatics studies performed by J. Bujnicki and colleagues led to the creation and considerable enrichment of present-day knowledge of RNA modifications and the enzymes implicated in such processes. The database created, which became known as *MODOMICS*, contains a lot of extremely valuable information on the systems biology of RNA modifications, the enzymes involved and their properties, and the biosynthetic pathways leading to the formation of complex hypermodified nucleotides [2, 3].

All ribosomal RNA modifications have been enrolled into a dedicated database (http://people.biochem.umass.edu/fournierlab/3dmodmap/) [23], which also allows the three-dimensional visualization of their positions in the ribosome. Since the majority of eukaryotic rRNA modifications are pseudouridine residues and 2′-O-methylations, information on the associated H/ACA and C/D snoRNA guides responsible for their biosynthesis can also be found in this database.

1.2
The Post-Transcriptional Nature of RNA Modifications

In contrast to the modified nucleotides in DNA – which may result from both enzymatic modifications and from spontaneous damage due to oxidative stress or spontaneous alkylation reactions – all of the known modified nucleotides in RNA molecules are formed by the specific action of RNA-modifying enzymes. Only recently, "damaged," nonenzymatically created nucleotides in RNAs were detected in certain pathological conditions (e.g., Alzheimer disease [24]; see also Sect. 6).

Most modifications – such as base methylation, pseudouridylation, deamination, reduction and many others – are produced by the single activity of one enzyme, whereas the formation of more complex structures, such as the synthesis of queuosine or archaeosine, requires several enzymes acting in a cascade of reactions [25–29] (Fig. 3).

The apposition of modifications in the sequences of the 18S, 5.8S and 25S rRNAs in the yeast *Saccharomyces cerevisiae* by the snoRNPs representing a singular class of RNA-modifying enzymes, is an early-stage

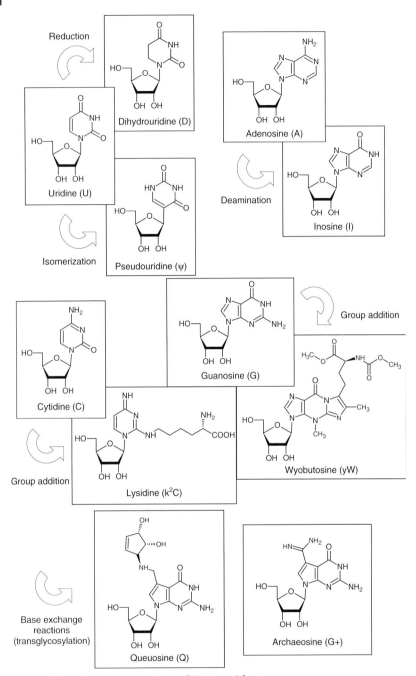

Fig. 3 Selected common reactions of RNA modification and the resulting modified residues, dihydrouridine, pseudouridine, inosine, lysidine, wyobutosine, queuosine, and archaeosine.

process that takes place in the dense fibrillar center of the nucleolus [30, 31]. It is unclear, however, whether all or only some of the modifications take place on the nascent pre-rRNA precursor via a cotranscriptional mechanism, or whether they are only generated post-transcriptionally. In the case of rRNAs, only a few modifications are added by protein enzymes during the late steps of pre-rRNA processing, or in the cytoplasm during the final protein-assembly steps on the ribosomes [32, 33].

1.3
Enzymes and Cofactors Implicated in RNA Modification: A Brief Overview

Although the properties and RNA-substrate recognition by RNA-modifying enzymes will be discussed below, the majority of these proteins also utilize a low-molecular-weight cosubstrate (frequently termed a *cofactor*). With only one exception, S-adenosyl-L-methionine (SAM; also known as AdoMet) is used as a cosubstrate for all RNA methylation reactions, while some amino acids (e.g., glycine, threonine, cysteine) or other small molecules (N^5,N^{10}-methylenetetrahydrofolate for m^5U in Gram-positive bacteria [34]; Δ^2-isopentenylpyrophosphate for i^6A [35]; flavin mononucleotide (FMN) for dihydrouridine [36]) are required for the formation of complex modifications (see Table 1). In some instances, such as the formation of 7-deaza derivatives of guanosine, a separate metabolic pathway is employed to form the precursor molecule, after which the precursor base replaces the encoded G residue. In the case of thiolation reactions, the sulfur atom must be activated by a separate enzymatic activity, which utilizes cysteine as a source of sulfur.

Tab. 1 Cofactors (substrates) used in RNA modification reactions.

Reaction	Cofactor (cosubstrate)	RNA modification
Methylation	S-Adenosyl-L-methionine (SAM, AdoMet)	All methylated nucleotides
Methylation	N^5,N^{10}-Methylenetetrahydrofolate	m^5U
Deamination	H_2O	Inosine (I)
Reduction	FMN	Dihydrouridine (D)
Isopentenylation	Δ^2-Isopentenylpyrophosphate	i^6A, io^6A, ms^2i^6A
Group addition	Threonine, HCO_3^-	t^6A, ms^2t^6A
Group addition	Lysine	Lysidine (k^2C)
Group addition	Taurine	tau m^5U, tau m^5s^2U
Group addition	–	Ribosyl-phosphate Gr(p), Ar(p)
Base-exchange	Queuine, Archaeine	All 7-deazaG derivatives
Thiolation	Cysteine, ATP	s^2U, s^4U, s^2C
Oxidation	O_2?	io^6A, ms^2io^6A
Methylthiolation	S-Adenosyl-L-methionine (SAM, AdoMet)	ms^2-group (ms^2i^6A, ms^2t^6A)

FMN, flavin mononucleotide.

2
Chemical Nature and Properties of Modified Nucleotides in RNA

Modified RNA nucleotides differ from the genetically encoded parental nucleotides in several aspects, including chemical reactivity, conformation and base-pairing properties. These differences are described in greater detail in the following sections.

2.1
Chemical Reactivity

Chemical groups, when added to the parental nucleobase, can cause a considerable change in the chemical reactivity of the latter toward a variety of chemical reagents. This provides an excellent basis for the selective chemical derivatization of modified nucleotides, and also their specific detection by employing physico-chemical and molecular biology analytical methods.

Methyl groups at nitrogen and oxygen atoms reduce the electron density and generally increase the reactivity of methylated nucleotides towards nucleophilic reagents. An example of this is the specific detection of m^5C by bisulfite ion – a reaction that is commonly used for the specific deamination of C, but not m^5C, residues in DNA and RNA [37, 38]. On the other hand, pseudouridine (Ψ, the uridine isomer) will react specifically with carbodiimides and other suitable chemicals [16, 19, 39]. Thiolated nucleobases react specifically via a sulfur atom, while free amino-group- or carboxylic-group-containing modified residues can be easily derivatized using amide bond formation. Details of these and other specific chemical reactions involving RNA-modified nucleotides are reviewed elsewhere [39].

2.2
Base-Pairing Properties and Specific Conformation and RNA Folding

Depending on the position concerned, the addition of a chemical group to a nucleobase frequently changes not only the general hydrophobicity of the nucleobase but also its capacity for H-bond formation; thus, such modified nucleotides may display an altered base-pairing capacity. In some instances, a modified nucleotide is completely unable to base-pair (m^1G, m^1A, m^2_2G, m^6_2A, etc.), whereas on other occasions base-pairing still may take place, but the partner nucleotide is changed (Inosine, s^2U, etc.) (Fig. 4). Only modifications at position 5 of pyrimidines and at positions 6, 7, and 8 of purines do not directly affect H-bond formation. However, the base-pairing capacities of these modified nucleotides may be still altered due to an inappropriate conformation (as for dihydrouridine, D) or an incapacity of a bulky chemical group to integrate the three-dimensional structure of the RNA helix. Such altered base-pairing behavior can be detected by primer extension using reverse transcriptase, and frequently serves for the detection of certain modifications.

In contrast, the Ψ residues may play a role in the stabilization of RNA–RNA or RNA–protein interactions. Indeed, nuclear magnetic resonance studies have shown that the additional imino group of Ψ stabilizes the interaction in RNA helices via a water molecule. This would favor the formation of unique RNA helices structures, which could not be obtained in the presence of uridine [40]. Compared to a U residue, a Ψ also increases the thermal stability of RNA–RNA duplexes [41, 42]. It has also been shown, using molecular dynamics simulations, that the presence of a Ψ decreased the mobility of the neighboring

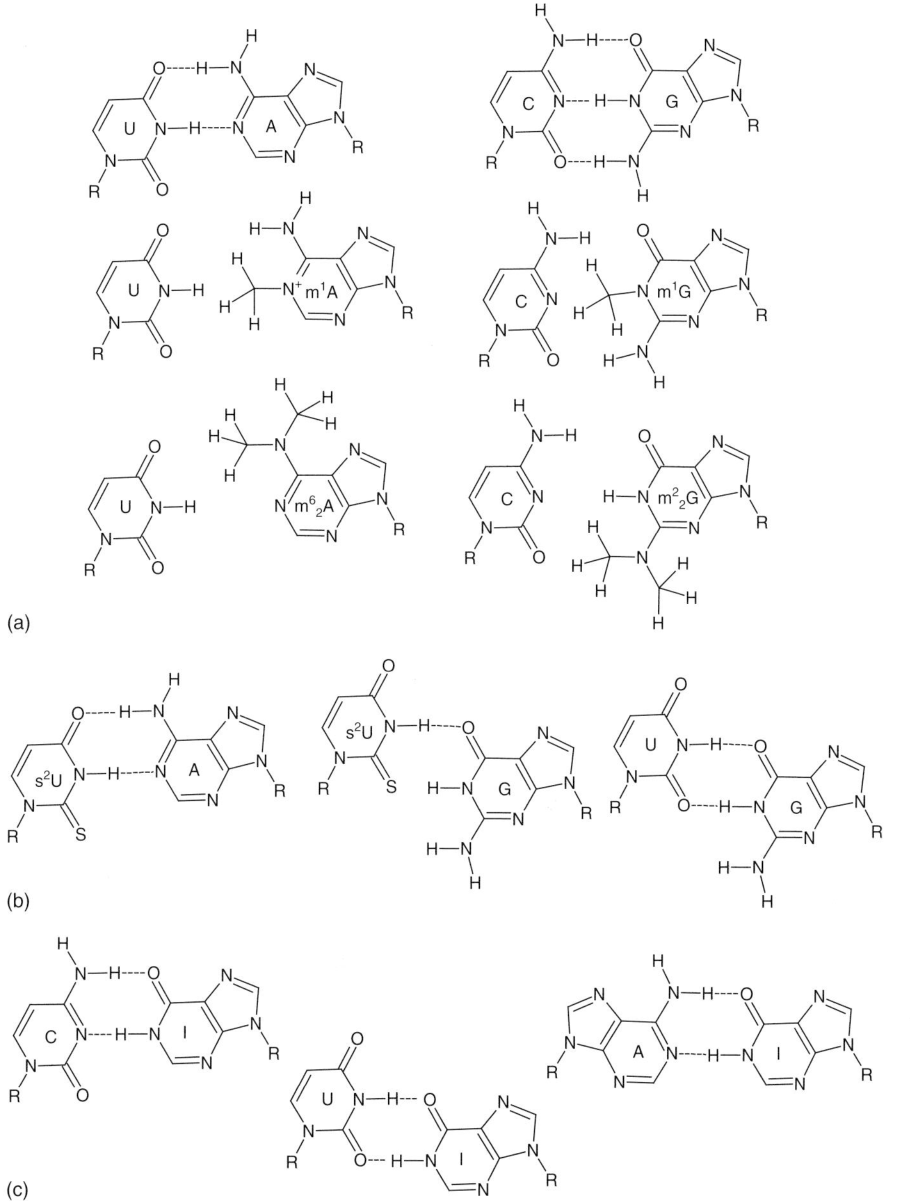

Fig. 4 Altered base-pairing properties of selected modified nucleotides. (a) Methylated adenosines (m¹A and m⁶₂A, left) and guanosines (m¹G and m²₂G, right) are unable to base-pair with the U and C residues due to the presence of the methyl groups at the Watson–Crick base positions; (b) Modulation of base-pairing by s²-group of s²U. Thiolated s²U normally base-pairs with A (left), but the Wobble base-pair with G is unstable as only one H-bond is formed (middle). The normal G*U base-pair is shown on the right; (c) Extended Wobbling with inosine, which base-pairs with C (like G), U (like G*U wobble base-pair) and with A.

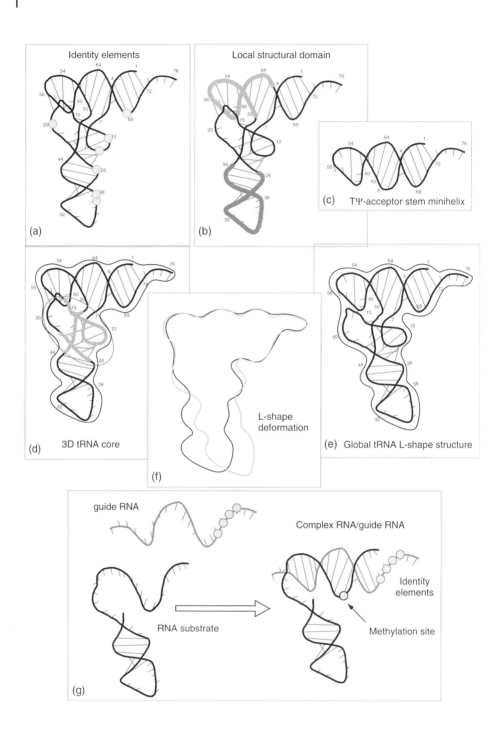

base by stiffening the ribose-phosphate backbone [43]. This was a consequence of water molecule coordination by the free imino group of Ψ, and the coordinated water in turn interacted with the adjacent phosphate [44]. Taken together, these unique properties of the Ψ residue could explain its prevalence in RNA structures. Another very frequent modification in RNA, namely the 2′-O-methylation of ribose, favors its 3′-endo conformation [45, 46]; this increases the rigidity of the RNA chain and facilitates not only the stacking of the nucleotides but also the base pairing and interactions of the RNA with associated proteins.

3
Enzymes and Enzymatic Mechanisms of RNA Modification

3.1
Two Concepts for Target Nucleotide Definition

Nature has developed two conceptually different mechanisms to define the positions of modified residues in RNA. First, unique or low-frequency residues are formed via a traditional approach, with the enzyme being specific to a given RNA sequence, or to a given local two- or three-dimensional structure. Depending on the strictness of this choice, the enzyme may modify a unique position in one single RNA species or, more commonly, it may act on a subset of RNAs having closely related two- and three-dimensional structures (such as tRNAs). In some cases, the specificity may be so relaxed that the same enzyme can modify different classes of cellular RNA molecules; however, this possibility is conceptually similar to many other situations in cellular metabolism.

A second way to define the target nucleotide for modification has emerged in the Archaea and Eukarya for the frequent modifications found in ribosomal rRNAs and some other RNA species. Indeed, these RNA molecules contain numerous Ψ residues and 2′-O-methylations. In these particular cases, the same catalytic subunit (Nop1/fibrillarin for 2′-O-methylations and Cbf5/dyskerin for Ψ) is responsible for the formation of modified nucleotides at dozens (or even hundreds) of sites. In these cases, the catalytic subunit provides only the required catalytic activity, while specificity is defined by an associated snoRNA guide (box C/D snoRNA/sRNA for 2′-O-methylations and box H/ACA snoRNA/sRNA for Ψ formation) [47–49]. A stable association of the catalytic subunit

Fig. 5 Possible RNA (tRNA) recognition strategies used by RNA modification enzymes. (a) Case of identity determinants for precise recognition, specific nucleotides located in the spatial proximity to the target residue is used for RNA substrate selection; (b) Recognition of the local structural domain (e.g., TΨ-loop or the anticodon loop, gray shaded); (c) Isolated structure of the TΨ-stem–loop which can be used as substrate by some tRNA-modifying enzymes for the formation of T54, ψ55, and m^{1}A58; (d) Recognition of the 3D core of the tRNA structure. The outline shows the global tRNA shape; (e) tRNA with disrupted interaction between TΨ- and D-loops, the L-shape is deformed; (f) Deformation of the angle between two highly ordered regions in tRNA (TΨ-stem/acceptor stem and anticodon stem/D-stem) due to perturbations in the 3D tRNA structure. Deformation of the L-shaped structure may be used for RNA substrate discrimination; (g) Use of the guide RNA (C/D or H/ACA snoRNA) to provide identity elements and recognize 3D RNA structure for modification. The case of 2′-O-methylation with C/D snoRNA guides is illustrated.

with the snoRNA guide is achieved during a complex process via the formation of snoRNP biogenesis, which in eukaryotes requires a specific assembly machinery [50] (Fig. 5).

3.2
Protein-Based RNA-Modifying Enzymes

Except for Ψ residues and 2′-O-methylations in eukaryotic rRNA, and some other small RNAs which are formed by snoRNP [sRNP/scaRNP (RNP containing small Cajal body-specific RNA)] complexes, the formation of all other modified nucleotides known to date is catalyzed by monopolypeptide RNA-modifying enzymes. Despite the importance of snoRNA-guide machineries, this group of protein stand-alone enzymes represents the great majority of RNA-modifying activities and is responsible for the majority of all known RNA modifications (Fig. 6). The main groups of RNA-modifying enzymes are briefly reviewed in the following subsections.

3.2.1 RNA-Methyltransferases

RNA-methyltransferases (RNA-MTases) belong to a vast group of MTases that catalyze the transfer of a CH_3-group (Me-group) from a methyl donor to a biomolecule [51, 52] (see Fig. 3). Depending on the presence of characteristic sequence motifs, and on the global organization of the protein fold, the MTases can be grouped into five distinct classes. Most DNA- and RNA-MTases belong to classes I (the most represented Rossmann fold MTases) [51] and IV (mostly SpoU type 2′-O-RNA-MTases and m^1G-tRNA-MTases) [53, 54]. RNA-MTases may be compared to other MTases acting on DNA, proteins and even on small molecules [55]. The almost universal methyl donor for all Me-transfer reactions in living organisms is SAM (AdoMet), with only a few MTases using other Me-donors, such as N^5,N^{10}-methylenetetrahydrofolate [56].

An analysis of the numerous AdoMet (SAM)-dependent MTases known to date highlighted the existence and relatively important conservation of AdoMet binding protein motifs. These motifs are present in almost any MTase, independently of its substrate specificity; some are known to stabilize AdoMet in the enzyme active site, but motifs IV, V and VI are involved in the pure catalytic step of the Me-transfer reaction [51, 52]. In addition to traditional RNA-MTases, a different catalytic mechanism for RNA methylation has been recently discovered. These RNA-MTases also use SAM as a source of the Me-group, but belong to the so-called "radical SAM family" which includes numerous enzymes catalyzing radical reactions using SAM as substrate; some of these proteins are also able to transfer Me-groups to biological substrates. Only a few radical SAM-MTases acting on RNA are known, the best characterized examples being the m^2A- and m^8A-generating enzymes RlmN and Cfr from antibiotic-resistant bacteria [57–59].

3.2.2 RNA-Pseudouridine (ψ)-Synthases

Numerous RNA-ψ-synthases from different organisms catalyzing the U → ψ conversion (see Fig. 3) have been characterized by three distinct approaches: biochemically; by sequence homology with RNA-ψ-synthases previously characterized; or by screening the gene products [60].

Known RNA-ψ-synthases can be allocated to six distinct families based on

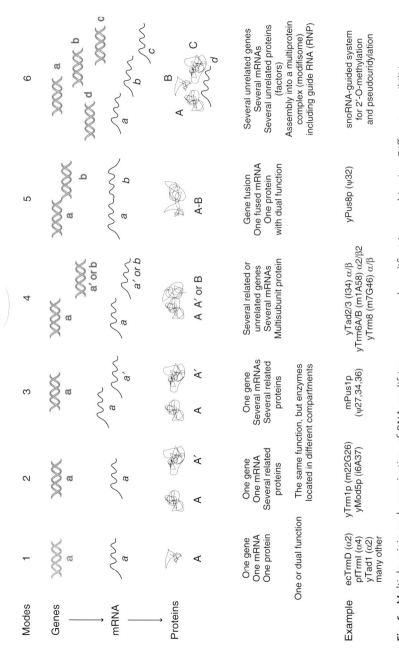

Fig. 6 Multiple origin and organization of RNA-modifying enzymes and modification machineries. Different possibilities are used to create the diversity of RNA-modifying enzymes in the cell and to ensure their correct cellular targeting in eukaryotes.

sequence similarity; such families are derived from TruA (tRNA PseudoUridine-synthase A), TruB, RluA (rRNA Large subunit PseudoUridine-synthase A) and RsuA (rRNA Small subunit PseudoUridine-synthase A), and TruD (tRNA Pseudo-Uridine-synthase D) [61] and archaeal Pus10 protein [62]. It is noteworthy that Pus10 is currently the only family which lacks representatives in bacteria [63].

The families RluA, RsuA, TruA, TruB, and TruA have three common conserved motifs (motifs I, II, and III), while additional conserved motifs, patterns IIa and IIIa, were also observed in some of families [64]. Motif II forms the binding site of the uridine and corresponds to the catalytic site [65]. The involvement of the conserved aspartic acid in catalysis has been demonstrated for the enzymes TruA, RluA/RluC, RsuA, and TruB [61, 66].

RNA-ψ-synthases recognize uridine only in the context of an RNA sequence. The maintenance of mechanisms for the accurate recognition of uridine within certain substrates is responsible for conservation during evolution of the location of some Ψ residues, especially in rRNAs, tRNAs, and snRNAs. Protein stand-alone enzymes are able to recognize their target in the absence of an accessory protein; these enzymes thus provide by themselves, in addition to catalysis, the recognition of one or more type(s) of RNA and access within the RNA to one or more specific positions. The recognition of U targets is often achieved by a base extrusion or flipping out mechanism [61]. This type of mechanism is also observed for enzymes that require access to a base in DNA during modification or repair processes, such as DNA-MTases or DNA-glycosylases [55].

The RNA-ψ-synthases feature three major different pattern of substrate specificity (see Fig. 5):

- Enzymes with site-specific recognition modify only one single position. Many RNA-ψ-synthases recognize RNA substrates in this way, and this very specific recognition is generally based on a unique sequence or a unique local RNA conformation.
- In the multiple-sites recognition mode, the enzymes have the capacity to recognize several uridine residues apart on the same RNA, or located on different RNAs. The modified residues are then present in structurally similar contexts. This is the case, for instance, for RluA which modifies U32 in tRNAs and U746 in the 23S rRNA of *E. coli* [67].
- Other enzymes modify several uridine residues present in close proximity one to another in an RNA sequence, a feature termed region-specificity. TruA modifies the positions 38, 39, and 40 in the helix of the tRNA anticodon arm [68]. This recognition mode allows stable helices to be distinguished, avoiding excessive stabilization from helices of low stability of which it is possible to enhance stability by converting U residues to ψ. This mode of action allows a fine adjustment of the stability of the anticodon of each tRNA arm.

3.2.3 Other Common Groups

Taking into account the important number of different RNA modifications, dozens of other enzymes participate in the modification of primary RNA transcripts. Thiolases catalyze the formation of sulfur-containing nucleotides, while transglycosylases are responsible for biosynthesis of deaza-G-derived modifications (see Fig. 3).

The reaction catalyzed by tRNA-guanine transglycosylases (TgTs) was discovered fortuitously by an observation that the radioactively labeled G base was incorporated into the tRNA fraction upon incubation in a cell-free extract. An understanding of the biological meaning of this phenomenon became clear much later, when deaza-guanosine derivatives (modified nucleotides of the queuosine family) were discovered in the anticodon loop of both bacterial and higher eukaryotic tRNA molecules. Indeed, the enzymes of the TgT family catalyze the exchange of an initially incorporated G residue by a deaza-guanosine base, which is synthesized via a separate metabolic pathway. The incorporated queuosine Q (or pre-queuosine preQ$_1$) base is further modified to become Q, ManQ, or GalQ [27, 29, 69]. A very similar reaction was also identified in Archaea, where the exchanged archaeosine residue is located at position 15 in the D-loop, except that the G residue is replaced by preQ$_0$ and further converted to archaeosine [70, 71].

In order to achieve the base replacement, TgT enzymes employ a precise cleavage of the C–N glycosidic bond (as do RNA-ψ-synthases), followed by the formation of a covalent reactive intermediate between the RNA substrate and enzyme. This reaction intermediate is further used to incorporate an external 7-deaza-guanosine derivative.

The biosynthesis of thiolated nucleotides in tRNA is catalyzed by a group of sulfurtransferases, responsible for the synthesis of s^{2}U, s^{4}U, s^{2}C, and also ms^{2}A and its derivatives. These thiolated nucleotides are important both for mRNA decoding during translation and, in addition, are implicated in bacterial photosensitivity to ultraviolet light [72]. The best-studied example of tRNA-sulfurtransferase is the enzyme ThiI, which is responsible for s^{4}U formation at position 8 in tRNA [73, 74]. In contrast, the enzymes (methylthiotransferase; MTTases) responsible for ms^2-group (methyl-thio) formation belong to the radical SAM enzyme family [75].

3.3
RNA-Guided RNA-Modifying Complexes

The main role of box C/D and box H/ACA snoRNPs (Fig. 7) is the addition of chemical modifications to specific sites in archaeal and eukaryotic pre-rRNA and certain other RNA substrates [76]. In addition, some snoRNPs are essential chaperones for the formation of pre-ribosomal complexes and the achievement of early cleavages of pre-rRNA processing. Unlike spliceosomal UsnRNPs, the snoRNPs are conserved in Archaea in the form of small ribonucleoprotein particles (sRNPs), suggesting that the appearance of the systems for modification guided by noncoding RNAs is an ancient event, appearing prior to two billion years ago and the separation between the Archaea and Eukarya. The composition, biogenesis and function of these particles have been the subject of extensive research during the past decade, and much structural information has been obtained, in particular for archaeal sRNPs.

3.3.1 C/D snoRNA-Guided RNA Methylation

The box C/D snoRNPs catalyze reactions of 2′-O-methylation. These particles form complex structures that typically consist of two RNPs, assembled respectively on K-turn motifs formed by the association of pairs of conserved C/D and C′/D′ boxes present on the snoRNA. Recognition of the substrate RNA is ensured by the snoRNA variable component of the complex, which

254 | RNA Modification

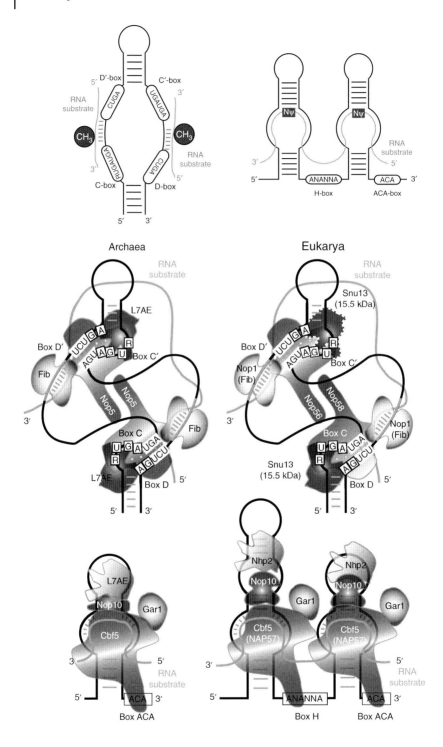

allows an adaptation of the specificity to a given RNA substrate sequence. These particles exert their function within the nucleolus in the maturation of the precursor of the rRNAs. The involvement of snoRNPs in the 2′-O-methylation of the U6 snRNA during its nucleolar maturation was also demonstrated in mammals and in *Xenopus* [79, 80]. Within the human brain, it has also been shown that the snoRNA MBII-52 could specifically modify an adenosine in the pre-mRNA of the 5-HT$_2$C receptor of serotonin [81]. In some eukaryotes, box C/D scaRNPs are also involved in the maturation of UsnRNAs in Cajal bodies [82]. Finally, in the Archaea some tRNAs are also the target of box C/D sRNPs.

The box C/D snoRNAs, which were first identified more than 30 years ago by the fractionation of RNAs isolated from the nucleolus, are small molecules that range in size from 70 to 300 nt. These RNAs are characterized by the presence of conserved sequences, termed *box C* (RUGAUGA 5′-3′) and *box D* (5′-CUGA-3′), which interact together to form a K-turn motif that is responsible for the recruitment of Snu13p/15.5K protein in eukaryotes and ribosomal protein L7Ae in Archaea. Boxes C and D are located respectively at the 5′ and 3′ ends of the snoRNA, while formation of the C/D motif leads to the establishment of a terminal stem–loop structure, which clamps both ends of the RNA. Most snoRNAs also have two additional internal sequences, boxes C′ and D′, but these are less conserved than boxes C and D [83]. In Archaea, the box C/D sRNAs are smaller, generally between 50 and 70 nt, and have a better conservation of the couple of boxes C′/D′ [49].

Base-pairing of the sequence located 5′ to the box D or D′ of the snoRNA with its RNA target results in the formation of a 10–20 bp RNA helix [84]. This represents essential structural information for the 2′-O-methylase Nop1p/Fibrillarin. It invariably targets the fifth base-paired nucleotide upstream of boxes D or D′ [85].

Ribosomal protein L7Ae in Archaea and proteins Snu13p/15.5K in eukaryotes represent, with the C/D and C′/D′ K-turn motifs, a nucleation center for the assembly of snoRNP [86]. In addition to these RNA-binding proteins, box C/D snoRNPs contain three additional proteins: the 2′-O-methylase Nop1p/Fibrillarin, which belongs to the family of class I SAM-dependent MTases [51]; Nop56p/NOP56; and Nop58p/NOP58. In Archaea, these last two factors are replaced by a single counterpart, the NOP5 protein.

3.3.2 H/ACA snoRNA-Guided Pseudouridylation

RNA-ψ-synthases belonging to the TruB family and named Cbf5p in *S. cerevisiae*, Dyskerin in humans and NAP57 in

Fig. 7 Top panel – Wire structure of C/D and H/ACA box snoRNAs. The conserved sequences (boxes C/D and C′/D′ in C/D snoRNA, as well as H and ACA boxes in H/ACA snoRNA) are indicated; Middle panel – Schematic representation of the assembled C/D snoRNPs from Archaea (left) and Eukarya (right). Conventional view of guide C/D snoRNA-mediated methylation machinery in Archaea and Eukarya. The sno(s)RNA guide is shown in black, the substrate RNA in gray. Conserved sequences of Boxes C, C′, D, D′ are indicated, protein 15.5 kDa may be present only in one copy in higher eukaryotes (indicated by dashed lane) (adapted from Ref. [77]. Bottom panel – Schematic representation of the H/ACA snoRNP particles in Archaea (left) and Eukarya (right); the expected locations of snoRNP proteins are indicated. NAP57 is a higher eukaryotic analog of Cbf5. Adapted from Ref. [78].

rodents [63, 78, 87], modify RNAs at many positions through the use of guide RNAs termed box H/ACA snoRNA in eukaryotes and box H/ACA sRNA in the Archaea. As with box C/D snoRNPs, some of the box H/ACA snoRNPs are also involved in the nucleolytic maturation of pre-rRNA.

In addition to the boxes H/ACA RNA guides, CBF5 enzyme and its counterparts require the intervention of three other proteins named *Nop10p*, *Gar1p* and *Nhp2p* in yeast, NOP10, GAR1 and Nhp2 in humans, and aNOP10, aGAR1 and L7Ae in Archaea. Interestingly, the archaeal aCBF5 protein, either alone or in combination with proteins aNOP10 and aGAR1, can catalyze the non-RNA-guided pseudouridylation at position 55 of tRNA [62, 88].

As is the case for the box C/D particles, the *in vitro* study of archaeal box H/ACA sRNP provided extensive information on the functioning of this RNA modification system. An approach allowing reconstitution of sRNP helped to define the key elements for the assembly of a catalytically active particle [89, 90].

In eukaryotes, the characterization of H/ACA box snoRNPs functions was initially undertaken in cell extracts [78], or from particles purified from the cells [91]. Experiments with proteins produced in reticulocyte lysates allowed an establishment of the interactions taking place between the components of the snoRNP. More recently, the *S. cerevisiae* H/ACA snoRNP particle was successfully reconstituted *in vitro* from recombinant proteins coexpressed in *E. coli* [92].

3.4
Catalytic Mechanisms of RNA Modification

Modification of the nucleotides in the RNA chain requires significant activation to increase the reactivity of different atoms in the base or ribose.

Independent of the exact chemical structure of the modified residue, at the molecular level the whole enzymatic reaction of RNA modification can be tentatively split into at least four elementary steps: RNA substrate and cofactor (cosubstrate) binding; conformational change of both RNA and protein to enter the target nucleotide into the active site; the "chemical" catalytic step; and product release.

At every such elementary step the amino acids of the enzyme enter into contact with the RNA chain via electrostatic, H-bonds, and van der Waals interactions. The electrostatic interactions between negatively charged phosphate moieties in RNA and Lys and Arg residues in the protein are known to stabilize the transient RNA–protein complex, and favor the subsequent conformational change of the RNA and enzyme. During this mutual structural adaptation between macromolecules, the target base is frequently stabilized in the active site by stacking interaction with aromatic Phe or Tyr residue. During the catalytic step of the reaction, the amino acids of the enzyme active site contribute to alter the distribution of the electronic density and thus to the activation of a given atom or a group for modification reaction. At the last step, the modified RNA chain must be released from the enzyme; this dissociation step may represent a rate-limiting step in the reaction.

3.4.1 Methylation

All RNA-MTases proceed with the direct or indirect transfer of a methyl group to a target atom, resulting in C-methylation, O-methylation, or N-methylation. In the great majority of cases, the methyl group is derived from AdoMet/SAM

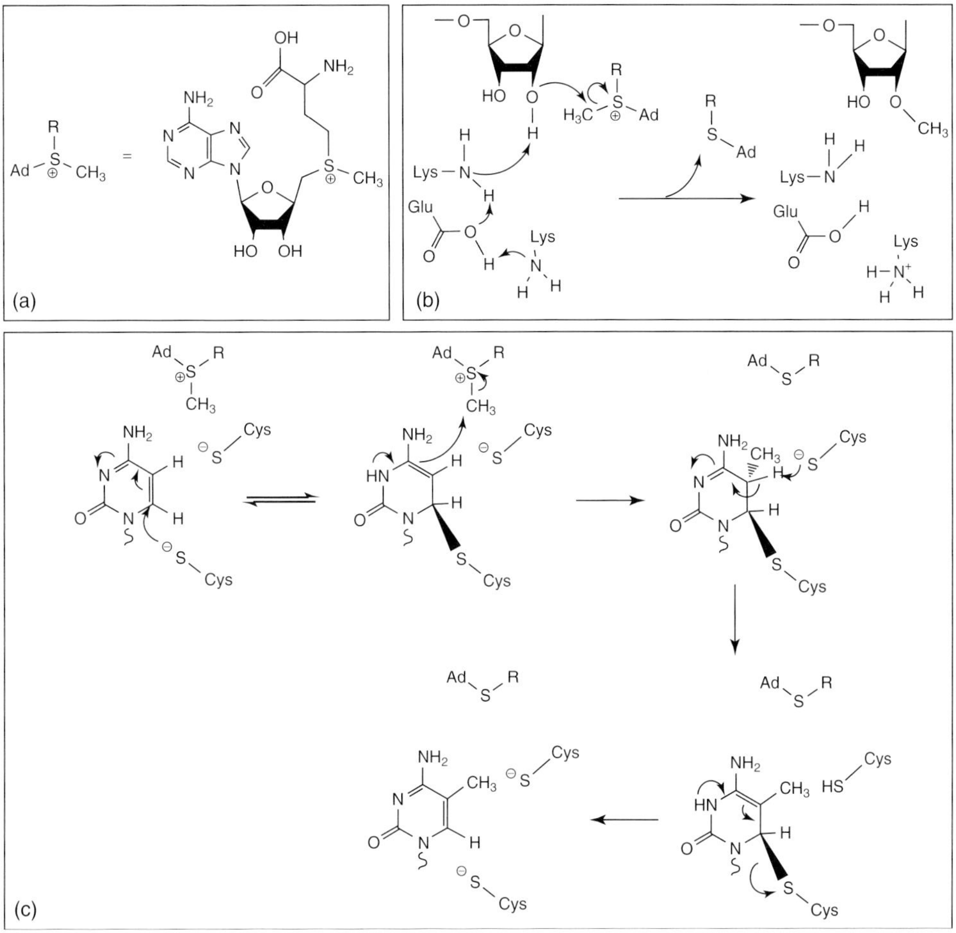

Fig. 8 Proposed mechanisms for O- and C-methylation reactions. (a) Chemical structure of AdoMet and its schematic representation; (b) 2′-O-methylation: Proposed catalytic mechanism of 2′-O-cap MTase nsp10. Adapted from Ref. [94]. The ε-amino group of the catalytic Lys, deprotonated by the concerted action of Glu and Lys residues, activates the 2′-OH group, which in turn can attack the electrophilic AdoMet methyl group; (c) Mechanism and catalysis of m^5C formation. The reaction proceeds through the nucleophilic attack at the C^6 of cytosine by the thiolate group of a catalytic Cys residue. Formation of the covalent Michael adduct leads to activated C^5 of the nucleoside and results in methyl transfer from the reactive methylthiol of the AdoMet cofactor. Another Cys residue acts as a general base in the β-elimination step. Adapted from Ref. [95].

(Fig. 8a), and the reaction resembles a S_N2 displacement mechanism involving the attack of a nucleophile from the RNA substrate on the electrophilic methyl group of SAM, with concomitant release of S-adenosyl-L-homocysteine (AdoHcy) [93]. A wide variety of mechanisms have evolved to activate the catalytic nucleophile atom, depending on its chemical properties. On the one hand, due to their intrinsic nucleophilic character, methylations at N and O atoms involve a direct transfer of the methyl group which can require a proton removal before, concurrent with, or after methyl transfer. The AdoMet conformation is also crucial to the mechanism of methyl transfer, which necessarily involves the precise juxtaposition of the electrophilic methyl group with a nucleophilic target atom. On the other hand, the sp^2-hybridized carbons display an electrophilic rather than a nucleophilic character, preventing a direct transfer mechanism for C-methylations. In these cases, two strategies have evolved: (i) a two-electron transfer reaction via a two-state mechanism with the formation of a covalent enzyme–RNA intermediate; or (ii) a one-electron transfer mechanism involving the intermediate methylation of a cysteine residue [93].

Exocyclic oxygens and nitrogens are generally sufficiently nucleophilic for methylation reactions using AdoMet as Me-donor, but the cycle carbons require significant activation. Even if the mechanism of N-methylation seems to be universal, the catalysis does not appear to be restricted to one model. Indeed, the chemical mechanism being presumed to entail an in-line $S_N 2$-type displacement mechanism with the incoming nitrogen nucleophile and AdoHcy leaving group occupying apical positions in a putative pentavalent transition state of the methyl carbon, catalysis may intervene at different levels, not necessarily independent of each other: chemical catalysis, substrate proximity, orientation, or transition-state stabilization.

In the case of 2′-O-methylation, the catalysis is essentially devoted to facilitate the removal of proton from the RNA OH-group, an essential step for this $S_N 2$-type displacement mechanism, via a general base-catalytic mechanism, as is the case for nsp10 and VP39, for example. Based on the results of structural and site-directed mutagenesis studies, it has been proposed that a Lys residue activates the 2′-OH group. The Lys ε-amino group is proposed to be itself activated by a concerted mechanism involving Glu and Lys (nsp10) or Asp and Arg (VP39) residues. The catalysis of nsp10 should also implicate an Asp residue probably being involved in the transition state stabilization via electrostatic interactions [94, 96] (Fig. 8b).

Fibrillarin activity in 2′-O-methylation is based on a catalytic triad, KDK, that is conserved and located in the substrate-binding pocket [97]. Aspartate plays an important role in catalysis, and could indeed be responsible for deprotonation of the 2′-OH group of ribose and/or stabilization of the cosubstrate SAM [98]. Interestingly, the fibrillarin is active only when integrated within an RNP, and no activity has indeed been observed in the isolated state.

Methylations at carbon atoms [most frequently at C^5 to produce 5-methyluridine (ribothymidine) and 5-methylcytosine] implicate a Michael addition at the C^6 ring position by a nucleophilic sulfhydryl group of the enzyme to generate a covalent intermediate. Formation of this covalent Michael adduct leads to activated C^5 of the nucleotide and results in methyl transfer from the SAM cofactor. The base abstraction of a proton from the C^5 then leads to product formation. In terms of catalysis, these reactions employ a general base to facilitate the β-elimination, which has been shown to be a carboxylic acid group for $m^5 U$ (Glu residue) or a thiolate for $m^5 C$ (Cys residue) [95] (Fig. 8c).

3.4.2 U → ψ Conversion: Pseudouridylation

Although the reaction mechanism of pseudouridylation is still not clearly defined, it has been well established that it is based

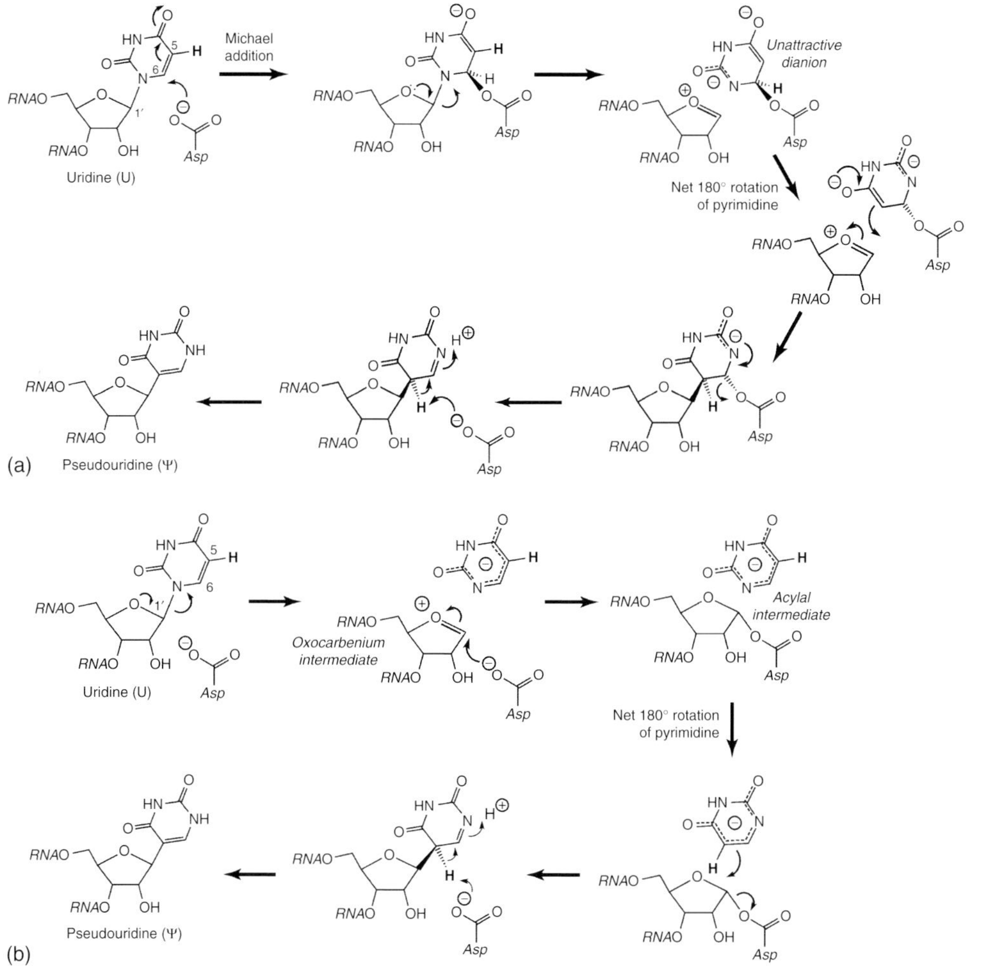

Fig. 9 Alternative mechanisms of pseudouridine formation. (a) ψ-generation via the formation of a Michael adduct by nucleophilic attack of the active site Asp residue on the C⁶ of U. The glycosidic bond breaks and the tethered uracil rotates about the ester bond and the new C-C bond forms. The aspartate then departs as a leaving group, the C⁵ is deprotonated, and the N¹ protonated. Adapted from Ref. [66]; (b) ψ-generation via formation of an acylal intermediate by nucleophilic attack of the active site Asp residue at $C^{1'}$ of U. The tightly bound uracilate anion then rotates about an axis perpendicular to the ring and the new C-C bond forms with the aspartate as the leaving group. The subsequent deprotonation of C⁵ and protonation of N¹ yield ψ. Adapted from Ref. [66].

on the reactivity of a catalytic aspartic acid. The reaction is perfectly conserved among different families of RNA-ψ-synthases [99]; however, two mechanisms have been proposed [66, 100].

The first mechanism (see Fig. 9) is based on "Michael addition," and involves a nucleophilic attack by the catalytic aspartate on the C⁶ of uracil. This reaction leads to the formation of a covalent link between the enzyme and the RNA. There follows a rupture of the N-glycosidic linkage between the base and the ribose, and a 180° rotation of uracil such that a new linkage is created between C^5 of the base and C^1 of the ribose. The driving force here is relocation of the electrons from the oxygen of C^4. The last two steps lead to a detachment of the base aspartate and finally to charge neutralization by the removal of a proton to form Ψ.

The second mechanism is based on nucleophilic substitutions S_N1 or S_N2, with the formation of an acyl intermediate. The catalytic aspartate accomplishes a nucleophilic attack on the $C^{1'}$ of ribose, and two options are possible. Either an S_N2 occurs where the N-glycosidic link between the base and ribose is broken in one step, or ribose is detached by a two-step S_N1 reaction. During a rotation of 180°, the base should be close to the ribose and is probably tightly retained in the active site of the enzyme, since uracil is not covalently bound to the aspartate as in the first discussed mechanism. The terminal steps of the isomerization take place in a manner similar to the first mechanism, leading to the formation of Ψ.

4
Functions of RNA Modifications

The presence of modified nucleotides in the RNA molecules alters the physico-chemical properties of the latter and may also influence their global two- and three-dimensional folding, as well as their interactions with other RNAs and with cellular protein partners. Hence, modified nucleotides are used for the adaptation and fine-tuning of RNA's properties with regards to the cellular processes in which these molecules are involved. At the molecular level, RNA modifications contribute to the acquisition of new conformational, chemical and physico-chemical properties of RNA species [30, 101–103].

4.1
Overall Stabilization of RNA Folding

The two- and three-dimensional folding of RNA is influenced by the presence of modified nucleotides in a different ways. First, many modified nucleotides do not base-pair with common partners or, in some instances, they may require a specific base-pairing partner or a particular conformation. Thus, the potentially complementary stretches of nucleotides do not interact together, favoring an alternative RNA conformation. The best-studied example of such modulation of the two-dimensional RNA structure is a folding of human mitochondrial with or without m^1A at position 9. By various methods, it was shown clearly that the single modified nucleotide would significantly favor the cloverleaf tRNA conformation, whereas in its absence the partially modified transcript would tend to adopt a long stem–loop structure, which is nonfunctional in translation [104–107]. Modification m^1A at position 9 seems to be particularly important for the correct folding of nonconventional tRNA species, such as certain mitochondrial tRNAs lacking the entire T-arm [108]. A similar role of a decisive position for correct cloverleaf tRNA folding was also proposed for another modified nucleotide in tRNA, namely the m^2_2G at position 26 [109]. It is almost certain that other non-base-pairing-modified RNA nucleotides may fulfill identical functions in tRNAs and other RNAs.

The stabilization of global three-dimensional RNA folding by modified residues is a characteristic feature of many cellular RNAs. In most cases, this modulation of three-dimensional structure is a result of conformational rigidity or the flexibility of modified nucleotides compared to their unmodified counterparts [77, 110]. Such fine-tuning of the three-dimensional RNA structure was assessed by using the chemical probing of nucleotides in tRNA structure, and also by measuring the hydrodynamic parameters of modified tRNA and unmodified transcript [111]. In both cases, the results indicated that the modified RNA had a more compact structure in solution, and therefore was less accessible for degradation by nucleolytic activities in the cell [110, 112]. This served as a basis for molecular dysfunctions characteristic for the loss of one (or indeed several) RNA modifications.

In some instances, modification of the base alters the capacity to form hydrogen bonds with external partners. Methylation at a different position of the base is certainly the most common reason for the loss of such bonds, although in rare cases the modification may create additional H-bond donors or acceptors, and in this situation the interaction with partners may be reinforced. Such a case was documented for Ψ, which has an additional imino (NH) group that provides an additional stabilizing H-bond [44].

4.2
Specific Effects of Chemical Reactivity or Conformation

The chemical reactivity of the modified RNA residues frequently differs from that of their unmodified counterparts. The presence of additional chemical groups on the base or ribose changes the electronic density on certain atoms and, as a consequence, the chemical reactivity towards nucleophilic or electrophilic reagents. The most common effect of this type is the protection of 2′-O-methylated RNA residues against nucleolytic degradation under alkaline conditions. The presence of the Me-group on the ribose prevents the nucleophilic attack of the 2′-oxygen to phosphodiester bond between nucleotides (Fig. 10). This feature of 2′-O-methylation is important in hyperthermophilic bacterial and archaeal organisms, to protect their RNAs against degradation at extreme growth temperatures [46].

The overall conformation of the nucleotides in the RNA chain also depends on the modifications. Additional chemical groups may modify the hydrophobicity of the molecule and favor a specific conformation of the ribose ring (2′-endo or 3′-endo sugar pucker). Some of the modified residues may freeze this conformational change and, in consequence, rigidify the RNA structure, whereas some others may favor flexibility of the RNA chain. For example, the modified nucleotide s^2T, which is found at position 54 in tRNA from hyperthermophic bacteria, favors a 3′-*endo*/*syn* conformation and thus increases the rigidity of the TΨ-stem that is required for tRNA stability at high growth temperatures [113, 114]. These subtle conformational effects may be also important for the efficient interaction of modified RNAs with other RNAs, or with proteins. Similar conformational effects may also exist for *syn*-/*anti*-rotation of the modified base around the C–N glycosidic bond.

Recent X-ray and NMR studies of codon–anticodon interactions formed by modified tRNA stem–loops have shown

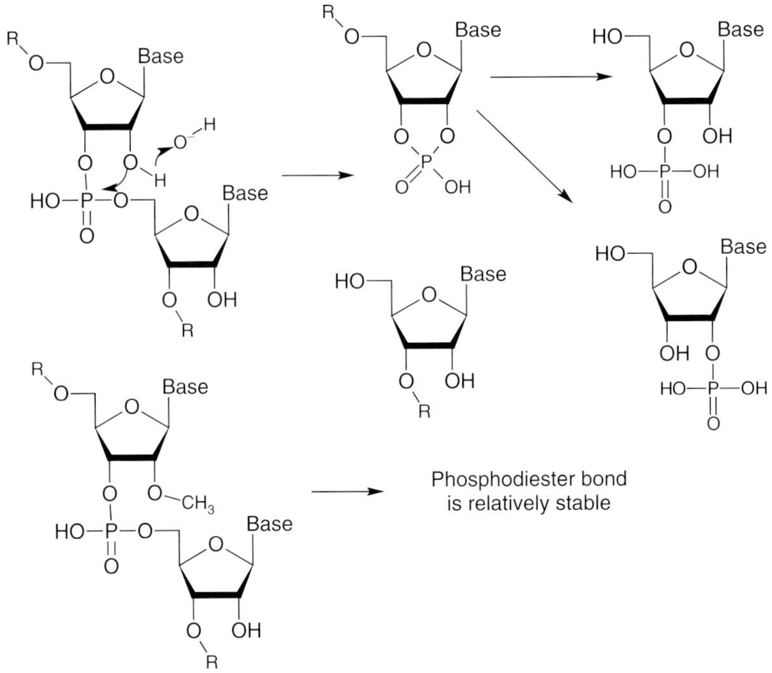

Fig. 10 RNA stabilization at the alkaline conditions by ribose 2′-O-methylation. The instability of the RNA chain in the presence of strong bases in aqueous solution is related to deprotonation of the ribose 2′-OH by the base (OH⁻), followed by the attack to the 3′-adjacent phosphodiester bond. The presence of a 2′-O-Me group hampers the deprotonation and significantly stabilizes the RNA chain under alkaline conditions.

that the presence of modifications in this extremely important region of tRNA can modulate the three-dimensional structure of the tRNA anticodon loop and create conformational effects by adjusting the base-pairing geometry. Indeed, these data now provide a rational explanation of the extended Wobble decoding provided by heavily modified U34 residues that are frequently present in the tRNAs and decoding four- and two-codon boxes in the genetic code [115–118].

Another level of structural modification is an improved stacking interactions provided by the modified bases, though this generally results in a stabilization of the RNA structure or heterologous RNA duplexes [119–121].

4.3
RNA Thermostability

The overall stabilization of RNA folding and conformational effects described above contribute to an increased thermal stability of the modified RNA molecules compared to unmodified RNA transcripts. This observation was made by comparing the melting curves of model RNA molecules *in vitro*, and also by observing that RNAs issued from thermophilic and hyperthermophilic organisms bear a much higher proportion of modified RNA residues compared to mesophilic species [113, 122, 123]. Such regulation of RNA modification as a function of growth temperature was observed for thermophilic bacteria cultivated under various conditions [124]. From an evolutionary viewpoint, this mechanism has contributed to bacterial species adapting to high temperatures, since nucleic acids – particularly RNA – are extremely unstable under these conditions.

4.4
Functions of Modified Nucleotides in mRNA Decoding

One of the most important and best-studied functions of modified nucleotides is a modulation of the codon–anticodon interaction during decoding of mRNA by the ribosome. As mRNA is only sparsely modified, the modified nucleotides in the tRNA anticodon loop are in charge of such fine-tuning.

mRNA decoding during translation is an extremely complex process that involves dozens of players, of both RNA and protein origin. In the decoding ribosome A-site, the short duplex between mRNA and the tRNA anticodon loop is stabilized by multiple interactions with the rRNA and ribosomal proteins (Fig. 11). Numerous functional and structural studies have shown that correct decoding cannot take place in the absence of a tRNA modification, or such decoding would be subject to errors both in reading frame maintenance and amino acid incorporation. Thus, the modified nucleotides in the tRNA anticodon loop are extremely important for the correct progressing of protein synthesis.

An analysis of the tRNA-modified nucleotides indicated that position 34 (wobble position) and position 37 of the tRNA molecules are both heavily modified, whereas positions 35 and 36 contain only a very limited number of modifications. It is also worth noting that position 34 base-pairs with a highly degenerated third codon nucleotide, while the tRNA nucleotide 37 stacks on the top of a short helix formed by the codon and anticodon in the ribosomal A-site [125]. This explains the importance of the RNA modifications in tRNA. The modified nucleotide at position 34 modulates the wobble base-pairing in order to adapt the geometry of interaction, whereas nucleotide 37 (which generally is a highly hydrophobic modified nucleotide) stabilizes the stacking of codon and anticodon bases in a short helix. Recent structural studies have provided an explanation for the extended wobble base-pairing which allows single tRNA molecule to decode all four codons in the codon box [102, 117, 118].

The modified nucleotide at position 37 may also prevent unproductive interaction with the complementary mRNA nucleotide and, in consequence, prevent loss of the reading frame (frameshifting) during decoding [126] (Fig. 12). At the cellular level, the frameshift suppression mediated by modified nucleotides in tRNA may affect the viral (phage) infection, which frequently uses the programmed ribosomal frameshifting for the expression of viral (phage) proteins [127].

The ribosomal RNA and its modified nucleotides also contribute to the decoding capacity of the A-site. It was noted earlier that the modified residues present in LSU and SSU rRNAs are clustered at functionally important regions of these long RNA species [18, 128, 129]. Moreover, the recent three-dimensional analysis of ribosomal subunits and the assembled ribosomes showed clearly that the majority of the modified residues were in fact present at both the A-site (responsible for decoding) and in the peptidyl transferase center (PTC), which ensures the catalysis

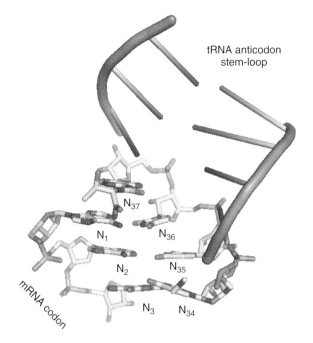

(a) mnm⁵U34-t⁶A37-tRNA^Lys UUU (PDB structure 1XMO)

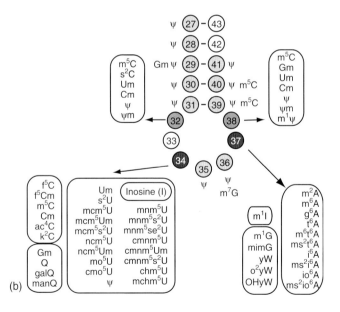

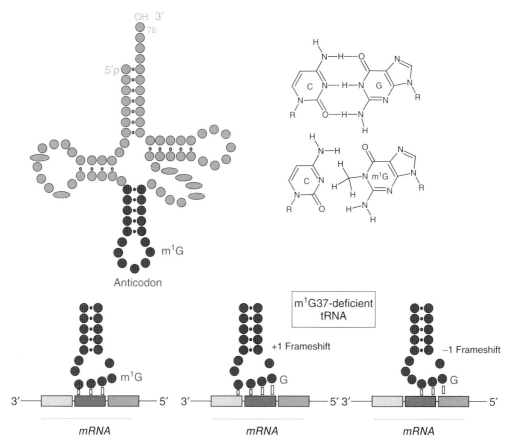

Fig. 12 Importance of m^{1}G37 for translational frameshifting. The modified nucleotide m^{1}G is frequently present at the position 37 in the tRNA anticodon loop (top left). This nucleotide is unable to base-pair with C (right) or any other nucleotide (top right), and this prevents unwanted interactions with other codons in mRNA. In the absence of m^{1}G37, the unmodified G nucleotide may interact with the nucleotides of mRNA, and, depending on the exact mRNA sequence, this may result in +1 or −1 frameshift.

←

Fig. 11 (a) Overview of the ribosome decoding site with the anticodon loop of tRNA (ASL). The anticodon region of tRNA^{Lys}UUU paired with AAA codon is used as example (Image taken from PDB structure 1XMO Ref. [117]). tRNA is shown in dark gray, mRNA codon nucleotides are in light gray, interacting bases are indicated. Codon nucleotides are numbered N1, N2 and N3 and tRNA anticodon nucleotides are labeled N34, N35, N36, The modified nucleotide at position 37 (t^{6}A in tRNA$_{Lys}$UUU) is also shown in sticks; (b) Types and localization of modified nucleotides in the anticodon loop of tRNAs. The greatest number and variety of modified nucleotides is found at positions 34 (Wobble position) and at position 37 (black circles), the intensity of gray shade corresponds to variety of modified nucleotides present at given position.

of peptide bond formation. It is, therefore, not surprising that an absence of these highly conserved rRNA modifications generally alters the efficiency of the protein synthesis [130, 131].

Nevertheless, these residues must act in synergy, as the absence of only one residue has no detectable impact *in vivo*. In yeast, only Ψ at position 2919 in the PTC [132] and the hypermodified $m^1acp^3\psi1191$ of the decoding center [133] showed any measurable effect on cell metabolism. However, an accumulation of mutations in a given rRNA region can lead to serious malfunctions of the ribosome. Such changes, which are located primarily at the heart of the decoding center and at the interface between the two subunits of the ribosome, have been shown as important in the production of 18S and 25S rRNAs, in the formation of polysomes, the activity of the ribosome, the fidelity of translation, and for cell growth [134, 135]. These recent data have revealed the collective importance of nucleotide modifications in the biogenesis and function of ribosomes. Moreover, in the absence of any strong individual impacts, their synergistic action and high conservation reflect the selective advantages that they have brought to optimizing the translation process.

4.5
rRNA Modifications and Antibiotic Resistance

Among other adaptation strategies, some antibiotic-resistant bacteria have developed an elegant means of escaping cytotoxic antibiotic actions by the selective modification (generally methylation) of key nucleotides in rRNA [136]. These additional methylated nucleotides are frequently located in close proximity to the housekeeping modifications in rRNA, and certainly alter the hydrophobicity of the binding site as well as the H-bond formation between the ribosome and antibiotic molecule. Various types of RNA methylation are used for this protection, namely m^1A, m^7G, m^1G, m^8A, and Am/Cm/Um in different bacteria [77]. This type of resistance is frequently used for aminoglycosides, which rely on electrostatic and H-bond interactions with the key nucleotides in the decoding and PTCs to affect ribosomal translation. Those genes which encode enzymes responsible for antibiotic-resistance methylations are shared between bacterial species via mobile genetic elements such as multiresistance plasmids, and thus are spread rapidly among the bacterial population.

It is noteworthy that the absence of a housekeeping modification may also confer antibiotic resistance. For example, a deficiency of Ksg RNA:MTase confers resistance to kasugamycin in *E. coli* [137]. Similarly, a loss of the RluC RNA-ψ-synthase responsible for rRNA modification in bacteria conferred an increased resistance towards some antibiotics [138].

4.6
Positive and Negative Determinants for RNA–Protein Recognition

One of the most spectacular examples of changes in tRNA aminoacylation specificity is the conversion from Met to Ile charging via a single RNA modification in the anticodon loop. In all bacteria, decoding of the rare AUA Ile codon is ensured by a specific $tRNA^{Ile}(k^2CAU)$, which results from the conversion of $tRNA^{Met/Ile}(CAU)$ primary transcript. The single modification of the C34

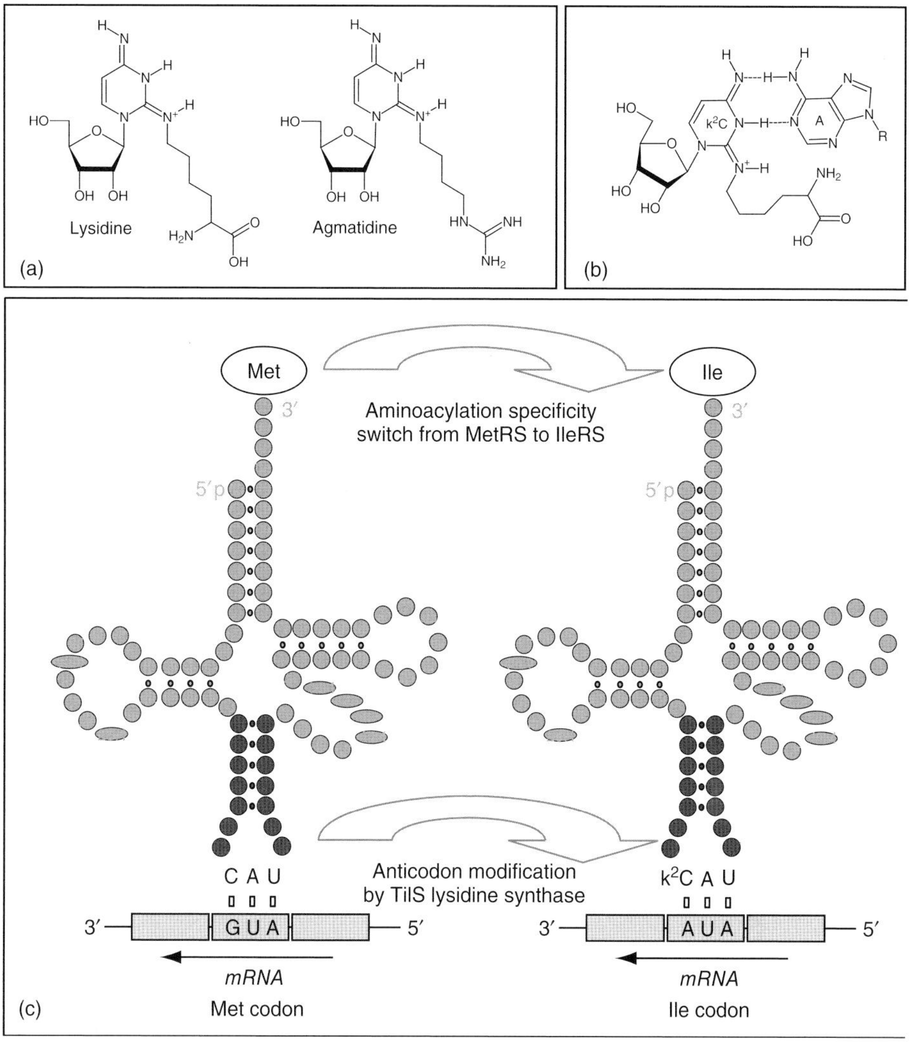

Fig. 13 Lysidine and agmatidine in AUA decoding. Panel A – Chemical structures of Lysidine (left) and agmatidine (right). Panel B – Base-pairing properties of lysidine, the lysidine (k^2C) – A base-pair is shown. Panel C – Lysidine modification at Wobble position in tRNA converts both the codon and amino acid specificities of tRNA^{Ile}. The precursor tRNA bearing C34 can be aminoacylated with methionine and decodes the AUG (Met) codon. After being modified to k^2C34 by TilS, tRNA bearing lysidine gains isoleucine-accepting activity and AUA-codon specificity. Adapted from Ref. [140].

in tRNA$^{Met/Ile}$(CAU) to k^2C not only changes the base-pairing from AUG to AUA codon, but also concomitantly converts the aminoacylation specificity from Met to Ile. The modified nucleotide in tRNA$^{Met/Ile}$(CAU) to k^2C k^2C (lysidine) then serves as a major identity determinant for IleRS and as the anti-determinant for MetRS, to avoid ambiguity in the reading of AUG and AUA codons [139, 140] (Fig. 13). In Archaea, another modified nucleotide, agmatidine, plays the role of lysidine in bacterial tRNA for AUA decoding [141]. A similar conversion between aminoacylation specificity of unmodified transcript and native tRNA was observed in the case of tRNA^{Arg}/tRNA^{Asp} par AspRS, where the modified nucleotide responsible for the specificity switch was a m^{1}G37 in the anticodon loop [142]. The modified nucleotide inosine was also found to be the positive determinant for tRNA^{Ile} aminoacylation by IleRS in the yeast *S. cerevisiae* [143].

One other case of an altered RNA recognition due to the presence of specific RNA modification involves recognition of the eukaryotic initiator tRNA^{Met}$_i$ by the initiation factor IF2. In contrast to bacteria, where the methionine fixed to the initiator tRNA^{fMet} is specifically formylated, eukaryotes employ a normal methionine residue to initiate translation. In order to be specifically recognized by eIF2, the tRNA^{Met}$_i$ has a number of specific sequence features, and the G64 residue is modified to Gr(p) [144]. This particular modification is found only in tRNA tRNA^{Met}$_i$ and serves as a negative identity element for EF-1α recognition [145].

It is not only translation factors and aminoacyl-tRNA synthetases that are sensitive to the modification status of the tRNA molecule. Rather, gamma-toxin from *Kluyveromyces lactis* specifically recognizes tRNA species bearing the complex modification mcm^{5}s^{2}U at the wobble position, formed in part by RNA-MTase Trm9. In the absence of the entire mcm^5 group, cleavage is considerably reduced, which explains the resistance to gamma-toxin seen in Trm9-deficient strains [146].

4.7
Other Effects Outside of Translation

RNA nucleotide modification plays an important role not only in "traditional" tRNA functions in protein translation. Several eukaryotic tRNAs, and namely tRNA^{Lys}$_3$, serve as primers for the reverse transcription of RNA-base retroviruses, such as HIV-1. As human tRNA^{Lys}$_3$ is heavily modified, the importance of these modified residues for tRNA^{Lys}$_3$ functions outside of translation was extensively studied [147–149]. The formation of the reverse-transcription initiation complex, as well as the switch to elongation mode, are stimulated by the presence of modifications in human tRNA^{Lys}$_3$. In addition, the strand transfer is also facilitated with a completely modified tRNA^{Lys}$_3$ primer.

The encapsidation of tRNA^{Lys}$_3$ in viral particles depends on the presence of certain modifications, namely mcm^{5}s^{2}U$_{34}$ and ms^{2}t^{6}A$_{37}$. Both of these modified nucleotides are located in the anticodon loop of the tRNA^{Lys}$_3$ and are required for efficient binding of the nucleocapsid protein Ncp7 [115].

4.8
Modulation of Immune Response Using RNA-Modified Nucleotides

The modification status of RNA is also used by the innate immune system to distinguish between self and nonself RNA, in particular bacterial and viral RNAs [150]. Although the exact molecular mechanism is still poorly understood, this response seems to include the TLR7 and TLR8 receptors, which are responsible for single-stranded RNA presentation. Methylated nucleotides (and specifically 2′-O-methylations) in RNA seem to play an important role in this immunosuppression [151], and these observations are considered to be a means of reducing the immunogenic response to small RNAs, such as synthetic siRNA [152, 153].

5
Static and Regulated Character of RNA Modification

For decades since the discovery of modified nucleotides in RNAs, these chemical alterations have been considered as constitutive, stable, and maintained during the entire life-time of RNA molecules. However, during the past four to five years this point of view has evolved significantly to the concept of "RNA epigenetics," which relates to the dynamic character of RNA modifications and their active implication in gene expression regulation [154–156].

5.1
Known Examples of Regulated RNA Modifications

The possibility of an enzymatic removal of RNA modification via oxidation reactions was first suggested for methylated nucleotides such as m^1A in RNA [157, 158]. Whilst this de-modification of RNA by AlkB and related proteins resembles the mechanism of DNA repair, these de-modification reactions were considered to be an exception. However, further studies allowed this observation to be extended to other modified nucleotides such as m^3U [159]. More recently, the regulation of mRNA modification m^6A by FTO-mediated oxidative demethylation reaction was documented in mouse and human cells [160–162] (Fig. 14).

Another member of the human AlkB-related proteins, ALKBH5, is also able to catalyze the oxidative demethylation of m^6A residues in mRNAs *in vitro* and *in vivo*. Alkbh5-deficient male mice showed an increased level of m^6A residues in mRNAs and had a reduced fertility due to the apoptotic death of spermatocytes. These data demonstrated that reversible m^6A mRNA modification has various functions in mammalian cells [163].

Pseudouridylation has long been perceived and described as a constitutive mechanism in the cell, and so could be neither induced nor reversible. Recently, however, it was shown that two positions in the U2 snRNA could be pseudouridylated according to the environmental conditions [164] – that is, following heat stress or under nutrient deficiency. The presence of these additional modifications reduced the functionality of U2 snRNA in splicing, which was consistent with a downregulation of the cellular metabolism under stress conditions. Thus, the mechanism of RNA pseudouridylation can also be regulated [165].

5.2
Regulation of mRNA Translation by Modified Residues

The first link between the tRNA modification profile and a translational regulation of specific mRNA was predicted for RNA-MTase Trm9, participating in the formation of mcm^5s^2U at position 34 in certain yeast tRNAs. This wobble modification ensures the decoding of AG(A/G) and GA(A/G) codons. The translation of certain yeast mRNAs is indeed affected *in vivo* by the absence of Trm9, while the translation of hundreds of other mRNAs with a similar codon usage is most likely modulated [166]. The

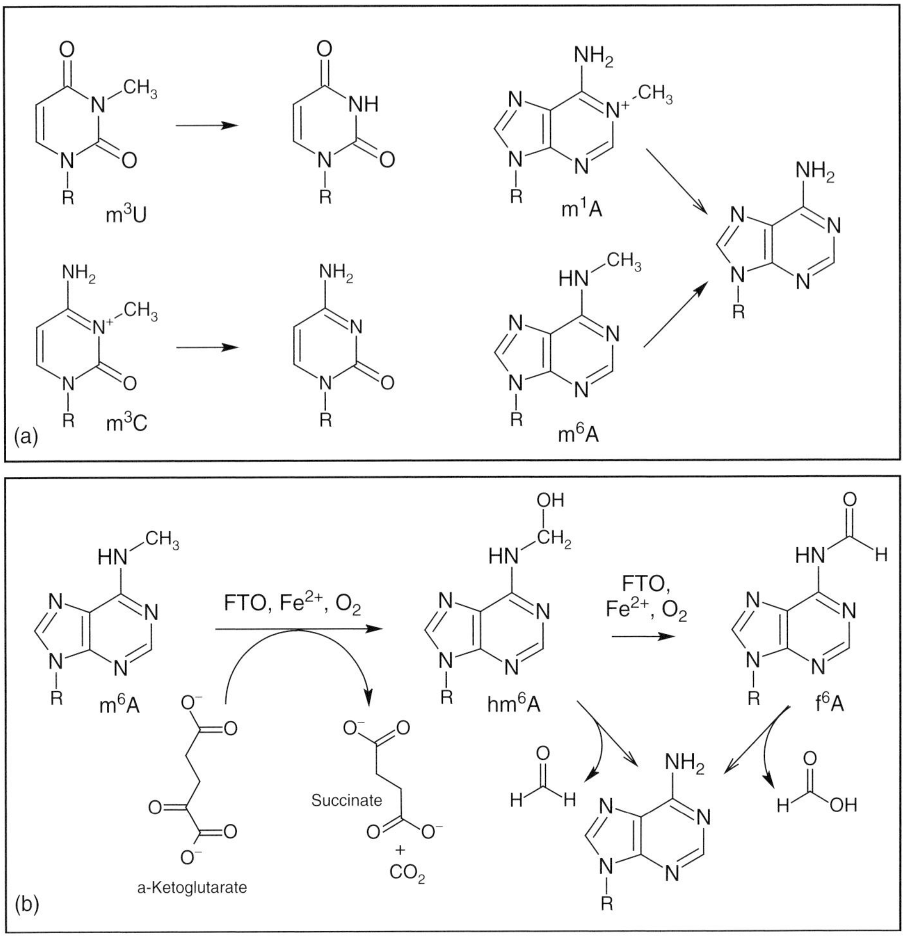

Fig. 14 Panel A – Modified RNA nucleotides which serve as substrates for oxidative RNA de-methylation (m^3U, m^3C, m^1A and m^6A). Panel B – Reactions of oxidative RNA demethylation of m^6A catalyzed by FTO enzyme. Reaction intermediates of oxidative de-methylation (hm^6A and f^6A) are shown, as well as low-molecular co-substrate (α-ketoglutarate). The reaction requires molecular oxygen O_2 and Fe^{2+} ions.

absence of Trm9 also increases the translational infidelity and thus promotes the accumulation of incorrectly folded protein isoforms, with a concomitant activation of stress response genes [167]. Interestingly, human tRNA methyltransferase 9-like protein (hTRM9L/KIAA1456) was found to be downregulated in different cancers, while its re-expression significantly suppressed tumor growth *in vivo*. The regulation of tumor growth seems to involve LIN9 and HIF1-α-dependent mechanisms [168]. A detailed analysis of ribosomal translation in TRM9-deficient yeast strains indicated that the translation of AAA, CAA, and GAA codons is indeed slowed down, but the major effect derives from an activation of the GCN4-mediated stress response. Thus, the loss on mcm^5s^2U modification affects

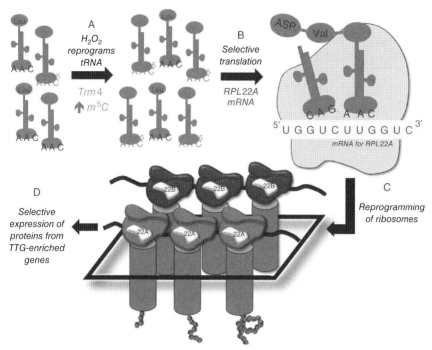

Fig. 15 Proposed mechanism by which an increase in m⁵C level regulates the translation of ribosomal protein paralogs and confers resistance to H$_2$O$_2$. Exposure to H$_2$O$_2$ leads to an elevation in the level of m^{5}C at the wobble position of the leucine tRNA for translating the codon UUG on mRNA (A), which enhances translation of the UUG-enriched *RPL22A* mRNA relative to its paralog *RPL22B* (B) and leads to changes in ribosome composition (C). This reprogramming of tRNA and ribosomes ultimately causes selective translation of proteins from genes enriched with the codon TTG (D). Adapted from Ref. [170].

global gene expression due to perturbed cell signaling [169].

Another recent example of stress response regulation is a selective reprogramming of the ribosomal translation under the conditions of oxidative stress [170]. The treatment of yeast cells with H$_2$O$_2$ increases the level of m^{5}C modification at the Wobble position of tRNA^{Leu}(CAA), and this causes a selective translation of TTG-enriched mRNAs (Fig. 15). This demonstrates the stress regulation of important tRNA modifications and, in consequence, the reprogramming of translational machinery to cell survival.

6
Molecular Pathologies Linked to RNA-Modification Defects

Although variations of RNA modification content in relation to human pathologies were first observed long ago [171], the exact molecular mechanisms linking RNA modification and disease were uncovered only recently in certain cases. As a general trend, RNA-modification-related pathologies may result either from the absence of a given modified nucleotide in one or several RNA species, or from the absence or deficiency of the RNA-modifying enzyme, which acts as an essential cellular protein. These two cases are considered below.

6.1 Mitochondrial Diseases

The mitochondrial compartment in a eukaryotic cell is frequently affected by deficiencies in RNA (mostly tRNA and rRNA) maturation. Only a few genes are directly expressed in the human mitochondrion, but these specific mitochondrial mRNAs encode the essential proteins required for oxidative phosphorylation. It is, therefore, not surprising that defects in tRNA maturation/modification/aminoacylation have direct effects on the production of ATP by mitochondria. Numerous human pathologies have mitochondrial deficiency as a major cellular and molecular cause [172, 173]. A lack of certain mitochondrial tRNA and rRNA modifications is the major cause of various diseases belonging to the family of mitochondrial myopathies, which are frequently fatal in humans.

6.1.1 Taurine Modification in Mitochondrial tRNA

One of the best-studied examples concerns the absence of the taurine group at uridine at position 34 in the anticodon of the mitochondrial tRNA^{Lys} or tRNA^{Leu}. Mutations in the mitochondrial genome observed in myoclonic epilepsy with ragged red fibers (MERRF) and mitochondrial encephalomyopathy, lactic acidosis, and stroke-like episodes (MELAS) patients are responsible for the synthesis of aberrant undermodified tRNA lacking taum^{5}U and taum^{5}s^{2}U [174, 175] (Fig. 16). At the translational level, the undermodified tRNAs are unable to efficiently decode certain mRNA codons and, depending on the deficient tRNA species, provoke the global or specific reduction of mRNA translation [176]. One other mitochondrial modification enzyme, the tRNA:thiolase MTU1 (TRMU), also contributes to taum^{5}s^{2}U synthesis, and its deficiency leads to the phenotypic features similar to those observed in MERRF [177]. However, the resulting phenotype may be also related to additional functions of MTU1 [178].

Fig. 16 Taurine modification in mitochondrial tRNA (mt tRNA) and its function in decoding. (a) Biosynthetic pathway that introduces the taum^{5}s^{2}U modification of mt tRNAs. *MSS1* and *MTO1* are involved in the initial step of taum^{5}s^{2}U synthesis on mt tRNAs. Mitochondrial taurine is subsequently incorporated into mt tRNAs by an unidentified transferase to build taum^{5}s^{2}U. The sulfur from cysteine is transferred to unknown sulfur mediators. MTU1 then acts as a mitochondria-specific 2-thiouridylase for taum^{5}s^{2}U by using the activated sulfur from the mediators. Adapted from Ref. [177]; (b) Cloverleaf structures of human mt tRNAs$^{Leu(UUR)}$ from wild-type (WT) cells (left) and from MELAS cytoplasmic hybrid cells with the A3243G mutation (center) or T3271C mutation (right). Encircled "U" indicates an unmodified wobble uridine. White letters on a square black background represent the respective point mutations. Adapted from Ref. [176]; (c) Possible mechanism of molecular pathogenesis caused by the wobble modification deficiency of the mutant tRNAs in MELAS (left) and MERRF (right) patients. A pathogenic point mutation (A3243G or U3271C) in mutant tRNA$^{Leu(UUR)}$ from MELAS patients causes a taum^{5}U-modification deficiency (indicated by gray shade), which results in a UUG codon-specific translational defect. The MERRF 8344 mutation in tRNA^{Lys} causes a taum^{5}s^{2}U-modification deficiency (indicated by gray shade) that results in a translational defect for both cognate codons (AAA and AAG). Adapted from Ref. [176]. The fragment of the universal genetic code indicating positions of Leu and Lys codons is shown on the right.

6.1.2 Pseudouridine Modification of tRNAs by hPUS1

Pseudouridine residues in tRNA represent another common modification contributing to mitochondrial pathologies. A missense mutation (R116W) in the active site of human pseudouridine synthase 1 (hPUS1) [179, 180] is associated with mitochondrial myopathy, lactic acidosis and sideroblastic anemia (MLASA), and the loss of pseudouridine residues in both cytoplasmic and mitochondrial tRNAs was confirmed. Another nonsense mutation of the hPUS1 gene was also reported that also led to MLASA [181]. hPUS1 has a broad site and substrate specificity, and is responsible for U → ψ conversion in tRNAs and U2 snRNA [182]. A comparative

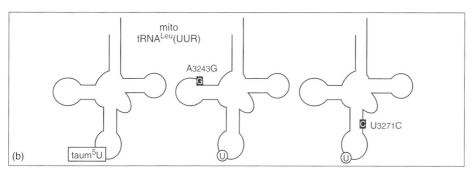

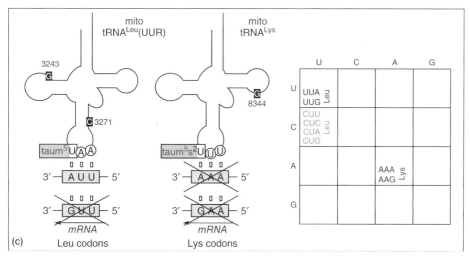

microarray analysis of the global human transcriptome showed that the expression of many genes is affected by the absence of pseudouridines in tRNAs, though the exact molecular mechanism involved in human pathology is yet to be resolved [183].

6.1.3 Mitochondrial 12S rRNA Mutation in Inherited Deafness

A synthetic phenotype between mitochondrial 12S rRNA mutation A1555G and mitochondrial tRNA modification was observed in the case of maternally inherited deafness. The defect in protein synthesis due to the 12S rRNA mutation can be compensated by an increased expression of ribosomal proteins, but additional defects in mitochondrial tRNA:thiolase MTU1 or in other factors important for tRNA transcription or maturation lead to hearing loss [184].

6.2 Link between Type II Diabetes and tRNA Modification

A recent analysis of diabetes type II-associated genes indicated that variations in the cyclin-dependent protein kinase 5 (Cdk5) regulatory associated protein 1-like 1 (Cdkal1) are associated with an impaired insulin production. Subsequently, Cdkal1 emerged as a mammalian thiotransferase responsible for ms^2-group addition in the ms^2t^6A biosynthetic pathway [185]. The ms^2t^6A modification is present in the tRNA^{Lys}(UUU) and is important for the global three-dimensional structure of the tRNA anticodon loop and for preventing translational errors during mRNA decoding. This defect will, in turn, affect insulin-producing cells and lead to a deregulation of insulin response in type II diabetes [186]. Another closely homologous radical SAM methylthiotransferase CDK5RAP1, acting in ms^2i^6A biosynthesis, is preferentially localized in the mitochondrion, although its activity seems not to be limited exclusively to mitochondrial tRNAs. This observation represented an unexpected connection between CDK5 which is critical for neuronal cell differentiation and anticodon tRNA modification [187].

6.3 Dyskeratosis Congenita

Dyskeratosis congenita (DC; also known as DKC or Zinsser–Cole–Engman syndrome) is a rare genetic disease with a prevalence estimated at one in one million [188, 189]. This polymorphic disease is characterized by alterations in tissues having high renewal rates, such as skin and bone marrow, and in some cases may lead to cancers [190]. There are three distinct forms of DC: an X-linked form, an autosomal dominant form, and an autosomal recessive form.

Mutations affecting some components of the H/ACA snoRNPs have been identified for each of these forms. Hence, the X-linked form involves mutations in the enzyme CBF5 [191], while the autosomal dominant form has been associated with mutations in genes encoding the telomerase reverse transcriptase (TERT) of the telomerase complex and its telomerase RNA component (TERC) RNA. In vertebrates, the 3′-region of TERC RNA carries a H/ACA motif that allows the recruitment of constituent proteins of the box H/ACA snoRNP. Although this box H/ACA particle has not yet been associated with any pseudouridylation reaction, it is essential for the maturation of RNA TERC and the activity of the telomerase complex [192, 193]. Mutations in the genes encoding NHP2, NOP10 and TERT have also been identified in cases of the autosomal recessive form of DC. Given that some of these mutations affect the box H/ACA snoRNPs, the box H/ACA scaRNPs and the telomerase complex, it was difficult to determine the contribution of each of the generated defects in the development of DC pathology. As most patients with DC show shortened telomeres, it is now well accepted that this disease is caused primarily by a defect in telomere maintenance due to decrease of telomerase activity and/or stability [194]. However, it is also very likely that the mutations located in the dyskerin, NOP10 and NHP2 proteins lead to a lower level of H/ACA snoRNPs and thus to defects in the ribosome biogenesis, which would increase the symptoms of the disease.

References

1 Limbach, P.A., Crain, P.F., McCloskey, J.A. (1994) Summary: the modified nucleosides of RNA. *Nucleic Acids Res.*, **22**, 2183–2196.
2 Czerwoniec, A., Dunin-Horkawicz, S., Purta, E., Kaminska, K.H. et al. (2009) MODOMICS: a database of RNA modification pathways. 2008 update. *Nucleic Acids Res.*, **37**, D118–D121.
3 Machnicka, M.A., Milanowska, K., Osman Oglou, O., Purta, E. et al. (2013) MODOMICS: a database of RNA modification pathways – 2013 update. *Nucleic Acids Res.*, **41**, D262–D267.
4 Motorin, Y., Grosjean, H. (2005) Transfer RNA Modification, in Encyclopedia of Life Sciences, John Wiley & Sons, Ltd, Chichester.
5 Motorin, Y., Grosjean, H. (1998) Chemical Structures and Classification of Posttranscriptionally Modified Nucleosides in RNA, in: Grosjean, H., Benne, R. (eds), *Modification and Editing of RNA*, ASM Press, Washington, DC, pp. 543–549.
6 Massenet, S., Mougin, A., Branlant, C. (1998) Posttranscriptional Modifications in the U Small Nuclear RNAs, in: Grosjean, H., Benne, R. (eds), *Modification and Editing of RNA*, ASM Press, Washington, DC, pp. 201–228.
7 Rozenski, J., Crain, P.F., McCloskey, J.A. (1999) The RNA Modification Database: 1999 update. *Nucleic Acids Res.*, **27**, 196–197.
8 Edelheit, S., Schwartz, S., Mumbach, M.R., Wurtzel, O. et al. (2013) Transcriptome-wide mapping of 5-methylcytidine RNA modifications in bacteria, archaea, and yeast reveals m5C within archaeal mRNAs. *PLoS Genet*, **9**, e1003602.
9 Anantharaman, V., Koonin, E.V., Aravind, L. (2002) Comparative genomics and evolution of proteins involved in RNA metabolism. *Nucleic Acids Res.*, **30**, 1427–1464.
10 Ferré-D'Amaré, A.R. (2003) RNA-modifying enzymes. *Curr. Opin. Struct. Biol.*, **13**, 49–55.
11 Mushegian, A.R., Koonin, E.V. (1996) A minimal gene set for cellular life derived by comparison of complete bacterial genomes. *Proc. Natl Acad. Sci. USA*, **93**, 10268–10273.
12 Björk, G.R. (1995) Genetic dissection of synthesis and function of modified nucleosides in bacterial transfer RNA. *Prog. Nucleic Acids Res. Mol. Biol.*, **50**, 263–338.
13 Lane, B.G. (1998) Historical Perspectives on RNA Nucleotide Modifications, in: Grosjean, H., Benne, R. (eds), *Modification and Editing of RNA*, ASM Press, Washington, DC, pp. 1–20.
14 Nishimura, S., Watanabe, K. (2006) The discovery of modified nucleosides from the early days to the present: a personal perspective. *J. Biosci.*, **31**, 465–475.
15 Sprinzl, M., Horn, C., Brown, M., Ioudovitch, A. et al. (1980) Compilation of tRNA sequences. *Nucleic Acids Res.*, **8**, r1–r22.
16 Bakin, A., Ofengand, J. (1993) Four newly located pseudouridylate residues in *Escherichia coli* 23S ribosomal RNA are all at the peptidyltransferase center: analysis by the application of a new sequencing technique. *Biochemistry*, **32**, 9754–9762.
17 Bakin, A., Ofengand, J. (1995) Mapping of the 13 pseudouridine residues in *Saccharomyces cerevisiae* small subunit ribosomal RNA to nucleotide resolution. *Nucleic Acids Res.*, **23**, 3290–3294.

18 Smith, J.E., Cooperman, B.S., Mitchell, P. (1992) Methylation sites in *Escherichia coli* ribosomal RNA: localization and identification of four new sites of methylation in 23S rRNA. *Biochemistry*, **31**, 10825–10834.

19 Motorin, Y., Muller, S., Behm-Ansmant, I., Branlant, C. (2007) Identification of modified residues in RNAs by reverse transcription-based methods. *Methods Enzymol.*, **425**, 21–53.

20 Douthwaite, S., Kirpekar, F. (2007) Identifying modifications in RNA by MALDI mass spectrometry. *Methods Enzymol.*, **425**, 3–20.

21 Kellner, S., Burhenne, J., Helm, M. (2010) Detection of RNA modifications. *RNA Biol.*, **7**, 237–247.

22 Cantara, W.A., Crain, P.F., Rozenski, J., McCloskey, J.A. et al. (2011) The RNA modification database, RNAMDB: 2011 update. *Nucleic Acids Res.*, **39**, D195–D201.

23 Piekna-Przybylska, D., Decatur, W.A., Fournier, M.J. (2008) The 3D rRNA modification maps database: with interactive tools for ribosome analysis. *Nucleic Acids Res.*, **36**, D178–D183.

24 Poulsen, H.E., Specht, E., Broedbaek, K., Henriksen, T. et al. (2012) RNA modifications by oxidation: a novel disease mechanism? *Free Radic. Biol. Med.*, **52**, 1353–1361.

25 Slany, R.K., Bösl, M., Crain, P.F., Kersten, H. (1993) A new function of S-adenosylmethionine: the ribosyl moiety of AdoMet is the precursor of the cyclopentenediol moiety of the tRNA wobble base queuine. *Biochemistry*, **32**, 7811–7817.

26 Van Lanen, S.G., Reader, J.S., Swairjo, M.A., de Crécy-Lagard, V. et al. (2005) From cyclohydrolase to oxidoreductase: discovery of nitrile reductase activity in a common fold. *Proc. Natl Acad. Sci. USA*, **102**, 4264–4269.

27 Vinayak, M., Pathak, C. (2010) Queuosine modification of tRNA: its divergent role in cellular machinery. *Biosci. Rep.*, **30**, 135–148.

28 Iwata-Reuyl, D. (2008) An embarrassment of riches: the enzymology of RNA modification. *Curr. Opin. Chem. Biol.*, **12**, 126–133.

29 Morris, R.C., Elliott, M.S. (2001) Queuosine modification of tRNA: a case for convergent evolution. *Mol. Genet. Metabol.*, **74**, 147–159.

30 Decatur, W.A., Fournier, M.J. (2003) RNA-guided nucleotide modification of ribosomal and other RNAs. *J. Biol. Chem.*, **278**, 695–698.

31 Hage, A.E., Tollervey, D. (2004) A surfeit of factors: why is ribosome assembly so much more complicated in eukaryotes than bacteria?. *RNA Biol.*, **1**, 10–15.

32 Lafontaine, D., Delcour, J., Glasser, A.L., Desgrès, J. et al. (1994) The DIM1 gene responsible for the conserved m6(2)Am6(2)A dimethylation in the 3'-terminal loop of 18 S rRNA is essential in yeast. *J. Mol. Biol.*, **241**, 492–497.

33 Lapeyre, B., Purushothaman, S.K. (2004) Spb1p-directed formation of Gm2922 in the ribosome catalytic center occurs at a late processing stage. *Mol. Cell*, **16**, 663–669.

34 Urbonavicius, J., Skouloubris, S., Myllykallio, H., Grosjean, H. (2005) Identification of a novel gene encoding a flavin-dependent tRNA:m5U methyltransferase in bacteria – evolutionary implications. *Nucleic Acids Res.*, **33**, 3955–3964.

35 Caillet, J., Droogmans, L. (1988) Molecular cloning of the *Escherichia coli* miaA gene involved in the formation of delta 2-isopentenyl adenosine in tRNA. *J. Bacteriol.*, **170**, 4147–4152.

36 Yu, F., Tanaka, Y., Yamashita, K., Suzuki, T. et al. (2011) Molecular basis of dihydrouridine formation on tRNA. *Proc. Natl Acad. Sci. USA*, **108**, 19593–19598.

37 Schaefer, M., Pollex, T., Hanna, K., Lyko, F. (2009) RNA cytosine methylation analysis by bisulfite sequencing. *Nucleic Acids Res.*, **37**, e12.

38 Pollex, T., Hanna, K., Schaefer, M. (2010) Detection of cytosine methylation in RNA using bisulfite sequencing. *Cold Spring Harb. Protoc.*, **10**, protocol 5505.

39 Behm-Ansmant, I., Helm, M., Motorin, Y. (2011) Use of specific chemical reagents for detection of modified nucleotides in RNA. *J. Nucleic Acids*, **2011**, 408053.

40 Newby, M.I., Greenbaum, N.L. (2002) Investigation of Overhauser effects between pseudouridine and water protons in RNA

41 Hall, K.B., McLaughlin, L.W. (1991) Properties of a U1/mRNA 5′ splice site duplex containing pseudouridine as measured by thermodynamic and NMR methods. *Biochemistry*, **30**, 1795–1801.
42 Newby, M.I., Greenbaum, N.L. (2001) A conserved pseudouridine modification in eukaryotic U2 snRNA induces a change in branch-site architecture. *RNA*, **7**, 833–845.
43 Auffinger, P., Westhof, E. (1996) H-bond stability in the tRNA(Asp) anticodon hairpin: 3 ns of multiple molecular dynamics simulations. *Biophys. J.*, **71**, 940–954.
44 Charette, M., Gray, M.W. (2000) Pseudouridine in RNA: what, where, how, and why. *IUBMB Life*, **49**, 341–351.
45 Kawai, G., Ue, H., Yasuda, M., Sakamoto, K. et al. (1991) Relation between functions and conformational characteristics of modified nucleosides found in tRNAs. *Nucleic Acids Symp. Ser.*, **25**, 49–50.
46 Kawai, G., Yamamoto, Y., Kamimura, T., Masegi, T. et al. (1992) Conformational rigidity of specific pyrimidine residues in tRNA arises from posttranscriptional modifications that enhance steric interaction between the base and the 2′-hydroxyl group. *Biochemistry*, **31**, 1040–1046.
47 Bachellerie, J.P., Cavaillé, J., Hüttenhofer, A. (2002) The expanding snoRNA world. *Biochimie*, **84**, 775–790.
48 Kiss, T. (2002) Small nucleolar RNAs: an abundant group of noncoding RNAs with diverse cellular functions. *Cell*, **109**, 145–148.
49 Omer, A.D., Ziesche, S., Decatur, W.A., Fournier, M.J. et al. (2003) RNA-modifying machines in archaea. *Mol. Microbiol.*, **48**, 617–629.
50 Boulon, S., Marmier-Gourrier, N., Pradet-Balade, B., Wurth, L. et al. (2008) The Hsp90 chaperone controls the biogenesis of L7Ae RNPs through conserved machinery. *J. Cell Biol.*, **180**, 579–595.
51 Schubert, H.L., Blumenthal, R.M., Cheng, X. et al. (2003) Many paths to methyltransfer: a chronicle of convergence. *Trends Biochem. Sci.*, **28**, 329–335.
52 Kozbial, P.Z., Mushegian, A.R. (2005) Natural history of S-adenosylmethionine-binding proteins. *BMC Struct. Biol.*, **5**, 19.
53 Watanabe, K. (2005) Roles of conserved amino acid sequence motifs in the SpoU (TrmH) RNA methyltransferase family. *J. Biol. Chem.*, **280**, 10368–10377.
54 Watanabe, K., Nureki, O., Fukai, S., Endo, Y. et al. (2006) Functional categorization of the conserved basic amino acid residues in TrmH (tRNA (Gm18) Methyltransferase) enzymes. *J. Biol. Chem.*, **281**, 34630–34639.
55 Cheng, X., Roberts, R.J. (2001) AdoMet-dependent methylation, DNA methyltransferases and base flipping. *Nucleic Acids Res.*, **29**, 3784–3795.
56 Hamdane, D., Argentini, M., Cornu, D., Myllykallio, H. et al. (2011) Insights into folate/FAD-dependent tRNA methyltransferase mechanism: role of two highly conserved cysteines in catalysis. *J. Biol. Chem.*, **286**, 36268–36280.
57 Vey, J.L., Drennan, C.L. (2011) Structural insights into radical generation by the radical SAM superfamily. *Chem. Rev.*, **111**, 2487–2506.
58 Hutcheson, R.U., Broderick, J.B. (2012) Radical SAM enzymes in methylation and methylthiolation. *Metallomics*, **4**, 1149–1154.
59 Grove, T.L., Benner, J.S., Radle, M.I., Ahlum, J.H. et al. (2011) A radically different mechanism for S-adenosylmethionine-dependent methyltransferases. *Science*, **332**, 604–607.
60 Ansmant, I., Motorin, I. (2001) Identification of RNA modification enzymes using sequence homology. *Mol. Biol. (Mosk.)*, **35**, 248–267.
61 Hamma, T., Ferré-D'Amaré, A.R. (2006) Pseudouridine synthases. *Chem. Biol.*, **13**, 1125–1135.
62 Roovers, M. Hale, C., Tricot, C., Terns, M.P. et al. (2006) Formation of the conserved pseudouridine at position 55 in archaeal tRNA. *Nucleic Acids Res.*, **34**, 4293–4301.
63 Watanabe, Y., Gray, M.W. (2000) Evolutionary appearance of genes encoding proteins associated with box H/ACA snoRNAs: cbf5p in *Euglena gracilis*, an early diverging eukaryote, and candidate Gar1p and Nop10p homologs in archaebacteria. *Nucleic Acids Res.*, **28**, 2342–2352.
64 Del Campo, M., Ofengand, J., Malhotra, A. (2004) Crystal structure of the catalytic

domain of RluD, the only rRNA pseudouridine synthase required for normal growth of *Escherichia coli*. *RNA*, **10**, 231–239.

65 Koonin, E.V. (1996) Pseudouridine synthases: four families of enzymes containing a putative uridine-binding motif also conserved in dUTPases and dCTP deaminases. *Nucleic Acids Res.*, **24**, 2411–2415.

66 Mueller, E.G., Ferré-D'Amaré, A.R. (2009) Pseudouridine Formation, The Most Common Transglycosylation in RNA, in: Grosjean, H. (Ed.), *DNA and RNA Modification Enzymes: Structure, Mechanism, Function and Evolution*, Landes Bioscience, Austin, TX, pp. 363–376.

67 Wrzesinski, J., Nurse, K., Bakin, A., Lane, B.G. et al. (1995) A dual-specificity pseudouridine synthase: an *Escherichia coli* synthase purified and cloned on the basis of its specificity for psi 746 in 23S RNA is also specific for psi 32 in tRNA(phe). *RNA*, **1**, 437–448.

68 Hur, S., Stroud, R.M. (2007) How U38, 39, and 40 of many tRNAs become the targets for pseudouridylation by TruA. *Mol. Cell*, **26**, 189–203.

69 Costa, A., de Barros, J.P.P., Keith, G., Baranowski, W.B. et al. (2004) Determination of queuosine derivatives by reverse-phase liquid chromatography for the hypomodification study of Q-bearing tRNAs from various mammal liver cells. *J. Chromatogr. B, Anal. Technol. Biomed. Life Sci.*, **801**, 237–247.

70 Iwata-Reuyl, D. (2003) Biosynthesis of the 7-deazaguanosine hypermodified nucleosides of transfer RNA. *Bioorg. Chem.*, **31**, 24–43.

71 Phillips, G., de Crécy-Lagard, V. (2011) Biosynthesis and function of tRNA modifications in Archaea. *Curr. Opin. Microbiol.*, **14**, 335–341.

72 Rajakovich, L.J., Tomlinson, J., Dos Santos, P.C. (2012) Functional analysis of *Bacillus subtilis* genes involved in the biosynthesis of 4-thiouridine in tRNA. *J. Bacteriol.*, **194**, 4933–4940.

73 Mueller, E.G., Palenchar, P.M., Buck, C.J. (2001) The role of the cysteine residues of ThiI in the generation of 4-thiouridine in tRNA. *J. Biol. Chem.*, **276**, 33588–33595.

74 Palenchar, P.M., Buck, C.J., Cheng, H., Larson, T.J. et al. (2000) Evidence that ThiI, an enzyme shared between thiamin and 4-thiouridine biosynthesis, may be a sulfurtransferase that proceeds through a persulfide intermediate. *J. Biol. Chem.*, **275**, 8283–8286.

75 Arragain, S., Handelman, S.K., Forouhar, F., Wei, F.Y. et al. (2010) Identification of eukaryotic and prokaryotic methylthiotransferase for biosynthesis of 2-methylthio-N6-threonylcarbamoyladenosine in tRNA. *J. Biol. Chem.*, **285**, 28425–28433.

76 Lui, L., Lowe, T. (2013) Small nucleolar RNAs and RNA-guided post-transcriptional modification. *Essays Biochem.*, **54**, 53–77.

77 Motorin, Y., Helm, M. (2011) RNA nucleotide methylation. *Wiley Interdiscip. Rev. RNA*, **2**, 611–631.

78 Wang, C., Meier, U.T. (2004) Architecture and assembly of mammalian H/ACA small nucleolar and telomerase ribonucleoproteins. *EMBO J.*, **23**, 1857–1867.

79 Ganot, P., Jady, B.E., Bortolin, M.L., Darzacq, X. et al. (1999) Nucleolar factors direct the 2′-O-ribose methylation and pseudouridylation of U6 spliceosomal RNA. *Mol. Cell. Biol.*, **19**, 6906–6917.

80 Tycowski, K.T., You, Z.H., Graham, P.J., Steitz, J.A. (1998) Modification of U6 spliceosomal RNA is guided by other small RNAs. *Mol. Cell*, **2**, 629–638.

81 Cavaillé, J., Seitz, H., Paulsen, M., Ferguson-Smith, A.C. et al. (2002) Identification of tandemly-repeated C/D snoRNA genes at the imprinted human 14q32 domain reminiscent of those at the Prader-Willi/Angelman syndrome region. *Hum. Mol. Genet.*, **11**, 1527–1538.

82 Jády, B.E., Darzacq, X., Tucker, K., Matera, A.G. et al. (2003) Modification of Sm small nuclear RNAs occurs in the nucleoplasmic Cajal body following import from the cytoplasm. *EMBO J.*, **22**, 1878–1888.

83 Tycowski, K.T., Smith, C.M., Shu, M.D., Steitz, J.A. (1996) A small nucleolar RNA requirement for site-specific ribose methylation of rRNA in *Xenopus*. *Proc. Natl Acad. Sci. USA*, **93**, 14480–14485.

84 Cavaillé, J., Nicoloso, M., Bachellerie, J.P. (1996) Targeted ribose methylation of RNA in vivo directed by tailored antisense RNA guides. *Nature*, **383**, 732–735.

85 Weinstein, L.B., Steitz, J.A. (1999) Guided tours: from precursor snoRNA to functional snoRNP. *Curr. Opin. Cell Biol.*, **11**, 378–384.

86 Watkins, N.J., Ségault, V., Charpentier, B., Nottrott, S. et al. (2000) A common core RNP structure shared between the small nucleolar box C/D RNPs and the spliceosomal U4 snRNP. *Cell*, **103**, 457–466.

87 Meier, U.T., Blobel, G. (1994) NAP57, a mammalian nucleolar protein with a putative homolog in yeast and bacteria. *J. Cell Biol.*, **127**, 1505–1514.

88 Muller, S., Fourmann, J.B., Loegler, C., Charpentier, B et al. (2007) Identification of determinants in the protein partners aCBF5 and aNOP10 necessary for the tRNA:Psi55-synthase and RNA-guided RNA:Psi-synthase activities. *Nucleic Acids Res.*, **35**, 5610–5624.

89 Charpentier, B., Fourmann, J.B., Branlant, C. (2007) Reconstitution of archaeal H/ACA sRNPs and test of their activity. *Methods Enzymol.*, **425**, 389–405.

90 Baker, D.L., Youssef, O.A., Chastkofsky, M.I., Dy, D.A. et al. (2005) RNA-guided RNA modification: functional organization of the archaeal H/ACA RNP. *Genes Dev.*, **19**, 1238–1248.

91 Karijolich, J., Stephenson, D., Yu, Y.T. (2007) Biochemical purification of box H/ACA RNPs involved in pseudouridylation. *Methods Enzymol.*, **425**, 241–262.

92 Li, S., Duan, J., Li, D., Yang, B. et al. (2011) Reconstitution and structural analysis of the yeast box H/ACA RNA-guided pseudouridine synthase. *Genes Dev.*, **25**, 2409–2421.

93 Boschi-Muller, S., Motorin, Y. (2013) Chemistry enters nucleic acids biology: enzymatic mechanisms of RNA modification. *Biochemistry (Mosc.), Special issue, Biol. Chem. Rev.*, **78** pp. 1392–1404 (in press).

94 Decroly, E., Debarnot, C., Ferron, F., Bouvet, M. et al. (2011) Crystal structure and functional analysis of the SARS-coronavirus RNA Cap 2′-O-methyltransferase nsp10/nsp16 complex. *PLoS Pathog.*, **7**, e1002059.

95 Foster, P.G., Nunes, C.R., Greene, P., Moustakas, D. et al. (2003) The first structure of an RNA m5C methyltransferase, Fmu, provides insight into catalytic mechanism and specific binding of RNA substrate. *Structure*, **11**, 1609–1620.

96 Hodel, A.E., Gershon, P.D., Quiocho, F.A. (1998) Structural basis for sequence-nonspecific recognition of 5′-capped mRNA by a cap-modifying enzyme. *Mol. Cell*, **1**, 443–447.

97 Feder, M., Pas, J., Wyrwicz, L.S., Bujnicki, J.M. (2003) Molecular phylogenetics of the RrmJ/fibrillarin superfamily of ribose 2′-O-methyltransferases. *Gene*, **302**, 129–138.

98 Aittaleb, M., Rashid, R., Chen, Q., Palmer, J.R. et al. (2003) Structure and function of archaeal box C/D sRNP core proteins. *Nat. Struct. Biol.*, **10**, 256–263.

99 Huang, L., Pookanjanatavip, M., Gu, X., Santi, D.V. (1998) A conserved aspartate of tRNA pseudouridine synthase is essential for activity and a probable nucleophilic catalyst. *Biochemistry*, **37**, 344–351.

100 Hamilton, C.S., Greco, T.M., Vizthum, C.A., Ginter, J.M. et al. (2006) Mechanistic investigations of the pseudouridine synthase RluA using RNA containing 5-fluorouridine. *Biochemistry*, **45**, 12029–12038.

101 Agris, P.F. (1996) The importance of being modified: roles of modified nucleosides and Mg^{2+} in RNA structure and function. *Prog. Nucleic Acids Res. Mol. Biol.*, **53**, 79–129.

102 Agris, P.F. (2004) Decoding the genome: a modified view. *Nucleic Acids Res.*, **32**, 223–238.

103 Gustilo, E.M., Vendeix, F.A., Agris, P.F. (2008) tRNA's modifications bring order to gene expression. *Curr. Opin. Microbiol.*, **11**, 134–140.

104 Helm, M. (2006) Post-transcriptional nucleotide modification and alternative folding of RNA. *Nucleic Acids Res.*, **34**, 721–733.

105 Helm, M., Giegé, R., Florentz, C. (1999) A Watson–Crick base-pair-disrupting methyl group (m1A9) is sufficient for cloverleaf folding of human mitochondrial tRNALys. *Biochemistry*, **38**, 13338–13346.

106 Helm, M., Brulé, H., Degoul, F., Cepanec, C. et al. (1998) The presence of modified nucleotides is required for cloverleaf folding of a human mitochondrial tRNA. *Nucleic Acids Res.*, **26**, 1636–1643.

107 Voigts-Hoffmann, F., Hengesbach, M., Kobitski, A.Y., van Aerschot, A. et al. (2007) A methyl group controls conformational equilibrium in human mitochondrial tRNA(Lys). *J. Am. Chem. Soc.*, **129**, 13382–13383.

108 Sakurai, M., Ohtsuki, T., Watanabe, K. (2005) Modification at position 9 with 1-methyladenosine is crucial for structure and function of nematode mitochondrial tRNAs lacking the entire T-arm. *Nucleic Acids Res.*, **33**, 1653–1661.

109 Steinberg, S., Cedergren, R. (1995) A correlation between N2-dimethylguanosine presence and alternate tRNA conformers. *RNA*, **1**, 886–891.

110 Motorin, Y., Helm, M. (2010) tRNA stabilization by modified nucleotides. *Biochemistry*, **49**, 4934–4944.

111 Serebrov, V., Vassilenko, K., Kholod, N., Gross, H.J. et al. (1998) Mg^{2+} binding and structural stability of mature and in vitro synthesized unmodified *Escherichia coli* tRNAPhe. *Nucleic Acids Res.*, **26**, 2723–2728.

112 Alexandrov, A., Chernyakov, I., Gu, W., Hiley, S.L. et al. (2006) Rapid tRNA decay can result from lack of nonessential modifications. *Mol. Cell*, **21**, 87–96.

113 Horie, N., Hara-Yokoyama, M., Yokoyama, S., Watanabe, K. et al. (1985) Two tRNAIle1 species from an extreme thermophile, *Thermus thermophilus* HB8: effect of 2-thiolation of ribothymidine on the thermostability of tRNA. *Biochemistry*, **24**, 5711–5715.

114 Shigi, N., Sakaguchi, Y., Suzuki, T., Watanabe, K. (2006) Identification of two tRNA thiolation genes required for cell growth at extremely high temperatures. *J. Biol. Chem.*, **281**, 14296–14306.

115 Bilbille, Y., Vendeix, F.A., Guenther, R., Malkiewicz, A. et al. (2009) The structure of the human tRNALys3 anticodon bound to the HIV genome is stabilized by modified nucleosides and adjacent mismatch base pairs. *Nucleic Acids Res.*, **37**, 3342–3353.

116 Cantara, W.A., Murphy, F.V. IV, Demirci, H., Agris, P.F. (2013) Expanded use of sense codons is regulated by modified cytidines in tRNA. *Proc. Natl Acad. Sci. USA*, **110**, 10964–10969.

117 Murphy, F.V. IV, Ramakrishnan, V., Malkiewicz, A., Agris, P.F. (2004) The role of modifications in codon discrimination by tRNA(Lys)UUU. *Nat. Struct. Mol. Biol.*, **11**, 1186–1191.

118 Vendeix, F.A.P., Murphy, F.V. IV, Cantara, W.A., Leszczyńska, G. et al. (2012) Human tRNA(Lys3)(UUU) is pre-structured by natural modifications for cognate and wobble codon binding through keto-enol tautomerism. *J. Mol. Biol.*, **416**, 467–485.

119 Cabello-Villegas, J., Nikonowicz, E.P. (2005) Solution structure of psi32-modified anticodon stem-loop of *Escherichia coli* tRNAPhe. *Nucleic Acids Res.*, **33**, 6961–6971.

120 Davis, D.R. (1995) Stabilization of RNA stacking by pseudouridine. *Nucleic Acids Res.*, **23**, 5020–5026.

121 Rife, J.P., Moore, P.B. (1998) The structure of a methylated tetraloop in 16S ribosomal RNA. *Structure*, **6**, 747–756.

122 Watanabe, K., Oshima, T., Iijima, K., Yamaizumi, Z. et al. (1980) Purification and thermal stability of several amino acid-specific tRNAs from an extreme thermophile, *Thermus thermophilus* HB8. *J. Biochem. (Tokyo)*, **87**, 1–13.

123 Yokoyama, S., Watanabe, K., Miyazawa, T. (1987) Dynamic structures and functions of transfer ribonucleic acids from extreme thermophiles. *Adv. Biophys.*, **23**, 115–147.

124 Kowalak, J.A., Dalluge, J.J., McCloskey, J.A., Stetter, K.O. (1994) The role of post-transcriptional modification in stabilization of transfer RNA from hyperthermophiles. *Biochemistry*, **33**, 7869–7876.

125 Yarian, C., Townsend, H., Czestkowski, W., Sochacka, E. et al. (2002) Accurate translation of the genetic code depends on tRNA modified nucleosides. *J. Biol. Chem.*, **277**, 16391–16395.

126 Hagervall, T.G., Ericson, J.U., Esberg, K.B., Li, J.N. et al. (1990) Role of tRNA modification in translational fidelity. *Biochim. Biophys. Acta*, **1050**, 263–266.

127 Maynard, N.D., Macklin, D.N., Kirkegaard, K., Covert, M.W. (2012) Competing pathways control host resistance to virus via tRNA modification and programmed ribosomal frameshifting. *Mol. Syst. Biol.*, **8**, 567.

128 Brimacombe, R., Mitchell, P., Osswald, M., Stade, K. et al. (1993) Clustering of modified nucleotides at the functional center of bacterial ribosomal RNA. *FASEB J.*, **7**, 161–167.

129 Lane, B.G., Ofengand, J., Gray, M.W. (1995) Pseudouridine and O2′-methylated nucleosides. Significance of their selective occurrence in rRNA domains that function in ribosome-catalyzed synthesis of the peptide bonds in proteins. *Biochimie*, **77**, 7–15.

130 Jenner, L., Melnikov, S., Garreau de Loubresse, N., Ben-Shem, A. et al. (2012) Crystal structure of the 80S yeast ribosome. *Curr. Opin. Struct. Biol.*, **22**, 759–767.

131 Melnikov, S., Ben-Shem, A., Garreau de Loubresse, N., Jenner, L. et al. (2012) One core, two shells: bacterial and eukaryotic ribosomes. *Nat. Struct. Mol. Biol.*, **19**, 560–567.

132 King, T.H., Liu, B., McCully, R.R., Fournier, M.J. (2003) Ribosome structure and activity are altered in cells lacking snoRNPs that form pseudouridines in the peptidyl transferase center. *Mol. Cell*, **11**, 425–435.

133 Liang, X.-H., Liu, Q., Fournier, M.J. (2009) Loss of rRNA modifications in the decoding center of the ribosome impairs translation and strongly delays pre-rRNA processing. *RNA*, **15**, 1716–1728.

134 Baudin-Baillieu, A., Fabret, C., Liang, X.H., Piekna-Przybylska, D. et al. (2009) Nucleotide modifications in three functionally important regions of the Saccharomyces cerevisiae ribosome affect translation accuracy. *Nucleic Acids Res.*, **37**, 7665–7677.

135 Baxter-Roshek, J.L., Petrov, A.N., Dinman, J.D. (2007) Optimization of ribosome structure and function by rRNA base modification. *PloS ONE*, **2**, e174.

136 Vester, B., Long, K.S. (2000) Antibiotic Resistance in Bacteria Caused by Modified Nucleosides in 23S Ribosomal RNA. http://www.ncbi.nlm.nih.gov/books/NBK6514/ (accessed 18 September 2013).

137 Chow, C.S., Lamichhane, T.N., Mahto, S.K. (2007) Expanding the nucleotide repertoire of the ribosome with post-transcriptional modifications. *ACS Chem. Biol.*, **2**, 610–619.

138 Toh, S.-M., Mankin, A.S. (2008) An indigenous posttranscriptional modification in the ribosomal peptidyl transferase center confers resistance to an array of protein synthesis inhibitors. *J. Mol. Biol.*, **380**, 593–597.

139 Soma, A., Ikeuchi, Y., Kanemasa, S., Kobayashi, K. et al. (2003) An RNA-modifying enzyme that governs both the codon and amino acid specificities of isoleucine tRNA. *Mol. Cell*, **12**, 689–698.

140 Suzuki, T., Miyauchi, K. (2010) Discovery and characterization of tRNAIle lysidine synthetase (TilS). *FEBS Lett.*, **584**, 272–277.

141 Mandal, D., Köhrer, C., Su, D., Russell, S.P. et al. (2010) Agmatidine, a modified cytidine in the anticodon of archaeal tRNA(Ile), base pairs with adenosine but not with guanosine. *Proc. Natl Acad. Sci. USA*, **107**, 2872–2877.

142 Grosjean, H. (Ed.) (2009) *DNA and RNA Modification Enzymes: Structure, Mechanism, Function and Evolution*, Landes Bioscience, Austin, TX.

143 Senger, B., Auxilien, S., Englisch, U., Cramer, F. et al. (1997) The modified wobble base inosine in yeast tRNAIle is a positive determinant for aminoacylation by isoleucyl-tRNA synthetase. *Biochemistry*, **36**, 8269–8275.

144 Glasser, A.L., Desgres, J., Heitzler, J., Gehrke, C.W. et al. (1991) O-ribosyl-phosphate purine as a constant modified nucleotide located at position 64 in cytoplasmic initiator tRNAs(Met) of yeasts. *Nucleic Acids Res.*, **19**, 5199–5203.

145 Kolitz, S.E., Lorsch, J.R. (2010) Eukaryotic initiator tRNA: finely tuned and ready for action. *FEBS Lett.*, **584**, 396–404.

146 Lu, J., Huang, B., Esberg, A., Johansson, M.J. et al. (2005) The *Kluyveromyces lactis* gamma-toxin targets tRNA anticodons. *RNA*, **11**, 1648–1654.

147 Isel, C., Marquet, R., Keith, G., Ehresmann, C. et al. (1993) Modified nucleotides of tRNA(3Lys) modulate primer/template loop-loop interaction in the initiation complex of HIV-1 reverse transcription. *J. Biol. Chem.*, **268**, 25269–25272.

148 Isel, C., Lanchy, J.M., Le Grice, S.F., Ehresmann, C. et al. (1996) Specific initiation and switch to elongation of human immunodeficiency virus type 1 reverse transcription require the post-transcriptional modifications of primer tRNA3Lys. *EMBO J.*, **15**, 917–924.

149 Auxilien, S., Keith, G., Le Grice, S.F., Darlix, J.L. (1999) Role of post-transcriptional modifications of primer tRNALys,3 in the fidelity and efficacy of plus strand DNA transfer during HIV-1 reverse transcription. *J. Biol. Chem.*, **274**, 4412–4420.

150 Marshak-Rothstein, A. (2006) Toll-like receptors in systemic autoimmune disease. *Nat. Rev. Immunol.*, **6**, 823–835.

151 Gehrig, S., Eberle, M.E., Botschen, F., Rimbach, K. et al. (2012) Identification of modifications in microbial, native tRNA

that suppress immunostimulatory activity. *J. Exp. Med.*, **209**, 225–233.

152 Robbins, M., Judge, A., MacLachlan, I. (2009) siRNA and innate immunity. *Oligonucleotides*, **19**, 89–102.

153 Judge, A., MacLachlan, I. (2008) Overcoming the innate immune response to small interfering RNA. *Hum. Gene Ther.*, **19**, 111–124.

154 Saletore, Y., Meyer, K., Korlach, J., Vilfan, I.D. et al. (2012) The birth of the Epitranscriptome: deciphering the function of RNA modifications. *Genome Biol.*, **13**, 175.

155 Saletore, Y., Chen-Kiang, S., Mason, C.E. (2013) Novel RNA regulatory mechanisms revealed in the epitranscriptome. *RNA Biol.*, **10**, 342–346.

156 Yi, C., Pan, T. (2011) Cellular dynamics of RNA modification. *Acc. Chem. Res.*, **44**, 1380–1388.

157 Falnes, P.Ø., Klungland, A., Alseth, I. (2007) Repair of methyl lesions in DNA and RNA by oxidative demethylation. *Neuroscience*, **145**, 1222–1232.

158 van den Born, E., Omelchenko, M.V., Bekkelund, A., Leihne, V. et al. (2008) Viral AlkB proteins repair RNA damage by oxidative demethylation. *Nucleic Acids Res.*, **36**, 5451–5461.

159 Jia, G., Yang, C.G., Yang, S., Jian, X. et al. (2008) Oxidative demethylation of 3-methylthymine and 3-methyluracil in single-stranded DNA and RNA by mouse and human FTO. *FEBS Lett.*, **582**, 3313–3319.

160 Fu, Y., Jia, G., Pang, X., Wang, R.N. et al. (2013) FTO-mediated formation of N6-hydroxymethyladenosine and N6-formyladenosine in mammalian RNA. *Nat. Commun.*, **4**, 1798.

161 Jia, G., Fu, Y., He, C. (2013) Reversible RNA adenosine methylation in biological regulation. *Trends Genet.*, **29**, 108–115.

162 Niu, Y., Zhao, X., Wu, Y.S., Li, M.M. et al. (2013) N6-methyl-adenosine (m6A) in RNA: an old modification with a novel epigenetic function. *Genomics Proteomics Bioinformatics*, **11**, 8–17.

163 Zheng, G., Dahl, J.A., Niu, Y., Fedorcsak, P. et al. (2013) ALKBH5 is a mammalian RNA demethylase that impacts RNA metabolism and mouse fertility. *Mol. Cell*, **49**, 18–29.

164 Wu, G., Xiao, M., Yang, C., Yu, Y.T. (2011) U2 snRNA is inducibly pseudouridylated at novel sites by Pus7p and snR81 RNP. *EMBO J.*, **30**, 79–89.

165 Meier, U.T. (2011) Pseudouridylation goes regulatory. *EMBO J.*, **30**, 3–4.

166 Begley, U., Dyavaiah, M., Patil, A., Rooney, J.P. et al. (2007) Trm9-catalyzed tRNA modifications link translation to the DNA damage response. *Mol. Cell*, **28**, 860–870.

167 Patil, A., Chan, C.T., Dyavaiah, M., Rooney, J.P. et al. (2012) Translational infidelity-induced protein stress results from a deficiency in Trm9-catalyzed tRNA modifications. *RNA Biol.*, **9**, 990–1001.

168 Begley, U., Sosa, M.S., Avivar-Valderas, A., Patil, A. et al. (2013) A human tRNA methyltransferase 9-like protein prevents tumour growth by regulating LIN9 and HIF1-α. *EMBO Mol. Med.*, **5**, 366–383.

169 Zinshteyn, B., Gilbert, W.V. (2013) Loss of a conserved tRNA anticodon modification perturbs cellular signaling. *PLoS Genet.*, **9**, e1003675.

170 Chan, C.T.Y., Pang, Y.L., Deng, W., Babu, I.R. et al. (2012) Reprogramming of tRNA modifications controls the oxidative stress response by codon-biased translation of proteins. *Nat. Commun.*, **3**, 937.

171 Dirheimer, G., Baranowski, W., Keith, G. (1995) Variations in tRNA modifications, particularly of their queuine content in higher eukaryotes. Its relation to malignancy grading. *Biochimie*, **77**, 99–103.

172 Nicholls, T.J., Rorbach, J., Minczuk, M. (2013) Mitochondria: mitochondrial RNA metabolism and human disease. *Int. J. Biochem. Cell Biol.*, **45**, 845–849.

173 Shutt, T.E., Shadel, G.S. (2010) Inventory of the human mitochondrial gene expression machinery with links to disease. *Environ. Mol. Mutagen.*, **51**, 360–379.

174 Suzuki, T., Suzuki, T., Wada, T., Saigo, K. et al. (2002) Taurine as a constituent of mitochondrial tRNAs: new insights into the functions of taurine and human mitochondrial diseases. *EMBO J.*, **21**, 6581–6589.

175 Yasukawa, T., Suzuki, T., Ishii, N., Ueda, T. et al. (2000) Defect in modification at the anticodon wobble nucleotide of mitochondrial tRNA(Lys) with the MERRF encephalomyopathy pathogenic mutation. *FEBS Lett.*, **467**, 175–178.

176 Kirino, Y., Yasukawa, T., Ohta, S., Akira, S. et al. (2004) Codon-specific translational defect caused by a wobble modification

176 deficiency in mutant tRNA from a human mitochondrial disease. *Proc. Natl Acad. Sci. USA*, **101**, 15070–15075.

177 Umeda, N., Suzuki, T., Yukawa, M., Ohya, Y. et al. (2005) Mitochondria-specific RNA-modifying enzymes responsible for the biosynthesis of the wobble base in mitochondrial tRNAs. Implications for the molecular pathogenesis of human mitochondrial diseases. *J. Biol. Chem.*, **280**, 1613–1624.

178 Sasarman, F., Antonicka, H., Horvath, R., Shoubridge, E.A. (2011) The 2-thiouridylase function of the human MTU1 (TRMU) enzyme is dispensable for mitochondrial translation. *Hum. Mol. Genet.*, **20**, 4634–4643.

179 Patton, J.R., Bykhocskaya, Y., Mengesha, E., Bertolotto, C. et al. (2005) Mitochondrial myopathy and sideroblastic anemia (MLASA): missense mutation in the pseudouridine synthase 1 (PUS1) gene is associated with the loss of tRNA pseudouridylation. *J. Biol. Chem.*, **280**, 19823–19828.

180 Zeharia, A., Fischel-Ghodsian, N., Casas, K., Bykhocskaya, Y. et al. (2005) Mitochondrial myopathy, sideroblastic anemia, and lactic acidosis: an autosomal recessive syndrome in Persian Jews caused by a mutation in the PUS1 gene. *J. Child Neurol.*, **20**, 449–452.

181 Fernandez-Vizarra, E., Berardinelli, A., Valente, L., Tiranti, V. et al. (2007) Nonsense mutation in pseudouridylate synthase 1 (PUS1) in two brothers affected by myopathy, lactic acidosis and sideroblastic anaemia (MLASA). *J. Med. Genet.*, **44**, 173–180.

182 Sibert, B.S., Patton, J.R. (2012) Pseudouridine synthase 1: a site-specific synthase without strict sequence recognition requirements. *Nucleic Acids Res.*, **40**, 2107–2118.

183 Bykhovskaya, Y., Mengesha, E., Fischel-Ghodsian, N. (2007) Pleiotropic effects and compensation mechanisms determine tissue specificity in mitochondrial myopathy and sideroblastic anemia (MLASA). *Mol. Genet. Metabol.*, **91**, 148–156.

184 Bykhovskaya, Y., Mengesha, E., Fischel-Ghodsian, N. (2009) Phenotypic expression of maternally inherited deafness is affected by RNA modification and cytoplasmic ribosomal proteins. *Mol. Genet. Metabol.*, **97**, 297–304.

185 Wei, F.-Y., Suzuki, T., Watanabe, S., Kimura, S. et al. (2011) Deficit of tRNA(Lys) modification by Cdkal1 causes the development of type 2 diabetes in mice. *J. Clin. Invest.*, **121**, 3598–3608.

186 Wei, F.-Y., Tomizawa, K. (2011) Functional loss of Cdkal1, a novel tRNA modification enzyme, causes the development of type 2 diabetes. *Endocr. J.*, **58**, 819–825.

187 Reiter, V., Matschkal, D.M., Wagner, M., Globisch, D. et al. (2012) The CDK5 repressor CDK5RAP1 is a methylthiotransferase acting on nuclear and mitochondrial RNA. *Nucleic Acids Res.*, **40**, 6235–6240.

188 Kirwan, M., Dokal, I. (2008) Dyskeratosis congenita: a genetic disorder of many faces. *Clin. Genet.*, **73**, 103–112.

189 Walne, A.J., Dokal, I. (2009) Advances in the understanding of dyskeratosis congenita. *Br. J. Haematol.*, **145**, 164–172.

190 Ruggero, D., Grisendi, S., Piazza, F., Rego, E. et al. (2003) Dyskeratosis congenita and cancer in mice deficient in ribosomal RNA modification. *Science*, **299**, 259–262.

191 Zucchini, C., Strippoli, P., Biolchi, A., Solmi, R. et al. (2003) The human TruB family of pseudouridine synthase genes, including the Dyskeratosis Congenita 1 gene and the novel member TRUB1. *Int. J. Mol. Med.*, **11**, 697–704.

192 Mitchell, J.R., Cheng, J., Collins, K. (1999) A box H/ACA small nucleolar RNA-like domain at the human telomerase RNA 3' end. *Mol. Cell. Biol.*, **19**, 567–576.

193 Mitchell, J.R., Wood, E., Collins, K. (1999) A telomerase component is defective in the human disease dyskeratosis congenita. *Nature*, **402**, 551–555.

194 Nelson, N.D., Bertuch, A.A. (2012) Dyskeratosis congenita as a disorder of telomere maintenance. *Mutat. Res.*, **730**, 43–51.

10
Regulation of Gene Expression*

Anil Kumar¹, Sarika Garg¹,², and Neha Garg¹,³
¹*Devi Ahilya University, School of Biotechnology, Khandwa Road, Indore 452001, India*
²*Max Planck Unit for Structural Molecular Biology, C/O DESY, Gebäude 25b, Notkestrasse 85, 22607 Hamburg, Germany*
Present address: University of Saskatchewan, Department of Psychiatry, Rm B45 HSB, 107 Wiggins Road, Saskatoon, SK S7N 5E5, Canada
³*Barkatullah University, Biotechnology Department, Bhopal, 462026, India*

1	Introduction	291
2	**Regulation of Gene Expression in Prokaryotes**	**292**
2.1	Induction and Repression	292
2.2	The Operon	293
2.2.1	The Lactose Operon (lac Operon)	294
2.2.2	The Histidine Operon	299
2.2.3	The Tryptophan Operon	299
2.2.4	The Arabinose Operon (ara Operon)	302
2.3	Positive and Negative Control	303
2.4	Attenuation: The Leader Sequence	303
2.5	Catabolite Repression	305
2.6	Cyclic AMP Receptor Protein	306
2.7	Guanosine-5′-Diphosphate,3′-Diphosphate	309
2.8	Riboswitch	309
2.9	Regulon	309
3	**Regulation of Gene Expression in Eukaryotes**	**311**
3.1	Transcriptionally Active Chromatin	316
3.2	Regulation of Gene Expression at the Initiation of Transcription	317
3.3	Regulation of Gene Expression in Chloroplasts	320
3.4	Regulation of Gene Expression in Mitochondria	321

*This is an updated version of a chapter previously published in: Meyers, R.A. (Ed.) Epigenetic Regulation and Epigenomics, 2013, ISBN 978-3-527-32682-2.

RNA Regulation: Advances in Molecular Biology and Medicine, First Edition. Edited by Robert A. Meyers.
© 2014 Wiley-VCH Verlag GmbH & Co. KGaA. Published 2014 by Wiley-VCH Verlag GmbH & Co. KGaA.

4	**RNA Splicing** 322
4.1	Nuclear Splicing 323
4.2	Splicing Pathways 325
4.2.1	Spliceosomal Introns 325
4.2.2	Spliceosome Formation and Activity 326
4.2.3	Self-Splicing 326
4.2.4	tRNA Splicing 328
4.3	cis- and trans-Splicing Reactions 330
4.4	Alternate Splicing 330

5	**Role of microRNAs (miRNAs) in the Regulation of Gene Expression** 332
6	**Chromatin Structure and the Control of Gene Expression** 333
7	**Epigenetic Control of Gene Expression** 335
8	**Gene Regulation by Hormonal Action** 336
9	**Post-Transcriptional Regulation of mRNA** 337
10	**Transport of Processed mRNA to the Cytoplasm** 338
11	**Regulation of Gene Expression at the Level of Translation** 339
	Acknowledgments 340
	References 341

Keywords

Alternate splicing
This occurs when after splicing, a single gene gives rise to more than one mRNA sequence. It may be due to the joining of exons in different series. Occasionally, HnRNA may splice differently (a portion of sequence may act as the intron in one case, and as the exon in another case).

Alzheimer's disease
A neurodegenerative disorder that leads to the irreversible loss of neurons and dementia. The apparent symptoms are progressive impairment in memory, judgment, decision making, orientation to physical surroundings, and language.

Attenuation
A mechanism that controls RNA polymerase to read through an attenuator, an intrinsic termination sequence that is present at the start of the transcription unit. This type of control is present in some prokaryotic operons.

Bromodomain
A protein domain of about 110 amino acids that recognizes acetylated lysine residues, such as those on the N-terminal tails of histones. This recognition is often a prerequisite

for protein–histone association and chromatin remodeling. It is found in a variety of mammalians, invertebrates, and yeast DNA-binding proteins.

CAAT box
A conserved sequence located about 75 nucleotides upstream of the start point of transcription units. It is found in eukaryotes, and is also known the -75 box sequence. It is recognized by certain transcription factors, and has the consensus sequence GGCAATCT. The CAAT box plays an important role in increasing the promoter strength.

Chromodomain
A protein structural domain of about 40–50 amino acid residues, commonly found in proteins associated with the manipulation of chromatin. The domain is highly conserved among both plants and animals, and is represented in a large number of different proteins in many genomes. Some chromodomain-containing genes have multiple alternative splicing isoforms.

Coffin–Lowry syndrome
An X-linked dominant genetic disorder that causes severe mental problems. It is sometimes also associated with abnormalities of growth, cardiac abnormalities, kyphoscoliosis, as well as auditory and visual abnormalities.

Cyclic AMP receptor protein (CRP or CAP)
A regulatory protein activated by $3',5'$-cyclic AMP (cAMP). In prokaryotes, the transcription of many genes is activated after binding of this protein (in the form of a CRP–cAMP complex) at a specific site in the DNA. Two molecules of cyclic AMP bind with one molecule of CRP.

Epigenetics
The study of changes in phenotype (appearance) or gene expression, caused by mechanisms other than changes in the underlying DNA sequence.

Epigenotype
The stable pattern of gene expression outside the actual base pair sequence of DNA.

Epigenetic regulation
Cells in multicellular organisms are genetically homogeneous, but structurally and functionally heterogeneous, due to differential expression of genes, mostly during development. This differential expression is subsequently retained through mitosis. Stable alterations of this type are termed epigenetic regulation.

Exon
A segment of interrupted gene having a coding region and present in mature mRNA.

Gratuitous inducer
A substance that induces the transcription of a gene(s), but is not a substrate for its enzyme protein product. Generally, it is an analog of the substrate, a normal inducer.

Intron or intervening sequence
A segment of interrupted gene found in eukaryotes. The intron is transcribed but does not code for a protein product. Intron sequence are removed during the maturation of primary transcript; this process is termed RNA splicing.

Inducer
A small molecule that triggers the biosynthesis of RNA by binding to the cytoplasmic repressor (the product of a regulatory gene). It is generally the substrate of the enzyme protein product of the structural gene.

Induction
The ability of bacteria or yeast to synthesize certain enzymes only when their substrates are present. The inducer binds to the cytoplasmic repressor, preventing it from binding to the operator region. If a cytoplasmic repressor is already bound with the operator; it becomes detached from the operator region after binding with the inducer.

Lariat
An intermediate that is formed during RNA splicing, where a circular structure with a tail is formed by a $5', 2'$ bond.

Leader sequence
A nontranslated sequence at the $5'$ end of mRNA, preceding the initiation codon.

Myoblast
A type of progenitor cell that gives rise to myocytes.

Operator
A DNA sequence to which a cytoplasmic repressor (the protein product of a regulatory gene) binds specifically.

Polyadenylation
The addition of Poly A sequence to the $3'$ end of an eukaryotic RNA (a post-transcriptional change).

Polycistronic mRNA
An mRNA having the information for more than one protein. It is formed after the transcription of more than one gene present in a cluster (operon).

Promoter
The region of DNA involved in the binding of RNA polymerase to start RNA biosynthesis.

Regulatory gene
This gene codes for an RNA or a protein that controls the expression of other genes.

Repression
The inhibition of enzyme biosynthesis by a product of the metabolic pathway. Generally, inhibition is at the level of transcription. The product of the regulatory gene (cytoplasmic repressor) and the product of the metabolic pathway (corepressor) complex bind to the operator region on the DNA.

Rett syndrome
A neurodevelopmental disorder classified as an autism spectrum disorder. It was first described by the Austrian pediatrician, Andreas Rett, in 1966. The clinical features include a deceleration of the rate of head growth, and small hands and feet. Repetitive hand movements such as mouthing or wringing are also noted.

Riboswitch
A riboswitch is an mRNA that senses the environment directly, shutting itself down in response to particular chemical cues.

Ribozyme
RNA as an enzyme. Some RNA molecules are capable of self-RNA splicing without the involvement of any protein. This type of RNA is called a ribozyme.

RITS (RNA-induced transcriptional silencing)
A form of RNA interference by which short RNA molecules (viz., small interfering RNAs; siRNAs) trigger the downregulation of transcription of a particular gene or genomic region. RITS is generally accomplished by the post-translational modification of histone tails (e.g., methylation of lysine 9 of histone H3), which target the genomic region for heterochromatin formation. The protein complex that binds to siRNAs and interacts with the methylated lysine 9 residue of histones H3 is the RITS complex.

SnRNAs (small nuclear RNAs)
These are small RNAs present in the nucleus. They are considered to be involved in RNA splicing/other processing reactions.

SnRNPs
These are small nuclear ribonucleoproteins. Within the SnRNPs, the SnRNAs are associated with proteins.

Splicing
The removal of introns and joining of exons in RNA.

TATA box
A conserved sequence found about 25 nucleotides upstream from the start point of the eukaryotic RNA polymerase II transcription unit. It is considered to be involved in positioning RNA polymerase II for correct initiation.

Telomerase
An enzyme resembling reverse transcriptase. The action of telomerase is to add telomeres to chromosome ends.

Telomere
A specialized structure at the ends of linear eukaryotic chromosomes. Telomeres generally have many tandem copies of a short oligonucleotide sequence, $T_a G_b$ in one strand, and $C_b A_a$ in the complementary strand (where a and b are on average 1–4).

Totipotent
Under appropriate conditions, a single cell divides and produces all of the differentiated cells in an organism. These cells are termed totipotent; the phenomenon is termed totipotency.

Tropomyosin
An actin-binding protein that regulates actin mechanics. It is important for muscle contraction. Tropomyosin, along with the troponin complex, associates with actin in muscle fibers and regulates muscle contraction by regulating the binding of myosin.

Upstream
The sequences found at the 5′ end of and beyond the region of expression.

Gene expression can be regulated at the stage of transcription, RNA processing (post-transcriptional changes), and translation. In prokaryotes, the on–off of transcription serves as the main regulatory control of the gene expression whereas, in eukaryotes, more complex regulatory mechanism of transcription takes place. In addition, RNA splicing also plays a major role in the regulation of gene expression. The primary transcript of DNA has complementary sequences of both exons and introns, and is termed heterogeneous RNA (HnRNA). The HnRNA is spliced by the removal of introns and the ligation of exons. The regulation of gene expression in both prokaryotes and eukaryotes is important, as it determines whether a particular protein should be synthesized, and in what quantity. The cells of a multicellular organism are genetically homogeneous, but structurally and functionally heterogeneous, owing to the differential expression of genes. Many of these differences in gene expression arise during development, and are subsequently retained through mitosis. Stable alterations of this type are termed *epigenetic*. These alterations are heritable in the short term, but do not involve mutations of the DNA itself. The main molecular mechanisms that mediate epigenetic phenomena are DNA methylation and histone modification(s).

1
Introduction

The central dogma of gene expression is that "DNA makes the RNA, a process called *transcription*; and RNA makes the protein, a process known as *translation*" [1, 2]. Whilst in prokaryotes the cells do not have a distinct, well-defined nucleus, in eukaryotes the cells have a distinct, well-defined nucleus. Examples of prokaryotes include bacteria and blue green algae, while eukaryotes include animals, plants, and fungi [3]. In prokaryotes, the RNA primary product may itself be the target of regulation, whereas in eukaryotic cells – because of compartmentation – the transport of mRNA from the nucleus to the cytoplasm may serve as an additional target for regulation. Bacterial mRNA is directly available for protein biosynthesis soon after its synthesis, while the regulation of transcription usually occurs at the stage of initiation. At this point, it would be pertinent to mention that eukaryotic genes have been found to have both coding and noncoding sequences. In fact, as per the recently acquired human genome sequence data, more than 50% of sequences are noncoding, the function of which is unclear. These sequences are referred to as *introns* or "junk sequences," while the coding sequences are known as *exons* [4, 5]. Of course, regulatory elements also exist.

In eukaryotes, the regulation of gene expression has been shown previously to occur mainly at the level of transcription. However, more recently such regulation has also been reported to occur significantly at the translational level.

In the past, the ability to control the expression of genes in mammalian cells exogenously has served as a powerful tool in biomedical research [6, 7]. Indeed, gene regulation technology has played a key role in the efforts to understand the role of specific gene products in fundamental biological processes, in both normal development and disease states [7]. Clearly, an understanding of the role of gene regulation within the context of an entire system, in relation to disease processes, would aid in the development of therapeutic approaches. Similarly, a knowledge of gene expression and promoter control to be able to improve gene and protein networks, in conjunction with a knowledge of signal transduction, might help in the treatment of complex diseases. Consequently, attempts have been made in the following sections to describe all of the important aspects of gene regulation.

2
Regulation of Gene Expression in Prokaryotes

2.1
Induction and Repression

In prokaryotes, induction and repression – especially in the case of enzyme proteins – represent the most prominent means of regulation at the level of transcription. While certain proteins are synthesized at a constant rate at all times, other proteins – especially enzymes – are often produced in larger amounts when certain other materials are present. This type of material, which generally is the substrate of the enzyme protein, will enhance the synthesis of the enzyme and is referred to as an *inducer* of that enzyme. Consequently, the enzyme is referred to as being an *inducible enzyme*, and the whole process is known as *enzyme induction*. It is not necessary for the inducer to be the substrate of the enzyme; in fact, the inducer may simply resemble the enzyme's natural substrate, and need not necessarily be affected by it. An inducer that is not the natural substrate of the enzyme is termed a *gratuitous inducer*. In addition, if the genes expressing more than one enzyme are arranged in cluster, then a single inducer may induce expression of all the genes.

Although an inducible enzyme is normally present only in trace amounts in a bacterial cell, its concentration can be rapidly increased (by 1000-fold or more) when its substrate is present in the medium. This is particularly the case when the substrate is the sole carbon source of the cell since, under these conditions, the induced enzyme is required to transform the substrate into a metabolite that can be utilized directly by the cell. One well-studied example of an inducible enzyme is β-galactosidase from *Escherichia coli*. Those *E. coli* cells with a wild-type β-galactosidase gene are unable to utilize lactose if glucose is also present in the medium. However, if only lactose is present as sole carbon source, or when the utilization of glucose is complete, the bacterial cells will synthesize the β-galactosidase enzyme and begin to utilize lactose within only a 1–2 min period. Simultaneously, the cells will synthesize the enzyme β-galactoside permease, which is required for the transfer of β-galactoside inside the bacterial cell, as well as β-thiogalactoside transacetylase. Subsequently, if the induced bacterial cells are transferred into a medium that is deficient in lactose, the synthesis of β-galactosidase (along with β-galactoside permease and β-thiogalactoside transacetylase) will cease immediately and the previously induced enzyme will then decline to normal levels. The induction of a group of related enzymes or proteins to the same extent by a single inducing agent is termed *coordinate induction*. The case just described for *E. coli*, involving the induction of β-galactosidase, β-galactoside permease and β-thiogalactoside transacetylase, is an excellent example of coordinate induction.

Previously, two hypotheses were proposed in an attempt to explain the mechanism of enzyme induction, using the β-galactosidase system:

1. That an activation of pre-existing protein occurred.
2. That there was a *de novo* synthesis of the protein.

The first possibility was discounted because, prior to induction, no protein could be detected which had antigenic properties similar to those of β-galactosidase. This observation suggested that the actual synthesis of the enzyme protein occurred following the addition of an inducer. At the time, it was realized that a specific protein would not be synthesized except in the presence of gene that would dictate its primary structure (i.e., the amino acid sequence). Based on these observations, it was clear that the *E. coli* cells carried a structural gene for the β-galactosidase enzyme protein; hence, the question remained as to why enzyme protein synthesis did not occur in the absence of an inducer. Two explanations were proposed for this:

1. The inducer might serve as a form of template that would trigger the enzyme protein synthesis (this may have been why the inducer resembled the substrate).
2. The synthesis of the enzyme protein may be inhibited by an unknown agent(s). In this case, the inducer itself might act as a form of inhibitor, which would in turn block the activity of the inhibiting agent(s). Although, initially, this possibility seemed very complex, subsequently obtained evidence supported this hypothesis, and today this sequence of events is known to be correct.

Similarly, in the presence of a material produced by a particular enzyme reaction, the synthesis of an enzyme protein may be reduced. This phenomenon is referred to as *enzyme repression*, and the material is known as a *corepressor* [1].

Induction and repression are two complementary phenomena. Generally, biosynthetic pathways have been found to be under the control of repression (e.g., the biosynthesis of many amino acids). For example, if histidine is added to the *E. coli* growth medium, all of the enzymes involved in its biosynthesis will no longer be produced as the cells do not need to synthesize histidine. Consequently, in the presence of histidine, all of the enzymes required for its biosynthesis – starting from ATP phosphoribosyl transferase, which catalyzes the first reaction in the histidine biosynthetic pathway, namely the biosynthesis of phosphoribosyl ATP from phosphoribosyl pyrophosphate (PRPP) and ATP – are repressed. This repression of the synthesis of a group of enzymes by a single corepressor – known as *coordinate repression* – is generally caused by the end product of the biosynthetic pathway (for this reason it may also be referred to as *end-product repression*).

2.2
The Operon

The concept of the *operon* was first proposed in 1961 by Jacob and Monod. The suggestion was that genes encoding proteins with functions that were related (e.g., consecutive gene proteins in a pathway) may be organized into a cluster that, in turn, would be transcribed into a polycistronic mRNA from a single operator. The control of this operator would then allow the expression of the entire structural genes in the operon to be regulated. This unit of regulation, which contained the structural gene, regulator gene(s), and the *cis*-acting elements, was referred to as the operon. Furthermore, the genes could be classified into two groups, depending on their coded protein functions:

- *Structural genes*, the protein products of which are directly involved in metabolic activity (when they act as enzymes), or they may serve as the constituents of an organelle.

- *Regulatory genes*, the protein products of which regulate the transcription of the structural genes.

The activity of an operon is controlled by the regulatory gene(s), the protein product(s) of which interact(s) with the control elements. Although many operons have been examined in detail, one of the best-studied examples is the *lac* (lactose) operon of *E. coli*. In this operon, *lac* i is the regulatory gene, the protein product of which (known as a *cytoplasmic repressor*) is involved solely in regulation, whereas *lac* Z, *lac* Y, and *lac* A are structural genes that code for the enzymes, β-galactosidase, β-galactoside permease, and β-thiogalactoside transacetylase, respectively. Details of the *lac* operon are provided in the following subsections.

2.2.1 The Lactose Operon (lac Operon)

If *E. coli* cells are grown in a medium containing lactose as a sole carbon and energy source, the cells will synthesize the enzymes, β-galactosidase (which catalyzes the hydrolysis of lactose into glucose and galactose), β-galactoside permease (which is involved in the entry of lactose into the bacterial cell), and β-thiogalactoside transacetylase (which catalyzes the transfer of an acetyl group from acetyl CoA onto the 6-position of β-thiogalactoside, to generate 6-acetyl-β-thiogalactoside). Whilst glucose is easily metabolized and enters the glycolytic pathway directly, galactose must first be converted into glucose before such entry can be made. Studies with the *lac* operon have shown it to consist of three adjacent structural genes, *lac* Z, *lac*Y, and *lac* A; preceding these genes are *lac* O, *lac* P, and *lac* i, which have regulatory roles. Typically, *lac* i codes for a protein that serves as a cytoplasmic repressor, *lac* P is the promoter site onto which RNA polymerase binds, and *lac* O is the controlling site onto which the cytoplasmic repressor binds. Following binding, the transcription of the structural genes is switched off (Fig. 1). All three structural genes are transcribed as a single polycistronic messenger RNA carrying genetic information for the three enzyme proteins. Besides lactose, many other analogs, including isopropyl-β-thiogalactoside (IPTG), methyl-β-thiogalactoside-, and mellibiose, also act as inducers. *In vitro*, IPTG is the most commonly used inducer of the *lac* operon [8].

The genes located in the *lac* operon are:

- **Gene Z$^+$ (*lac* Z$^+$):** in mutant condition, this gene results in a loss of the ability to synthesize active β-galactosidase, either in the presence or absence of an inducer.
- **Gene Y$^+$ (*lac* Y$^+$):** in mutant condition, this gene results in a loss of the ability to synthesize active β-galactoside permease, either in the presence or absence of inducer.
- **Gene A$^+$ (*lac* A$^+$):** in mutant condition, this gene results in a loss of the ability to synthesize active β-thiogalactoside transacetylase, either in the presence or absence of inducer.
- **Gene i$^+$ (*lac* i$^+$):** this gene causes changes in the influence of the inducer on the synthesis of β-galactosidase, β-galactoside permease, and β-thiogalactoside transacetylase.

Many i mutants synthesize large amounts of the enzymes in the absence of an inducer. By using different combinations of wild-type and mutant genes of *lac* Z, *lac* Y, and *lac* i genes, different types of genetic structure of the *lac* region of the *E. coli* chromosome may be assumed (Table 1).

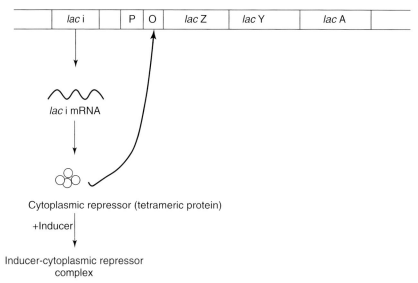

Fig. 1 Line diagram of the lac operon. P, promoter; O, operator; *lac* i, regulatory gene; *lac* Z, Y, and A, structural genes for β-galactosidase, β-galactoside permease, and β-thiogalactoside transacetylase, respectively. RNA polymerase binds on the promoter site, while the cytoplasmic repressor binds on the operator site and represses transcription of the structural genes (negative control). In the presence of the inducer, there is formation of an inducer–cytoplasmic repressor complex, which is not capable of binding on the operator site. If the cytoplasmic repressor is already bound to the operator site, it becomes detached from there after forming a complex with the inducer, and this results in transcription of the structural genes.

Regulatory gene(s) need not necessarily be located close to the structural genes; rather, the regulatory action(s) is (are) due to the biosynthesis of various intracellular substance(s). The study of mutations of the regulatory genes have provided insights into the main mechanism of induction and repression. The *E. coli* cells containing *lac* i$^+$ produce β-galactosidase only in the presence of an inducer, whereas cells containing the mutated *lac* i (referred to as *lac* i$^-$) can produce β-galactosidase both in the presence and absence of an inducer. These findings indicate that *lac* i$^+$ is dominant, while *lac* i$^-$ is recessive.

The regulatory gene, *lac* i$^+$, codes for a cytoplasmic repressor (also called lac repressor). The *lac* repressor was first isolated by Walter Gilbert and Beno Muller Hill in 1966, and is a tetrameric protein with four identical subunits each of molecular weight ca. 37 000 Da that binds specifically to the operator region. Each of the subunits is formed by a chain of 347 amino acids, in which the N-terminal amino acid is methionine and the C-terminal amino acid is glutamine. The tetrameric protein may be dissociated in the presence of sodium dodecyl sulfate. Each of the subunits has one binding site for the inducer; the subunits are also able to bind to an inducer, but not to the operator region. The binding constant for IPTG has been calculated as approximately 10^{-6} M. The cytoplasmic repressor binds very

Tab. 1 Various genotypes of the *E. coli* lactose system.

Genotype	Noninducible		Inducible	
	β-Galactosidase	β-Galactoside permease	β-Galactosidase	β-Galactoside permease
Z⇓← Y⇓← i⇓	–	–	+	+
Z•← Y⇓← i⇓	–	–	–	+
Z⇓← Y•← i⇓	–	–	+	–
Z•← Y•← i⇓	–	–	–	–
Z⇓← Y⇓← i•	+	+	+	+
Z•← Y⇓← i•	–	+	–	+
Z⇓← Y•–← i•	+	–	+	–
Z•← Y•← i•	–	–	–	–

tightly to the operator region, the equilibrium constant of the complex being about 10^{-13} M; the rate constants for association and dissociation are 7×10^9 M^{-1} s^{-1} and 6×10^{-4} s^{-1}, respectively. Although the content of the various amino acids is normal, the tryptophan content is comparatively low, with only two tryptophanyl residues among the 347 amino acids of each subunit. The OD$_{280}^{1 \text{ mg ml}^{-1}}$ is 0.59 (a comparatively low value due to the low tryptophan content) [9].

By employing techniques of circular dichroism (CD) and optical rotary dispersion, estimates have been made of the α-helix content (ca. 33–40%) and β-structure (18–42%) of the subunits. Based on the Chou–Fasman model, and the primary sequence, predictions of 37% α-helix content and 35% β-structure have been made for the subunits.

The use of electron microscopy (after negative staining) and powder X-ray diffraction (XRD) analysis indicated the tetramer to have an asymmetric dumbbell shape, with dimensions of about 45×60 Å, and with four tetramers being contained in one unit cell of 91×117 Å. Subsequent powder XRD analyses indicated third unit cell dimension (which is not seen in electron micrographs) to be 140 Å. Four tetramers could be easily packed into this cell, in a manner which accounted for the stain distribution observed on electron microscopy. The molecule was shown to extend the full length of the 140 Å cell, and to cause the tetramer to have an elongated shape with molecular dimensions of approximately $140 \times 60 \times 45$ Å. On the basis of these data, a model was proposed in which the subunits are related by 222 symmetry and placed at the corners of a rectangular plane. This suggests the existence of two operator binding sites per tetramer, if the repressor were to maintain perfect 222 symmetry [9].

The shape of the *lac* repressor in solution appears quite different, however, with the tetramer appearing as a square structure with dimensions of approximately 105×95 Å. Moreover, neither the shape nor the dimensions of the molecule were changed in the presence of IPTG. Although the subunits could be distinguished within the tetrameric structure, the poor resolution of the system meant that no decision could be made on the geometry of their arrangement.

By using X-ray crystallography, the *lac* repressor has been shown to consist of three distinct regions: (i) a core region which binds allolactose; (ii) a tetramerization region which joins four monomers in an α-helix bundle; and (iii) a DNA-binding region having a helix–turn–helix structural region that binds the operator site. The tetrameric lac repressor may be viewed as two dimers, each of which is capable of binding to a single *lac* operator. In turn, the two subunits each bind to a slightly separated major groove region of the operator.

It would appear that two different types of binding site should be present in the tetrameric structure – the first for the low-molecular-weight effectors, and the second for the *lac* operator. There are indications that both inducers and anti-inducers bind to the same site, or at least to overlapping sites; *O*-nitrophenyl β-D-fucoside has been found to act as anti-inducer for the *lac* operon. Similarly, the operator binding site involves the same region of the tetrameric repressor that binds nonoperator DNA. Each repressor subunit seems to have an effector binding site, besides contributing some interactions with operator DNA. Based on experimental evidence, it has been concluded that the effector binding site and operator binding site are distinct and nonoverlapping. Likewise, based on the results of trypsin-limited digestion studies (which split 59 N-terminal residues and 20 C-terminal residues of each subunit, leaving a tetrameric trypsin-resistant core protein composed of subunits having 60 to 327 residues), it has been shown that the N- and C-terminal residues are not required for either the binding of an inducer, nor for folding of the subunits into a correct tetrameric conformation. However, it could not be confirmed whether the interaction between the subunits of the core was identical to the interaction of the subunits in the native repressor. Whilst there are indications that terminal regions are involved in operator binding, there are also indications that effector binding changes the affinity of the repressor for its operator. Both, the N- and C-terminal regions have hydroxyl group containing amino acids (threonyl residues at positions 5, 19, 34, 315, 316, 321, and 323; seryl residues at positions 16, 21, 28, 31, 309, 325, 332, 341, and 345) which could contribute their hydroxyl groups of the side chains to form hydrogen bonds with specific groups of the bases in the *lac* operator. Among a total of eight tyrosyl residues in the chain, there were four in the 59-residue N-terminal region (at positions 7, 12, 17, and 47), which offers the possibility of interactions with DNA either by providing additional hydroxyl groups or by intercalation between the bases. In addition, eight positively charged amino acids have been found in the 59-residue N-terminal region, and six positively charged amino acids in the 36-residue C-terminal region. Thus, a higher content of positively charged amino acids is present in the terminal regions (14.7%) compared to the total percentage in the molecule (10.7%). It has been considered that a combination of the N- and C-terminal regions constitutes a basic region which, through electrostatic interactions, makes contact with the negatively charged DNA and contributes to its specific binding to the operator region. The region between 215 and 324 amino acyl residues has few charged residues (only three each positive and negatively charged residues), and is enriched in hydrophobic residues; hence, it may be involved in the stabilization of both the native tetramer and its tryptic core. This region may serve as a hydrophobic nucleus that is resistant to trypsin attack [9].

The *lac* repressor can initiate four types of interaction:

1. Specific interaction between the *lac* repressor and its operator.
2. Nonspecific interaction between the *lac* repressor and any DNA.
3. Specific interaction between the *lac* repressor and its low-molecular-weight effectors, which include inducers and anti-inducers.
4. The effector (inducer) may also interact with the *lac* repressor that is already bound to the operator, to form an intermediate ternary complex.

One molecule of the inducer IPTG was found to be sufficient to release the *lac* repressor from its specific operator. After binding of the inducer, an almost 1000-fold decrease was apparent in the affinity of the repressor for its operator. The *lac* repressor shows a single emission maximum at 338 nm in the fluorescence spectrum, which is characteristic of tryptophan. However, on the addition of an inducer at saturating concentrations, a shift in the emission maximum to 330 nm occurs, but with no change in the peak shape or fluorescence intensity. The change in emission maximum suggests that at least one tryptophanyl residue per subunit has become less accessible to the solvent upon inducer binding. The absence of any change in the shape of the fluorescence spectrum may indicate that either both tryptophanyl residues of the subunit are similarly affected, or that only one contributes to the change in the emission maximum. At least two sequential steps appear to be involved in the binding of IPTG to the *lac* repressor: (i) a bimolecular step, which is much slower than would be expected for a diffusion-controlled reaction; and (ii) a monomolecular step, which may be attributed to a conformational change in the protein. Subsequent CD studies showed that no major changes had occurred in the overall geometry of the peptide backbone of the repressor upon binding of the inducer, while sedimentation coefficient studies indicated a compactness in the protein molecule upon inducer binding. Glycerol perturbation spectra indicated that fewer aromatic residues are available to the solvent in the presence of the inducer than in the protein alone, or in the presence of anti-inducers. It has been predicted that the repressor undergoes a conformational change upon binding to its operator, and a major change in induction may be taking place upon interaction of the inducer with the repressor–operator complex.

When, subsequently, the *lac* repressor became available in pure form, it was employed to isolate the operator region. For this, DNA with a *lac* region was first fragmented into units each of almost 1000 nucleotides; the *lac* repressor was then added to the fragmented mixture and, after incubation, the reaction mixture was filtered through a cellulose nitrate membrane. Those DNA fragments without the bound *lac* repressor passed through the filter, whereas those with the bound *lac* repressor remained tightly bound to the filter membrane. The bound DNA was released from the filter by adding IPTG, after which the *lac* repressor was added and the fragments treated with deoxyribonuclease (DNase). The operator region was protected against digestion by DNase after binding the *lac* repressor, whereupon a nucleotide sequence determination revealed that the repressor had protected a total of 27 nucleotides. Moreover, this sequence has a *dyad* symmetry that is important for the specific binding of the *lac* repressor with its operator. The symmetrical sequence of the *lac* operator region is as follows:

5′TGGAATTGTGAGCGGATAACAATT3′
3′ACCTTAACACTCGCCTATTGTTAA5′

It has also been shown that allolactose may serve as an inducer for the *lac* operon. As β-galactosidase is able to convert lactose into allolactose, the genes necessary for this conversion are under control of the *lac* promoter. It has been shown that, if the number of repressor molecules in a bacterium is sufficiently low, a small proportion of the cells will have insufficient cytoplasmic repressor to inhibit the transcription. Consequently, with time, an increasing number of cells in the culture will (transiently) have no *lac* cytoplasmic repressor and will express the *lac* operon such that, under these conditions, lactose will be converted into allolactose. The latter material will then bind to the cytoplasmic repressor, resulting in an increase in the expression of the genes of the *lac* operon. Moreover, this induced state is epigenetic and somewhat heritable [1, 9, 10].

2.2.2 The Histidine Operon

The histidine operon is an example of enzyme repression. In *Salmonella*, most of the structural genes encoding the enzymes required for histidine biosynthesis are arranged in the same order as the sequence of chemical reactions catalyzed by the respective enzymes, except in one or two cases. The sequence of the genes in the histidine operon is as follows, where O denotes its operator:

EIFAHBCDGO

In total, there are nine genes that specify the structure of nine proteins involved in histidine biosynthesis. The biosynthesis of histidine, starting from PRPP and ATP, along with the genes coding for the enzymes involved, is shown in Fig. 2. Two enzymes in the pathway have been found to be bifunctional. As written above, with the exception of one or two genes, the arrangement of genes on the chromosome is related to their position in the pathway *in vivo*. This indicates that the chromosome contains a remarkable amount of information, not only of the sequence of the amino acids in the enzyme proteins but also regarding the metabolic pathway catalyzed by these enzymes [1].

2.2.3 The Tryptophan Operon

The tryptophan operon consists of five structural genes that code for the enzymes involved in the biosynthesis of tryptophan. The latter process, in addition to enzyme repression, is also controlled by feedback inhibition, with tryptophan inhibiting the activity of the first enzyme that is unique to the tryptophan biosynthetic pathway. However, the tryptophan operon is also controlled by attenuation; a line diagram of the tryptophan operon, showing the structural genes, regulatory genes and other regulatory elements, is shown in Fig. 3. The regulatory gene – which is referred to as *trp R* – produces the cytoplasmic repressor, a dimeric protein with two identical subunits each of 107 amino acids and a molecular weight of almost 12 500 Da. As in all cases of enzyme repression, in the absence of tryptophan (as corepressor) the cytoplasmic repressor does not bind with the operator region. In the presence of tryptophan, a cytoplasmic repressor–corepressor complex is formed that binds with the operator region which, in turn, is partially

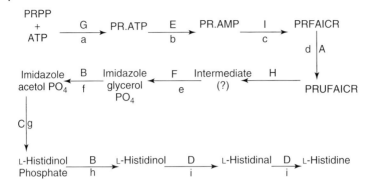

PRPP – Phosphoribosyl pyrophosphate
PRATP – Phosphoribosyl adenosine triphosphate
PRAMP – Phosphoribosyl adenosine monophosphate
PRFAICR – Phosphoribosyl formimino amino imidazole carboxamide ribonucleotide
PRUFAICR – Phosphoribulosyl formimino amino imidazole carboxamide ribonucleotide

a - ATP phosphoribosyl transferase
b - Pyrophosphohydrolase
c - Phosphoribosyl adenosine monophosphate cyclohydrolase
d - Phosphoribosyl formimino 5-amino-imidazole-4-carboxamide ribonucleotide isomerase
e - Glutamine amido transferase
f - Imidazole glycerol-3-phosphate dehydratase
g - L-Histidinol phosphate amino transferase
h - Histidinol phosphate phosphatase
i - Histidinol dehydrogenase

Fig. 2 The histidine biosynthetic pathway. The capital letters on the arrows denote the genes that encode the enzyme catalyzing the biochemical steps.

overlapped with the promoter region. The points contacted by the cytoplasmic repressor lie symmetrically and occupy the region from positions −23 to −3. The operator has a region of *dyad* symmetry, which also includes the consensus sequence of the promoter at −10. As a result, RNA polymerase is incapable of binding with the promoter region, thereby repressing transcription of the structural genes. It is clear that the different needs of induction and repression are accomplished in an almost similar manner, the difference being that the effector molecule modulates the operator binding specificity of the cytoplasmic repressor in a different way [1, 3, 11].

In the case of the tryptophan operon, a deprivation of tryptophan results in an approximately 70-fold increase in the frequency of initiation events at the tryptophan promoter. Moreover, even under repressing conditions transcription of the structural genes remains at a low level. In the case of the *lac* operon, the basal level of synthesis is only about one-thousandth of the induced level. This indicates that the efficiency of repression in the tryptophan operon is much lower than that seen in the *lac* operon [1].

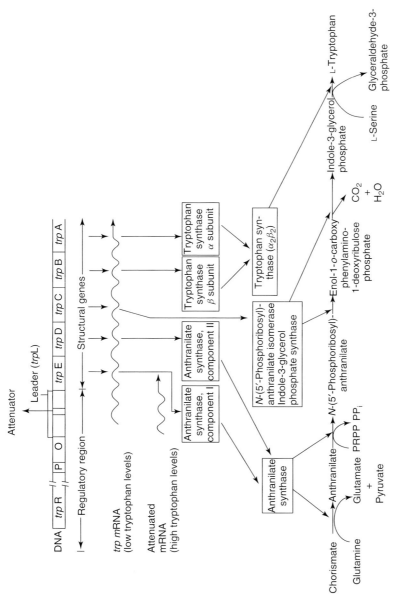

Fig. 3 Line diagram of the tryptophan operon. *trp* R, regulatory gene; P, promoter; O, operator. Between O and *trp* E is the leader sequence used in attenuation control; *trp* E codes for anthranilate synthase component I, and *trp* D for component II. The components I and II, on combination, form active anthranilate synthase. *trp* C codes for N-(5'-phosphoribosyl)-anthranilate isomerase-indole-3-glycerol phosphate synthase; *trp* B codes for the β subunit of tryptophan synthase; and *trp* A for the α subunit of tryptophan synthase. The $\alpha_2\beta_2$ complex forms active tryptophan synthase.

2.2.4 The Arabinose Operon (ara Operon)

The arabinose (ara) operon in E. coli consists of three structural genes that code for the enzymes involved in the utilization of arabinose (the bacterium can utilize arabinose as a carbon source). The ara operon is an example of both positive and negative control; a line diagram of the operon, showing the structural genes, regulatory gene, and regulatory elements, is shown in Fig. 4. The product of the ara C (regulatory) gene is referred to as *Ara C protein*, the biosynthesis of which is self-regulatory after binding with the ara O_1 operator and repressing ara C gene transcription. In general, the cell contains about 40 copies of the Ara C protein, and it acts as a positive and negative regulator for transcription of the structural genes, ara B, ara A, and ara D, which in turn code for L-ribulose kinase, L-arabinose isomerase, and L-ribulose-5-phosphate epimerase, respectively. Some regulatory DNA sequences exert their effect from a distance; these sequences are not always contiguous with the promoters, with distant DNA sequences being made closer via DNA looping mediated by specific protein–protein and protein–DNA interactions [3].

When glucose is present and arabinose absent, the Ara C protein binds to both ara O_2 and ara I, forming a DNA loop of about 210 nucleotides. Under these conditions, transcription of the structural genes is repressed. In contrast, if glucose is absent and arabinose present, then cyclic AMP (cAMP) and cyclic AMP receptor protein (CRP) become abundant, such that a complex of cAMP and CRP binds to its site adjacent to ara I. Arabinose then binds with the Ara C protein, altering its conformation; this binding causes the DNA loop to be opened, while the Ara C protein bound to ara I acts as activator and, in concurrence

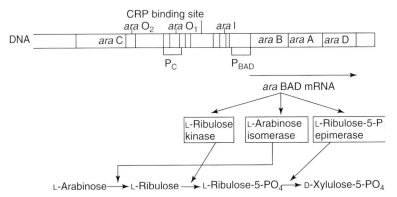

Fig. 4 Line diagram of the arabinose operon. ara C, regulatory gene; ara O_2, ara O_1 and ara I, regulatory elements to which the ara C gene product may bind; P_c, promoter for ara C gene; P_{BAD}, promoter for BAD genes; ara B, ara A and ara D, structural genes for L-ribulose kinase, L-arabinose isomerase, and L-ribulose-5-phosphate epimerase, respectively. The ara C protein regulates its own synthesis after binding on ara O_1, resulting in a repression of transcription of the ara C gene. The ara C protein acts as positive as well as negative regulator for ara BAD genes. If arabinose is absent and glucose present, ara C protein binds with ara O_2 as well as ara I to form a DNA loop, and there is repression of the ara BAD genes. If arabinose is present, cAMP–CRP becomes abundant and binds to the site adjacent to the ara I site (CRP-binding site). Arabinose also binds to the ara C protein, altering its conformation, the DNA loop is opened; the ara C protein bound on ara I then acts as an activator for the transcription of the ara BAD genes.

with the cAMP–CRP complex, induces transcription of the structural genes. Finally, if arabinose and glucose are both present, then a repression of transcription will occur, possibly due to catabolite repression caused by glucose [1, 3].

2.3
Positive and Negative Control

Positive and negative control systems can be distinguished on the basis of the mode of action of the cytoplasmic repressor. Genes under the negative control are unable to be transcribed in the presence of the product of the regulatory gene (cytoplasmic repressor), but will be transcribed in its absence. This indicates that the cytoplasmic repressor switches off the transcription, either by binding to the DNA to prevent RNA polymerase from initiating transcription, or by binding to the mRNA to prevent a ribosome from initiating translation. In fact, such negative control provides a fail-safe mechanism. The *lac* operon, as described above, represents an example of negative control.

The tryptophan operon described above represents another example of negative control, as the level of tryptophan in the cell regulates both the activity and generation of the tryptophan-synthesizing enzymes. Moreover, as tryptophan inhibits the activity of the first enzyme of the synthetic pathway, it will also inhibit the synthesis of further tryptophan. Tryptophan may also act as corepressor that activates the product of the *trp* R gene. In the presence of tryptophan, the tryptophan operon is repressed by binding of the cytoplasmic repressor (the product of the *trp* R gene) – tryptophan complex to the operator region.

Genes under the positive control are expressed only when an active regulatory protein is present. This regulatory protein acts to switch on transcription, and is thus an activator protein; such activator proteins are also referred to as *positive control factors*. The regulatory protein interacts with DNA and with RNA polymerase to assist the initiation. A positive control factor that responds to a small molecule is known as an *activator*. Unfortunately, the activator alone cannot bind to the operon; rather, it requires another molecule to be bound to the activator protein, which in turn increases the DNA-binding ability. An example of this is cAMP-activated CRP which activates the arabinose operon, which is an example of both negative and positive control [1, 11, 12].

2.4
Attenuation: The Leader Sequence

In the regulation of amino acid operons, it is generally the end product (amino acid) that acts as a corepressor to repress transcription of the structural genes. On the basis of the mechanism of enzyme repression, it was considered originally that a regulatory gene-deleted operon or operon having a mutant regulatory gene should not be under transcriptional repression. However, in already derepressed *trp* R⁻ mutants, tryptophan synthesis can be stimulated by the deprival of tryptophan, and also by an internal deletion of the region between the operator and the first structural gene. Based on these findings, a second mechanism of regulation that involved a variable, premature termination of transcription in this region was elucidated; this process was termed *attenuation*. On analysis of the early mRNA of the tryptophan operon, a part of the sequence was found to code for a short leader peptide, with any variation in translation of the leader peptide being dependent on the supply of tryptophan. The latter material influences the frequency of termination of transcription at the attenuator site, which lies still further ahead.

The process of attenuation controls the ability of the RNA polymerase to read through an *attenuator* – an intrinsic terminator located at the beginning of a transcription unit. The common feature of attenuator systems from different operons is that some external event controls the formation of the hairpin required for intrinsic termination. Typically, if the hairpin is allowed to form, then termination will prevent RNA polymerase transcribing the structural genes. However, if the hairpin is prevented from forming, then RNA polymerase will elongate through the terminator such that the structural genes are expressed. Control by attenuation requires a precise timing of the events that control termination. For example, translation of the leader peptide must occur at exactly the same time that RNA polymerase approaches the terminator site. The RNA polymerase will then remain paused until translation of the leader peptide occurs on the ribosome. Subsequently, the RNA polymerase is released and moves toward the attenuation site. In providing a mechanism to sense the inadequacy of the supply of Trp-tRNA, attenuation is able to respond directly to the needs of the cell for tryptophan in protein biosynthesis, and also employs attenuation as a control mechanism [1].

In the case of the tryptophan operon, the attenuator lies within the transcribed leader sequence of 162 nucleotides that precedes the initiation codon of *trp* E. It has a *rho*-independent termination site, and is a barrier for transcription, while a short GC-rich palindrome sequence is followed by eight successive U residues. RNA polymerase terminates at this site, producing only a 140 nucleotide mRNA. The leader region sequence contains a ribosome-binding site, and has an open-reading frame for coding a peptide of 14 amino acids called a *leader peptide*, which is unstable and has the following sequence:

Met – Lys – Ala – Ile – Phe – Val – Leu – Lys – Gly – Trp – Trp – Arg – Thr – Ser

It is clear from the sequence of the leader peptide that, among the 14 amino acids present, two are tryptophan. As tryptophan is considered to be a rare amino acid in proteins, its abundance in the leader peptide has a certain significance. For example, when the amount of tryptophan in the cell is deficient, the biosynthesis of the leader peptide on the ribosome will be stopped when the *trp* codons are reached. The sequence of the mRNA suggests that this "ribosome stalling" may in turn influence termination at the attenuator. Pairing of the regions generates the hairpin that precedes the oligo U sequence, which is a termination signal for transcription. The position of ribosome can determine which structure is formed; for example, when tryptophan is deficient in the cell the ribosomes will stall at the trp codons, which form part of region 1. Consequently, region 1 will be sequestered within the ribosome and cannot base-pair with region 2. Under these conditions region 2 will base-pair with region 3, thus compelling region 4 to remain in single-stranded form. In the absence of a terminator hairpin, RNA polymerase will continue transcription after the attenuator. When tryptophan is present in the cell, the biosynthesis of the leader peptide occurs through the *trp* codons and continues along the leader section of the mRNA to the UGA codon, which lies between regions 1 and 2.

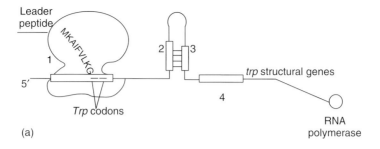

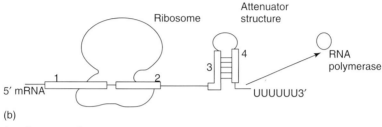

Fig. 5 Ribosomal stalling. (a) When tryptophan levels are low, biosynthesis of the leader peptide on the ribosome becomes paused at the *trp* codons in region 1. The region 1 sequence is then sequestered within the ribosome and cannot base-pair with region 2; therefore, region 2 base-pairs with region 3, compelling region 4 to remain in a single-stranded form; (b) When tryptophan levels are high, biosynthesis of the leader peptide occurs through *trp* codons. Synthesis continues along the leader section of the mRNA to the UGA codon present between regions 1 and 2. The ribosome extends over region 2 and prevents it from base-pairing with region 3; region 3 then base-pairs with region 4, generating a termination hairpin.

Under these conditions, the ribosomes will extend over region 2, preventing it from base-pairing with region 3. At this point, region 3 remains available to base-pair with region 4, thus generating a hairpin that results in a termination of transcription at the attenuator (Fig. 5) [1, 3].

Regulation via an attenuation mechanism has been identified in many amino acid operons of *E. coli*, for example, *His*, *Phe*, *Leu*, *Thr*, and *ilv*.

2.5
Catabolite Repression

Glucose is the most easily utilizable sugar for energy purposes, and is therefore preferred by *E. coli* as a carbon source. If the bacterial cells are grown in a medium containing both glucose and lactose, there is no induction of the *lac* operon. However, when the utilization of glucose is complete, the cells will begin to utilize lactose such that the *lac* operon will be induced to initiate the biosynthesis of β-galactosidase. In the presence of glucose, the lactose operon will not be induced; in this case, the inhibitory molecule is not glucose itself but rather is an unknown catabolite that is derived from glucose and functions by preventing the expression of several operons, including those of lactose, galactose, and arabinose. Collectively, this effect is referred to as *catabolite repression/carbon catabolite repression*.

Catabolite repression is generally mediated by several mechanisms which can either affect the synthesis of catabolic enzymes via global or specific regulators, or inhibit the uptake of a carbon source and result in a decline of the corresponding inducer. The phosphoenolpyruvate (PEP):carbohydrate phosphotransferase system (PTS) and protein phosphorylation play a major role in catabolite repression. The PTS components form a protein phosphorylation cascade which uses PEP as the phosphoryl donor. Most of the PTS-mediated catabolite repression mechanisms respond to the phosphorylation level of a PTS protein that is controlled by the metabolic state of the cell.

In *E. coli*, an important enzyme of the PTS system – enzyme IIA (EIIA) – plays an important role in this mechanism. In *E. coli*, EIIA is specific for glucose transport such that, when glucose levels are high inside the bacterial cell the enzyme is present mostly in its nonphosphorylated form, and this leads to an inhibition of adenylyl cyclase. In contrast, recently acquired genetic data have suggested that the adenylyl cyclase enzyme is stimulated by phosphorylated EIIA^{Glc}. Indeed, a direct correlation has been observed in the levels of phosphorylated EIIA^{Glc} and secreted cAMP. However, other evidence has indicated that an additional factor is required for the phosphorylated EIIA-mediated stimulation of cAMP secretion.

The non-phosphorylated EIIA interacts with proteins of several non-PTS sugar-transport systems (e.g., lactose permease and maltose permease), and inhibits their activities, which leads to a non-transportation of lactose inside the bacterial cell. In the Firmicutes, it is the histidine protein (HPr) that exerts this role, with HPr being phosphorylated not only at His15 in a PEP-dependent reaction but also at Ser46 in an ATP-requiring reaction. Notably, the HPr exists in four different forms, all of which exert different regulatory functions.

Whereas, catabolite repression has been studied extensively only in the Enterobacteriaceae and Firmicutes, evidence exists in certain other pathogens of a relationship between carbon metabolism and virulence. The mechanisms that are operative in carbon catabolite repression appear also to control virulence gene regulators, cell adhesion, and pili formation. Indeed, various studies have shown that the expression of the *pil*T and *pil*D genes of *Clostridium perfringens*, and of the multiple gene regulator (*mga*) gene of *Streptococcus pyogenes*, all of which encode a virulence regulator, are controlled by catabolite-controlled protein A (CcpA) [1, 13].

2.6
Cyclic AMP Receptor Protein

Cyclic AMP plays an important role in controlling the catabolic activity of both prokaryotic and eukaryotic cells. In prokaryotes, cAMP modulates transcription through CRP (also known as *CAP*), whereas in eukaryotes cAMP modulates the enzyme activity via covalent modulation through cAMP-dependent protein kinases.

Typically, CRP does not bind to DNA without the prior binding of cAMP. Among the genes that are activated in bacteria in response to an increase in cAMP are those that encode the enzymes for the catabolism of lactose, arabinose, galactose, and maltose. The presence of cAMP is necessary for the activation of transcription in bacteria, a situation that has been demonstrated by mutating the gene coding for adenylyl cyclase (*cya*$^-$), which converts ATP into cAMP. If cAMP is added externally to such a system, then an activation of the transcription will occur. Promoters of the operons – the expression of which depends on cAMP and CRP – contain specific sites for binding the cAMP–CRP complex. The *in vitro* transcription of DNA fragments containing cAMP–CRP-dependent promoters is also activated by cAMP–CRP. In some cases, mutant promoters have been isolated at which cAMP–CRP is unable to bind; in this case, the cAMP–CRP fails to activate transcription, both in intact cells and *in vitro*.

More recently, it has been shown that one CRP dimer (after binding two cAMP molecules) binds at the specific site in the operon where transcription is activated by cAMP–CRP. Aided by the results of CRP protection experiments to monitor "chewing" by DNase, it was shown that approximately 25 base pairs are protected by cAMP–CRP against chemical attack, and that the mutations which prevent CRP binding are located within these sequences. The results of the experiments also indicated that CRP forms major contacts in two successive grooves of the DNA, with the most conserved sequence to bind CRP being 5'TGTGA3'. Other evidence has indicated that the 5'TGTGA3' sequence is critical for CRP binding. Point mutations that are known to prevent stable CRP binding are located at the *gal* and *lac* sites whilst, at the *ara* site, the results of deletion experiments highlighted the importance of this sequence for CRP binding. Another sequence 6 bp downstream of the TGTGA motif had an inverted repeat although, in many cases, this was not an exactly inverted repeat sequence. Irrespective of the symmetry of the sequence, this second motif has also been shown necessary for efficient CRP binding. The results of many types of experiment have indicated that the two subunits of CRP recognize two zones of sequences, separated by 6 bp. As noted above, the first of these zones contains the sequence 5'TGTGA3', while the second zone contains either a symmetrically arranged version of the sequence or another type of sequence. However, the affinity of CRP for DNA appears to be greater when the 6 bp downstream sequence of 5'TGTGA3' is symmetrical rather than non-symmetrical. The distance between the transcription start point and the CRP binding site is different for the various promoters. For some promoters, such as those for the *ara* operon and the *mal* operon, an additional protein – the Ara C protein or the Mal T protein – is required to activate the transcription (these activator proteins also bind to the promoter). In some cases (e.g., *lac*, *cat*) two CRP binding sites have been found, with the secondary sites binding CRP less tightly and assisting in the quest for CRP at the primary sites [14, 15].

Although, the main function of CRP is to activate transcription, in some cases the binding of CRP has been shown to repress transcription. Two promoters, P1 and P2, are located at the *gal* promoter. The binding of CRP at P1 causes the transcription to be activated, whereas CRP binding at P2 causes it to be repressed. This situation occurs because, at P2, the CRP binds close to the -35 region and blocks the binding of RNA polymerase at the P2 promoter. CRP also acts as a repressor of transcription of its own promoter *in vitro*. It also inhibits transcription of the gene for the major outer membrane protein, Omp A, again by binding close to the -35 region of the promoter [14].

CRP is a dimeric protein with two identical subunits, each containing 210 amino acids, the complete sequence of which has been deduced from the nucleotide sequence of the gene. The results of equilibrium dialysis studies have indicated that two molecules of cAMP can bind per CRP dimer, while CRP has a two-domain structure, as confirmed by the high-resolution crystal structure of CRP when complexed with cAMP. A large N-terminal domain that extends from residue 1 to 135 is separated by a cleft from a smaller C-terminal domain (CTD) that extends from residue 136 to 210. The N-terminal domain of each subunit contains one cAMP molecule buried in the interior of the protein, while residues from both subunits are involved in the binding of cAMP. Typically, the 6-amino group of the adenine ring in cAMP interacts with Thr127 on one subunit, and with Ser128 on the other subunit.

The N-terminal domain of CRP in the region of residues 30–89 exhibits sequence homology with the regulatory subunit of the protein kinase of eukaryotes. The regulatory subunit of protein kinase also has two cAMP-binding sites.

The CTDs of the two CRP subunits consist of three α-helices connected by short, β-sheet structures. On each subunit, one of the α-helices protrudes from the surface of the CRP dimer; these two α-helices are considered to be involved in DNA binding. The other DNA-binding proteins, such as Cro and cI proteins, also have α-helices but these are located at the N-terminal region. All of the DNA-binding proteins have been shown to have a helix–turn–helix domain that is essential for interactions with DNA. However, the E. coli fnr protein, which is essential for the anaerobic respiratory metabolism, also has a helix–turn–helix domain in the C-terminal region. Additional homology is also found in the N-terminal regions of the two proteins. Although the fnr protein does not bind to cAMP, it has a somewhat similar function as CRP, serving as a pleiotropic activator for a series of genes that are turned on under limiting aerobic conditions. Subsequent sequence comparisons indicated that the *fnr* gene might have been derived as a result of duplication either from the CRP gene itself, or from a common ancestor.

Interactions between RNA polymerase and promoters may be described as a two-step event. In the first step, the enzyme binds to the promoter to form a closed complex; this binding is reversible and characterized by an association constant, K_B. In the second step, the closed complex isomerizes to give rise to an open complex; this isomerization includes a localized unwinding of the DNA over a distance of approximately 12 bp near the transcription start, it is generally irreversible, and the corresponding rate constant, K_f, is slow. Strong promoters have high values of both K_B and K_f, whereas weak promoters have low values for both constants. The addition of cAMP and CRP has two effects on the *lac* promoter: (i) it enhances the rate of open complex formation by increasing the value of K_B without affecting K_f; and (ii) the presence of cAMP–CRP increases the binding of RNA polymerase on the P1 promoter. This latter increase is due to an inhibition of RNA polymerase binding on other secondary sites in the presence of cAMP–CRP. The structure of the CRP–DNA complex is interesting in that the DNA has a bend, and the proteins may distort the double-helical structure of DNA when they bind, while several regulatory proteins may induce a bend in the axis. Consequently, a dramatic change occurs in the organization of the DNA double helix following the binding of CRP [14].

2.7
Guanosine-5′-Diphosphate,3′-Diphosphate

The *rel* A gene, which is required for the synthesis of guanosine-5′-diphosphate, 3′-diphosphate (ppGpp), has been shown to enhance the transcription of the *lac* Z and *glg* genes (*glg* genes code for glycogen biosynthetic enzymes). It has been indicated that ppGpp interacts directly with RNA polymerase to alter the transcription of various genes and, indeed, a small protein has been shown to mediate the effect of ppGpp on the *lac* Z gene under certain conditions. Nitrogen regulatory proteins C and A (Ntr C and Ntr A) have also been shown to activate the *gln* promoter. The *ntr* C and *ntr* A genes encode a specific DNA-binding protein and an alternate sigma factor for RNA polymerase, respectively. However, neither Ntr C nor Ntr A increased the synthesis of glycogen biosynthetic enzymes [16–19].

2.8
Riboswitch

Previously, various research groups were of the opinion that the regulation of gene activity in response to environmental cues was mediated only by proteins. In fact, in the classical model of gene regulation, the cells monitor their environment through a variety of specialized sensor proteins that are deployed either on their surfaces or internally. Today, riboswitches have been demonstrated as mRNAs that sense the environment directly and shut themselves down in response to particular chemical cues. Recently, it was shown that bacterial genes for enzymes which direct the synthesis of vitamin B_{12} employ a riboswitch. In this case, the mRNAs transcribed from these genes were shown to fold into a specialized shape, creating a binding pocket for coenzyme B_{12}. Following B_{12} binding, the mRNA would alter its shape in such a way as to mask a nearby sequence that otherwise would instruct the ribosomes to start reading at that point. Consequently, when coenzyme B_{12} is abundant these sequences are hidden and the enzymes for B_{12} synthesis are no longer produced. Many other riboswitches have been reported in bacteria, including those that control the synthesis of vitamins B_1 and B_2, and guanine nucleotides. There is also some evidence that riboswitches are present in plants and fungi [20, 21].

2.9
Regulon

Regulon are also referred to as *multigene systems* or *global regulatory systems*. In contrast to operons, the coordinately regulated genes of a regulon are located physiologically at different parts of the chromosome, and are controlled by their own promoters, but are regulated by the same mechanisms. One of the most well-known examples of a regulon is the production of heat shock proteins (Hsps) in *E. coli* which, as a mesophile, exhibits normal growth at between 20 and 37 °C. The bacterium responds to an abrupt increase in temperature, from 30 to 42 °C, by producing a set of almost 30 different proteins, termed collectively Hsps; in fact, when the temperature is raised from 30 to 42 °C, Hsp production by *E. coli* is increased almost 10-fold within a 5 min period. Subsequently, the Hsp level decreases slightly to a steady-state level, which is maintained while the cells remain at the elevated temperature. If the temperature is then decreased from 42 to 30 °C, the levels of Hsps decrease abruptly almost 10-fold within the same, 5-min period. In addition to a change in temperature, however, other agents (viz.

organic solvents or other DNA-damaging agents) can induce heat shock gene expression. Clearly, the heat shock regulon deals with the variety of cellular damage that may occur in many different ways [22].

Many of the Hsp genes encode either proteases or chaperones. The proteases degrade any abnormal proteins, including incompletely synthesized and misfolded proteins, whereas the chaperones bind to abnormal proteins, causing them to unfold and then attempt to re-fold into an active configuration. Although the genes encoding Hsps are scattered around the chromosome, they are coordinately regulated and therefore they constitute a regulon. The regulator of the heat shock response is an alternative sigma factor, named sigma 32 (σ^{32}); this protein has a molecular weight of 32 kDa, is involved in the initiation of the transcription of heat shock genes by recognizing the heat shock promoters, and is coded by a gene known as *rpoH* (RNA polymerase subunit heat shock). The heat shock promoters have different −10 (CCCCAT) and −35 (CTTGAAA) consensus sequences than do promoters that are recognized by a normal sigma factor with a molecular weight of 70 kDa. Typically, σ^{32} is unstable at low temperatures, with a half-life of about 1 min, but is almost fivefold more stable at a higher temperature. Regulation of the *rpoH* gene occurs at the translational level; indeed, although a significant amount of *rpoH* mRNA may be detected in cells at low temperatures, it is not translated. At high temperatures, the inhibition of translation is relieved and the synthesis of σ^{32} occurs. Previously, it has been shown that two regions of the *rpoH* mRNA are required for translational inhibition – one region close to the +1 site, and another at between +150 and +250 in the mRNA. These two regions form a stem-loop structure, which may prevent binding at the ribosome binding site and thus inhibit translation, as well as possibly increasing the stability of this mRNA [22].

The σ^{32} protein is degraded by a specific protease, termed *Hfl B*. The degradation of σ^{32} at 30 °C also requires a chaperone composed of three proteins, termed *DnaK*, *DnaJ*, and *GrpE*. The degradation of σ^{32} at 30 °C is decreased almost 10-fold by mutations in any one of the genes that code the Hfl B, DnaK, DnaJ, and GrpE proteins.

Evidence is available that the interactions between DnaK and σ^{32} are temperature-dependent and occur only at low temperatures. At higher temperatures, σ^{32} is capable of interacting with RNA polymerase, but is unable to interact with DnaK. It has been assumed that this temperature-dependent interaction between DnaK and σ^{32} brings stability to σ^{32}, although when the temperature falls from 42 to 30 °C a translational inhibition of mRNA returns and σ^{32} again becomes sensitive to degradation. Such temperature-sensitive properties allows the heat shock response to be turned on and off very quickly.

A second heat-induced regulon is controlled by another sigma factor, known as *sigma E* (σ^E). The σ^E-controlled promoters are much more active at about 50 °C; in fact, deletions of the gene encoding σ^E have been shown to be temperature-sensitive at 42 °C, whereas deletions of the gene encoding σ^{32} are temperature-sensitive at 20 °C. The σ^E promoter responds to misfolded outer membrane proteins, whereas σ^{32} responds to misfolded cytoplasmic proteins. The σ^{32} gene also has a σ^E promoter, so that all of the Hsps are induced by a cascade effect when σ^E-regulated genes are expressed. Typically, the σ^E regulon provides proteins which are required under more extreme conditions [22].

Another important example of the regulons is the SOS regulon, which becomes activated in response to extensive DNA damage. Previously, Weigle *et al.* were the first to demonstrate the induction of DNA-repairing genes in case of reactivated ultraviolet light-irradiated lambda (λ) phage. Similar to the heat shock regulon, the SOS regulon also has a mechanism for signaling the "on" and "off" of the regulon. In prokaryotes, the SOS system is regulated by two main proteins, namely LexA and RecA. The transcription of almost 48 genes has been shown to be regulated by the LexA protein, a homodimer that acts as a transcriptional repressor after binding with a sequence near the promoter/operator region in these proteins, called the SOS box. In *E. coli*, the SOS boxes are almost 20 nucleotide-long sequences with a palindromic structure and a conservative sequence. In other prokaryotes, however, the sequence of the SOS boxes varies considerably, with different lengths and compositions. Nonetheless, in all cases the sequence is conservative and is one of the strongest short signals in the genome. Those SOS promoters that are bound by LexA are unable to initiate transcription although, upon DNA damage the LexA is inactivated and removed, which results in an expression of the SOS genes. Previously, it has been shown that, upon exposure to DNA-damaging agents, large amounts of single-stranded DNA become accumulated such that single-stranded DNA will bind to the RecA protein, which is involved in homologous recombination and postreplication DNA repair. At the time when the DNA damage occurs, the RecA-bound single-stranded DNA will bind to LexA and induce the latter to cleave itself (autocleavage). The autocleavage of LexA has been shown to take place between two specific amino acids that separate the repressor into two domains – the DNA-binding domain and the dimerization domain. As a result of this disruption of dimerization, LexA is removed from the SOS box, after which the SOS genes are expressed at high levels. Subsequent to the SOS response, the amount of RecA that is complexed to single-stranded DNA will be decreased due to DNA repair, while LexA fails to undergo autocleavage; this results in a return of the regulon to the uninduced state [22–26].

During the SOS response, cell division is also halted, so that any damaged chromosomes do not become segregated into the daughter cells. Consequently, during the SOS response, in addition to the DNA-repair enzymes a cell division-inhibitory protein is also expressed, at high levels.

3
Regulation of Gene Expression in Eukaryotes

In comparison to prokaryotes, the eukaryotes have a much more complex regulatory mechanism of transcription, with RNA splicing also playing an important role in the regulation of gene expression. In addition to the activation of gene structure, the polyadenylation, capping, transport to the cytoplasm, and translation of mRNA represent potent control points in the process of regulating gene expression. Five potential control points for regulating gene expression in eukaryotes are shown in Fig. 6. The most important method of control is to regulate the initiation of transcription (i.e., the interaction of RNA polymerase with the promoter region), which may be demonstrated using a technique known as

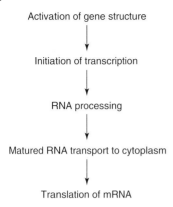

Fig. 6 Regulation of gene expression. Gene regulation may take place in a gene-specific manner at any of the several sequential steps. However, there are five potential control steps.

run-off transcription. In this case, the nuclei are first isolated from the cells and then incubated with radiolabeled nucleoside triphosphates. Under suitable conditions, unfinished transcripts will be completed, but no new transcripts will be synthesized; consequently, the RNA that is labeled by using this method will have been derived from those genes that started transcription at the time the nuclei were isolated. Subsequently, when the labeled RNA is used to probe DNA from a clone of genes under investigation, an absence of hybridization between the labeled RNA and the cloned DNA indicates that the DNA was not transcribed in the tissue. The use of this technique to examine several genes has led to the realization that an absence of gene expression does, indeed, result from an absence of transcription [27].

Nowadays, *DNA microarray technology* – which is more commonly known as *"gene chip technology"* – is also used widely to identify the presence of complementary sequences of DNA. In fact, microarray technology can be regarded very much as a modern-day 'genetic revolution,' and comparable to the development of microprocessors in the computer revolution of 30 years ago. Today, with the advent of microarray technology, the task of screening genetic information has become an automatic routine that exploits the tendency for a molecule, that is carrying a template for synthesizing mRNA and protein, to bind to the very DNA that produces it. Currently, microarrays incorporate many thousands of probes, each of which is imbued with a different nucleic acid from known and unknown genes, to bind with mRNA. Subsequently, the resulting bonded molecules will fluoresce under different colors of laser light, thus demonstrating which complementary sequence is present. In this way, these microarrays can be used to measure the incidence of genes and their expression.

More recently, following the determination of the human genome sequence, the importance of the single nucleotide polymorphism (SNP) has also been realized. The SNPs represent minor variations in DNA that define the differences that occur among people, that may predispose a person to disease, and that may influence a patient's response to a drug. Consequently, with the genetic make-up of humankind now broadly known, it is possible to create microarrays that are capable of targeting individual SNP variations, and thereby to make much greater comparisons across the genome. Taken together, the results of these studies may help to identify the roots of many diseases, especially when combined with specific software that has been developed to design microarrays incorporating very large numbers of probes [28].

The real-time-polymerase chain reaction (RT-PCR) has also been used to quantify the level of gene expression. This technology, which is both highly sensitive and convenient, includes approaches that serve as a natural complement of transcriptome analysis, either when the tuning of array results is necessary or when an array sensitivity limit is reached for low-level transcripts of interest. In RT-PCR, the sensitive quantification of PCR products relies on the detection of a fluorescent signal that is proportional to the amount of product. Typically, PCR products can be measured in real time by using a dye that will bind with double-stranded, but not single-stranded, DNA, or with labeled oligonucleotides that can bind specifically to the PCR products.

The cells of multicellular organisms are genetically homogeneous but structurally and functionally heterogeneous, because of the differential expression of genes. Many of these differences in gene expression occur during development, and are retained through mitosis. Stable alterations of this type are referred to as being *epigenetic*, because they are heritable in the short term but do not involve mutations of the DNA itself. The term "epigenetics" is used to define the mechanism by which changes in the pattern of inherited gene expression occur in the absence of alterations or changes in the nucleotide composition of a given gene. In the past, research investigations have been focused on two molecular mechanisms that mediate epigenetic phenomena, namely DNA methylation and histone modifications. Previously, it has been shown that epigenetic effects via DNA methylation have an important role in development, but can also arise stochastically as animals age. The identification of proteins that mediate these effects has been helpful in elucidating the epigenetic effect which, when perturbed, may result in disease. Typically, external factors that apply to epigenetic processes are associated with the diet in long-term diseases, such as cancer. Indeed, it has been proposed that epigenetic mechanisms might allow an organism to respond to the environment through changes in gene expression [29, 30].

The fact that many genes are transcribed in one tissue or organ, but not in others, may explain the need for cell differentiation in eukaryotic organisms, whereby some genes are expressed under the influence of certain signaling agents, such as the substrates of specific enzymes, hormones, and regulatory nucleotides. Gene expression under the influence of certain signaling agents has also been considered as the phenomenon of *induction*, which is less prominent among eukaryotes than in prokaryotes. Typically, in eukaryotes more time is required for induction, and the extent of stimulation may be only 10- to 20-fold; this contrasts greatly with bacteria, where many thousand-fold levels of stimulation may occur as a result of induction. Since, in eukaryotes, monocistronic mRNAs are generally found, compared to polycistronic RNAs in prokaryotes, coordinate induction has not been reported in eukaryotes. Many years ago, the so-called "Britten–Davidson model" was proposed to explain the induction phenomena in eukaryotic genes, according to which the eukaryotic genome contains a large number of sensor sites that recognize specific molecular signaling agents such as hormones and the substrates of specific enzymes. Each sensor site is adjacent to an integrator gene such that, when a sensor site is activated following the binding of a signaling agent, the integrator gene is transcribed to form its complementary

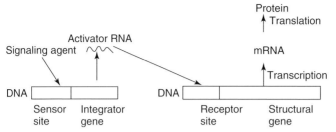

Fig. 7 The Britten–Davidson model of gene regulation. When a signaling agent such as a hormone binds to a sensor site, transcription of the integrator gene occurs, such that complementary RNA (the activator RNA) is formed. The activator RNA is recognized by the receptor site located elsewhere in the genome. When the activator RNA binds to the receptor site, the adjacent structural gene is transcribed and subsequently translated.

RNA, termed an *activator RNA*. The latter, in turn, is recognized by one or more receptor sites that are located elsewhere in the genome and may be on the same or another chromosome. It is when the activator RNA binds to the receptor site that an adjacent structural gene is transcribed [1, 31, 32] (Fig. 7).

Although, initially, the phenomenon of coordinate expression in eukaryotes was considered nonexistent owing to the presence of monocistronic mRNAs, the use of DNA microarrays and global expression analysis has illustrated the highly coordinate expression of genes that function in common processes in eukaryotes. This process, which has been termed *"syn-expression"* in eukaryotes, has been considered comparable to the role of operons in prokaryotes. Moreover, it has also been proposed as a key determinant facilitating evolutionary change leading to animal diversity. By using DNA microarrays, the simultaneous monitoring of thousands of transcripts is possible, and this has in turn provided global insights into gene expression. Ultimately, however, the expression data have revealed a high degree of order in the genetic program, and a tight coordination of the expression of groups of genes that function in a common process [33].

It was while working with mammalian cells infected with SV40 virus that Frenster first suggested the existence of a de-repressor model for gene regulation in eukaryotes. Based on experimental data indicating the ability of exogenous DNA or RNA to de-repress specific loci on the host cellular genome, this model suggested a close relationship to the normal mechanisms of gene regulation in animal cells, which may be subverted to allow the re-expression of otherwise repressed embryonic information. This derepression model accounted for a selective gene transcription that was locus- and strand-specific, but failed to discuss gene–gene interaction.

Subsequently, Frenster proposed a Mated Model of gene regulation in eukaryotes, according to which a derepressor RNA (dRNA) binds to the anticoding strand of an operator locus, thus permitting the transcription of operator and structural gene loci. The dRNA of an operon is complementary in base sequence to its operator portion of the direct transcription product. Following this, the direct transcription product would be split into mRNA and operator RNA (oRNA), with cleavage occurring either directly or after the formation of a heterometric duplex RNA by the base-pairing of dRNA with the operator portion. Thereafter, the dRNA would be removed selectively from the operator locus, providing a feedback inhibition of transcription of the operon. Following the consumption of mRNA and degradation of the oDNA, the dRNA would be released from the duplex, providing a positive feedback derepression of transcription of the operon. As the different structural genes may share operators with common base sequences, they would be equally sensitive to given species of dRNA, during both transcription of the gene and its selective inhibition [34], (Fig. 8).

In eukaryotes, cell division is normally highly regulated, aided by growth factors that cause the cells to undergo cell division and, in some cases, cell differentiation. Among these growth factors, some are specific for certain types of cell, due to specific receptors present at the cell surface, while others are general rather than specific in their effects. Other growth factors which control cell division include epidermal growth factor, nerve growth factor, platelet-derived growth factor, fibroblast growth factor, and lymphokines. Occasionally, the failure of growth factors to control cell division may lead to the creation of tumors.

Although, in prokaryotes, the negative control of transcription plays an important role (e.g., fail-safe mechanism), a positive control in eukaryotes is even more important, for the simple reason that, in a large genome, such an approach is more efficient. If a large number of genes is to be negatively controlled, then each cell would need to synthesize the same number of different repressors in sufficient amounts as to permit the specific binding of each. In addition, the nonspecific DNA-binding of regulatory proteins (repressors) is especially important in much larger genomes of the higher eukaryotes, as the

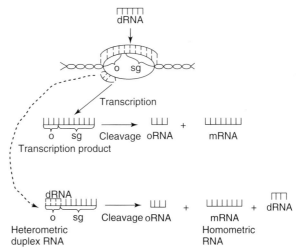

Fig. 8 Mated model for gene regulation. o, operator; sg, structural gene; dRNA, derepressor RNA. The derepressor RNA (dRNA) binds to the anticoding strand of the operator locus, permitting transcription of the operator and structural gene loci. The transcription product is then split into mRNA and operator RNA (oRNA). Cleavage may occur either directly, or after the formation of heterometric duplex RNA.

chance of a specific-binding sequence being present at random at inappropriate sites would also increase with genome size [35].

3.1
Transcriptionally Active Chromatin

The major part of the eukaryotic genome is sequestered in the nucleus, where it is surrounded by a nuclear membrane to safeguard it against exposure to the cytoplasm. As the transcription of genes also occurs within the nucleus, but translation occurs mainly in the cytoplasm, the two processes cannot be coupled. Typically, the chromosomes of eukaryotes are more complex than those of bacteria, with each containing a double-helix DNA molecule that may be more than 20-fold larger than that of a bacterial chromosome. In eukaryotes, the DNA is tightly complexed with histone proteins that are thought to have structural and protective functions; other loosely bound nonhistone proteins are also generally present, albeit in smaller quantities than the histones. Although the functions of the nonhistone proteins are not clear, they may have role(s) in transcription and/or replication. Among other DNAs present in the cells, mitochondrial and chloroplast DNAs are both small, double-stranded molecules. Typically, the mitochondrial DNA of plants is larger than that of animals, while all plants appear to have similarly sized chloroplast DNAs. The mitochondrial and chloroplast DNAs resemble bacterial DNA but, unlike eukaryotic nuclear DNA, are not associated with histones [36].

It has been shown that those chromosomal regions which have been activated for transcription are more sensitive to DNase degradation, which is indicative of their lesser degree of protection by histones. The actively transcribed regions have also been found to include sequences with a high sensitivity to DNase, termed *hypersensitive sites*. The latter are generally no longer than 200 bp, and are found within the 1000 bp that flank the 5′ ends of the transcribed genes. In some cases, these hypersensitive sites may be located farther from the 5′ end, close to the 3′ end, or even within the gene itself. Many hypersensitive sites have been found to serve as binding sites for the regulatory proteins [37].

The *telomeres* are specialized structures which are located at the ends of linear eukaryotic chromosomes, and which generally have many tandem copies of a short oligonucleotide sequence (T_aG_b) in one strand, with C_bA_a in the complementary strand (where a and b are 1 to 4). The structure of the telomere poses a biological problem, however, in that DNA replication requires a primer, but in linear DNA molecules it is impossible to synthesize an RNA primer starting at the end nucleotide. However, this problem is resolved by employing *telomerase*, an enzyme which resembles reverse transcriptase and catalyzes the addition of telomeres to the chromosome ends. Within its structure, telomerase has both protein and RNA regions; the RNA portion is about 150 nucleotides in length and has about 1.5 copies of the C_bA_a telomere repeat that serves as a template for the synthesis of the T_aG_b strand. The telomerase-like reverse transcriptase synthesizes only a segment of DNA that is complementary to an internal RNA molecule [38].

The DNA in transcriptionally active chromatin has been found to be methylated to a lesser degree; moreover, nucleosomes have not been found in the transcriptionally active regions (at least in some cases). Chromatin has been classified into two groups: *heterochromatin*, a highly condensed chromatin which is transcriptionally inert; and *euchromatin*, a loosely packed chromatin which is transcriptionally active.

3.2
Regulation of Gene Expression at the Initiation of Transcription

Although the regulation of gene expression at the initiation stage of transcription (i.e., the binding of RNA polymerase to the promoter) has been demonstrated, there is at present no evidence for the control of gene expression at the subsequent stages of transcription. Three RNA polymerases have been identified in eukaryotic cells as being involved in the biosynthesis of different classes of RNAs. For example, the biosynthesis of heterogeneous RNA (HnRNA) occurs in the presence of RNA polymerase II, while the initiation of transcription by RNA polymerase II is regulated by a series of DNA elements that may be divided into the core promoter elements consisting of the TATA box, the transcription initiation site, and upstream activating sequencess (UASs). The UASs are generally located upstream of the core promoter sequence, although in some cases they have been found downstream of the transcription start site (Fig. 9). In this case, a specific protein is bound to each UAS, and this results in a positive or negative effect on the core promoter activity. The TATA box is found generally 25 bp upstream from the transcription initiation site and, although it is common in eukaryotic genes, very few genes have been shown to be expressed without the TATA box. In addition to the TATA box, another sequence – referred to as the *CAAT box* – has been found at the −75 position from the initiation site. The CAAT box (which has the consensus sequence GGCAATCT) functions in either orientation, and plays an important role in increasing the promoter strength. A GC box at the −90 position from the initiation site has also been found; this may occur in either orientation, and is a common component of the promoters of housekeeping genes. The GC box has the consensus sequence GGGCCGGG.

Transcription factor II D (TFIID) for RNA polymerase II has been shown to play an important role in the initiation of transcription, by binding to the TATA box sequences. Such binding of TFIID facilitates the binding of the RNA polymerase II on to the promoter. The assembly of an initiation complex and RNA polymerase at the promoter is a complex process that requires the participation of many other initiation factors. Typically, TFIID has two components – the TATA box binding protein (TBP) and another protein termed the TBP-associated factor (TAF). Whilst TAF is important for regulating the transcription, TBP – which is also referred to as the *"commitment factor"* – binds to DNA in the minor groove. The inner surface of the TBP binds to DNA, while the outer surface is available to extend contacts to other proteins. The DNA-binding sites of TBP consist of sequences that are conserved between species, with the variable N-terminal tail being exposed to interact with other proteins. Normally, TBP is the only transcription factor to make contacts with the specific sequences in the DNA. The activity of TFIID has also been shown to be regulated by inhibitory proteins that interact with TBP; these inhibitory proteins may serve an important regulatory role by maintaining any genes that have been removed from inactive chromatin in a repressed, but rapidly inducible, state [39–41].

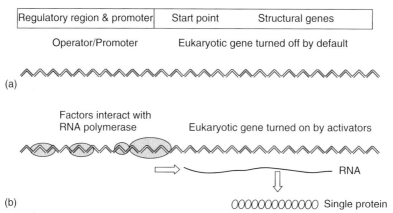

Fig. 9 Regulation of eukaryotic gene expression at the transcriptional level. Eukaryotic control is often positive: trans-acting factors bind to cis-acting sites in order for RNA polymerase to initiate transcription at the promoter. (a) Transcription is turned off by default (if the correct initiation factors are not bound in the regulatory region); (b) Transcription becomes turned on after the binding of initiation factors on the regulatory region, after which RNA polymerase binds to the promoter region.

A number of other factors also play important roles in the regulation of initiation of transcription. For example, TFIIA, which joins the initiation complex after TFIID, has two subunits in yeast and three in mammals. Following the joining of TFIIA, the TFIID is able to protect a region that extends further upstream, while the addition of TFIIB provides further protection to the region of the template in the vicinity of the start point, from -1 to $+10$ bp. TFIIF is a dimeric protein, in which the larger subunit has an ATP-dependent DNA helicase activity that may be involved in opening the DNA during initiation, while the smaller subunit has been found to be equivalent to the sigma factor of *E. coli*. The TFIIF brings RNA polymerase II to the assembling transcription complex, and also provides the means for its binding; interaction with TFIIB may be important when TFIIF joins the complex. Polymerase binding extends the sites that are protected downstream to $+15$ on the template strand and $+20$ on the nontemplate strand. TFIIE binds at the upstream boundary. Two more factors – TFIIH and TFIIJ – also join the complex after TFIIE; TFIIH has kinase activity that may phosphorylate the CTD tail of RNA polymerase II, which consists of multiple repeats of the consensus sequence, Tyr-Ser-Pro-Thr-Ser-Pro-Ser, that is unique to RNA polymerase II. Phosphorylation of the tail (at either seryl or threonyl residues) is required to release the enzyme from the transcription factors, so that it can leave promoter region to start elongation. The TATA box determines the location of the start point, while the general initiation process of transcription is the same as in bacteria. The enzyme RNA polymerase generates a closed complex, and subsequently is converted into an open complex where the DNA strands become separated. The removal of TFIIE occurs during the process of open complex formation.

The CAAT box is recognized by the proteins of the CCAAT-binding transcription factor (CTF) family, which are generated by the alternative splicing from a single gene. The CAAT box-binding protein 1

(CP1) factor binds to the CAAT boxes of α-globin, while CP2 binds the CAAT box in a β-fibrinogen gene. Other proteins also bind to the CAAT boxes; for example, the albumin CCAAT factor (ACF) protein binds CAAT in the albumin promoter. The CAAT box may also serve as a regulatory point; in embryonic tissues, a protein referred to as the CAAT displacement protein (CDP) binds to the CAAT boxes, preventing the transcription factors from recognizing them. In the testes, the promoter is bound by transcription factors at the TATA box, CAAT box, and the octamer sequences. In embryonic tissues, the exclusion of a CAAT binding factor from the promoter prevents a transcription complex from being assembled. This behavior is analogous to the effect of a bacterial repressor.

The GC box is recognized by the factor SP1, a monomeric protein which makes contacts on one strand of the DNA over a -20 bp binding site, including the GC box. In the SV40 promoter, the multiple boxes between -70 and -110 all bind this factor, thus protecting the whole region. However, in the thymidine kinase promoter, SP1 interacts with a factor at the CAAT box on one side, and with TFIID bound at the TATA box on the other side.

Additional regulatory sequence elements are enhancers in higher eukaryotes, and UASs in yeast. In the case of the enhancer, the location and orientation of sequences relative to the transcription start site are relatively unimportant. Typically, the enhancers exert their regulatory effects even when moved experimentally, and they may occur naturally thousands of base pairs away from the gene which is being regulated. The enhancers have no promoter activity of their own, but may stimulate transcription over considerable distances. Moreover, the enhancers may be involved in the regulation of gene expression during the development of the organism, such as the immunoglobulin enhancer that only functions in B lymphocytes.

A regulatory role of enhancer activity has been identified in the transcription of genes that are responsive to steroid hormones. In this case, the steroid is bound onto a soluble protein, which in turn binds the enhancers for the steroid-responsive genes. Transcriptional activation is also accompanied by a decondensation of the chromatin in the regions containing the genes; this is evident from the fact that the region becomes more sensitive to DNase I digestion and subsequent binding of the transcriptional factors to the promoter regions. An enhancer may also provide an entry site, a point at which RNA polymerase and/or other essential protein associates with chromatin. This involves the same type of interaction with the basal apparatus as the interactions promoted by the upstream promoter elements.

The UASs found in yeast are analogous to the enhancers of higher eukaryotes, and are located upstream of the gene in a region having two identical sequences of 72 bp, each repeated at tandem 200 bp upstream of the initiation start site. The -72 bp repeat is located within a hypersensitive site of chromatin.

RNA polymerase I transcribes the genes for ribosomal RNA from a single type of promoter. The promoters for RNA polymerase I have the least diversity in the eukaryotic genome. The promoter, which has been found located 70 bp downstream of a control element called the upstream control element (UCE), consists of a bipartite sequence in the region preceding the start site which, in turn, surrounds the start site extending from -45 to $+20$ bp, and is able to start the transcription. The UCE located at −180 to −107 bp increases the efficiency of the promoter. Both regions are enriched in terms of GC content, and two initiation factors are required for the initiation of transcription by RNA polymerase I. The upstream binding factor 1 (UBF 1) binds in sequence-specific manner to related sequences in the core promoter and UCE. Another factor, termed the spliced leader 1 (SL1) binds cooperatively to UBF 1 to extend the region of DNA. Following the binding of both the factors, RNA polymerase is able to initiate transcription after first binding with the promoter region. Notably, the SL1 is species-specific; for example, mouse SL1 cannot function on human DNA. The SL1, which consists of four proteins (including one known as *TBP*, that is required for the initiation of transcription by RNA polymerase II and III), has been considered analogous to the sigma factor of bacteria.

RNA polymerase III transcribes the DNA coding for 5S rRNA, tRNA, and many small nuclear RNAs (snRNAs). The DNAs transcribed by RNA polymerase III are all smaller in size, generally less than 300 nucleotides. Studies of the regulation of oocyte 5S rRNA synthesis have shown that it requires three transcription factors for initiation, known as TF IIIA, TF IIIB, and TF IIIC. TF IIIA is a member of the zinc finger proteins, whereas TF IIIB consists of TBP and two other proteins, and TF IIIC is still used as a partially purified preparation containing five subunits. TF IIIA has been shown to be specific for oocyte 5S rRNA, whereas TF IIIB and TF IIIC are required for the transcription of all DNAs by RNA polymerase III.

Promoters recognized by RNA polymerase III are of two types, lying upstream and downstream of the initiation site, and are recognized by different initiation factors. Typically, the promoters for 5S rRNA and tRNA genes lie downstream of the start site, whereas promoters for the snRNA gene are located upstream of the start point. The promoters for the 5S rRNA gene are located between −55 and +80 bp within the gene.

The promoters for RNA polymerase III have a bipartite structure, with the two short sequences being separated by a variable sequence. The type I promoter consists of box A sequence separated from a box C sequence, while the type II promoter consists of a box A sequence separated from a box B sequence. In type II promoters, TF IIIC recognizes box B but binds to a region involving box A, as well as box B. In the type I promoters, TF IIIA binds on box C. In promoters of both types I and II, the binding of TF IIIC facilitates the binding of TF IIIB to the sequence surrounding the start site. Recently, TF IIIB has been shown to be the main initiation factor for RNA polymerase III, whereas both TF IIIA and TF IIIC help TF IIIB to bind at the correct site. The efficiency of the transcription by RNA polymerase is found to be increased by the presence of the proximal sequence element (PSE). All the transcription factors for RNA polymerase bind at the promoter region, forming a preinitiation complex before the binding of RNA polymerase III onto the promoter [42–44].

3.3
Regulation of Gene Expression in Chloroplasts

In contrast to the nuclear genome, chloroplasts have their own genetic system which has certain prokaryotic as well as eukaryotic features. Many chloroplast genes are also organized as operons and, in contrast to nuclear transcription (where monocistronic RNA is transcribed), polycistronic RNA formation occurs in the chloroplasts. In addition, chloroplast gene expression more closely resembles the prokaryotic systems, as it has σ^{70}-type promoters. The plastid operons are transcribed as polycistronic units by at least two distinct RNA polymerases – a plastid-encoded RNA polymerase (PERP), and the nuclear-encoded RNA polymerase (NERP). The PERP resembles the bacterial RNA polymerase, and consists of four different subunits, α, β, β', and β'', which are encoded on the plastid genome by the *rpoA*, *rpoB*, *rpoC1*, and *rpoC2* genes, respectively. The activity of the PERP core enzyme is regulated by sigma-like transcription factors that play a role in promoter selection in a similar manner to the RNA polymerase from *E. coli*. These primary transcripts are processed into smaller RNAs, which are further modified to generate functional RNAs. Although, in general, the RNA-processing mechanisms are unknown, they represent an important step in the control of chloroplast gene expression. Such mechanisms include RNA cleavage, stabilization, intron splicing, and RNA editing. Some nuclear-encoded proteins that participate in diverse plastid RNA processing have been characterized, and most of these appear to belong to the pentatricopeptide repeat (PPR) protein family that is implicated in many crucial functions, including organelle biogenesis and plant development. The PPR proteins seem to bind to specific chloroplast transcripts, thus modulating their expression with other general factors, and also appear to be involved in the control of post-transcriptional gene expression in chloroplasts, including transcript processing, stabilization, editing, and translation. Efforts are required to identify and study interacting enzymes to understand the role of the PPR proteins in post-transcriptional activities, such as splicing, stabilization, editing, and the translation of diverse transcripts in chloroplasts [45–47].

In the case of translation, chloroplasts have 70S ribosomes much like prokaryotes, and have also been shown to possess Shine–Dalgarno-like sequences. In contrast, chloroplast genes have the characteristics of nuclear systems, including the presence of introns and highly stable mRNAs. Generally, however, the transcription rates and steady-state mRNA levels are not comparable, which suggests that post-transcriptional RNA processing and stabilization are decisive steps in the control of gene expression in chloroplasts.

3.4
Regulation of Gene Expression in Mitochondria

In similar fashion to chloroplasts, the mitochondria in eukaryotes possess an independent genetic system. On average, a mitochondrion will include at least one ribosomal protein gene, together with the rRNA and tRNA required for the mitochondrial translation system. In plant mitochondria, genes may be present that are responsible for coding the proteins involved in the electron transport and ATPase complexes. The size of the mitochondrial genome has been reported to be larger, despite only 30 coded proteins have been identified using polyacrylamide gel electrophoresis.

Mitochondrial DNA is unusual, as it is neither wholly prokaryotic nor eukaryotic in nature. Rather, some similarities to bacterial protein synthesis have been found, such as a sensitivity to antibiotics, the sequence homology of rRNAs, and the use of *N*-formyl methionine for protein chain initiation. Notably, the diversity of mitochondrial tRNAs and their structures differ from those in prokaryotes, and from the eukaryotic cytoplasm or chloroplasts. For example, the sizes of the mitochondrial ribosomes range from 55S to 77S, compared to 70S for chloroplast ribosomes and 80S for cytoplasmic ribosomes. The major difference between the mitochondrial genetic system and all other systems, however, is that the mitochondria employ a slightly altered genetic code. Although, in general, the genetic code is considered to be universal, in animal and yeast mitochondria UGA serves as a codon for tryptophan, whereas in plant mitochondria it is used as a stop codon. In plant mitochondria there also appears to be a strong bias towards the use of codons ending in T, whereas in yeast the preference is for those ending in A or T, and in animals for those ending in A or C. It is very difficult, therefore, to express nuclear or chloroplast genes in the mitochondrial system.

As in the case of chloroplast genes, mitochondrial genes often produce a complex set of transcripts. Processing occurs at the ends of the tRNAs, which are inserted much like punctuation marks at the ends of the structural genes. Polyadenylation occurs in neither yeast nor plant mitochondria, and the transcripts do not include the entire genome. Comparatively little is known about the regulation of gene expression in mitochondria; likewise, little is known about the splicing mechanism, except that splicing is known to depend on the RNA secondary structure rather than on specific splicing signals, which is unlike nuclear and chloroplast RNA splicing. On the other hand it is well known that, although the amount of mitochondrial DNA in a plant is less than 1% of the total cellular DNA, it plays an important role in the development and reproduction of the plant [48, 49].

4
RNA Splicing

Most eukaryotic genes have been found to include noncoding sequences (introns) in addition to coding sequences (exons). The introns, which are present in DNA and in the primary transcription product of the gene (HnRNA), are removed by RNA splicing before the mature mRNA is transported to the cytoplasm. The number of introns varies between the genes; for example, the *dystrophin* gene has 70 introns, whereas the α-*interferon* gene has no intron. The size of the intron also varies from almost 100 to 200 000 nucleotides.

At least four types of reaction of RNA splicing have been identified, namely the splicing of nuclear introns, of group I and II introns, and of tRNA introns. Each reaction carries a change within the individual RNA molecule, and therefore is considered to be a cis-acting event. In RNA splicing, only very short consensus sequences are required, and these are located as the end sequences of the intron, GT-AG. In yeast, a branch sequence UACUAAC is also required, which is a less-conserved sequence in mammals. The ends of the introns are identified by RNA–RNA base pairing between the HnRNA and uridine-rich small nuclear ribonucleoprotein particles (snRNPs). As the conserved splice site sequences are short, are not precisely conserved between introns, and occur frequently in the primary sequences of many HnRNAs, this allows the spliceosomes to combine different 5′ and 3′ splice sites in the HnRNA to produce several alternatively spliced mRNAs from a single nuclear gene. Consequently, due to the process of alternate splicing, multiple proteins with different primary amino acid sequences and biological activities may be produced from a single gene. The alternatively spliced mRNAs have been shown to be regulated in either temporal, developmental, or tissue-specific manner in many cases. An alternative RNA splice site choice has been shown to regulate the expression of a somatic sex-determination pathway in *Drosophila*. In this case, the sex-lethal and transformer 1 proteins were shown to control the maintenance of gender by regulating *Drosophila* gene expression at the level of alternative RNA splicing.

RNA splicing starts with a 5′ splicing site, while the formation of a lariat occurs by joining the GU end of the intron to the A position of the branch sequence, via a 5′, 2′ linkage. Subsequently, the 3′-OH end of the exon attacks the 3′ splicing site in such a way that the ligation of exons occurs, releasing the intron as a lariat. Both reactions involve *trans*-esterification, in which the bonds are conserved. At several stages, ATP hydrolysis occurs, most likely to fuel the conformational changes occurring in the RNA and/or proteins. The lariat formation is responsible for the 3′ splicing site, while nuclear splicing requires the formation of a *spliceosome*, which contains various snRNPs and splicing factors. The snRNPs recognize consensus sequences, and also share some interacting proteins. Typically, the U1, U2, and U5 snRNPs each contain a single snRNA and several proteins, whereas the U4/U6 snRNP contains two snRNAs and several proteins. The U1 snRNP base-pairs with the 5′ splicing site, U2 base-pairs with the branch sequence, and U5 snRNP acts at the 5′ splicing site. From U4/U6, there is cleavage of U4, after which U6 base-pairs with U2 to create the catalytic center for splicing. The Group I and Group II introns perform RNA splicing as a self-catalyzed property of RNA. In Group I introns, the hydroxyl group required for attack at the 5′ exon-intron junction is provided by a free guanine nucleotide, whereas in Group II introns the internal 2′-OH position serves as the source. Although these introns also follow the GT-AG rule, they form a characteristic secondary structure that holds the splice sites in the appropriate position. tRNA splicing in yeast has been shown to involve separate endonuclease and ligase reactions, whereby the endonuclease recognizes the secondary structure of the pre-tRNA and cleaves both ends of the intron. The two halves of the tRNAs released by the removal of introns can be ligated, using the enzyme RNA ligase, in the presence of ATP [1, 50–52].

4.1
Nuclear Splicing

In eukaryotes, the majority of genes have introns. However, because of the presence of such introns (as noncoding sequences) in the gene there is much discrepancy in size between the nuclear genes and their corresponding mRNAs. The average size and complexity of the nuclear RNA (HnRNA) was found to be much greater than for mRNA. The HnRNP has also been found to be a ribonucleoprotein in which HnRNA is bound by proteins, such that it resembles a bead connected by a fiber. The "beads" are in fact globular-shaped RNAs associated with six common proteins, A1, A2, B1, B2, C1, and C2, which are referred to as *core proteins*, with sizes ranging from 34 to 120 kDa. The exact structure of the HnRNP and the function of RNAs packaging in the form of beads are not clear.

The RNA splicing and other post-transcriptional changes occur in the nucleus, the substrate for these processes is HnRNP. In this process, the transcript is capped at the 5' end, the introns are removed, and polyadenylation occurs at the 3' end; collectively, these reactions are referred to as "RNA processing." After processing, the RNA is transported through the nuclear pores to the cytoplasm, where it is available for translation.

Currently, many types of splicing system have been identified (Fig. 10):

1. Introns are removed from the nuclear RNAs, using the spliceosome. This reaction requires a large splicing system.
2. The excision of certain introns is an autonomous property of the RNA itself. The ability of RNA to act as an enzyme is seen in the self-cleavage of viroid RNAs, and in the catalytic activity of RNase P.
3. The removal of introns in yeast pre-tRNAs involves endonuclease and RNA ligase, whose dealings with pre-tRNA seem to resemble those of the RNA-processing enzymes. A critical feature here is the conformation of the pre-tRNA.

Nuclear RNA splicing junctions are interchangeable, but are read in pairs. There is no extensive homology or complementarity between the two ends (5' GU-AG 3') of an intron and, as written above, the junctions have well-conserved consensus sequences. The really high conservation is found only within the introns at the presumed junctions. The 5' and 3' end dinucleotide sequences define the left (or 5') and right (or 3') splicing sites; these are also referred to as *donor* and *acceptor* sites. Although it has been shown that there is a common mechanism for nuclear HnRNA splicing, the consensus is not applied to the introns of mitochondria, chloroplasts, and pre-tRNA introns.

In order to ensure splicing of the correct pairs of junctions, the following two points may be applicable:

1. It may be an intrinsic property of the RNA to connect the sites at the ends of a particular intron, because of the base-pairing involving these regions.
2. All of the 5' sites may be functionally equivalent, and all 3' sites may be similarly indistinguishable. The splicing could follow rules which ensure that the 5' site is always connected to the 3' site, which locates next in the RNA.

The splicing sites are generic; they do not have specific individual RNA precursors and besides, the apparatus (spliceosome) for splicing is not tissue-specific. The RNA may be spliced by any cell, and the conformation of the RNA will influence the accessibility of the splicing sites. The reaction does not proceed sequentially along the precursor, and the RNA splicing is also independent of any modifications to the RNA. *In vitro*, a cut is first made at the 5' end of the intron separating the left exon and the right intron–exon molecule. In this case, the left exon takes the form of a linear molecule, whereas the right intron–exon is not a linear molecule. The 5' terminus generated at the left end of the intron becomes linked by a 5'-2' bond to the A in the branch site located 30 nucleotides upstream of the 3' end of the intron. This linkage keeps the intron in the form of a structure called a "lariat." Subsequently, cutting at the 3' end releases the free intron in a lariat form, while the right exon becomes ligated with the left exon. The lariat is then debranched to

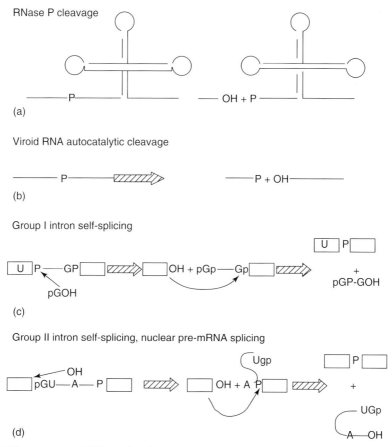

Fig. 10 (a–d) Different types of RNA-catalyzed intron splicing reactions.

provide a linear excised intron, which is rapidly degraded [1, 53].

4.2
Splicing Pathways

Whilst several methods of RNA splicing occur in Nature, the type of splicing will depend on the structure of the spliced intron and the catalysts required for splicing.

4.2.1 Spliceosomal Introns

Spliceosomal introns often reside in eukaryotic protein-coding genes, and within the intron a 3′ splice site, 5′ splice site, and branch site are required for splicing. The 5′ splice site or splice donor site includes an almost invariant sequence GU at the 5′ end of the intron, within a larger, less highly conserved consensus region. The 3′ splice site or splice acceptor site terminates the intron with an almost invariant AG sequence. Upstream from the AG, there is a region enriched in pyrimidines (C and U), or polypyrimidine tract. Upstream from the polypyrimidine tract is the branch point, which includes an adenine nucleotide. Point mutations in the underlying DNA or errors during transcription, can activate a "cryptic splice site" in part of the transcript that usually is not spliced. This results in a mature mRNA with a missing section of an exon. In this way, a point mutation, which usually only affects a single amino acid, can manifest as a deletion in the final protein.

4.2.2 Spliceosome Formation and Activity

Many small RNAs have been found in the nucleus and cytoplasm of eukaryotic cells; these may be referred to as *small nuclear RNAs* (snRNAs) or "snurps" and small cytoplasmic RNAs (scRNAs) or *"scrps"*, respectively. A snRNP generally contains one snRNA and about 10 proteins, some of which are common in all snRNPs, while some are unique to a particular snRNP. The common proteins are recognized by an autoimmune antiserum (anti-Sm), and are considered to be involved in the autoimmune reaction. Many snRNPs are involved in RNA splicing. The snRNAs present in these snRNPs have sequences complementary to the 5' or 3' splicing sites, or to the branching sequence. It is considered that base-pairing between snRNA and HnRNA or between snRNAs plays an important role in splicing.

The spliceosome consists of many snRNPs and many additional proteins that often are referred to as *splicing factors* (Fig. 11). The snRNPs are U1, U2, U5, and U4/U6, and are named according to the snRNPs present in the spliceosome. The snRNPs present in spliceosome together incorporate about 40 proteins, some of which may be directly involved in splicing, while others may have structural roles for assembly or for interaction with the snRNPs.

In the U1 snRNP, a region of 5' terminal 11 nucleotides that is single-stranded and has a stretch which is complementary to the consensus sequence at the 3' site of the exon, is considered to be directly involved in splicing. The intact U1 snRNP can bind to a 5' splicing site *in vitro*; only the snRNA of U1 cannot bind with the 5' splicing site. The U1 RNP first binds at the 5' splice site, and then also binds to the branch site, although how the U1 RNP recognizes the branch site is not known. The U2 RNA binds to the branch site by recognizing the base-pairing interaction; however, for the binding of U2 RNP, a prior binding of U1 RNP is essential. Although interaction with U1 snRNP is responsible for recognizing the splicing site, this does not control the cleavage. Initially, the U5 snRNA binds close to exon sequences at the 5' splice site, but it then changes its position to the vicinity of the intron. Based on the results obtained, it has been suggested that the snRNA components of snRNPs interact both among themselves and with the substrate RNA by base-pairing interactions, and that these interactions allow for changes in structure that may bring reaction groups into opposition, thereby creating a catalytic center [54, 55].

A series of loci containing genes which may potentially code for splicing factors were originally thought to code RNA, but are now known to encode pre-RNA processing proteins (PRPs). Some of the PRPs are components of snRNPs, while others may function as independent factors. One protein, PRP16 (an ATP-dependent helicase), has been shown to be involved in the second catalytic step of RNA splicing. Another protein, PRP22 (yet another ATP-dependent helicase), has been shown to be required to release mRNA from the spliceosome [56, 57].

4.2.3 Self-Splicing

Self-splicing occurs rarely in RNA; this type of RNA is referred to as a *ribozyme*.

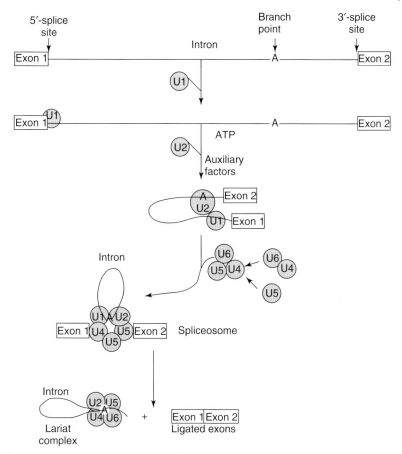

Fig. 11 Pre-mRNA splicing using the spliceosome. Spliceosome formation involves the interaction of a component that recognizes the consensus sequences. U1, U2, U3, U4, U5, and U6 are different small ribonuclear proteins.

Two types of self-splicing intron have been identified, termed *Group I* and *Group II*. These introns perform splicing similar to the spliceosome, but without any requirement for protein. Such similarity indicates that the Group I and II introns may be evolutionarily related to the spliceosome. Self-splicing may have existed in an "RNA world" that was present before protein. Although, tRNA splicing requires other enzymes (viz. endonuclease and RNA ligase), it has been shown that only the RNA part of ribonuclease P (an enzyme protein having RNA in its structure) may cut the pre-tRNA molecule at a specific site. In general, splicing is a cis-reaction, although trans-splicing has also been reported [1, 4].

Group I and Group II Introns Group I introns (where the hydroxyl group is provided by a free guanine nucleotide) are more common than Group II introns (where the hydroxyl group is provided by an internal 2'-OH position). Both, Group I and II introns can perform the splicing by themselves, without a need for enzymic activities to be provided by the proteins. The Group II mitochondrial introns have splicing sites that resemble the nuclear introns, and they are also spliced by the same mechanism as nuclear HnRNA. Two transesterification reactions can be performed by the Group II introns although, as the number of phosphodiester bonds is conserved in the reaction, there is no need for an external energy supply – which may have been an important feature in the evolution of splicing. In autocatalytic splicing, the RNA folds into a specific conformation or series of conformations, and splicing occurs in *cis*-conformation. In contrast, the snRNAs act in *trans*-form upon HnRNA.

Previously, Cech and colleagues, while working with *Tetrahymena thermophila* for the first time, showed that RNA molecules were capable of self-RNA splicing, without the involvement of any protein. This led Cech *et al.* to coin the term *ribozyme*, meaning RNA as an enzyme, and they subsequently showed that RNA could indeed catalyze its own splicing. In the self-splicing of RNA by *T. thermophila* (as shown in Fig. 12), the enzymes act on molecules other than on themselves – hence the term ribozyme. The same group later showed that ribozyme could act on a slightly different form of the same RNA and was, therefore, an enzyme in the true sense. It was also suggested that, because RNA can serve as both a catalyst and an informational molecule, at the time when life on Earth first began RNA may have functioned alone, in the absence of DNA or proteins [4].

4.2.4 tRNA Splicing

All the genes that code for tRNAs do not have noncoding sequences (introns) in their structures. In fact, only about 40 of almost 400 nuclear tRNA gene products in yeast are known to be interrupted, with only one intron having been found present just one nucleotide beyond the 3' side of the anticodon. The size of these introns varies from 14 to 46 nucleotides, and no consensus sequence has been found within them. RNA splicing in the primary transcript of the tRNA gene may occur in a different fashion, there being separate cleavage and ligation reactions (Fig. 13). The same mode of splicing as occurs in yeast has also been reported to occur in the nuclear tRNA gene products of plants, amphibians, and mammals. All of the introns in the tRNA gene products have a sequence which is complementary to the anticodon of the tRNA; this is an alternative conformation for the anticodon arm, in which the anticodon is base-paired to form an extension of the usual arm. The splicing of a tRNA gene product depends primarily on the recognition of a common secondary structure in tRNA, although to date no common sequences within the introns have been reported.

In tRNA gene product splicing, there is a cleavage of the phosphodiester bond, assisted by an endonuclease, but the hydrolysis of ATP is not required as an energy source. Subsequently, an enzyme – RNA ligase – is required for bond formation, with the ligase-catalyzed reaction requiring energy via the hydrolysis of ATP. The generation of a 2', 3'-cyclic phosphate bond also occurs during splicing in plants and mammals. On cleavage of the phosphodiester bond as a result of the endonuclease reaction, there is first a generation of 2', 3'-cyclic phosphate and 5'-OH termini, after which the cyclic phosphate is opened to form 3'-OH and 2'-phosphate groups, and the 5'-OH is phosphorylated. Following release of the intron, the tRNA half-molecules are folded into a tRNA-like

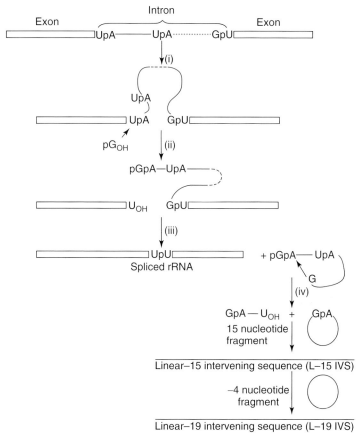

Fig. 12 Self-cleavage of the rRNA intron. The intron of pre-rRNA of *Tetrahymena* is cleaved by autocatalytic splicing. (i) Folding of RNA to act as ribozyme; (ii) A hydroxyl group attached to GTP attacks the 5′ end phosphate of the intron, such that the phosphodiester bond between exon and intron is broken and a new bond is formed between the guanine nucleotide and the intron; (iii) The hydroxyl group at the left exon attacks at the 3′ end of the intron. The bond is broken and the exons are ligated, releasing the intron; (iv) A similar reaction enables the intron to form a circle, snipping 15 nucleotides from its end in the process. The circle opens into a linear molecule, and then closes with the removal of four nucleotides. The final open form is called L-19 IVS (linear minus 19 intervening sequence).

structure that has a 3′-OH, 5′-phosphate which is sealed by the action of RNA ligase.

Ribonuclease P, a tRNA-processing enzyme that is found in both bacteria and eukaryotes, is a nucleoprotein. It was noted earlier that both the RNA and protein components are required for the nuclease to cut the tRNA precursor at a specific point. However, it was shown subsequently that only the RNA part of ribonuclease P can cut the pre-tRNA molecule at a specific site. In contrast, the protein part of the enzyme alone could not do this, which indicates that RNA catalyzes the splicing of the pre-tRNA molecule [4, 58–60].

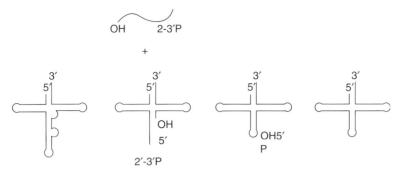

Fig. 13 Splicing of tRNA. This involves cleavage of the exon–intron boundaries by endonuclease to generate 2'-3'-cyclic phosphate and 5'-OH termini. The cyclic phosphate is opened to generate 3'-OH and 2'-phosphate groups. The 5'-OH is phosphorylated and, after releasing the intron, the tRNA half molecule folds into a tRNA-like structure that now has a 3'-OH and 5'-phosphate, which is then joined by RNA ligase.

4.3
cis- and trans-Splicing Reactions

Splicing occurs generally as an intramolecular cis-reaction in which a controlled deletion of the introns takes place. When the introns are removed from the RNA molecule, this allows the exons of the RNA molecule to be spliced together. An inter-molecular splicing also occurs, whereby the exons present in different RNA molecules can be spliced (ligated) into one molecule; these reactions, which are referred to as *trans-splicing*s (Fig. 14), are rare and never occur between pre-mRNA transcripts of the same gene.

The trans-splicing occurs *in vivo* under certain special conditions. In trypanosomes, a 35-nucleotide leader sequence is present at the end of many mRNAs; such RNA, which is known as a spliced leader (SL) RNA, donates the 5' exon required for trans-splicing. The SL RNAs that are found in certain species of trypanosomes and nematodes have common features; notably, they fold into a common secondary structure having three stem loops and a single-stranded region, resembling snRNPs. Typically, trypanosomes possess the U2, U4, and U6 snRNAs, but not the U1 and U3 snRNAs. The SL RNA functions without recognition of the 3' splicing site, and depends directly on RNA.

Some chloroplast genes are also trans-spliced. For example, the *psa* gene of the *Chlamydomonas* chloroplast has three widely separated exons, with Exon 1 being located 50 kb away from Exon 2, and Exon 2 being 90 kb away from Exon 3. Although many other genes lie between these exons, they cannot be transcribed as a common transcript, since Exon 1 is in reversed orientation from Exon 2 and Exon 3. In addition, as several other genes are required for one or the other of trans-splicing reactions, the process of splicing this mRNA together is quite complex [1].

4.4
Alternate Splicing

When a single gene provides more than one mRNA sequence, the situation is referred to as *alternate splicing*. In some cases, the use of a different start point (5' splicing site) and/or 3' splice site will

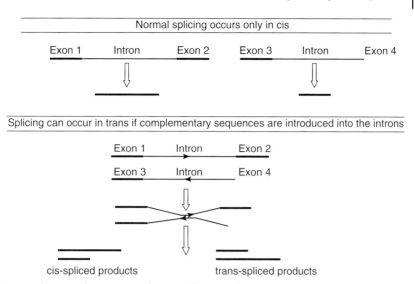

Fig. 14 Schematic diagram showing cis-splicing and trans-splicing reactions.

alter the pattern of splicing although, as noted above, this occurs only when dinucleotide sequences are present at the ends of the introns. As a result, it is possible that the same portion of the gene may act as an exon in one mRNA, and as an intron in another mRNA. Alternate splicing also occurs following the substitution, addition, or deletion of internal exons. For example, if a gene has exon number 1, 2, 3, 4, 5, 6, 7, and/or 8, these may be ligated in different ways, such as 1, 2, 3, 4; 2, 3, 5, 7; 1, 2, 4, 6; 1, 2, 3, 7; and 1, 3, 5, 6. These multiple products are created in the same cell, but in other cells the process may be regulated so that particular splicing patterns will occur only under certain conditions. In some cases, proteins that intervene to bias the use of alternate splice sites have been identified. In the recently sequenced human genome, the number of genes has been estimated to be in the range of 25 000, whilst the total number of proteins at different stages of development has been estimated at almost 500 000. This situation has been explained on the basis of alternate splicing, it having been estimated that, under different conditions, almost 20 proteins may be coded from one gene.

In *Drosophila melanogaster*, native splicing may be caused by mutations in the genes. In the case of the T/t antigens, the 5' site for the T antigen removes a termination codon present in the t antigen mRNA – which is why the T antigen is much larger than the t antigen. In E1A transcripts, one of the 5' sites is connected to the last exon in a different open-reading frame, which then causes a change in the C-terminal region of the protein.

Drosophila flies with one X chromosome and two sets of autosomes (A) are male, while those with an equal number of chromosomes and sets of autosomes are female. The X : A ratio activates the so-called *Sxl* gene, which exerts a positive control on its own expression as well as that of three other genes. This provides a mechanism so that a single X chromosome of males is transcribed into as much RNA as the two X chromosomes of females. The

Sxl gene produces either male- or female-specific spliced transcripts which have identical 5′ ends but differ in the presence or absence of a small male-specific exon that inserts a stop codon into the transcript. The protein encoded by the *Sxl* gene has an 80-amino acid RNA-binding domain; the same motif has been reported in many other RNA-binding proteins and perhaps provides a clue as to the control of its own processing, and that of further proteins in the regulatory cascade.

Another important example of alternate splicing is the tropomyosin genes of *Drosophila* and vertebrates. The tropomyosins are a family of closely related proteins that mediate the interactions between actin and troponin, and help in the regulation of muscle contraction. Different tissues – both muscle and non-muscle – are characterized by the presence of different tropomyosin isoforms. It is generally considered that many of these isoforms are produced from the same gene, via alternate splicing [1].

5
Role of microRNAs (miRNAs) in the Regulation of Gene Expression

During recent years, it has been argued that a majority of RNA molecules represent the principal actors in the largely unexplored networks of gene regulation. Indeed, it has been suggested that an understanding of RNA-based gene regulatory networks might provide the key to explaining the difference between a yeast cell and a fruitfly, and/or between a fruitfly and a human. According to John Mattick (University of Queensland in Brisbane, Australia), complexity is hidden in the noncoding output of the genome. Recently, a new class of noncoding RNAs has been reported – microRNAs (miRNAs) – which have been predicted to regulate the production of proteins from other genes. The genomes of higher organisms may have up to almost 98% noncoding DNA sequence, much of which is never read at all, although some of it may be transcribed to RNA; in this case, it is considered that the genes contain noncoding introns between the exons. When a HnRNA is transcribed from a gene, the introns are cut out and the exons ligated. Many sequences outside protein-coding genes are also transcribed into RNA [61]. According to Mattick, noncoding RNAs interact with one another, with mRNAs, with DNA, and also with proteins, to form networks that can regulate gene activity with almost infinite potential complexity. This is a very convincing suggestion, as a straightforward comparison of gene numbers cannot explain the difference between simple and complex organisms. As humans appear not to have a much larger number of genes than do simple organisms (e.g., the nematode *Caenorhabditis elegans*), it would appear that higher organisms bolster their complexity by "mixing and matching" the protein domains, so as to generate new combinations (although other ploys may be required to explain the complexity of humans and other vertebrates). Mattick compared the RNA-based networks with a computer, as the controlling software of which allows the processor to be easily reconfigured for a new task by changing the control codes. Evidence became available that, in human chromosomes, many more (up to 10-fold) sequences were being transcribed than was predicted, and therefore, the role(s) of the noncoding sequences is (are) of great importance. Whilst the noncoding sequences have been shown to be common, it cannot be proved on this basis that most are involved in networks of gene regulation. The subsequent discovery of miRNAs strengthened this concept, however. With a length of approximately 22 nucleotides, the first genes coding for miRNAs, lin-4 and let-7, were identified in C. elegans. The miRNAs are known to be cut from longer hairpin-shaped RNAs that are transcribed from lin-4 and let-7, and bind to specific target mRNAs, thus blocking their translation to proteins. Evidence has also been provided for the presence of miRNAs in a diverse range of species, including vertebrates and plants. An intriguing link to a gene-silencing mechanism – RNA interference (RNAi) – that is considered to defend cells from viruses and jumping genes, has also been identified. The role of RNAi begins when the cell detects an unusual RNA with paired strands; an enzyme known as *Dicer* then cleaves the offending double-stranded RNA into fragments of 21–25 nucleotides that are referred to as small interfering RNAs (siRNAs). Single strands from these fragments then bind to further copies of the original RNA, targeting them for destruction. RNAi has also been used experimentally to silence a cell's own genes by adding double-stranded RNA sequences that match the gene's mRNA [62–64].

Previously, miRNAs have been considered as being regulators for all types of biological systems, one such role being to convert proliferating oligodendrocyte precursor cells into mature, myelinating oligodendrocytes. As noted above, Dicer 1 was found to be involved in the processing of larger RNA precursors into smaller, active, 20- to 24-nucleotide miRNAs, and subsequent knockout studies conducted with Dicer 1 in an oligodendrite precursor cell lineage in mice led to the creation of animals without myelin. Along the same lines, attempts have been made to correlate miRNAs with the demyelination of dendrocytes, as occurs in Alzheimer's disease [64–66].

6
Chromatin Structure and the Control of Gene Expression

As noted above, two forms of chromatin structure have been identified, namely heterochromatin and euchromatin, the original designation being based on cytological observations of how darkly the two regions could be stained. Heterochromatin is more densely packed than euchromatin, is often located close to the centromeres of the chromosomes, and is generally transcriptionally inactive. In contrast, euchromatin is more loosely packed and transcriptionally active.

Whilst it is possible to predict the transcriptionally active regions of chromatin, based on cytological assays, more modern investigations have defined the molecular basis for chromatin structure in the context of the regulation of gene expression. Two primary mechanisms exist that alter chromatin structure and, consequently, affect gene expression: (i) the methylation of cytidine residues in the DNA, located in the dinucleotide CG (this is most often written as a CpG dinucleotide); and (ii) histone modification(s).

While previous observations have suggested that over 90% of methyl-C is located in the dinucleotide, CpG, not all CpG dinucleotides will have a methylated C residue. It has also been shown that the promoter regions of genes contain 10- to 20-fold more CpGs than the remainder of the genome. In general, there is an inverse relationship between methylation and transcription; for example, when cells undergo differentiation, the transcriptionally active genes have been shown to exhibit a reduction in methylation level compared to that prior to activation, and that such under-methylation persists when the transcription has ceased. The role of DNA methylation in the control of gene transcription was first demonstrated by treating cells in culture with the cytidine analog, 5-azacytidine (5-azaC), which has a nitrogen instead of a carbon at position 5 of the pyrimidine ring and so cannot serve as a substrate for methylation. When fibroblasts were grown in the presence of 5-azaC, and then differentiated into myoblasts, such differentiation was shown to have resulted from an under-methylation and activation of the *MyoD* gene (a master regulator of muscle differentiation).

The methylation of DNA is catalyzed by several different DNA methyltransferases. The critical role of DNA methylation in controlling developmental fate was observed in mice by inactivating either DNA methyltransferase 3a or 3b, whereby the loss of either gene resulted in animal death shortly after birth. When cells divide, the newly formed DNA will contain one strand of parental DNA, and one newly replicated DNA strand. However, if the DNA contains methylated cytidines in the CpG dinucleotides, then the newly replicated DNA strand should be methylated in order to maintain the parental pattern of methylation. Such "maintenance" methylation is catalyzed by DNA methyltransferase 1 (also referred to as *maintenance methylase*).

Today, many proteins have been identified that bind to methylated, but not unmethylated, CpGs. One such protein example is methyl CP binding protein 2 (MeCP2) which, when bound to methylated CpG dinucleotides, causes the DNA to take on a closed chromatin structure, with the subsequent repression of transcription. The ability of MeCP2 to bind methylated CpGs is controlled by its phosphorylation and dephosphorylation states. Although, phosphorylated MeCP2 has a lesser affinity for methylated CpGs, its binding leads to the DNA acquiring a more open chromatin state. The importance of MeCP2 in regulating chromatin structure and, consequently, the transcription process, has been confirmed by the fact that a deficiency in this protein results in Rett syndrome. This neurodevelopmental disorder occurs almost exclusively in females, and manifests as mental retardation, seizures, microcephaly, arrested development, and loss of speech.

Those histone proteins that remain bound to DNA also undergo a number of modifications that affect the chromatin structure. In fact, it has been shown that if the histone is acetylated then the chromatin structure will be more open, and such modified histones will be located in regions of transcriptionally active chromatin. A direct correlation between histone acetylation and transcriptional activity has been confirmed by the fact that protein complexes, known previously as *transcriptional activators*, demonstrate histone acetylase activity, whereas transcriptional repressor complexes possess histone deacetylase activity. Other proteins that interact with acetylated lysines in histones together form a more open chromatin structure. Those proteins that bind to acetylated histones incorporate a so-called *bromodomain*, which is composed of a bundle of four α-helices and is involved in protein–protein interactions in several cellular systems, in addition to acetylated histone binding and chromatin structure modification.

Both, acetylation and methylation in histones have been shown to affect chromatin structure, although no direct correlation between histone methylation and a specific effect on transcription has yet been observed. The methylation of histone H4 on arginine at position 4 promotes an open chromatin structure, and consequently accelerates transcriptional activation. The methylation of histone H3 on lysine at positions 4 and 79 has also been shown to accelerate transcriptional activation. In contrast, the methylation of histone H3 on lysine at positions 9 and 27 has been shown to result in transcriptionally inactive genes. The binding of some specific proteins on methylated histones may result in the formation of a more compact chromatin. Those proteins that bind to methylated lysines present in histones incorporate a so-called *"chromodomain"*, which consists of a conserved stretch of 40–50 amino acids and is found in many proteins involved in chromatin remodeling complexes. Chromodomain proteins are also found in the RNA-induced transcriptional silencing (RITS) complex, which involves the siRNA- and miRNA-mediated downregulation of transcription.

The histone proteins may also be modified by binding a small protein, *ubiquitin*, though this occurs only with histones H2A and H2B (typically, only a small percentage of histone H2A is ubiquitinated). Whilst ubiquitinated H2A is involved in the repression of transcription, ubiquitinated histone H2B causes the stimulation of gene expression. The ubiquitinated histone H2B has also been shown to promote the methylation of histone H3 at lysine at positions 4 and 79 such that, in turn, the methylated histone H3 promotes an open chromatin structure.

The phosphorylation of histones has also been reported, based on outside signals such as growth factor stimulation, or stress inducers such as heat shock. The binding of phosphorylated histones causes the genes to become transcriptionally active, an effect that becomes apparent in patients with Coffin–Lowry syndrome, a disease which results from defects in the *RSK2* gene that encodes the histone-phosphorylating enzyme. Coffin–Lowry syndrome, a rare form of X-linked mental retardation, is characterized by skeletal malformations, growth retardation, hearing deficit, paroxysmal movement disorders, and cognitive impairment in affected males [29].

7
Epigenetic Control of Gene Expression

Originally, the term "epigenetics" was coined by Conrad Waddington in 1939, to define the unfolding of the genetic program during development. In addition, the term *epigenotype* was coined to define " ... the total developmental system consisting of interrelated developmental pathways through which the adult form of an organism is realized." Nowadays, the term epigenetics is used to define the mechanism by which changes in the pattern of inherited gene expression occur in the absence of any alterations or changes in the nucleotide composition of a given gene. Epigenetics can also be explained as being " ... in addition to changes in genome sequence."

It may help to explain epigenetics through the example of a fertilized egg which, at the moment of fertilization is totipotent; that is, as the egg divides the daughter cells will ultimately differentiate into all of the different cells of the organism. The only differences between the various cells of the resultant organism are the consequences of differential gene expression; they are not due to any differences in the sequences of the genes themselves.

To date, several different types of epigenetic event have been identified, among which DNA methylation is likely to be the most important for controlling and maintaining the pattern of gene expression during development. Other DNA-modifying events that are also known to affect the epigenetic phenomenon include the acetylation, methylation, phosphorylation, ubiquitylation, and sumoylation of histone proteins. Consequently, the same events that affect chromatin structure can be considered also as epigenetic events. Notably, the control of gene expression by siRNAs is also considered to be an epigenetic event.

Epigenesis plays an important role in the regulation and maintenance of gene expression, and may result in many differentiation states of cells within an organism. Recently acquired evidence has demonstrated a connection between epigenetic processes and diseases, the most significant of which is the link between epigenesis and cancer (epigenesis has been suggested as a contributory factor in many different types of cancer). In particular, a correlation has been observed between changes in the methylation status of the tumor suppressor genes and the development of many types of cancer. Epigenetic effects on immune system function have also been reported, as has a correlation between epigenetic processes and mental health [29].

8
Gene Regulation by Hormonal Action

It has been shown that signals originate from various glands and/or secretory cells that stimulate the target tissues or cells to carry out dramatic changes in their metabolic patterns, including altered patterns of differentiation. As peptide hormones are generally larger molecules, and are generally unable to enter the cell, they exert their effects by binding to cell-surface receptors, with subsequent activation of the protein enzyme transcription factors via a mechanism of phosphorylation.

In contrast, steroid hormones (e.g., estrogens) are smaller molecules that can readily penetrate the plasma membrane. Following entry, these molecules become tightly bound to specific receptor proteins that are present only in the cytoplasm of the target cells.

Hormone–receptor protein complexes may activate the transcription of specific genes in two different ways:

- The hormone–receptor protein complex activates the transcription of target genes by binding to specific DNA sequences present in the *cis*-acting regulatory regions of genes.
- The hormone–receptor protein complex interacts with specific nonhistone chromosomal proteins, after which the complex stimulates transcription of the correct genes.

In the past, it has been considered that nonhistone chromosomal proteins play an important role in the regulation of gene expression in eukaryotes. However, further evidence suggests that the hormone–receptor protein complexes may activate gene expression by interacting directly with specific DNA sequences present within the enhancer or promoter regions, that regulate transcription of the target genes [67, 68].

In addition, the possibility exists that histone modifications or nonhistone chromosomal proteins are involved in some aspects of hormone-regulated gene expression.

9
Post-Transcriptional Regulation of mRNA

Although the regulation of gene expression in eukaryotes at the level of initiation of transcription is considered to be very important, regulation at the level of post-transcription has also been noted in many cases. Although capping at the 5′ end of the eukaryotic mRNA is considered essential, polyadenylation at the 3′ end has not been identified in all mRNAs. Whether the inhibition of polyadenylation is used specifically to block the expression of particular genes is not known, although some genes have multiple putative polyadenylation sites that may be used for alternate splicing (i.e., the formation of more than one mRNA from one gene). The choice of polyadenylation site may also vary during the development of a cell, with the switching of splicing patterns occurring in a developmentally significant manner.

Polyadenylation does not occur only at the extreme 3′ end of the mRNA; rather, between 10 and 30 nucleotides may be transcribed that precede the polyadenylation signal, which has the sequence 5′AAUAAA3′, or a variant of it. These terminal nucleotides are cleaved with the assistance of an endonuclease, thereby producing an intermediate 3′ end to which the polyA tail is subsequently attached by the enzyme polyA polymerase. For polyadenylation, there is a requirement for a specificity factor that also recognizes the 5′AAUAAA3′ sequence. This specificity factor incorporates three subunits which, together, will bind specifically to RNA containing the sequence 5′AAUAAA3′. The polyA polymerase first synthesizes almost 10 residue oligo-As at the 3′ end of the mRNA, in the presence of a specificity factor; subsequently, this oligo-A tail is extended to almost 200 residues in the presence of another factor which recognizes the oligo-A tail and directs polyA polymerase to catalyze its extension. As noted above, the polyadenylation of mRNA is not essential for further translation; rather, it is considered that it may affect the stability of the mRNA in the cell. The polyA tail is associated with a particular protein termed the polyA binding protein (PABP); it is believed that the binding of polyA with PABP is essential to protect mRNA against degradation by nucleases [1, 69, 70].

As the stability of mRNA may be regulated in the cytoplasm, this may result in changes in its concentration. In fact, it has been found that estrogen not only induces transcription of the vitellogenin gene but also increases the stability of its mRNA in the cytoplasm, increasing its half-life from 16 h to 300 h.

Among the eukaryotes, all mRNAs have been shown to possess a 5′ cap. Although the exact significance of capping is unclear, it is thought to serve as a recognition point for the attachment of a ribosome at the outset of translation. This is considered equivalent to the Shine–Dalgarno sequence (GGAGGC), which is found in prokaryotes and is the sequence to which the small subunit of the ribosome attaches in order to commence protein biosynthesis. The ribosome recognizes the cap structure as its binding site and, after becoming attached, migrates along the mRNA until it reaches the initiation codon. Those mRNAs that are translated on cytoplasmic ribosomes have also been shown to be capped, but no capping has been identified on mitochondrial and chloroplast mRNAs.

In eukaryotes, a mRNA after splicing at the 5′ end has the structure 5′pppPuNp----3′, where Pu is a purine residue, N is the sugar component of the nucleotide, and p represents a phosphate group. However, mature mRNA (after post-transcriptional changes) has at the 5′ end 5′-7-mGpppPuNp-----3′, where 7mG (7-methyl guanosine) is attached after transcription and is known as a *cap*. During capping, cleavage of the terminal phosphate group of the first nucleotide occurs, catalyzed by a phosphohydrolase. Subsequently, a guanylyl residue is transferred to the 5′ end from GTP by the enzyme, guanylyl transferase, and thereafter modified to a 7-methyl guanylyl residue by the enzyme mRNA, guanylyl-7-methyltransferase in the presence of S-adenosyl methionine (SAM), which acts as a methyl donor. In this case, the newly added guanylyl residue is in the reverse orientation compared to other nucleotides present in the mRNA. A cap containing a single methyl group is known as *cap o*; however, if there is an addition of another methyl group on the second nucleotide (which was in fact the first nucleotide in the original mRNA), this is referred to as *cap 1* (though this occurs only if it is an adenine nucleotide). The methyl group is added on the N^6 position of the adenine nucleotide.

In some cases, another methyl group may be added to the third nucleotide; the substrate for this reaction is cap 1 mRNA, and the acceptor of the methyl group is the ribosyl moiety at the 2′ position, and this is referred to as *cap 2*. This reaction is catalyzed by the enzyme 2′-O-methyltransferase, while the methyl group donor (SAM) is unchanged. The number of caps is considered characteristic of the organism, while a low frequency of internal methylation (1 in 1000 nucleotides) is also known to occur in the mRNA of higher eukaryotes [1].

In prokaryotic mRNA, post-transcriptional changes do not generally occur. Rather, because of an absence of compartmentation there may be a coupled translation whereby, as soon as mRNA biosynthesis occurs (or is in progress), it may bind with the ribosome to begin translation.

10
Transport of Processed mRNA to the Cytoplasm

Following any post-transcriptional modifications, the matured mRNA is transported from the nucleus in a very rapid process and, on entering the cytoplasm, becomes bound to the cytoplasmic ribosomes in readiness for translation. The latter process occurs within only 1–5 min after the mRNA has left the nucleus. It has been suggested that specific proteins exist to assist the transportation of mature mRNA from the nucleus, though the exact process involved is not presently understood.

Evidence indicates that the mRNA transport process is not restricted to the simple passage of mRNA through the nuclear pore complex, which spans the nuclear envelope; rather, it is embedded into the gene-expression pathway. During transcription, the message is capped, spliced, and polyadenylated, while the mRNA export factors are loaded onto the nascent transcript. This maturation and assembly of the mRNA to form a messenger ribonucleoprotein particle (mRNP) is controlled by nuclear surveillance systems; the nuclear exosome and the Mlp1–2 system combine to prevent the escape of aberrant transcripts to the cytoplasm. As a consequence, only correctly assembled mRNPs are transported through the nuclear pore to the cytoplasm by the mRNA export receptor Mex67–Mtr2/Tap–p15, which is attached to the mRNA by interaction with the mRNP-bound transcription export (TREX) complex and splicing reporter (SR) proteins [71].

11
Regulation of Gene Expression at the Level of Translation

The majority of the regulation of gene expression takes place at the level of transcription. The production of mRNA involves many steps, several of which – such as promoter utilization, RNA splicing, and polyadenylation – are known to be regulated. Whilst pre-mRNA stability and the transport of mRNA from the nucleus to the cytoplasm provide a very rapid control of gene expression, on occasion the level of translation may be manipulated by changing the essential components of the translational machinery of the cell. In this regard, phosphorylation of the ribosomal components (particularly 5S rRNA in the 40S ribosomal subunit) has been correlated with higher polysome levels in the presence of different growth factors in mammalian cells. A similar situation occurs in the brine shrimp egg where, upon fertilization, a previously absent translational initiation factor that is involved in polysome formation suddenly appears [1, 72].

In mammalian reticulocytes the control of protein synthesis by hemin is mediated via the formation from a presynthesized precursor (prorepressor) of a potent inhibitor of a polypeptide chain initiator, the hemin controlled repressor (HCR). Despite these cells having lost their nuclei, they retain high levels of stable mRNAs that encode mostly hemoglobin chains. In reticulocyte lysates, protein synthesis occurs at high rate but declines rapidly in the absence of hemin. Within the cells, hemin synthesis occurs in the mitochondria, but these are absent from the lysate. In fact, HCR is activated in the absence of heme, but inhibited in its presence. Although the mode of action of HCR was a mystery for many years, it has now been shown to act as a specific kinase for phosphorylation of the α subunit of translation initiation factor 2. The presence of the eukaryotic initiation factor 2 (eIF2), GTP–GDP exchange cycle leads to the phosphorylation of even a small fraction of eIF2 being sufficient to block the initiation of protein synthesis. Indeed, it appears that all of the eIF2B (which is present in lesser amounts than eIF2) is sequestered into eIF2–eIF2B complexes, such that it is no longer available to recycle the remaining unphosphorylated eIF2 [1, 73].

A translational inhibitor, which is present in Friend leukemia cells and has been characterized as a heat-labile, sulfhydryl reagent-insensitive protein of molecular weight almost 214 kDa, inhibits the initiation of protein synthesis by preventing the initiation factor-dependent binding of methionyl-tRNA to the 40S ribosomal subunit. However, this does not interfere with the formation of a ternary complex between eIF2, methionyl-tRNA, and GTP. Rather, the inhibitor functions as a protein kinase which phosphorylates the α subunit of eIF2, and has been considered analogous to the HCR of reticulocyte cells [1, 74].

A phosphoprotein phosphatase enzyme capable of releasing the phosphate group from the phosphorylated α subunit of eIF2 has been reported in rabbit reticulocytes, that could restore the activity of eIF2 lost after phosphorylation. The activity of this enzyme was stimulated almost threefold by an optimal concentration of Mn^{2+} ions, but not by Ca^{2+} or Mg^{2+} ions. In contrast, the enzyme activity was greatly inhibited by Fe^{2+} ions and purine nucleoside diphosphates [1, 75].

During post-translational modifications, many proteins are modified by processes such as phosphorylation, acetylation, and hydroxylation at the side chains of the amino acids. In many proteins, there is also conjugation of nonprotein component(s).

Recently, the post-translational regulation of transcription factors has been shown to play an important role in the control of gene expression in eukaryotes. The mechanisms of regulation include not only factor modifications, but also regulated protein–protein interaction, protein degradation, and intracellular partitioning. In plants, the basic-region leucine zipper (bZIP) transcription factors contribute to many transcriptional response pathways. It has been suggested that plant bZIP factors are under the control of various partially signal-induced and reversible post-translational mechanisms that are crucial for the control of their function. However, only a few plant bZIPs have yet been investigated with respect to post-translational regulation [76].

Oct4 is a key component of the molecular circuitry which regulates embryonic stem cell proliferation and differentiation. It is essential for the maintenance of undifferentiated, pluripotent cell populations, and binds with DNA in multiple heterodimeric and homodimeric configurations. At present, very little is known regarding the regulation of the formation of these complexes, and of the mechanisms by which Oct4 proteins respond to complex extracellular stimuli that regulate pluripotency. However, a phosphorylation-based mechanism has been proposed for the regulation of specific Oct4 homodimer conformations, whereby the point mutations of a putative phosphorylation site might specifically abrogate the transcriptional activity of a specific homodimer assembly, with minimal effect on other configurations. It has also been shown that altering the Oct4 protein levels has an effect on the transcription of Oct4 target genes, with several signaling pathways having been identified that may mediate this phosphorylation and act in combination to regulate Oct4 transcriptional activity and protein stability [77].

Other strategies that act either at or before the translation initiation step include alterations to the inherent variability in the life span of eukaryotic mRNA, and mRNA stability in response to certain agents.

Acknowledgments

The authors acknowledge the facilities of the Distributed Bioinformatics Sub-Centre of the Department of Biotechnology, Government of India, New Delhi within the School of Biotechnology, Devi Ahilya University, Indore used in the preparation of this chapter. A.K. thanks Horizon Scientific Press, Wymondham, England for granting permission to use some material and figures from the author's own book, *Advanced Topics in Molecular Biology* (copyright is with the publisher).

References

1. Kumar, A., Srivastava, A.K. (Eds) (2001) *Advanced Topics in Molecular Biology*, Horizon Scientific Press, Wymondham.
2. Alberts, B., Bray, D., Lewis, J., Raff, M., Roberts, K., Watson, J. (Eds) (1994) *Molecular Biology of the Cell*, Garland, New York.
3. Lehninger, A.L., Nelson, D.L., Cox, M.M. (Eds) (1993) *Principles of Biochemistry*, CBS Publishers & Distributors, Delhi.
4. Cech, T.R., Bass, B.L. (1986) Biological catalysis by RNA. *Ann. Rev. Biochem.*, **55**, 599–629.
5. Kumar, A., Garg, N. (2006) *Genetic Engineering*, Novascience Publishers Inc., New York.
6. Gossen, M., Bujard, H. (2002) Studying gene function in eukaryotes by conditional gene inactivation. *Annu. Rev. Genet.*, **36**, 153–173.
7. Yen, L., Magnier, M., Weissleder, R., Stockwell, B.R., Mulligen, R.C. (2006) Identification of inhibitors of ribozyme self cleavage in mammalian cells via high-throughput screening of chemical libraries. *RNA*, **12** (5), 797–806.
8. Miller, J.H., Reznikoff, W.S. (1978) *The Operon*, Cold Spring Harbor Laboratory Press, New York.
9. Bourgeois, S., Pfahl, M. (1976) Repressor. *Adv. Protein Chem.*, **30**, 1–99.
10. Lewis, M. (2005) The lac repressor. *C. R. Biol.*, **328**, 521–548.
11. Lewin, B. (1994) *Genes V*, Oxford University Press, Oxford.
12. Raiband, O., Schwartz, M. (1984) Positive control of transcription initiation in bacteria. *Annu. Rev. Genet.*, **18**, 173–206.
13. Deutscher, J. (2008) The mechanisms of carbon catabolite repression in bacteria. *Curr. Opin. Microbiol.*, **11**, 87–93.
14. Crombrugghe, B.D., Busby, S., Buc, H. (1984) Cyclic AMP receptor protein: role in transcription activation. *Science*, **224**, 831–838.
15. Gesteland, R.F., Atkins, J.F. (1993) *The RNA World*, Cold Spring Harbor Laboratory Press, New York.
16. Romeo, T., Preiss, J. (1989) Genetic regulation of glycogen biosynthesis in *Escherichia coli*: in vitro effects of cyclic AMP and guanosine-5'-diphosphate-3'-diphosphate and analysis of in vitro transcripts. *J. Bacteriol.*, **171**, 2773–2782.
17. Abound, M., Pastan, I. (1975) Activation of transcription by guanosine-5'-diphosphate-3'-diphosphate, transfer ribonucleic acid and a novel protein from *Escherichia coli*. *J. Biol. Chem.*, **250**, 2189–2195.
18. Leckie, M.P., Tieber, V.L., Porter, S.E., Roth, W.G., Dietzler, D.N. (1985) Independence of cyclic AMP and rel A gene stimulation of glycogen synthesis in intact *Escherichia coli* cells. *J. Bacteriol.*, **161**, 133–140.
19. Primakoff, P., Artz, S.W. (1979) Positive control of lac operon expression *in vitro* by guanosine-5'-diphosphate-3'-diphosphate. *Proc. Natl Acad. Sci. USA*, **76**, 1726–1730.
20. Blount, K.F., Breaker, R.R. (2006) Riboswitches as antibacterial drug targets. *Nat. Biotechnol.*, **24**, 1558–1564.
21. Knight, J. (2003) Switched on to RNA. *Nature*, **425**, 232–233.
22. Trun, N., Trempy, J. (2004) Gene expression and regulation, in: *Fundamental Bacterial Genetics*, Blackwell Publishing, Malden, USA, Chapt. 12, pp. 191–212.
23. Janion, C. (2001) Some aspects of the SOS response system – a critical survey. *Acta Biochim. Pol.*, **48**, 599–610.
24. Erill, I., Campoy, S., Barbe, J. (2007) Aeons of distress: an evolutionary perspective on the bacterial SOS response. *FEMS Microbiol. Rev.*, **31**, 637–656.
25. Schlacher, K., Pham, P., Cox, M.M., Goodman, M.F. (2006) Roles of DNA polymerase V and RecA protein in SOS damage induced mutation. *Chem. Rev.*, **106**, 406–419.
26. DNA repair. http://en.wikipedia.org/wiki/DNA_repair (accessed 11 February 2010).
27. Collins, S., Bolanowski, M.A., Caron, M.G., Lefkowitz, R.J. (1989) Genetic regulation of β-adrenergic receptors. *Annu. Rev. Physiol.*, **51**, 203–215.
28. Anonymous (2002) The race to computerize biology, in *The Economist*, December 12, 2002.
29. Epigenetic control of gene expression. http://themedicalbiochemistrypage.org/gene-regulation.html (accessed 22 March 2011).
30. Jaenisch, R., Bird, A. (2003) Epigenetic regulation of gene expression: how the genome integrates intrinsic and environmental signals. *Nat. Genet.*, **33**, 245–254.

31 http://www.molecular-plant-biotechnology.info/regulation-of-gene-expression/mechanisms-of-gene-regulation-in-eukaryotes.htm (accessed 22 March 2011).
32 Britten, R.J., Davidson, E.H. (1969) Gene regulation for higher cells: a theory, new facts regarding the organization of the genome provide clues to the nature of gene regulation. *Science*, **165**, 349–357.
33 Niehrs, C., Pollet, N. (1999) Synexpression groups in eukaryotes. *Nature*, **402**, 483–487.
34 Herstein, P.R., Frenster, J.H. (1972) Mated Models of Gene Regulation in Eukaryotes, in: Anderson, N.G., Coggin, J.H. (Eds) *Embryonic and Fetal Antigens in Cancer*, Vol. 2, National Technical Information Service, U.S. Department of Commerce, Springfield, pp. 5–7.
35 Adams, M.D., Rudner, D.Z., Rio, D.C. (1996) Biochemistry and regulation of pre-mRNA splicing. *Curr. Opin. Cell Biol.*, **8**, 331–339.
36 Godde, J.S., Ura, K. (2009) Dynamic alterations of linker histone variants during development. *Int. J. Dev. Biol.*, **53**, 215–229.
37 Lehninger, A.L., Nelson, D.L., Cox, M.M. (2005) Regulation of gene expression, in: *Principles of Biochemistry. Part IV: Information Pathways*, 2nd edn, CBS Publishers and Distributors, Delhi, India, Chapt. 27, pp. 973–977.
38 Osterhage, J.L., Friedman, K.L. (2009) Chromosome end maintenance by telomerase. *J. Biol. Chem.*, **284**, 16061–16065.
39 Singer, M., Berg, P. (1991) *Genes and Genome*, University Science Press, Mill Valley.
40 Latchman, D.S. (1993) *Eukaryotic Transcription Factors*, Academic Press, New York.
41 Wasyl, K.B. (1988) Enhancers and transcription factors in control of gene expression. *Biochim. Biophys. Acta*, **951**, 17–35.
42 Maldonado, E., Reinberg, D. (1995) News on initiation and elongation of transcription by RNA polymerase II. *Curr. Opin. Cell Biol.*, **7**, 352–361.
43 Mcknight, S.L., Yamamoto, K.R. (1992) *Transcriptional Regulation*, Cold Spring Harbor Laboratory Press, New York.
44 Nagai, K. (1996) RNA-protein complexes. *Curr. Opin. Struct. Biol.*, **6**, 53–61.
45 del Campo, E.M. (2009) Post-transcriptional control of chloroplast gene expression. *Gene Regul. Syst. Biol.*, **3**, 31–47.
46 Liere, K., Borner, T. (2007) Transcription of Plastid Genes, in: Grasser, K.D. (Ed.) *Regulation of Transcription in Plants*, Blackwell Publishing, Oxford, pp. 184–224.
47 Little, M.C., Hallick, R.B. (1988) Chloroplast *rpo*A, *rpo*B and *rpo*C genes specify at least three components of a chloroplast DNA dependent RNA polymerase active in tRNA and mRNA transcription. *J. Biol. Chem.*, **263**, 14302–14307.
48 Mitochondrial gene regulation. http://www.molecular-plant-biotechnology.info/mitochondrial-genome/gene-content-structure-and-expression-of-mitochondrial-genome.htm (accessed 22 March 2011).
49 Smart, C.J., Moneger, F., Leaver, C.J. (1994) Cell-specific regulation of gene expression in mitochondria during anther development in sunflower. *Plant Cell*, **6**, 811–825.
50 Sharp, P.A. (1987) Splicing of mRNA precursors. *Science*, **253**, 766–771.
51 Smith, H.C., Snowden, M.P. (1996) Base modification mRNA editing through deamination – the good, the bad, and the unregulated. *Trends Genet.*, **12**, 418–424.
52 Tijan, R. (1995) Molecular machines that control genes. *Sci. Am.*, **272**, 38–45.
53 Mattick, J.S. (2004) RNA regulation: a new genetics? *Nat. Rev. Genet.*, **5**, 316–323.
54 Akhtar, A. (2003) Dosage compensation: an intertwined world of RNA and chromatin remodelling. *Curr. Opin. Genet. Dev.*, **13**, 161–169.
55 Darnell, J. (1982) Variety in the level of gene control in eukaryotic cell. *Nature*, **297**, 365–371.
56 Morrissey, J.P., Tollervery, D. (1995) Birth of the snoRNPs – the evolution of RNase MRP and the eukaryotic pre-rRNA-processing system. *Trends Biochem. Sci.*, **20**, 78–82.
57 Ross, J. (1996) Control of mRNA stability in higher eukaryotes. *Trends Genet.*, **12**, 171–175.
58 Scott, W.G., Klug, A. (1996) Ribozymes: structure and mechanism of RNA catalysis. *Trends Biochem. Sci.*, **21**, 220–224.
59 Tuschl, T., Thomson, J.B., Eckstein, F. (1995) RNA cleavage by small catalytic RNAs. *Curr. Opin. Struct. Biol.*, **5**, 296–302.
60 Padgett, R.A. (1985) Splicing messenger RNA precursor: branch site and lariat RNAs. *Trends Biochem. Sci.*, **10**, 154–157.

61 Dennis, C. (2002) The brave new world of RNA. *Nature*, **418**, 122–124.

62 Mattick, J.S. (2001) Non-coding RNAs: the architects of eukaryotic complexity. *EMBO Rep.*, **2**, 986–991.

63 Mattick, J.S. (2003) Challenging the dogma: the hidden layer of non-protein-coding RNAs in complex organisms. *BioEssays*, **24**, 930–939.

64 Zhao, X., He, X., Han, X., Yu, Y., Ye, F., Chen, Y., Hoang, T.N., Xu, X., Mi, Q.-S., Xin, M., Wang, F., Appel, B., Lu, Q.R. (2010) MicroRNA-mediated control of oligodendrocyte differentiation. *Neuron*, **65**, 612–626.

65 Dugas, J.C., Cuellar, T.L., Scholze, A., Ason, B., Ibrahim, A., Emery, B., Zamanian, J.L., Foo, L.C., McManus, M.T., Barres, B.A. (2010) Dicer 1 and miR-219 are required for normal oligodendrocyte differentiation and myelination. *Neuron*, **65**, 597–611.

66 Nave, K.A. (2010) Oligodendrocytes and the "Micro Brake" of progenitor cell proliferation. *Neuron*, **65**, 577–579.

67 Yamamoto, K.R. (1985) Steroid receptor regulated transcription of specific genes and gene networks. *Annu. Rev. Genet.*, **19**, 209–252.

68 Gorski, J., Seyfred, M.A., Kladde, M.P., Meier, D.A., Murdoch, F.E. (1990) Steroid hormone regulation of gene expression. *J. Anim. Sci.*, **68**, 18–27.

69 Wahle, E., Keller, W. (1992) Polyadenylation. *Annu. Rev. Biochem.*, **61**, 419–440.

70 Wahle, E., Keller, W. (1996) The biochemistry of polyadenylation. *Trends Biochem. Sci.*, **21**, 247–250.

71 Rother, S., Strasser, K. (2009) mRNA Export- An Integrative Component of Gene Expression, in: Kehlenbach, R.H. (Ed.) *Nuclear Transport*, Landes Bioscience, Austin, Texas.

72 Laemmli, U.K., Tijan, R. (1996) Nucleus and gene expression – a nuclear traffic jam – unraveling multicomponent machines and compartments. *Curr. Opin. Cell Biol.*, **8**, 299–302.

73 Gross, M. (1979) Control of protein synthesis by hemin: evidence that the human controlled translational repressor inhibits formation of 80S initiation complexes from 48S intermediate initiation complexes. *J. Biol. Chem.*, **254**, 2370–2377.

74 Pinphanichakarn, P., Kramer, G., Hardesty, B. (1977) Partial purification and characterization of a translational inhibitor from Friend leukemia cells. *J. Biol. Chem.*, **252**, 2106–2112.

75 Grankawski, N., Lehmusvirta, D., Kramer, G., Hardesty, B. (1980) Partial purification and characterization of reticulocyte phosphatase with activity for phosphorylated peptide initiation factor-2. *J. Biol. Chem.*, **255**, 310–317.

76 Schutze, K., Harter, K., Chaban, C. (2008) Post-translational regulation of plant bZIP factors. *Trends Plant Sci.*, **13**, 247–255.

77 Saxe, J.P., Tomilin, A., Scholer, H.R., Plath, K., Huang, J. (2009) Post-translational regulation of Oct4 transcriptional activity. *PLoS ONE*, **4**, e4467.

11
RNA Silencing in Plants

Charles W. Melnyk[1] *and C. Jake Harris*[2]
[1]*The Sainsbury Laboratory, University of Cambridge, Bateman Street, Cambridge CB2 1LR, UK*
[2]*Department of Plant Sciences, University of Cambridge, Downing Street, Cambridge, CB2 3EA, UK*

1	Introduction	348
2	An Historical Overview of RNA Silencing	348
3	The Core Mechanism	350
4	Double-Stranded RNA	351
5	Dicers	351
6	Argonautes	352
7	Signal Amplification	354
8	**MicroRNAs**	**354**
8.1	The Biogenesis of miRNAs	355
8.2	miRNA Targeting Mechanisms and Prediction	355
8.3	Origins and Evolution	357
8.4	The Function of miRNAs	357
9	**Small Interfering RNAs**	**358**
9.1	Trans-Acting siRNAs	358
9.2	Natural Antisense siRNAs	360
9.3	Heterochromatin-Associated siRNAs	360
9.3.1	The Biogenesis of hc-siRNAs	360
9.3.2	RNA-Directed DNA Methylation	363
9.3.3	The Function of hc-siRNAs	364
10	**Viruses and RNA Silencing**	**364**
10.1	Plant Viruses: Genomes, Replication Strategies, and Detection by Host RNA Silencing Machinery	365

RNA Regulation: Advances in Molecular Biology and Medicine, First Edition. Edited by Robert A. Meyers.
© 2014 Wiley-VCH Verlag GmbH & Co. KGaA. Published 2014 by Wiley-VCH Verlag GmbH & Co. KGaA.

10.2	Host Factors Involved in Antiviral Silencing 365
10.3	Viral Silencing Repressors 367
10.4	Crosstalk: RNA Silencing and the Innate Immune System 370

11	**Mobile Small RNAs 371**
11.1	Cell-to-Cell Movement of an RNA Silencing Signal 371
11.2	Long-Distance Movement of an RNA Silencing Signal 372
11.3	Functions of sRNA Movement 374

| **12** | **Conclusion 375** |

Acknowledgments 376

References 376

Keywords

Argonaute (AGO)
A protein that binds small RNAs and uses these molecules as a guide to seek out, by Watson–Crick base-pairing, complementary RNA. Binding of the AGO-small RNA complex to the target RNA can cause RNA cleavage, degradation, translation repression, or recruit factors that direct epigenetic marks such as DNA methylation or histone modifications.

Dicer-like (DCL)
A ribonuclease protein that cleaves double-stranded RNA to generate sRNA duplexes. Known as *Dicer* in many Eukaryotes, and *Dicer-like* in plants.

MicroRNA (miRNA)
A small RNA molecule, typically 20–22 nt in length, which binds complementary mRNA to direct cleavage or translational repression. miRNAs are formed from an imperfect hairpin-like sequence and, for a given miRNA gene, a single type of mature duplex is generated.

Natural antisense small interfering RNA (nat-siRNA)
Small RNA molecules formed by the DCL-mediated cleavage of double-stranded RNA that arises from overlapping mRNA transcripts.

Post-transcriptional gene silencing (PTGS)
Gene silencing that occurs after RNA has been transcribed. Typically, PTGS involves cleavage of the RNA transcript, RNA destabilization or repressing translation by preventing ribosomes from performing protein synthesis.

RNA-directed DNA Methylation (RdDM)
The process by which small RNAs direct DNA methylation, an epigenetic mark often associated with gene silencing.

Small RNA deep sequencing
A technique whereby the small RNA population of a sample is sequenced using high-throughput sequencing techniques such as the 454 or Illumina platforms. Typically, millions of small RNAs are sequenced, providing in-depth information of their abundance.

RNA-dependent RNA polymerase (RDR)
The enzyme that converts single-stranded RNA to double-stranded RNA. This enzyme is present in plants, yeast, and *Caenorhabditis elegans* among other species.

RNA-induced silencing complex (RISC)
A complex comprised of the sRNA loaded into the Argonaute protein. The complex may include other proteins required for the AGO-sRNA to direct gene silencing. The RISC is also referred to as the *"effector complex."*

Small interfering RNA (siRNA)
20–24 nt small RNAs generated from perfectly or near-perfectly matching double-stranded RNA. siRNAs are generated along the length of the region of double-strandedness. siRNAs are commonly produced from transposable elements, regions of heterochromatin, transgenes, and viruses. Those associated with transgenes and heterochromatin are referred to as *heterochromatin-associated small interfering RNAs* (hc-siRNAs). siRNAs can regulate gene expression via PTGS or TGS.

Small RNA (sRNA)
An RNA species that is typically 18–30 nt in length.

Trans-acting small interfering RNA (tasiRNA)
Small RNAs produced via a miRNA-initiated cleavage of an RNA transcript. This cleavage product is acted upon by an RDR to produced double-stranded RNA that is subsequently diced to produce tasiRNAs. tasiRNAs occur roughly every 21 nt along the length of the template RNA. They are referred to as *trans acting* because they target other RNA transcripts *in trans* to silencing gene expression.

Transcriptional gene silencing (TGS)
A gene-silencing phenomenon that blocks transcription of the RNA. TGS is associated with epigenetic marks, typically DNA methylation or changes in histone marks, which block transcription from occurring. TGS is directed by small RNAs often through the actions of AGO4 and DRM2, among other proteins.

The discovery of RNA silencing has greatly expanded the understanding of gene regulation. Present in nearly all Eukaryotes, RNA silencing has emerged as a potent method to regulate gene expression at the transcriptional and post-transcriptional levels. Some of the earliest observations and discoveries associated with RNA silencing were made in plants, and plant research is continuing to further the understanding of RNA silencing in other organisms. The role of RNA silencing in regulating plant development, stress responses, and adaptation is becoming increasing clear. An overview of the various plant RNA silencing pathways and their functions is presented in this chapter, and some of the latest findings in this field will be discussed.

1
Introduction

Gene regulation is a fundamental process of living organisms. Cells regulate gene expression through a number of mechanisms and a recently identified regulatory principle is RNA silencing. RNA silencing or RNA interference (RNAi) is the process by which long double-stranded RNA (dsRNA) is cleaved into 18 to 30 nt small RNAs (sRNAs) that regulate gene expression by binding complementary RNA to cause RNA degradation, transcriptional inhibition, or translational repression. Numerous biological phenomena caused by RNA silencing have been reported in the literature for over 80 years, but it was not until several pivotal discoveries during the late 1990s that research groups began to understand the key mechanism of gene silencing: the downregulation of gene expression by sRNAs derived from dsRNA [1–3]. For their discovery of gene silencing by dsRNA [1], Andrew Fire and Craig Mello were awarded the 2006 Nobel Prize in Physiology or Medicine. Although the understanding of RNA silencing has progressed rapidly since this discovery, many investigations predated Fire and Mello's findings.

2
An Historical Overview of RNA Silencing

Before RNA silencing was discovered, several earlier observations were difficult to reconcile for research groups who worked with transgenic plants, based on the knowledge available at the time. When plants were transformed with transgenes, these transgenes often failed to express [4], a phenomenon that came to be known as *silencing*. This silencing correlated to the number, orientation, and position of the transgene insertions [5–10]. Antisense transgenes could suppress endogenous gene expression through an unknown mechanism [11, 12] but, interestingly, transgenes in the sense orientation could have a similar effect and downregulate the expression of both transgene and complementary endogenous genes [13, 14]. One well-described example involved a petunia that developed novel pigmentation patterns following the introduction of transgenes that affected the expression of petal pigments. In some transgenic plants, the sense transgene reduced both the endogenous and transgenic chalcone synthase mRNA levels and produced petals that had no anthocyanins (Fig. 1) [13]. The term *"cosuppression"* was

Fig. 1 Cosuppression in petunia. Transformation of a wild-type petunia (a) with the chalcone synthase transgene can lead to silencing of the endogenous gene, which produces patches of white due to the loss of pigment (b, c) [13]. This gene silencing phenomenon is known as *cosuppression*. (Images (b) and (c) are adapted from Ref. [17].)

coined to describe when both the sense transgene and complementary endogenous gene were downregulated. A similar phenomenon occurred in the fungus *Neurospora crassa* [15] and in the nematode *Caenorhabditis elegans* [16], indicating that these "cosuppression" or "quelling" phenomena were common to plants, fungi, and animals.

Gene silencing is triggered not only when transgenes and endogenous genes share a complementary transcribed region, but also when transgenes are complementary to a promoter DNA. In one report, tobacco plants were transformed with two transgenes encoding different genes but that shared the same promoter [18]. One transgene directed DNA methylation at the promoter region of the other transgene that resulted in suppressed gene expression [18]. This finding showed that RNA silencing could also occur at the DNA level, although for many years this phenomenon was not associated with "cosuppression."

Studies with viruses have also revealed a connection between complementary nucleic acid sequences and RNA silencing. When transgenes containing viral DNA were transformed into a plant, they conferred resistance against viruses that had a similar nucleic acid sequence [19, 20]. Conversely, viruses could silence endogenous plant genes if they contained the corresponding nucleotide sequence [20, 21], a mechanism termed "Virus-Induced Gene Silencing" (VIGS); indeed, this mechanism is now frequently used to silence endogenous plant genes. Thus, viral infection could induce a phenomenon that resembled transgene silencing [22].

Several notable properties of RNA silencing were identified before the core mechanism became known. First, nuclear run-on assays revealed that cosuppression occurred at the post-transcriptional level [5, 7], and this phenomenon was therefore referred to as post-transcriptional gene silencing (PTGS). A pivotal observation was that viroid transgenes transformed into tobacco became methylated when RNA was replicated, indicating that *de novo* DNA methylation required RNA transcription [23]. This observation was the first to link RNA to chromatin regulation, and demonstrated that RNA was the specificity determinant for DNA methylation [23]. Gene silencing could also move from cell-to-cell and over long distances; indeed, grafting and transformation experiments revealed that a gene silencing signal could move from silenced tissue to nonsilenced tissue [24, 25]. The mobile silencing signal only targeted genes that had a complementary sequence to the silencing trigger, suggesting that a nucleic acid, most likely RNA, was the mobile species [24, 25].

3
The Core Mechanism

The core mechanism of RNA silencing was elucidated through several key experiments. First, Fire et al. sought to explain cosuppression, quelling, and sense mRNA degradation, and hypothesized that these effects were related to the production of aberrant or dsRNA [1]. Fire et al. purified single-stranded RNA and dsRNA and demonstrated that dsRNA was 10- to 100-fold more efficient at silencing mRNA than single-stranded RNA in the sense or antisense orientation [1]. A second fundamental study identified 25 nt sRNA species that were complementary to a gene undergoing PTGS in plants and marked the discovery of sRNAs [2]. Shortly afterwards, in Drosophila, 21 to 23 nt sRNAs were identified that cofractionated with cell extracts undergoing RNA silencing [3, 26]. These observations suggested that sRNAs were the produced from dsRNA in the RNA silencing pathway.

Another important discovery was the nuclease that generated the sRNAs. This enzyme, termed *Dicer*, was identified and characterized through a candidate gene approach in Drosophila, biochemical fractionation, and dsRNA cleavage assays [27]. Homologous Dicer genes were soon characterized in plants, animals, nematodes, several fungi, and algae [28–34]. The final core component to be identified was "slicer," the enzyme responsible for binding sRNAs to target complementary RNA. The slicer enzyme, a member of the P-element Induced Wimpy testis (PIWI)/Argonaute (AGO)/Zwille family, was initially identified in Arabidopsis and later associated with RNA silencing through a screen for C. elegans mutants deficient in RNA silencing [35, 36]. Biochemical purifications of the complex associated with mRNA cleavage found an AGO protein [37]. Subsequent analyses coprecipitated sRNAs and found an AGO protein that bound sRNAs [38]. Crystallization studies of the *Pyrococcus furiosus* AGO identified a conserved catalytic triad and a domain, piwi, containing similarity to ribonuclease H [39]. Mutations in this regions and *in vitro* cleavage assays confirmed that AGO was the enzyme responsible for loading the sRNAs to direct complementary RNA cleavage [40–42].

Briefly, the core mechanism of RNA silencing can be summarized as follows. First, dsRNA is generated from DNA by RNA polymerases and is acted upon by the endonuclease Dicer to generate double-stranded sRNAs. These species are loaded into the AGO protein where one strand is discarded and the other used as a guide to target complementary RNA though Watson–Crick base pairing. The target RNA can be cleaved, degraded, translationally repressed, or used as a scaffold to direct epigenetic marks such as DNA methylation or histone modifications [43–45]. When sRNAs target mRNAs for degradation or translational repression, the phenomenon is termed *post-transcriptional gene silencing* (PTGS), whereas when sRNAs direct epigenetic marks to DNA which silences gene expression, the phenomenon is known as *transcriptional gene silencing* (TGS).

RNA silencing occurs in most eukaryotes, including mammals, flies, nematodes, plants, algae, and some yeast [44]. Interestingly, some prokaryotes have an analogous system that bears a resemblance to eukaryotic RNA silencing [46, 47]. There are variations to the core mechanism in eukaryotes as most RNA silencing genes are encoded by multigene families. These variations account for multiple types of RNA silencing, the diverse phenotypes associated with RNA silencing mutants, and the different size classes of sRNAs. High-throughput sequencing of sRNAs has been extremely informative about the abundance and types of sRNAs [48]. Together with genetic analyses, high-throughput sequencing has revealed that multiple RNA silencing pathways exist in plants, including the microRNA (miRNA) pathway and the small interfering RNA (siRNA) pathway.

4
Double-Stranded RNA

dsRNA is the precursor molecule from which the majority of sRNAs are generated, an exception being Piwi-Associated RNAs (piRNAs) found in animals. There are three main sources of dsRNA. First, inverted-repeat (IR) formation by gene duplications and inversions is a common source of duplex RNA and occurs at endogenous loci and at transgenes. Second, overlapping or antisense RNA transcripts are a source of dsRNA, and approximately 7% of genes in *Arabidopsis*, 17% of genes in *Drosophila*, and 26% of genes in humans overlap [49, 50]. Although overlapping genes have the potential to generate overlapping transcripts, such transcripts are not abundant in *Arabidopsis* [51], perhaps due to their proposed role in regulating gene expression under stress conditions or in specific developmental contexts [52]. Lastly, RNA-dependent RNA polymerases (RDRs) can generate dsRNA from single-stranded RNA.

A role for RDRs in gene silencing was first identified in *Neurospora crassa* using mutants that were deficient in quelling [53]. Later, these enzymes were found to affect gene silencing in plants, *Saccharomyces pombe* (fission yeast), *Tetrahymena thermophila*, and *C. elegans* [54–58]. *C. elegans* has four RDRs while *S. pombe* has one RDR [59]. In *Arabidopsis*, there are six putative RDRs that exist in four genomic clusters. At least three of these genes, *RDR1*, *RDR2* and *RDR6*, are functional and have specialized roles in sRNA production [60]. The RDRs from these diverse organisms share a common sequence motif named the double-psi β-barrel domain and the catalytic active amino acid sequence DxDGD [61], suggesting a common ancestor for RDRs. RDR activity may also exist in insects and mammals. An RNA Pol II elongator complex with RDR activity has been identified in *Drosophila* [62] and in humans, a subunit of the telomerase complex has RDR activity at one locus [63]. Canonical RDRs do not appear to be present in mammals or insects.

5
Dicers

Dicers are enzymes that cleave dsRNA to form sRNAs. They contain several features including two RNase-III motifs, a Piwi-Argonaute-Zwille (PAZ) domain, and an RNA helicase domain [32]. The distance between the PAZ and RNase-III domains determines the length of the sRNA produced [64]. Dicer proteins generate sRNAs that range in size from 18 to 30 nt but, in *Arabidopsis* the most abundant sizes are 21–24 nt in length [65, 66]. Mammals, *C. elegans* and *S. pombe* have one Dicer gene, and *Drosophila* has two Dicers [32]. In plants, *Arabidopsis* has four *DICER-LIKE* (*DCL*) genes, rice has eight DCLs [67], and poplar trees have five *DCLs* [33]. Plant DCL proteins have undergone significant expansion and are thought to have evolved and diversified in response to viruses, diseases, and endogenous elements such as transposons [33, 68]. In *Arabidopsis*, the DCLs possess some redundant and overlapping functions. For example, *DCL1* or *DCL2* can function in place of *DCL4* for several processes, including viral defense when *DCL4* is mutated [66, 69–71]. The *Arabidopsis* DCLs use different substrates and, based on genetic analyses, generate different-sized products. In general, DCL1 generates 20–22 nt miRNAs, DCL2 generates 22 nt viral-derived small interfering RNAs (vsiRNAs), DCL3 generates 23–24 nt heterochromatin-associated small interfering RNAs (hc-siRNAs), and DCL4 generates 21 nt *trans*-acting small interfering RNAs (tasiRNAs) and 21 nt vsiRNAs. Notably, there are some exceptions to the substrates acted upon and the sRNA sizes generated.

Dicer-independent sRNA production occurs in *Drosophila*, mammals, and fungi [44], but has yet to be discovered in plants. A group of repeat-associated RNAs known as *piRNAs* are generated by a dicer-independent mechanism in *Drosophila* [72–74]. In mammals, miRNA 451 requires AGO catalytic activity but does not require Dicer to generate the mature sRNA [75, 76], whereas in *S. pombe* primal RNAs are generated from aberrant RNA transcripts in a Dicer-independent manner and may amplify RNA silencing and heterochromatin formation [77].

6
Argonautes

Argonaute (AGO) proteins play key roles in RNA silencing as they bind the sRNA and use this molecule as a guide to direct gene silencing. For this reason, they are often referred to as the *"effector"* or *"effector complex"* of RNA silencing. When bound to an sRNA, AGOs can cleave RNA, destabilize RNA, or direct epigenetic marks. There are two main components of the effector complex: the AGO protein and sRNA. Together, these make up the minimal RNA-Induced Silencing Complex (RISC) [78, 79] or, in fission yeast, they form part of the RNA-induced initiation of transcriptional gene silencing (RITS) complex [80]. AGO proteins are present in eukaryotes and some archaea and bacteria, which suggests a common ancestor [68]. They are named AGOs because mutant *ago1 Arabidopsis* plants resemble a small squid [35]. These proteins are grouped into two types: AGO and Piwi proteins [81]. Phylogenetically, the Piwi proteins are a distinct clade specific to the animals.

AGO proteins load the Dicer-generated sRNA duplex and then cleave (siRNAs) or discard (miRNAs) one of the two sRNA strands [82]. The discarded strand is typically the one with the higher 5′ thermodynamic stability, termed the *passenger strand*. The other strand is used as a guide to seek out complementary RNA [83, 84]. In animals, the sRNA–AGO complex binds to the complementary mRNA, generally at the 3′ end, and in the majority of cases causes mRNA destabilization [85, 86]. In some instances, it is thought that AGOs cause translational repression by increasing ribosome drop-off or by reducing translation initiation [85, 87, 88]. In plants, mRNA cleavage by the AGO–sRNA complex is a common targeting mechanism.

There are fewer examples of translational repression in plants, but this phenomenon is thought to be widespread [89, 90]. In S. pombe and plants, AGO proteins can bind noncoding RNA transcripts to recruit chromatin-modifying enzymes that direct histone marks or DNA methylation, which can cause gene silencing at the transcriptional level [45, 91]. AGOs also generate or help to produce the mature sRNA; they influence which strand of the double-stranded sRNA is discarded [82] and several sRNAs – including Dicer independent-piRNAs and some miRNAs – require AGOs for their biogenesis [73–76]. In one well-described example, the *Neurospora* AGO homolog QUELLING-DEFECTIVE-2 (QDE-2) generates single-stranded siRNAs along with QDE-2-INTERACTING PROTEIN (QIP), a QDE-2 interacting protein and exonuclease [92].

AGO proteins contain PAZ, mid, PIWI, and N terminal domains [93]. The PAZ domain binds sRNAs, whereas the PIWI domain has RNase H activity that cleaves or binds target RNA transcripts [39]. The function of the mid domain is not well understood, but it is required for correct cleavage function and contains a 5′ phosphate-binding pocket [94, 95]. Several proteins interact with the AGO PIWI domain. One of these proteins, GW182, has an AGO hook motif with WG/GW amino acid repeats that interacts with the AGO1 PIWI domain in *Drosophila* [96]. The Tas3 protein from *S. pombe*, KOW Domain-Containing Transcription Factor 1 (KTF1) and NRPE1 proteins from *Arabidopsis*, and the P38 viral suppressor of silencing have similar WG/GW motifs that bind AGO proteins [97–101].

In *S. pombe* there is one AGO protein (Ago1), and this is a member of the RITS complex along with Chp1 and Tas3 [80]. This complex contains sRNAs and initiates and maintains heterochromatin [80]. In other organisms, AGOs have diversified and expanded to perform multiple roles. For example, *C. elegans* has 27 AGO genes, humans have eight, and rice has 19 AGOs [67, 93, 102]. *Arabidopsis* has 10 AGO genes that are associated with various functions. AGO1 is involved in miRNA-regulated gene expression. Null *ago1* mutants have strong pleotropic effects and are infertile due to the requirements of miRNA-regulated gene expression for plant growth and development [35]. AGO1 was the first *Arabidopsis* AGO demonstrated to have *in vitro* slicing activity [41, 42]. Experiments performed to purify the AGO1–sRNA complex revealed that AGO1 was associated primarily with 21 nt miRNAs and to a lesser extent with 21 and 24 nt siRNAs [41, 103]. Notably, AGO1 transcripts are targeted by miRNA168 bound by AGO1. This allows for AGO1 to regulate its own mRNA level through a negative feedback loop, and this homeostasis is required for correct *AGO1* function [104]. Interestingly, AGO10 acts as a miRNA-sequestering protein by binding to miR165/miR166 and preventing their AGO1 loading and subsequent targeting of homeodomain/leucine-zipper (HD ZIP) transcription factor mRNAs [105]. AGO10 also binds to a lesser extent other miRNAs, and may act as a slicer to cleave target mRNA in these instances [106].

Several *Arabidopsis* AGOs, including AGO4, are associated with epigenetic modifications and heterochromatin. The function of AGO4 was identified in a mutant screen as *ago4* mutants reactivated transposable element transcription and reduced DNA methylation and siRNAs, indicating that AGO4 acted at the TGS level [107]. Consistent with this role, AGO4 binds primary 24 nt sRNAs to maintain heterochromatin and silence transposable elements, transgenes, and repeat elements [108]. AGO4 also has slicing activity that is required at some sRNA-generating and transposable element-associated regions [108]. AGO6 is related to AGO4 and may reinforce heterochromatin and share partial functional redundancy with AGO4 [109]. A third related AGO, AGO9, binds 24 nt siRNAs derived from transposable elements and controls female gamete formation [110, 111]. Rice encodes an AGO that, when mutated, produces sterile plants which are unable to complete meiosis [112]; this suggests that AGOs might play a conserved role in gamete formation in plants.

AGO proteins in *Arabidopsis* have binding size preferences: AGO1 and AGO7 bind primarily 21 nt sRNAs, AGO2 binds primarily 21 and 22 nt sRNAs, whereas AGO4, AGO6, and AGO9 bind primarily 24 nt sRNAs [41, 111, 113, 114]. They also have a bias for the first nucleotide composition of the sRNA. AGO2, AGO4, AGO6, and AGO9 primarily bind sRNAs starting with a 5′ adenosine, AGO5 primarily binds those starting with a 5′ cytosine, and AGO1 those with a 5′ uridine [111, 113–115]. Changing the first nucleotide causes an sRNA to load into a different AGO complex [114].

The large number of AGO proteins in *Arabidopsis* may result from functional diversification. AGO4, AGO6, and AGO9 bind sRNAs of the same length and first base-pair composition, which suggests that additional specificity determinants are important as these AGOs are nonredundant [109, 110]. AGO4, AGO6, and AGO9 can bind hc-siRNAs from the same locus, but are expressed in different regions of the plant [111]. Furthermore, AGO4, AGO6, and AGO9 differ in the epigenetic modifications they direct, suggesting that these proteins have different interacting proteins or modifiers [111].

7
Signal Amplification

One important feature of RNA silencing is the ability to amplify the primary RNA silencing signal. Primary sRNA biogenesis is thought to be RDR-independent and produces both miRNAs and siRNAs [116–118]. These primary sRNAs function in either *cis* or *trans* to cleave an RNA transcript that can be then acted upon by RDRs. RDRs amplify the cleaved products, generating dsRNA that becomes a source of secondary siRNAs [117, 118]. Secondary siRNA production is important for viral silencing, transgene silencing, and the amplification of mobile silencing signals. In addition, this amplification process is required for transitivity [119], the process by which RNA silencing is amplified downstream of where primary sRNAs have targeted an RNA, and for the generation of secondary siRNAs from tasiRNA genes [120–122].

8
MicroRNAs

Plants contain two types of sRNAs: miRNAs and siRNAs. miRNAs are potent and specific regulators of post-transcriptional gene expression. They are distinct from other sRNAs because they are excised as discrete species from imperfect stem–loop precursors, whereas siRNAs are derived as populations from perfect or near-perfect dsRNA molecules [43]. miRNAs target mRNAs with full or partial complementarity to induce mRNA cleavage, destabilization, or translational repression [123]. A vast array of biological processes are regulated by miRNAs, ranging from developmental programming [124, 125] to stress responses [123, 126].

Prior to the discovery of dsRNA-mediated gene regulation [1], miRNAs had been identified in *C. elegans*. Two reports suggested that the *lin-4* RNA, now known to be a miRNA, regulated *lin-14* mRNAs levels through antisense RNA base-pairing [127, 128]. Twenty-two-nucleotide sRNAs were produced at the *lin-4* locus but the active species and regulation mechanism were unknown [127]. Later, the identification of a second miRNA, *let 7*, in animals and invertebrates demonstrated that miRNA gene regulation was not unique to a single locus, and hinted that miRNAs might be a widespread gene regulation phenomenon [129, 130]. Soon after, miRNAs were cloned from humans, *C. elegans* and *Drosophila* [131–133]. miRNAs were also identified in plants, and *Arabidopsis* contains at least 155 miRNA-generating loci [134–136]. miRNAs were also found in moss and single-celled algae, which suggests that they were present since the evolution of plants [137, 138].

8.1
The Biogenesis of miRNAs

In plants, miRNAs are transcribed by RNA Polymerase II to form a hairpin molecule that has imperfect complementarity [139]. DCL1 performs three endonucleolytic cleavages on this transcript in the nucleus to generate the pri-miRNA, pre-miRNA, and finally the mature 20–22 nt miRNA duplex [140, 141]. *DCL1* produces miRNAs but can also generate 20–22 nt siRNAs from IRs and noncoding RNA transcripts [66, 142]. Several proteins act with *DCL1* to generate miRNAs. *SERRATE* (*SE*) and *HYPONASTIC LEAVES 1* (*HYL1*) are required for processing miRNA precursors (Fig. 2) [141, 143]. *DAWDLE* (*DDL*) interacts with DCL1 and facilitates miRNA biogenesis [144]. *HUA-ENHANCER 1* (*HEN1*) methylates the 3′ end of miRNAs and siRNAs to prevent degradation [145]. A recently identified RNA-binding protein, TOUGH (TGH), also associates with the DCL1–HYL1–SE complex, and is required for correct miRNA processing [146]. A homolog of the human pre-miRNA transporter Exportin-5, *HASTY* (*HST*), may be responsible for transporting mature miRNAs and siRNAs duplexes from the nucleus to the cytoplasm [147–150]. After export to the cytoplasm, one strand of the miRNA duplex is selected as the guide strand, and is loaded into the AGO silencing complex, most commonly AGO1 [104]. The complementary strand of the miRNA duplex, known as the *miRNA** *(star)*, is subsequently degraded [82–84, 151].

8.2
miRNA Targeting Mechanisms and Prediction

The miRNA-bound AGO1 complex is sufficient for directing mRNA cleavage to regulate gene expression [41]. Some miRNAs cause translational repression or mRNA destabilization in *Arabidopsis*, similar to the mechanism observed for the majority of animal miRNAs [85, 89, 90]. The cleavage of miRNA targets occurs between nucleotides 9–11 from the 5′ end of the miRNA, and complementarity at these bases is required for this slicing activity to occur [123]. Interestingly, a loss of base-pairing in this region can cause an increase in translational repression of

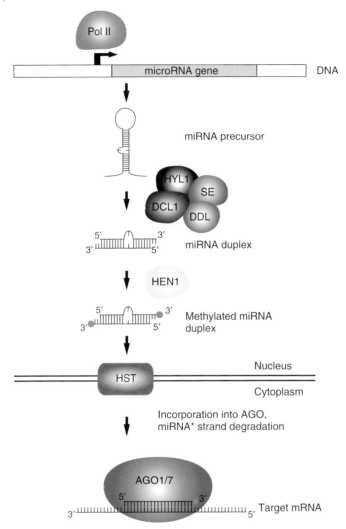

Fig. 2 An overview of the miRNA pathway in *Arabidopsis thaliana*. miRNA genes are transcribed by RNA polymerase II and acted upon by DCL1 along with HYL1, SE, and DDL to generate a duplex 21 nt miRNA. The miRNA is methylated by HEN1 and exported into the cytoplasm by HST. The majority of double-stranded miRNAs load into AGO1 where one strand is discarded, termed the *miRNA** (star), and the other used to target complementary mRNA for degradation or translational repression.

the mRNA target for at least one miRNA [152]. However, perfect complementarity at this region can lead to mRNA cleavage or translational repression [89], and it remains unclear what determines the extent of slicing versus translational repression for a given miRNA–mRNA interaction [123].

Mature miRNAs in plants typically exhibit a high degree of complementarity with their targets [153, 154], and this specificity of miRNAs allows target prediction to be reasonably accurate. Conversely, animal miRNAs can display as little as seven-nucleotide complementarity with their targets, which makes target prediction challenging [155]. This region of complementarity in animals is called the *seed sequence*, and is located from nucleotides 2–7/8 from the 5′ end of the miRNA [156].

8.3
Origins and Evolution

Although both plant and animal miRNAs are generated from long stem–loop precursors and act in *trans* to target complementary mRNAs, they are believed to have evolved convergently [157]. Animal and plant miRNAs vary in their biogenesis mechanism, precursor sequences and the mode of mRNA regulation [158]. Furthermore, no known miRNAs are conserved between plants and animals [158]. These observations argue for at least two independent origins for miRNAs from an ancestral sRNA pathway [157].

In plants, miRNAs may have arisen from invert duplications of endogenous genes [159]. These DNA hairpins would initially have been a source of siRNAs and, with time, would have mutated to lose perfect complementarity to instead produce a single sRNA species characteristic of a miRNA locus. These progenitor miRNA genes may be generated frequently, but are often selected against becoming miRNAs [160]. Another model proposes that random foldback sequences found throughout the genome occasionally evolve to generate a miRNA [161]. It is probable that both models contributed to the evolution and large number of miRNAs in plants.

8.4
The Function of miRNAs

In *Arabidopsis*, miRNAs are key regulators of gene expression and, similar to AGO1 mutants, DCL1 mutants have strong developmental defects [32, 124]. Before the mechanism of RNA silencing was understood, many *DCL1* mutant alleles had been identified in mutant screens searching for embryo lethality, short integuments or floral meristem defects – processes that are now known to be regulated by miRNAs [32, 162]. Some of the most deeply conserved plant miRNA families include miR156, miR160, miR319, and miR390 [163]. These miRNAs are also present in mosses, and target conserved transcription factors controlling developmental and hormonal regulation [163].

miRNAs also play important roles in plant development. During early embryogenesis, miR156 promotes embryonic patterning by preventing the precocious expression of SPL10/11 transcription factors [124]. Also during early embryogenesis, miR165/166 controls several HD ZIP transcription factors to maintain stem cells and leaf polarity [164]. Later in development, decreased miR156 levels cause an increased expression of the miR156 targets, SPL3/4/5, to promote the transition from early to late vegetative growth [125]. As well as controlling developmental transitions, miRNAs also regulate plant morphology. For instance, miR164 constrains the expression of *CUP-SHAPED COTYLEDON2* to control leaf margin serration [165]. miRNAs can also regulate plant responses to stress. For instance, miR393 is induced upon perception of a pathogenic threat [126]. miR393 targets F-box Auxin Receptors causing the repression of auxin signaling and thereby restricting pathogen growth.

9
Small Interfering RNAs

Plants contain a second type of sRNAs known as *small interfering RNAs*. siRNAs are typically 20–24 nt in length and are associated with both TGS and PTGS. They were initially identified associated with silenced transgenes, and later, with transposable elements and heterochromatic regions of the genome [2, 166]. siRNAs are formed from perfect or near-perfect complementary dsRNA transcripts and, after DCL cleavage, accumulate from both strands and throughout the region of complementarity. Typically, a dsRNA transcript will produce many species of overlapping siRNAs from both sense and antisense strands. siRNAs are categorized into those that are natural antisense small interfering RNAs (nat-siRNAs), tasiRNAs, or hc-siRNAs [43, 167].

9.1
Trans-Acting siRNAs

One group of siRNAs known as *trans*-acting siRNAs (tasiRNAs or TASs) have a unique biogenesis mechanism. Production is initiated by a miRNA cleavage of a TAS RNA transcript that is then acted upon by RDR6 and SUPPRESSOR OF GENE SILENCING 3 (SGS3) to generate dsRNA [168, 169] (Fig. 3). This double-stranded TAS RNA transcript is cleaved by DCL4 to produce tasiRNAs every 21 nt in a phased register [168, 169]. The tasiRNAs are then loaded into an AGO complex, either AGO1 or AGO7, and target complementary mRNAs for cleavage *in trans* and to direct PTGS (Fig. 3) [113, 170, 171]. The cleavage products can be further sources of tasiRNAs that result in a cascade of siRNAs generated from multiple loci [122, 172]. There are four well-described TAS gene families in *Arabidopsis*, namely *TAS1*, *TAS2*, *TAS3*, and *TAS4*, along with several other regions that behave similarly to the TAS genes [122]. TAS genes are also present in a wide variety of plants including tomato, legumes, and moss [121, 173, 174].

Several miRNAs are associated with tasiRNA production. miR173 targets the *TAS1* and *TAS2* transcripts, whereas miR390 targets *TAS3* transcripts [168, 169]. *TAS1* and *TAS2* each require one miRNA-directed cleavage by AGO1 for tasiRNA production, and are referred to as *one-hit loci* [121, 171]. Notably, 22 nt miRNAs have been identified as the triggers for TAS transcript cleavage and tasiRNA formation [175, 176]. Although 21 and 22 nt species load into AGO1, only the 22 nt form can direct cleavage to trigger RDR6 activity and the production of secondary tasiRNAs [175]. *TAS3* does not have this requirement [176], possibly because it is targeted at both the 3′ and 5′ ends by the 21 nt miR390 [121]. As a result, *TAS3* is termed a *two-hit locus* though, notably, only the 3′ target is cleaved [121]. AGO7 binds primarily miRNA390, and the main role of AGO7 may be to regulate *TAS3* RNA levels [113].

Many siRNA mutants do not produce obvious morphological phenotypes in *Arabidopsis*, but the tasiRNA mutants are exceptional. Mutations in *RDR6*, *DCL4*, *SGS3*, and *AGO7* affect the juvenile to mature leaf transition and affect leaf polarity, which produces *Arabidopsis* with a downward-curling leaf phenotype [170, 177]. These genes produce *TAS3* siRNAs [178, 179], which are expressed on the upper side of the leaf and move to the lower side to create a concentration gradient that regulates *Auxin Response Factor 3* (*ARF3*) levels, a gene required for leaf

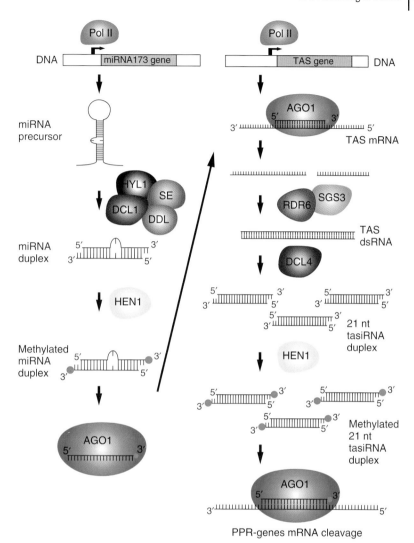

Fig. 3 An overview of the tasiRNA pathway in *Arabidopsis thaliana*. miRNA173, generated by the miRNA pathway (see Fig 2), targets TAS1 or TAS2 RNAs that are transcribed, presumably by RNA polymerase II, from *TAS1* or *TAS2* genes. After AGO1-mediated miRNA cleavage, the TAS RNA is acted upon by RDR6 and SGS3 to generated double-stranded RNA. DCL4 cleavage generates 21 nt tasiRNAs that are in phased 21 nt registers and act *in trans* to target mRNAs, such mRNAs containing pentatricopeptide repeats (PPR), for degradation.

development [180, 181]. The *TAS3* locus is highly conserved among plants, which suggests that the *TAS3–ARF* module plays an important developmental role [121]. Although *TAS1* and *TAS2* siRNAs control the expression of endogenous genes, including several pentatricopeptide repeat (PPR) -containing genes, no morphological phenotype is associated with mutant *tas1* or *tas2* genes or the loss of these siRNAs [167, 168]. In addition, the TAS1 and TAS2 genes are not well conserved among plant species [170, 177].

9.2
Natural Antisense siRNAs

One mechanism to generate dsRNA is via the production of overlapping and complementary mRNA transcripts. The phenomenon occurs at approximately 7% of *Arabidopsis* genes [49], yet only 84 pairs of overlapping *cis* transcripts that generate sRNAs have been identified in *Arabidopsis* [52]. nat-siRNAs are associated with stress responses and plant development. During pathogen stress, *ATGB2* increases in expression and downregulates the overlapping Pentatricopeptide Repeat Protein-Like gene (*PPRL*), a gene thought to negatively regulate host resistance genes, via the production of 22 nt siRNAs [182] (Fig. 4). These nat-siRNAs are dependent on *DCL1*, *HEN1*, *HYL1*, *RDR6*, the largest subunit of Pol IV (*NRPD1*), and *SGS3* (Fig. 4). In the absence of 22 nt nat-siRNAs, PPRL transcript levels are unaffected and plants are hypersusceptible to the pathogen *Pseudomonas syringae* [182]. Another stress-responsive nat-siRNAs in *Arabidopsis* is associated with the Δ^1-pyrroline-5-carboxylate dehydrogenase (*P5CDH*) gene. Overlapping transcripts at the *P5CDH* locus generate nat-siRNAs that repress this gene and increases salt stress tolerance [183]. Overlapping transcript production at a sperm-specific locus (*ARI14*) generates nat-siRNAs that downregulate *ARI14* mRNAs and allow fertilization to occur [184]. It appears that most nat-siRNA loci generate both 21 and 24 nt siRNAs, but little else is presently known about the 84 nat-siRNA loci [52]. Further nat-siRNAs will need to be identified and characterized to better understand their biogenesis and functional significance in *Arabidopsis*.

9.3
Heterochromatin-Associated siRNAs

In plants, hc-siRNAs are endogenous sRNAs that associate with IRs, direct repeats, transposable elements, and methylated regions of the genome [185, 186]. sRNA deep-sequencing techniques have identified over 4200 regions in the *Arabidopsis* genome that produce hc-siRNAs [186]. The hc-siRNAs affect TGS by modifying chromatin, but this process is only partially understood. Heterochromatin, the condensed form of DNA, has modifications to the DNA and histone proteins that associate with DNA. The histone proteins are post-transcriptionally modified, typically with methylation and acetylation at the lysine residues of the histone tails [187]. Another type of epigenetic modification, DNA methylation, is typically an inactivating mark that occurs at cytosine residues (C) in three sequence contexts: CG, CHG, or CHH, where H is any nucleotide other than guanine (G). Gene body methylation appears to have a minimal effect on transcription, which suggests that RNA Pol II is differentially sensitive to DNA methylation during transcriptional initiation versus elongation. When present in the promoters of genes, DNA methylation generally inhibits gene transcription [188].

9.3.1 The Biogenesis of hc-siRNAs

The genes required for hc-siRNA biogenesis were identified through forward and reverse genetic screens that assayed for changes in siRNA abundance, or for changes in gene silencing [56, 57, 189–191]. *Arabidopsis* plants deficient in gene silencing revealed the requirements

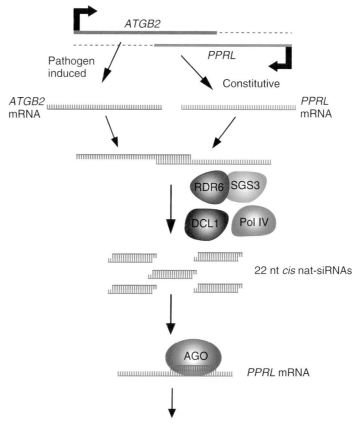

Fig. 4 An overview of the nat-siRNA pathway in *Arabidopsis thaliana*. *PPRL* is expressed, but upon *Pseudomonas syringae* infection, the overlapping *ATGB2* transcript is also expressed [182]. These transcripts generate overlapping mRNAs that are recognized by Pol IV, RDR6, SGS3, and DCL1, among other factors to generate 22 nt nat-siRNAs. These nat-siRNAs load into the RISC complex to target the *PPRL* transcript leading to mRNA degradation and resistance to *Pseudomonas syringae*.

for two noncanonical DNA-dependent RNA polymerases, RNA Polymerase IV (Pol IV) and RNA polymerase V (Pol V) [192–194]. These polymerases are unique to plants, and the largest subunit of Pol IV and Pol V has homology with RNA polymerase II (Pol II). Pol IV and Pol V share at least six subunits with Pol II, but also contain subunits that are specific to Pol IV and/or Pol V [98, 101]. Interestingly, Pol IV and Pol V appear to function independently from one another.

Pol II transcription appears to recruit Pol IV at heterochromatic regions and, based on genetic evidence, Pol IV is thought to transcribe noncoding RNAs to generate a template for hc-siRNA biogenesis [192, 193, 195]. Copurification experiments revealed that a putative chromatin remodeling factor, CLASSY1

(CLSY1), a homeodomain-containing protein SAWADEE HOMEODOMAIN HOMOLOG 1 (SHH) 1, and a putative transcription factor, RNA-DIRECTED DNA METHYLATION 4 (DMS4), associated with Pol IV [196, 197]. CLSY1, SHH1, and DMS4 are required for correct hc-siRNA biogenesis and might recruit Pol IV to loci [196, 197]. The biogenesis of 24 nt hc-siRNAs also requires the activity of DCL3 and RDR2 [189]. RDR2 physically associates with Pol IV and acts on the Pol IV transcript to generate dsRNA that is then diced by DCL3 to generate 24 nt hc-siRNAs (Fig. 5) [196, 198]. These hc-siRNAs are exported from the nucleus to the cytoplasm where they bind AGO proteins [151, 198]. In *Arabidopsis*, various effort complexes can load 24 nt hc-siRNAs to mediate gene silencing. Initially, AGO4 was identified as an effector of transcription gene silencing as it relieved the silencing of a methylated form of the *SUPERMAN* locus [107]. Later, AGO6 and AGO9 were also identified as proteins that load 24 nt hc-siRNAs and direct TGS, though they have a weaker effect on gene silencing than AGO4 [109–111]. After loading, the AGO–hc-siRNA complex moves back into the nucleus to direct gene silencing [151].

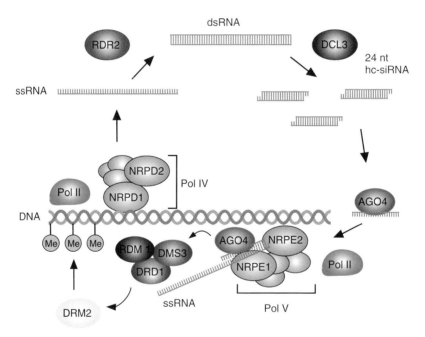

Fig. 5 An overview of the RNA-directed DNA methylation pathway in *Arabidopsis thaliana*. The Pol IV complex is thought to act with Pol II to transcribe single-stranded RNA from transposable elements, transgenes, and regions of heterochromatin. The single-stranded RNA is acted upon by RDR2 to from dsRNA that is cleaved into 24 nt sRNAs by DCL3. These species are loaded into AGO4. Independently, the Pol V complex transcribes RNA, likely with Pol II, to form a single-stranded RNA transcript. The AGO4-sRNA complex binds the Pol V transcript and associates with RDM1, DMS3, and DRD1. These proteins are thought to facilitate or be required for DNA methylation. This complex associates with the DNA methyltransferase DRM2 to direct DNA methylation.

Surprisingly, 24 nt hc-siRNAs are not present in gymnosperms (conifers and other "naked" seed-bearing plants), most likely due to the absence of *DCL3*. This siRNA size class may have been lost, as both angiosperms (flowering plants) and pteridophytes (mosses and ancestrally ancient plants) have 24 nt siRNAs [199]. The loss of 24 nt siRNAs suggests that these hc-siRNAs are dispensable, and that 21 nt hc-siRNAs maintain heterochromatin and silence transposable elements in gymnosperms [199].

9.3.2 RNA-Directed DNA Methylation

RNA polymerase V plays a key role in RNA-directed DNA methylation (RdDM) [194]. RNA Polymerase II transcription is thought to recruit Pol V to intergenic and heterochromatic regions [195]. Pol V requires a nucleosome-remodeling complex named DDR for activity [200, 201]. The DDR complex contains a SWI2/SNF2 chromatin remodeler, DEFECTIVE IN RNA-DIRECTED DNA METHYLATION 1 (DRD1), a structural maintenance of chromosome hinge-domain protein, DEFECTIVE IN MERISTEM SILENCING 3 (DMS3), and a methyl-DNA binding protein, RNA-DIRECTED DNA METHYLATION 1 (RDM1) [201, 202]. Genetic evidence indicates that Pol V transcribes intergenic and heterochromatic regions of the genome independently of Pol IV [200]. The transcripts generated by Pol V serve as a scaffold to recruit the AGO4–hc-siRNA complex. This AGO4–hc-siRNA complex interacts physically with the Pol V-dependent RNA through complementary base-pairing between the hc-siRNA and RNA [203]. Through this interaction, chromatin-modifying enzymes are recruited that direct DNA methylation or histone modifications to silence gene expression [91]. One such enzyme is the methyltransferase DOMAINS REARRANGED METHYLTRANSFERASE 2 (DRM2). Although no direct interaction between AGO4 and DRM2 has been shown, AGO4 and DRM2 colocalize in nuclear foci, and both proteins copurify with RDM1 [198, 202, 204]. DRM2 catalyzes *de novo* DNA methylation in all sequence contexts (CG, CHG, CHH) [190]. Therefore, the targeting of loci by Pol V and binding of hc-siRNAs to the AGO4 complex to recruit DRM2 appears essential to direct *de novo* DNA methylation in all sequence contexts (CG, CHG, CHH) (Fig. 5) [205, 206].

Once DNA methylation is directed, several pathways exist to maintain this epigenetic mark. METHYLTRANSFERASE 1 (MET1) is an ortholog of the mammalian DNA methyltransferase, DNMT1, and maintains CG DNA methylation [207, 208]. A loss of *MET1* reduces viability and fertility and causes lethality over generations. A second protein, CHROMOMETHYLASE 3 (CMT3), maintains CHG methylation [209]. Together, MET1 and CMT3 propagate DNA methylation from newly replicated hemi-methylated DNA onto the nascent DNA strand [91]. The maintenance and propagation of CHH is thought to present a challenge because CHH methylation is only present on one strand and cannot be used as a template for both nascent DNA strands. This phenomenon contrasts CG and CHG sites where methylation is present on both strands and can be used as a template to direct DNA methylation to each nascent DNA strand. One hypothesis for methylating nascent DNA at CHH sites is that sRNAs and DRM2 maintain CHH methylation by persistently directing *de novo* methylation [91]. This hypothesis is supported by the observations that plants which lose DNA methylation over several generations require the RNA-silencing machinery to restore DNA methylation to wild-type levels [206]. Another contributing factor may be a self-reinforcing loop between DNA methylation and histone modifications to maintain CHH methylation during replication [210].

9.3.3 The Function of hc-siRNAs

The primary role of the hc-siRNAs is thought to be the silencing of transposable elements and the maintenance of heterochromatin. An absence of the Pol IV largest subunit, *NRPD1*, changes histone 3 Lysine 9 (H3 K9) distribution, nucleolar structure, DNA methylation and transposable element-associated siRNA levels [193]. The loss of members of the hc-siRNA pathway also re-reactivates certain transposable elements [189, 192]. Interestingly, hc-siRNAs may also play a role in protecting again transposable element misregulation in subsequent generations. Heat-stressed *Arabidopsis* that are deficient in the Pol IV complex show a high number of new transposon insertions in their progeny, whereas unstressed or wild-type *Arabidopsis* do not show this phenomenon [211]. Together, these observations indicate that hc-siRNAs are important for protecting the genome against transposable and repeat DNA elements.

Mutations in the genes responsible for hc-siRNA biogenesis do not produce obvious morphological phenotypes in *Arabidopsis*, but they appear to play a role in controlling plant flowering. Under short-day conditions, mutations in the hc-siRNA pathway delay flowering time substantially, most likely due to a targeting of the flowering time genes *FWA* and *FLC* by hc-siRNAs [212].

In *Arabidopsis*, hc-siRNAs are most abundant in flowering tissues and developing seeds [213]. In particular, the pollen vegetative cell, which supports the sperm cells, reactivates specific transposable elements and increases hc-siRNAs levels. These hc-siRNAs are thought to move from the vegetative nucleus to the sperm cell to silence transposable element expression [214]. Similarly, the endosperm, which supports the developing embryo, undergoes demethylation and becomes enriched in maternally expressed 24 nt hc-siRNAs [213, 215, 216]. Maternally expressed hc-siRNAs play a role in silencing transposable elements and genes in the endosperm, but it remains unknown where these maternally expressed hc-siRNAs could move to the embryo to affect gene expression [217]. Hc-siRNAs are also important for long-distance mobile RNA silencing (as discussed below in Sect. 11).

Although *Arabidopsis* mutants that lack hc-siRNAs have no obvious morphological defects, this is not true for maize plants. Mutants in maize *RDR2*, *NRPD1*, and *NRPD2* homologs produce stunted plants and delay flowering [218–221]. *RDR2*, *NRPD1*, and *NRPD2* are also required for paramutation, the epigenetic transfer of information between alleles that results in a heritable change in one allele [222]. It remains unknown whether these morphological defects are due to the role of the Pol IV pathway in maize miRNA biogenesis [223], heterochromatin maintenance, or transposable element reactivation and insertion.

10
Viruses and RNA Silencing

In 1928, it was observed that the upper growing leaves of tobacco plants could recover from a viral infection and become resistant to subsequent infection by the same virus. It is now known that this resistance is likely to be due to the antiviral action of RNA silencing [43]. As a defense mechanism, RNA silencing is likely to have ancient evolutionary origins with examples from animals, fungi, and plants [43, 224]. The core mechanism involves the production of vsiRNAs from a dsRNA precursor. These vsiRNAs are then loaded into a host-encoded effector complex, which guides sequence specific silencing of the viral genome. Three main lines of evidence support an antiviral role for RNA silencing in plants. First, vsiRNAs accumulate to high levels during viral infection. Second, plants deficient in RNA silencing are often hypersusceptible to viral infection. Third, all plant viruses studied to date encode proteins called viral silencing repressors (VSRs) that repress RNA silencing.

10.1
Plant Viruses: Genomes, Replication Strategies, and Detection by Host RNA Silencing Machinery

Plant viruses contain either RNA or DNA genomes and can be divided into those with double-stranded (ds) or single-stranded (ss) genomes. ssRNA viruses are further subdivided into positive sense (+) or negative sense (−). The vast majority of plant viruses (~65%) encode (+)ssRNA genomes [225].

The replication of RNA viruses occurs in the cytoplasm using a viral-encoded replicase protein. During replication, the replicase protein transcribes genomic RNA from a complementary viral RNA template, forming dsRNA intermediates. Although it is possible for these perfect duplexes to trigger the RNA silencing machinery initiated by DCL cleavage, all available evidence suggests that structured viral genomic RNAs are the primary source of vsiRNAs [226–230] (Fig. 6).

Geminiviruses are the predominant form of plant DNA viruses. They encode single-stranded circular DNA genomes and utilize host machinery to replicate in the nucleus [231]. Overlapping sense and antisense geminiviral transcripts can provide a source of dsRNA, which triggers RNA silencing [231]. However, because the geminiviral genome is DNA-based, it is targeted not only by RNA cleavage-associated PTGS but also by DNA methylation-associated TGS [232].

10.2
Host Factors Involved in Antiviral Silencing

Of the four DCL proteins encoded by *Arabidopsis*, DCL4 and DCL2 are implicated in defense against RNA viruses [69, 233]. DCL4 is the primary antiviral DCL enzyme, cleaving viral dsRNAs to produce abundant 21 nt vsiRNAs during (+)RNA viral infections. In the absence of DCL4, DCL2-dependent 22 nt vsiRNAs become predominant; however, these 22 nt vsiRNAs do not fully compensate for the absence of DCL4 as *dcl4* plants produce a less-efficient antiviral response [69, 234]. Plants lacking both DCL4 and DCL2, are more hypersusceptible to all RNA viruses examined to date, including *Tobacco rattle virus* (TRV), *Turnip crinkle virus* (TCV), and *Cucumber mosaic virus* (CMV) [69, 235–237]. *Potato Virus X* (PVX) does not normally infect *Arabidopsis*. However, *Arabidopsis* plants lacking DCL2, DCL3 and DCL4 become susceptible hosts for PVX [238]. These observations illustrate the importance of dicers in basal antiviral defense and in determining pathogen host range.

In contrast to RNA viruses, all four dicers are required for defense against

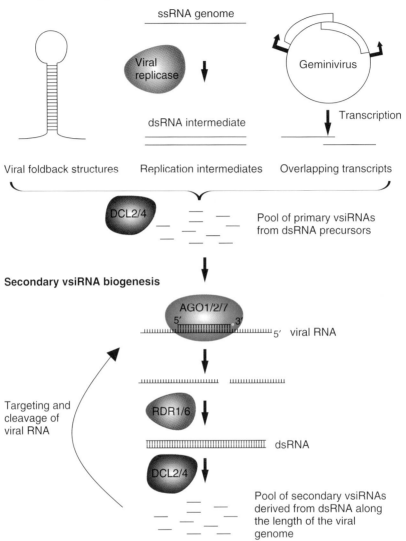

Fig. 6 Sources of primary and secondary vsiRNAs. Viral siRNAs are derived from double-stranded viral RNA in the form of structured viral RNA foldbacks, replication intermediates, or overlapping sense/antisense viral transcripts. These RNAs are processed by host DCLs (DCL2/4 in the case of RNA viruses, DCL1/2/3/4 for DNA viruses) to generate primary vsiRNAs. The primary vsiRNAs are loaded into AGO1/2/7, which directs cleavage of complementary viral mRNA and primes RDR1/6 to synthesize dsRNA along the length of the viral transcript. This dsRNA is processed by DCLs to generate secondary vsiRNAs. Secondary vsiRNAs can be loaded into AGO1/2/7 to direct cleavage of viral transcripts, or in the case of DNA viruses, loaded into AGO4 to induce gene silencing by DNA methylation of the viral genome.

nuclear-replicating DNA viruses [70, 239]. DCL3-dependent 24 nt vsiRNAs are the most abundant vsiRNA species produced during *Cauliflower mosaic virus* (CaMV) infection. These 24 nt vsiRNAs appear to direct DNA methylation at viral intergenic regions. High methylation levels at these regions correlate with low levels of viral DNA and reduced severity of symptoms, suggesting that the 24 nt vsiRNAs can direct antiviral TGS [240]. CaMV viral titers are increases in the *dcl2/3/4* triple mutant, but not in *dcl* single or double mutant combinations, consistent with a degree of redundancy in the DCL proteins [239]. DCL1 might be involved in cleaving viral miRNA precursor-like foldback structures, as it localizes to the nucleus with replicating CaMV, and is required for accumulating CaMV vsiRNAs. Cytoplasmically replicating RNA viruses are spatially separated from DCL1 and are consequently unaffected by this mechanism.

In principle, the action of DCL processing on viral RNA is sufficient to prevent viral replication without a requirement for the RISC complex. However, *ago1*, *ago2*, and *ago7* mutants are hypersusceptible to RNA viruses such as TCV and CMV [237, 241, 242]. AGO1, AGO2, and AGO5 bind vsiRNAs in infected *Arabidopsis* tissues [115, 243], and vsiRNA-loaded RISC can efficiently cleave viral RNA *in vivo* [244]. These observations suggest that the AGO–vsiRNA complex targets viral RNA for degradation and plays an important role in the antiviral response. Because DNA viruses are targeted by AGO4-mediated TGS, a requirement for RISC is expected. Consistent with this hypothesis, *AGO4* is required for recovery from geminiviral infection [245].

RNA silencing can be amplified by host-encoded RDRs that can produce dsRNA along the length of the viral genome. Primary vsiRNAs are derived from replication intermediates, secondary structures, or overlapping transcripts and bind to complementary viral templates. This binding primes RDR-mediated complementary strand synthesis, generating long dsRNA along the length of the viral genome. The dsRNA can be cleaved by DCLs, generating secondary vsiRNAs (Fig. 6) [43]. Of the six putative RDRs in *Arabidopsis*, RDR1 and RDR6 appear to play a key role in antiviral defense [246]. RDR1 is responsible for the majority of vsiRNAs in virus-infected plants [247]. However, *Nicotiana benthamiana* plants with reduced levels of RDR6 are hypersusceptible to PVX [248] and *Arabidopsis rdr6* mutants are hypersusceptible to a form of CMV that lacks the 2b VSR [234]. RDR6 is also essential for long-range movement of silencing in *Arabidopsis* [249], and prevents viral invasion of the apical meristem in *N. benthamiana* [248]. Although RDR2 is involved in the hc-siRNA pathway, evidence of an antiviral role for RDR2 against DNA viruses is currently lacking [250].

The core protein components of the canonical RNA silencing pathway – namely DCLs, AGOs and RDRs – all play essential roles in the antiviral silencing process. Because this mechanism generates a sequence-specific defense, it is a potent barrier to any potential viral threat. Viruses have therefore evolved a means to dampen host RNA silencing by producing VSRs.

10.3
Viral Silencing Repressors

VSR proteins from different viral families exhibit no conserved sequence similarities, and are likely the result of independent evolution across many viral lineages [43]. VSRs can act at multiple stages of the antiviral RNA silencing pathway (Fig. 7). Consistent with this idea, the constitutive expression of VSRs in plants phenotypically resemble mutations in the RNA silencing components [251]. Although the molecular mechanisms of VSR action are only just beginning to be understood [230], general strategies are becoming apparent. These strategies include the perturbation of vsiRNA availability, a direct inhibition of host proteins and, in the case of DNA viruses, the prevention of methylation-mediated TGS.

One of the most widely adopted VSR strategies is to sequester or inactivate the vsiRNAs after their biogenesis so that they cannot be loaded into the RISC complex. P19, from *Tomato bushy stunt virus*, is the archetypal example of this mechanism, binding specifically to 21 nt dsRNA. The crystal structure of P19 reveals that it forms a head-to-tail homodimer cradling the RNA duplex in a concave interface [252]. The HC-Pro silencing suppressor from Potyviruses prevents 2'-O-methylation at the 3' end of vsiRNAs, which is required for their stabilization and loading into the RISC [253]. The *Sweet potato chlorotic stunt virus* encodes a suppressor protein with an RNase III domain that processes 21, 22, and 24 nt vsiRNAs into 14 nt dsRNAs that are unable to load into the RISC [254].

The P38 and 2b VSRs from TCV and CMV, respectively, function to block RISC assembly by directly binding to AGO1 [100, 243]. P38 encodes a glycine/tryptophan (GW) repeat domain. This domain mimics host-encoded GW repeat domains that are required for RISC formation, facilitating a direct association between P38 and AGO1. P38 prevents siRNA, but not miRNA, loading into AGO1 [255]. This blockage perturbs the antiviral-targeting siRNA pathway, whereas the miRNA pathway – which

Fig. 7 RNA silencing-based antiviral immunity in a plant infected with a +ssRNA virus and a geminivirus. Plant viruses can enter cells through a wound. +ssRNA viruses disassemble in the cytoplasm and begin replication to form dsRNA intermediates. These dsRNAs are recognized and cleaved by DCL4 and/or DCL2 and methylated by HEN1 to generate primary vsiRNAs. vsiRNAs are loaded into AGO1/2 or AGO7 to target viral mRNA which is acted upon by RDR6/1 to synthesize dsRNA along the length of the viral transcript. This dsRNA is processed into secondary vsiRNAs. Secondary vsiRNAs are loaded into the RISC complex to guide further silencing of viral RNA. Viral silencing repressors (VSRs) P19 and HC-Pro prevent vsiRNA incorporation into AGO1/2 or AGO7 by sequestration or removal of 3'-O-methylation, respectively. The Sweet potato chlorotic stunt virus VSR, RNase3, processes vsiRNAs into 14 bp dsRNAs that are unable to load into AGO. P19, HC-Pro, and RNase3 can act on both primary and secondary vsiRNAs pools, and are shown to act on separate pools for display purposes only. The VSRs 2b and P38 perturb RISC function by directly binding to AGO1, whereas P0 marks AGO1 for degradation. Geminiviruses utilize host machinery to replicate in the nucleus. Overlapping sense and antisense geminiviral mRNA transcripts can be processed by all four DCLs (DCL1/2/3/4) and methylated by HEN1 to generate vsiRNAs. The vsiRNAs are loaded into AGO4 to direct DNA methylation of the geminiviral genome, inducing TGS. The AL2 and L2 VSRs inhibit the action of ADK, a positive regulator of RdDM, whereas C2 stabilizes SAMDC1, a negative regulator of RdDM.

RNA Silencing in Plants | 369

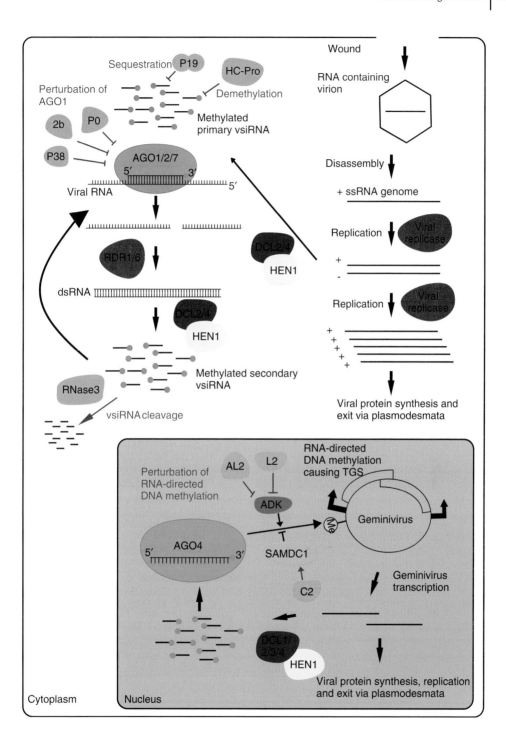

is responsible for plant development and growth – remains largely unaffected. The P0 VSR from poleroviruses also targets AGO1, but instead of preventing RISC assembly, it promotes its degradation by interacting with the SCF family of E3-ligases by means of its minimal F-box motif [256, 257]. Interestingly, this degradation is proteasome-independent, as proteasome inhibitors have no effect on AGO1 degradation. In addition to a well-established role in vsiRNA sequestration, P19 also indirectly promotes the degradation of AGO1 by causing the upregulation of a miRNA (miR168) that targets *AGO1* for translational repression [258]. Therefore, the targeting of AGO1 appears to be a common strategy employed by viruses, and provides compelling evidence of an essential role for RISC activity in antiviral defense.

DNA viruses are targets for silencing by the RNA-directed DNA methylation (RdDM) pathway. *In vitro*-methylated *Tomato golden mosaic virus* (TGMV) is unable to replicate in protoplasts, suggesting that DNA methylation of viral genomes is a valid defensive strategy [259]. TGMV and *Beet curly top virus* (BCTV) encoded VSRs, AL2, and L2, respectively, interact with and inhibit adenosine kinase (ADK) [260, 261]. ADK is an essential enzyme in the methyl cycle, required for *S*-adenosyl methionine (SAM) biosynthesis. Thus, ADK is required for the *de novo* methylation of viral genomes and provides a target for VSRs. In support of this hypothesis, a synthetic ADK inhibitor suppresses silencing in a similar manner to the geminiviral L2 protein [260]. The VSR protein, C2, from *Beet severe curly top virus* (BSCTV) interacts with and stabilizes *S*-adenosine-methionine decarboxylase1 (SAMDC1) [262]. An overaccumulation of SAMDC1 likely prevents the availability of free methyl groups for the RdDM pathway. Consistent with this observation, BSCTV is hypermethylated in *samdc1* knockout plants, and accumulates to lower levels [262].

In addition to viruses, other plant pathogens also perturb the RNA silencing machinery. Pathogenic *Pseudomonas* bacteria secrete proteins molecules called *effectors* that block plant defense responses. In particular, some *Pseudomonas* effectors block the biogenesis or transcriptional activation of miRNAs associated with pathogen attack response, such as miR393 and miR396 [263]. Thus, similar to viruses, bacteria – and likely other pathogens – also suppress RNA silencing to induce disease.

10.4
Crosstalk: RNA Silencing and the Innate Immune System

RNA silencing is not the only type of antiviral defense found in plants. Viral proteins can also be recognized by classical receptor-based pattern-triggered immunity (PTI) and effector-triggered immunity (ETI) [264]. The ETI receptors primarily encode proteins from the nucleotide binding leucine-rich repeat (NB-LRR) class, while PTI receptors encode transmembrane LRR kinases. Traditionally, the ETI/PTI and RNA silencing pathways were thought to be entirely separate. However, accumulating evidence suggests that there is a degree of cooperative crosstalk between RNA silencing and ETI/PTI. For instance, the defense hormone salicylic acid, induced by ETI and responsible for systemic acquired resistance, causes an upregulation of RDR1 [265, 266]. Rx, which is an NB-LRR receptor for PVX, activates a defense response that requires an AGO4-mediated translational repression of the virus in

Nicotiana benthamiana [267]. These results suggest that ETI/PTI pathways can prime the RNA silencing machinery. However, RNA silencing can also recruit ETI/PTI. For instance, miR482 targets a highly conserved region of the NB-LRR pathogen receptors, of which there are more than 100 target loci in *Solanaceous* plant species [173, 174, 268]. In response to pathogen challenge, tomato rapidly downregulates miR482, causing a corresponding upregulation in the NB-LRR defense protein transcripts [173]. This crosstalk provides a robust antiviral defense, as any VSR deployed by the virus would likely perturb miR482-mediated silencing and increase the accumulation of NB-LRR pathogen receptors.

11
Mobile Small RNAs

One interesting aspect of RNA silencing is the ability of the silencing signal to move from cell-to-cell and over long distances. This phenomenon has so far been observed in invertebrates such as *C. elegans* and in plants. Early experiments using tobacco ring spot virus produced necrotic plants that showed symptoms in the lower leaves but, with time, newly emerging tissues were resistant to the virus [269]. It was likely that an RNA signal had moved through the phloem to prime viral silencing in the leaf primordia that consequently became resistant to the virus. Later, RNA silencing spread was observed in plants undergoing spontaneous transgene silencing [270–273]. Since these initial observations, genetic screens and RNA deep sequencing has greatly improved the present knowledge of RNA silencing spread.

11.1
Cell-to-Cell Movement of an RNA Silencing Signal

An RNA silencing signal initially moves from cell-to-cell before entering the vascular tissue and moving systemically through the plant. The phloem is thought to transport the silencing signal as the signal typically moves from source to sink [24, 25], consistent with phloem movement, and no endogenous RNAs have been detected in the xylem [274]. In plants, cell-to-cell movement can either occur around cell membranes, termed *apoplastic movement*, or through membrane channels (called *plasmodesmata*) that connect plant cells, a process referred to as *symplastic movement*. Plasmodesmata have a size selectivity of around 27 kDa [275], although molecules such as viral movement proteins [276] and the KNOTTED transcription factor [277] can increase this size limit. It is thought that a cell-to-cell silencing signal moves symplastically through the plasmodesmata, but there is no reason to believe that apoplastic movement is not possible [278].

Genetic screens have shed light on the process of cell-to-cell movement and the identity of the mobile RNA silencing signal. The basis for these screens was the genetic separation of a silencer transgene from a reporter locus that produces a visual phenotype when silenced [279, 280]. Mutating genes required for RNA silencing movement perturbed the silencing of the reporter locus and uncovered the requirement for *NRPD1*, *RDR2*, and *DCL4* in cell-to-cell silencing movements [280–282]. Surprisingly, *DCL3* and *AGO4* were not required for RNA silencing spread [280, 281]. These analyses implicated the Pol IV pathway in the spread of a silencing signal, though whether NRPD1 and RDR2 act in the production, movement or targeting aspects of mobile silencing are not yet known. The identification of *DCL4* as a factor required for RNA silencing spread suggested that *DCL4*-dependent 21 nt sRNAs are the mobile cell-to-cell species [280, 282]. Consistent with this observation, DCL4 expression in only the tissue where the silencing trigger is expressed is sufficient to restore the mobile silencing phenotype, indicating that DCL4 produces the 21 nt silencing signal [283]. Vascular-specific expression of the P19 viral suppressor of silencing, which binds specifically 21 nt duplexes, also prevented the movement of a silencing signal from the vasculature. Bombarded fluorescently labeled double-stranded 21 nt siRNAs moved through several cells in *Arabidopsis* leaves, but single-stranded species were unable to move, consistent with the mobile species being double-stranded [283]. Cell-to-cell movement is not restricted to 21 nt siRNA species. Bombarded fluorescently labeled double-stranded 24 nt siRNAs also moved from cell-to-cell [150], suggesting that both 21 and 24 nt siRNAs are mobile as duplexes.

A cell-to-cell RNA silencing signal can move 10–15 cells [279], although certain transgenic miRNAs are restricted to move only one cell layer [154]. The 10–15 cell limit is likely to due to a concentration gradient created by the dilution of siRNAs from their source. To move beyond 10–15 cells, an amplification of the silencing signal by RDRs is required. This amplification is important for both cell-to-cell spread and long-distance movement as it can compensate for the dilution effect. Consistent with this idea, the amplification of a long-distance silencing signal by RDR2 and RDR6 allows for cell-to-cell spread of the silencing signal in the recipient tissue [248, 284].

11.2
Long-Distance Movement of an RNA Silencing Signal

Cell-to-cell and long-distance movement have several notable differences, which suggests that they may involve different mechanisms. Viral suppressors that reduced 24 nt siRNAs inhibited the long-distance movement but not the cell-to-cell movement of a silencing signal [166, 279]. In addition, low concentrations of cadmium prevented systemic but not cell-to-cell movement of a silencing signal due to the upregulation of a glycine-rich protein in the cell walls of the vascular tissue [285, 286]. Mechanistic and physiological differences are also apparent as root-to-shoot (sink-to-source) movement of a silencing signal in *Arabidopsis* proceeds by cell-to-cell movement, whereas reciprocal shoot-to-root (source-to-sink) movement spreads through the phloem [287].

Evidence for the long-distance transport of a silencing signal came from experiments that physically separated the RNA silencing trigger from the reporter of silencing. Transient expression of the gene encoding the Green Fluorescent Protein (GFP) in a leaf of *Nicotiana* stably expressing *GFP* caused *GFP* expression to become silenced in the leaf within several days, and after five weeks the GFP silencing had spread throughout the entire plant [25, 288]. A second line of evidence used grafting to physically separate a *Nicotiana* rootstock silencing the *GUS* (β-glucuronidase) reporter gene from a shoot expressing *GUS*. Upon grafting, a mobile signal spread from the silenced rootstock to the shoot, which caused GUS silencing [24]. In both examples, silencing was specific to the transgene, which indicated that the silencing signal included a nucleic acid complementary to the sequence of the transgene [24, 25].

Initial approaches to identify the mobile signal used direct sampling from the phloem to detect the species present, but did not assay for their mobility. Analyses of cucurbit phloem sap revealed the presence of 21–24 nt endogenous and transgene-derived sRNAs [289]. Interestingly, this study also detected a 21 kDa protein, sRNA BINDING PROTEIN1, that preferentially bound 25 nt single-stranded RNAs. Sampling phloem from *Brassica napus* revealed the presence of 21–24 nt sRNAs and 32 known miRNAs [274]. Three miRNAs – miR395, miR398, and miR399 – increased their abundance in the phloem in response to sulfate, copper, and phosphate stress, respectively [274].

A second approach used *Arabidopsis* grafting to physically separate the source of sRNAs from the recipient tissue. Overexpressing miR399 in the shoots increased miR399 levels in grafted wild-type roots [290]. Mobile miR399 downregulated *PHO2*, a negative regulator of phosphate uptake and was the first miRNA to be identified as moving systemically [290]. Invert-repeat (IR) siRNAs were also demonstrated to be mobile using a similar strategy. IR71 siRNAs produced in the shoot moved to grafted roots that were deficient in IR71 siRNAs [150].

Another grafting strategy attempted to identify the mobile signal and factors required in the source and recipient tissues for mobile RNA silencing. *GFP*-expressing shoots were grafted to a rootstock, thereby silencing the *GFP* transgene, and movement of the silencing signal across the graft junction was monitored. *AGO4*, *DCL3*, *NRPD1*, *RDR2*, and *RDR6* were each required for the reception of a *GFP*-silencing signal, but not its production [284]. When the *dcl1* mutant and the *dcl2*, *dcl3*, *dcl4* triple mutant were introduced to the silencing source, they did not affect the RNA silencing movement, which was consistent with the hypothesis that long RNAs were the mobile species [284]. However, this system may be mechanistically different from classical spreading systems as the leaves emerge completely silenced, rather than silencing spreading through the vasculature. Furthermore, some sRNAs may have been produced since DCLs are functionally redundant and low levels of siRNAs are generated in the weak *dcl1* background.

A whole-transcriptome analysis of mobile siRNAs was performed using a grafting approach and RNA deep sequencing techniques. *Arabidopsis* shoots competent to produce siRNAs were grafted to roots deficient in 22–24 nt siRNAs (the *dcl2*, *dcl2*, *dcl4* triple mutant), and siRNAs were measured in the roots at five weeks after grafting. Of the 20 679 loci that produced sRNAs, approximately 35% of the loci produced mobile sRNAs [291]. Furthermore, mobile 24 nt siRNAs could direct DNA methylation at several endogenous loci and transgenic 24 nt siRNAs could direct DNA methylation and reduce gene expression at transgene promoters in meristemic root tissue [291, 292]. These experiments demonstrated that a substantial number of 22–24 nt siRNAs move systemically (Fig. 8), though to what extent 21 nt sRNAs are mobile over long distances remains unknown.

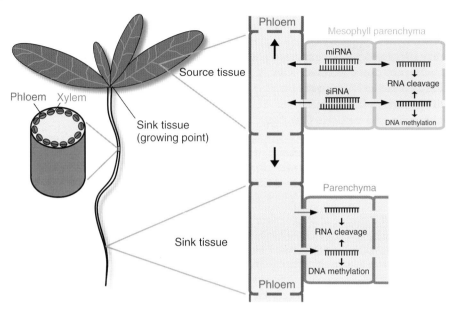

Fig. 8 Small RNA movement in *Arabidopsis thaliana*. sRNAs produced in photosynthetic source tissues such as the leaf mesophyll parenchyma may initially move from cell-to-cell through plasmodesmata until reaching the phloem companion cells, where they load into the enucleated sieve elements. Once in the phloem, sRNAs move long distances to photosynthetic sinks and growing points. There, sRNAs unload at source tissues and move from cell-to-cell to meristems and other tissues [292]. Certain 21 nt miRNAs move either long distances or from cell-to-cell to regulate gene expression by mRNA cleavage [290, 293]. Mobile 21 nt tasiRNAs move from the top of the leaf to the bottom to cleave auxin response factor mRNAs [180, 181], whereas 24 nt hc-siRNAs move long distances to direct DNA methylation. The parenchyma represents the bulk of the plant tissue. (Figure modified from Ref. [278].)

11.3 Functions of sRNA Movement

Mobile sRNAs are implicated in several biological functions, including plant development. In addition to miR399, which moves from shoots to roots to regulate phosphate uptake, another mobile miRNA, miR165/miR166, is produced in the root endodermis and moves from cell-to-cell to the stele. The miR165/miR166 miRNA gradient degrades several transcription factor mRNAs and determines xylem cell identity [293]. tasiRNAs from the *TAS3* locus also diffuse down a concentration gradient from the upper to lower leaf surface to regulate ARF3 mRNAs. The resulting ARF3 expression gradient influences leaf polarity and development [180, 181].

Virus resistance may be another important property conferred by mobile sRNAs. vsiRNAs are thought to move ahead of the viral infection and prime defense systems against invading viruses. In support of this idea, RDR6 is required in distant tissues for viral resistance [248] and studies using recombinant viruses that are unable to move have shown that they can initiate a silencing signal that moves 10–15 cells from the virus-infected cells [279, 294].

One interesting aspect of mobile RNA silencing is its ability to move between organisms. An RNA silencing signals produced in a plant can silence complementary genes in pathogens that are feeding on the plant. To date, silencing movement from plants to fungi, insects, and nematodes has been demonstrated [295–299]. Parasitic plants can also transmit silencing signals. Lettuce plants silencing a transgene encoding the *GUS* enzyme transmitted a silencing signal to a lettuce plant stability expressing *GUS* through an intermediate wild-type parasitic plant infecting both lettuce plants [300]. Transmission of the RNA silencing signal caused *GUS* silencing, and presents an interesting strategy for genetic information to move between plants.

12 Conclusion

The discovery of RNA silencing has been pivotal to the present understanding of gene regulation, and explained phenomena that previously were difficult to reconcile, such as systemic viral resistance and transgene silencing. During the past 15 years, a comprehensive understanding of the various RNA silencing pathways in *Arabidopsis* has emerged. These mechanisms are not rigid, as proteins from one pathway can substitute in other pathways, and in many cases proteins can act in multiple pathways. For instance, the PHD-finger-containing protein NEEDED FOR RDR2-INDEPENDENT DNA METHYLATION (NERD) binds both histones and an AGO protein associated with virus resistance, AGO2. Mutations in NERD or AGO2 affect epigenetic marks, indicating that the TGS and PTGS pathways are functionally linked [301]. Furthermore, there is evidence that sRNAs can substitute for one another, such as miRNAs directing DNA methylation in rice [302]. The dynamic nature of RNA silencing is also illustrated by feedback loops. Both AGO1 and DCL1 mRNAs are targeted by a miRNA, allowing a feedback loop to regulate the strength and extent of miRNA-mediated gene silencing [104, 303]. This feedback loop may be important for viral defense in which viral suppressors sequester miRNAs or inactivate AGO1. Feedback may also occur in the hc-siRNA pathway through the interaction between histone marks and DNA methylation. hc-siRNAs can modify histone marks, either directly, or more likely using a feedback loop linking DNA methylation to histone marks through the actions of the CMT3 and KRYPTONITE (KYP) proteins [91, 210]. Such a loop could be important for propagating DNA methylation and histone marks during mitosis and meiosis. Although the understanding of RNA silencing has progressed rapidly, there remain many unknown aspects in the field. Fully identifying sRNAs and their roles should be a priority; in particular, how sRNAs are spatially and temporally regulated throughout development, and in response to stress. Interestingly, a novel group of sRNAs, termed *double-strand break-induced small RNAs* (diRNAs) might be important in repairing double-strand breaks [304]. New classes of siRNA are also being identified, such as the 30–40 nt long small interfering RNAs (lsiRNAs) that are upregulated upon pathogen infection [305].

Hc-siRNAs are the most abundant sRNA in plants, yet their loss produces few discernible phenotypes. Instead, the role of hc-siRNAs is to protect the genome from transposable elements and to maintain heterochromatin in parallel with other pathways [306, 307]. Hc-siRNAs might also have important roles in adaptation. There is evidence that stress can trigger the formation of epigenetic marks. Interestingly, in some cases these epigenetic marks are inherited to the progeny where they increase stress-resistance or affect the rate of homologous recombination [211, 308–312]. There is evidence to suggest that hc-siRNAs play a role in this phenomenon, but to what extent is not yet known. Ultimately, harnessing trans-generational stress adaptation and using sRNAs to engineering plants that are better adapted to pathogens and abiotic stress will be of great interest to science and of benefit to agriculture.

Acknowledgments

The authors thank Ian Henderson and Raymond Wightman for their critical reading of the manuscript, and Richard Jorgensen for permission to reproduce Fig. 1b,c. C.W.M. is supported by a Junior Research Fellowship from Clare College, Cambridge, UK. C.J.H. is supported by the Biotechnology and Biological Sciences Research Council.

References

1 Fire, A., Xu, S., Montgomery, M.K., Kostas, S.A. et al. (1998) Potent and specific genetic interference by double-stranded RNA in *Caenorhabditis elegans*. *Nature*, **391**, 806–811.

2 Hamilton, A.J., Baulcombe, D.C. (1999) A species of small antisense RNA in post-transcriptional gene silencing in plants. *Science*, **286**, 950–952.

3 Zamore, P.D., Tuschl, T., Sharp, P.A., Bartel, D.P. (2000) RNAi: double-stranded RNA directs the ATP-dependent cleavage of mRNA at 21 to 23 nucleotide intervals. *Cell*, **101**, 25–33.

4 Flavell, R.B. (1994) Inactivation of gene expression in plants as a consequence of specific sequence duplication. *Proc. Natl Acad. Sci. USA*, **91**, 3490–3496.

5 de Carvalho, F., Gheysen, G., Kushnir, S., Van Montagu, M. et al. (1992) Suppression of β-1,3-glucanase transgene expression in homozygous plants. *EMBO J.*, **11**, 2595–2602.

6 Assaad, F.F., Tucker, K.L., Signer, E.R. (1993) Epigenetic repeat-induced gene silencing (RIGS) in *Arabidopsis*. *Plant Mol. Biol.*, **22**, 1067–1085.

7 Elmayan, T., Vaucheret, H. (1996) Expression of single copies of a strongly expressed 35S transgene can be silenced post-transcriptionally. *Plant J.*, **9**, 787–797.

8 Que, Q.D., Wang, H.Y., English, J.J., Jorgensen, R.A. (1997) The frequency and degree of cosuppression by sense chalcone synthase transgenes are dependent on transgene promoter strength and are reduced by premature nonsense codons in the transgene coding sequence. *Plant Cell*, **9**, 1357–1368.

9 Van Blokland, R., Van der Geest, N., Mol, J.N.M., Kooter, J.M. (1994) Transgene-mediated suppression of chalcone synthase expression in *Petunia hybrida* results from an increase in RNA turnover. *Plant J.*, **6**, 861–877.

10 Stam, M., de Bruin, R., Kenter, S., van der Hoorn, R.A.L. et al. (1997) Post-transcriptional silencing of chalcone synthase in Petunia by inverted transgene repeats. *Plant J.*, **12**, 63–82.

11 Ecker, J.R., Davis, R.W. (1986) Inhibition of gene-expression in plant-cells by expression of antisense RNA. *Proc. Natl Acad. Sci. USA*, **83**, 5372–5376.

12 Smith, C.J.S., Watson, C.F., Ray, J., Bird, C.R. et al. (1988) Antisense RNA inhibition of polygalacturonase gene expression in transgenic tomatoes. *Nature*, **334**, 724–726.

13 Napoli, C., Lemieux, C., Jorgensen, R.A. (1990) Introduction of a chimeric chalcone synthase gene into Petunia results in reversible co-suppression of homologous genes *in trans*. *Plant Cell*, **2**, 279–289.

14 van der Krol, A.R., Mur, L.A., Beld, M., Mol, J.N.M. *et al.* (1990) Flavonoid genes in petunia: addition of a limited number of gene copies may lead to a suppression of gene expression. *Plant Cell*, **2**, 291–299.

15 Romano, N., Macino, G. (1992) Quelling: transient inactivation of gene expression in *Neurospora crassa* by transformation with homologous sequences. *Mol. Microbiol.*, **6**, 3343–3353.

16 Guo, S., Kemphues, K. (1995) par-1, a gene required for establishing polarity in *C. elegans* embryos, encodes a putative Ser/Thr kinase that is asymmetrically distributed. *Cell*, **81**, 611–620.

17 Que, Q., Wang, H.-Y. and Jorgensen, R.A. (1998) Distinct patterns of pigment suppression are produced by allelic sense and antisense chalcone synthase transgenes in petunia flowers. *Plant J.*, **13**, 401–409.

18 Matzke, M.A., Primig, M., Trnovsky, J., Matzke, A.J.M. (1989) Reversible methylation and inactivation of marker genes in sequentially transformed tobacco plants. *EMBO J.*, **8**, 643–649.

19 Abel, P.P., Nelson, R.S., De, B., Hoffmann, N. *et al.* (1986) Delay of disease development in transgenic plants that express the tobacco mosaic virus coat protein gene. *Science*, **232**, 738–743.

20 Lindbo, J.A., Silva-Rosales, L., Proebsting, W.M., Dougherty, W.G. (1993) Induction of a highly specific antiviral state in transgenic plants: implications for regulation of gene expression and virus resistance. *Plant Cell*, **5**, 1749–1759.

21 Kumagai, M.H., Donson, J., Della-Cioppa, G., Harvey, D. *et al.* (1995) Cytoplasmic inhibition of carotenoid biosynthesis with virus-derived RNA. *Proc. Natl Acad. Sci. USA*, **92**, 1679–1683.

22 Ratcliff, F., Harrison, B.D., Baulcombe, D.C. (1997) A similarity between viral defense and gene silencing in plants. *Science*, **276**, 1558–1560.

23 Wassenegger, M., Heimes, S., Riedel, L., Sanger, H.L. (1994) RNA-directed de novo methylation of genomic sequences in plants. *Cell*, **76**, 567–576.

24 Palauqui, J.-C., Elmayan, T., Pollien, J.-M., Vaucheret, H. (1997) Systemic acquired silencing: transgene-specific post-transcriptional silencing is transmitted by grafting from silenced stocks to non-silenced scions. *EMBO J.*, **16**, 4738–4745.

25 Voinnet, O., Baulcombe, D.C. (1997) Systemic signalling in gene silencing. *Nature*, **389**, 553.

26 Hammond, S.M., Bernstein, E., Beach, D., Hannon, G. (2000) An RNA-directed nuclease mediates post-transcriptional gene silencing in *Drosophila* cell extracts. *Nature*, **404**, 293–296.

27 Bernstein, E., Caudy, A.A., Hammond, S.M., Hannon, G.J. (2001) Role for a bidentate ribonuclease in the initiation step of RNA interference. *Nature*, **409**, 363–366.

28 Grishok, A., Pasquinelli, A.E., Conte, D., Li, N. *et al.* (2001) Genes and mechanisms related to RNA interference regulate expression of the small temporal RNAs that control *C. elegans* developmental timing. *Cell*, **106**, 23–34.

29 Ketting, R.F., Fischer, S.E.J., Bernstein, E., Sijen, T. *et al.* (2001) Dicer functions in RNA interference and in synthesis of small RNA involved in developmental timing in *C. elegans*. *Genes Dev.*, **15**, 2654–2659.

30 Knight, S.W., Bass, B.L. (2001) A role for the RNase III enzyme DCR-1 in RNA interference and germ line development in *Caenorhabditis elegans*. *Science*, **293**, 2269–2271.

31 Hutvagner, G., McLachlan, J., Pasquinelli, A.E., Balint, E. *et al.* (2001) A cellular function for the RNA-interference enzyme Dicer in the maturation of the *let-7* small temporal RNA. *Science*, **293**, 834–838.

32 Schauer, S.E., Jacobsen, S.E., Meinke, D.W., Ray, A. (2002) DICER-LIKE1: blind men and elephants in *Arabidopsis* development. *Trends Plant Sci.*, **7**, 487–491.

33 Margis, R., Fusaro, A.F., Smith, N.A., Curtin, S.J. *et al.* (2006) The evolution and diversification of Dicers in plants. *FEBS Lett.*, **580**, 2442–2450.

34 Jacobsen, S.E., Running, M.P., Meyerowitz, E.M. (1999) Disruption of an RNA helicase/RNase III gene in *Arabidopsis* causes unregulated cell division in floral meristems. *Development*, **126**, 5231–5243.

35 Bohmert, K., Camus, I., Bellini, C., Bouchez, D. *et al.* (1998) *AGO1* defines a novel locus of *Arabidopsis* controlling leaf development. *EMBO J.*, **17**, 170–180.

36 Tabara, H., Sarkissian, M., Kelly, W.G., Fleenor, J. *et al.* (1999) The rde-1 gene,

RNA interference and transposon silencing in *C. elegans*. *Cell*, **99**, 123–132.

37 Hammond, S.M., Boettcher, S., Caudy, A.A., Kobayashi, R. *et al.* (2001) Argonaute2, a link between genetic and biochemical analyses of RNAi. *Science*, **293**, 1146–1150.

38 Martinez, J., Patkaniowska, A., Urlaub, H., Luhrmann, R. *et al.* (2002) Single-stranded antisense siRNAs guide target RNA cleavage in RNAi. *Cell*, **110**, 563–574.

39 Song, J.J., Smith, S.K., Hannon, G.J., Joshua-Tor, L. (2004) Crystal structure of Argonaute and its implications for RISC slicer activity. *Science*, **305**, 1434–1437.

40 Liu, J., Carmell, M.A., Rivas, F.V., Marsden, C.G. *et al.* (2004) Argonaute2 is the catalytic engine of mammalian RNAi. *Science*, **305**, 1437–1441.

41 Baumberger, N., Baulcombe, D.C. (2005) Arabidopsis ARGONAUTE1 is an RNA Slicer that selectively recruits micro RNAs and short interfering RNAs. *Proc. Natl Acad. Sci. USA*, **102**, 11928–11933.

42 Qi, Y., Denli, A.M., Hannon, G.J. (2005) Biochemical specialization within *Arabidopsis* RNA silencing pathways. *Mol. Cell.*, **19**, 421–428.

43 Baulcombe, D. (2004) RNA silencing in plants. *Nature*, **431**, 356–363.

44 Ghildiyal, M., Zamore, P.D. (2009) Small silencing RNAs: an expanding universe. *Nat. Rev. Genet.*, **10**, 94–108.

45 Moazed, D. (2009) Small RNAs in transcriptional gene silencing and genome defence. *Nature*, **457**, 413–420.

46 Barrangou, R., Fremaux, C., Deveau, H., Richards, M. *et al.* (2007) CRISPR provides acquired resistance against viruses in prokaryotes. *Science*, **315**, 1709.

47 Brouns, S.J., Jore, M.M., Lundgren, M., Westra, E.R. *et al.* (2008) Small CRISPR RNAs guide antiviral defense in prokaryotes. *Science*, **321**, 960–964.

48 Mardis, E.R. (2008) Next-generation DNA sequencing methods. *Annu. Rev. Genomics Hum. Genet.*, **9**, 387–402.

49 Henz, S.R., Cumbie, J.S., Kasschau, K.D., Lohmann, J.U. *et al.* (2007) Distinct expression patterns of natural antisense transcripts in *Arabidopsis*. *Plant Physiol.*, **144**, 1247–1255.

50 Zhang, Y., Liu, X.S., Liu, Q.R., Wei, L. (2006) Genome-wide in silico identification and analysis of cis natural antisense transcripts (cis-NATs) in ten species. *Nucleic Acids Res.*, **34**, 3465–3475.

51 Wang, X.J., Gaasterland, T., Chua, N.H. (2005) Genome-wide prediction and identification of cis-natural antisense transcripts in *Arabidopsis thaliana*. *Genome Biol.*, **6**, R30.

52 Zhang, X., Xia, J., Lii, Y.E., Barrera-Figueroa, B.E. *et al.* (2012) Genome-wide analysis of plant nat-siRNAs reveals insights into their distribution, biogenesis and function. *Genome Biol.*, **13**, R20.

53 Cogoni, C., Macino, G. (1999) Gene silencing in *Neurospora crassa* requires a protein homologous to RNA-dependent RNA polymerase. *Nature*, **399**, 166–169.

54 Volpe, T., Kidner, C., Hall, I.M., Teng, G. *et al.* (2002) Regulation of heterochromatic silencing and histone H3 lysine-9 methylation by RNAi. *Science*, **297**, 1833–1837.

55 Smardon, A., Spoerke, J.M., Stacey, S.C., Klein, M.E. *et al.* (2000) EGO-1 is related to RNA-directed RNA polymerase and functions in germ-line development and RNA interference in *C. elegans*. *Curr. Biol.*, **10**, 169–178.

56 Dalmay, T., Hamilton, A.J., Rudd, S., Angell, S. *et al.* (2000) An RNA-dependent RNA polymerase gene in *Arabidopsis* is required for posttranscriptional gene silencing mediated by a transgene but not by a virus. *Cell*, **101**, 543–553.

57 Mourrain, P., Beclin, C., Elmayan, T., Feuerbach, F. *et al.* (2000) *Arabidopsis SGS2* and *SGS3* genes are required for posttranscriptional gene silencing and natural virus resistance. *Cell*, **101**, 533–542.

58 Lee, S.R., Collins, K. (2007) Physical and functional coupling of RNA-dependent RNA polymerase and Dicer in the biogenesis of endogenous siRNAs. *Nat. Struct. Mol. Biol.*, **14**, 604–610.

59 Wassenegger, M., Krczal, G. (2006) Nomenclature and functions of RNA-directed RNA polymerases. *Trends Plant Sci.*, **11**, 142–151.

60 Wang, X.B., Wu, Q., Ito, T., Cillo, F. *et al.* (2010) RNAi-mediated viral immunity requires amplification of virus-derived siRNAs in *Arabidopsis thaliana*. *Proc. Natl Acad. Sci. USA*, **107**, 484–489.

61 Iyer, L.M., Koonin, E.V., Aravind, L. (2003) Evolutionary connection between the catalytic subunits of DNA-dependent

RNA polymerases and eukaryotic RNA-dependent RNA polymerases and the origin of RNA polymerases. *BMC Struct. Biol.*, **3**, 1–23.

62 Lipardi, C., Paterson, B.M. (2009) Identification of an RNA-dependent RNA polymerase in *Drosophila* involved in RNAi and transposon suppression. *Proc. Natl Acad. Sci. USA*, **106**, 15645–15650.

63 Maida, Y., Yasukawa, M., Furuuchi, M., Lassmann, T. *et al.* (2009) An RNA-dependent RNA polymerase formed by TERT and the RMRP RNA. *Nature*, **461**, 230–235.

64 MacRae, I.J., Zhou, K., Li, F., Repic, A. *et al.* (2006) Structural basis for double-stranded RNA processing by dicer. *Science*, **311**, 195–195.

65 Rajagopalan, R., Vaucheret, H., Trejo, J., Bartel, D.P. (2006) A diverse and evolutionarily fluid set of microRNAs in *Arabidopsis thaliana*. *Genes Dev.*, **20**, 3407–3425.

66 Henderson, I.R., Zhang, X., Lu, C., Johnson, L. *et al.* (2006) Dissecting *Arabidopsis thaliana* DICER function in small RNA processing, gene silencing and DNA methylation patterning. *Nat. Genet.*, **38**, 721–725.

67 Kapoor, M., Arora, R., Lama, T., Nijhawan, A. *et al.* (2008) Genome-wide identification, organization and phylogenetic analysis of Dicer-like, Argonaute and RNA-dependent RNA Polymerase gene families and their expression analysis during reproductive development and stress in rice. *BMC Genomics*, **9**, 451.

68 Cerutti, H. Casas-Mollano, J.A. (2006) On the origin and functions of RNA-mediated silencing: from protists to man. *Curr. Genet.*, **50**, 81–99.

69 Deleris, A., Gallego-Bartolome, J., Bao, J., Kasschau, K.D. *et al.* (2006) Hierarchical action and inhibition of plant dicer-like proteins in antiviral defense. *Science*, **313**, 68–71.

70 Blevins, T., Rajeswaran, R., Shivaprasad, P.V., Beknazariants, D. *et al.* (2006) Four plant Dicers mediate viral small RNA biogenesis and DNA virus induced silencing. *Nucleic Acids Res.*, **34**, 6233–6246.

71 Gasciolli, V., Mallory, A.C., Bartel, D.P., Vaucheret, H. (2005) Partially Redundant Functions of *Arabidopsis* DICER-like Enzymes and a Role for DCL4 in Producing *trans*-Acting siRNAs. *Curr. Biol.*, **15**, 1494–1500.

72 Vagin, V.V., Sigova, A., Li, C., Seitz, H. *et al.* (2006) A distinct small RNA pathway silences selfish genetic elements in the germline. *Science*, **313**, 320–324.

73 Gunawardane, L.S., Saito, K., Nishida, K.M., Miyoshi, K. *et al.* (2007) A slicer-mediated mechanism for repeat-associated siRNA 5′ end formation in Drosophila. *Science*, **315**, 1587–1590.

74 Brennecke, J., Aravin, A.A., Stark, A., Dus, M. *et al.* (2007) Discrete small RNA-generating loci as master regulators of transposon activity in *Drosophila*. *Cell*, **128**, 1–15.

75 Cifuentes, D., Xue, H., Taylor, D.W., Patnode, H. *et al.* (2010) A novel miRNA processing pathway independent of Dicer requires Argonaute2 catalytic activity. *Science*, **328**, 1694–1698.

76 Cheloufi, S., Dos Santos, C.O., Chong, M.M., Hannon, G.J. (2010) A dicer-independent miRNA biogenesis pathway that requires Ago catalysis. *Nature*, **465**, 584–589.

77 Halic, M., Moazed, D. (2010) Dicer-independent primal RNAs trigger RNAi and heterochromatin formation. *Cell*, **140**, 504–516.

78 Miyoshi, K., Tsukumo, H., Nagami, T., Siomi, H. *et al.* (2005) Slicer function of *Drosophila* Argonautes and its involvement in RISC formation. *Genes Dev.*, **19**, 2837–2848.

79 Rivas, F.V., Tolia, N.H., Song, J.-J., Aragon, J.P. *et al.* (2005) Purified Argonaute2 and an siRNA form recombinant human RISC. *Nat. Struct. Mol. Biol.*, **12**, 340–349.

80 Verdel, A., Jia, S., Gerber, S., Sugiyama, T. *et al.* (2004) RNAi-mediated targeting of heterochromatin by the RITS complex. *Science*, **303**, 672–676.

81 Carmell, M.A., Xuan, Z., Zhang, M., Hannon, G.J. (2002) The Argonaute family: tentacles that reach into RNAi, developmental control, stem cell maintenance, and tumorigenesis. *Genes Dev.*, **16**, 2733–2742.

82 Matranga, C., Tomari, Y., Shin, C., Bartel, D.P. *et al.* (2005) Passenger-strand cleavage facilitates assembly of siRNA into

Ago2-containing RNAi enzyme complexes. *Cell*, **123**, 1–14.

83 Schwarz, D.S., Hutvagner, G., Du, T., Xu, Z. et al. (2003) Asymmetry in the assembly of the RNAi enzyme complex. *Cell*, **115**, 199–208.

84 Khvorova, A., Reynolds, A., Jayasena, S.D. (2003) Functional siRNAs and miRNAs exhibit strand bias. *Cell*, **115**, 209–216.

85 Guo, H., Ingolia, N.T., Weissman, J.S., Bartel, D.P. (2010) Mammalian microRNAs predominantly act to decrease target mRNA levels. *Nature*, **466**, 835–840.

86 Bagga, S., Bracht, J., Hunter, S., Massirer, K. et al. (2005) Regulation by let-7 and lin-4 miRNAs results in target mRNA degradation. *Cell*, **122**, 553–563.

87 Pillai, R.S., Bhattacharyya, S.N., Artus, C.G., Zoller, T. et al. (2005) Inhibition of translational initiation by Let-7 MicroRNA in human cells. *Science*, **309**, 1573–1576.

88 Chendrimada, T.P., Finn, K.J., Ji, X., Baillat, D. et al. (2007) MicroRNA silencing through RISC recruitment of elF6. *Nature*, **447**, 823.

89 Brodersen, P., Sakvarelidze-Achard, L., Bruun-Rasmussen, M., Dunoyer, P. et al. (2008) Widespread translational inhibition by plant miRNAs and siRNAs. *Science*, **320**, 1185–1190.

90 Lanet, E., Delannoy, E., Sormani, R., Floris, M. et al. (2009) Biochemical evidence for translational repression by *Arabidopsis* microRNAs. *Plant Cell*, **21**, 1762–1768.

91 Law, J.A., Jacobsen, S.E. (2010) Establishing, maintaining and modifying DNA methylation patterns in plants and animals. *Nat. Rev. Genet.*, **11**, 204–220.

92 Maiti, M., Lee, H.C., Liu, Y. (2007) QIP, a putative exonuclease, interacts with the *Neurospora* Argonaute protein and facilitates conversion of duplex siRNA into single strands. *Genes Dev.*, **21**, 590–600.

93 Hutvagner, G., Simard, M.J. (2008) Argonaute proteins: key players in RNA silencing. *Nat. Rev. Mol. Cell Biol.*, **9**, 22–32.

94 Parker, J.S., Roe, S.M., Barford, D. (2005) Structural insights into mRNA recognition from a PIWI domain-siRNA guide complex. *Nature*, **434**, 663–666.

95 Ma, J.-B., Yuan, Y.-R., Meister, G., Pei, Y. et al. (2005) Structural basis for 5′-end-specific recognition of guide RNA by the *A. fulgidus* Piwi protein. *Nature*, **434**, 666–670.

96 Behm-Ansmant, I., Rehwinkel, J., Doerks, T., Stark, A. et al. (2006) mRNA degradation by miRNAs and GW182 requires both CCR4:NOT deadenylase and DCP1:DCP2 decapping complexes. *Genes Dev.*, **20**, 1885–1898.

97 Till, S., Lejeune, E., Thermann, R., Bortfeld, M. et al. (2007) A conserved motif in Argonaute-interacting proteins mediates functional interactions through the Argonaute PIWI domain. *Nat. Struct. Mol. Biol.*, **14**, 897–903.

98 Huang, L.F., Jones, A.M.E., Searle, I., Patel, K. et al. (2009) An atypical RNA polymerase involved in RNA silencing shares small subunits with RNA polymerase II. *Nat. Struct. Mol. Biol.*, **16**, 91–93.

99 El-Shami, M., Pontier, D., Lahmy, S., Braun, L. et al. (2007) Reiterated WG/GW motifs form functionally and evolutionarily conserved ARGONAUTE-binding platforms in RNAi-related components. *Genes Dev.*, **21**, 2539–2544.

100 Azevedo, J., Garcia, D., Pontier, D., Ohnesorge, S. et al. (2010) Argonaute quenching and global changes in Dicer homeostasis caused by a pathogen-encoded GW repeat protein. *Genes Dev.*, **24**, 904–915.

101 Ream, T.S., Haag, J.R., Wierzbicki, A.T., Nicora, C.D. et al. (2009) Subunit compositions of the RNA-silencing enzymes Pol IV and Pol V reveal their origins as specialized forms of RNA polymerase II. *Mol. Cell*, **33**, 192–203.

102 Yigit, E., Batista, P.J., Bei, Y., Pang, K.M. et al. (2006) Analysis of the *C. elegans* Argonaute family reveals that distinct argonautes act sequentially during RNAi. *Cell*, **127**, 747–757.

103 Wang, H., Zhang, X., Liu, J., Kiba, T. et al. (2011) Deep sequencing of small RNAs specifically associated with *Arabidopsis* AGO1 and AGO4 uncovers new AGO functions. *Plant J.*, **67**, 292–304.

104 Vaucheret, H., Vazquez, F., Crete, P., Bartel, D.P. (2004) The action of ARGONAUTE1 in the miRNA pathway and its regulation by the miRNA pathway are crucial for plant development. *Genes Dev.*, **18**, 1187–1197.

105 Zhu, H., Hu, F., Wang, R., Zhou, X. et al. (2011) *Arabidopsis* Argonaute10 specifically

sequesters miR166/165 to regulate shoot apical meristem development. *Cell*, **145**, 242–256.
106 Ji, L., Liu, X., Yan, J., Wang, W. *et al.* (2011) ARGONAUTE10 and ARGONAUTE1 regulate the termination of floral stem cells through two microRNAs in *Arabidopsis*. *PLoS Genet.*, **7**, e1001358.
107 Zilberman, D., Cao, X., Jacobsen, S.E. (2003) *ARGONAUTE4* control of locus specific siRNA accumulation and DNA and histone methylation. *Science*, **299**, 716–719.
108 Qi, Y., He, X.H., Wang, X., Kohany, O. *et al.* (2006) Distinct catalytic and non-catalytic roles of ARGONAUTE4 in RNA-directed DNA methylation. *Nature*, **443**, 1008–1012.
109 Zheng, X., Zhu, J., Kapoor, A., Zhu, J.-K. (2007) Role of Arabidopsis AGO6 in siRNA accumulation, DNA methylation and transcriptional gene silencing. *EMBO J.*, **26**, 1–11.
110 Olmedo-Monfil, V., Duran-Figueroa, N., Arteaga-Vazquez, M., Demesa-Arevalo, E. *et al.* (2010) Control of female gamete formation by a small RNA pathway in *Arabidopsis*. *Nature*, **464**, 628–632.
111 Havecker, E.R., Wallbridge, L.M., Hardcastle, T.J., Bush, M.S. *et al.* (2010) The *Arabidopsis* RNA-directed DNA methylation argonautes functionally diverge based on their expression and interaction with target loci. *Plant Cell*, **22**, 321–334.
112 Nonomura, K., Morohoshi, A., Nakano, M., Eiguchi, M. *et al.* (2007) A germ cell specific gene of the ARGONAUTE family is essential for the progression of premeiotic mitosis and meiosis during sporogenesis in rice. *Plant Cell*, **19**, 2583–2594.
113 Montgomery, T.A., Howell, M.D., Cuperus, J.T., Li, D. *et al.* (2008) Specificity of ARGONAUTE7-miR390 interaction and dual functionality in TAS3 trans-acting siRNA formation. *Cell*, **133**, 128–141.
114 Mi, S.J., Cai, T., Hu, Y.G., Chen, Y. *et al.* (2008) Sorting of small RNAs into *Arabidopsis* argonaute complexes is directed by the 5′ terminal nucleotide. *Cell*, **133**, 116–127.
115 Takeda, A., Iwasaki, S., Watanabe, T., Utsumi, M. *et al.* (2008) The mechanism selecting the guide strand from small RNA duplexes is different among argonaute proteins. *Plant Cell Physiol.*, **49**, 493–500.

116 Daxinger, L., Kanno, T., Bucher, E., van der Winden, J. *et al.* (2009) A stepwise pathway for biogenesis of 24-nt secondary siRNAs and spreading of DNA methylation. *EMBO J.*, **28**, 48–57.
117 Sijen, T., Steiner, F.A., Thijssen, K.L., Plasterk, R.H.A. (2007) Secondary siRNAs result from unprimed RNA synthesis and form a distinct class. *Science*, **315**, 244–247.
118 Pak, J., Fire, A. (2007) Distinct populations of primary and secondary effectors during RNAi in *C. elegans*. *Science*, **315**, 241–244.
119 Vaistij, F.E., Jones, L., Baulcombe, D.C. (2002) Spreading of RNA targeting and DNA methylation in RNA silencing requires transcription of the target gene and a putative RNA-dependent RNA polymerase. *Plant Cell*, **14**, 857–867.
120 Voinnet, O. (2008) Use, tolerance and avoidance of amplified RNA silencing by plants. *Trends Plant Sci.*, **13**, 317–328.
121 Axtell, M.J., Jan, C., Rajagopalan, R., Bartel, D.P. (2006) A two-hit trigger for siRNA biogenesis in plants. *Cell*, **127**, 565–577.
122 Howell, M.D., Fahlgren, N., Chapman, E.J., Cumbie, J.S. *et al.* (2007) Genome-wide analysis of the RNA-DEPENDENT RNA POLYMERASE6/DICER-LIKE4 pathway in *Arabidopsis* reveals dependency on miRNA- and tasiRNA-directed targeting. *Plant Cell*, **19**, 926–942.
123 Voinnet, O. (2009) Origin, biogenesis, and activity of plant microRNAs. *Cell*, **136**, 669–687.
124 Nodine, M.D., Bartel, D.P. (2010) MicroRNAs prevent precocious gene expression and enable pattern formation during plant embryogenesis. *Genes Dev.*, **24**, 2678–2692.
125 Wu, G., Poethig, R.S. (2006) Temporal regulation of shoot development in *Arabidopsis thaliana* by miR156 and its target SPL3. *Development*, **133**, 3539–3547.
126 Navarro, L., Dunoyer, P., Jay, F., Arnold, B. *et al.* (2006) A plant miRNA contributes to antibacterial resistance by repressing auxin signaling. *Science*, **312**, 436.
127 Lee, R.C., Feinbaum, R.L., Ambros, V. (1993) The *C. elegans* heterochronic gene lin-4 encodes small RNAs with antisense complementarity to lin-14. *Cell*, **75**, 843–854.
128 Wightman, B., Ha, I., Ruvkun, G. (1993) Posttranscriptional regulation of the heterochronic gene lin-14 by lin-4 mediates

temporal pattern-formation in *C. elegans. Cell*, **75**, 855–862.

129 Pasquinelli, A.E., Reinhart, B.J., Slack, F., Martindale, M.Q. et al. (2000) Conservation of the sequence and temporal expression of *let-7* heterochronic regulatory RNA. *Nature*, **408**, 86–89.

130 Reinhart, B.J., Slack, F.J., Basson, M., Pasquinelli, A.E. et al. (2000) The 21-nucleotide let-7 RNA regulates developmental timing in *Caenorhabditis elegans. Nature*, **403**, 901–906.

131 Lagos-Quintana, M., Rauhut, R., Lendeckel, W., Tuschl, T. (2001) Identification of novel genes coding for small expressed RNAs. *Science*, **294**, 853–858.

132 Lee, R., Ambros, V. (2001) An extensive class of small RNAs in *Caenorhabditis elegans. Science*, **294**, 862–864.

133 Lau, N.C., Lim, L.P., Weinstein, E.G., Bartel, D.P. (2001) An abundant class of tiny RNAs with probable regulatory roles in *Caenorhabditis elegans. Science*, **294**, 858–862.

134 Reinhart, B.J., Weinstein, E.G., Rhoades, M., Bartel, B. et al. (2002) MicroRNAs in plants. *Genes Dev.*, **16**, 1616–1626.

135 Llave, C., Kasschau, K.D., Rector, M.A., Carrington, J.C. (2002) Endogenous and silencing-associated small RNAs in plants. *Plant Cell*, **14**, 1605–1619.

136 Lu, C., Kulkarni, K., Souret, F.F., MuthuValliappan, R. et al. (2006) MicroRNAs and other small RNAs enriched in the *Arabidopsis* RNA-dependent RNA polymerase-2 mutant. *Genome Res*, **16**, 1276–1288.

137 Molnar, A., Schwach, F., Studholme, D.J., Thuenemann, E.C. et al. (2007) miRNAs control gene expression in the single-cell alga *Chlamydomonas reinhardtii. Nature*, **447**, 1126–U1115.

138 Axtell, M.J., Bartel, D.P. (2005) Antiquity of MicroRNAs and their targets in land plants. *Plant Cell*, **17**, 1658–1673.

139 Mallory, A.C., Elmayan, T., Vaucheret, H. (2008) MicroRNA maturation and action--the expanding roles of ARGONAUTEs. *Curr. Opin. Plant Biol.*, **11**, 560–566.

140 Kurihara, Y., Watanabe, Y. (2004) Arabidopsis micro-RNA biogenesis through Dicer-like 1 protein functions. *Proc. Natl Acad. Sci. USA*, **101**, 12753–12758.

141 Dong, Z., Han, M.H., Fedoroff, N. (2008) The RNA-binding proteins HYL1 and SE promote accurate in vitro processing of pri-miRNA by DCL1. *Proc. Natl Acad. Sci. USA*, **105**, 9970–9975.

142 Hirsch, J., Lefort, V., Vankersschaver, M., Boualem, A. et al. (2006) Characterization of 43 non-protein-coding mRNA genes in *Arabidopsis*, including the MIR162a-derived transcripts. *Plant Physiol.*, **140**, 1192–1204.

143 Lobbes, D., Rallapalli, G., Schmidt, D.D., Martin, C. et al. (2006) SERRATE: a new player on the plant microRNA scene. *EMBO Rep.*, **7**, 1052–1058.

144 Yu, B., Bi, L., Zheng, B., Ji, L. et al. (2008) The FHA domain proteins DAWDLE in *Arabidopsis* and SNIP1 in humans act in small RNA biogenesis. *Proc. Natl Acad. Sci. USA*, **105**, 10073–10078.

145 Yang, Z., Ebright, Y.W., Yu, B., Chen, X. (2006) HEN1 recognizes 21-24 nt small RNA duplexes and deposits a methyl group onto the 2′ OH of the 3′terminal nucleotide. *Nucleic Acids Res.*, **34**, 667–675.

146 Ren, G., Xie, M., Dou, Y., Zhang, S. et al. (2012) Regulation of miRNA abundance by RNA binding protein TOUGH in *Arabidopsis. Proc. Natl Acad. Sci. USA*, **109**, 12817–12821.

147 Yi, R., Doehle, B.P., Qin, Y., Macara, I.G. et al. (2005) Overexpression of Exportin 5 enhances RNA interference mediated by short hairpin RNAs and microRNAS. *RNA*, **11**, 220–226.

148 Yi, R., Qin, Y., Macara, I.G., Cullen, B.R. (2003) Exportin-5 mediates the nuclear export of pre-microRNAs and short hairpin RNAs. *Genes Dev.*, **17**, 3011–3016.

149 Lund, E., Guttinger, S., Calado, A., Dahlberg, J.E. et al. (2004) Nuclear Export of MicroRNA Precursors. *Science*, **303**, 95–98.

150 Dunoyer, P., Brosnan, C.A., Schott, G., Wang, Y. et al. (2010) An endogenous, systemic RNAi pathway in plants. *EMBO J.*, **29**, 1699–1712.

151 Ye, R., Wang, W., Iki, T., Liu, C. et al. (2012) Cytoplasmic assembly and selective nuclear import of *Arabidopsis* Argonaute4/siRNA complexes. *Mol. Cell*, **46**, 859–870.

152 Dugas, D.V., Bartel, B. (2008) Sucrose induction of Arabidopsis miR398 represses

two Cu/Zn superoxide dismutases. *Plant Mol. Biol.*, **67**, 403–417.

153 Molnar, A., Bassett, A., Thuenemann, E., Schwach, F. et al. (2009) Highly specific gene silencing by artificial microRNAs in the unicellular alga *Chlamydomonas reinhardtii*. *Plant J.*, **58**, 165–174.

154 Schwab, R., Ossowski, S., Riester, M., Warthmann, N. et al. (2006) Highly specific gene silencing by artificial microRNAs in *Arabidopsis*. *Plant Cell*, **18**, 1121–1133.

155 Brennecke, J., Stark, A., Russell, R.B., Cohen, S.M. (2005) Principles of microRNA-target recognition. *PLoS Biol.*, **3**, e85.

156 Lewis, B.P., Shih, I.H., Jones-Rhoades, M.W., Bartel, D.P. et al. (2003) Prediction of mammalian microRNA targets. *Cell*, **115**, 787–798.

157 Axtell, M.J. (2008) Evolution of microRNAs and their targets: are all microRNAs biologically relevant? *Biochim. Biophys. Acta*, **1779**, 725–734.

158 Axtell, M.J., Bowman, J.L. (2008) Evolution of plant microRNAs and their targets. *Trends Plant Sci.*, **13**, 343–349.

159 Allen, E., Xie, Z., Gustafson, A.M., Sung, G.-H. et al. (2004) Evolution of microRNA genes by inverted duplication of target gene sequences in *Arabidopsis thaliana*. *Nat. Genet.*, **36**, 1282–1290.

160 Fahlgren, N., Howell, M.D., Kasschau, K.D., Chapman, E.J. et al. (2007) High-throughput sequencing of *Arabidopsis* microRNAs: evidence for frequent birth and death of *MIRNA* genes. *PLoS ONE*, **2**, e219.

161 Felippes, F.F., Schneeberger, K., Dezulian, T., Huson, D.H. et al. (2008) Evolution of *Arabidopsis thaliana* microRNAs from random sequences. *RNA*, **14**, 2455–2459.

162 Golden, T.A., Schauer, S.E., Lang, J.D., Pien, S. et al. (2002) Short Integuments1/Suspensor1/Carpel factory, a Dicer homolog, is a maternal effect gene required for embryo development in *Arabidopsis*. *Plant Physiol.*, **130**, 808–822.

163 Garcia, D. (2008) A miRacle in plant development: role of microRNAs in cell differentiation and patterning. *Semin. Cell Dev. Biol.*, **19**, 586–595.

164 Kidner, C.A., Martienssen, R. (2004) Spatially restricted microRNA directs leaf polarity through ARGONAUTE1. *Nature*, **428**, 81–84.

165 Nikovics, K., Blein, T., Peaucelle, A., Ishida, T. et al. (2006) The balance between the MIR164A and CUC2 genes controls leaf margin serration in *Arabidopsis*. *Plant Cell*, **18**, 2929–2945.

166 Hamilton, A.J., Voinnet, O., Chappell, L., Baulcombe, D.C. (2002) Two classes of short interfering RNA in RNA silencing. *EMBO J.*, **21**, 4671–4679.

167 Vaucheret, H. (2006) Post-transcriptional small RNA pathways in plants: mechanisms and regulations. *Genes Dev.*, **20**, 759–771.

168 Yoshikawa, M., Peragine, A., Park, M.Y., Poethig, R.S. (2005) A pathway for the biogenesis of trans-acting siRNAs in *Arabidopsis*. *Genes Dev.*, **19**, 2164–2175.

169 Allen, E., Xie, Z., Gustafson, A.M., Carrington, J.C. (2005) microRNA-directed phasing during trans-acting siRNA biogenesis in plants. *Cell*, **121**, 207–221.

170 Vazquez, F., Vaucheret, H., Rajagopalan, R., Lepers, C. et al. (2004) Endogenous trans-acting siRNAs regulate the accumulation of *Arabidopsis* mRNAs. *Mol. Cell*, **16**, 69–79.

171 Montgomery, T.A., Yoo, S.J., Fahlgren, N., Gilbert, S.D. et al. (2008) AGO1-miR173 complex initiates phased siRNA formation in plants. *Proc. Natl Acad. Sci. USA*, **105**, 20055–20062.

172 Chen, H.M., Li, Y.H., Wu, S.H. (2007) Bioinformatic prediction and experimental validation of a microRNA-directed tandem trans-acting siRNA cascade in *Arabidopsis*. *Proc. Natl Acad. Sci. USA*, **104**, 3318–3323.

173 Shivaprasad, P.V., Chen, H.M., Patel, K., Bond, D.M. et al. (2012) A microRNA superfamily regulates nucleotide binding site-leucine-rich repeats and other mRNAs. *Plant Cell*, **24**, 859–874.

174 Zhai, J., Jeong, D.H., De Paoli, E., Park, S. et al. (2011) MicroRNAs as master regulators of the plant NB-LRR defense gene family via the production of phased, trans-acting siRNAs. *Genes Dev.*, **25**, 2540–2553.

175 Cuperus, J.T., Carbonell, A., Fahlgren, N., Garcia-Ruiz, H. et al. (2010) Unique functionality of 22-nt miRNAs in triggering RDR6-dependent siRNA biogenesis from

target transcripts in *Arabidopsis*. *Nat. Struct. Mol. Biol.*, **17**, 997–1003.

176 Chen, H.-M., Chen, L.-T., Patel, K., Li, Y.-H. et al. (2010) 22-nucleotide RNAs trigger secondary siRNA biogenesis in plants. *Proc. Natl Acad. Sci. USA*, **107**, 15269–15274.

177 Peragine, A., Yoshikawa, M., Wu, G., Albrecht, H.L. et al. (2004) *SGS3* and *SGS2/SDE1/RDR6* are required for juvenile development and the production of *trans*-acting siRNAs in *Arabidopsis*. *Genes Dev.*, **18**, 2368–2379.

178 Adenot, X., Elmayan, T., Laursergues, D., Boutet, S. et al. (2006) DRB4-dependent TAS3 trans-acting siRNAs control leaf morphology through AGO7. *Curr. Biol.*, **16**, 927–932.

179 Fahlgren, N., Montgomery, T.A., Howell, M.D., Allen, E. et al. (2006) Regulation of *AUXIN RESPONSE FACTOR3* by *TAS3* ta-siRNA affects developmental timing and patterning in *Arabidopsis*. *Curr. Biol.*, **16**, 939–944.

180 Chitwood, D.H., Nogueira, F.T.S., Howell, M.D., Montgomery, T.A. et al. (2009) Pattern formation via small RNA mobility. *Genes Dev.*, **23**, 549–554.

181 Schwab, R., Maizel, A., Ruiz-Ferrer, V., Garcia, D. et al. (2009) Endogenous TasiRNAs mediate non-cell autonomous effects on gene regulation in *Arabidopsis thaliana*. *PLoS ONE*, **4**, e5980.

182 Katiyar-Agarwal, S., Morgan, R.A., Dahlbeck, D., Borsani, O. et al. (2006) A pathogen-inducible endogenous siRNA in plant immunity. *Proc. Natl Acad. Sci. USA*, **103**, 18002–18007.

183 Borsani, O., Zhu, J., Verslues, P.E., Sunkar, R. et al. (2005) Endogenous siRNAs derived from a pair of natural *cis*-antisense transcripts regulate salt tolerance in *Arabidopsis*. *Cell*, **123**, 1279–1291.

184 Ron, M., Alandete Saez, M., Eshed Williams, L., Fletcher, J.C. et al. (2010) Proper regulation of a sperm-specific cis-nat-siRNA is essential for double fertilization in *Arabidopsis*. *Genes Dev.*, **24**, 1010–1021.

185 Zhang, X., Henderson, I.R., Lu, C., Green, P.J. et al. (2007) Role of RNA polymerase IV in plant small RNA metabolism. *Proc. Natl Acad. Sci. USA*, **104**, 4536–4541.

186 Mosher, R.A., Schwach, F., Studhollme, D., Baulcombe, D.C. (2008) PolIVb influences RNA-directed DNA-methylation independently of its role in siRNA biogenesis. *Proc. Natl Acad. Sci. USA*, **105**, 3145–3150.

187 Grewal, S.I., Moazed, D. (2003) Heterochromatin and epigenetic control of gene expression. *Science*, **301**, 798–802.

188 Zhang, X., Yazaki, J., Sundaresan, A., Cokus, S. et al. (2006) Genome-wide high resolution mapping and functional analysis of DNA methylation in *Arabidopsis*. *Cell*, **126**, 1–13.

189 Xie, Z., Johansen, L.K., Gustafson, A.M., Kasschau, K.D. et al. (2004) Genetic and functional diversification of small RNA pathways in plants. *PLoS Biol.*, **2**, 642–652.

190 Cao, X.F., Jacobsen, S.E. (2002) Role of the *Arabidopsis* DRM methyltransferases in de novo DNA methylation and gene silencing. *Curr. Biol.*, **12**, 1138–1144.

191 Chan, S.W.-L., Zilberman, D., Xie, Z., Johansen, L.K. et al. (2004) RNA silencing genes control *de novo* DNA methylation. *Science*, **303**, 1336.

192 Herr, A.J., Jensen, M.B., Dalmay, T., Baulcombe, D. (2005) RNA polymerase IV directs silencing of endogenous DNA. *Science*, **308**, 118–120.

193 Onodera, Y., Haag, J.R., Ream, T., Nunes, P.C. et al. (2005) Plant Nuclear RNA polymerase IV mediates siRNA and DNA methylation-dependent heterochromatin formation. *Cell*, **120**, 613–622.

194 Kanno, T., Huettel, B., Mette, M.F., Aufsatz, W. et al. (2005) Atypical RNA polymerase subunits required for RNA-directed DNA methylation. *Nat. Genet.*, **37**, 761–765.

195 Zheng, B., Wang, Z., Li, S., Yu, B. et al. (2009) Intergenic transcription by RNA polymerase II coordinates Pol IV and Pol V in siRNA-directed transcriptional gene silencing in *Arabidopsis*. *Genes Dev.*, **23**, 2850–2860.

196 Law, J.A., Vashisht, A.A., Wohlschlegel, J.A., Jacobsen, S.E. (2011) SHH1, a homeodomain protein required for DNA methylation, as well as RDR2, RDM4, and chromatin remodeling factors, associate with RNA polymerase IV. *PLoS Genet.*, **7**, e1002195.

197 Kanno, T., Bucher, E., Daxinger, L., Huettel, B. et al. (2010) RNA-directed DNA methylation and plant development require

an IWR1-type transcription factor. *EMBO Rep.*, **11**, 65–71.
198 Pontes, O., Li, C.F., Nunes, P.C., Haag, J.R. et al. (2006) The *Arabidopsis* chromatin-modifying nuclear siRNA pathway involves a nucleolar RNA processing center. *Cell*, **126**, 79–92.
199 Dolgosheina, E.V., Morin, R.D., Aksay, G., Sahinalp, S.C. et al. (2008) Conifers have a unique small RNA silencing signature. *RNA*, **14**, 1508–1515.
200 Wierzbicki, A.T., Haag, J.R., Pikaard, C.S. (2008) Noncoding transcription by RNA polymerase Pol IVb/Pol V mediates transcriptional silencing of overlapping and adjacent genes. *Cell*, **135**, 635–648.
201 Law, J.A., Ausin, I., Johnson, L.M., Vashisht, A.A. et al. (2010) A protein complex required for polymerase V transcripts and RNA-directed DNA methylation in *Arabidopsis*. *Curr. Biol.*, **20**, 951–956.
202 Gao, Z., Liu, H.L., Daxinger, L., Pontes, O. et al. (2010) An RNA polymerase II- and AGO4-associated protein acts in RNA-directed DNA methylation. *Nature*, **465**, 106–109.
203 Wierzbicki, A.T., Ream, T.S., Haag, J.R., Pikaard, C.S. (2009) RNA polymerase V transcription guides ARGONAUTE4 to chromatin. *Nat. Genet.*, **41**, 630–634.
204 Li, C.F., Pontes, O., El-Shami, M., Henderson, I.R. et al. (2006) An ARGONAUTE4-containing nuclear processing center colocalized with cajal bodies in *Arabidopsis thaliana*. *Cell*, **126**, 93–106.
205 Zilberman, D., Cao, X., Johansen, L.K., Xie, Z. et al. (2004) Role of *Arabidopsis ARGONAUTE4* in RNA-directed DNA methylation triggered by inverted repeats. *Curr. Biol.*, **14**, 1214–1220.
206 Teixeira, F.K., Heredia, F., Sarazin, A., Roudier, F. et al. (2009) A role for RNAi in the selective correction of DNA methylation defects. *Science*, **323**, 1600–1604.
207 Finnegan, E.J., Kovac, K.A. (2000) Plant DNA methyltransferases. *Plant Mol. Biol.*, **43**, 189–201.
208 Aufsatz, W., Mette, M.F., Matzke, A.J.M., Matzke, M. (2004) The role of MET1 in RNA-directed de novo and maintenance methylation of CG dinucleotides. *Plant Mol. Biol.*, **54**, 793–804.
209 Lindroth, A.M., Cao, X.F., Jackson, J.P., Zilberman, D. et al. (2001) Requirement of CHROMOMETHYLASE3 for maintenance of CpXpG methylation. *Science*, **292**, 2077–2080.
210 Johnson, L.M., Bostick, M., Zhang, X., Kraft, E. et al. (2007) The SRA methyl-cytosine-binding domain links DNA and histone methylation. *Curr. Biol.*, **17**, 379–384.
211 Ito, H., Gaubert, H., Bucher, E., Mirouze, M. et al. (2011) An siRNA pathway prevents transgenerational retrotransposition in plants subjected to stress. *Nature*, **472**, 115–119.
212 Pikaard, C.S., Haag, J.R., Ream, T., Wierzbicki, A.T. (2008) Roles of RNA polymerase IV in gene silencing. *Trends Plant Sci.*, **13**, 390–397.
213 Mosher, R.A., Melnyk, C.W., Kelly, K.A., Dunn, R.M. et al. (2009) Uniparental expression of PolIV-dependent siRNAs in developing endosperm of *Arabidopsis*. *Nature*, **460**, 283–U151.
214 Slotkin, R.K., Vaughn, M., Borges, F., Tanurdzic, M. et al. (2009) Epigenetic reprogramming and small RNA silencing of transposable elements in pollen. *Cell*, **136**, 461–472.
215 Gehring, M., Bubb, K.L., Henikoff, S. (2009) Extensive demethylation of repetitive elements during seed development underlies gene imprinting. *Science*, **324**, 1447–1451.
216 Hsieh, T.F., Ibarra, C.A., Silva, P., Zemach, A. et al. (2009) Genome-wide demethylation of *Arabidopsis* endosperm. *Science*, **324**, 1451–1454.
217 Lu, J., Zhang, C., Baulcombe, D.C., Chen, Z.J. (2012) Maternal siRNAs as regulators of parental genome imbalance and gene expression in endosperm of *Arabidopsis* seeds. *Proc. Natl Acad. Sci. USA*, **109**, 5529–5534.
218 Alleman, M., Sidorenko, L., McGinnis, K.M., Seshardri, V. et al. (2006) An RNA-dependent RNA polymerase is required for paramutation in maize. *Nature*, **442**, 295–295.
219 Erhard, K.F., Stonaker, J.L., Parkinson, S.E., Lim, J.P. Jr et al. (2009) RNA polymerase IV functions in paramutation in *Zea mays*. *Science*, **323**, 1201–1205.
220 Stonaker, J.L., Lim, J.P., Erhard, K.F., Hollick, J.B. Jr (2009) Diversity of Pol IV

function is defined by mutations at the maize rmr7 locus. *PLoS Genet.*, **5**, e1000706.
221. Sidorenko, L., Dorweiler, J.E., Cigan, A.M., Arteaga-Vazquez, M. et al. (2009) A dominant mutation in mediator of paramutation2, one of three second-largest subunits of a plant-specific RNA polymerase, disrupts multiple siRNA silencing processes. *PLoS Genet.*, **5**, e1000725.
222. Arteaga-Vazquez, M.A., Chandler, V.L. (2010) Paramutation in maize: RNA mediated trans-generational gene silencing. *Curr. Opin. Genet. Dev.*, **20**, 156–163.
223. Arteaga-Vazquez, M., Sidorenko, L., Rabanal, F.A., Shrivistava, R. et al. (2010) RNA-mediated trans-communication can establish paramutation at the b1 locus in maize. *Proc. Natl Acad. Sci. USA*, **107**, 12986–12991.
224. Ghabrial, S.A., Suzuki, N. (2009) Viruses of plant pathogenic fungi. *Annu. Rev. Phytopathol.*, **47**, 353–384.
225. Lewsey, M.G., Carr, J.P. (2009) Plant Pathogens: RNA Viruses, in: Schaechter, M. (Ed.) *Encyclopedia of Microbiology*, Elsevier, Oxford, pp. 443–458.
226. Szittya, G., Moxon, S., Pantaleo, V., Toth, G. et al. (2010) Structural and functional analysis of viral siRNAs. *PLoS Pathog.*, **6**, e1000838.
227. Molnar, A., Csorba, T., Lakatos, L., Varallyay, E. et al. (2005) Plant virus-derived small interfering RNAs originate predominantly from highly structured single-stranded viral RNAs. *J. Virol.*, **79**, 7812–7818.
228. Szittya, G., Molnar, A., Silhavy, D., Hornyik, C., et al. (2002) Short defective interfering RNAs of tombusviruses are not targeted but trigger post-transcriptional gene silencing against their helper virus. *Plant Cell*, **14**, 359–372.
229. Du, Q.S., Duan, C.G., Zhang, Z.H., Fang, Y.Y. et al. (2007) DCL4 targets Cucumber mosaic virus satellite RNA at novel secondary structures. *J. Virol.*, **81**, 9142–9151.
230. Burgyan, J., Havelda, Z. (2011) Viral suppressors of RNA silencing. *Trends Plant Sci.*, **16**, 265–272.
231. Vanitharani, R., Chellappan, P., Fauquet, C.M. (2005) Geminiviruses and RNA silencing. *Trends Plant Sci.*, **10**, 144–151.
232. Wang, M.B., Masuta, C., Smith, N.A., Shimura, H. (2012) RNA silencing and plant viral diseases. *Mol. Plant Microbe. Interact.*, **25**, 1275–1285.
233. Ding, S.W. (2010) RNA-based antiviral immunity. *Nat. Rev. Immunol.*, **10**, 632–644.
234. Wang, X.B., Jovel, J., Udomporn, P., Wang, Y. et al. (2011) The 21-nucleotide, but not 22-nucleotide, viral secondary small interfering RNAs direct potent antiviral defense by two cooperative argonautes in *Arabidopsis thaliana*. *Plant Cell*, **23**, 1625–1638.
235. Bouche, N., Lauressergues, D., Gasciolli, V., Vaucheret, H. (2006) An antagonistic function for *Arabidopsis* DCL2 in development and a new function for DCL4 in generating viral siRNAs. *EMBO J.*, **25**, 3347–3356.
236. Diaz-Pendon, J.A., Li, F., Li, W.X., Ding, S.W. (2007) Suppression of antiviral silencing by cucumber mosaic virus 2b protein in *Arabidopsis* is associated with drastically reduced accumulation of three classes of viral small interfering RNAs. *Plant Cell*, **19**, 2053–2063.
237. Qu, F., Ye, X.H., Morris, T.J. (2008) *Arabidopsis* DRB4, AG01, AG07, and RDR6 participate in a DCL4-initiated antiviral RNA silencing pathway negatively regulated by DCL1. *Proc. Natl Acad. Sci. USA*, **105**, 14732–14737.
238. Jaubert, M., Bhattacharjee, S., Mello, A.F.S., Perry, K.L. et al. (2011) ARGONAUTE2 mediates RNA-silencing antiviral defenses against potato virus X in *Arabidopsis*. *Plant Physiol.*, **156**, 1556–1564.
239. Moissiard, G., Voinnet, O. (2006) RNA silencing of host transcripts by cauliflower mosaic virus requires coordinated action of the four *Arabidopsis* Dicer-like proteins. *Proc. Natl Acad. Sci. USA*, **103**, 19593–19598.
240. Yadav, R.K., Chattopadhyay, D. (2011) Enhanced viral intergenic region-specific short interfering RNA accumulation and DNA methylation correlates with resistance against a geminivirus. *Mol. Plant Microbe Interact.*, **24**, 1189–1197.
241. Morel, J.B., Godon, C., Mourrain, P., Beclin, C. et al. (2002) Fertile hypomorphic ARGONAUTE (ago1) mutants impaired in post-transcriptional gene silencing and virus resistance. *Plant Cell*, **14**, 629–639.
242. Harvey, J.J.W., Lewsey, M.G., Patel, K., Westwood, J. et al. (2011) An antiviral defense role of AGO2 in plants. *PLoS ONE*, **6**, e14639.

243 Zhang, X., Yuan, Y.-R., Pei, Y., Lin, S.S. et al. (2006) Cucumber mosaic virus-encoded 2b suppressor inhibits *Arabidopsis* Argonaute1 cleavage activity to counter plant defense. *Genes Dev.*, **20**, 3255–3268.

244 Pantaleo, V., Szittya, G., Burgyan, J. (2007) Molecular bases of viral RNA targeting by viral siRNA programmed RISC. *J. Virol.*, **81**, 3797–3806.

245 Raja, P., Sanville, B.C., Buchmann, R.C., Bisaro, D.M. (2008) Viral genome methylation as an epigenetic defense against geminiviruses. *J. Virol.*, **82**, 8997–9007.

246 Garcia-Ruiz, H., Takeda, A., Chapman, E.J., Sullivan, C.M. et al. (2010) *Arabidopsis* RNA-dependent RNA polymerases and dicer-like proteins in antiviral defense and small interfering RNA biogenesis during Turnip Mosaic Virus infection. *Plant Cell*, **22**, 481–496.

247 Qu, F. (2010) Antiviral role of plant-encoded RNA-dependent RNA polymerases revisited with deep sequencing of small interfering RNAs of virus origin. *Mol. Plant Microbe Interact.*, **23**, 1248–1252.

248 Schwach, F., Vaistij, F.E., Jones, L., Baulcombe, D.C. (2005) An RNA-dependent RNA-polymerase prevents meristem invasion by Potato virus X and is required for the activity but not the production of a systemic silencing signal. *Plant Physiol.*, **138**, 1842–1852.

249 Voinnet, O. (2005) Non-cell autonomous RNA silencing. *FEBS Lett.*, **579**, 5858–5871.

250 Ding, S.W., Voinnet, O. (2007) Antiviral immunity directed by small RNAs. *Cell*, **130**, 413–426.

251 Voinnet, O. (2005) Induction and suppression of RNA silencing: insights from viral infections. *Nat. Rev. Genet.*, **6**, 206–220.

252 Ye, K., Malinina, L., Patel, D.J. (2003) Recognition of small interfering RNA by a viral suppressor of RNA Silencing. *Nature*, **426**, 874–878.

253 Ebhardt, H.A., Thi, E.P., Wang, M.-B., Unrau, P.J. (2005) Extensive 3′ modification of plant small RNAs is modulated by helper component-proteinase expression. *Proc. Natl Acad. Sci. USA*, **102**, 13398–13403.

254 Cuellar, W.J., Kreuze, J.F., Rajamaki, M.L., Cruzado, K.R. et al. (2009) Elimination of antiviral defense by viral RNase III. *Proc. Natl Acad. Sci. USA*, **106**, 10354–10358.

255 Schott, G., Mari-Ordonez, A., Himber, C., Alioua, A. et al. (2012) Differential effects of viral silencing suppressors on siRNA and miRNA loading support the existence of two distinct cellular pools of ARGONAUTE1. *EMBO J.*, **31**, 2553–2565.

256 Bortolamiol, D., Pazhouhandeh, M., Marrocco, K., Genschik, P. et al. (2007) The polerovirus F box protein P0 targets ARGONAUTE1 to suppress RNA silencing. *Curr. Biol.*, **17**, 1615–1621.

257 Baumberger, N., Tsai, C.-H., Lie, M., Havecker, E. et al. (2007) The Polerovirus silencing suppressor P0 targets ARGONAUTE proteins for degradation. *Curr. Biol.*, **17**, 1609–1614.

258 Varallyay, E., Valoczi, A., Agyi, A., Burgyan, J. et al. (2010) Plant virus-mediated induction of miR168 is associated with repression of ARGONAUTE1 accumulation. *EMBO J.*, **29**, 3507–3519.

259 Bisaro, D.M. (2006) Silencing suppression by geminivirus proteins. *Virology*, **344**, 158–168.

260 Wang, H., Buckley, K.J., Yang, X.J., Buchmann, R.C. et al. (2005) Adenosine kinase inhibition and suppression of RNA silencing by geminivirus AL2 and L2 proteins. *J. Virol.*, **79**, 7410–7418.

261 Wang, H., Hao, L.H., Shung, C.Y., Sunter, G. et al. (2003) Adenosine kinase is inactivated by geminivirus AL2 and L2 proteins. *Plant Cell*, **15**, 3020–3032.

262 Zhang, Z.H., Chen, H., Huang, X.H., Xia, R. et al. (2011) BSCTV C2 attenuates the degradation of SAMDC1 to suppress DNA methylation-mediated gene silencing in *Arabidopsis*. *Plant Cell*, **23**, 273–288.

263 Navarro, L., Jay, F., Nomura, K., He, S.Y. et al. (2008) Suppression of the microRNA pathway by bacterial effector proteins. *Science*, **321**, 964–967.

264 Jones, J.D., Dangl, J.L. (2006) The plant immune system. *Nature*, **444**, 323–329.

265 Xie, Z., Fan, B., Chen, C.H., Chen, Z. (2001) An important role of an inducible RNA-dependent RNA polymerase in plant antiviral defense. *Proc. Natl Acad. Sci. USA*, **98**, 6516–6521.

266 Yu, D., Fan, B., MacFarlane, S.A., Chen, Z. (2003) Analysis of the involvement of an inducible *Arabidopsis* RNA-dependent RNA polymerase in antiviral defense. *Mol. Plant Microbe Interact.*, **16**, 206–216.

267 Bhattacharjee, S., Zamora, A., Azhar, M.T., Sacco, M.A. et al. (2009) Virus resistance induced by NB-LRR proteins involves Argonaute4-dependent translational control. *Plant J.*, **58**, 940–951.

268 Li, F., Pignatta, D., Bendix, C., Brunkard, J.O. et al. (2012) MicroRNA regulation of plant innate immune receptors. *Proc. Natl Acad. Sci. USA*, **109**, 1790–1795.

269 Wingard, S.A. (1928) Hosts and symptoms of ring spot, a virus disease of plants. *J. Agric. Res.*, **37**, 127–153.

270 Kunz, C., Hanspeter, S., Stam, M., Kooter, J.M. et al. (1996) Developmentally regulated silencing and reactivation of tobacco chitinase transgene expression. *Plant J.*, **10**, 437–450.

271 Hart, C.M., Fischer, B., Neuhaus, J.M., Meins, F. Jr (1992) Regulated inactivation of homologous gene expression in transgenic *Nicotiana sylvestris* plants containing a defense-related tobacco chitinase gene. *Mol. Gen. Genet.*, **235**, 179–188.

272 Boerjan, W., Bauw, G., Van Montagu, M., Inzé, D. (1994) Distinct phenotypes generated by overexpression and suppression of S-adenosyl-L-methionine synthetase reveal developmental patterns of gene silencing in tobacco. *Plant Cell*, **6**, 1401–1414.

273 Palauqui, J.C., De Borne, F.D., Elmayan, T., Crete, P. et al. (1996) Frequencies, timing, and spatial patterns of co-suppression of nitrate reductase and nitrite reductase in transgenic tobacco plants. *Plant Physiol.*, **112**, 1447–1456.

274 Buhtz, A., Springer, F., Chappell, L., Baulcombe, D.C. et al. (2008) Identification and characterization of small RNAs from the phloem of *Brassica napus*. *Plant J.*, **53**, 739–749.

275 Imlau, A., Truernit, E., Sauer, N. (1999) Cell-to-cell and long distance trafficking of the green fluorescent protein in the phloem and symplastic unloading of the protein into sink tissues. *Plant Cell*, **11**, 309–322.

276 Carrington, J.C., Kasschau, K.D., Mahajan, S.K., Schaad, M.C. (1996) Cell-to-cell and long-distance transport of viruses in plants. *Plant Cell*, **8**, 1669–1681.

277 Lucas, W.J., Bouché-Pillon, S., Jackson, D.P., Nguyen, L. et al. (1995) Selective trafficking of KNOTTED1 homeodomain proteins and its mRNA through plasmodesmata. *Science*, **270**, 1980–1983.

278 Melnyk, C.W., Molnar, A., Baulcombe, D.C. (2011) Intercellular and systemic movement of RNA silencing signals. *EMBO J.*, **30**, 3553–3563.

279 Himber, C., Dunoyer, P., Moissiard, G., Ritzenthaler, C. et al. (2003) Transitivity-dependent and -independent cell-to-cell movement of RNA silencing. *EMBO J.*, **22**, 4523–4533.

280 Smith, L.M., Pontes, O., Searle, L., Yelina, N. et al. (2007) An SNF2 protein associated with nuclear RNA silencing and the spread of a silencing signal between cells in *Arabidopsis*. *Plant Cell*, **19**, 1507–1521.

281 Dunoyer, P., Himber, C., Ruiz-Ferrer, V., Alioua, A. et al. (2007) Intra- and intercellular RNA interference in *Arabidopsis thaliana* requires components of the microRNA and heterochromatic silencing pathways. *Nat. Genet.*, **39**, 848–856.

282 Dunoyer, P., Himber, C., Voinnet, O. (2005) DICER-LIKE 4 is required for RNA interference and produces the 21-nucleotide small interfering RNA component of the plant cell-to-cell silencing signal. *Nat. Genet.*, **37**, 1356–1360.

283 Dunoyer, P., Schott, G., Himber, C., Meyer, D. et al. (2010) Small RNA duplexes function as mobile silencing signals between plant cells. *Science*, **328**, 912–916.

284 Brosnan, C.A., Mitter, N., Christie, M., Smith, N.A. et al. (2007) Nuclear gene silencing directs reception of long-distance mRNA silencing in *Arabidopsis*. *Proc. Natl Acad. Sci. USA*, **104**, 14741–14746.

285 Ueki, S., Citovsky, V. (2001) Inhibition of systemic onset of post-transcriptional gene silencing by non-toxic concentrations of cadmium. *Plant J.*, **28**, 283–291.

286 Ueki, S., Citovsky, V. (2002) The systemic movement of a tobamovirus is inhibited by a cadmium-ion-induced glycine-rich protein. *Nat. Cell Biol.*, **4**, 478–486.

287 Liang, D., White, R.G., Waterhouse, P.M. (2012) Gene silencing in *Arabidopsis* spreads from the root to the shoot, through a gating barrier, by template-dependent, nonvascular, cell-to-cell movement. *Plant Physiol.*, **159**, 984–1000.

288 Voinnet, O., Vain, P., Angell, S., Baulcombe, D.C. (1998) Systemic spread of sequence-specific transgene RNA degradation is initiated by localised

introduction of ectopic promoterless DNA. *Cell*, **95**, 177–187.

289 Yoo, B.-C., Kragler, F., Varkonyi-Gasic, E., Haywood, V. *et al.* (2004) A systemic small RNA signaling system in plants. *Plant Cell*, **16**, 1979–2000.

290 Pant, B.D., Buhtz, A., Kehr, J., Scheible, W.R. (2008) MicroRNA399 is a long-distance signal for the regulation of plant phosphate homeostasis. *Plant J.*, **53**, 731–738.

291 Molnar, A., Melnyk, C.W., Bassett, A., Hardcastle, T.J. *et al.* (2010) Small silencing RNAs in plants are mobile and direct epigenetic modification in recipient cells. *Science*, **328**, 872–875.

292 Melnyk, C.W., Molnar, A., Bassett, A., Baulcombe, D.C. (2011) Mobile 24 nt small RNAs direct transcriptional gene silencing in the root meristems of *Arabidopsis thaliana*. *Curr. Biol.*, **21**, 1678–1683.

293 Carlsbecker, A., Lee, J.Y., Roberts, C.J., Dettmer, J. *et al.* (2010) Cell signalling by microRNA165/6 directs gene dose-dependent root cell fate. *Nature*, **465**, 316–321.

294 Voinnet, O., Lederer, C., Baulcombe, D.C. (2000) A viral movement protein prevents spread of the gene silencing signal in *Nicotiana benthamiana*. *Cell*, **103**, 157–167.

295 Huang, G., Allen, R., Davis, E.L., Baum, T.J. *et al.* (2006) Engineering broad root-knot resistance in transgenic plants by RNAi silencing of a conserved and essential root-knot nematode parasitism gene. *Proc. Natl Acad. Sci. USA*, **103**, 14302–14306.

296 Baum, J.A., Bogaert, T., Clinton, W., Heck, G.R. *et al.* (2007) Control of coleopteran insect pests through RNA interference. *Nat. Biotechnol.*, **25**, 1322–1326.

297 Mao, Y.B., Cai, W.J., Wang, J.W., Hong, G.J. *et al.* (2007) Silencing a cotton bollworm P450 monooxygenase gene by plant-mediated RNAi impairs larval tolerance of gossypol. *Nat. Biotechnol.*, **25**, 1307–1313.

298 Tinoco, M.L., Dias, B.B., Dall'Astta, R.C., Pamphile, J.A. *et al.* (2010) In vivo trans-specific gene silencing in fungal cells by in planta expression of a double-stranded RNA. *BMC Biol.*, **8**, 27.

299 Nowara, D., Gay, A., Lacomme, C., Shaw, J. *et al.* (2010) HIGS: host-induced gene silencing in the obligate biotrophic fungal pathogen *Blumeria graminis*. *Plant Cell*, **22**, 3130–3141.

300 Tomilov, A.A., Tomilova, N.B., Wroblewski, T., Michelmore, R. *et al.* (2008) Trans-specific gene silencing between host and parasitic plants. *Plant J.*, **56**, 389–397.

301 Pontier, D., Picart, C., Roudier, F., Garcia, D. *et al.* (2012) NERD, a plant-specific GW protein, defines an additional RNAi-dependent chromatin-based pathway in *Arabidopsis*. *Mol. Cell*, **48**, 121–132.

302 Wu, L., Zhou, H., Zhang, Q., Zhang, *et al.* (2010) DNA methylation mediated by a microRNA pathway. *Mol. Cell*, **38**, 465–475.

303 Xie, Z., Kasschau, K.D., Carrington, J.C. (2003) Negative feedback regulation of *Dicer-Like1* in *Arabidopsis* by microRNA-guided mRNA degradation. *Curr. Biol.*, **13**, 784–789.

304 Wei, W., Ba, Z., Gao, M., Wu, Y. *et al.* (2012) A role for small RNAs in DNA double-strand break repair. *Cell*, **149**, 101–112.

305 Katiyar-Agarwal, S., Gao, S., Vivian-Smith, A., Jin, H. (2007) A novel class of bacteria-induced small RNAs in *Arabidopsis*. *Genes Dev.*, **21**, 3123–3134.

306 Baubec, T., Dinh, H.Q., Pecinka, A., Rakic, B. *et al.* (2010) Cooperation of multiple chromatin modifications can generate unanticipated stability of epigenetic States in *Arabidopsis*. *Plant Cell*, **22**, 34–47.

307 Henderson, I.R., Jacobsen, S.E. (2008) Tandem repeats upstream of the *Arabidopsis* endogene SDC recruit non-CG DNA methylation and initiate siRNA spreading. *Genes Dev.*, **22**, 1597–1606.

308 Luna, E., Bruce, T.J., Roberts, M.R., Flors, V., *et al.* (2012) Next-generation systemic acquired resistance. *Plant Physiol.*, **158**, 844–853.

309 Whittle, C.A., Otto, S.P., Johnston, M.O., Krochko, J.E. (2009) Adaptive epigenetic memory of ancestral temperature regime in *Arabidopsis thaliana*. *Botany*, **87**, 650–657.

310 Molinier, J., Ries, G., Zipfel, C., Hohn, B. (2006) Transgeneration memory of stress in plants. *Nature*, **442**, 1046–1049.

311 Lang-Mladek, C., Popova, O., Kiok, K., Berlinger, M. *et al.* (2010) Transgenerational inheritance and resetting of

stress-induced loss of epigenetic gene silencing in *Arabidopsis*. *Mol. Plant*, **3**, 594–602.

312 Boyko, A., Kathiria, P., Zemp, F.J., Yao, Y. et al. (2007) Transgenerational changes in the genome stability and methylation in pathogen-infected plants: (virus-induced plant genome instability). *Nucleic Acids Res.*, **35**, 1714–1725.

12
mRNA Stability

Ashley T. Neff, Carol J. Wilusz, and Jeffrey Wilusz
Colorado State University, 1682 Campus Delivery, Fort Collins, CO 80523-1682, USA

1	Introduction	393
2	**mRNA Decay Factors and Enzymes**	**394**
2.1	Deadenylation	394
2.2	Exosome-Mediated Decay	397
2.3	Decapping and 5′ → 3′ Exoribonucleolytic Decay	398
2.4	Deadenylation-Independent Decapping	399
2.5	Endoribonucleolytic Cleavage	400
3	**Regulation of mRNA Decay through *cis*-Acting Elements and *trans*-Acting Factors**	**401**
4	**AU-, GU-, and C-Rich Elements and Their Associated RNA-Binding Proteins**	**401**
4.1	MicroRNAs	403
4.2	Modulation of *trans*-Acting Factors	403
4.2.1	Modulation of Abundance	403
4.2.2	Modulation of Activity	404
4.2.3	Modulation of Availability	405
4.3	RNA Regulons	405
4.4	Interplay between *trans*-Acting Factors	406
5	**mRNA Surveillance Pathways**	**406**
5.1	Nonsense-Mediated Decay	407
5.2	Non-Stop Decay	407
5.3	No-Go Decay	408
6	**Conclusions and Future Questions**	**408**

RNA Regulation: Advances in Molecular Biology and Medicine, First Edition. Edited by Robert A. Meyers.
© 2014 Wiley-VCH Verlag GmbH & Co. KGaA. Published 2014 by Wiley-VCH Verlag GmbH & Co. KGaA.

Acknowledgments 410

References 410

Keywords

cis-Acting elements
Sequence motifs or structural elements found within an mRNA that serve to either recruit the decay machinery or protect it from decay through binding of RNA-binding proteins or miRNAs.

Deadenylase
An enzyme that catalyzes the removal of adenosine nucleotides from RNAs in the $3' \rightarrow 5'$ direction.

Decapping
Removal of the 5' 7-methylguanosine cap structure from an mRNA.

Endoribonuclease
An enzyme that catalyzes the hydrolysis of ester linkages to create internal RNA breaks.

Exoribonuclease
An enzyme that catalyzes the removal of mononucleotides from either the 5' or 3' end of an RNA molecule.

Messenger ribonucleoprotein (mRNP)
A complex comprised of an mRNA bound by at least one RNA-binding protein.

MicroRNAs (miRNAs)
Short (21–25 nt) noncoding RNAs (functional RNA molecules not translated into proteins) that target mRNAs for decay or translation repression through the binding of complementary seed sequences.

mRNA decay
A mechanism used by the cell to degrade normal and aberrant messenger RNA transcripts.

mRNA surveillance pathway
A quality control mechanism used by the cell to identify defective mRNAs and rapidly target them for decay.

Processing bodies (P-bodies)
Cytoplasmic concentrations of RNA decay factors related to miRNA-mediated silencing and the 5′ → 3′ mRNA decay pathway.

RNA-binding proteins (RBPs)
Proteins able to bind RNAs to influence their processing, stability, localization, and/or translation status.

Transcriptome
The entire population of mRNAs present within a given cell.

The regulated stability of eukaryotic mRNAs plays a major role in determining the levels of gene expression. Global analyses have shown that ~20–50% of changes in cellular gene expression in various experimental scenarios can be attributed to alterations in mRNA stability rather than to transcription rates. In addition, RNA stability plays a major role in determining the quality of overall gene expression in the cell. Damaged, malformed, or mutated transcripts are often subject to rapid degradation to preserve the integrity of the cellular transcriptome. Thus, a detailed understanding of the processes of mRNA decay is vital for the study and interdisciplinary applications of molecular and cellular biology. The aim of this chapter is to provide the reader with a detailed overview of the processes, enzymes, and regulatory factors that influence the stability of an mRNA in eukaryotic cells.

1
Introduction

Eukaryotic gene expression is largely dependent on the abundance and quality of mRNAs available for translation. Importantly, these two criteria are determined in part by the rate and fidelity of synthesis (i.e., transcription and RNA processing), but equally through the rate of mRNA decay. Just as mRNA synthesis is modulated by a wide array of transcription factors and chromatin modifiers, mRNA decay is a tightly controlled process that is highly dependent on RNA-binding proteins (RBPs) and small noncoding RNAs (ncRNAs) (Fig. 1). Moreover, in order to achieve an appropriate control of gene expression, the transcription and mRNA decay must be coordinated. Such coordination is evident – and essential – in many biological events that require rapid global changes in gene expression including differentiation, circadian rhythms, the immune response, and cell cycle progression.

Although the contribution of mRNA degradation pathways has been previously underappreciated, exciting studies continue to illustrate the complexity and importance of regulated mRNA decay in supporting accurate gene expression. In this chapter the many mechanisms that target both normal and aberrant mRNAs for decay in eukaryotic cells will be reviewed, and the vast array of factors required for this vital last step in the mRNA life-cycle briefly outlined.

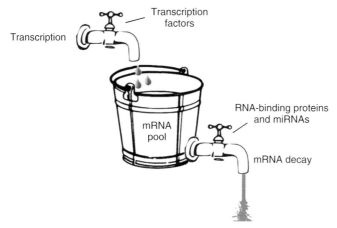

Fig. 1 Bucket model for gene expression, illustrating that both transcription and decay contribute to establishing mRNA abundance.

2 mRNA Decay Factors and Enzymes

The rate and mechanism of decay for each mRNA is largely determined at the transcript's birth in the nucleus. During transcription and processing, each transcript undergoes modifications that result in the deposition of proteins on the RNA body. Together, the mRNA and its associated *trans*-acting factors form a messenger ribonucleoprotein (mRNP) that allows the various subsequent steps of mRNA metabolism (export, localization, translation, and decay) to proceed correctly. Nuclear processing leaves almost all mRNAs with a 5′,7-methylguanosine cap and a 3′ poly(A) tail comprised of approximately 200 adenosine nucleotides. These universal features are recognized by several factors that facilitate mRNA functions, namely cap- and poly(A) tail-binding proteins. As will be seen below, the cap and the poly(A) tail are essential determinants of mRNA stability, and their removal is tightly controlled.

In the cytoplasm, the eukaryotic translation initiation factor eIF4E binds the 5′ cap, while the cytoplasmic poly(A)-binding protein (PABPC) coats the poly(A) tail. An interaction between eIF4E and PABPC, bridged by eIF4G, leads to transcript circularization (Fig. 2). This closed-loop conformation serves to promote translation and prevent cytoplasmic exonucleases from accessing the 5′ and 3′ ends of the transcript. In most cases, decay is initiated through removal of the poly(A) tail (deadenylation), which eventually disrupts the circular configuration and leaves the mRNA vulnerable to 3′ → 5′ exoribonucleolytic decay that is mediated by the exosome and/or decapping and 5′ → 3′ exoribonucleolytic decay by XRN1. Many of the factors and enzymes involved in this major deadenylation-dependent mRNA decay pathway (Fig. 3) have been identified, and their roles are further described below.

2.1 Deadenylation

The poly(A) tail is added cotranscriptionally by poly(A) polymerase (PAP) in the nucleus, and is bound by poly(A)-binding

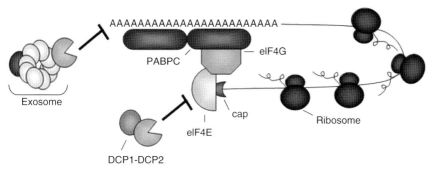

Fig. 2 Closed-loop conformation of mRNA formed by an interaction between PABPC on the 3' poly(A) tail and eIF4E on the 5' cap bridged by eIFAG enhances translation and protects mRNA termini from exonucleases.

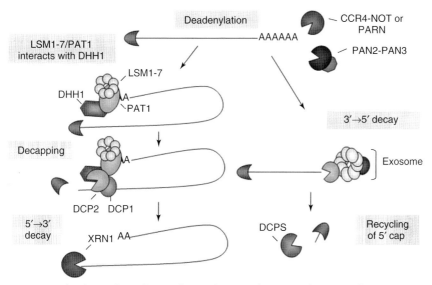

Fig. 3 Deadenylation-dependent pathway of mRNA decay in eukaryotic cells.

proteins (PABPs) in both compartments [1]. Particularly in higher eukaryotes, the poly(A) tail lengths can vary significantly between transcripts, as well as in the same transcript under different conditions, which indicates the importance of tail length as a modulator of mRNA function. The stability of many mRNAs is regulated by *trans*-acting factors that recruit deadenylases for rapid poly(A) tail removal [2]. Although poly(A) tails can be renewed by cytoplasmic PAPs to enhance translation initiation [3, 4], deadenylation frequently serves as the first step in eukaryotic deadenylation-dependent mRNA decay.

The best-characterized eukaryotic deadenylases include PAN2–PAN3 [5, 6], CCR4–NOT [7–9], nocturnin [10], and poly(A)-specific ribonuclease (PARN) [11], though several additional poly(A) shortening enzymes are present in mammalian

Tab. 1 Deadenylases involved in eukaryotic regulated RNA decay.

Protein	Also known as	Superfamily
2'PDE	2' Phosphodiesterase	EEP
ANGEL	–	EEP
CAF1Z	CCR4-associated factor 1Z	RNase D
CNOT6 (CCR4)	CCR4–NOT complex subunit 6	EEP
CNOT6L	CCR4–NOT complex subunit 6-like	EEP
CNOT7 (POP2)	CCR4–NOT complex subunit 7	RNase D
CNOT8	CCR4–NOT complex subunit 8	RNase D
Nocturnin	–	EEP
PAN2	Poly(A)-specific ribonuclease subunit	RNase D
PARN	Poly(A)-specific ribonuclease	RNase D

genomes (Table 1). Although not formally demonstrated, it seems likely that there is some redundancy in function as the depletion of individual deadenylase enzymes has surprisingly small effects on cell viability [8]. The catalytic activity of deadenylases is Mg^{2+}-dependent and functions to hydrolyze adenosine nucleotides from the 3' end of RNAs, releasing 5'-adenosine monophosphate as the product [12]. All deadenylases identified to date are members of either the RNase D or exonuclease-endonuclease-phosphatase (EEP) superfamilies. RNase D-type proteins are members of the DEDD superfamily, and share three sequence motifs containing four invariant Asp and Glu amino acids surrounded by several conserved residues, while EEP-type deadenylases have conserved catalytic Asp and His residues within their exonuclease domains [13]. Structural information is available for several deadenylases, including CNOT6L [14] and CNOT7 [15] of the CCR4–NOT complex, as well as PARN [16].

PAN2 is a member of the RNase D exonuclease subfamily that drives the hydrolytic activity of PAN2–PAN3 to trim the poly(A) tails of nascent mRNAs, from lengths of approximately 200 nt to about 80 nt [6]. Subsequent deadenylation is performed by CCR4–NOT, a complex that contains two catalytic subunits – an EEP-type exonuclease (either CNOT6 or CNOT6L) and a CCR4-associated factor (CAF) (CNOT7 or CNOT8) [17, 18]. The results of several studies have suggested that the CCR4–NOT complex plays a major role in the deadenylation of mRNAs in many eukaryotes [19, 20].

PARN, which also is a member of the RNase D family, is unique in that it is able to interact with both mRNA termini [16, 21]. The exonuclease activity of PARN is stimulated through an interaction with the 5' cap, and is inhibited by cap-binding protein 80 (CBP80) and eIF4E [22]. Thus, removal of the translation initiation complex allows PARN access to the cap structure, leading to deadenylation. Although PARN is an important, developmentally regulated deadenylase in higher eukaryotes, there is no homolog in either *Saccharomyces cerevisiae* or *Drosophila melanogaster*.

As poly(A) tail shortening is a highly regulated event, deadenylase activity is influenced through interactions with other proteins. The binding of PABPC along the poly(A) tail of mRNAs can serve to either inhibit or recruit decay machinery [23]. While the activities of CCR4 and PARN are inhibited by PABPC [24], PAN2 deadenylation is stimulated by its presence [25]. In order to overcome inhibition by PABPC, the deadenylation complexes interact with enhancing cofactors. It has been shown that CAF1, a component of the CCR4–NOT complex of which the human homologs are CNOT7 and CNOT8, is recruited to the 3′ end of mRNAs by the transducer of ERBB2 (TOB) family of antiproliferative proteins. The TOB proteins interact with CAF1 and bind PABPC [26]. The RNA stability factor tristetraprolin (TTP) also recruits the CCR4–NOT complex to mRNAs [27]. In cytoplasmic foci containing aggregates of repressed mRNAs and components of the decay machinery – known as processing bodies (P-bodies) [28, 29] – the Argonaute-interacting GW182 protein has been shown to serve as a docking platform to direct the PAN2–PAN3 and CCR4–NOT deadenylases to microRNA (miRNA) targets [30, 31]. Finally, cytoplasmic polyadenylation element binding protein (CPEB) interacts with PARN to stimulate its activity [32]. Undoubtedly, additional proteins and factors that regulate the assembly of deadenylases on RNA substrates will be elucidated in the future.

Deadenylation leaves the mRNA with an oligo(A) tail that is the key to any subsequent complete degradation of the transcript. At this point, both ends of the mRNA are open to attack by exonucleases.

2.2
Exosome-Mediated Decay

Decay in the 3′ → 5′ direction is carried out by a large complex known as the *exosome*. The exosome requires a region of unstructured RNA and a 3′OH in order to initiate decay [33], and in many cases this "toehold" may be provided by the remnants of the poly(A) tail. The huge exosome complex (~2 MDa) contains three S1-family RBPs (EXOSC1-EXOSC3), six RNase PH-domain proteins (EXOSC4-EXOSC9), and two RNase D-like enzymes (EXOSC10 and DIS3), all of which function in concert to promote the highly processive unwinding and hydrolysis of RNA substrates [34–36]. The nine-subunit, ring-shaped exosome core consists of catalytically inactive proteins, and requires at least one additional catalytic factor to become active. The core of the exosome forms a channel through which the mRNA is thought to thread as it is being degraded [37]. A tenth subunit to assemble, DIS3, confers both exo- and endonuclease activities, and is homologous to bacterial RNase II; this region of the protein is responsible for much of the exonuclease activity of the exosome [38]. Endonucleolytic reactions are achieved through the DIS3 PIN domain [39]. Final assembly of the exosome differs between the nucleus and cytoplasm. For example, in yeast the nuclear exosome contains the EXOSC10-homolog Rrp6p which also has exonuclease activity, while the cytoplasmic exosome includes Ski7p [40]. Human exosome complexes are similar to yeast but differ in that RRP6/EXOSC10 is present in both the nucleus and cytoplasm; the nuclear enzyme also has an additional subunit, M-phase phosphoprotein 6 (MPP6) [41] (Fig. 4). Exonuclease activity in the cytoplasm requires recruitment of the exosome to the 3′ end of mRNAs targeted for degradation through an interaction with SKI7, a component of the SKI

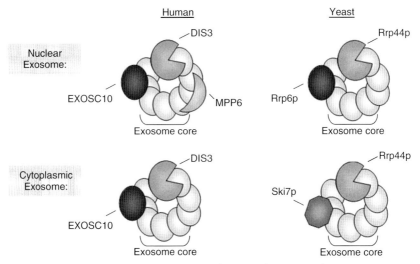

Fig. 4 Nuclear and cytoplasmic exosome subunits in humans and yeast.

complex which also contains the RNA helicase SKI2 [42, 43]. Following 3′ → 5′ exonucleolytic digestion by the exosome, when only a few 5′ nucleotides remain on the mRNA substrate, the scavenger decapping enzyme DCPS cleaves and recycles the 5′ cap [44]. Interestingly, as well as finalizing decay from the 3′ end, the yeast scavenger decapping enzyme Dcs1p is able to influence decay from the 5′ direction through its action as a cofactor for the XRN1 5′ → 3′ exonuclease [45].

2.3
Decapping and 5′ → 3′ Exoribonucleolytic Decay

The body of deadenylated mRNAs may alternatively be degraded starting from the 5′ end with decapping [46, 47]. Interestingly, decapping at the 5′ end also requires recognition of the deadenylated 3′ end of the transcript. The ring-shaped LSM1-7/PAT1 complex associates with the 3′ end of mRNAs by binding to the drastically shortened poly(A) tail that remains after deadenylation [48]. This 3′ complex, along with the DEAD box helicase DHH1 [49], then recruits the DCP1–DCP2 decapping complex to the 5′ end of the transcript while simultaneously preventing 3′ exonucleolytic trimming [50]. The decapping complex is well characterized, and is comprised of catalytically active decapping protein 2 (DCP2), a member of the Nudix superfamily of hydrolases, along with coactivator DCP1 [51, 52]. An alternative decapping enzyme, nudix-motif containing (NUDT) 16, is also present in cells and appears to regulate the decapping of a select subset of mRNAs [53]. Decapping generates a 5′ monophosphorylated transcript that is vulnerable to 5′ → 3′ exonucleolytic decay by the highly processive XRN1 enzyme [47], which represents the core cytoplasmic 5′ → 3′ decay apparatus in eukaryotic cells.

A multitude of factors that influence decapping have been identified. Perhaps not surprisingly, cap-binding protein eIF4E and poly(A) tail-binding proteins negatively regulate decapping activity [22, 54]. Conversely, enhancers of DCPs EDC3 and EDC4/HEDLS/Ge-1, as well as the

LSM14-homolog SCD6, interact with the DCP1–DCP2 complex to favor an active conformation [52, 55].

The enzymes described above, namely deadenylases, decapping enzymes, and exonucleases, represent three of the four major activities involved in removing unwanted mRNAs from the cytoplasm in eukaryotic cells. Together, these enzymes comprise the deadenylation-dependent decay pathway, though other minor pathways exist that include endonucleolytic decay; these are described briefly in the following section.

2.4
Deadenylation-Independent Decapping

As described above, one way to recruit the decapping machinery to an mRNA is through association of the LSM1-7/PAT1 complex with the shortened oligo(A) tail that remains following deadenylation [48]. However, this is not the only means of targeting an mRNA for decapping [56]. Perhaps the best example of deadenylation-independent decapping can be found in metazoan histone mRNAs, which do not have a 3′ poly(A) tail and thus cannot undergo deadenylation. Rather, a stem-loop structure found at the 3′ end of all histone mRNAs recruits a variety of *trans*-acting factors to modulate translation and decay [57]. Surprisingly, despite never undergoing deadenylation, the histone mRNAs are decapped through a very similar mechanism to deadenylated mRNAs. The 3′ histone stem-loop and the associated stem-loop-binding protein (SLBP) recruit a terminal uridylyl transferase (TUTase) to stimulate the addition of short uridine tracts to the end of the histone message [58]. This oligo(U) tract permits binding of the LSM1-7/PAT1 complex [59] which recruits DCP1–DCP2 and triggers 5′ → 3′ decay [60]. Exosome-mediated 3′ → 5′ decay is also activated to ensure the rapid removal of all histone transcripts (Fig. 5).

Deadenylation can also be bypassed through direct recruitment of decapping activators. For example, the S. cerevisiae

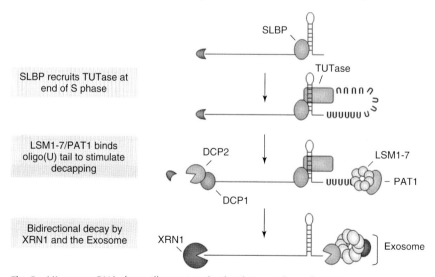

Fig. 5 Histone mRNA decay illustrating deadenylation-independent decay.

Rps28b protein binds to a stem-loop in the 3′UTR (untranslated region) of its own mRNA, and recruits Edc3p to activate 5′ cap removal and subsequent degradation [56]. Certain aberrant mRNAs can also be targeted for deadenylation-independent decapping through mRNA surveillance pathways (as discussed later).

2.5 Endoribonucleolytic Cleavage

Endonucleolytic cleavage is a decay mechanism that targets actively translating mRNAs and results in two mRNA fragments with unprotected ends subject to 5′ → 3′ decay by XRN1 and exosome-mediated 3′ → 5′ degradation (Fig. 6). In contrast to deadenylation, which gradually inactivates mRNA over time, endonucleases instantly incapacitate their mRNA substrates. In addition, endonucleases exhibit some sequence specificity and thus generally target only a subset of transcripts.

Many endoribonucleases, some of which are listed in Table 2, exist in mammalian cells. A noteworthy endo-nuclease is the Argonaute 2 (AGO 2) protein, which serves as an integral component of the miRNA machinery involved in modulating mRNA decay and translation [61, 62]. Interestingly, as well as endonucleolytically cleaving its mRNA targets, the AGO2 protein can also instigate biphasic deadenylation through interaction with the GW182 protein to recruit PAN2–PAN3 and CCR4–NOT deadenylases [63]. The SMG6 protein of the nonsense-mediated decay (NMD) pathway (as discussed below) is another interesting endonuclease that cleaves mRNAs close to the sites of premature termination codons (PTCs) [64]. Other endonucleases, such as RNase L and ZC3H12A, play vital roles during immune responses. For example, RNase L is activated during viral infection, and targets both cellular and viral RNAs for rapid destruction [65], often resulting in apoptosis. ZC3H12A is also induced in response to infection, and targets mRNAs encoding cytokines such as interleukin (IL) -6 for decay to restrain

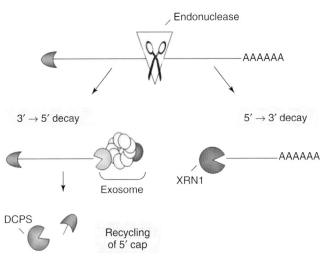

Fig. 6 Endonucleolytic cleavage resulting in two RNA fragments subject to exosome-mediated decay 3′ → 5′, and 5′ → 3′ decay by XRN1.

Tab. 2 Endoribonucleases involved in eukaryotic regulated RNA decay.

Protein	Also known as	Substrate
AGO2	Argonaute 2	miRNA targets
DICER	Ribonuclease type III	Pre-miRNA
IRE1	Inositol-requiring enzyme-1	Endoplasmic reticulum mRNA
PMR1	P-type ATPase, Ca^{2+}-transporting	Translating poly(A) mRNA
SMG6	Nonsense-mediated decay factor	Nonsense-mediated decay targets
ZC3H12A	Zinc finger CCCH-type containing 12A	Poly(A) mRNA

the immune response [66]. Finally, the polysome-associated PMR1 endonuclease plays a key role in the regulation of several transcripts, and has been implicated in the regulation of cell mobility [67]. In summary, endonucleases can clearly have a significant impact on mRNA stability in a variety of important biological contexts.

3
Regulation of mRNA Decay through cis-Acting Elements and trans-Acting Factors

Although almost all mRNAs have a cap and poly(A) tail, and the vast majority undergo decay through the deadenylation-dependent pathway, the mRNA half-lives may vary drastically from as little as a few minutes to over 24 h. The rate of decay is specified through sequence elements (cis-acting elements) embedded within the mRNA, most commonly in the 3′-UTR [68], but there is some evidence that they can also be found in the 5′UTR and coding region. These sequence elements associate with trans-acting factors (RBPs and miRNAs), which in turn recruit or impede the mRNA decay machinery [2] (Fig. 7). Such trans-acting factors act most often at the level of deadenylation, as this is the first rate-limiting step in the mRNA decay pathway. Importantly, the rate of decay for an individual mRNA can vary widely in different cell types and conditions, due largely to fluctuations in the array of active factors available in response to cellular signals. In addition, sequence elements present within the transcript can change as a result of alternative processing. For example, the use of a downstream poly(A) site can add extra regulatory elements, allowing the recruitment of new RBPs and miRNAs [69, 70].

4
AU-, GU-, and C-Rich Elements and Their Associated RNA-Binding Proteins

Over the years, several sequence elements that influence mRNA decay have come to light. The first to be uncovered – and consequently the best studied – is the AU-rich element (ARE). Typically, AREs comprise 50 to 150 base sequences in the 3′UTR that contains one or more AUUUA motifs within a U-rich context, and are found in up to 9% of cellular mRNAs [71]. These elements serve as binding sites for proteins to target mRNAs for rapid deadenylation, and are most often found in transcripts that require transient and rapid expression, such as cytokines

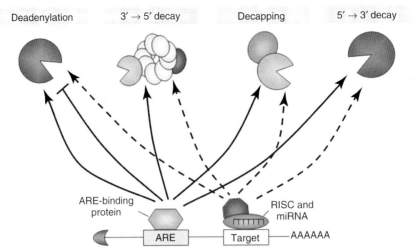

Fig. 7 Binding of *trans*-acting factors such as AU-rich elements (AREs) and miRNA target binding sites influence stability by either recruiting or inhibiting decay machinery.

[72]. A tight control of proinflammatory factors such as these is necessary because prolonged expression can damage tissue; consequently, a rapid destabilization to halt protein production is critical once the signal eliciting response has ceased. AREs have also been identified in mRNAs encoding transcription factors, cell cycle modulators, and oncogenes [73].

The activity of an ARE depends on several factors, such as sequence context, the secondary structure of the mRNA, and the binding of various RBPs and miRNAs within the vicinity. Most AU-rich element-binding proteins (ARE-BPs) are destabilizers that recruit deadenylases such as CCR4–NOT and PARN to the transcript [71, 74]. Well-characterized destabilizing ARE-BPs include the TTP family of CCCH-type zinc finger proteins (ZFP36, ZFP36L1, and ZFP36L2) [75], KHSRP, and AUF1 (HNRNPD) [76]. A few ARE-BPs act as stabilizers, including HuR (ELAVL1) [77], TIA1, and TIAR [78]. Such stabilization mechanisms are relatively poorly understood, but may involve competing a destabilizing RBP off the mRNA or relocalizing the mRNA to a cytoplasmic location that is protected from the decay machinery.

GU-rich elements (GREs) function very similarly to AREs, in that they recruit RBPs that destabilize mRNAs, but they are recognized and bound by different proteins. Generally, GREs consist of UG-repeats or U-rich regions interspersed with Gs (e.g., UGUUUGU) [79]. GREs are found in genes involved in regulating transcription, nucleic acid metabolism, developmental processes, and neurogenesis [80]. Functional analyses have shown that at least 5% of human genes contain GREs [81]. CELF1, the most prolific GRE-binding protein, has been shown to regulate many transcripts harboring this instability element [82, 83]. However, HuR (ELAVL1) also has similar binding preferences and may counteract CELF1 in certain situations. One way that CELF1 mediates mRNA decay is through a direct association with the PARN deadenylase [84].

The final mRNA stability element to be considered here is the pyrimidine-rich or C-rich tract that can serve as binding sites for members of the poly(C)-binding protein (PCBP) family [85, 86]. Unlike AREs and GREs, C-rich elements are generally found in relatively stable mRNAs such as α-globin [87] and collagen-α1 [88]. The influence of PCBPs on mRNA stability was first demonstrated in α-globin mRNA, where they clearly stabilized the transcript in erythroid cells [87]. Other studies have shown that PCBP2 functions as a cytoplasmic polyadenylation factor in *Xenopus* embryos [89]; increasing the poly(A) tail length would most likely stabilize mRNAs, but the effects of cytoplasmic polyadenylation on mRNA decay in this case have not been formally tested.

4.1
MicroRNAs

In addition to RBPs, small ncRNAs known as *miRNAs* (and the related piRNAs and siRNAs) also play an important role in regulating mRNA stability and/or translation for up to 30% of human genes [90]. miRNAs are approximately 21–25 nt in length and are incorporated into a cytoplasmic RNP complex known as the RNA-induced silencing complex (RISC), which contains the core proteins AGO2 and GW182 [91]. To date, hundreds of miRNAs have been identified in mammalian cells, with each miRNA family recognizing a different set of mRNA targets through its "seed" sequence. The miRNA guides the RISC to a complementary sequence within the target mRNA, most often within the 3′UTR but, on occasion, within the coding region and 5′UTR [92]. The level of complementarity of the miRNA to the mRNA determines whether the mRNAs is degraded or translationally repressed. A perfect complementarity will activate the AGO2 component of RISC to induce endonucleolytic cleavage, and this will result in a rapid destruction of the transcript via XRN1 and exosome activities. Mismatches between the miRNA and the mRNA result in deadenylation and translation repression [93]. In this pathway, the GW182 component of the RISC is able to independently interact with PAN2–PAN3 and CCR4–NOT complexes to stimulate deadenylation and mRNA silencing [30, 94] (Fig. 8).

4.2
Modulation of *trans*-Acting Factors

As noted above, mRNA decay rates can be changed dramatically in response to cellular signals. Such changes are achieved by directly altering the sequence elements present in the mRNA through alternative processing choices, and also by modulating the abundance, activity, and/or availability of RBPs and miRNAs.

4.2.1 Modulation of Abundance

The most obvious way in which the cell can regulate the activity of *trans*-acting factors is by simply increasing or decreasing their abundance. As for all genes, such changes are achieved through alterations of transcription or processing efficiency and mRNA/miRNA decay rates. For RBPs, changes in translation efficiency are also a factor. Levels of the mRNA destabilizing protein TTP, for example, are specifically induced by B cells treated with tumor necrosis factor [95], while estrogen upregulates AUF1 (HNRNPD) expression [96]. The regulation of abundance applies to both RBPs and miRNAs, although for miRNAs it is a predominant means of control, as their activity depends primarily on the RISC complex which interacts

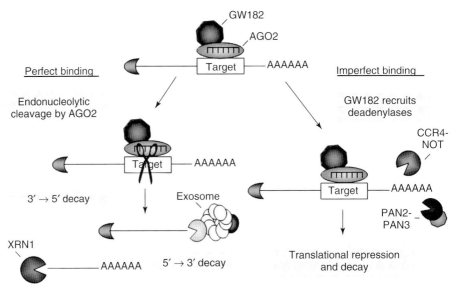

Fig. 8 miRNA binding complementarity determines whether mRNA is degraded or repressed.

with all miRNAs, preventing any isolated control of individual miRNAs.

4.2.2 Modulation of Activity

The activity of RBPs can be rapidly and dramatically changed by post-translational modifications that include phosphorylation, sumoylation, and methylation. Such modifications can instantly alter the RBP conformation and profoundly influence its function. One possible outcome is an altered binding affinity for the substrate, either repression or enhancement. For example, the GRE-binding protein CELF1 exhibits reduced RNA-binding following phosphorylation upon T-cell activation [97]. Sumoylation (i.e., conjugation to a small, ubiquitin-like modification) has been observed in the RBPs HNRNPA1, HNRNPF, and HNRNPK at the sites of RNA recognition [98]. The phosphorylation of *Drosophila* RBP HOW affects both protein–protein and RNA–protein interactions to increase substrate binding. Although dimerization of the HOW protein is not required for RNA binding, phosphorylation-induced dimerization by mitogen-activated protein kinase (MAPK)/extracellular signal-regulated kinase (ERK) leads to an increased binding of targets compared to monomeric HOW [99]. Finally, the methylation of arginine residues, catalyzed by protein arginine methyltransferases (PRMTs), is a form of post-translational modification to which many groups of RBPs are subject, including hnRNP, Sm, and HuD protein families [100, 101]. This activity of RBP methylation is evolutionarily conserved and can either enhance or disrupt protein–protein interactions [102]. Modifications to the miRNA termini can also lead to changes in activity, although the mechanisms involved are less well understood. The addition of a 3′ uridine nucleotide to miRNA sequences blocks processing by DICER, while the 3′ adenylation of animal miRNAs reduces activity possibly by preventing uptake by RISC [103].

4.2.3 Modulation of Availability

The cellular localization of decay enzymes, RBPs, and miRNAs influences their ability to act on targets. Factors sequestered away from mRNAs cannot perform regulated degradation or translation repression to modulate gene expression. In fact, a variety of ways exists by which the cell can regulate the availability of *trans*-acting factors to their targets, including the relocalization of RBPs between compartments or mRNAs to specialized foci within the cytoplasm. Another recently observed, but not well understood, mechanism of regulation is the expression of ncRNAs known as *sponges*, which compete for miRNA binding [104]. Both of these methods serve to alter the half-life of target transcripts by preventing interactions with *trans*-acting factors.

The relocalization of proteins and mRNAs in response to cellular conditions demonstrates the importance of spatial organization in maintaining correct expression patterns. HuR and other RBPs relocalize from the nucleus to the cytoplasm under a variety of stress conditions [105]. The nuclear retention of RBPs prevents them from associating with target mRNAs in the cytoplasm to influence stability or translation. However, the transcripts themselves can be directed towards concentrated cytoplasmic structures for regulation. P-bodies are constitutively present cytoplasmic aggregates of translationally repressed mRNAs, and also decay intermediates associated with components of repression and decay machinery. These foci are thought to serve as sites of mRNA decay [28, 29, 106]. Conditions of stress that require global translational silencing induce the formation of another type of mRNA silencing foci, known as stress granules (SGs). The latter contain translationally silenced mRNAs which are aggregated either for temporary storage or to be targeted for degradation. The colocalization of SGs with P-bodies suggests that these mRNA-silencing foci interact and may exchange transcripts [107, 108]. The sequestration of mRNAs away from *trans*-acting factors available in the cytoplasm provides another way in which the cell can modulate transcript stability and translation status that does not require changes in abundance.

Although the endogenous expression of miRNA sponges has only recently been observed, it offers an exciting insight to miRNA-mediated regulation pathways [105]. A pseudogene resulting from a mutation in the start codon of PTEN, known as *PTENP1*, fails to produce protein and instead competes for miRNA binding through seed sequences shared with those found in the 3′UTR of PTEN [109]. miRNAs bound to RNA sponges are unable to bind their protein-encoding transcript targets, and this leads to an increased stabilization or translation of the mRNA. Further studies regarding the role(s) of these features in miRNA regulation will be necessary to fully understand their importance and function in gene expression.

4.3 RNA Regulons

One currently emerging theme in the mRNA decay field is that of the RNA regulon. This idea postulates that the expression of functionally related genes is coordinated in part through a recruitment of shared *trans*-acting factors to modulate post-transcriptional events, including translation and mRNA decay [110]. For example, in *S. cerevisiae*, a group of mRNAs encoding mitochondrial protein mRNAs are all associated with Puf3p, and this facilitates their localization to mitochondria for translation. These mRNAs all bear a Puf3p-binding site that allows their coordinated regulation [111]. Such an RNA regulon can act in a combinatorial fashion to synchronize global gene expression changes that occur in response to various stimuli from cell cycle progression to differentiation to stress. In principle, through regulating the abundance or activity of a single RBP or miRNA, the cell can either upregulate or downregulate an entire array of mRNAs encoding factors needed for a specific response. miRNA-126, for example, has been shown recently to downregulate an RNA regulon that dictates endothelial cell recruitment and metastasis in breast cancer cells [112]. A variety of developmental RNA regulons that are controlled by RBPs (e.g., TbZFP3) have also been described in trypanosomes [113]. Thus, coordinate control of gene expression in specific pathways can be obtained by post-transcriptional mechanisms.

4.4
Interplay between *trans*-Acting Factors

A single transcript can contain several different *cis*-acting elements targeted by multiple *trans*-acting factors. Although the expression of many factors is often spatially or temporally regulated, an additional level of control derives from the interplay between *trans*-acting factors [114]. For example, miRNA function can be repressed through the binding of RBPs adjacent to, or overlapping with, their target site, which leads to a decreased miRNA accessibility. The seed sequence may be obscured by the RBP, or binding may cause conformational changes in the RNA structure that prevent miRNA annealing. ARE-BP HuR and miR-122 show this type of interaction on the CAT1 mRNA.

Under resting conditions in hepatoma cells, CAT1 expression is repressed by miR-122 through translational silencing. In times of stress, however, HuR binds CAT1 mRNA to displace miR-122 binding [115]. Additionally, RBPs that compete for the same binding site can also interact to influence binding capacity. HuR and another ARE-BP, AUF1 (HNRNPD) target many common transcripts, and can competitively bind a single ARE to confer either stabilization or destabilization, respectively [116, 117]. While the exact nature of their interaction is unclear, there is evidence to show that these proteins interact closely and have the potential to heterodimerize on an ARE to form a single complex [118]. This observed interplay adds a layer of complexity to understanding global regulation of mRNA stability.

5
mRNA Surveillance Pathways

Due to the large impact that the transcriptome has on gene expression, it comes as no surprise that the cell has established conserved quality control mechanisms to ensure the integrity and accuracy of mRNAs translated into proteins. The targets of these pathways can either be newly synthesized or undergoing translation. While canonical bulk decay is cytoplasmic, surveillance mechanisms police both the nucleus and the cytoplasm for defective transcripts such as those harboring premature stop codons, mRNAs that lack a termination codon, and strong secondary structures that inhibit complete translation. Nuclear mRNA surveillance has been studied almost exclusively in *S. cerevisiae*, and investigations performed in this area have been reviewed by Houseley and Tollervey [119]. At this point, attention is focused on cytoplasmic mRNA surveillance, and the specialized factors and pathways that recognize defective mRNAs and facilitate their rapid degradation through mechanisms crucial for correct gene expression, are discussed.

5.1
Nonsense-Mediated Decay

The NMD pathway is the most well understood of the mRNA surveillance mechanisms, and is estimated to regulate 10–20% of normal mRNAs [120, 121]. In mammalian cells, the targets of this pathway may be restricted to newly exported transcripts that contain a PTC. There are two major models of how eukaryotic cells determine that a termination codon is "premature," rather than appropriately positioned. The first model involves the exon junction complex (EJC) which is deposited about 24 bases upstream of a spliced junction during nuclear processing in metazoans. Importantly, the 3′ UTR region downstream of a normal termination codon is rarely spliced in mammalian cells, and thus generally contains no EJCs. EJCs contain the surveillance factor up-shift protein 3 (UPF3), attached to a related NMD protein, UPF2 [122]. During the first or "pioneer" round of translation, EJCs are removed from the mRNAs due to the transit of a translating ribosome. The translation of mRNAs containing a PTC more than 50–55 nt upstream of an EJC bound exon–exon junction triggers the initiation of NMD, since the EJC is not removed from the transcript [120]. In the second model, the distinction between translation termination at a normal stop codon versus a premature one is determined by the physical distance between a terminating ribosome and the poly(A) tail/PABP complex [123].

The encounter of a ribosome with a PTC triggers a cascade of events that, ultimately, leads to destruction of the mRNA through the assembly of a decay-promoting complex. The stalled ribosome induces binding to the SURF complex containing SMG1, UFP1, and peptide-releasing eRF1 and eRF3. In the EJC-promotion model of NMD, CBP80 promotes the interaction of UPF1 in the SURF complex with UPF2 bound to the EJC [124]. This interaction between UPF proteins elicits NMD, as normal mRNAs are not bound by the SURF complex. SGM1 phosphorylates UPF1, which leads to the dissociation of eRF1 and eRF3 and binding of the heterodimer SMG5/SMG7 [125] (Fig. 9). The transcript is then degraded by a number of mechanisms, included deadenylation-independent decapping and endonucleolytic cleavage by SMG6. This results in the dramatic instability observed for defective mRNAs containing PTCs.

5.2
Non-Stop Decay

Non-stop decay (NSD) is initiated when an mRNA that lacks an in-frame termination codon allows a ribosome to move all the way to the 3′ end of the transcript. One way in which these non-stop transcripts can arise is through the aberrant usage of upstream "premature" polyadenylation signals during nuclear pre-mRNA processing, though they can also be generated by breaks in the mRNA. Once at the 3′ end, the ribosome is stalled and unable to translate normal mRNAs until it is released from the defective transcript. In a mechanism conserved in yeast through

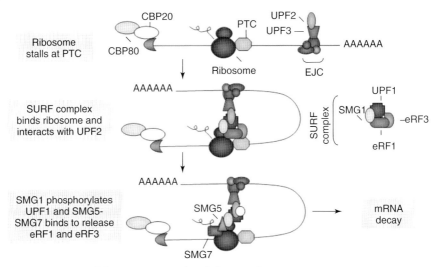

Fig. 9 Initiation of the nonsense-mediated decay pathway by a stalled ribosome at an mRNA premature termination codon.

humans, the stalled ribosome is recognized by either the adaptor protein SKI7, which binds its empty A site to release the mRNA, or a complex comprising the endonuclease Pelota (Dom34p in yeast) and HBS1, a SKI7 paralog [126, 127]. Following ribosome ejection, the mRNA is then subjected to 3′ → 5′ decay by the cytoplasmic exosome [128]. In the absence of SKI7, the non-stop transcript is subject to decapping and 5′ → 3′ decay by exoribonuclease XRN1 following the displacement of PABPC (Fig. 10). NSD serves not only to remove incorrect mRNAs from translating populations but also to free ribosomes, so that they can continue to function in active translation.

5.3
No-Go Decay

Like NSD, no-go decay (NGD) is initiated by a stalled ribosome on a defective mRNA. Here, the translating ribosome cannot elongate due to strong secondary features in the coding region that disrupts codon–anticodon pairing. Similar to the NSD pathway, the stalled ribosome is recognized by the Pelota(Dom34)–HBS1 complex, promoting cleavage of the transcript near the stall site [129, 130]. Once cleaved, the ribosome is released and the resulting mRNA fragments are subjected to 3′ → 5′ and 5′ → 3′ decay pathways (Fig. 11).

6
Conclusions and Future Questions

While the mechanistic details regarding mRNA stability and its regulation have become much clearer over the past two decades, there remains much to explore in this area. Consequently, six interesting aspects can be proposed for future study:

- A variety of experiments have revealed intimate connections between mRNA stability and other processes of gene

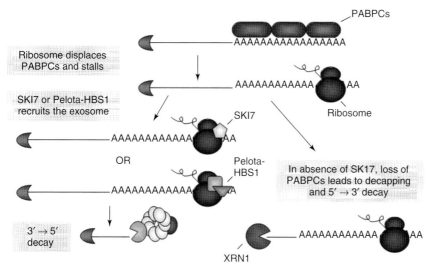

Fig. 10 Non-stop decay leading to degradation of mRNAs with ribosomes stalled at the 3′ end, either by 3′ → 5′ decay or by decapping and 5′ → 3′ decay.

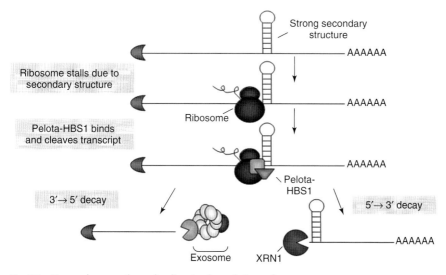

Fig. 11 No-go decay pathway leading to degradation of mRNAs harboring stalled ribosomes due to strong secondary structures.

expression, including translation and transcription. The underlying principles of these interconnections, as well as how they contribute to the regulation of gene expression in the context of cell developmental, growth, and differentiation, need to be better established.

- Mutations in RNA decay factors and in sequence elements that modulate mRNA decay are associated with a wide

range of human diseases. In addition, alterations in RNA decay enzymes and factors may serve as biomarkers for a variety of pathologies.
- Many aspects of how ncRNAs (in addition to the highly studied miRNAs) influence, and are themselves influenced by, regulated RNA decay remains to be investigated. This should be a particularly interesting area for future experimentation as the biological relevance of the "dark matter" of the transcriptome becomes clear.
- While recent experiments have shed some light on how viruses interface with and avoid the cellular RNA decay machinery during infection, this area should likely reveal additional insights and surprises in the future.
- It can readily be anticipated that a significant amount of additional progress will be made in identifying mechanisms and the biological impact of RNA stability through advancements in technology and methods such as efficient deep sequencing and single cell/molecule analyses. In particular, genome-wide studies should reveal significant new insights into the global interplay of RNA stability and gene expression.
- Finally, a key remaining challenge in the field is to understand the true nature and dynamics of mRNPs, and how they relate to mRNA stability. It is important to remember that the substrate of regulated mRNA stability is the mRNA particle, not the naked, unstructured mRNA nor an mRNA with a single protein attached. It is likely that a full understanding of mRNP dynamics will be necessary to truly understand the combinatorial control of mRNA stability. Undoubtedly, there are exciting times ahead in this field.

Acknowledgments

Studies of the mechanisms of mRNA stability in eukaryotic cells, conducted in the Wilusz laboratories, are supported by NIH grants (R01GM072481 and AR059247) to J.W. and C.J.W., respectively.

References

1 Chekanova, J.A., Belostotsky, D.A. (2003) Evidence that poly(A) binding protein has an evolutionarily conserved function in facilitating mRNA biogenesis and export. *RNA*, **9**, 1476–1490.

2 Glisovic, T., Bachorik, J.L., Yong, J., Dreyfuss, G. (2008) RNA-binding proteins and post-transcriptional gene regulation. *FEBS Lett.*, **582**, 1977–1986.

3 Lasko, P. (2011) Posttranscriptional regulation in *Drosophila* oocytes and early embryos. *Wiley Interdiscip. Rev.: RNA*, **2**, 408–416.

4 Coll, O., Villalba, A., Bussotti, G., Notredame, C. *et al.* (2010) A novel, noncanonical mechanism of cytoplasmic polyadenylation operates in *Drosophila* embryogenesis. *Genes Dev.*, **24**, 129–134.

5 Boeck, R., Tarun, S.J., Rieger, M., Deardorff, J.A. *et al.* (1996) The yeast Pan2 protein is required for poly(A)-binding protein-stimulated poly(A)-nuclease activity. *J. Biol. Chem.*, **271**, 432–438.

6 Brown, C.E., Tarun, S.Z., Boeck, R., Sachs, A.B. (1996) PAN3 encodes a subunit of the Pab1p-dependent poly(A) nuclease in *Saccharomyces cerevisiae. Mol. Cell. Biol.*, **16**, 5744–5753.

7 Schwede, A., Ellis, L., Luther, J., Carrington, M. *et al.* (2008) A role for Caf1 in mRNA deadenylation and decay in trypanosomes and human cells. *Nucleic Acids Res.*, **36**, 3374–3388.

8 Mittal, S., Aslam, A., Doidge, R., Medica, R. *et al.* (2011b) The Ccr4a (CNOT6) and Ccr4b (CNOT6L) deadenylase subunits of the human Ccr4–NOT complex contribute to the prevention of cell death and senescence. *Mol. Biol. Cell*, **22**, 748–758.

9 Temme, C., Zhang, L., Kremmer, E., Ihling, C. *et al.* (2010) Subunits of the *Drosophila* CCR4-NOT complex and their

roles in mRNA deadenylation. *RNA*, **16**, 1356–1370.

10 Baggs, J.E., Green, C.B. (2003) Nocturnin, a deadenylase in *Xenopus laevis* retina: a mechanism for posttranscriptional control of circadian-related mRNA. *Curr. Biol.*, **13**, 189–198.

11 Copeland, P.R., Wormington, M. (2001) The mechanism and regulation of deadenylation: identification and characterization of *Xenopus* PARN. *RNA*, **7**, 875–886.

12 Beese, L.S., Steitz, T.A. (1991) Structural basis for the 3′-5′ exonuclease activity of *Escherichia coli* DNA polymerase I: a two metal ion mechanism. *EMBO J.*, **10**, 25–33.

13 Zuo, Y., Deutscher, M.P. (2001) Exoribonuclease superfamilies: structural analysis and phylogenetic distribution. *Nucleic Acids Res.*, **29**, 1017–1026.

14 Wang, H., Morita, M., Yang, X., Suzuki, T. et al. (2010) Crystal structure of the human CNOT6L nuclease domain reveals strict poly(A) substrate specificity. *EMBO J.*, **29**, 2566–2576.

15 Bartlam, M., Yamamoto, T. (2010) The structural basis for deadenylation by the CCR4-NOT complex. *Protein Cell*, **1**, 443–452.

16 Wu, M., Reuter, M., Lilie, H., Liu, Y. et al. (2005) Structural insight into poly(A) binding and catalytic mechanism of human PARN. *EMBO J.*, **24**, 4082–4093.

17 Tucker, M., Staples, R.R., Valencia-Sanchez, M.A., Muhlrad, D. et al. (2002) Ccr4p is the catalytic subunit of a Ccr4p/Pop2p/Notp mRNA deadenylase complex in *Saccharomyces cerevisiae*. *EMBO J.*, **21**, 1427–1436.

18 Thore, S., Mauxion, F., Séraphin, B., Suck, D. (2003) X-ray structure and activity of the yeast Pop2 protein: a nuclease subunit of the mRNA deadenylase complex. *EMBO Rep.*, **4**, 1150–1155.

19 Ito, K., Takahashi, A., Morita, M., Suzuki, T. et al. (2011) The role of the CNOT1 subunit of the CCR4–NOT complex in mRNA deadenylation and cell viability. *Protein Cell*, **2**, 755–763.

20 Liu, H.Y., Badarinarayana, V., Audino, D.C., Rappsilber, J. et al. (1998) The NOT proteins are part of the CCR4 transcriptional complex and affect gene expression both positively and negatively. *EMBO J.*, **17**, 1096–1106.

21 Wilusz, C.J., Wormington, M., Peltz, S.W. (2001) The cap-to-tail guide to mRNA turnover. *Nat. Rev. Mol. Cell Biol.*, **2**, 237–246.

22 Ramirez, C.V., Vilela, C., Berthelot, K., McCarthy, J.E.G. (2002) Modulation of eukaryotic mRNA stability via the cap-binding translation complex eIF4F. *J. Mol. Biol.*, **318**, 951–962.

23 Bernstein, P., Ross, J. (1989) Poly(A), poly(A) binding protein and the regulation of mRNA stability. *Trends Biochem. Sci.*, **14**, 373–377.

24 Khanna, R., Kiledjian, M. (2004) Poly(A)-binding-protein-mediated regulation of hDcp2 decapping in vitro. *EMBO J.*, **23**, 1968–1976.

25 Uchida, N., Hoshino, S., Katada, T. (2004) Identification of a human cytoplasmic poly(A) nuclease complex stimulated by poly(A)-binding protein. *J. Biol. Chem.*, **279**, 1383–1391.

26 Ezzeddine, N., Chen, C.-Y.A., Shyu, A.-B. (2012) Evidence providing new insights into TOB-promoted deadenylation and supporting a link between TOB's deadenylation-enhancing and antiproliferative activities. *Mol. Cell. Biol.*, **32**, 1089–1098.

27 Sandler, H., Kreth, J., Timmers, H.T.M., Stoecklin, G. (2011) Not1 mediates recruitment of the deadenylase Caf1 to mRNAs targeted for degradation by tristetraprolin. *Nucleic Acids Res.*, **39**, 4373–4386.

28 Cougot, N., Babajko, S., Séraphin, B. (2004) Cytoplasmic foci are sites of mRNA decay in human cells. *J. Cell Biol.*, **165**, 31–40.

29 Parker, R., Sheth, U. (2007) P bodies and the control of mRNA translation and degradation. *Mol. Cell*, **25**, 635–646.

30 Fabian, M.R., Cieplak, M.K., Frank, F., Morita, M. et al. (2011) miRNA-mediated deadenylation is orchestrated by GW182 through two conserved motifs that interact with CCR4–NOT. *Nat. Struct. Mol. Biol.*, **18**, 1211–1217.

31 Braun, J.E., Huntzinger, E., Fauser, M., Izaurralde, E. (2011) GW182 proteins directly recruit cytoplasmic deadenylase complexes to miRNA targets. *Mol. Cell*, **44**, 120–133.

32 Kim, J.H., Richter, J.D. (2006) Opposing polymerase-deadenylase activities regulate

cytoplasmic polyadenylation. *Mol. Cell*, **24**, 173–183.
33 Schmid, M., Jensen, T.H. (2008) The exosome: a multipurpose RNA-decay machine. *Trends Biochem. Sci.*, **33**, 501–510.
34 Brouwer, R., Allmang, C., Raijmakers, R., Van Aarssen, Y. et al. (2001) Three novel components of the human exosome. *J. Biol. Chem.*, **276**, 6177–6184.
35 Chlebowski, A., Tomecki, R., López, M.E.G., Séraphin, B. et al. (2011) Catalytic properties of the eukaryotic exosome. *Adv. Exp. Med. Biol.*, **702**, 63–78.
36 Mitchell, P., Petfalski, E., Shevchenko, A., Mann, M. et al. (1997) The exosome: a conserved eukaryotic RNA processing complex containing multiple $3'{\rightarrow}5'$ exoribonucleases. *Cell*, **91**, 457–466.
37 Tsanova, B., van Hoof, A. (2010) Poring over exosome structure. *EMBO Rep.*, **11**, 900–901.
38 Dziembowski, A., Lorentzen, E., Conti, E., Séraphin, B. (2007) A single subunit, Dis3, is essentially responsible for yeast exosome core activity. *Nat. Struct. Mol. Biol.*, **14**, 15–22.
39 Lebreton, A., Tomecki, R., Dziembowski, A., Séraphin, B. (2008) Endonucleolytic RNA cleavage by a eukaryotic exosome. *Nature*, **456**, 993–996.
40 Graham, A.C., Kiss, D.L., Andrulis, E.D. (2006) Differential distribution of exosome subunits at the nuclear lamina and in cytoplasmic foci. *Mol. Biol. Cell*, **17**, 1399–1409.
41 Schilders, G., Raijmakers, R., Raats, J.M.H., Pruijn, G.J.M. (2005) MPP6 is an exosome-associated RNA-binding protein involved in 5.8S rRNA maturation. *Nucleic Acids Res.*, **33**, 6795–6804.
42 Araki, Y., Takahashi, S., Kobayashi, T., Kajiho, H. et al. (2001) Ski7p G protein interacts with the exosome and the Ski complex for 3′-to-5′ mRNA decay in yeast. *EMBO J.*, **20**, 4684–4693.
43 Houseley, J., LaCava, J., Tollervey, D. (2006) RNA-quality control by the exosome. *Nat. Rev. Mol. Cell Biol.*, **7**, 529–539.
44 Liu, H., Rodgers, N.D., Jiao, X., Kiledjian, M. (2002) The scavenger mRNA decapping enzyme DcpS is a member of the HIT family of pyrophosphatases. *EMBO J.*, **21**, 4699–4708.
45 van Dijk, E., Hir, H.L., Séraphin, B. (2003) DcpS can act in the $5'{\rightarrow}3'$ mRNA decay pathway in addition to the $3'{\rightarrow}5'$ pathway. *Proc. Natl Acad. Sci. USA*, **100**, 12081–12086.
46 Coller, J., Parker, R. (2004) Eukaryotic mRNA decapping. *Annu. Rev. Biochem.*, **73**, 861–890.
47 Muhlrad, D., Decker, C.J., Parker, R. (1994a) Deadenylation of the unstable mRNA encoded by the yeast MFA2 gene leads to decapping followed by $5'{\rightarrow}3'$ digestion of the transcript. *Genes Dev.*, **8**, 855–866.
48 Tharun, S. (2009) Lsm1-7-Pat1 complex: a link between 3′ and 5′-ends in mRNA decay? *RNA Biol.*, **6**, 228–232.
49 Coller, J.M., Tucker, M., Sheth, U., Valencia-Sanchez, M.A. et al. (2001) The DEAD box helicase, Dhh1p, functions in mRNA decapping and interacts with both the decapping and deadenylase complexes. *RNA*, **7**, 1717–1727.
50 Fischer, N., Weis, K. (2002) The DEAD box protein Dhh1 stimulates the decapping enzyme Dcp1. *EMBO J.*, **21**, 2788–2797.
51 Li, Y., Song, M.-G., Kiledjian, M. (2008) Transcript-specific decapping and regulated stability by the human Dcp2 decapping protein. *Mol. Cell. Biol.*, **28**, 939–948.
52 Ling, S.H.M., Qamra, R., Song, H. (2011) Structural and functional insights into eukaryotic mRNA decapping. *Wiley Interdiscip. Rev.: RNA*, **2**, 193–208.
53 Song, M.-G., Li, Y., Kiledjian, M. (2010) Multiple mRNA decapping enzymes in mammalian cells. *Mol. Cell*, **40**, 423–432.
54 Vilela, C., Velasco, C., Ptushkina, M., McCarthy, J.E. (2000) The eukaryotic mRNA decapping protein Dcp1 interacts physically and functionally with the eIF4F translation initiation complex. *EMBO J.*, **19**, 4372–4382.
55 Harigaya, Y., Jones, B.N., Muhlrad, D., Gross, J.D. et al. (2010) Identification and analysis of the interaction between Edc3 and Dcp2 in *Saccharomyces cerevisiae*. *Mol. Cell. Biol.*, **30**, 1446–1456.
56 Badis, G., Saveanu, C., Fromont-Racine, M., Jacquier, A. (2004) Targeted mRNA degradation by deadenylation-independent decapping. *Mol. Cell*, **15**, 5–15.
57 Marzluff, W.F., Wagner, E.J., Duronio, R.J. (2008) Metabolism and regulation of canonical histone mRNAs: life without a poly(A) tail. *Nat. Rev. Genet.*, **9**, 843–854.

58 Mullen, T.E., Marzluff, W.F. (2008) Degradation of histone mRNA requires oligouridylation followed by decapping and simultaneous degradation of the mRNA both 5′ to 3′ and 3′ to 5′. *Genes Dev.*, **22**, 50–65.

59 Song, M.-G., Kiledjian, M. (2007) 3′ Terminal oligo U-tract-mediated stimulation of decapping. *RNA*, **13**, 2356–2365.

60 Herrero, A.B., Moreno, S. (2011) Lsm1 promotes genomic stability by controlling histone mRNA decay. *EMBO J.*, **30**, 2008–2018.

61 Liu, J., Carmell, M.A., Rivas, F.V., Marsden, C.G. et al. (2004) Argonaute2 is the catalytic engine of mammalian RNAi. *Science*, **305**, 1437–1441.

62 Meister, G., Landthaler, M., Patkaniowska, A., Dorsett, Y. et al. (2004) Human Argonaute2 mediates RNA cleavage targeted by miRNAs and siRNAs. *Mol. Cell*, **15**, 185–197.

63 Behm-Ansmant, I., Rehwinkel, J., Doerks, T., Stark, A. et al. (2006) mRNA degradation by miRNAs and GW182 requires both CCR4:NOT deadenylase and DCP1:DCP2 decapping complexes. *Genes Dev.*, **20**, 1885–1898.

64 Eberle, A.B., Lykke-Andersen, S., Mühlemann, O., Jensen, T.H. (2008) SMG6 promotes endonucleolytic cleavage of nonsense mRNA in human cells. *Nat. Struct. Mol. Biol.*, **16**, 49–55.

65 Davidson, B.L., McCray, P.B. (2011) Current prospects for RNA interference-based therapies. *Nat. Rev. Genet.*, **12**, 329–340.

66 Schoenberg, D.R. (2011) Mechanisms of endonuclease-mediated mRNA decay. *Wiley Interdiscip. Rev.: RNA*, **2**, 582–600.

67 Bremer, K.A., Stevens, A., Schoenberg, D.R. (2003) An endonuclease activity similar to *Xenopus* PMR1 catalyzes the degradation of normal and nonsense-containing human beta-globin mRNA in erythroid cells. *RNA*, **9**, 1157–1167.

68 Matoulkova, E., Michalova, E., Vojtesek, B., Hrstka, R. (2012) The role of the 3′-untranslated region in post-transcriptional regulation of protein expression in mammalian cells. *RNA Biol.*, **9**, 563–576.

69 Hughes, T.A. (2006) Regulation of gene expression by alternative untranslated regions. *Trends Genet.*, **22**, 119–122.

70 Mayr, C., Bartel, D.P. (2009) Widespread shortening of 3′UTRs by alternative cleavage and polyadenylation activates oncogenes in cancer cells. *Cell*, **138**, 673–684.

71 Chen, C.-Y., Gherzi, R., Ong, S.-E., Chan, E.L. et al. (2001) AU binding proteins recruit the exosome to degrade ARE-containing mRNAs. *Cell*, **107**, 451–464.

72 Anderson, P. (2008) Post-transcriptional control of cytokine production. *Nat. Immunol.*, **9**, 353–359.

73 Gingerich, T.J., Feige, J.-J., LaMarre, J. (2004) AU-Rich elements and the control of gene expression through regulated mRNA stability. *Anim. Health Res. Rev.*, **5**, 49–63.

74 Stoecklin, G., Mayo, T., Anderson, P. (2006) ARE-mRNA degradation requires the 5′-3′ decay pathway. *EMBO Rep.*, **7**, 72–77.

75 Lai, W.S., Parker, J.S., Grissom, S.F., Stumpo, D.J. et al. (2006) Novel mRNA targets for tristetraprolin (TTP) identified by global analysis of stabilized transcripts in TTP-deficient fibroblasts. *Mol. Cell. Biol.*, **26**, 9196–9208.

76 Li, H., Chen, W., Zhou, Y., Abidi, P. et al. (2009) Identification of mRNA binding proteins that regulate the stability of LDL receptor mRNA through AU-rich elements. *J. Lipid Res.*, **50**, 820–831.

77 Dean, J.L.E., Wait, R., Mahtani, K.R., Sully, G. et al. (2001) The 3′ untranslated region of tumor necrosis factor alpha mRNA is a target of the mRNA-stabilizing factor HuR. *Mol. Cell. Biol.*, **21**, 721–730.

78 Gueydan, C., Droogmans, L., Chalon, P., Huez, G. et al. (1999) Identification of TIAR as a protein binding to the translational regulatory AU-rich element of tumor necrosis factor alpha mRNA. *J. Biol. Chem.*, **274**, 2322–2326.

79 Rattenbacher, B., Beisang, D., Wiesner, D.L., Jeschke, J.C. et al. (2010) Analysis of CUGBP1 targets identifies GU-repeat sequences that mediate rapid mRNA decay. *Mol. Cell. Biol.*, **30**, 3970–3980.

80 Vlasova, I.A., Bohjanen, P.R. (2008) Post-transcriptional regulation of gene networks by GU-rich elements and CELF proteins. *RNA Biol.*, **5**, 201–207.

81 Halees, A.S., Hitti, E., Al-Saif, M., Mahmoud, L. et al. (2011) Global assessment of GU-rich regulatory content

and function in the human transcriptome. *RNA Biol.*, **8**, 681–691.

82 Vlasova-St Louis, I., Bohjanen, P.R. (2011) Coordinate regulation of mRNA decay networks by GU-rich elements and CELF1. *Curr. Opin. Genet. Dev.*, **21**, 444–451.

83 Vlasova, I.A., Tahoe, N.M., Fan, D., Larsson, O. *et al.* (2008) Conserved GU-rich elements mediate mRNA decay by binding to CUG-binding protein 1. *Mol. Cell*, **29**, 263–270.

84 Moraes, K.C.M., Wilusz, C.J., Wilusz, J. (2006) CUG-BP binds to RNA substrates and recruits PARN deadenylase. *RNA*, **12**, 1084–1091.

85 Du, Z., Fenn, S., Tjhen, R., James, T.L. (2008) Structure of a construct of a human poly(C)-binding protein containing the first and second KH domains reveals insights into its regulatory mechanisms. *J. Biol. Chem.*, **283**, 28757–28766.

86 Kang, D.-H., Song, K.Y., Choi, H.S., Law, P.-Y. *et al.* (2012) Novel dual-binding function of a poly (C)-binding protein 3, transcriptional factor which binds the double-strand and single-stranded DNA sequence. *Gene*, **501**, 33–38.

87 Kong, J., Sumaroka, M., Eastmond, D.L., Liebhaber, S.A. (2006) Shared stabilization functions of pyrimidine-rich determinants in the erythroid 15-lipoxygenase and alpha-globin mRNAs. *Mol. Cell. Biol.*, **26**, 5603–5614.

88 Czyzyk-Krzeska, M.F., Bendixen, A.C. (1999) Identification of the poly(C) binding protein in the complex associated with the 3′ untranslated region of erythropoietin messenger RNA. *Blood*, **93**, 2111–2120.

89 Vishnu, M.R., Sumaroka, M., Klein, P.S., Liebhaber, S.A. (2011) The poly(rC)-binding protein αCP2 is a noncanonical factor in *X. laevis* cytoplasmic polyadenylation. *RNA*, **17**, 944–956.

90 Neilson, J.R., Zheng, G.X.Y., Burge, C.B., Sharp, P.A. (2007) Dynamic regulation of miRNA expression in ordered stages of cellular development. *Genes Dev.*, **21**, 578–589.

91 Gregory, R.I., Chendrimada, T.P., Cooch, N., Shiekhattar, R. (2005) Human RISC couples microRNA biogenesis and posttranscriptional gene silencing. *Cell*, **123**, 631–640.

92 Schnall-Levin, M., Rissland, O.S., Johnston, W.K., Perrimon, N. *et al.* (2011) Unusually effective microRNA targeting within repeat-rich coding regions of mammalian mRNAs. *Genome Res.*, **21**, 1395–1403.

93 Pillai, R.S. (2005) MicroRNA function: multiple mechanisms for a tiny RNA? *RNA*, **11**, 1753–1761.

94 Fabian, M.R., Sonenberg, N., Filipowicz, W. (2010) Regulation of mRNA translation and stability by microRNAs. *Annu. Rev. Biochem.*, **79**, 351–379.

95 Frasca, D., Landin, A.M., Alvarez, J.P., Blackshear, P.J. *et al.* (2007) Tristetraprolin, a negative regulator of mRNA stability, is increased in old B cells and is involved in the degradation of E47 mRNA. *J. Immunol.*, **179**, 918–927.

96 Arao, Y., Kikuchi, A., Ikeda, K., Nomoto, S. *et al.* (2002) A+U-rich-element RNA-binding factor 1/heterogeneous nuclear ribonucleoprotein D gene expression is regulated by oestrogen in the rat uterus. *Biochem. J.*, **361**, 125–132.

97 Beisang, D., Rattenbacher, B., Vlasova-St Louis, I.A., Bohjanen, P.R. (2012) Regulation of CUG-binding protein 1 (CUGBP1) binding to target transcripts upon T cell activation. *J. Biol. Chem.*, **287**, 950–960.

98 Li, T., Evdokimov, E., Shen, R.-F., Chao, C.-C. *et al.* (2004) Sumoylation of heterogeneous nuclear ribonucleoproteins, zinc finger proteins, and nuclear pore complex proteins: a proteomic analysis. *Proc. Natl Acad. Sci. USA*, **101**, 8551–8556.

99 Nir, R., Grossman, R., Paroush, Z., Volk, T. (2012) Phosphorylation of the *Drosophila melanogaster* RNA-binding protein HOW by MAPK/ERK enhances its dimerization and activity. *PLoS Genet.*, **8**, e1002632.

100 Blackwell, E., Ceman, S. (2012) Arginine methylation of RNA-binding proteins regulates cell function and differentiation. *Mol. Reprod. Dev.*, **79**, 163–175.

101 Liu, Q., Dreyfuss, G. (1995) In vivo and in vitro arginine methylation of RNA-binding proteins. *Mol. Cell. Biol.*, **15**, 2800–2808.

102 Bachand, F. (2007) Protein arginine methyltransferases: from unicellular eukaryotes to humans. *Eukaryot. Cell*, **6**, 889–898.

103 Burroughs, A.M., Ando, Y., de Hoon, M.J.L., Tomaru, Y. et al. (2010) A comprehensive survey of 3′ animal miRNA modification events and a possible role for 3′ adenylation in modulating miRNA targeting effectiveness. *Genome Res.*, **20**, 1398–1410.

104 Ebert, M.S., Sharp, P.A. (2010) Emerging roles for natural microRNA sponges. *Curr. Biol.*, **20**, R858–R861.

105 Kim, H.H., Gorospe, M. (2008) Phosphorylated HuR shuttles in cycles. *Cell Cycle*, **7**, 3124–3126.

106 Franks, T.M., Lykke-Andersen, J. (2008) The control of mRNA decapping and P-body formation. *Mol. Cell*, **32**, 605–615.

107 Kedersha, N., Anderson, P. (2007) Mammalian Stress Granules and Processing Bodies, *Translation Initiation: Cell Biology, High-Throughput Methods, and Chemical-Based Approaches*, Academic Press, pp. 61–81.

108 Kedersha, N., Stoecklin, G., Ayodele, M., Yacono, P. et al. (2005) Stress granules and processing bodies are dynamically linked sites of mRNP remodeling. *J. Cell Biol.*, **169**, 871–884.

109 Poliseno, L., Salmena, L., Zhang, J., Carver, B. et al. (2010) A coding-independent function of gene and pseudogene mRNAs regulates tumour biology. *Nature*, **465**, 1033–1038.

110 Keene, J.D. (2007) RNA regulons: coordination of post-transcriptional events. *Nat. Rev. Genet.*, **8**, 533–543.

111 Saint-Georges, Y., Garcia, M., Delaveau, T., Jourdren, L. et al. (2008) Yeast mitochondrial biogenesis: a role for the PUF RNA-binding protein Puf3p in mRNA localization. *PLoS ONE*, **3**, e2293.

112 Png, K.J., Halberg, N., Yoshida, M., Tavazoie, S.F. (2012) A microRNA regulon that mediates endothelial recruitment and metastasis by cancer cells. *Nature*, **481**, 190–194.

113 Walrad, P.B., Capewell, P., Fenn, K., Matthews, K.R. (2011) The post-transcriptional trans-acting regulator, TbZFP3, co-ordinates transmission-stage enriched mRNAs in *Trypanosoma brucei*. *Nucleic Acids Res.*, **40** (7), 2869–2883.

114 Mittal, N., Scherrer, T., Gerber, A.P., Janga, S.C. (2011) Interplay between posttranscriptional and posttranslational interactions of RNA-binding proteins. *J. Mol. Biol.*, **409**, 466–479.

115 Lewis, A.P., Jopling, C.L. (2010) Regulation and biological function of the liver-specific miR-122. *Biochem. Soc. Trans.*, **38**, 1553.

116 Barker, A., Epis, M.R., Porter, C.J., Hopkins, B.R. et al. (2012) Sequence requirements for RNA binding by HuR and AUF1. *J. Biochem.*, **151**, 423–437.

117 Lal, A., Mazan-Mamczarz, K., Kawai, T., Yang, X. et al. (2004) Concurrent versus individual binding of HuR and AUF1 to common labile target mRNAs. *EMBO J.*, **23**, 3092–3102.

118 David, P.S., Tanveer, R., Port, J.D. (2007) FRET-detectable interactions between the ARE binding proteins, HuR and p37AUF1. *RNA*, **13**, 1453–1468.

119 Houseley, J., Tollervey, D. (2008) The nuclear RNA surveillance machinery: the link between ncRNAs and genome structure in budding yeast? *Biochim. Biophys. Acta*, **1779**, 239–246.

120 Nicholson, P., Yepiskoposyan, H., Metze, S., Zamudio Orozco, R. et al. (2010) Nonsense-mediated mRNA decay in human cells: mechanistic insights, functions beyond quality control and the double-life of NMD factors. *Cell. Mol. Life Sci.*, **67**, 677–700.

121 González, C.I., Wang, W., Peltz, S.W. (2001) Nonsense-mediated mRNA decay in *Saccharomyces cerevisiae*: a quality control mechanism that degrades transcripts harboring premature termination codons. *Cold Spring Harbor Symp. Quant. Biol.*, **66**, 321–328.

122 He, F., Jacobson, A. (2001) Upf1p, Nmd2p, and Upf3p regulate the decapping and exonucleolytic degradation of both nonsense-containing mRNAs and wild-type mRNAs. *Mol. Cell. Biol.*, **21**, 1515–1530.

123 Kerényi, Z., Mérai, Z., Hiripi, L., Benkovics, A. et al. (2008) Inter-kingdom conservation of mechanism of nonsense-mediated mRNA decay. *EMBO J.*, **27**, 1585–1595.

124 Maquat, L.E., Hwang, J., Sato, H., Tang, Y. (2010) CBP80-promoted mRNP rearrangements during the pioneer round of translation, nonsense-mediated mRNA decay, and thereafter. *Cold Spring Harbor Symp. Quant. Biol.*, **75**, 127–134.

125 Kashima, I., Yamashita, A., Izumi, N., Kataoka, N. et al. (2006) Binding of a novel SMG-1-Upf1-eRF1-eRF3 complex (SURF) to the exon junction complex triggers Upf1 phosphorylation and nonsense-mediated mRNA decay. *Genes Dev.*, **20**, 355–367.

126 Kobayashi, K., Kikuno, I., Kuroha, K., Saito, K. et al. (2010) Structural basis for mRNA surveillance by archaeal pelota and GTP-bound EF1α complex. *Proc. Natl Acad. Sci. USA*, **107**, 17575–17579.

127 Atkinson, G.C., Baldauf, S.L., Hauryliuk, V. (2008) Evolution of nonstop, no-go and nonsense-mediated mRNA decay and their termination factor-derived components. *BMC Evol. Biol.*, **8**, 290.

128 Schaeffer, D., van Hoof, A. (2011) Different nuclease requirements for exosome-mediated degradation of normal and nonstop mRNAs. *Proc. Natl Acad. Sci. USA*, **108**, 2366–2371.

129 Tsuboi, T., Kuroha, K., Kudo, K., Makino, S. et al. (2012) Dom34:hbs1 plays a general role in quality-control systems by dissociation of a stalled ribosome at the 3′ end of aberrant mRNA. *Mol. Cell*, **46**, 518–529.

130 Graille, M., Chaillet, M., van Tilbeurgh, H. (2008) Structure of yeast Dom34: a protein related to translation termination factor Erf1 and involved in No-Go decay. *J. Biol. Chem.*, **283**, 7145–7154.

13
Visualization of RNA and RNA Interactions in Cells

Natalia E. Broude
Boston University Department of Biomedical Engineering, 44 Cummington Mall Boston, MA 02215, USA

1	**Introduction** 418	
2	**Methods for Studying RNA and RNA Interactions** 420	
2.1	General Remarks 420	
3	**Fixed Cells** 420	
3.1	FISH 420	
3.2	Rolling Circle Amplification and Proximity Ligation Assay 422	
4	**Live Cells** 424	
4.1	General Remarks 424	
4.2	Molecular Beacons, 2′-O-Methyl Riboprobes, MTRIPs 424	
4.3	Labeling Methods Relying on Tagging RNAs with the Aptamers 427	
4.3.1	Introduction 427	
4.3.2	MS2 Method and Similar Methods Employing Full-Size FPs 427	
4.4	Spinach Method 428	
4.4.1	RNA Labeling Methods Based on Protein Complementation 429	
4.5	Detection of Endogenous Specific mRNAs in Live Cells 430	
5	**Conclusions** 431	
	Acknowledgments 431	
	References 431	

RNA Regulation: Advances in Molecular Biology and Medicine, First Edition. Edited by Robert A. Meyers.
© 2014 Wiley-VCH Verlag GmbH & Co. KGaA. Published 2014 by Wiley-VCH Verlag GmbH & Co. KGaA.

Keywords

Fluorescence *in-situ* hybridization (FISH)
A method employed to visualize nucleic acids in fixed cells using fluorescently labeled antisense oligonucleotides

Proximity ligation assay (PLA)
A method used to visualize protein–protein and RNA–protein interactions in fixed cells. To visualize RNA, the method includes four major components: reverse trauscription; padlock probes; ligation; and rolling circle amplification

Molecular beacons (MBs)
These are single-stranded RNA and DNA oligonucleotides capable of folding into stem–loop structures and labeled with a fluorophore and a quencher at the two ends. In the absence of a target, the MBs are in a stem–loop form and do not fluoresce due to quenching of the fluorophore by the quencher. In the presence of a target, the MBs hybridize to the target, the fluorophore and the quencher become spatially separated, and the MB becomes fluorescent

Aptamers
Single-stranded RNA and DNA oligonucleotides capable of folding into stem–loop structures that bind various ligands with high affinity

Protein complementation
A method used to study protein–protein and nucleic acid–protein interactions; it has been adapted to study RNA localizations in live cells.

The increasing diversity of biological pathways in which RNA molecules play a role requires different approaches to RNA studies. Among other methods, RNA studies on a single-cell level offer the unique possibility not only to measure transcription kinetics but also to obtain information on time- and space-determined RNA dynamics. The past decade has witnessed significant progress in the development of new methods allowing RNA studies on the level of single cells. These methods, and the results obtained, are discussed in this chapter.

1
Introduction

The very chemical nature of the RNA molecule determines its versatility and high propensity in forming different types of bonding – whether covalent, complementary, van der Vaals, salt, and/or combinations of the above. This chemical versatility of RNA led several of the most visionary scientists to suggest that the RNA molecule is the primordial molecule and to formulate the "RNA world" hypothesis [1–4]. However, in the general perception the functional importance of RNA has long been underestimated and restricted to the roles of messenger, transport, and structural molecules. Yet, the new studies that revealed the enzymatic and regulatory roles of small RNAs changed this perception only gradually, and it was the sequencing of large genomes and discovery of a new world of noncoding regulatory RNAs where proportions correlate with the complexity of the genome that led to RNA molecules taking center-stage [5]. Today, RNA is recognized as a pervasive regulatory molecule, the functional complexity of which greatly exceeds the functional landscape initially ascribed to these molecules by the central dogma (for recent reviews on this topic, see Refs [5, 6]).

Thus, RNA functional versatility is determined by its structure, and the sources of RNA functional complexity can be found on different levels. First, it is the primary structure of the eukaryotic mRNAs that is subject to a splicing mechanism that creates an enormous variety of transcripts necessary for both time- and space-determined gene expression. A second level of functional diversity is provided by the secondary and tertiary RNA structures. In this case, the RNA molecule might fold into alternative conformations that would result in different functionalities. Different conformers might depend on different temperatures or salt concentrations, on the presence of a small molecule or other molecular effectors. Examples of such a type of RNA functionality are represented by riboswitches, thermosensors, attenuators, and RNzymes (for a review, see Ref. [6]).

There are, of course, overlapping cases when very small variations in the primary structures (e.g., single-nucleotide polymorphisms; SNPs) are responsible for new RNA functional properties. An example of such variation is the μ-opioid receptor mRNA, which adopts different conformations depending on one SNP, which in turn results in different opiate responses [7].

Another level of functional complexity is presented by the multifunctional RNAs, which are capable of fulfilling different roles. In this case, the RNAs usually contain different modules [8]. In many cases the RNAs have short open reading frames (ORFs) that can be translated as short peptides. The peptide can, in turn, either play a role independent of RNA or its function can be related to the function of the maternal RNA. One such example is the bacterial SgrS RNA, a small bacterial regulatory RNA that is expressed under phospho-sugar stress and forms a complementary bonding with the *ptsG* mRNA, which encodes the major phospho-sugar transporter. This hybridization inhibits the PtsG translation that helps to overcome phospho-sugar stress [9–12]. Recently, it has been discovered that SgrS RNA contains a short ORF encoding a small peptide SgrT that is also involved in the inhibition of phospho-sugar stress [12, 13].

Other examples of multifunctional RNAs include an RNA that is a messenger and a regulatory molecule (bacterial tmRNA), or ribosomal 23S RNA that serves as a backbone for the assembly of ribosomal proteins while at the same time being a ribozyme capable of catalytic peptidyl transferase activity. It is also assumed that many eukaryotic regulatory RNAs play structural roles in maintaining the cellular cytoskeletal structure [8, 14–16].

2
Methods for Studying RNA and RNA Interactions

2.1
General Remarks

The same properties of RNA molecules that make them ubiquitous regulators bestow a challenge to researchers involved in RNA studies. Indeed, RNA is the most flexible, mobile, and transient molecule in the cell, it has multiple protein partners that change continuously with the progression of the cell cycle and with RNA movements and localizations. These RNA–protein interactions – but mostly the secondary and tertiary structures of RNA molecules – make finding accessible sites for efficient hybridization probes a difficult task. Moreover, the short lifetime of RNAs also determines the need to consider kinetic factors: the interaction of the probe with RNA should be rapid and reversible in order to reflect on the RNA dynamics [6].

Currently, the methods aimed at studying RNA and RNA–protein interactions are keeping pace with the discovery of new RNA functions, and are being developed at an amazing rate. RNA studies (and the corresponding methods) can be categorized by applying different principles, and include: (i) studies *in vitro* versus studies *in vivo*; (ii) studies in bulk versus studies on a single-cell level; and (iii) studies in fixed cells versus studies in live cells. In order to make this chapter more manageable, the discussions here will be restricted to RNA and RNA–protein studies conducted at the level of single cells, and a further distinction will be made between studies in fixed versus live cells. Today, multiple excellent reviews are available that consider the different aspects of these RNA studies [17–21].

Single-cell RNA studies are complementary to studies of large cell populations, or to *in vitro* studies on the biochemical level. Indeed, such studies provide unique information that cannot be obtained by other approaches. The analysis of single cells allows the fine resolution of spatial and/or temporal changes in RNA dynamics, in addition to measuring rates of transcription. Subsequently, cell-to-cell variability (transcriptional noise) may offer insights into the mechanisms of gene expression regulation. Finally, the ultimate goal of performing RNA studies to achieve single-molecule resolution and sensitivity may also become possible, by using single cells in conjunction with highly sensitive probes.

3
Fixed Cells

3.1
FISH

Fluorescence *in-situ* hybridization (FISH) is a technique used for the detection nucleic acids in fixed cells by employing hybridization with complementary probes (for recent reviews, see Refs [18, 22, 23]). The two major challenges with RNA FISH involve finding accessible sites in a folded and protein-covered RNA within the ribonucleoprotein (RNP) complexes, and obtaining a signal that exceeds background, given the low RNA concentrations. In one of the first studies aimed at the visualization of individual mRNA molecules, the authors targeted a relatively abundant β-actin mRNA and used five 50 nt labeled probes, each containing five labels that would

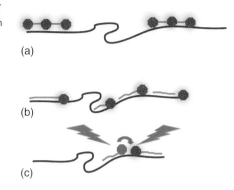

Fig. 1 Schematic representation of different probe designs employed in FISH. (a) Several multiply labeled probes [24, 25]; (b) Multiple singly labeled probes [27–30]; (c) Pairs of fluorescent probes capable of Forster resonance energy transfer (FRET) [31].

hybridize to adjacent sequences within the transcript [24] (Fig. 1a). The authors reported single-molecule detection, and verified this by two-color labeling using differently labeled probes that hybridized the same molecules. The results of the study also demonstrated a simultaneous detection of different transcripts (β- and γ-actins) with two different labels. Finally, the authors were able to capture the kinetics of β-actin mRNA transcription by using probes which hybridized mRNA at sites that were at different distances from the transcription start [24]. This method was subsequently used to label 11 different transcripts in the nuclei of the in vitro-cultured cells [25]. In another study, RNA FISH with multiply labeled probes was used to study the patterning of six different transcripts in Drosophila embryos [26]. In this case, it was possible to genotype homozygous and heterozygous transcripts and to resolve RNAs separated by only 20 kb [26].

An alternative FISH method is based on using a large number of singly labeled probes [27–30] (Fig. 1b). In one application of this approach, a total of 48 Alexa 594-labeled probes was targeted to the coding region of mRNA encoding enhanced green fluorescent protein (EGFP). Additionally, the 3′ untranslated region (UTR) contained 32 to 80 nt-long repeats that were targeted with four oligonucleotides labeled with tetramethylrhodamine [29]. By using this approach, the authors were able to visualize three different mRNAs in a variety of cells. Ultimately, this approach was applied in several studies that included RNA labeling in both bacterial [32] and mammalian cells [33]. Montero Llopis and coworkers used FISH with 38 Cy3-labeled probes to study the localization of bacterial mRNAs in Caulobacter crescentus cells, and found mRNAs to be localized at focal points close to their transcription sites [32]. Later, Shih et al. applied FISH to mammalian cells to study the localization of the reporter luc2 mRNA subjected to microRNA (miRNA) translational repression. By using FISH with 65 labeled probes against luc2 mRNA combined with the live cell protein labeling, it was possible to visualize single mRNAs and to colocalize these with the miRNA-associated proteins, Argonaute2, Dcp1a, hedls, and Rck [33].

In some examples, RNA FISH has been successful when using a single RNA probe containing multiple labels [34], or with a single probe directed towards multiple repeats introduced into the RNA transcript [32]. If RNA is abundant enough, then a single probe with an increased affinity to RNA such as locked nucleic acid (LNA) can be successful [32]. Thus, Taniguchi et al.

reported a single-molecule sensitivity detection when analyzing bacterial mRNAs using FISH with a 20 nt-long probe labeled with Atto594 [35], while Russell and Keiler localized a small regulatory tmRNA in *C. crescentus* cells using FISH with a single 31 nt-long oligonucleotide labeled with Cy3 [36].

Several approaches have been described in attempts to increase FISH efficiency. For example, a combination of the Cy3- and Cy5-labeled peptide nucleic acid (PNA) probes that enabled Forster resonance energy transfer (FRET) was used to analyze mRNA splice variants [31] (Fig. 1c). Chemically modified oligonucleotides that included locked, morpholino-, negatively charged PNA, caged oligos, and multiply labeled tetravalent RNA imaging probes (MTRIPs) [37, 38] were used in both fixed and live cells. Comprehensive reviews on the use of chemically modified oligonucleotides for FISH can be found in recent publications [19, 21].

It would seem that the most robust – but not the cheapest – approach to visualizing single molecules of unmodified mRNAs when using FISH is to employ about 50 fluorescent labels per mRNA molecule, which would require the design of about 50 singly labeled complementary probes covering the target.

3.2
Rolling Circle Amplification and Proximity Ligation Assay

Rolling circle amplification (RCA) is an isothermal DNA amplification method based on the ability of DNA polymerases to amplify single-stranded DNA circles, so as to yield products containing multiple repeats of the circular sequence [39]. A combination of RCA with antibody-coupled signal detection (immuno-RCA) has been applied to detect β-actin mRNA *in situ* in rat liver T6 stellate cells [40]. Later, a similar approach was combined with the proximity ligation assay (PLA) for sensitive RNA detection *in situ*.

The PLA method was developed by Ulf Landegren and colleagues for conducting *in situ* gene, transcript and protein studies (for reviews, see Refs [41–43]). The principle of the method is based on the question of whether two molecules are localized in close proximity or, in other words, whether they interact with each other. In order to measure the proximity (interaction) of the two proteins, they are first tagged with the two oligonucleotides. If the two proteins are close to each other, the two tags will hybridize head-to-tail to a specially designed linear padlock probe that is complementary to the oligonucleotide tags. If complementarity is perfect, the padlock probe can then be ligated into a circle by applying DNA ligase, an enzyme that is extremely sensitive to mismatches and will ligate only perfectly double-stranded nicked DNA. Thus, the proximity of the two protein molecules can be transformed into the DNA circle that is amplified using a rolling circle mechanism. Typically, PLA involves three major steps: (i) the padlock probe hybridization; (ii) circularization with the ligase; and (iii) the RCA [41, 42]. In order to analyze the RNA transcripts, an additional step has been shown necessary; for this, RNA should be reverse-transcribed into cDNA [44], which is then hybridized with a padlock probe that is complementary to the target and then ligated into a circle. The circle is then amplified by RCA using Φ29 DNA polymerase (Fig. 2a). The RCA product is a single-stranded DNA polymer collapsing into a random coil that is detected by hybridization with a

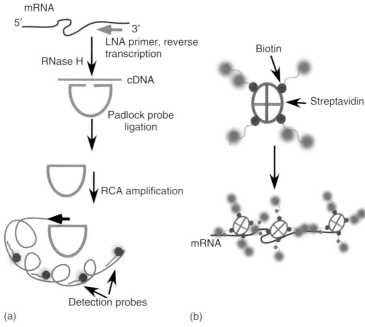

Fig. 2 Schematic representation of proximity ligation assay (PLA) for RNA labeling and MTRIPs probes. (a) Fragment of mRNA is hybridized with the LNA primer and reverse-transcribed into cDNA. cDNA is circularized with the help of the padlock probe and ligase, and amplified by the rolling circle amplification (RCA) mechanism. The product is detected by hybridization with the labeled probes binding the RCA product [44]; (b) MTRIPs are composed of the 2′-O-ethylRNA/DNA chimera with amino-modified thymidines for conjugation with fluorophore and biotinylation for binding to streptavidin [37].

fluorescent oligonucleotide. This method has been used successfully to detect β-actin mRNA in human cells and tissues, to differentiate between α-, β-, and γ-actins, and to perform multiplex detection of the transcripts in human and mouse cell lines and tissues [44].

A further development of PLA has recently been reported in which the authors visualized simultaneously the proteins of the signaling cascade (platelet-derived growth factor receptor, beta subunit; PDGFRb) and mRNA (DUSP6 mRNA) synthesized as a result of the signal [45].

In another study, PLA has been combined with the peptide-modified MTRIPs to study the interaction between a viral RNA (human respiratory syncytial virus) and a nucleocapsid protein N [46]. MTRIPs are composed of the 2′-O-methylRNA/DNA chimera with amino-modified thymidines for conjugation with fluorophore and biotinylation for binding to streptavidin (Fig. 2b). Ultimately, the probe is about 200–300 kDa and two to three such probes would allow single-molecule detection [38]. When combined with PLA this method was used to study the effect of actinimycin D on the interactions of the β-actin mRNA and polyA+ mRNA with the RNA-binding protein HuR [46]. Hence, this method allows the simultaneous visualization of RNA and its protein partners, with single-molecule resolution.

The strength of PLA is based on the extremely high specificity of the ligation reaction, which forms the major discriminatory step in the entire procedure. Thus, PLA represents an alternative method to FISH that is especially suited to analyzing highly similar sequences such as splice or allelic variants. At the same time, the design of PLA is much more complex than that of FISH, and the multiple steps and enzymatic reactions involved when performing PLA require certain skills and experience. In addition, the efficiency of the protein and RNA detection is relatively low [44, 45].

4
Live Cells

4.1
General Remarks

Studies of RNA in live cells represent an additional challenge as compared to studies in fixed cells. Indeed, issues such as cell viability, RNA stability, rate of signal development and related questions should all be taken into account. At the same time, live-cell studies offer the possibility of studying the movements and dynamics of RNA and RNA–protein complexes in real time, as opposed to the snapshot images obtained when using FISH or by other methods employing fixed cells.

The choice of methods for live-cell RNA imaging depends mostly on the type of cell, with the main difference being between large and small cells. In the case of large cells, the probes can either be delivered to the cell or synthesized by the cell. In case of small cells, active delivery is impossible and the remaining options are either to label the RNA with small molecules that can easily penetrate the cell membrane, or to use probes synthesized by the cell itself.

4.2
Molecular Beacons, 2′-O-Methyl Riboprobes, MTRIPs

There are several major challenges that should be overcome when using oligonucleotide probes for RNA detection in live cells. Notably, it is necessary to: (i) identify accessible sites in the target; (ii) deliver the probe into the cell and into the correct compartment; and (iii) protect the probe against nucleases. As each of these challenges can have a direct effect on the signal-to-background ratio, the achievement of a high signal over background requires the efficient solution of the above-mentioned problems.

One meticulous study aimed at identifying accessible sites in mRNAs for oligonucleotide hybridization resulted in the conclusion that there is no correlation between the sites accessible to antisense oligonucleotides and to siRNAs [47]. Most likely, this reflects differences between the two mechanisms of complex formation: pure RNA/DNA hybridization versus RNP complex formation with the participation of proteins in case of RNA interference. The results of this study confirmed the earlier conclusion that the 5′- and 3′-ends of mRNAs are more accessible to hybridization than are the internal sequences [47].

As noted above, oligonucleotide probes require delivery into the cell, and this can be achieved by microinjection, transfection with cationic lipids, or by using electroporation or membrane permeabilization with streptolysin-O. Delivery with the assistance of membrane-permeating peptides has been also described (for

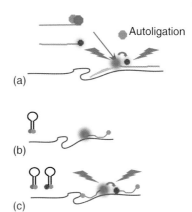

Fig. 3 Schematic representation of autoligation probes and molecular beacons. (a) Two oligonucleotides are hybridized to RNA head-to-tail. One oligonucleotide has a fluorophore and quencher. Hybridization of the two oligonucleotides triggers autoligation that displaces the quencher. The remaining fluorophore becomes fluorescent [52, 53]; (b) Molecular beacons are dim in a closed stem–loop form, but become fluorescent upon hybridization to the target that because of the opening of the stem–loop structure [54]; (c) Two MBs with different fluorophores can be designed to hybridize to the two adjacent sites on RNA that enables FRET [55, 56].

reviews of these delivery methods, see Refs [18, 19]).

Another difficulty when using oligonucleotide probes is their degradation by nucleases, which reduces the efficiency of labeling the RNA and results in an increased background. Oligonucleotide stability can be improved by employing chemically modified oligonucleotides that are not digested by nucleases (for details, see Ref. [21]). Some modifications not only increase the stability of the probes but also increase the affinity binding to RNA, as is the case with the 2'-O-methyl riboprobes [21, 22]. One further complication with negatively charged oligonucleotides is their accumulation in the nucleus, and to overcome this problem oligonucleotides have been conjugated with streptavidin, quantum dots, or tRNA. This makes them bulkier and results in a reduced accumulation in the nucleus [18, 48–51].

In order to increase the signal-to-background ratio, many studies have employed hybridization probes that change their fluorescent properties upon binding to RNA. These probes include molecular beacons (MBs), as well as linear probes and beacons that form FRET pairs (Fig. 3).

One such approach involved the use of linear oligonucleotides that were capable of "quenched autoligation" [52, 53]. Here, one of the oligonucleotides contained both a fluorophore and a quencher; this oligonucleotide was also capable of autoligation with the other oligonucleotide if it was hybridized to target mRNA in close proximity. As a result of the ligation the quencher separated from the fluorophore, after which FRET took place between the donor and acceptor oligonucleotides (Fig. 3a) [52, 53].

Typically, MBs are stem–loop oligonucleotides that bear a fluorophore and a quencher at the ends. In the closed stem–loop form they do not fluoresce because of the contact-mediated quenching,

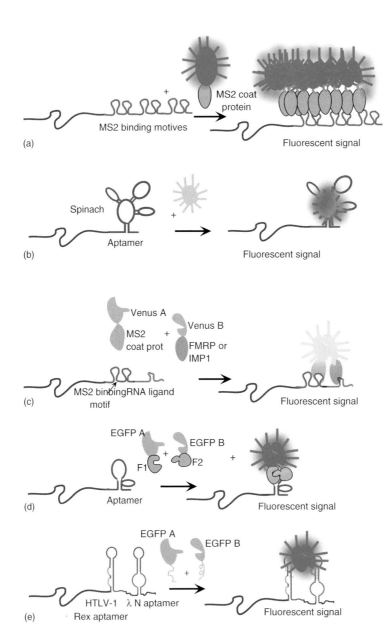

but upon hybridization to their target RNAs the MBs unwind and become fluorescent [18, 54]. Recently, MBs have become the most widely used oligonucleotide probes for mRNA studies in live cells [18, 54, 57]. Their performance has been improved by using not only nuclease-resistant backbones but also pairs of MB hybridizing side-by-side on mRNA and forming a FRET pair [55, 56]. The sensitivity of MBs is relatively low, however, and to achieve single-molecule detection level the target mRNA must be modified by tagging with multiple repeats of the sequences binding the MBs [58]. Recently introduced MTRIPs (as mentioned above) have been employed as probes for RNA detection in fixed cells; these may also be applied to live-cell imaging with single-molecule sensitivity [37, 46].

4.3
Labeling Methods Relying on Tagging RNAs with the Aptamers

4.3.1 Introduction

All of the methods discussed above require the probe to be delivered into the cell, to a specific compartment. Consequently, much effort has been invested in developing methods that rely on the cells synthesizing the probes by themselves. A vast majority of current methods in which the probes are synthesized intracellularly employ RNA aptamer–protein interactions.

4.3.2 MS2 Method and Similar Methods Employing Full-Size FPs

The first and the most widely used method for studying RNA localization and movements in live cells is based on the natural high-affinity binding of the MS2 phage coat protein with a short RNA sequence (aptamer) found in the phage RNA. In order to label the target RNA, it should first be tagged with an array of these RNA sequences binding coat protein. The MS2 coat protein (hereafter referred to as *MS2*) should be expressed as a fusion with the intact fluorescent protein (FP). Each of the aptamers can bind two copies of the fusion protein. When the tagged RNA and fusion proteins are coexpressed in the cell, the fusion proteins will bind cognate aptamers and so render the target RNA fluorescent

Fig. 4 Schematics of the methods based on using RNA aptamers. (a) MS2 method [59, 60]. Target RNA is tagged with an array of aptamers binding MS2 coat protein with high affinity. MS2 coat protein is expressed in the cell as a fusion with a fluorescent protein. When the fusion protein binds cognate aptamers, RNA becomes fluorescent; (b) A Spinach approach [75, 76]. RNA is tagged with an aptamer that binds a low-molecular-weight compound that easily penetrates the cell and becomes fluorescent upon RNA binding; (c) Protein complementation applied to a study of RNA–protein interactions [77]. MS2-binding motives are introduced into the target RNA. Fluorescent protein (FP) is split into two inactive fragments. One fragment is fused with the MS2 coat protein; another fragment is fused with the protein for which interaction with RNA is in question. If this protein binds RNA, two fragments of the split FP will come together, re-associate, and become fluorescent; (d) Fluorescent protein complementation applied to RNA detection and localization [78, 79]. An RNA-binding protein eIF4A is split into two fragments and fused with the fragments of the split FP. RNA is tagged with an aptamer raised against the eIF4A protein. In the presence of RNA protein, complementation takes place and the RNA becomes fluorescently labeled; (e) Same principle as in (c), but two different aptamers binding two viral peptides drive protein complementation [80].

(Fig. 4a). The MS2 method was originally developed to study the localization of *Ash1* mRNA in yeast [59, 60], but later applied to kinetic and localization studies of different mRNAs in a variety of cells [61–69], including bacteria [32, 34, 70–73]. As the MS2 method employs an intact FP, it inherently displays a strong background fluorescence. In eukaryotic cells, this problem can be partly mitigated by directing the bulk of the FP to compartments other than the expected site of RNA localization. For example, if nuclear RNA is studied, the fluorescent fusion protein can be directed to the cytoplasm, or *vice versa* [74]. Unfortunately, this approach is not applicable in bacteria, where alternative modifications of the method have been used to perform RNA kinetics and localization studies [32, 34, 70–73].

The MS2 coat protein is not the only RNA-binding protein used to label RNA with FPs. An alternative system uses the RNA-binding protein U1Ap and the corresponding RNA aptamer to tag target RNA [81–83]. The lambda phage N peptide has also been used to label RNA in live cells [84]. By using a combination of the MS2 protein and the lambda N peptide fused with different FPs, Lange and coworkers were able to label two different RNAs simultaneously and to reveal the details of RNA movement in yeast [85].

Recently, a modification of the MS2 method has been described in an attempt to increase the signal-to-background ratio [86]. In the first approach, a single-chain tandem repeat of the MS2 coat protein was used that showed a better signal-to-background ratio than a monomer of the MS2 coat protein. In a second approach, the PP7 coat protein was used together with the corresponding aptamer, and this also displayed a better signal-to-background ratio than the original MS2 method. By using these improved labeling methods, the authors were able to measure the diffusion of single mRNA molecules in the nucleus and cytoplasm [86].

4.4
Spinach Method

A high fluorescent background is typical of methods employing full-size FPs, and serves as a substantial "roadblock" for quantitative RNA studies; hence, different approaches have been explored over the years to mitigate this problem. One currently emerging approach for labeling aptamer-tagged RNAs with a higher sensitivity is based on new non-toxic dyes that have a low molecular weight, can penetrate cell membranes, and change fluorescence upon binding to RNA aptamers [87–89]. Recently, an alternative method has been developed that employed low-molecular-weight chemical compounds reminiscent of the chromophore in FPs [75]. These authors synthesized several imidazolinones (aromatic, nonfluorescent heterocycles) and then used SELEX (Systematic Evolution of Ligands by Exponential Enrichment) to isolate RNA aptamers that would bind these compounds with a high affinity and render them fluorescent. Several chosen chromophore/aptamer pairs displayed fluorescence upon binding; of these, the most promising combination was dubbed "Spinach" because of its green fluorescence [75, 76] (Fig. 4b).

Low-molecular-weight chromophores that become fluorescent upon RNA binding represent a new and promising approach to labeling RNA in live cells. Moreover, they offer several advantages over other methods, in that they are practically dim without aptamers, which yields a high signal-to-background ratio. In addition, they can easily penetrate the cell membrane and be used to study both prokaryotic and eukaryotic cells. The labeling complex is relatively small, as the protein fusion commonly used to label RNA is substituted by a low-molecular-weight aromatic heterocycle. Another benefit is that the chromophore is pre-synthesized, which implies that the fluorescent signal will develop much faster on binding to RNA than the time required for the chromophore to oxidize and mature in most FPs. The Spinach approach does have several limitations, however. Notably, it brightness on binding to RNA is only 80% that of EGFP, and its K_d is much higher (500 nM) than for other aptamer/ligand pairs (when it is often in the low nanomolar range). Hence, further studies must be conducted before this approach will be capable of detecting low concentrations of RNAs in live cells.

4.4.1 RNA Labeling Methods Based on Protein Complementation

Another approach towards reducing the fluorescent background of FPs is based on the use of protein complementation (PC). Here, a FP is split into two fragments that are fluorescently inactive; however, when the inactive fragments bind to two ligands that interact with each other, the FP fragments are re-associated and this results in the development of fluorescence [76, 90–93].

Originally, PC was developed to study protein–protein interactions [90–94], but later adapted by Rackham and Brown to study RNA–protein interactions in mammalian cells [77] (Fig. 4c). These authors tagged mRNAs with a cassette of MS2-aptamers, while the MS2 coat protein was fused to one fragment of the split yellow FP, Venus. The complementing fragment of Venus was fused to an RNA-binding candidate protein whose RNA-binding properties were in question (e.g., Fragile X Mental Retardation Protein; FMRP). If the protein candidate does bind mRNA, then the two fragments of Venus will be brought into close proximity so they can re-associate and form a FP (Fig. 4c). Subsequently, the authors showed that isoform 18 of the FMRP protein and the human zip-code transport protein IMP1 (insulin-like growth factor II mRNA-binding protein) would interact with common mRNAs to form granules in the cytoplasm. These findings suggested a link between mRNA transport and translational repression [77]. This method was used later to visualize interactions between the viral HIV-1 Gag protein and cellular protein Staufen1 [95].

The PC technique was later adapted for the direct labeling of different RNAs in bacterial [78, 79] and mammalian [96] cells. In both cases, different RNA-binding proteins were used as connectors between the RNA and the split fluorescent reporter.

In order to label RNA in bacteria, Valencia-Burton et al. used eukaryotic initiation factor 4A (eIF4A) as an RNA-binding protein [78, 79]. For this, the protein eIF4A was split into two fragments that corresponded to the two protein domains, with each domain being fused to a fragment of the split EGFP. In the presence of RNA tagged with the eIF4A-specific aptamer, the two fragments of the eIF4A protein bound the aptamer and brought together the fragments of the split EGFP, which re-assembled and rendered RNA fluorescent (Fig. 4d). By using this method the authors were able to visualize *LacZ* mRNA and 5S ribosomal RNA in live *Escherichia coli* cells [78].

Recently, an alternative method for RNA visualization based on PC has been described [80] that is based on the interactions between two aptamers and two viral peptides that trigger complementation of the split FP (Fig. 4e). The same principle has been applied creatively by Silver's group to build bacterial cells with RNA-nanostructures overproducing hydrogen, H_2 [97].

In all studies using PC, spurious protein–protein interactions may take place which contribute to the background, although this background will be substantially lower than that observed when using intact (i.e., non-split) FPs. Nevertheless, appropriate controls to verify the dependence of the signal on target RNA are necessary to ensure the absence of any artifacts [80].

These RNA labeling methods allow live bacterial cell imaging with the possibility of measuring RNA in real time, to study subcellular RNA localization, and to reveal transcriptional variabilities between cells. Yet, these methods also have their limitations, mostly because they require RNA tagging with an aptamer. A recent study from Vogel's laboratory showed that aptamer-tagging can affect the expression level and the functional activity of small non-coding RNAs [98]. Additionally, RNAs in complexes with RNA-binding proteins can display a higher stability (dubbed "immortalization") than unmodified RNAs, which might compromise kinetic studies [71]. This limitation can be overcome in some cases, however. For example, it was shown recently that an assembly-defective MS2 coat protein, MS2 dIFG, does not "immortalize" RNA, in contrast to the MS2 dimer protein, MSd [32].

4.5
Detection of Endogenous Specific mRNAs in Live Cells

As noted above, one important limitation of all methods using RNA tagged with aptamers is that they require a genetic modification of the RNA to introduce extra-sequences with strong secondary structures. Unfortunately, this modification can affect RNA recognition by the cellular machinery, and may also result in unexpected adverse effects on the cell as a whole. In practical applications, the need to modify the RNA limits the value of these methods in both diagnostic and therapeutic studies.

Therefore, the labeling of endogenous non-engineered RNAs with probes synthesized by the cell itself remains a major challenge, and a universal method for labeling unmodified RNAs would be highly desirable.

The only noninvasive method for labeling eudogeuous RNA described to date has involved the use of a pumilio homology domain of human pumilio 1 protein (PUM-HD) [96]. PUM-HD protein recognizes a 8 nt sequence within RNA [99, 100], and therefore many RNAs contain sequences similar to PUM-HD recognition sites. In one variant of this method, two PUM-HD proteins were mutagenized to directly recognize two specific 8 nt-long sites in the mitochondrial mRNA encoding NADH dehydrogenase subunit 6 (ND6). These mutants were then fused with the fragments of a split FP, after which PC was triggered by the presence of the ND6 mRNA [96] or the β-actin mRNA in NIH3T3 cells [101]; in the latter case, the authors achieved single- molecule detection level [97]. Recently, another variant of the method has been described that used a protein probe consisting of a full-size

EGFP flanked by the two PUM-HD proteins that recognized and bound two neighboring sites in the β-globin mRNA [102]. The advantage of this variant was that only one protein construct had to be expressed in the cell, although expression of the full-size FP undoubtedly led to an increased fluorescence background. Unfortunately, the most restrictive part of this approach was the need to mutagenize the PUM-HD proteins to increase protein affinity for every RNA target.

5
Conclusions

The past 10–15 years have witnessed great progress in improvements of the "old" methods and development of "new" methods to study RNAs in the cellular environment. Yet, this area of research is a moving target and each month brings new ideas and new data. The results of these studies are equally impressive. Studies on a single-cell level have led to an understanding of the role of cell variability that takes place both during transcription and translation [27, 35]; they have also resulted in the discovery and studies of transcription bursting [27, 66, 69, 71, 103]; and they have made possible the imaging and measurement of RNA movements in the cell [28, 32, 38, 47–49, 53, 58–60, 86, 92]. In microbiology, these methods have revealed that some RNAs might be localized [32, 34–36, 78], and they have allowed the visualization of single mRNA molecules [29, 30, 35, 37, 44, 64, 86] and their complexes with different protein partners [24, 25, 33, 77, 94]. Recently, the first studies on a single-cell level addressed localization and protein interactions of the long noncoding RNAs [104]. Yet, these are just a few examples illustrating the great progress in RNA studies on the level of single cells; there is no doubt that many more insights into RNA biology will follow in the near future.

Acknowledgments

The author thanks Maxim Frank-Kamenetskii and the reviewers for their valuable comments on the chapter manuscript, and also apologizes to everybody whose work has not been mentioned in this chapter. These studies have been supported in part by NSF grant 0244045099 to NEB.

References

1 Woese, C.R. (1967) *The Genetic Code: The Molecular Basis for Genetic Expression* Harper & Row, New York, p. 186.
2 Orgel, L.E. (1968) Evolution of the genetic apparatus. *J. Mol. Biol.*, **38**, 381–393.
3 Crick, F.H. (1968) The origin of the genetic code. *J. Mol. Biol.*, **38**, 367–379.
4 Gilbert, W. (1986) Origin of life – the RNA world. *Nature*, **319**, 618.
5 Mattick, J.S. (2011) The double life of RNA. *Biochimie*, **93**, viii–viix.
6 Dethoff, E.A., Chugh, J., Mustoe, A.M., Al-Hashimi, H.M. (2012) Functional complexity and regulation through RNA dynamics. *Nature*, **482**, 322–330.
7 Shabalina, S.A., Zaykin, D.V., Gris, P., Ogurtsov, A.Y. et al. (2009) Expansion of the human μ-opioid receptor gene architecture: novel functional variants. *Hum. Mol. Genet.*, **18**, 1037–1051.
8 Guttman, M., Rinn, J.L. (2012) Modular regulatory principles of large non-coding RNAs. *Nature*, **482**, 339–346.
9 Vanderpool, C.K., Gottesman, S. (2004) Involvement of a novel transcriptional activator and small RNA in post-transcriptional regulation of the glucose phosphoenolpyruvate phosphotransferase system. *Mol. Microbiol.*, **54**, 1076–1089.
10 Kawamoto, H., Morita, T., Shimizu, A., Inada, T. et al. (2005) Implication of

membrane localization of target mRNA in the action of a small RNA: mechanism of post-transcriptional regulation of glucose transporter in *Escherichia coli*. *Genes Dev.*, **19**, 328–338.

11 Kawamoto, H., Koide, Y., Morita, T., Aiba, H. (2006) Base-pairing requirement for RNA silencing by a bacterial small RNA and acceleration of duplex formation by Hfq. *Mol. Microbiol.*, **61**, 1013–1022.

12 Vanderpool, C.K., Balasubramanian, D., Lloyd, C.R. (2011) Dual-function RNA regulators in bacteria. *Biochimie*, **93**, 1943–1949.

13 Wadler, C.S., Vanderpool, C.K. (2007) A dual function for a bacterial small RNA: SgrS performs base pairing-dependent regulation and encodes a functional polypeptide. *Proc. Natl Acad. Sci. USA*, **104**, 20454–20459.

14 Kloc, M., Foreman, V., Reddy, S.A. (2011) Binary function of mRNA. *Biochimie*, **93**, 1955–1961.

15 Ulveling, D., Francastel, C., Hubé, F. (2011) When one is better than two: RNA with dual functions. *Biochimie*, **93**, 633–644.

16 Dinger, M.E., Gascoigne, D.K., Mattick, J.S. (2011) The evolution of RNAs with multiple functions. *Biochimie*, **93**, 2013–2018.

17 Rodriguez, A.J., Condeelis, J., Singer, R.H., Dichtenberg, J.B. (2007) Imaging mRNA movement from transcription sites to translation sites. *Sem. Cell Dev. Biol.*, **18**, 202–208.

18 Tyagi, S. (2009) Imaging intracellular RNA distribution and dynamics in living cells. *Nat. Methods*, **6**, 331–338.

19 Santangelo, P.J., Alonas, E., Jung, J., Lifland, A.W. et al. (2012) Probes for intracellular RNA imaging in live cells. *Methods Enzymol.*, **505**, 383–399.

20 Itzkovitz, S., van Oudenaarden, A. (2011) Validating transcripts with probes and imaging technology. *Nat. Methods*, **8** (Suppl. 4), S12–S19.

21 Shestopalov, I.A., Chen, J.K. (2010) Oligonucleotide-based tools for studying zebrafish development. *Zebrafish*, **7**, 31–40.

22 Dirks, R.W., Tanke, H.J. (2006) Advances in fluorescent tracking of nucleic acids in living cells. *Biotechniques*, **40**, 489–496.

23 Armitage, B.A. (2011) Imaging of RNA in live cells. *Curr. Opin. Chem. Biol.*, **15**, 806–812.

24 Femino, A.M., Fay, F.S., Fogarty, K., Singer, R.H. (1998) Visualization of single RNA transcripts in situ. *Science*, **280**, 585–590.

25 Levsky, J.M., Shenoy, S.M., Pezo, R.C., Singer, R.H. (2002) Single-cell gene expression profiling. *Science*, **297**, 836–840.

26 Kosman, D., Mizutani, C.M., Lemons, D., Cox, W.G. et al. (2004) Multiplex detection of RNA expression in *Drosophila* embryos. *Science*, **305**, 846.

27 Raj, A., Peskin, C.S., Tranchina, D., Vargas, D.Y. et al. (2006) Stochastic mRNA synthesis in mammalian cells. *PLoS Biol.*, **4**, e309.

28 Raj, A., van den Bogaard, P., Rifkin, S.A., van Oudenaarden. A. et al. (2008) Imaging individual mRNA molecules using multiple singly labeled probes. *Nat. Methods*, **5**, 877–879.

29 Batish, M., Raj, A., Tyagi, S. (2011) Single molecule imaging of RNA in situ. *Methods Mol. Biol.*, **714**, 3–13.

30 Jakt, L.M., Moriwaki, S., Nishikawa, S. (2013) A continuum of transcriptional identities visualized by combinatorial fluorescent in situ hybridization. *Development*, **140** (1), 216–225.

31 Blanco, A.M., Rausell, L., Aguado, B., Perez-Alonso, M. et al. (2009) A FRET-based assay for characterization of alternative splicing events using peptide nucleic acid fluorescence in situ hybridization. *Nucleic Acids Res.*, **37**, e116.

32 Montero Llopis, P.M., Jackson, A.F., Sliusarenko, O., Surovtsev, I. et al. (2010) Spatial organization of the flow of genetic information in bacteria. *Nature*, **466**, 77–81.

33 Shih, J.D., Waks, Z., Kedersha, N., Silver, P. (2011) Visualization of single mRNA reveals temporal association of proteins with microRNA-regulated mRNA. *Nucleic Acids Res.*, **39**, 7740–7749.

34 Nevo-Dinur, K., Nussbaum-Shochat, A., Ben-Yehuda, S., Amster-Choder, O. (2011) Translation-independent localization of mRNA in *E. coli*. *Science*, **331**, 1081–1084.

35 Taniguchi, Y., Choi, P.J., Li, G.W., Chen, H. et al. (2010) Quantifying *E. coli* proteome and transcriptome with single-molecule

sensitivity in single cells. *Science*, **329**, 533–538.

36 Russell, J.H., Keiler, K.C. (2009) Subcellular localization of a bacterial regulatory RNA. *Proc. Natl Acad. Sci. USA*, **106**, 16405–16409.

37 Santangelo, P.J., Lifland, A.W., Curt, P., Sasaki, Y. *et al.* (2009) Single molecule-sensitive probes for imaging RNA in live cells. *Nat. Methods*, **6**, 347–349.

38 Lifland, A,W,, Zurla, C., Yu, J., Santangelo, P.J. (2011) Dynamics of native β-actin mRNA transport in the cytoplasm. *Traffic*, **12**, 1000–1011.

39 Fire, A., Xu, S.Q. (1995) Rolling replication of short DNA circles. *Proc. Natl Acad. Sci. USA*, **92**, 4641–4645.

40 Zhou, Y., Calciano, M., Hamann, S., Leamon, J.H. *et al.* (2001) In situ detection of messenger RNA using digoxigenin-labeled oligonucleotides and rolling circle amplification. *Exp. Mol. Pathol.*, **70**, 281–288.

41 Landegren, U., Dahl, F., Nilsson, M., Fredriksson, S. *et al.* (2003) Padlock and proximity probes for in situ and array-based analyses: tools for the post-genomic era. *Comp. Funct. Genomics*, **4**, 525–530.

42 Landegren, U., Nilsson, M., Gullberg, M., Söderberg, O. *et al.* (2004) Prospects for *in situ* analyses of individual and complexes of DNA, RNA, and protein molecules with padlock and proximity probes. *Methods Cell Biol.*, **75**, 787–797.

43 Nilsson, M., Malmgren, H., Samiotaki, M., Kwiatkowski, M. *et al.* (1994) Padlock probes: circularizing oligonucleotides for localized DNA detection. *Science*, **265**, 2085–2088.

44 Larsson, C., Grundberg, I., Söderberg, O., Nilsson, M. (2010) *In situ* detection and genotyping of individual mRNA molecules. *Nat. Methods*, **7**, 395–397.

45 Weibrecht, I., Grundberg, I., Nilsson, M., Söderberg. O. (2011) Simultaneous visualization of both signaling cascade activity and end-point gene expression in single cells. *PLoS ONE*, **6**, e20148.

46 Jung, J., Lifland, A.W., Zurla, C., Alonas, E.J. *et al.* (2013) Quantifying RNA-protein interactions in situ using modified-MTRIPs and proximity ligation. *Nucleic Acids Res.*, **41** (1), e12.

47 Rhee, W.J., Santangelo, P.J., Jo, H., Bao, G. (2008) Target accessibility and signal specificity in live-cell detection of BMP-4 mRNA using molecular beacons. *Nucleic Acids Res.*, **36**, e30.

48 Tsuji, A., Koshimoto, H., Sato, Y., Hirano, M. *et al.* (2000) Direct observation of specific messenger RNA in a single living cell under a fluorescence microscope. *Biophys. J.*, **78**, 3260–3274.

49 Tyagi, S., Alsmadi, O. (2004) Imaging native beta-actin mRNA in motile fibroblasts. *Biophys. J.*, **87**, 4153–4162.

50 Mhlanga, M.M., Vargas, D.Y., Fung, C.W., Kramer, F.R. *et al.* (2005) tRNA-linked molecular beacons for imaging mRNAs in the cytoplasm of living cells. *Nucleic Acids Res.*, **33**, 1902–1912.

51 Chen, A.K., Behlke, M.A., Tsourkas, A. (2008) Efficient cytosolic delivery of molecular beacon conjugates and flow cytometric analysis of target RNA. *Nucleic Acids Res.*, **36**, e69.

52 Sando, S., Kool, E.T. (2002) Imaging of RNA in bacteria with self-ligating quenched probes. *J. Am. Chem. Soc.*, **124**, 9686–9687.

53 Abe, H., Kool, E.T. (2006) Flow cytometric detection of specific RNAs in native human cells with quenched autoligating FRET probes. *Proc. Natl Acad. Sci. USA*, **103**, 263–268.

54 Tyagi, S., Kramer, F.R. (1996) Molecular beacons: probes that fluoresce upon hybridization. *Nat. Biotechnol.*, **14**, 303–308.

55 Santangelo, P.J., Nix, B., Tsourkas, A., Bao, G. (2004) Dual FRET molecular beacons for mRNA detection in living cells. *Nucleic Acids Res.*, **32**, e57.

56 Bratu, D.P., Cha, B.J., Mhlanga, M.M., Kramer, F.R. *et al.* (2003) Visualizing the distribution and transport of mRNAs in living cells. *Proc. Natl Acad. Sci. USA*, **100**, 13308–13313.

57 Broude, N. E. (2002) Stem-loop oligonucleotides: a robust tool for molecular biology and biotechnology. *Trends Biotechnol.*, **20**, 249–256.

58 Vargas, D.Y., Raj, A., Marras, S.A., Kramer, F.R. *et al.* (2005) Mechanism of mRNA transport in the nucleus. *Proc. Natl Acad. Sci. USA*, **102**, 17008–17013.

59 Bertrand, E., Chartrand, P., Schaefer, M., Shenoy, S.M. *et al.* (1998) Localization of

ASH1 mRNA particles in living yeast. *Mol. Cell*, **2**, 437–445.

60 Beach, D.L., Salmon, E.D., Bloom, K. (1999) Localization and anchoring of mRNA in budding yeast. *Curr. Biol.*, **9**, 569–578.

61 Rook, M.S., Lu, M., Kosik, K.S. (2000) CaMKIIalpha 3′ untranslated region-directed mRNA translocation in living neurons: visualization by GFP linkage. *J. Neurosci.*, **20**, 6385–6393.

62 Shestakova, E.A., Singer, R.H., Condeelis, J. (2001) The physiological significance of beta -actin mRNA localization in determining cell polarity and directional motility. *Proc. Natl Acad. Sci. USA*, **98**, 7045–7050.

63 Forrest, K.M., Gavis, E.R. (2003) Live imaging of endogenous RNA reveals a diffusion and entrapment mechanism for nanos mRNA localization in *Drosophila*. *Curr. Biol.*, **13**, 159–1168.

64 Shav-Tal, Y., Darzacq, X., Shenoy, S.M., Fusco, D. *et al.* (2004) Dynamics of single mRNPs in nuclei of living cells. *Science*, **304**, 1797–1800.

65 Janicki, S.M., Tsukamoto, T., Salghetti, S.E., Tansey, W.P. *et al.* (2004) From silencing to gene expression: real-time analysis in single cells. *Cell*, **116**, 683–698.

66 Chubb, J.R., Trcek, T., Shenoy, S.M., Singer, R.H. (2006) Transcriptional pulsing of a developmental gene. *Curr. Biol.*, **16**, 1018–1025.

67 Hüttelmaier, S., Zenklusen, D., Lederer, M., Dictenberg, J. *et al.* (2005) Spatial regulation of beta-actin translation by Src-dependent phosphorylation of ZBP1. *Nature*, **438**, 512–515.

68 Zimyanin, V.L., Belaya, K., Pecreaux, J., Gilchrist, M.J. *et al.* (2008) In vivo imaging of oskar mRNA transport reveals the mechanism of posterior localization. *Cell*, **134**, 843–853.

69 Karpova, T.S., Kim, M.J., Spriet, C., Nalley, K. *et al.* (2008) Concurrent fast and slow cycling of a transcriptional activator at an endogenous promoter. *Science*, **319**, 466–469.

70 Golding, I., Cox, E.C. (2004) RNA dynamics in live *Escherichia coli* cells. *Proc. Natl Acad. Sci. USA*, **101**, 11310–11315.

71 Golding, I., Paulsson, J., Zawilski, S.M., Cox, E.C. (2005) Real-time kinetics of gene activity in individual bacteria. *Cell*, **123**, 1025–1036.

72 Le, T.T., Harlepp, S., Guet, C.C., Dittmar, K. *et al.* (2005) Real-time RNA profiling within a single bacterium. *Proc. Natl Acad. Sci. USA*, **102**, 9160–9164.

73 Guet, C.C., Bruneaux, L., Min, T.L., Siegal-Gaskins, D. *et al.* (2008) Minimally invasive determination of mRNA concentration in single living bacteria. *Nucleic Acids Res.*, **36**, e73.

74 Singer, R.H. (2003) RNA localization: visualization in real-time. *Curr. Biol.*, **13**, R673–R675.

75 Paige, J.S., Wu, K.Y., Jaffrey, S.R. (2011) RNA mimics of green fluorescent protein. *Science*, **333**, 642–646.

76 Paige, J.S., Nguyen-Duc, T., Song, W., Jaffrey, S.R. (2012) Fluorescence imaging of cellular metabolites with RNA. *Science*, **335**, 1194.

77 Rackham, O., Brown, C.M. (2004) Visualization of RNA-protein interactions in living cells: FMRP and IMP1 interact on mRNAs. *EMBO J.*, **23**, 3346–3355.

78 Valencia-Burton, M.A., McCullough, R.M., Cantor, C.R., Broude, N.E. (2007) RNA visualization in live bacterial cell using split fluorescent protein approach. *Nat. Methods*, **4**, 421–427.

79 Valencia-Burton, M., Shah, A., Sutin, J., Borogovac, A. *et al.* (2009) Spatiotemporal patterns and transcription kinetics of induced RNA in single bacterial cells. *Proc. Natl Acad. Sci. USA*, **106**, 16399–16404.

80 Yiu, H.W., Demidov, V.V., Toran, P., Cantor, C.R. *et al.* (2011) RNA detection in live bacterial cells using fluorescent protein complementation triggered by Interaction of two RNA aptamers with two RNA-binding peptides. *Pharmaceuticals*, **4** (3), 494–508.

81 Takizawa, P.A., Vale, R.D. (2000) The myosin motor, Myo4p, binds Ash1 mRNA via the adapter protein, She3p. *Proc. Natl Acad. Sci. USA*, **97**, 5273–5278.

82 Brodsky, A.S., Silver, P.A. (2000) Pre-mRNA processing factors are required for nuclear export. *RNA*, **6**, 1737–1749.

83 Brengues, M., Teixeira, D., Parker, R. (2005) Movement of eukaryotic mRNAs between polysomes and cytoplasmic processing bodies. *Science*, **310**, 486–489.

84 Daigle, N., Ellenberg, J. (2007) Lambda N-GFP: an RNA reporter system for live-cell imaging. *Nat. Methods*, **4**, 633–636.

85 Lange, S., Katayama, Y., Schmid, M., Burkacky, O. et al. (2008) Simultaneous transport of different localized mRNA species revealed by live-cell imaging. *Traffic*, **9**, 1256–1267.

86 Wu, B., Chao, J.A., Singer, R.H. (2012) Fluorescence fluctuation spectroscopy enables quantitative imaging of single mRNAs in living cells. *Biophys. J.*, **102**, 2936–2944.

87 Babendure, J.R., Adams, S.R., Tsien, R.Y. (2003) Aptamers switch on fluorescence of triphenylmethane dyes. *J. Am. Chem. Soc.*, **125**, 14716–14617.

88 Sparano, B.A. Koide, K. (2007) Fluorescent sensors for specific RNA: a general paradigm using chemistry and combinatorial biology. *J. Am. Chem. Soc.*, **129**, 4785–4794.

89 Kolpashchikov, D.M. (2005) Binary malachite green aptamer for fluorescent detection of nucleic acids. *J. Am. Chem. Soc.*, **127**, 12442–12443.

90 Michnick, S.W. (2003) Protein fragment complementation strategies for biochemical network mapping. *Curr. Opin. Biotechnol.*, **14**, 610–617.

91 Kerppola, T.K. (2006) Visualization of molecular interactions by fluorescence complementation. *Nat. Rev. Mol. Cell Biol.*, **7**, 449–456.

92 Shyu, Y.J., Hu, C.D. (2008) Fluorescence complementation: an emerging tool for biological research. *Trends Biotechnol.*, **26**, 622–630.

93 Ozawa, T. (2009) Protein reconstitution methods for visualizing biomolecular function in living cells. *Yakugaku Zasshi*, **129**, 289–295.

94 Hu, C.D., Chinenov, Y., Kerppola, T.K. (2002) Visualization of interactions among bZIP and Rel family proteins in living cells using bimolecular fluorescence complementation. *Mol. Cell*, **9**, 789–798.

95 Milev, M.P., Brown, C.M., Mouland, A.J. (2010) Live cell visualization of the interactions between HIV-1 Gag and the cellular RNA-binding protein Staufen1. *Retrovirology*, **7**, 41.

96 Ozawa, T., Natori, Y., Sato, M. Umezawa, Y. (2007) Imaging dynamics of endogenous mitochondrial RNA in single living cells. *Nat. Methods*, **4**, 413–419.

97 Delebecque, C.J., Lindner, A.B., Silver, P.A., Aldaye, F.A. (2011) Organization of intracellular reactions with rationally designed RNA assemblies. *Science*, **333**, 470–474.

98 Said, N., Rieder, R., Hurwitz, R., Deckert, J. et al. (2009) In vivo expression and purification of aptamer-tagged small RNA regulators. *Nucleic Acids Res.*, **37**, e133.

99 Wang, X., McLachlan, J., Zamore, P.D., Hall, T.M. (2002) Modular recognition of RNA by a human pumilio-homology domain. *Cell*, **110**, 501–512.

100 Cheong, C.G., Hall, T.M. (2006) Engineering RNA sequence specificity of Pumilio repeats. *Proc. Natl Acad. Sci. USA*, **103**, 13635–13639.

101 Yamada, T., Yoshimura, H., Inaguma, A., Ozawa, T. (2011) Visualization of nonengineered single mRNAs in living cells using genetically encoded fluorescent probes. *Anal. Chem.*, **83**, 5708–5714.

102 Yoshimura, H., Inaguma, A., Yamada, T., Ozawa, T. (2012) Fluorescent probes for imaging endogenous β-actin mRNA in living cells using fluorescent protein-tagged pumilio. *ACS Chem. Biol.*, **7**, 999–1005.

103 Muramoto, T., Cannon, D., Gierlinski, M, Corrigan, A. et al. (2012) Live imaging of nascent RNA dynamics reveals distinct types of transcriptional pulse regulation. *Proc. Natl Acad. Sci. USA*, **109**, 7350–7355.

104 Chakraborty, D., Kappei, D., Theis, M., Nitzsche, A. et al. (2012) Combined RNAi and localization for functionally dissecting long noncoding RNAs. *Nat. Methods*, **9**, 360–362.

14
RNA as a Regulator of Chromatin Structure

Yota Murakami
Hokkaido University, Department of Chemistry, Faculty of Science, N10 W8 Kita-ku, Sapporo, 001-0010 Japan

1	**Introduction to Chromatin Structure** 439	
1.1	Chromatin 439	
1.2	Histone Modifications and DNA Methylation in Chromatin Structure 440	
2	**RNA as a Regulator of Chromatin** 441	
3	**Small RNA-Mediated Regulation of Chromatin** 442	
3.1	RNA Interference 442	
3.2	RNAi-Dependent Heterochromatin Assembly in Fission Yeast 442	
3.2.1	Fundamental Features of RNAi-Mediated Heterochromatin Formation 444	
3.2.2	Organization of the RNAi-Dependent Heterochromatin-Formation Pathway 444	
3.2.3	Mechanism of Silencing at Heterochromatin 446	
3.2.4	Regulation of Heterochromatic ncRNA Transcription during the Cell Cycle 447	
3.2.5	Dynamics of ncRNA in RNAi-Dependent Heterochromatin Formation 447	
3.2.6	Other Systems of RNA-Dependent Heterochromatin Formation 449	
3.2.7	Heterochromatin Islands in Euchromatin 450	
3.3	Small RNA-Mediated DNA Methylation in Plants 451	
3.4	piRNA and Chromatin Regulation in Drosophila and Mammals 453	
4	**Long RNA-Mediated Chromatin Regulation** 453	
4.1	Cis-Acting Long RNAs: Lessons from Xist RNA 454	
4.2	Other Cis-Acting Long RNAs 455	
4.3	Trans-Acting Long RNAs 457	
4.4	Structural Roles of Long RNAs in Chromatin Organization 458	

RNA Regulation: Advances in Molecular Biology and Medicine, First Edition. Edited by Robert A. Meyers.
© 2014 Wiley-VCH Verlag GmbH & Co. KGaA. Published 2014 by Wiley-VCH Verlag GmbH & Co. KGaA.

5   Summary   460

References   460

Keywords

Heterochromatin
A condensed chromatin structure in which transcription and other DNA transactions are restricted.

Euchromatin
A relaxed chromatin structure in which transcription or other DNA transactions are allowed to take place.

Noncoding RNA (ncRNA)
RNA that does not encode protein.

Long noncoding RNA (lncRNA)
Noncoding RNA more than 200 nt in length.

RNA interference (RNAi)
A system that evolved to prevent RNA-virus infection or propagation of transposons via the degradation of target RNA.

Small inhibitory RNA (siRNA)
Small (~20–30 nt) RNA produced by RNAi factors, which are incorporated into a complex containing an Argonaute-family protein; Argonaute and its homologs are important RNAi factors.

PIWI-interacting RNA (piRNA)
Small (~20–30 nt) RNA that associates with Piwi, an Argonaute-family protein expressed in the germline cells of higher eukaryotes.

Xist RNA
Noncoding RNA transcribed from the inactive X chromosome in mammals; Xist covers the inactive X chromosome and is essential for X-chromosome inactivation.

Chromatin structure plays a major role in the epigenetic regulation of gene expression, and is strictly regulated by histone modifications and DNA methylation. Recent studies have identified the molecular basis of the connections between these modifications and chromatin structure; however, the mechanisms underlying the spatiotemporal regulation of chromatin structure remain obscure. Knowledge of these mechanisms is essential for a complete understanding of the epigenetic regulation of gene expression in the context of development and disease. In contrast, technical advances in whole-genome analysis – such as microarrays and deep sequencing – have revealed that an unexpectedly large number of noncoding RNAs (ncRNAs) and small RNAs are produced from many genomic loci. Emerging evidence also indicates that some of these ncRNAs and small RNAs are directly involved in regulating higher-order chromatin structure. Hence, an improvement in knowledge relating to the roles of these RNAs in chromatin regulation is necessary if the epigenetic regulation of genome function is to be fully understood.

1
Introduction to Chromatin Structure

1.1
Chromatin

Eukaryotic chromosomes – long molecules of DNA that contain the genes which govern the development and physiological function of the organism – are packaged into a relatively small cellular compartment called the *nucleus*. For example, in a human cell, approximately 2 m of DNA is packaged into a nucleus that has a diameter of about 10μm. This extensive condensation is achieved by multilayered systems involved in DNA folding. The most fundamental folded structure, the *nucleosome*, consists of DNA in complex with proteins referred to as *histones*: H2A, H2B, H3, and H4. In a nucleosome, two molecules of each of the four histones form an octamer, which is wrapped twice by a 147-bp segment of DNA. Genomic DNA packaged into nucleosomes is folded into a higher-order structure called *chromatin*. Based on their nucleosome density, chromatin structures can be categorized into two structures: heterochromatin and euchromatin. Heterochromatin has a more condensed structure, in which the nucleosomes are regularly positioned and packed tightly. By contrast, *euchromatin* has a looser structure in which the extent of nucleosome packaging varies from region to region and is dynamically regulated.

Heterochromatin can be further subdivided into two classes: constitutive and facultative heterochromatin. Constitutive heterochromatin is found in almost all eukaryotic cells and is localized at repeated sequences and transposons. Because repeated sequences (including transposons) are generally enriched at centromeres and in subtelomeric regions, large domains of constitutive heterochromatin are generally observed in these regions, where they are thought to play important roles.

Like constitutive heterochromatin, facultative heterochromatin is stable and suppresses the transcription and recombination of the embedded DNA sequences; however, the formation of facultative heterochromatin is regulated spatiotemporally. Facultative heterochromatin stably represses the transcription of protein-encoding genes over many cell generations. Many of the genes contained within facultative heterochromatin are developmental genes; consequently, the formation of facultative heterochromatin must be developmentally regulated.

1.2 Histone Modifications and DNA Methylation in Chromatin Structure

The chromatin structures of both heterochromatin and euchromatin are primarily determined by histone modifications, the so-called "histone code" [1]. The N termini of histones, which have disordered structures, protrude from the nucleosome and become targets for various chemical modifications, such as acetylation, methylation, and phosphorylation. Each modification serves as a specific recognition signal for particular protein domains, thereby providing a platform for chromatin-binding proteins that contain those domains ("code-readers"). Once associated with histones, these chromatin-binding proteins recruit various proteins ("effectors"), which then implement various chromatin-related functions such as activating and repressing transcription (Fig. 1). Thus, different combinations of histone modifications and code-readers determine chromatin structure and function [1]. For example, constitutive heterochromatin contains particular histone modifications, namely, the di- or tri-methylation of lysine 9 of histone H3 (H3K9me2 or 3). These modifications are recognized by the chromodomain of heterochromatin protein (HP) family proteins [2–4]. HP1 recruits various effector proteins, including repressors of transcription and cohesin, which connects sister chromatids

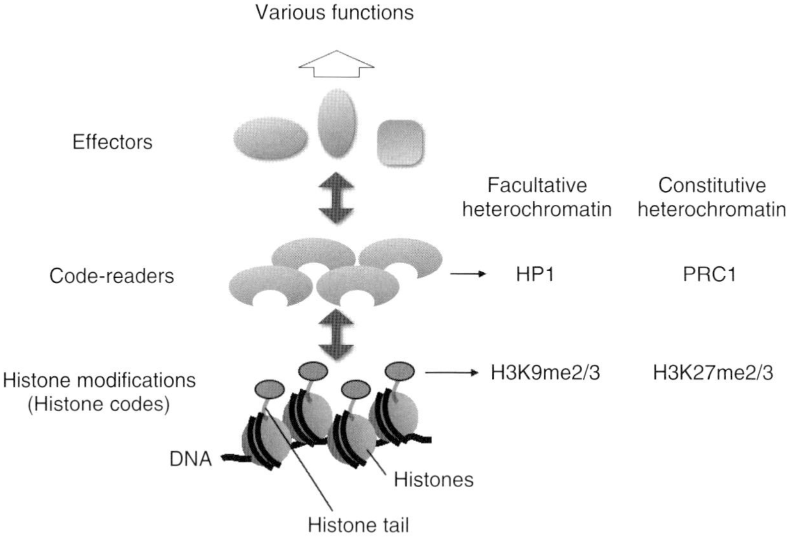

Fig. 1 Histone modifications and higher-order chromatin structures. Histone tails are subjected to various covalent modifications. The pattern of histone modification ("histone codes") is specifically recognized by "code-readers," which contain domain(s) that bind to specific histone codes. Code-readers interact with effector proteins ("effectors"), which promote various functions, including transcriptional regulation. Examples of two higher-order chromatin structures, facultative and constitutive heterochromatin, are indicated.

after chromosome replication (Fig. 1). Similarly, H3K27me2/3 and its binding partner, Polycomb repressive complex (PRC), define facultative heterochromatin [5–9].

In higher eukaryotic cells, DNA methylation forms another layer of regulation of chromatin structure. In mammals, methylation of the cytosines within CpG dinucleotides serves as a marker for heterochromatin [10, 11]. Methylated CpG provides a binding site for methyl-CpG-binding proteins (MBPs), which in turn recruit the histone-modifying enzymes that promote heterochromatin formation (Fig. 2) [10, 11]. Plant cells contain several distinct types of cytosine methylation (non-CpG methylation); these modifications are thought to recruit different types of MBP, resulting in different types of chromatin structure [12, 13]. However, the functional outputs of the different types of methylation are not yet fully understood.

2 RNA as a Regulator of Chromatin

Recent analyses of mammalian genomes show that protein-coding sequences occupy only 1–2% of the genome; however, more than 50% of the genome is transcribed to produce various types of noncoding RNAs (ncRNAs) [14–17]. Whilst the functions of these ncRNAs have been extensively investigated, and many of their functions remain unknown, some ncRNAs are involved in the regulation of multiple cellular processes, including chromatin regulation. ncRNAs can be categorized into two classes: long noncoding RNAs (>200 nt) (lncRNAs) and small (<200 nt) ncRNAs. Some ncRNAs belonging to each category are intimately involved in the regulation of chromatin structure; consistent with this, many ncRNAs associate with chromatin [18, 19]

In contrast to DNA, which functions as a medium for storing genetic information, RNA plays multiple roles within the cell, particularly in processes related to the expression of genetic information. Messenger RNAs (mRNAs) store genetic information temporarily and transfer it to ribosomes in the cytoplasm, while ribosomal RNAs (rRNAs) not only function as scaffolds for ribosomal proteins but also contribute to the enzymatic function of the ribosomes. Transfer RNAs (tRNAs) connect each amino acid with its corresponding codon to interpret the genetic code of mRNAs. Similarly, ncRNAs play various roles in chromatin regulation. Some lncRNAs are structural components of chromatin, whereas others serve as platforms for complexes that regulate chromatin structure. In addition, some types of small RNAs and lncRNAs connect chromatin-modifying enzymes with chromatin.

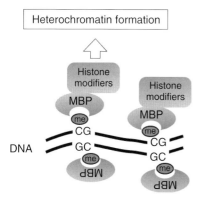

Fig. 2 DNA methylation and heterochromatin. Methylated CpG sequences are recognized by specific proteins that contain methyl-CpG binding domains (MBPs). MBPs recruit histone-modifying proteins that promote heterochromatin formation.

At this stage, the roles played by ncRNAs in chromatin regulation are still under extensive investigation and are only partially characterized. Some typical examples of the way in which ncRNA regulates chromatin are described in the following sections.

3
Small RNA-Mediated Regulation of Chromatin

RNA interference (RNAi) was first identified as a system of post-transcriptional gene silencing (PTGS) [20, 21], which evolved as a defense mechanism against "selfish" genetic elements such as viruses and transposons. The discovery of this system prompted a search for other functional small RNAs, and resulted in the identification of different types of small RNAs, including PIWI-interacting RNA (piRNA; for details, see Sect. 3.4) [22]. Subsequent functional analyses of small RNAs revealed that they can mediate heterochromatin formation [23, 24]. In the next section, the fundamental features of RNAi are first detailed, followed by a detailed description of small RNA-mediated chromatin regulation in fission yeast, plants, and vertebrates.

3.1
RNA Interference

The core features of RNAi are illustrated in Fig. 3. Conventional RNAi is triggered by the presence of double-stranded RNAs (dsRNAs) in cells; such dsRNAs are presumably derived from infecting RNA viruses, or the intermediate products of retroviruses, retroposons, and bidirectional transcripts. dsRNAs are processed into small dsRNAs (ca. 20–30 bp) by the enzyme Dicer, which belongs to the RNase III family. The small RNAs produced by Dicer are called small inhibitory RNAs (siRNAs) [25–28]. An individual siRNA is captured by Argonaute, a catalytic subunit of the effector complex RNA induced silencing complex (RISC) [29]. Argonaute has RNase activity, referred to as *"slicer" activity* [30, 31], by which it selectively digests and discards one strand of the double-stranded siRNA, so as to leave a single-stranded siRNA in the RISC complex [32–34]. RISC binds to target RNAs via complementary interactions between the target RNA and the siRNA within the complex [27, 28], thereby inhibiting the function of the target RNA (i.e., its translation into a protein) by degrading it via Argonaute's slicer activity [30, 31].

In plants and *Caenorhabditis elegans* (in which RNAi was first identified), RNAs that trigger the RNAi system are efficiently amplified by an RNA-dependent RNA polymerase (RdRNAP), which synthesizes dsRNA from single-stranded RNA (ssRNA) or dsRNA [35, 36]. The amplification of dsRNA is thought to magnify the signal triggering RNAi. By contrast, no RdRNAP polymerase has been identified in mammals or in *Drosophila*; thus, RNAi may function without the amplification step in mammals and flies.

3.2
RNAi-Dependent Heterochromatin Assembly in Fission Yeast

The existence of small RNA-mediated heterochromatin formation was first suggested in *Arabidopsis thaliana* [24]. As described in the next section, however, plant genomes encode many paralogs of RNAi proteins, which complicates analysis. By contrast, the fission yeast *Schizosaccharomyces pombe*

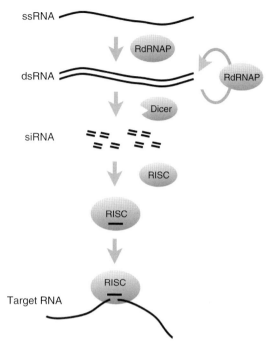

Fig. 3 Fundamental features of RNA interference (RNAi). RNAi is triggered by double-stranded DNA (dsRNA), which is degraded by the RNase Dicer, resulting in the production of 20–24-nt siRNAs. One strand of the siRNA is incorporated into the effector complex, RISC. RISC, which contains an Argonaute-family protein, binds to the target RNA using the siRNA as a guide molecule, and then cuts the target RNA via the slicer activity of the Argonaute protein, thereby inactivating it. RNA-dependent RNA polymerase (RdRNAP), produces dsRNA from single-stranded RNA (ssRNA) or dsRNA to amplify the signal for RNAi.

has only one set of genes encoding the major RNAi proteins Argonaute (*ago1*), RNA-dependent RNA polymerase (*rdp1*), and Dicer (*dcr1*). Furthermore, many *S. pombe* heterochromatin-related genes are nonessential [23]; this situation simplifies the functional analysis of the genes because there is no need to generate conditional mutations of each gene. In addition, the chromatin regulation system in fission yeast is similar to – but simpler than – the homologous system in higher eukaryotic cells. These features make fission yeast a good model system in which to analyze chromatin regulation, though one potential drawback of using fission yeast as a model organism for chromatin studies is the absence of DNA methylation; however, this absence makes the system much simpler and easier to understand. As will be seen in the following sections, the RNAi-mediated heterochromatin system is already sufficiently complex, even in fission yeast.

Because siRNA-mediated heterochromatin formation is one of the most extensively studied forms of RNA-mediated chromatin regulation, and provides many clues to understanding the other systems, this system will be discussed in detail in the following sections.

3.2.1 Fundamental Features of RNAi-Mediated Heterochromatin Formation

In fission yeast, heterochromatin exists at pericentromeric, subtelomeric and mating-type loci, occupying a total of about 5% of the genome. As in other eukaryotic cells, heterochromatin in fission yeast is defined by H3K9me2/3 and its binding partners Swi6 and Chp2 [4, 37, 38], both of which are HP1 homologs that recognize H3K9me2/3 via their chromodomains [38].

The current model of RNAi-mediated heterochromatin formation is illustrated in Fig. 2. This system is triggered by the transcription of ncRNA from heterochromatic repeats by RNA polymerase II (RNAPII) [39, 40]. These ncRNAs become the targets for the RNA-induced transcriptional silencing (RITS) complex, which comprises three proteins (the Argonaute homolog Ago1, Tas3, and the chromodomain protein Chp1) and siRNA derived from an ncRNA [41, 42]. The RITS complex binds to the ncRNA via complementary interactions between siRNA and ncRNA, and then recruits two complexes – Clr4-containing complex (CLRC) and RNA-dependent RNA polymerase complex (RDRC) – onto chromatin [41–44].

CLRC consists of five subunits: Clr4, Rik1, Dos1, Dos2 and Cul4 [45, 46]. Clr4 is a histone H3K9-specific methyltransferase [4] that is highly conserved among eukaryotes. The recruitment of CLRC induces H3K9 methylation and subsequent heterochromatin formation.

RDRC consists of three subunits: Rdp1, Hrr1, and Cid12 [43, 44]. Rdp1 is an RdRNAP that produces dsRNA, which becomes a substrate for siRNA production [43, 44]. Hrr1 contains an RNA-helicase motif and is thought to function in dsRNA production by removing secondary structures from template RNA. Cid12 is a poly-A polymerase [43]. Because there are multiple poly-A polymerases, and poly-A addition seems to be coupled to different steps during RNA processing, polyadenylation by Cid12 may be coupled to dsRNA synthesis by Rdp1 [43]. The dsRNAs produced by RDRC are processed into ds-siRNA by Dcr1, a homolog of Dicer [47]. ds-siRNA is first incorporated into ARC, which consists of Ago1, Arb1, and Arb2, and then subsequently is transferred to the RITS complex [48]. During the transfer, ds-siRNA is converted to ss-siRNA, probably via the slicer activity of Ago1 [48]. Heterochromatin formation and siRNA production are interdependent and form a self-reinforcing loop, resulting in the stable maintenance of heterochromatin [44].

3.2.2 Organization of the RNAi-Dependent Heterochromatin-Formation Pathway

As noted in Sect. 3.2.1, RNAi-dependent heterochromatin formation includes several steps and is supported by many factors (Fig. 4). Like other systems in cells, each reaction in RNAi-dependent heterochromatin formation is coupled to the subsequent reaction, and the overall reaction pathway seems to be well coordinated on heterochromatin. Some of the coupling between reactions is based on direct interactions between factors. For example, RDRC and RITS interact physically, and this interaction is important for recruiting RDRC to ncRNA, which is necessary for efficient dsRNA production [43]. In addition, a few proteins couple specific reactions (Fig. 5). The Lim-domain protein Stc1 connects RITS with CLRC to recruit CLRC for heterochromatin formation; this is demonstrated by the observation that artificial recruitment of Stc1

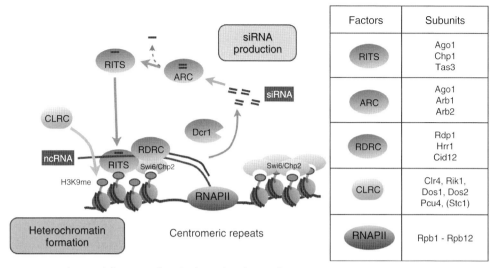

Fig. 4 Fundamental features of RNAi-dependent heterochromatin formation in fission yeast. For details, see Sect. 3.2.1.

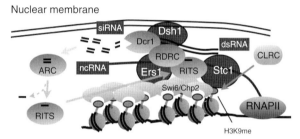

Fig. 5 Organization of the RNAi-dependent heterochromatin-formation pathway. Several factors (shown by white on black) organize RNAi-dependent heterochromatin formation. Stc1 binds to the RITS complex and recruits CLRC to promote H3K9 methylation. Ers1 connects RDRC with Swi6 to promote dsRNA synthesis on heterochromatin. Dsh1 connects RDRC and Dcr1 to couple dsRNA synthesis with siRNA production. Because Dcr1 and Dsh1 colocalize at the inner nuclear periphery, the overall reaction is organized in this part of the cell. For details, see Sect. 3.2.2.

onto euchromatin also recruits CLRC, ultimately resulting in heterochromatin formation [49]. Ers1 was first isolated as a factor required for RNAi-dependent heterochromatin formation [50]. Ers1 was subsequently shown to bind to both Swi6 and Hrr1 (a component of RDRC) and enhance RDRC–RITS interaction, which promotes efficient dsRNA synthesis on heterochromatin [51]. Another coupling factor, Dsh1, forms a tight complex with Dcr1, interacts with RDRC, and also promotes the association of RDRC with heterochromatin. These interactions couple dsRNA synthesis and processing, resulting in a robust synthesis of siRNA from small amounts of RNA on heterochromatin [52]. Dcr1 associates with the nuclear-pore complex, and that association seems to be important for the function of Dcr1 in RNAi-dependent heterochromatin formation [53]. Dsh1 is also localized at the nuclear periphery [52]. Thus, production of siRNA and subsequent heterochromatin formation seems to occur in a compartment at the inner nuclear periphery that is organized by Dsh1. Because heterochromatin is usually localized at the nuclear periphery in almost all eukaryotic cells (including yeasts), and this localization is thought to be important for gene silencing [54], these observations may explain the functional significance of the localization of heterochromatin to the nuclear periphery.

3.2.3 Mechanism of Silencing at Heterochromatin

Fission-yeast heterochromatin is primarily defined by methylation at H3K9me and the ensuing binding of the HP1 homologs Swi6 and Chp2; such heterochromatin is refractory to transcription. Both, Swi6 and Chp2 form homodimers [38] that are thought to induce the compact structure of heterochromatin by connecting neighboring nucleosomes. However, the formation of a compact structure is not sufficient to establish transcriptionally silent chromatin. In order to establish a silent state, a complex called *SHREC* must be recruited via the interaction with Swi6/Chp2 [55, 56]. SHREC contains the histone deacetylase (HDAC) Clr3 and the chromatin-remodeling protein Mit1, which respectively induce the hypoacetylation of histones and silent chromatin structures [55]. The localization of SHREC at heterochromatin requires phosphorylation of Sw6/Chp2 by casein kinase 2, which suggests that the recruitment of SHREC (i.e., the silent state) is dynamically regulated by Swi6/Chp2 phosphorylation [57]. In addition to SHREC, other HDACs – including Clr6 and the Sir2 family – also seem to be involved in regulating heterochromatic silencing [58–61].

Another factor, Epe1, is also recruited to heterochromatin by interacting with Swi6 [62]. Epe1 was initially identified as a factor that is required to establish boundaries between euchromatin and heterochromatin, and it negatively affects heterochromatic silencing [63]. Epe1 contains a JmjC domain, a signature of histone demethylases [64–66]. Epe1 can activate transcription in heterochromatin by increasing the recruitment of RNAPII [62], but it remains unknown whether Epe1 is an active demethylase [62, 66, 67]. The localization of Epe1 at heterochromatin competes with SHREC [57], which again suggests a dynamic regulation of transcription in heterochromatin.

Gene silencing in heterochromatin (as described above) occurs at the transcriptional level. In addition, RNA transcribed in heterochromatin is subjected to post-transcriptional by RNAi. Heterochromatic ncRNA becomes a target for RNAi (see Sect. 3.1). Furthermore, transcripts from genes residing in heterochromatin also become targets for RNAi, and siRNAs derived from these transcripts can be detected [68]; these siRNAs contribute to the RNAi-mediated spreading of heterochromatin into the these genes [69, 70]. Another form of PTGS at heterochromatin is RNA elimination by RNase complexes known as *exosomes*. Cid14 is a catalytic subunit of the TRAMP polyadenylation complex, which promotes the degradation of aberrant transcripts by exosomes [71–73]. The deletion of *cid14* causes a marked increase in the level of transcripts derived from heterochromatin, despite the preservation of heterochromatic structure; this indicates that heterochromatic transcripts are degraded in a TRAMP-mediated fashion [68]. Swi6 binds to RNA, which competes with its binding to H3K9me [74]. This RNA-binding activity is required for a heterochromatin-specific checkpoint that involves the capture and priming of heterochromatin RNAs for the RNA-degradation machinery [74].

3.2.4 Regulation of Heterochromatic ncRNA Transcription during the Cell Cycle

Although heterochromatin usually represses transcription, RNAi-dependent heterochromatin formation depends strictly on the transcription of ncRNA in heterochromatin. The question is, how do cells resolve this apparent paradox? Detailed analyses of the transcription and behavior of heterochromatic proteins in fission yeast have shown that ncRNA transcription from heterochromatic repeats occurs in a narrow window of the cell cycle, early S phase [75, 76]. In fission yeast, heterochromatin also replicates in early S phase; this is in contrast to other organisms in which heterochromatin is generally replicated in late S phase [77]. Therefore, it seems that the RNAi-dependent formation of heterochromatin, induced by ncRNA transcription in early S phase, re-establishes heterochromatin on newly replicated chromosomes. Further experiments are required to clarify the generality of this coupling between ncRNA transcription and heterochromatin formation in other organisms.

How do cells regulate transcription in a cell cycle-dependent manner? According to one proposed model (Fig. 6) [75], when cells enter M phase, Aurora B kinase phosphorylates Ser10 of histone H3 (H3S10), which may simultaneously inhibit the binding of Swi6 to H3K9me [78] and provide binding sites for condensin, which induces chromosome condensation [79]. Condensin binding and/or chromosome condensation seem to maintain gene silencing during M phase [75] (Fig. 4). In anaphase, condensin is released from chromatin when the level of H3S10 phosphorylation decreases due to the inactivation of Aurora B kinase. Because fission yeast has a very short (or no) G_1 phase, cells enter S phase immediately after the end of mitosis and initiate the replication of heterochromatic chromatin that bears decreased levels of Swi6; this allows ncRNA transcription and induces heterochromatin formation on replicated chromatin via the RNAi pathway. On the re-established heterochromatin, Swi6 recruits SHREC, which then shuts down transcription, as mentioned above. Based on these observations, ncRNA transcription from heterochromatic repeats seems to be coupled to heterochromatin replication in early S phase, which allows the re-establishment of heterochromatic structures on newly replicated chromatin [75].

3.2.5 Dynamics of ncRNA in RNAi-Dependent Heterochromatin Formation

In the RNAi-dependent heterochromatin-formation system, ncRNA appears to remain on the chromatin after it is transcribed. There, it serves as a platform for the RNAi machinery, as shown in Figs 4 and 5. Thus, heterochromatic ncRNA has a completely different fate from that of mRNA, even though both types of RNA are transcribed by RNAPII. The molecular mechanisms that distinguish between mRNA and heterochromatic ncRNA remain obscure. Transcription within heterochromatin may be crucial for discriminating the fates of transcribed RNA; transcripts from euchromatic genes inserted in heterochromatin are not translated into protein, but instead become

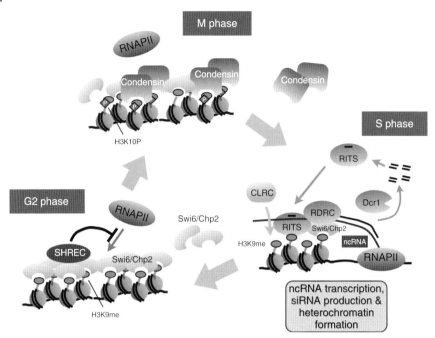

Fig. 6 Cell-cycle regulation of ncRNA transcription during RNAi-dependent heterochromatin formation. ncRNA transcription, which triggers RNAi-dependent heterochromatin formation, occurs during a small window of the cell cycle (S phase), and promotes heterochromatin formation on the newly replicated DNA. At other phases of the cell cycle, ncRNA transcription is repressed by distinct mechanisms. For details, see Sect. 3.2.4.

targets for the RNAi machinery, as in the case of heterochromatic ncRNA [74]. One clue to understanding the mechanism was provided by the discovery of a specific RNAPII mutant that compromises siRNA production from heterochromatic ncRNA but does not affect mRNA transcription [40]. In addition, other factors involved in regulating transcription, such as Spt6 and Mediator, are involved in heterochromatin formation [80–82]. These results suggest that RNAPII, together with other transcription factors, actively determines the fate of transcribed RNA. Accordingly, siRNA-containing Ago1 binds to nascent transcripts during the elongation phase of transcription and inhibits RNAPII transcription via an unknown mechanism, which in turn induces the release of RNAPII from the template [83]. This observation further supports the idea that the RNAi machinery forms heterochromatin cotranscriptionally. In addition, recent observations that some RNA-splicing factors are involved in RNAi-dependent heterochromatin formation have indicated that splicing factors may also be involved in the fate of heterochromatic ncRNA [84, 85].

One of the most important features of heterochromatic ncRNA is its association with chromatin. The results of a recent study showed that some heterochromatic ncRNA cotranscriptionally forms a DNA–RNA hybrid in order to remain on chromatin after transcription [86]. Decreased levels of DNA–RNA hybrids upon the overproduction of RNase H (which

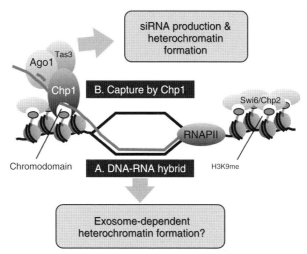

Fig. 7 Mechanisms by which ncRNA is tethered onto heterochromatin. Two hypothetical mechanisms for the association of ncRNA with heterochromatin are indicated. Mechanism A: Cotranscriptional formation of a DNA–RNA hybrid between template DNA and ncRNA. Mechanism B: Capture of ncRNA by the chromodomain of Chp1, which binds H3K9me and RNA cooperatively. These mechanisms are not mutually exclusive. For details, see Sect. 3.2.6.

specifically degrades DNA–RNA hybrids) causes a reduction in H3K9me at heterochromatin; likewise, increased levels of hybrids upon the depletion of RNase H results in an decrease in H3K9me. This observation suggests that DNA–RNA hybrid formation contributes to heterochromatin formation (Fig. 7). Mutations, or the depletion of RNA-transport or RNA-processing factors, result in the abnormal formation of DNA–mRNA hybrids in yeast and mammals [87–90], and suggests the existence of an active mechanism that regulates DNA–RNA hybrid formation on heterochromatin. Chromatin also associates with ncRNA when RNA is tethered by RNA-binding chromatin factors. In this context, Swi6 associates with RNA, but its RNA-binding activity competes with the H3K9me-binding activity required for chromatin association (see Sect. 3.2.3) [74]. Therefore, Swi6 cannot mediate the association of ncRNA with chromatin. Another chromodomain protein, Chp1, a subunit of the RITS complex, also binds RNA. In this case, the RNA-binding activity of the chromodomain of Chp1 (which binds to H3K9me) is strengthened by H3K9me binding, which is necessary for heterochromatin formation [91]. Therefore, Chp1, which binds to H3K9me at heterochromatin, might capture nascent RNAs, resulting in the stable association of ncRNA on chromatin, where it can serve as a platform for RNAi factors (Fig. 7).

3.2.6 Other Systems of RNA-Dependent Heterochromatin Formation

In addition to the RNAi-dependent system of heterochromatin formation, fission yeast also has an RNAi-independent system. This system depends on the DNA-binding transcription factor Atf1/Pcr1, whose targets include stress-response genes [92]. The heterochromatic region in the mating-type locus contains a sequence that is very similar (ca. 98% identical) to centromeric repeats (*cenH*)

[93], and ncRNA transcribed from this sequence triggers RNAi-dependent heterochromatin formation [41, 92]. In addition to *cenH*, the mating-type locus contains an Atf1/Pcr1 binding site that nucleates heterochromatin independently of RNAi but is dependent on the HDAC, Clr3. Thus, disrupting either RNAi or Atf1/Pcr1 at the mating-type locus has no effect on heterochromatin; however, disrupting both systems results in the complete loss of heterochromatin [92].

The Atf1-dependent pathway does not function at centromeres. However, in RNAi mutants substantial heterochromatin (i.e., chromatin with high levels of H3K9me and Swi6) is retained at the centromeres [94], suggesting the existence of additional heterochromatin-formation systems. Until recently, however, the nature of such systems had not been investigated in detail. As described in Sect. 3.2.3, exosomes degrade heterochromatic RNAs and function in both the cytoplasm and nucleus; nuclear exosomes contain Rrp6 as a catalytic subunit [95]. Disrupting *rrp6* in fission yeast reduces the levels of H3K9me at the centromeric repeats (as in RNAi mutants harboring deletions of *ago1* or *dcr1*) [96]. When *rrp6* and *ago1* are simultaneously disrupted, the H3K9me retained at the centromeric repeats in each single mutant is almost completely abolished [96]. These results clearly indicate that the RNAi-dependent and exosome-dependent systems function in parallel at centromeric heterochromatin. Furthermore, recent analyses have shown that DNA–RNA hybrid formation contributes to the cotranscriptional degradation of RNA by nuclear exosomes, which play an important role in the quality control of RNA [89, 90]. Thus, it is quite possible that DNA–RNA hybrid formation at heterochromatin (see Sect. 3.2.6) functions during exosome-dependent heterochromatin formation. Because exosomes function cotranscriptionally, exosome-dependent heterochromatin could also assemble cotranscriptionally.

3.2.7 Heterochromatin Islands in Euchromatin

The heterochromatin marker, H3K9me, is predominantly localized to facultative heterochromatin: the pericentromere, telomeres, and the mating-type locus in fission yeast. However, close examination shows that there are several small "islands" of H3K9me on which Swi6 is also localized in an H3K9me-dependent manner [97–99]. The formation of some of these islands, including those at meiotic genes such as *mei4* and *ssm4*, depends on the meiotic gene-specific RNA-degradation system [97–99]. Importantly, these heterochromatin islands form in vegetative cells but diminish upon sexual differentiation, demonstrating that the islands represent facultative heterochromatin [97–99] (Fig. 8). Some meiotic genes become targets for nuclear exosome-dependent RNA degradation via the action of the RNA-binding protein, Mim1, which binds to a specific RNA sequence (DSR) contained in its target mRNAs and recruits Rrp6 for RNA degradation [100, 101]. The heterochromatin islands formed around meiotic genes are abolished by deleting *mim1*, *rrp6*, or *red1*; the Red1 protein interacts with Mim1 and is essential for Mim1-dependent RNA elimination [98, 99, 102]. Mim1 recruits the RITS complex to meiotic genes [97], but RNAi is dispensable for the formation of facultative heterochromatin, as demonstrated by the observation that deletion of *ago1* or *dcr1*

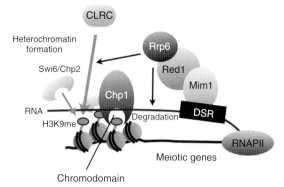

Fig. 8 Facultative heterochromatin formation at meiotic genes in fission yeast. Facultative heterochromatin is formed at several meiotic genes, such as *mei4* and *ssm6*, in vegetative cells. This heterochromatin-formation system depends on the RNA-degradation system for meiotic genes, which includes Mim1, Red1, Rrp6, and the RNA element DSR. Chp1, but not other RNAi factors, plays an important role in this system. For details, see Sect. 3.2.7.

has no effect on heterochromatin assembly [98]. Among the different components of the RITS complex, Chp1 seems to play an important role [97, 98]; in particular, the H3K9me- and RNA-binding activities of Chp1 (see Sect. 3.2.6) might contribute to the tethering of RNA at the target loci. Mim1 directs heterochromatin formation irrespective of its RNA-degradation activity, as shown by the observation that a *mim1* mutant that cannot form facultative heterochromatin still retains the ability to degrade RNA [98]. Moreover, in addition to DSR, another *cis*-element, which must be transcribed in order to function, plays a role in heterochromatin formation by recruiting Red1 at the *mei4* locus [98]. Although the molecular mechanism underlying Mim1-dependent H3K9 methylation remains unknown, the system may share mechanistic aspects with the exosome-dependent heterochromatin formation (as described in Sect. 3.2.6).

In addition to the heterochromatin islands described above, facultative heterochromatin can be formed under specific conditions at diverse loci, including sexual-differentiation genes and retrotransposons that are silenced by the exosome- and RNAi-dependent heterochromatin systems [103]. Transcripts from these loci are usually degraded by exosomes but, in the absence of exosomes, they are processed by the RNAi factors to generate siRNAs, resulting in heterochromatin formation [103].

Cooperation between RNAi and exosomes is also observed in *Drosophila*, suggesting that RNA degradation and processing are closely coordinated with chromatin regulation in a wide range of eukaryotes [103].

3.3 Small RNA-Mediated DNA Methylation in Plants

In higher eukaryotic cells, both DNA methylation and histone modifications mediate chromatin regulation. Methylation at the fifth carbon of cytosine residues has been widely observed and extensively studied. Usually, cytosine methylation promotes heterochromatin formation and is important for the suppression of mobile elements (transposons) and for the developmental regulation of gene expression. Cytosine methylation is limited almost exclusively to CG sequences in mammals [104], but non-CG methylation (CHG and CHH; H = A, C, T) is observed in plant cells [12].

In plants, RNA-dependent DNA methylation (RdDM) induces *de novo* cytosine methylation in all sequence contexts (CG, CHG, and CHH) [24, 105], and is required for the suppression of mobile elements as well as for plant development [99, 105–107]. RdDM depends on the RNAi machinery and, like RNAi-dependent heterochromatin formation in fission yeast, can be divided into two steps: (i) the generation of 24-nt siRNAs; and (ii) the subsequent conversion of siRNA signals into *de novo* DNA methylation at the target loci. In fission yeast, both steps depend on transcripts generated by a single RNAP, RNAPII, whereas in plants each step requires a plant-specific RNAP; the enzymes involved are RNAPIV and RNAPV [106] (Fig. 9).

RNAPII, IV, and V have similar subunit compositions (12 subunits); the holoenzymes share small subunits, but each polymerase has specific large subunits, including the largest subunits [13, 106, 108]. These differences in subunit composition may reflect functional differences among the polymerases, but the details are not yet fully understood.

RNAPIV is essential for siRNA production, and is assumed to transcribe the target region to produce the substrates for siRNA generation [109–111], although the RNAP activity of RNAPIV and the presence of its transcripts in cells have not yet been demonstrated. One foundation of this assumption is the observation that RNAPIV interacts physically with RNA-dependent RNA polymerase 2 (RDR2), which is required for siRNA generation [112]. The RNAPIV transcripts

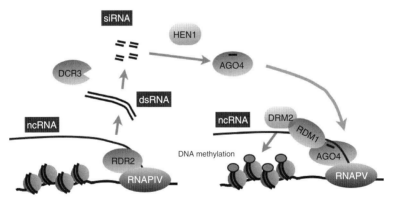

Fig. 9 RNA-dependent DNA methylation in plants. In plants, ncRNAs and siRNAs cooperate to promote DNA methylation, a process analogous to RNAi-dependent heterochromatin formation in fission yeast. Two different RNA polymerases, RNA polymerase IV and V (RNAPIV and RNAPV), function at different steps. RNAPV produces ncRNAs that serve as substrates for siRNA production by RNA-dependent RNA polymerase (RDR2) and a Dicer homolog (DCR3). RNAPIV transcribes target loci to produce ncRNA, which serves as a target for the effector AGO4 in complex with siRNA. AGO4 also interacts with RNAPV and recruits the DNA methyltransferase, DRM2, via RDM1 to methylate DNA. HEN1 modifies the 5′ ends of siRNAs as they mature. For details, see Sect. 3.3.

seem to be immediately converted to dsRNAs and subsequently degraded into 24-nt siRNAs by DICER-LIKE 3 (DCL3) [113]. After the maturation process, which involves the addition of methyl groups to the 3' ends of siRNAs by HUA ENHANCER (HEN1) [99], the siRNAs are loaded onto ARGONAUTE4 (AGO4) to form a silencing–effector complex [114].

The AGO4-containing silencing–effector complex is recruited to target loci by RNAPV [114–117]. RNAPV transcribes the target loci, and the resulting ncRNAs accumulate at those loci. Because AGO4 interacts with the C-terminal domain of the largest subunit of RNAPV, RNAPV is thought to promote the binding of the AGO4–siRNA complex to ncRNA accumulated at the target loci [118–120]. The AGO4-containing effector complex recruits the DNA methylase DRM2, probably via the bridging protein, RDM1 [121].

Many factors in addition to those described above play roles in both the first and second steps via complex protein–protein interaction networks [13]. At some target loci, RNAPII transcribes the region and is then assumed to be required for the recruitment of RNAPIV and V [122]. The mechanism by which RdDM identifies the target loci remains unknown.

3.4
piRNA and Chromatin Regulation in *Drosophila* and Mammals

As described in Sect. 3.1, small-RNA pathways in metazoans primarily function during PTGS. In both *Drosophila* and mammals, complexes comprising siRNAs and Argonaute-family proteins silence the expression of transposable elements and viral genes in somatic cells via a post-transcriptional mechanism [123, 124]. Piwi, a member of the Argonaute family, associates with another class of 25–29-nt small RNAs called *Piwi-interacting RNAs*, and represses mobile genetic elements in the germline cells of both *Drosophila* and mammals [125]. Two of three Piwi-family proteins in *Drosophila* – AUB and AGO3 – function in the cytoplasm to repress transposons by cleaving transcripts. By contrast, Piwi resides in the nucleus, and the elimination of Piwi causes changes in histone modifications on transposable elements [126, 127]. Note that *Drosophila* lacks DNA methylation. Piwi targets transposable elements and represses transcription by promoting H3K9 methylation and the recruitment of HP1 [128, 129], a mechanism that may be analogous to that of RNAi-dependent heterochromatin formation in fission yeast.

In mammals, the Piwi protein Miwi2, which associates with piRNAs derived from transposable elements, is expressed during the period of *de novo* DNA methylation in germline cells and resides in the nucleus. The elimination of Miwi2 correlates with a reduction in DNA methylation at transposable elements [130–132], suggesting that, as in plants, a piRNA-dependent DNA-methylation system exists in mammals. However, the mechanistic details of this system remain to be elucidated.

4
Long RNA-Mediated Chromatin Regulation

In a recent study, it was reported that lncRNAs play a role in chromatin regulation, generally functioning as a platform for chromatin-modifying enzymes, thereby recruiting the latter enzymes to target loci. This is analogous to the function of ncRNA during

RNAi-dependent heterochromatin formation in fission yeast, in which RNAi factors such as the RITS complex and RDRC (see Fig. 5) assemble on ncRNA. Chromatin-related lncRNAs can be further categorized according to whether they regulate chromatin in *cis* or in *trans*. Examples of each category are described in the following sections.

4.1
Cis-Acting Long RNAs: Lessons from Xist RNA

One classic example of *cis*-acting lncRNA is X-inactive-specific transcript (Xist) RNA, which functions during X-chromosome inactivation [133]. In mammals, one of the X chromosomes in females is almost entirely inactivated to compensate for the difference in X chromosome number between males and females. Xist RNA is 17–20 kb in length and is essential for X-chromosome inactivation. Xist RNA is transcribed from a specific region of the inactive X chromosome, called the X-chromosome inactivation center (XIC), and coats the entire X chromosome in *cis* [134, 135] (Fig. 10a). Xist RNA then recruits Polycomb complex 2 (PRC2), which contains EZH2, an H3K27-specific methyltransferase. EZH2 induces

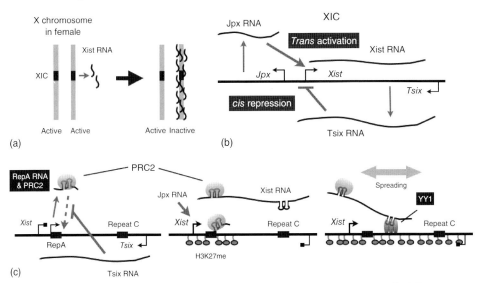

Fig. 10 X-chromosome inactivation by Xist RNA. (a) X-chromosome inactivation. Xist RNA is transcribed from the X-chromosome inactivation center (XIC) of the X chromosome to be inactivated. Xist RNA associates with XIC and then covers the entire X chromosome in *cis*; (b) Regulation of Xist RNA transcription. An antisense RNA of Xist, Tsix RNA, is transcribed to repress *Xist* transcription on the active X. The Jpx ncRNA, transcribed from the upstream region of *Xist*, activates *Xist* expression in *trans*; (c) Functions of RepA RNA and YY1 during X-chromosome inactivation. The short RepA RNA is transcribed from within the *Xist* locus and binds to RPC2 via its stem–loop structure. Tsix RNA prevents the RepA/PRC2 complex from associating with XIC. When Tsix is repressed, RepA/PRC2 associates with XIC and promotes the methylation of H3K27. Xist RNA also associates with PRC2. The transcription factor, YY1, binds to specific sites in *Xist* and simultaneously associates with Xist RNA via its C-repeat sequence. Xist RNA then spreads throughout the X chromosome to inactivate it. For details of panels (a) to (c), see Sect. 4.1.

H3K27 methylation and the subsequent formation of Polycomb-dependent facultative heterochromatin throughout the chromosome [136, 137]. Thus, in this system, Xist RNA connects chromatin and histone-modifying enzymes.

Two other ncRNAs, Tsix and Jpx, regulate the expression of Xist (Fig. 10b). Tsix RNA, which is an antisense transcript of Xist expressed from the active X chromosome, inhibits the function of Xist [136, 138] in several ways [139]. One of these inhibitory mechanisms involves recruiting DNA methyltransferase (Dnmt3) to silence the *Xist* gene [140, 141], which is analogous to the mechanism observed in plant RdDM (Fig. 9). Jpx RNA, which is transcribed upstream of Xist, is required to activate Xist on the inactive X chromosome. Expression of Jpx RNA from an autosomal locus rescues the deletion of *Jpx*, indicating that Jpx RNA is diffusible and acts in *trans* [142].

The mechanism by which Xist RNA recruits PRC2 is complicated (Fig. 10c). At the initiation of X-chromosome inactivation, PRC2 first binds to RepA RNA, which is a short (ca. 1.6 kb) ncRNA transcribed from the first exon of *Xist*. RepA RNA is required to recruit PRC2 to the *Xist* promoter [137]. RepA RNA contains a 28-nt repeated sequence that forms a stem–loop structure that is recognized by PRC2 via its EZH2 subunit. Docking of the RepA–PRC2 complex to the *Xist* promoter is precluded by Tsix RNA expression. When Tsix transcription is repressed during development, the RepA–PRC2 complex associates with XIC and deposits H3K27 methylation. It is not yet clear how the complex is tethered to XIC.

After inducing full-length Xist, PRC2 binds to the RepA sequence of Xist RNA and the Xist–PRC2 complex is then tethered to XIC, from which it spreads across the entire X chromosome [137]. A key factor involved in tethering Xist RNA was identified recently (Fig. 10c). The transcription factor, YY1, functions to both activate and repress transcription, and YY1-binding sites are present in XIC [143–147]. The depletion of YY1 from female cells causes a release of Xist RNA from the inactive X, indicating that YY1 is essential for the recruitment of Xist RNA [148]. Detailed analyses have shown that YY1 forms a bridge between the inactive X and Xist RNA through its bivalent activity; namely, sequence-specific DNA and RNA binding. YY1 binds to three sites in XIC of the inactive X as well as to a type of repeated sequence in Xist RNA, called the *C repeats*. Binding of YY1 to the active X XIC seems to be blocked (via an unknown mechanism) by one of the mechanisms which ensures that Xist RNA acts only in *cis* [148]. Although YY1 explains the mechanism by which Xist RNA binds to XIC, the mechanism by which Xist RNA spreads along the inactive X remains unclear.

4.2 Other *Cis*-Acting Long RNAs

Chromatin-regulation by cis-acting lncRNA is observed in budding yeast, a well-known model organism. Antisense lncRNAs repress PHO84 gene through histone deacetylation by recruiting HDACs, though the mechanism underlies the recruitment is not clear [149]. Since antisense lncRNAs are destabilized by Rrp6-dependent RNA surveillance system, PHO84 is expressed in the normal condition, but the degradation is suppressed during chronological aging, causing the repression of PHO84 [149].

LncRNA-mediated chromatin regulation (which is similar to that of Xist

RNA) has been observed during the silencing of various imprinted genes in mammals. Some gene clusters in either the maternal or paternal chromosomes are silenced [150]. Recent investigations have shown that lncRNAs are transcribed from these loci and play roles in silencing [129, 151–153]. For example, genes within the locus containing insulin-like growth factor receptor 2 (*Igfr2*) are paternally repressed in mouse placenta. An antisense lncRNA, called *Air*, is paternally expressed from the second intron of Igf2r and induces epigenetic silencing [154]. Air binds not only to the promoter regions of *Igf2r* but also to those of its neighboring genes; the mechanism of binding is not yet understood, although by analogy to Xist RNA it can be imagined that the secondary structure of Air RNA and its binding partner contribute to this binding. Promoter-bound Air recruits G9a, a H3K9-specific methyltransferase, thereby forming a repressive chromatin structure [155]. Similarly, the Kcnqlot1 lncRNA is paternally transcribed from the imprinted gene *Kcnq1*. This lncRNA silences *Kcnq1* and its neighboring genes in *cis* by recruiting PRC2 and G9a, resulting in a deposition of the repressive histone modifications H3K27me and H3K9me [156, 157]. The basic role of lncRNA in imprinting is quite similar to that of Xist RNA in X-chromosome inactivation; however, spreading of the repressive chromatin structure is limited to the imprinted region, the extent of which is related to the positions of various insulator sequences [158, 159].

LncRNA-dependent chromatin regulation has also been observed in the epigenetic regulation of autosomal loci. The *INK4b/ARF/INK4a* locus encodes two cyclin-dependent kinase inhibitors, p15INK4b and p16INK4a, and a regulator of the p53 pathway, ARF, all of which are tumor-suppressor genes that are down-regulated in several types of tumor cell [160]. ANR1L is an antisense lncRNA, which is transcribed from this locus [161]; ANR1L expression is inversely correlated with p15INK4b expression in some leukemias, and it mediates the silencing of p15INK4b and p16INK4a through DNA methylation and heterochromatin formation [162]. In another study, ANR1L was shown to interact with CBX7, a component of PRC1. In cancer cells, ANR1L colocalizes with EZH2 and CBX7 at the p16INK4a promoter, thereby suppressing the expression of p16INK4a [163]. PRC2 undergoes a similar interaction with lncRNA to regulate the cyclin-dependent kinase inhibitor p21, an important tumor-suppressor gene. Antisense lncRNA, transcribed at the p21 locus, represses p21 gene transcription via the deposition of the H3K27me3 marker [164]. Recent studies have provided additional examples of lncRNA involvement in the regulation of autosomal genes via the deposition of repressive histone markers (H3K27me and H3K9me) and/or DNA methylation [151, 165].

As described above, many types of lncRNAs associate with PRC2 and recruit the complex to target loci for silencing. In addition, short RNAs may also contribute to PRC2 targeting. H3K27me3 associates with the transcription-initiation sites of PRC2 target genes [166–168]. In fact, short RNAs, of approximately 50 to 200 nt, are transcribed by RNAPII around the initiation site of PRC2 target genes in the absence of mRNA transcription [169]. These short RNAs form stem–loop structures that resemble the PRC2-binding sites in *Xist* RepA RNA and interact with PRC2 via SUZ12. Based on the observation that these short RNAs are depleted from PRC2 target genes when the genes are activated during cell differentiation, they appear to play a role in the association of PRC2 with the target genes it represses [169].

NcRNAs frequently localize at the promoter region of genes. These ncRNAs are relatively short (50–200 nt) and are thought to be abortive transcripts made by paused RNAPII. However, at least some of these short RNAs seem to play a role in regulating gene expression by facilitating the recruitment of transcriptional repressors (as described above for lncRNA). For example, short ncRNAs transcribed at the *CCND1* promoter, which encodes the cell-cycle regulator cyclin D1, tether the transcriptional repressor TLS and allosterically modify the repressor to silence transcription [170]

In addition to lncRNA-dependent repression, lncRNA-dependent activation of some autosomal genes may occur. The anti-sense lncRNAs, Evx1as and Hox5/6as, are transcribed from the *Evx* and *Hoxb5/6* genes during a specific stage of mouse embryonic stem cell differentiation [171]. These lncRNAs associate with chromatin-containing H3K4me3, which marks active chromatin. Moreover, Evx1as and Hox5/6as RNAs also associate with MLL1, the mammalian Trithorax protein responsible for H3K4 methylation [171]. Similarly, the mouse forebrain-specific lncRNA, Evf2, which is transcribed from an enhancer sited between *Dlx5* and *Dlx6*, forms a complex with the DLX2 homeodomain protein to activate the enhancer and promote *Dlx5/Dlx6* expression [172]. Furthermore, the results of a recent study have suggested that Evf2 also functions to repress *Dlx5* and *Dlx6* via antisense interference and the recruitment of a repressor complex, respectively, at a different stage of differentiation [173]. Taken together, the results of these studies highlight the importance of lncRNAs during development.

Similarly, the lncRNA HOTTIP is involved in activating target genes. HOTTIP is transcribed in the opposite orientation from the 5′ end of the *HOX-A* locus [174]. The knockdown of HOTTIP in fibroblasts and chick embryos decreases *HOX-A* expression, which is correlated with a loss of H3K4me2/3 across the affected genes. The reduction in H3K4me2/3 acts in *cis* and depends on the distance from the *HOTTIP* gene; genes closer to *HOTTIP* show a greater decrease. Because HOTTIP associates with WDR5, a component of the core complex involved in H3K4 methylation, it is likely that HOTTIP recruits this complex to target loci [174].

4.3
Trans-Acting Long RNAs

In budding yeast, antisense lcnRNA control Ty1 retroposon mobility by repressing Ty1 transcription through histone deacetylation [175], which is resemble to the control of PHO84 (Sect 4.3). However, the lcnRNA is degraded by the cytoplasmic 5′-3′ Xrn1 exonuclease, which suggest the lcnRNA acts in trans. Recent genome wide study shows that many Xrn1-sensitive unstable transcripts (XUTs) are antisense to open reading frames. In many cases, the expression of genes with XUTs is suppressed at the transcription level. Thus, XUTs that escape from Xrn1-dependent degradation may form silencing complex for transcriptional gene silencing [176].

Drosophila, which like mammals has X and Y chromosomes for gender determination, employs a different strategy to compensate for the difference in the number of X chromosomes: a twofold upregulation of all the genes on the single male X chromosome [177]. To achieve this upregulation, the Dosage Compensation Complex [DDC; also called the

Male-Specific Lethal (MSL) complex] localizes to many different binding sites on the male X chromosome. Two lncRNAs, roX1 and roX2, which are transcribed from the male X chromosome and components of DDC, play an essential role in targeting DDC to the male X chromosome. Based on the observation that roX1 still coats the male X chromosome when its locus is translocated onto an autosome, this RNA can function in *trans*, unlike Xist RNA [178].

As in the case of Xist, the *cis*-acting lncRNAs described above are often transcribed as anti-sense RNAs of their target loci. The cotranscriptional tethering of lncRNAs to chromatin, followed by binding to histone modifiers, might explain the *cis*-acting nature of these lncRNAs. However, many eukaryotic genomes produce large numbers of large intergenic non-coding RNAs (lincRNAs) [179]. In human cells, EZH2, a component of PRC2 and a H3K27-specific histone methyltransferase, binds a large number (more than 200) of lincRNAs [180]. This observation strongly suggests that lincRNA also functions in the epigenetic regulation of gene expression by altering chromatin structure. Indeed, some lincRNAs act as chromatin regulators, and can regulate chromatin structure at distant loci – that is, they exert their functions in *trans* [180].

One example of lincRNA-dependent chromatin regulation is HOTAIR RNA (Fig. 11). HOTAIR is encoded in an antisense orientation in the *HOX-C* cluster on chromosome 12 [181]. HOX genes are involved in differentiation; in mouse fibroblast cells, in which HOTAIR is expressed, the *HOX-D* cluster on chromosome 2 is repressed by deposition of the H3K27me repressive marker. In addition to H3K27me, the *HOX-D* cluster is also associated with PRC2. HOTAIR associates with SUZ12, a component of PRC2, and depletion of HOTAIR causes a de-repression of *HOX-D* (concomitant with a reduction in H3K27me and PRC2 at that locus), indicating that HOTAIR forms a complex with PRC2 and targets it to the *HOX-C* cluster to repress expression [181]. Further analysis has revealed that the repressed *HOX-D* contains low levels of H3K4me2/3, a marker of active chromatin [182]. Accordingly, HOTAIR associates with LSD1, an H3K4me-specific demethylase, which coordinates the methylation of H3K27 and the demethylation of H3K4 to form repressive chromatin [183]. A genome-wide analysis has revealed that more than 700 repressed genes in fibroblasts associate with high levels of SUZ12 and LSD1. About 40% of the genes occupied by SUZ12 and LSD1 lost both factors upon depletion of HOTAIR, suggesting that HOTAIR is involved in another LSD1/SUZ12-dependent silencing pathway through the coordinated recruitment of both complexes [183]. Although HOTAIR is required for the association of the PRC2/LSD1 complex with target loci, the mechanism underlying this recruitment is not known. The HOTAIR-dependent SUZ12- and LSD1-binding sequences share common motifs, which may indicate the existence of specific binding partners that are involved in tethering the HOTAIR–PRC2/LSD1 complex [183].

4.4
Structural Roles of Long RNAs in Chromatin Organization

In some cases, lncRNAs are integral components of chromatin and play roles in its structural organization. In mouse nuclei, constitutive heterochromatin – which is assembled on major satellite repeats at pericentromeric regions – forms large foci

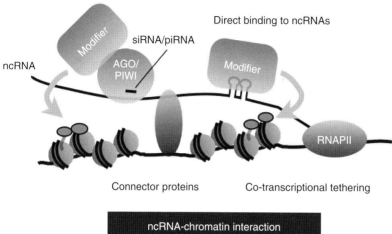

Fig. 11 Schematic showing the functions of RNAs during chromatin regulation. For details, see Sect. 5.

that can be visualized by DAPI staining. These loci colocalize with H3K9me and HP1α. RNase treatment of cells causes the loss of DAPI-dense foci, although the total amounts of H3K9me and HP1α are unaffected; this indicates that RNA brings histones bearing the repressive marker together in a specific configuration within the nucleus [184]. Furthermore, lncRNAs derived from the major satellite repeats are present at the pericentromere [185]. HP1α binds these RNAs, and this binding activity, together with its H3K9me-binding activity, is required for the localization of HP1α on heterochromatin [186]. HP1α (bound to satellite lncRNAs) is modified at its hinge domain by SUMO; both the hinge domain and SUMOylation are critical for the initial targeting of HP1α to pericentromeres, but they are not required to maintain HP1α at these loci [185].

Another example of the structural role of lncRNA is Xist RNA. The human inactive X chromosome forms a compact structure called the *Barr body*, and repressive histone markers (H3K9me- and H3K27me-rich domains) are interspersed throughout the inactive X [187, 188]. The H3K27me domains are associated with Xist RNA, whereas H3K9me domains are associated with HP1α [187–189]. The structural maintenance of chromosomes hinge domain-containing protein 1 (SMCHD1) localizes to H3K27me domains by associating with Xist RNA, whereas HBiX1, an HP1-binding protein, localizes at H3K9me domains by binding to HP1α. SMCHD1 interacts tightly with HBiX1, which brings the H3K27me and H3K9me domains together to form a compact structure. The depletion of Xist RNA, as well as depleting either SMCHD1 or HBiX1, causes a decompaction of the Barr body.

Therefore, Xist RNA, together with SMCHD1, HBiX1, and HP1α, contribute to the organization of the compact Barr body [188].

Recently, lncRNA is shown to play an active role in recognition of homologous chromosomes for pairing in meiosis [190]. This finding further expands the function of lncRNA from the regulation of gene expression to chromosome function.

5
Summary

In this chapter, the ncRNAs have been classified as short and/or long, and their roles during chromatin regulation have been described; however, the mechanisms underlying chromatin regulation by both types of ncRNA share common features (as summarized in Fig. 11). ncRNAs bring chromatin/DNA modifiers to the target loci by interacting with chromatin/DNA modifiers and by associating with their target loci (Fig. 11).

First, ncRNAs associate with chromatin/DNA modifiers, such as histone-modification proteins or Dnmt. In some cases (e.g., PRC2 and Xist) chromatin regulators directly recognize an ncRNA, based on its secondary structure. During siRNA/piRNA-mediated heterochromatin formation, complexes that contain Argonaute-family proteins and small RNAs connect ncRNAs with chromatin regulators via complementary binding between the small RNAs and their target RNAs.

Second, ncRNAs associate with target loci via distinct mechanisms that are not yet fully understood. One such mechanism is cotranscriptional, in which case chromatin regulators interact with nascent RNAs as they are transcribed, and this results in the recruitment of chromatin regulators to targets in *cis*. The pausing or stalling of RNAPs, or the cotranscriptional formation of DNA–RNA hybrids, may promote the stable association of transcribed ncRNAs with their target loci. Another mechanism involves connector proteins, such as YY1 for Xist/RepA RNA, which mediate the association between ncRNAs and target loci. Although, to date, few connectors have been described, further analyses should identify additional proteins that connect RNA and chromatin. The functional characterization of these proteins should then help to explain the targeting mechanisms of *trans*-acting RNAs, such as HOTAIR.

Clearly, studies of RNA-mediated chromatin regulation are still in their infancy. However, future analyses – including genome-wide profiling techniques – should further reveal the importance of RNAs in the epigenetic regulation of chromatin.

References

1 Jenuwein, T., Allis, C.D. (2001) Translating the histone code. *Science*, **293**, 1074–1080.
2 Bannister, A.J., Zegerman, P., Partridge, J.F., Miska, E.A. *et al.* (2001) Selective recognition of methylated lysine 9 on histone H3 by the HP1 chromo domain. *Nature*, **410**, 120–124.
3 Lachner, M., O'Carroll, D., Rea, S., Mechtler, K. *et al.* (2001) Methylation of histone H3 lysine 9 creates a binding site for HP1 proteins. *Nature*, **410**, 116–120.
4 Nakayama, J., Rice, J.C., Strahl, B.D., Allis, C.D. *et al.* (2001) Role of histone H3 lysine 9 methylation in epigenetic control of heterochromatin assembly. *Science*, **292**, 110–113.
5 Cao, R., Wang, L., Wang, H., Xia, L. *et al.* (2002) Role of histone H3 lysine 27 methylation in Polycomb-group silencing. *Science*, **298**, 1039–1043.
6 Czermin, B., Melfi, R., McCabe, D., Seitz, V. *et al.* (2002) Drosophila enhancer

of Zeste/ESC complexes have a histone H3 methyltransferase activity that marks chromosomal Polycomb sites. *Cell*, **111**, 185–196.

7 Kuzmichev, A., Nishioka, K., Erdjument-Bromage, H., Tempst, P. *et al.* (2002) Histone methyltransferase activity associated with a human multiprotein complex containing the Enhancer of Zeste protein. *Genes Dev.*, **16**, 2893–2905.

8 Lichtneckert, R., Muller, P., Schmid, V., Reber-Muller, S. (2002) Evolutionary conservation of the chromatin modulator Polycomb in the jellyfish *Podocoryne carnea*. *Differentiation*, **70**, 422–428.

9 Muller, J., Hart, C.M., Francis, N.J., Vargas, M.L. *et al.* (2002) Histone methyltransferase activity of a *Drosophila* Polycomb group repressor complex. *Cell*, **111**, 197–208.

10 Craig, J.M. (2005) Heterochromatin – many flavours, common themes. *BioEssays*, **27**, 17–28.

11 Sarraf, S.A., Stancheva, I. (2004) Methyl-CpG binding protein MBD1 couples histone H3 methylation at lysine 9 by SETDB1 to DNA replication and chromatin assembly. *Mol. Cell*, **15**, 595–605.

12 Cokus, S.J., Feng, S., Zhang, X., Chen, Z. *et al.* (2008) Shotgun bisulphite sequencing of the *Arabidopsis* genome reveals DNA methylation patterning. *Nature*, **452**, 215–219.

13 Saze, H., Tsugane, K., Kanno, T., Nishimura, T. (2012) DNA methylation in plants: relationship to small RNAs and histone modifications, and functions in transposon inactivation. *Plant Cell Physiol.*, **53**, 766–784.

14 Birney, E., Stamatoyannopoulos, J.A., Dutta, A., Guigo, R. *et al.* (2007) Identification and analysis of functional elements in 1% of the human genome by the ENCODE pilot project. *Nature*, **447**, 799–816.

15 Carninci, P., Kasukawa, T., Katayama, S., Gough, J. *et al.* (2005) The transcriptional landscape of the mammalian genome. *Science*, **309**, 1559–1563.

16 Clark, M.B., Amaral, P.P., Schlesinger, F.J., Dinger, M.E. *et al.* (2011) The reality of pervasive transcription. *PLoS Biol.*, **9**, e1000625; discussion die1001102.

17 Katayama, S., Tomaru, Y., Kasukawa, T., Waki, K. *et al.* (2005) Antisense transcription in the mammalian transcriptome. *Science*, **309**, 1564–1566.

18 Mondal, T., Rasmussen, M., Pandey, G.K., Isaksson, A. *et al.* (2010) Characterization of the RNA content of chromatin. *Genome Res.*, **20**, 899–907.

19 Rodriguez-Campos, A., Azorin, F. (2007) RNA is an integral component of chromatin that contributes to its structural organization. *PLoS ONE*, **2**, e1182.

20 Fire, A., Xu, S., Montgomery, M.K., Kostas, S.A. *et al.* (1998) Potent and specific genetic interference by double-stranded RNA in *Caenorhabditis elegans*. *Nature*, **391**, 806–811.

21 Montgomery, M.K., Xu, S., Fire, A. (1998) RNA as a target of double-stranded RNA-mediated genetic interference in *Caenorhabditis elegans*. *Proc. Natl Acad. Sci. USA*, **95**, 15502–15507.

22 Thomson, T., Lin, H. (2009) The biogenesis and function of PIWI proteins and piRNAs: progress and prospect. *Annu. Rev. Cell Dev. Biol.*, **25**, 355–376.

23 Volpe, T.A., Kidner, C., Hall, I.M., Teng, G. *et al.* (2002) Regulation of heterochromatic silencing and histone H3 lysine-9 methylation by RNAi. *Science*, **297**, 1833–1837.

24 Wassenegger, M., Heimes, S., Riedel, L., Sanger, H.L. (1994) RNA-directed de novo methylation of genomic sequences in plants. *Cell*, **76**, 567–576.

25 Bernstein, E., Caudy, A.A., Hammond, S.M., Hannon, G.J. (2001) Role for a bidentate ribonuclease in the initiation step of RNA interference. *Nature*, **409**, 363–366.

26 Hamilton, A.J., Baulcombe, D.C. (1999) A species of small antisense RNA in post-transcriptional gene silencing in plants. *Science*, **286**, 950–952.

27 Hammond, S.M., Bernstein, E., Beach, D., Hannon, G.J. (2000) An RNA-directed nuclease mediates post-transcriptional gene silencing in *Drosophila* cells. *Nature*, **404**, 293–296.

28 Tuschl, T., Zamore, P.D., Lehmann, R., Bartel, D.P. *et al.* (1999) Targeted mRNA degradation by double-stranded RNA in vitro. *Genes Dev.*, **13**, 3191–3197.

29 Hammond, S.M., Boettcher, S., Caudy, A.A., Kobayashi, R. *et al.* (2001) Argonaute2, a link between genetic and biochemical analyses of RNAi. *Science*, **293**, 1146–1150.

30 Liu, J., Carmell, M.A., Rivas, F.V., Marsden, C.G. et al. (2004) Argonaute2 is the catalytic engine of mammalian RNAi. *Science*, **305**, 1437–1441.

31 Song, J.J., Smith, S.K., Hannon, G.J., Joshua-Tor, L. (2004) Crystal structure of Argonaute and its implications for RISC slicer activity. *Science*, **305**, 1434–1437.

32 Matranga, C., Tomari, Y., Shin, C., Bartel, D.P. et al. (2005) Passenger-strand cleavage facilitates assembly of siRNA into Ago2-containing RNAi enzyme complexes. *Cell*, **123**, 607–620.

33 Miyoshi, K., Tsukumo, H., Nagami, T., Siomi, H. et al. (2005) Slicer function of *Drosophila* Argonautes and its involvement in RISC formation. *Genes Dev.*, **19**, 2837–2848.

34 Rand, T.A., Petersen, S., Du, F., Wang, X. (2005) Argonaute2 cleaves the anti-guide strand of siRNA during RISC activation. *Cell*, **123**, 621–629.

35 Lipardi, C., Wei, Q., Paterson, B.M. (2001) RNAi as random degradative PCR: siRNA primers convert mRNA into dsRNAs that are degraded to generate new siRNAs. *Cell*, **107**, 297–307.

36 Sijen, T., Fleenor, J., Simmer, F., Thijssen, K.L. et al. (2001) On the role of RNA amplification in dsRNA-triggered gene silencing. *Cell*, **107**, 465–476.

37 Nakayama, J., Klar, A.J., Grewal, S.I. (2000) A chromodomain protein, Swi6, performs imprinting functions in fission yeast during mitosis and meiosis. *Cell*, **101**, 307–317.

38 Sadaie, M., Kawaguchi, R., Ohtani, Y., Arisaka, F. et al. (2008) Balance between distinct HP1 proteins controls heterochromatin assembly in fission yeast. *Mol. Cell. Biol.*, **28**, 6973–6988.

39 Djupedal, I., Portoso, M., Spahr, H., Bonilla, C. et al. (2005) RNA Pol II subunit Rpb7 promotes centromeric transcription and RNAi-directed chromatin silencing. *Genes Dev.*, **19**, 2301–2306.

40 Kato, H., Goto, D.B., Martienssen, R.A., Urano, T. et al. (2005) RNA polymerase II is required for RNAi-dependent heterochromatin assembly. *Science*, **309**, 467–469.

41 Noma, K., Sugiyama, T., Cam, H., Verdel, A. et al. (2004) RITS acts in cis to promote RNA interference-mediated transcriptional and post-transcriptional silencing. *Nat. Genet.*, **36**, 1174–1180.

42 Verdel, A., Jia, S., Gerber, S., Sugiyama, T. et al. (2004) RNAi-mediated targeting of heterochromatin by the RITS complex. *Science*, **303**, 672–676.

43 Motamedi, M.R., Verdel, A., Colmenares, S.U., Gerber, S.A. et al. (2004) Two RNAi complexes, RITS and RDRC, physically interact and localize to noncoding centromeric RNAs. *Cell*, **119**, 789–802.

44 Sugiyama, T., Cam, H., Verdel, A., Moazed, D. et al. (2005) RNA-dependent RNA polymerase is an essential component of a self-enforcing loop coupling heterochromatin assembly to siRNA production. *Proc. Natl Acad. Sci. USA*, **102**, 152–157.

45 Hong, E.J., Villén, J., Gerace, E.L., Gygi, S.P. et al. (2005) A cullin E3 ubiquitin ligase complex associates with Rik1 and the Clr4 histone H3-K9 methyltransferase and is required for RNAi-mediated heterochromatin formation. *RNA Biol.*, **2**, 106–111.

46 Jia, S., Kobayashi, R., Grewal, S.I. (2005) Ubiquitin ligase component Cul4 associates with Clr4 histone methyltransferase to assemble heterochromatin. *Nat. Cell Biol.*, **7**, 1007–1013.

47 Colmenares, S.U., Buker, S.M., Buhler, M., Dlakić, M. et al. (2007) Coupling of double-stranded RNA synthesis and siRNA generation in fission yeast RNAi. *Mol. Cell*, **27**, 449–461.

48 Buker, S.M., Iida, T., Bühler, M., Villén, J. et al. (2007) Two different Argonaute complexes are required for siRNA generation and heterochromatin assembly in fission yeast. *Nat. Struct. Mol. Biol.*, **14**, 200–207.

49 Bayne, E.H., White, S.A., Kagansky, A., Bijos, D.A. et al. (2010) Stc1: a critical link between RNAi and chromatin modification required for heterochromatin integrity. *Cell*, **140**, 666–677.

50 Rougemaille, M., Shankar, S., Braun, S., Rowley, M. et al. (2008) Ers1, a rapidly diverging protein essential for RNA interference-dependent heterochromatic silencing in *Schizosaccharomyces pombe*. *J. Biol. Chem.*, **283**, 25770–25773.

51 Hayashi, A., Ishida, M., Kawaguchi, R., Urano, T. et al. (2012) Heterochromatin protein 1 homologue Swi6 acts in concert with Ers1 to regulate RNAi-directed heterochromatin assembly. *Proc. Natl Acad. Sci. USA*, **109**, 6159–6164.

52. Kawakami, K., Hayashi, A., Nakayama, J., Murakami, Y. (2012) A novel RNAi protein, Dsh1, assembles RNAi machinery on chromatin to amplify heterochromatic siRNA. *Genes Dev.*, **26**, 1811–1824.

53. Emmerth, S., Schober, H., Gaidatzis, D., Roloff, T. et al. (2010) Nuclear retention of fission yeast dicer is a prerequisite for RNAi-mediated heterochromatin assembly. *Dev. Cell*, **18**, 102–113.

54. Towbin, B.D., Meister, P., Gasser, S.M. (2009) The nuclear envelope – a scaffold for silencing? *Curr. Opin. Genet. Dev.*, **19**, 180–186.

55. Sugiyama, T., Cam, H.P., Sugiyama, R., Noma, K. et al. (2007) SHREC, an effector complex for heterochromatic transcriptional silencing. *Cell*, **128**, 491–504.

56. Yamada, T., Fischle, W., Sugiyama, T., Allis, C.D. et al. (2005) The nucleation and maintenance of heterochromatin by a histone deacetylase in fission yeast. *Mol. Cell*, **20**, 173–185.

57. Shimada, A., Dohke, K., Sadaie, M., Shinmyozu, K. et al. (2009) Phosphorylation of Swi6/HP1 regulates transcriptional gene silencing at heterochromatin. *Genes Dev.*, **23**, 18–23.

58. Freeman-Cook, L.L., Sherman, J.M., Brachmann, C.B., Allshire, R.C. et al. (1999) The *Schizosaccharomyces pombe* hst4(+) gene is a SIR2 homologue with silencing and centromeric functions. *Mol. Biol. Cell*, **10**, 3171–3186.

59. Grewal, S.I., Bonaduce, M.J., Klar, A.J. (1998) Histone deacetylase homologs regulate epigenetic inheritance of transcriptional silencing and chromosome segregation in fission yeast. *Genetics*, **150**, 563–576.

60. Nicolas, E., Yamada, T., Cam, H.P., Fitzgerald, P.C. et al. (2007) Distinct roles of HDAC complexes in promoter silencing, antisense suppression and DNA damage protection. *Nat. Struct. Mol. Biol.*, **14**, 372–380.

61. Shankaranarayana, G.D., Motamedi, M.R., Moazed, D., Grewal, S.I. (2003) Sir2 regulates histone H3 lysine 9 methylation and heterochromatin assembly in fission yeast. *Curr. Biol.*, **13**, 1240–1246.

62. Zofall, M., Grewal, S.I. (2006) Swi6/HP1 recruits a JmjC domain protein to facilitate transcription of heterochromatic repeats. *Mol. Cell*, **22**, 681–692.

63. Ayoub, N., Noma, K., Isaac, S., Kahan, T. et al. (2003) A novel jmjC domain protein modulates heterochromatization in fission yeast. *Mol. Cell. Biol.*, **23**, 4356–4370.

64. Klose, R.J., Kallin, E.M., Zhang, Y. (2006) JmjC-domain-containing proteins and histone demethylation. *Nat. Rev. Genet.*, **7**, 715–727.

65. Trewick, S.C., McLaughlin, P.J., Allshire, R.C. (2005) Methylation: lost in hydroxylation? *EMBO Rep.*, **6**, 315–320.

66. Tsukada, Y., Fang, J., Erdjument-Bromage, H., Warren, M.E. et al. (2006) Histone demethylation by a family of JmjC domain-containing proteins. *Nature*, **439**, 811–816.

67. Trewick, S.C., Minc, E., Antonelli, R., Urano, T. et al. (2007) The JmjC domain protein Epe1 prevents unregulated assembly and disassembly of heterochromatin. *EMBO J.*, **26**, 4670–4682.

68. Bühler, M., Haas, W., Gygi, S.P., Moazed, D. (2007) RNAi-dependent and -independent RNA turnover mechanisms contribute to heterochromatic gene silencing. *Cell*, **129**, 707–721.

69. Irvine, D.V., Zaratiegui, M., Tolia, N.H., Goto, D.B. et al. (2006) Argonaute slicing is required for heterochromatic silencing and spreading. *Science*, **313**, 1134–1137.

70. Simmer, F., Buscaino, A., Kos-Braun, I.C., Kagansky, A. et al. (2010) Hairpin RNA induces secondary small interfering RNA synthesis and silencing in trans in fission yeast. *EMBO Rep.*, **11**, 112–118.

71. LaCava, J., Houseley, J., Saveanu, C., Petfalski, E. et al. (2005) RNA degradation by the exosome is promoted by a nuclear polyadenylation complex. *Cell*, **121**, 713–724.

72. Vanacova, S., Wolf, J., Martin, G., Blank, D. et al. (2005) A new yeast poly(A) polymerase complex involved in RNA quality control. *PLoS Biol.*, **3**, e189.

73. Wyers, F., Rougemaille, M., Badis, G., Rousselle, J.C. et al. (2005) Cryptic pol II transcripts are degraded by a nuclear quality control pathway involving a new poly(A) polymerase. *Cell*, **121**, 725–737.

74. Keller, C., Adaixo, R., Stunnenberg, R., Woolcock, K.J. et al. (2012) HP1(Swi6) mediates the recognition and destruction of heterochromatic RNA transcripts. *Mol. Cell*, **47**, 215–227.

75 Chen, E.S., Zhang, K., Nicolas, E., Cam, H.P. et al. (2008) Cell cycle control of centromeric repeat transcription and heterochromatin assembly. *Nature*, **451**, 734–737.

76 Kloc, A., Zaratiegui, M., Nora, E., Martienssen, R. (2008) RNA interference guides histone modification during the S phase of chromosomal replication. *Curr. Biol.*, **18**, 490–495.

77 Kim, S.M., Dubey, D.D., Huberman, J.A. (2003) Early-replicating heterochromatin. *Genes Dev.*, **17**, 330–335.

78 Hirota, T., Lipp, J.J., Toh, B.H., Peters, J.M. (2005) Histone H3 serine 10 phosphorylation by Aurora B causes HP1 dissociation from heterochromatin. *Nature*, **438**, 1176–1180.

79 Giet, R., Glover, D.M. (2001) *Drosophila* aurora B kinase is required for histone H3 phosphorylation and condensin recruitment during chromosome condensation and to organize the central spindle during cytokinesis. *J. Cell Biol.*, **152**, 669–682.

80 Carlsten, J.O., Szilagyi, Z., Liu, B., Lopez, M.D. et al. (2012) Mediator promotes CENP-a incorporation at fission yeast centromeres. *Mol. Cell. Biol.*, **32**, 4035–4043.

81 Kiely, C.M., Marguerat, S., Garcia, J.F., Madhani, H.D. et al. (2011) Spt6 is required for heterochromatic silencing in the fission yeast *Schizosaccharomyces pombe*. *Mol. Cell. Biol.*, **31**, 4193–4204.

82 Thorsen, M., Hansen, H., Venturi, M., Holmberg, S. et al. (2012) Mediator regulates non-coding RNA transcription at fission yeast centromeres. *Epigenetics & Chromatin*, **5**, 19.

83 Zaratiegui, M., Castel, S.E., Irvine, D.V., Kloc, A. et al. (2011) RNAi promotes heterochromatic silencing through replication-coupled release of RNA Pol II. *Nature*, **479**, 135–138.

84 Bayne, E.H., Portoso, M., Kagansky, A., Kos-Braun, I.C. et al. (2008) Splicing factors facilitate RNAi-directed silencing in fission yeast. *Science*, **322**, 602–606.

85 Chinen, M., Morita, M., Fukumura, K., Tani, T. (2010) Involvement of the spliceosomal U4 small nuclear RNA in heterochromatic gene silencing at fission yeast centromeres. *J. Biol. Chem.*, **285**, 5630–5638.

86 Nakama, M., Kawakami, K., Kajitani, T., Urano, T. et al. (2012) DNA-RNA hybrid formation mediates RNAi-directed heterochromatin formation. *Genes Cells*, **17**, 218–233.

87 Huertas, P., Aguilera, A. (2003) Cotranscriptionally formed DNA:RNA hybrids mediate transcription elongation impairment and transcription-associated recombination. *Mol. Cell*, **12**, 711–721.

88 Li, X., Manley, J.L. (2005) Inactivation of the SR protein splicing factor ASF/SF2 results in genomic instability. *Cell*, **122**, 365–378.

89 Mischo, H.E., Gomez-Gonzalez, B., Grzechnik, P., Rondon, A.G. et al. (2011) Yeast Sen1 helicase protects the genome from transcription-associated instability. *Mol. Cell*, **41**, 21–32.

90 Skourti-Stathaki, K., Proudfoot, N.J., Gromak, N. (2011) Human senataxin resolves RNA/DNA hybrids formed at transcriptional pause sites to promote Xrn2-dependent termination. *Mol. Cell*, **42**, 794–805.

91 Ishida, M., Shimojo, H., Hayashi, A., Kawaguchi, R. et al. (2012) Intrinsic nucleic acid-binding activity of Chp1 chromodomain is required for heterochromatic gene silencing. *Mol. Cell*, **47**, 228–241.

92 Jia, S., Noma, K., Grewal, S.I. (2004) RNAi-independent heterochromatin nucleation by the stress-activated ATF/CREB family proteins. *Science*, **304**, 1971–1976.

93 Grewal, S.I., Klar, A.J. (1997) A recombinationally repressed region between mat2 and mat3 loci shares homology to centromeric repeats and regulates directionality of mating-type switching in fission yeast. *Genetics*, **146**, 1221–1238.

94 Sadaie, M., Iida, T., Urano, T., Nakayama, J. (2004) A chromodomain protein, Chp1, is required for the establishment of heterochromatin in fission yeast. *EMBO J.*, **23**, 3825–3835.

95 Lykke-Andersen, S., Tomecki, R., Jensen, T.H., Dziembowski, A. (2011) The eukaryotic RNA exosome: same scaffold but variable catalytic subunits. *RNA Biol.*, **8**, 61–66.

96 Reyes-Turcu, F.E., Zhang, K., Zofall, M., Chen, E. et al. (2011) Defects in RNA quality control factors reveal RNAi-independent nucleation of heterochromatin. *Nat. Struct. Mol. Biol.*, **18**, 1132–1138.

97 Hiriart, E., Vavasseur, A., Touat-Todeschini, L., Yamashita, A.

et al. (2012) Mmi1 RNA surveillance machinery directs RNAi complex RITS to specific meiotic genes in fission yeast. *EMBO J.*, **31**, 2296–2308.

98 Tashiro, S., Asano, T., Kanoh, J., Ishikawa, F. (2013) Transcription-induced chromatin association of RNA surveillance factors mediates facultative heterochromatin formation in fission yeast. *Genes Cells*, **18**, 327–339.

99 Zofall, M., Yamanaka, S., Reyes-Turcu, F.E., Zhang, K. et al. (2012) RNA elimination machinery targeting meiotic mRNAs promotes facultative heterochromatin formation. *Science*, **335**, 96–100.

100 Harigaya, Y., Tanaka, H., Yamanaka, S., Tanaka, K. et al. (2006) Selective elimination of messenger RNA prevents an incidence of untimely meiosis. *Nature*, **442**, 45–50.

101 Yamanaka, S., Yamashita, A., Harigaya, Y., Iwata, R. et al. (2010) Importance of polyadenylation in the selective elimination of meiotic mRNAs in growing *S. pombe* cells. *EMBO J.*, **29**, 2173–2181.

102 Sugiyama, T., Sugioka-Sugiyama, R. (2011) Red1 promotes the elimination of meiosis-specific mRNAs in vegetatively growing fission yeast. *EMBO J.*, **30**, 1027–1039.

103 Yamanaka, S., Mehta, S., Reyes-Turcu, F.E., Zhuang, F. et al. (2013) RNAi triggered by specialized machinery silences developmental genes and retrotransposons. *Nature*, **493**, 557–560.

104 Lister, R., Pelizzola, M., Dowen, R.H., Hawkins, R.D. et al. (2009) Human DNA methylomes at base resolution show widespread epigenomic differences. *Nature*, **462**, 315–322.

105 Law, J.A., Jacobsen, S.E. (2010) Establishing, maintaining and modifying DNA methylation patterns in plants and animals. *Nat. Rev. Genet.*, **11**, 204–220.

106 Haag, J.R., Pikaard, C.S. (2011) Multisubunit RNA polymerases IV and V: purveyors of non-coding RNA for plant gene silencing. *Nat. Rev. Mol. Cell Biol.*, **12**, 483–492.

107 He, X.J., Chen, T., Zhu, J.K. (2011) Regulation and function of DNA methylation in plants and animals. *Cell Res.*, **21**, 442–465.

108 Ream, T.S., Haag, J.R., Wierzbicki, A.T., Nicora, C.D. et al. (2009) Subunit compositions of the RNA-silencing enzymes Pol IV and Pol V reveal their origins as specialized forms of RNA polymerase II. *Mol. Cell*, **33**, 192–203.

109 Herr, A.J., Jensen, M.B., Dalmay, T., Baulcombe, D.C. (2005) RNA polymerase IV directs silencing of endogenous DNA. *Science*, **308**, 118–120.

110 Pontes, O., Li, C.F., Costa Nunes, P., Haag, J. et al. (2006) The *Arabidopsis* chromatin-modifying nuclear siRNA pathway involves a nucleolar RNA processing center. *Cell*, **126**, 79–92.

111 Zhang, X., Henderson, I.R., Lu, C., Green, P.J. et al. (2007) Role of RNA polymerase IV in plant small RNA metabolism. *Proc. Natl Acad. Sci. USA*, **104**, 4536–4541.

112 Law, J.A., Vashisht, A.A., Wohlschlegel, J.A., Jacobsen, S.E. (2011) SHH1, a homeodomain protein required for DNA methylation, as well as RDR2, RDM4, and chromatin remodeling factors, associate with RNA polymerase IV. *PLoS Genet.*, **7**, e1002195.

113 Xie, Z., Johansen, L.K., Gustafson, A.M., Kasschau, K.D. et al. (2004) Genetic and functional diversification of small RNA pathways in plants. *PLoS Biol.*, **2**, E104.

114 He, X.J., Hsu, Y.F., Zhu, S., Wierzbicki, A.T. et al. (2009) An effector of RNA-directed DNA methylation in *Arabidopsis* is an ARGONAUTE 4- and RNA-binding protein. *Cell*, **137**, 498–508.

115 Kanno, T., Huettel, B., Mette, M.F., Aufsatz, W. et al. (2005) Atypical RNA polymerase subunits required for RNA-directed DNA methylation. *Nat. Genet.*, **37**, 761–765.

116 Lahmy, S., Pontier, D., Cavel, E., Vega, D. et al. (2009) PolV(PolIVb) function in RNA-directed DNA methylation requires the conserved active site and an additional plant-specific subunit. *Proc. Natl Acad. Sci. USA*, **106**, 941–946.

117 Pontier, D., Yahubyan, G., Vega, D., Bulski, A. et al. (2005) Reinforcement of silencing at transposons and highly repeated sequences requires the concerted action of two distinct RNA polymerases IV in *Arabidopsis*. *Genes Dev.*, **19**, 2030–2040.

118 El-Shami, M., Pontier, D., Lahmy, S., Braun, L. et al. (2007) Reiterated WG/GW motifs form functionally and evolutionarily conserved ARGONAUTE-binding platforms in RNAi-related components. *Genes Dev.*, **21**, 2539–2544.

119 Wierzbicki, A.T., Haag, J.R., Pikaard, C.S. (2008) Noncoding transcription by RNA polymerase Pol IVb/Pol V mediates transcriptional silencing of overlapping and adjacent genes. *Cell*, **135**, 635–648.

120 Wierzbicki, A.T., Ream, T.S., Haag, J.R., Pikaard, C.S. (2009) RNA polymerase V transcription guides ARGONAUTE4 to chromatin. *Nat. Genet.*, **41**, 630–634.

121 Gao, Z., Liu, H.L., Daxinger, L., Pontes, O. et al. (2010) An RNA polymerase II- and AGO4-associated protein acts in RNA-directed DNA methylation. *Nature*, **465**, 106–109.

122 Zheng, B., Wang, Z., Li, S., Yu, B. et al. (2009) Intergenic transcription by RNA polymerase II coordinates Pol IV and Pol V in siRNA-directed transcriptional gene silencing in *Arabidopsis*. *Genes Dev.*, **23**, 2850–2860.

123 Chung, W.J., Okamura, K., Martin, R., Lai, E.C. (2008) Endogenous RNA interference provides a somatic defense against *Drosophila* transposons. *Curr. Biol.*, **18**, 795–802.

124 Ghildiyal, M., Seitz, H., Horwich, M.D., Li, C. et al. (2008) Endogenous siRNAs derived from transposons and mRNAs in *Drosophila* somatic cells. *Science*, **320**, 1077–1081.

125 Siomi, M.C., Sato, K., Pezic, D., Aravin, A.A. (2011) PIWI-interacting small RNAs: the vanguard of genome defence. *Nat. Rev. Mol. Cell Biol.*, **12**, 246–258.

126 Klenov, M.S., Sokolova, O.A., Yakushev, E.Y., Stolyarenko, A.D. et al. (2011) Separation of stem cell maintenance and transposon silencing functions of Piwi protein. *Proc. Natl Acad. Sci. USA*, **108**, 18760–18765.

127 Poyhonen, M., de Vanssay, A., Delmarre, V., Hermant, C. et al. (2012) Homology-dependent silencing by an exogenous sequence in the *Drosophila* germline. *G3 Bethesda*, **2**, 331–338.

128 Le Thomas, A., Rogers, A.K., Webster, A., Marinov, G.K. et al. (2013) Piwi induces piRNA-guided transcriptional silencing and establishment of a repressive chromatin state. *Genes Dev.*, **27**, 390–399.

129 Wang, S.H., Elgin, S.C. (2011) *Drosophila* Piwi functions downstream of piRNA production mediating a chromatin-based transposon silencing mechanism in female germ line. *Proc. Natl Acad. Sci. USA*, **108**, 21164–21169.

130 Aravin, A.A., Sachidanandam, R., Bourc'his, D., Schaefer, C. et al. (2008) A piRNA pathway primed by individual transposons is linked to de novo DNA methylation in mice. *Mol. Cell*, **31**, 785–799.

131 Carmell, M.A., Girard, A., van de Kant, H.J., Bourc'his, D. et al. (2007) MIWI2 is essential for spermatogenesis and repression of transposons in the mouse male germline. *Dev. Cell*, **12**, 503–514.

132 Kuramochi-Miyagawa, S., Watanabe, T., Gotoh, K., Totoki, Y. et al. (2008) DNA methylation of retrotransposon genes is regulated by Piwi family members MILI and MIWI2 in murine fetal testes. *Genes Dev.*, **22**, 908–917.

133 Brown, C.J., Hendrich, B.D., Rupert, J.L., Lafreniere, R.G. et al. (1992) The human XIST gene: analysis of a 17 kb inactive X-specific RNA that contains conserved repeats and is highly localized within the nucleus. *Cell*, **71**, 527–542.

134 Clemson, C.M., McNeil, J.A., Willard, H.F., Lawrence, J.B. (1996) XIST RNA paints the inactive X chromosome at interphase: evidence for a novel RNA involved in nuclear/chromosome structure. *J. Cell Biol.*, **132**, 259–275.

135 Wutz, A. (2011) Gene silencing in X-chromosome inactivation: advances in understanding facultative heterochromatin formation. *Nat. Rev. Genet.*, **12**, 542–553.

136 Lee, J.T., Davidow, L.S., Warshawsky, D. (1999) Tsix, a gene antisense to Xist at the X-inactivation centre. *Nat. Genet.*, **21**, 400–404.

137 Zhao, J., Sun, B.K., Erwin, J.A., Song, J.J. et al. (2008) Polycomb proteins targeted by a short repeat RNA to the mouse X chromosome. *Science*, **322**, 750–756.

138 Lee, J.T., Lu, N. (1999) Targeted mutagenesis of Tsix leads to nonrandom X inactivation. *Cell*, **99**, 47–57.

139 Lee, T.L., Xiao, A., Rennert, O.M. (2012) Identification of novel long noncoding RNA transcripts in male germ cells. *Methods Mol. Biol.*, **825**, 105–114.

140 Sado, T., Hoki, Y., Sasaki, H. (2005) Tsix silences Xist through modification of chromatin structure. *Dev. Cell*, **9**, 159–165.

141 Sun, B.K., Deaton, A.M., Lee, J.T. (2006) A transient heterochromatic state in Xist preempts X inactivation choice without RNA stabilization. *Mol. Cell*, **21**, 617–628.

142 Tian, D., Sun, S., Lee, J.T. (2010) The long noncoding RNA, Jpx, is a molecular switch for X chromosome inactivation. *Cell*, **143**, 390–403.

143 Hariharan, N., Kelley, D.E., Perry, R.P. (1991) Delta, a transcription factor that binds to downstream elements in several polymerase II promoters, is a functionally versatile zinc finger protein. *Proc. Natl Acad. Sci. USA*, **88**, 9799–9803.

144 Kim, J.D., Kang, K., Kim, J. (2009) YY1's role in DNA methylation of Peg3 and Xist. *Nucleic Acids Res.*, **37**, 5656–5664.

145 Park, K., Atchison, M.L. (1991) Isolation of a candidate repressor/activator, NF-E1 (YY-1, delta), that binds to the immunoglobulin kappa 3 enhancer and the immunoglobulin heavy-chain mu E1 site. *Proc. Natl Acad. Sci. USA*, **88**, 9804–9808.

146 Seto, E., Shi, Y., Shenk, T. (1991) YY1 is an initiator sequence-binding protein that directs and activates transcription in vitro. *Nature*, **354**, 241–245.

147 Shi, Y., Seto, E., Chang, L.S., Shenk, T. (1991) Transcriptional repression by YY1, a human GLI-Kruppel-related protein, and relief of repression by adenovirus E1A protein. *Cell*, **67**, 377–388.

148 Jeon, Y., Lee, J.T. (2011) YY1 tethers Xist RNA to the inactive X nucleation center. *Cell*, **146**, 119–133.

149 Camblong, J., Iglesias, N., Fickentscher, C., Dieppois, G. et al. (2007). Antisense RNA stabilization induces transcriptional gene silencing via histone deacetylation in S. cerevisiae. *Cell*, **131**, 706–717.

150 Verona, R.I., Mann, M.R., Bartolomei, M.S. (2003) Genomic imprinting: intricacies of epigenetic regulation in clusters. *Annu. Rev. Cell Dev. Biol.*, **19**, 237–259.

151 Lee, J.T. (2012) Epigenetic regulation by long noncoding RNAs. *Science*, **338**, 1435–1439.

152 Mohammad, F., Mondal, T., Kanduri, C. (2009) Epigenetics of imprinted long noncoding RNAs. *Epigenetics*, **4**, 277–286.

153 Wan, L.B., Bartolomei, M.S. (2008) Regulation of imprinting in clusters: noncoding RNAs versus insulators. *Adv. Genet.*, **61**, 207–223.

154 Sleutels, F., Zwart, R., Barlow, D.P. (2002) The non-coding Air RNA is required for silencing autosomal imprinted genes. *Nature*, **415**, 810–813.

155 Nagano, T., Mitchell, J.A., Sanz, L.A., Pauler, F.M. et al. (2008) The Air noncoding RNA epigenetically silences transcription by targeting G9a to chromatin. *Science*, **322**, 1717–1720.

156 Pandey, R.R., Mondal, T., Mohammad, F., Enroth, S. et al. (2008) Kcnq1ot1 antisense noncoding RNA mediates lineage-specific transcriptional silencing through chromatin-level regulation. *Mol. Cell*, **32**, 232–246.

157 Terranova, R., Yokobayashi, S., Stadler, M.B., Otte, A.P. et al. (2008) Polycomb group proteins Ezh2 and Rnf2 direct genomic contraction and imprinted repression in early mouse embryos. *Dev. Cell*, **15**, 668–679.

158 Gaszner, M., Felsenfeld, G. (2006) Insulators: exploiting transcriptional and epigenetic mechanisms. *Nat. Rev. Genet.*, **7**, 703–713.

159 Kim, T.H., Abdullaev, Z.K., Smith, A.D., Ching, K.A. et al. (2007) Analysis of the vertebrate insulator protein CTCF-binding sites in the human genome. *Cell*, **128**, 1231–1245.

160 Popov, N., Gil, J. (2010) Epigenetic regulation of the INK4b-ARF-INK4a locus: in sickness and in health. *Epigenetics*, **5**, 685–690.

161 Pasmant, E., Laurendeau, I., Heron, D., Vidaud, M. et al. (2007) Characterization of a germ-line deletion, including the entire INK4/ARF locus, in a melanoma-neural system tumor family: identification of ANRIL, an antisense noncoding RNA whose expression coclusters with ARF. *Cancer Res.*, **67**, 3963–3969.

162 Yu, W., Gius, D., Onyango, P., Muldoon-Jacobs, K. et al. (2008) Epigenetic silencing of tumour suppressor gene p15 by its antisense RNA. *Nature*, **451**, 202–206.

163 Yap, K.L., Li, S., Munoz-Cabello, A.M., Raguz, S. et al. (2010) Molecular interplay of the noncoding RNA ANRIL and methylated histone H3 lysine 27 by polycomb CBX7 in transcriptional silencing of INK4a. *Mol. Cell*, **38**, 662–674.

164 Morris, K.V., Santoso, S., Turner, A.M., Pastori, C. et al. (2008) Bidirectional transcription directs both transcriptional gene

activation and suppression in human cells. *PLoS Genet.*, **4**, e1000258.
165. Magistri, M., Faghihi, M.A., St Laurent, G. III, Wahlestedt, C. (2012) Regulation of chromatin structure by long noncoding RNAs: focus on natural antisense transcripts. *Trends Genet.*, **28**, 389–396.
166. Azuara, V., Perry, P., Sauer, S., Spivakov, M. et al. (2006) Chromatin signatures of pluripotent cell lines. *Nat. Cell Biol.*, **8**, 532–538.
167. Bernstein, B.E., Mikkelsen, T.S., Xie, X., Kamal, M. et al. (2006) A bivalent chromatin structure marks key developmental genes in embryonic stem cells. *Cell*, **125**, 315–326.
168. Roh, T.Y., Cuddapah, S., Cui, K., Zhao, K. (2006) The genomic landscape of histone modifications in human T cells. *Proc. Natl Acad. Sci. USA*, **103**, 15782–15787.
169. Kanhere, A., Viiri, K., Araujo, C.C., Rasaiyaah, J. et al. (2010) Short RNAs are transcribed from repressed polycomb target genes and interact with polycomb repressive complex-2. *Mol. Cell*, **38**, 675–688.
170. Wang, X., Arai, S., Song, X., Reichart, D. et al. (2008) Induced ncRNAs allosterically modify RNA-binding proteins in cis to inhibit transcription. *Nature*, **454**, 126–130.
171. Dinger, M.E., Amaral, P.P., Mercer, T.R., Pang, K.C. et al. (2008) Long noncoding RNAs in mouse embryonic stem cell pluripotency and differentiation. *Genome Res.*, **18**, 1433–1445.
172. Feng, J., Bi, C., Clark, B.S., Mady, R. et al. (2006) The Evf-2 noncoding RNA is transcribed from the Dlx-5/6 ultraconserved region and functions as a Dlx-2 transcriptional coactivator. *Genes Dev.*, **20**, 1470–1484.
173. Bond, A.M., Vangompel, M.J., Sametsky, E.A., Clark, M.F. et al. (2009) Balanced gene regulation by an embryonic brain ncRNA is critical for adult hippocampal GABA circuitry. *Nat. Neurosci.*, **12**, 1020–1027.
174. Wang, K.C., Yang, Y.W., Liu, B., Sanyal, A. et al. (2011) A long noncoding RNA maintains active chromatin to coordinate homeotic gene expression. *Nature*, **472**, 120–124.
175. Berretta, J., Pinskaya, M., Morillon, A. (2008). A cryptic unstable transcript mediates transcriptional trans-silencing of the Ty1 retrotransposon in S. cerevisiae. *Genes Dev.*, **22**, 615–626.
176. van Dijk, E.L., Chen, C.L., d'Aubenton-Carafa, Y. et al. (2011). XUTs are a class of Xrn1-sensitive antisense regulatory non-coding RNA in yeast. *Nature*, **475**, 114–117.
177. Straub, T., Becker, P.B. (2011) Transcription modulation chromosome-wide: universal features and principles of dosage compensation in worms and flies. *Curr. Opin. Genet. Dev.*, **21**, 147–153.
178. Ilik, I., Akhtar, A. (2009) roX RNAs: non-coding regulators of the male X chromosome in flies. *RNA Biol.*, **6**, 113–121.
179. Guttman, M., Amit, I., Garber, M., French, C. et al. (2009) Chromatin signature reveals over a thousand highly conserved large non-coding RNAs in mammals. *Nature*, **458**, 223–227.
180. Khalil, A.M., Guttman, M., Huarte, M., Garber, M. et al. (2009) Many human large intergenic noncoding RNAs associate with chromatin-modifying complexes and affect gene expression. *Proc. Natl Acad. Sci. USA*, **106**, 11667–11672.
181. Rinn, J.L., Kertesz, M., Wang, J.K., Squazzo, S.L. et al. (2007) Functional demarcation of active and silent chromatin domains in human HOX loci by noncoding RNAs. *Cell*, **129**, 1311–1323.
182. Fanti, L., Perrini, B., Piacentini, L., Berloco, M. et al. (2008) The trithorax group and Pc group proteins are differentially involved in heterochromatin formation in *Drosophila*. *Chromosoma*, **117**, 25–39.
183. Tsai, M.C., Manor, O., Wan, Y., Mosammaparast, N. et al. (2010) Long noncoding RNA as modular scaffold of histone modification complexes. *Science*, **329**, 689–693.
184. Maison, C., Bailly, D., Peters, A.H., Quivy, J.P. et al. (2002) Higher-order structure in pericentric heterochromatin involves a distinct pattern of histone modification and an RNA component. *Nat. Genet.*, **30**, 329–334.
185. Maison, C., Bailly, D., Roche, D., Montes de Oca, R. et al. (2011) SUMOylation promotes de novo targeting of HP1alpha to pericentric heterochromatin. *Nat. Genet.*, **43**, 220–227.
186. Muchardt, C., Guilleme, M., Seeler, J.S., Trouche, D. et al. (2002) Coordinated methyl

and RNA binding is required for heterochromatin localization of mammalian HP1alpha. *EMBO Rep.*, **3**, 975–981.

187 Chadwick, B.P., Willard, H.F. (2004) Multiple spatially distinct types of facultative heterochromatin on the human inactive X chromosome. *Proc. Natl Acad. Sci. USA*, **101**, 17450–17455.

188 Nozawa, R.S., Nagao, K., Igami, K.T., Shibata, S. *et al.* (2013) Human inactive X chromosome is compacted through a PRC2-independent SMCHD1-HBiX1 pathway. *Nat. Struct. Mol. Biol.*, **20**, 566–573.

189 Chadwick, B.P. (2007) Variation in Xi chromatin organization and correlation of the H3K27me3 chromatin territories to transcribed sequences by microarray analysis. *Chromosoma*, **116**, 147–157.

190 Ding, D.Q., Okamasa, K., Yamane, M. *et al.* (2012). Meiosis-specific noncoding RNA mediates robust pairing of homologous chromosomes in meiosis. *Science*, **336**, 732–736.

15
Intracellular RNA Localization and Localized Translation

Florence Besse[1,2,3]
[1] *University of Nice Sophia Antipolis, Institute of Biology Valrose, Parc Valrose, 06108 Nice, France*
[2] *UMR7277 CNRS, Institute of Biology Valrose, Parc Valrose, 06108 Nice, France*
[3] *U1091 Inserm, Institute of Biology Valrose, Parc Valrose, 06108 Nice, France*

1	**Localized mRNAs: an Expanding Number of Destinations and Functions**	473
1.1	A Historical Perspective	473
1.2	A Variety of Localization Patterns	473
1.3	A Variety of Functions	474
1.3.1	Early Embryo Patterning	474
1.3.2	Asymmetric Cell Division	474
1.3.3	Epithelium Polarization	475
1.3.4	Cell Migration and Guidance	475
1.3.5	Synaptic Plasticity and Long-Term Memory	476
2	**Why Localize mRNAs Rather Than Proteins?**	476
3	**Different Mechanisms to Localize mRNAs**	477
3.1	Localized Protection from Degradation	477
3.2	Diffusion and Entrapment	477
3.3	Directed Transport	478
4	**Mechanisms Underlying the Directed Transport of mRNAs**	478
4.1	Assembly of RNP Complexes	478
4.1.1	Localization Elements	478
4.1.2	trans-Acting Binding Factors	481
4.1.3	Stepwise Assembly and Remodeling of RNPs	483
4.1.4	How Many RNA Molecules per Transport Particle?	485
4.2	Recruitment of Molecular Motors and Transport along the Cytoskeleton	486
4.2.1	Cytoskeletal Elements and Molecular Motors	486

4.2.2	Regulation of mRNA Transport 487
4.3	Anchoring of mRNAs at the Final Destination 488

5	**Coupling mRNA Localization with Translational Control** 489
5.1	Translational Repression of Localizing mRNAs 489
5.2	Translational Derepression upon Arrival 490
5.3	Translational Derepression in Response to Specific Signals 490

6	**Integrated Examples of Localizing mRNAs** 491
6.1	mRNA Localization in the Budding Yeast: Focus on ASH1 mRNA 492
6.2	mRNA Localization in Drosophila Oocyte: Focus on oskar mRNA 492
6.3	mRNA Localization in Mammalian Dendrites: Focus on CaMKII mRNA 495

7	**Conclusions and Perspectives** 498
	Acknowledgments 498
	References 498

Keywords

Ribonucleoprotein particles (RNPs)
Multimolecular complexes composed of mRNA molecules and associated proteins. Transport RNPs typically contain RNA-binding proteins, components and regulators of the protein synthesis machinery, as well as molecular motors.

Localization elements
cis-Regulatory RNA elements usually (but not exclusively) located in 3′UTR, and directing mRNAs for transport to specific subcellular locations. Localization elements are selectively recognized by *trans*-acting localizing factors.

***trans*-Acting localizing factors**
Factors (usually RNA-binding proteins) recognizing specific elements in mRNAs to be targeted, and coupling these mRNAs to the transport machinery.

Directed mRNA transport
Active mechanism mediated by molecular motors and generating a net displacement of mRNA molecules. Transport directionality is dictated by the polarity of cytoskeletal tracks, as well as by the oriented activity of molecular motors.

Intracellular targeting of mRNAs, when coupled to local translation, represents a very efficient means to regulate gene expression in space and time, and to asymmetrically segregate proteins within polarized cells. While mRNAs with specific distributions have been known for 30 years, the prevalence of mRNA localization has emerged only recently, with the advent of genome-wide studies. Mechanistically, mRNA localization can be achieved via selective degradation or diffusion coupled to entrapment at specific cellular sites, but appears to be mediated in multiple cell types and species by directed transport. Directed transport relies on the recognition of RNA *cis*-regulatory elements by *trans*-acting binding factors, and on the further recruitment of molecular motors. This is coupled with precise translational control, such that localized mRNAs are translated specifically at their destination sites. Current challenges in the field include understanding how mRNA localization and local translation are regulated in response to external signals or developmental programs, and starting to predict mRNA fate based on primary sequence.

1
Localized mRNAs: an Expanding Number of Destinations and Functions

1.1
A Historical Perspective

Some 30 years ago, Brodeur and colleagues reported the asymmetric distribution of β-*actin* mRNA in ascidian eggs, providing the first evidence for the cytoplasmic localization of specific mRNAs [1]. Shortly thereafter, localized mRNAs were identified in different species and cell types, including *Xenopus* [2] and *Drosophila* [3, 4] oocytes, chicken embryonic fibroblasts [5], or mammalian oligodendrocytes [6] and neurons [7]. These initial studies, soon completed by functional studies, established that the intracellular localization of mRNAs, when coupled to local protein synthesis, would provide polarized cells with an efficient means to target proteins to specific cellular sites. Despite the following identification of various mRNAs with asymmetric subcellular distribution, the total number of cytoplasmically localized mRNAs listed in a comprehensive review from the late 1990s was still limited to 75 [8]. Thus, mRNA localization has long been considered as a mechanism used in specific cell types, for a restricted number of mRNAs. As described in the next section, recent genome-wide analyses have dramatically changed this view, revealing that mRNA localization may rather be a prevalent mechanism used by most polarized cells to generate functional asymmetries (for recent reviews on mRNA localization, see Refs [9–11]).

1.2
A Variety of Localization Patterns

As demonstrated by recent high-throughput studies based on visual screens or expression profiling experiments, it now appears that a very large number of transcripts are asymmetrically distributed in differentiated cells, and that mRNAs can be targeted to a variety of cellular compartments [12]. Indeed, mRNAs associating with cellular structures as diverse as cell–cell junctions, mitotic apparatus, endoplasmic reticulum, or chromatin foci have been discovered in a high resolution *in-situ* hybridization screen performed in *Drosophila* embryos [13]. Moreover, a striking 70% of the 2314 expressed transcripts analyzed in this screen appear to localize with specific patterns. Complementary studies performed in mammalian cells, and relying on the purification of specific subcellular compartments, have also led to the discovery that dozens to thousands of mRNAs are present in the pseudopodia of migrating cells [14], and in the dendrites [15, 16] or axons [17–19] of neuronal cells.

Importantly, an asymmetric distribution of transcripts has been observed in a variety of organisms, ranging from higher eukaryotes (including plants [20] and animals [11]) to fungi [21, 22] and bacteria [23], further suggesting that mRNA localization is a widespread mechanism.

1.3
A Variety of Functions

mRNA localization, long known for its role in early embryo patterning and neuronal functions, has more recently been shown to underlie a variety of cellular and developmental processes [11]. As a comprehensive review of the processes regulated by mRNA localization is beyond the scope of this section, only selected examples illustrating the range of functions associated with mRNA targeting will be mentioned at this point.

1.3.1 Early Embryo Patterning
Translation of localized mRNAs is particularly important during early development, as many organisms rely on proteins synthesized from maternally deposited mRNAs to control embryonic germ layer specification, axis formation, and/or germline determination [24]. An asymmetric distribution of mRNAs has been observed in a wide range of organisms, including cnidarians, ascidians, insects, amphibians, or fish, and is used as a means to control gene expression in space and time prior to the onset of zygotic transcription (for more comprehensive reviews, see Refs [24–28]).

In Cnidarian eggs, for example, *CheFrizzled1* and *CheFrizzled3* mRNAs are localized respectively at the animal and vegetal hemispheres, where they induce oral and aboral fates [29, 30]. In *Drosophila*, specification of the embryonic anteroposterior axis depends on the accumulation of *bicoid* and *oskar* mRNAs at the oocyte anterior and posterior poles, respectively [3, 31–33]. Furthermore, posterior accumulation of *nanos* mRNA in late stage oocytes is essential for the determination of abdominal structures and germ cells [34, 35]. Finally, establishment of the dorsoventral axis relies on the localization of *gurken* mRNA at the anterodorsal oocyte corner [36]. In *Xenopus*, the targeting of maternal mRNAs is essential for germ layer patterning, as an accumulation of *Vg1* and *VegT* mRNAs at the vegetal pole of oocytes underlies the specification of endodermal and mesodermal structures during embryogenesis [37–40]. Furthermore, targeting of *Xdazl* and *Nanos1* mRNAs to the vegetally localized germ plasm is required for the differentiation of primordial germ cells in this system [41, 42].

1.3.2 Asymmetric Cell Division
A key feature of asymmetric cell division is to produce daughter cells with different fates, a process mediated by differential segregation of key determinants. While these determinants can be segregated as proteins, examples of determinants inherited asymmetrically as mRNAs have been described.

In the budding yeast, for example, *ASH1* mRNA is transported to the bud tip during anaphase, which results in the asymmetric sorting of Ash1p protein to the daughter cell nucleus, the subsequent repression of mating type switching, and the generation of two cells with different mating types [43, 44]. In cleavage-stage embryos of organisms ranging from sea snail to mouse, unequal inheritance of determinant-encoding mRNAs appears to underlie lineage diversification [45–47]. The trophectoderm determinant-encoding *cdx2* mRNA, for example, is inherited exclusively by outside daughter cells upon division of eight-cell mouse blastomeres [45]. Asymmetric inheritance of determinant-encoding mRNAs has also been observed in the context of stem cell divisions in *Drosophila*, where *prospero* and *inscuteable* mRNAs are targeted to the basal and apical poles of dividing neural precursors, respectively [48–50].

1.3.3 Ephithelium Polarization

As described in different contexts, mRNAs encoding proteins with a wide range of functions (secreted morphogens, transmembrane receptors, components of cell–cell junctions, cytoskeletal regulators, transcription factors, etc.) distribute asymmetrically along the apicobasal axis of epithelial cells [11, 13]. Furthermore, functional studies have demonstrated that apical or basal targeting of specific mRNAs is critical for epithelial polarization, maintenance, or secretory activity [51–57]. In the *Drosophila* embryo, for example, an apical localization of *wingless* (*wg*) transcripts is essential for apical secretion of the Wg morphogen, and appears to potentiate Wg activity by preventing it from diffusing away [56]. In mammals, transcripts encoding the tight-junction protein zonula occludens-1 (ZO-1) are targeted to the apical side of mammary epithelial cells, a process required for the assembly of tight-junctions and for cell polarization [54].

1.3.4 Cell Migration and Guidance

Migrating cells exhibit a clear front to rear polarity axis, accumulating proteins involved in cue sensing and cell motility at their front. As shown by different studies, mRNAs encoding cytoskeletal elements or regulators accumulate in the protrusions of migrating cells, suggesting that their targeting may promote migration [5, 14, 58]. Consistent with this view, mRNAs encoding β-actin or the Arp2/3 complex subunit Arp2 accumulate at the leading edge of chicken embryonic fibroblasts, and their targeting is essential for persistent and directional cell migration [59, 60]. Notably, mRNA localization may also regulate cell migration in pathological conditions, as changes in the subcellular distribution of transcripts encoding actin regulators correlate with an increased ability of human breast carcinoma cell lines to invade surrounding tissues [61].

The leading tip of migrating axons, also named the *growth cone*, is a specialized structure translating external guidance cues into appropriate turning behavior. As revealed by expression profiling experiments, this structure appears to contain up to hundreds of mRNAs encoding proteins with various functions [18, 19, 62]. Furthermore, local translation of axonally localized mRNAs is induced in response to attractive or repulsive cues, and is essential for axon turning or collapse [63, 64]. Indeed, the transport of *β-actin* mRNA to growth cones, and local translation in response to attractive cues such as Netrin-1 or Brain-derived neurotrophic factor (BDNF), are essential for axon turning in *Xenopus* cultured neurons [65, 66]. Furthermore, axonal synthesis of the actin cytoskeleton regulator RhoA is required for Semaphorin3A-induced growth cone collapse in mammalian cultured neurons [67].

1.3.5 Synaptic Plasticity and Long-Term Memory

A precise modulation of the efficacy of individual synapses within neuronal networks is required for the formation and storage of memory traces, but represents a challenging task if it is considered that dendritic processes commonly extend millimeters away from the cell body, and that a given neuron can establish up to 10 000 contacts *in vivo*. In this context, the localization of mRNAs to dendrites coupled to local translation at synapses, appears to provide an efficient means to differentially tag synapses and to regulate protein content locally, in response to synaptic activity [68–70].

Indeed, hundreds to thousands of mRNAs are found in dendrites [15, 16], and various dendritically localized mRNAs including *CaMKIIα*, *Arc*, *Glutamate receptor 1* and *2*, or *post-synaptic density-95* (*PSD-95*) are recruited and/or translated in response to synaptic stimulation [71–76]. Furthermore, as demonstrated using live-imaging, mRNA translation can be restricted to individual stimulated synapses, and regulated in a stimulus-specific manner [77]. Finally, consistent with a functional requirement of dendritic translation in synaptic plasticity and learning and memory processes *in vivo*, mice with mislocalized *CaMKIIα* mRNA show a reduced late-phase long-term potentiation (L-LTP) as well as a defective long-term memory [78]. Similarly, mistargeting of dendritically localized *BDNF* transcripts appears to impair long-term potentiation [79].

2
Why Localize mRNAs Rather Than Proteins?

Several advantages are associated with localizing mRNAs rather than proteins, the most obvious one being a reduction in transport costs, as one mRNA molecule can generate many copies of proteins through successive rounds of translation. Producing proteins on site can also protect cells from the deleterious activity of proteins that would otherwise have to travel across the cell before reaching their destination. Indeed, mislocalization of maternal determinants disrupts embryonic patterning [24], and ectopic accumulation of the Myelin Basic Protein (MBP) is associated with the ectopic appearance of myelin-like membranes [80]. Furthermore, freshly produced proteins, in contrast to pre-existing protein copies transported from perinuclear regions, lack post-translational modifications modulating their properties. Nascent β-actin proteins, for example, are neither arginylated nor glutathionylated, and thus exhibit polymerization properties distinct from those of mature β-actin [81, 82]. By restricting protein synthesis to specific microdomains, mRNA localization may in addition generate locally high protein concentrations required for the assembly of multimolecular complexes. In particular, the local translation of β-actin and components of the Arp2/3 complex in the protrusions of migrating cells or axons has been suggested to promote actin nucleation, which usually is the rate-limiting step for actin filament assembly [58, 59].

Finally, a main advantage of mRNA localization is to allow a precise regulation of gene expression in space and time. Indeed, mRNAs can be transported to specific cellular compartments and kept silenced, their translation being activated in response to very specific stimuli. As mainly demonstrated in neurons, such a mechanism provides a very efficient means to decentralize the control of gene expression and rapidly respond to external signals. In *Aplysia*, for example, *sensorin* mRNA localizes to neuronal processes, and is translated at activated synapses specifically in response to the spaced application of serotonin [77]. In mammalian neurons, different isoforms of *BDNF* transcripts are targeted differentially along dendritic processes, allowing a selective accumulation of BDNF along the proximodistal axis, and a spatially restricted modulation of dendritic arborization [83].

3
Different Mechanisms to Localize mRNAs

Three main mechanisms contribute to the asymmetric localization of mRNAs within the cytoplasm: localized protection from degradation; diffusion coupled to local anchoring; and directed transport along a polarized cytoskeleton. Although these mechanisms are molecularly distinct, they can be combined to allow efficient mRNA localization.

3.1
Localized Protection from Degradation

The initial uniform distribution of transcripts, followed by selective degradation of nonlocalized mRNAs, has been extensively studied in *Drosophila* embryos where maternal *hsp83* transcripts, first distributed throughout fertilized eggs, are later restricted to the posterior germ plasm [84]. The degradation of *hsp83* transcripts in the bulk of the embryo is mediated by *cis*-regulatory elements recognized by the RNA-binding protein Smaug, which recruits the CCR4/NOT deadenylase complex, thereby triggering *hsp83* mRNA decay [85, 86]. Similarly, a posterior accumulation of *nanos* mRNA in the *Drosophila* embryo also partly results from Smaug-dependent destabilization of nonlocalized molecules [87], which suggests that Smaug may promote the selective degradation of various maternal transcripts. Consistent with this hypothesis, genome-wide expression profiling experiments have revealed that a vast number of maternal transcripts are abnormally stabilized in *smaug* mutants [88]. Although the proportion of Smaug direct targets selectively stabilized at the posterior pole of embryos is still unknown, it is likely that many of them will be enriched posteriorly [13, 88]. How Smaug target mRNAs become stabilized at the posterior pole remains unclear, although this might be mediated by competitive binding of Smaug to posteriorly localized proteins [87].

3.2
Diffusion and Entrapment

Local entrapment of mRNA molecules diffusing throughout the cytoplasm is another mechanism supporting the accumulation of transcripts at specific cellular sites. As demonstrated by live-imaging experiments, indeed, localization of *Xcat2* (*Nanos1*) and *Xdazl* transcripts in early *Xenopus* oocytes does not require an intact cytoskeleton, and is mediated by the association of freely diffusing mRNA molecules with an endoplasmic reticulum network concentrated in the vegetally localized mitochondrial cloud [89]. Similarly, the dynamic visualization of fluorescently tagged mRNAs has revealed that the accumulation of *nanos* mRNA at the posterior pole of late *Drosophila* oocytes is largely microtubule MT-independent, and relies on diffusion followed by entrapment [90]. In these examples, however, the identity of the molecular anchors involved in the entrapment process remains to be determined.

3.3
Directed Transport

In contrast to the previously described mechanisms, directed mRNA transport requires a polarized cytoskeleton, and the association of mRNA molecules with molecular motors. Active and directed transport of mRNAs has been observed in a wide range of cell types and organisms, and is the most extensively described (and probably the predominant) mechanism used to direct mRNA localization [91, 92]. The following section is thus dedicated to the cellular and molecular mechanisms underlying this process.

4
Mechanisms Underlying the Directed Transport of mRNAs

4.1
Assembly of RNP Complexes

The subcellular destination of transcripts is encoded by *cis*-acting elements present in localized mRNAs, and acting as platforms for the recruitment of *trans*-acting factors and the assembly of large ribonucleoprotein particles (RNPs). These RNA transport particles, often called granules, are visualized upon the injection of fluorescently labeled RNAs (for seminal studies, see Ref. [93]) or by *in-situ* hybridization, and can be biochemically isolated based on their sedimentation in heavy fractions [94–97]. As revealed by systematic proteomic analyses, they contain a large number of associated proteins, including RNA-binding proteins, components, and regulators of the protein synthesis machinery, as well as molecular motors [94–96, 98–100]. Interestingly, complexes with distinct molecular compositions have been purified, suggesting the coexistence of multiple RNP species within differentiated cells [94, 95]. Whereas, some components are found in different transport complexes and might therefore constitute a core module regulating the assembly and the regulatory properties of various RNPs, some *trans*-acting binding factors rather associate with unique sets of target mRNAs through the recognition of specific motifs.

4.1.1 Localization Elements

Definition RNA localization elements (also named *zipcodes*) are *cis*-acting elements that direct mRNAs for transport to specific subcellular locations [101, 102]. They range in length from few nucleotides to over 1 kb, and can drive the localization of heterologous RNAs when incorporated into reporter constructs. Although some localization elements have been identified in 5′UTR (untranslated regions) [103, 104], coding regions [52, 105–109], or even intronic sequences [110], they are usually located in 3′UTRs of targeted mRNAs [111]. A 54 nt zipcode mapped to the proximal part of β-*actin* 3′UTR, for example, is sufficient to support the accumulation of transcripts in the peripheral cytoplasm of chicken embryonic fibroblasts [112]. Similarly, a 11 nt sequence found in the 3′UTR of mammalian MBP mRNA is necessary and sufficient for the transport of injected

RNAs to oligodendrocyte peripheral processes [113, 114]. In *Drosophila*, a 44 nt sequence located within the 3′UTR of *K10* mRNA is both necessary and sufficient for the localization of *K10* transcripts to oocytes [115].

Mutational analyses performed on the few RNA localization elements minimized sufficiently to allow detailed studies have revealed that they are defined by both primary sequence and structural requirements [101, 102]. The recognition of β-actin mRNA by the transport machinery, for example, depends primarily on two nucleotide stretches forming a bipartite RNA sequence located in the 5′-most region of the β-actin zipcode [116, 117]. In addition, many localization elements are predicted to form stem–loop structures and can accommodate base changes, provided that compensatory mutations are introduced to preserve the secondary structures [104–106, 115, 118–121]. The three-dimensional configuration adopted by *cis*-acting element has recently emerged as a key determinant of their functionality [121–123]. Indeed, as demonstrated by a recent study combining NMR spectroscopy and functional localization assays, the key determinants for *Drosophila* K10 localization element activity are two double-stranded RNA helices adopting an unusual A′-form conformation with widened major grooves [122].

Localization Elements Are Modular An important feature emerging from the dissection of localization elements is that most of them are composed of different modules acting either independently to regulate the different steps of the localization program, or in a concerted manner to ensure robust targeting. The entire signal required for the localization of *bicoid* mRNA to the anterior pole of the *Drosophila* oocyte, for example, resides within a 625 nt fragment of the 3′UTR [124]. While a 53 nt sub-fragment named *bicoid* localization element 1 (BLE1) supports the early stages of localization, additional stem–loop structures spread over the 3′UTR are required for later steps, including anchoring of the transcript at the anterior pole [124, 125]. Noteworthy, localization elements with complementary functions can distribute over both 5′ and 3′UTR sequences. *Aplysia sensorin* mRNA, for example, contains both a 3′UTR-located element targeting transcripts to distal neurites, and a 66 nt regulatory element in the 5′UTR required for concentration of mRNA molecules at synapses [104].

Examples of *cis*-acting elements acting redundantly to promote localization have also been described [112, 126–128]. In *Xenopus*, *Vg1* mRNA is targeted to the oocyte vegetal pole via a 340 nt localization signal located within the 3′UTR of the mRNA [129]. This element contains a series of redundant repeated motifs (the so-called VM1 and E2 motifs) that act synergistically to promote *Vg1* mRNA localization [127, 130]. Although necessary, clusters of E2 motifs are not sufficient to promote localization, and need to be associated with overlapping clusters of VM1 motifs, suggesting that clusters of E2 and VM1 motifs act in concert for efficient *Vg1* mRNA targeting [131, 132]. Another example of synergistic action is given by the four *cis*-acting elements (named E1, E2A, E2B, and E3) mediating *ASH1* mRNA targeting to the yeast bud tip. Although each single element supports the localization of reporter RNAs [105], the presence of all four elements is required for an efficient localization [126]. Further suggesting that the multimerization of localization elements may promote transport, the artificial duplication of sub-elements with weak localization activity has been shown in different contexts to drive efficient mRNA targeting [118, 125, 127].

Regulation of mRNA Targeting by Differential Incorporation of Localization Elements Recent studies have revealed that the inclusion of localization elements may not be a "default" process, but rather a highly regulated process allowing the differential targeting of specific isoforms. 5′UTR variants of *BDNF* transcripts, for example, are targeted to specific destinations within hippocampal neurons, ranging from proximal to distal dendritic compartments [83]. Furthermore, the apical localization of *stardust* transcripts in *Drosophila* follicular epithelial cells is controlled by a developmentally regulated alternative splicing event that ensures inclusion of a localization element-containing exon specifically in early stages of oogenesis [52]. The use of alternative polyadenylation sites, and a selective inclusion of distal localization elements in extended 3′UTRs, has also been described [17, 79]. For example, two *Impa1* isoforms with 3′UTRs of different lengths have been discovered in rat sympathetic neurons, yet only the long 3′UTR supports the localization of reporter mRNAs to axons [17]. Given the striking stage- and tissue-specific patterns of alternative cleavage and polyadenylation revealed by recent genome-wide studies [133–136], such a mechanism might commonly be used to modulate mRNA distribution in response to external signals or developmental programs.

Localization Elements Confer Specific Distributions in Various Cell Types and Different Species Strikingly, localization elements have been shown to confer specific subcellular distribution in multiple cell types, suggesting that they may be recognized by common transport machineries. *β-actin* zipcode-containing mRNAs, indeed, localize to the leading edge of migrating chicken embryonic fibroblasts [5, 112], and are also transported to the growth cones of elongating axons [137, 138], or to differentiated dendrites of cultured neurons [139]. Furthermore, reporter mRNAs containing the A2 response element (A2RE) sequence responsible for *MBP* localization to murine oligodendrocyte processes [24] are transported to dendrites when injected in cultured hippocampal neurons [140]. In *Drosophila*, the 44 nt *cis*-acting element mediating all aspects of *K10* localization in oocytes is sufficient to drive peripheral apical localization upon injection into syncytial blastoderm embryos [51].

Predicting Localization Elements Genome-wide studies have led to the identification of groups of mRNAs targeted to specific cell compartments, and thus potentially recognized by the same machinery. Despite the recent development of computational tools aimed at searching consensus RNA signals [141–143], it has proven difficult to identify *de novo* any *cis*-acting elements shared by RNAs with similar distributions [144]. Improvements in tertiary structure prediction methods, together with better refinements of the molecular requirements for *trans*-acting factor binding, are however likely to help in implementing computational methods [121]. Indeed, a recent biochemical and structural characterization of the RNA elements required for the recognition of *β-actin* zipcode by the transport machinery has provided selective parameters for a computational search of mRNAs

Tab. 1 Best-studied examples of *cis*-acting localization elements and their associated *trans*-acting localizing factors.

Organism	RNA	Localization element(s)	Subcellular destination	Binding protein(s)	References
Yeast	*ASH1*	E1, E2A, E2B, E3	Bud tip	She2/3	[105, 107, 151–153]
Drosophila	*K10*	TLS (44 nt stem-loop in the 3'UTR)	Oocyte anterior	Egl/BicD	[115, 154]
Xenopus	*Vg1*	E2 sites (5 nt)	Oocyte vegetal pole	Vg1RBP/Vera	[130, 149]
		VM1 sites (6 nt)	Oocyte vegetal pole	hnRNP I	[127, 148]
Chicken	*β-actin*	54 nt zipcode, 3'UTR	Fibroblast leading edge	ZBP1	[112, 145]
Rat	*MBP*	A2RE (11 nt in the 3'UTR)	Oligo-dendrocyte processes	hnRNP A2	[113, 114]

regulated similarly to β-actin [117]. Although a restricted number of mRNAs identified with this approach has been validated so far, promising candidates have been discovered.

4.1.2 *trans*-Acting Binding Factors

Identification of *trans*-Acting Localizing Factors As revealed by biochemical studies, RNA localization elements are bound by RNA-binding proteins that promote mRNA localization through recruitment of the localization machinery. Several of these *trans*-acting localizing factors have been identified based on the following criteria:

- They bind specifically to localizing elements and colocalize with their target mRNAs.
- Mutations in the localizing element that abrogate binding of these factors also interfere with RNA localization.
- Inactivation of these factors (or interference with their ability to bind target mRNAs) impairs mRNA localization *in vivo*.

The Zipcode-Binding Protein 1 (ZBP1), for example, was identified as a protein selectively binding to β-actin zipcode, and the strength of the zipcode–ZBP1 interaction *in vitro* was shown to directly correlate with the localization activity of reporter mRNAs *in vivo* [145]. Furthermore, the injection of zipcode-antisense oligonucleotides disrupts both β-actin–ZBP1 complexes and β-actin accumulation at the periphery of migrating fibroblasts [146], identifying ZBP1 as a key *trans*-acting binding factor promoting β-actin mRNA localization *in vivo*. As summarized in Table 1, several other *trans*-acting binding factors have been shown to associate specifically with discrete localization elements, and to promote mRNA transport upon binding. Heterogeneous nuclear ribonucleoprotein (hnRNP) A2, for example, was shown to associate selectively with the A2RE *MBP* localization element [147], and to promote the targeting of *MBP* mRNA to oligodendrocyte processes [114]. Vg1 RNA-binding protein (Vg1RBP)/Vera and hnRNP I were each shown to bind, respectively, to VM1 and E2 sites present in the *Vg1* localization element, thereby mediating *Vg1* mRNA accumulation to the vegetal pole of *Xenopus* oocytes [148, 149]. In yeast, She2p and She3p proteins are required for the localization of *ASH1* transcripts to the bud tip [43, 150], and directly interact with its localization elements [151–153].

Cooperative Binding for Increased Specificity Although RNA-binding proteins required for mRNA localization are very specifically associated with their target mRNAs *in vivo*, they often bind with relatively low affinity to localization elements *in vitro*. Recent reports have revealed that the cooperative interaction of *trans*-acting factors may generate complexes of high affinity and specificity. In yeast, for example, She2p and She3p proteins bind RNA rather unspecifically on their own, yet associate very selectively with *ASH1* mRNA when cocomplexed [153]. Furthermore, an interaction of BicD with the RNA-binding protein Egalitarian appears to improve the selective recognition of RNA localization signals by Egalitarian in *Drosophila* [154]. These results support a previously proposed model in which a selective binding of protein complexes to target mRNAs results from the summing of multiple low-affinity and/or low-specificity interactions. As shown in *Drosophila*, none of the numerous proteins specifically associating with *bicoid* stem IV/V localization element in the context of an intact recognition complex can bind selectively to this element on its own [98].

Building-Up RNP Complexes and Recruiting Molecular Motors The binding of *trans*-acting localizing factors promotes the assembly of transport RNPs and their coupling to the cytoskeleton. How this coupling is molecularly achieved has been elucidated in few cases where a direct link between RNA-binding proteins associating with localization elements and recruitment of molecular motors has been established. In yeast, the She2p:She3p cocomplex has been proposed to directly link *ASH1* mRNA to the myosin motor Myo4p, as it binds directly to *ASH1* mRNA and Myo4p, and artificial tethering of She3p to a reporter RNA induces its localization to the bud tip [151–153]. In *Drosophila*, the RNA-binding protein Egalitarian has also been shown to directly connect target mRNAs to the localization machinery through physical interactions with RNA localization signals on one hand, and Dynein Light Chain and the dynein/dynactin cofactor BicD on the other hand [154–156]. In *Xenopus* oocytes, the double-stranded RNA-binding protein Staufen is required for vegetal localization of *Vg1* mRNA, and interacts with the kinesin I motor [157]. Finally, the RNA-binding protein fragile X mental retardation protein (FMRP) mediates the association of various dendritically localized transcripts with the KIF5 motor in rat cortical neurons, and interacts with the KIF5 regulatory chain kinesin light chain (KLC) via its C-terminal domain [158]. Notably, FMRP may promote the recruitment of different molecular motors to RNP complexes, as it is also associating with the minus-end directed motor dynein [158] and the dynein-interacting BicD protein [159].

Zipcode-binding proteins may also promote association of RNP complexes with the cytoskeleton indirectly. Vg1RBP, for example, is required for *Vg1* mRNA localization in *Xenopus* oocytes, and promotes the association of *Vg1* mRNA with appropriately organized MTs [160]. However, a direct binding of Vg1RBP to MT-associated protein(s) has not been demonstrated so far. As revealed by a recent structural study performed on the Vg1RBP ortholog IGF-II mRNA-binding protein-1 (IMP-1), this protein might in fact facilitate the recruitment of MT-associated molecular motors by regulating the architecture of RNP complexes and modulating mRNA accessibility. The binding of IMP-1 K homology (KH) domains 3 and 4 to target RNA, indeed, induces a looping of the transcript, and brings sequences that are distant from one another into closer proximity, thereby potentially creating binding sites for additional factors [116]. Similar properties have been described for Polypyrimidine Tract Binding protein (PTB), another RNA-binding protein associating with localized mRNAs and regulating the architecture of RNP complexes [161–163].

Conservation of *trans*-Acting Localizing Factors Strikingly, studies performed in various cell types and organisms have revealed that localizing factors present in transport RNPs belong to protein families with highly conserved functions. The best-described members known for their role in mRNA localization include cytoplasmic polyadenylation element binding (CPEB) protein, hnRNP A/B, Egalitarian, PTB Staufen or VICKZ (Vg1 RBP/Vera, IMP-1,2,3, CRD-BP, KOC, ZBP-1) family members [164–167]; these are listed in Table 2, together with their target mRNAs.

4.1.3 Stepwise Assembly and Remodeling of RNPs

From the Nucleus to Specific Cytoplasmic Destinations: Maturation of mRNA Transport Complexes The recognition of mRNAs to be targeted, and the assembly of localizing complexes, are initiated in the nucleus with the cotranscriptional loading of *trans*-acting binding factors [177, 178]. In yeast, for example, She2p has been shown to shuttle into the nucleus [179, 180], and to bind cotranscriptionally to *ASH1* mRNA through interaction with the RNA polymerase II-associated transcription elongation factors Spt4 and Spt5 [181]. Whether the interaction of She2p with nascent RNAs is specific to localizing mRNAs is still under debate [153, 181]. Other examples of localizing factors loaded in the nucleus include Vg1RBP and hnRNP I that associate with *Vg1* and *VegT* in the nucleus of *Xenopus* oocytes [182], and ZBP1 and ZBP2 that bind to β-actin mRNA in the nucleus of cultured chicken fibroblasts [146, 183].

Upon export to the cytoplasm, core complexes formed in the nucleus are remodeled, and additional factors are recruited to generate highly specific and transport-competent RNPs. As demonstrated in yeast, the strictly cytoplasmic protein She3p is recruited upon nuclear export to the She2p:*ASH1* mRNA complex, thereby forming a very selective and affine tertiary complex further recruiting the Myo4p motor [153]. In *Xenopus* oocytes, the RNA-binding proteins XStaufen and Prrp have been shown to associate with *Vg1* and *VegT* mRNAs specifically in the cytoplasm [182].

At the destination site, translation activation of localized mRNAs induces another remodeling step in which translational repressors are released from the complexes,

Tab. 2 Families of RNA-binding proteins with function in mRNA localization conserved across species and/or cell types.

Family	Species	Specific name	Cell type (destination)	Target mRNA	Reference(s)
CPEB	Rat	–	Hippocampal neurons (dendrites)	MAP2	[168]
		–		CaMKIIα	[168]
	Mouse	–	Mammary epithelial cells (apical)	ZO-1	[54]
Egl	Drosophila	–	Early oocyte	K10	[51]
			Blastoderm embryo (apical)	Hairy	[51]
			Neuroblast (apical)	Inscuteable	[49]
			Epithelial polar cells (apical)	Unpaired	[57]
hnRNP A/B	Drosophila	Squid	Oocyte (anterodorsal corner)	Gurken	[169]
		Hrp48	Oocyte (posterior pole)	Oskar	[170]
			Oocyte (anterodorsal corner)	Gurken	[171]
	Rat	hnRNP A2	Oligodendrocytes (distal processes)	MBP	[114]
			Hippocampal neurons (dendrites)	–	[140]
PTB	Xenopus	PTB/hnRNP I	Oocyte (vegetal pole)	Vg1	[148]
	Rat	PTB	PC12 cells (neurites)	β-actin	[162]
Staufen	Drosophila	Staufen	Oocyte (posterior pole)	Oskar	[172]
			Oocyte (anterior pole)	Bicoid	[172]
			Neuroblast (apical cortex)	Prospero	[48, 50]
	Xenopus	XStaufen	Oocyte (vegetal pole)	Vg1	[157]
	Mouse	Staufen 1	Hippocampal neurons (dendrites)	β-actin	[173]
VICKZ	Xenopus	Vg1RBP/Vera	Oocyte (vegetal pole)	Vg1	[149]
				VegT	[174]
			Retinal ganglion cells (growing axons)	β-actin	[65]
	Chicken	ZBP1	Embryonic fibroblasts (leading edge)	β-actin	[145, 146]
			Forebrain neurons (axons)	β-actin	[138]
	Rat	ZBP1	Cortical neurons (axons)	β-actin	[66, 175]
			Hippocampal neurons (dendrites)	β-actin	[176]
	Human	IMP-1	Non-metastatic breast carcinoma cells (cell–cell contacts)	β-actin	[61]
				E-cadherin	[61]
			Metastatic breast carcinoma cells (leading edge)	α-actinin	[61]
				Arp16	[61]

and ribosomal components associate into functional units on the mRNA (see Sect. 5).

Functional Importance of the Nuclear History As revealed by functional assays, a loading of *trans*-acting factors in the nucleus is essential for the correct targeting of mRNAs in the cytoplasm. Indeed, blocking the nuclear import of yeast She2p does not affect the association of She2p with *ASH1* mRNA, yet prevents the recruitment of the translation repressors Loc1p and Puf6p to *ASH1* mRNA, thereby impairing *ASH1* localization to the bud tip [179, 180]. The export of localizing complexes from the nucleus may also be a regulated and functionally important step, as *ASH1* and other bud tip localizing mRNAs are selectively retained into the nucleus upon inactivation of the nonessential nucleoporin protein Nup60p [184]. Under these conditions, *ASH1* molecules exported to the cytoplasm fail to localize correctly, consistent with a key function of nuclear events on the cytoplasmic fate of mRNAs.

Another striking example of the importance of mRNA nuclear history is given by *oskar* mRNA in *Drosophila* oocytes. Indeed, the localization of *oskar* transcripts to the oocyte posterior pole relies both on components of the Exon Junction Complex (EJC), a complex deposited upstream of exon–exon junctions upon nuclear splicing [185–187], and on the splicing of *oskar* first intron [188]. As revealed by a recent structure–function study, *oskar* splicing creates a 28 nt localization element that is essential for the motility of *oskar* particles, and is composed of stretches of nucleotides flanking intron1 and assembling into a stem–loop structure upon splicing [106]. Thus, the nuclear splicing of *oskar* would serve two functions: (i) recruiting components of the EJC to the mRNA; and (ii) generating a *cis*-acting element required for localization.

Although these examples suggest that the nuclear assembly of RNP complexes is essential for an efficient recruitment of motors in the cytoplasm [106], or for efficient translation repression of localizing mRNAs [180], this may not be required for all localizing mRNAs. Indeed, many transcripts can localize normally when directly injected into the cytoplasm [93, 189–192]. Further studies are thus required to determine the extent to which nuclear recruitment of *trans*-acting factors is needed.

4.1.4 How Many RNA Molecules per Transport Particle?

Although biochemical purifications have revealed that localized mRNAs are packaged into large "granules" containing various proteins [94–96], the precise stoichiometry of RNP complexes remains to be determined. In particular, how many copies of mRNA molecules are present in a single RNP is still an open question in the field. Supporting a model where localizing mRNAs are coassembled into high-order RNP complexes containing multiple copies of mRNA molecules, *Drosophila bicoid* mRNA is oligomerizing *in vitro*, and intermolecular RNA–RNA interactions are required for the assembly of transport granules upon injection into early embryos [120]. Also consistent with this model, *Drosophila oskar* mRNA is multimerizing *in vitro* and *in vivo* [99, 193], and localizing incompetent *oskar* transcripts can hitchhike on endogenous *oskar* molecules for their transport to the oocyte posterior pole [188]. The cotransport of two different mRNA species has also been described in yeast by dynamic imaging of bud tip-localizing *ASH1* and *IST2* mRNAs [194]. The copackaging of several mRNA molecules may present several advantages, including a reduction in transport costs and a coordinated regulation of mRNA translation. Furthermore, the multimerization of mRNA molecules has been suggested to create specific structures required for the binding of *trans*-acting factors [120], and to promote the translation repression of transported mRNAs [99, 161, 195], perhaps by interfering with the recruitment of the translation machinery.

The presence of several mRNA molecules within single RNPs, however, may not be a general rule, as several recent reports have suggested that mRNAs are transported independently of each other in *Drosophila* embryos and mammalian neurons. By using quantitative *in-situ* hybridization methods, Bullock and colleagues have indeed shown in *Drosophila* blastoderm embryos that apically localizing RNPs contain single molecules of *hairy*, and that *hairy* is not cotransported with *even-skipped*, another mRNA localizing apically [196]. Similar conclusions have been reached in cultured neurons where individual copies of dendritically localizing mRNAs are present in most localizing RNPs [197, 198].

Further studies are thus required to uncover the structural constraints and the functional requirements associated with the assembly of multimolecular complexes.

4.2
Recruitment of Molecular Motors and Transport along the Cytoskeleton

The live imaging of fluorescently tagged mRNAs has revealed that transport RNPs undergo active directional movement relying on intact actin and/or MT networks, and are regulated by a combination of molecular motors.

4.2.1 Cytoskeletal Elements and Molecular Motors

Localizing mRNAs associate with – and are transported along – cytoskeletal elements that, depending on the cell types and/or the mRNA, consist of actin filaments or MTs. The targeting of bud tip-localized mRNAs in yeast or *β-actin* in chicken fibroblasts, for example, relies on intact actin filaments [44, 199]. In contrast, the targeting of maternal transcripts in *Xenopus* and *Drosophila* oocytes, or of *β-actin* mRNA in vertebrate axons, is MT-dependent [200–202]. Importantly, an overall polarization of the cytoskeletal tracks mediating mRNA transport is required for an asymmetric accumulation of localized mRNAs, although the extent of cytoskeleton polarization may depend on cell types. While a highly polarized array of MTs has been suggested to underlie the localization of pair-rule transcripts in *Drosophila* blastoderm embryo [203], a mild overall polarization of the MT cytoskeleton appears to support the localization of *oskar* mRNA to the posterior pole of *Drosophila* oocytes [204, 205].

The molecular motors recruited to transport RNPs link the localized mRNAs to cytoskeletal elements (actin filaments for myosin motors and MTs for kinesins or dynein), and interpret cytoskeleton polarity by transporting their cargoes unidirectionally [91, 206]. In yeast, the type V myosin motor Myo4p associates with *ASH1* mRNA and is required for its transport to the bud tip [44, 207]. In *Drosophila*, the MT minus end-directed motor dynein is recruited to various maternal mRNAs and pair-rule transcripts, and is responsible for their transport within developing egg chambers and in the embryo, respectively [51, 190, 192]. In addition, kinesin family members, that move toward the plus-ends of MTs, are responsible for the transport of maternal mRNAs within *Xenopus* and *Drosophila* oocytes [106, 205, 208].

As will be further described in Sect. 4.2.2, the localization of many mRNAs has been shown to depend on different types of cytoskeletal motors, suggesting a tight regulation of their recruitment and/or activity.

4.2.2 Regulation of mRNA Transport

Properties of Directed mRNA Transport
Quite surprisingly, live-imaging experiments have demonstrated that both asymmetrically distributing and uniformly distributing transcripts undergo rapid, MT-, and motor-dependent movements [189, 196, 209]. However, a key property of localized mRNA movement is that, although bidirectional, it is nonrandom and exhibits a net bias. As shown by quantitative analyses of mRNA trajectories, this net bias depends on the presence and the number of functional RNA localization elements, and results from an increased frequency and duration of directed transport events [189, 196].

Increased Recruitment of Molecular Motors
Localization signals appear to increase the ability of localized mRNAs to move persistently by facilitating the recruitment of multiple copies of the same motors. In yeast, the association of multiple copies of monomeric, nonprocessive Myo4p [210] to *ASH1* mRNA may be required for persistent movement. Indeed, a single *ASH1* localization element is capable of recruiting four copies of She2p:She3p:Myo4p complex, thus creating a tetrameric motor that can take multiple steps on microfilaments before falling of the track [211]. Furthermore, increasing the number of bound Myo4 motors by inserting additional localization elements improves the localization of *ASH1* mRNA to the bud tip [211]. In *Drosophila*, stepwise photobleaching experiments performed in a reconstituted *in vitro* system have revealed that a functional *K10* localization element increases the number of dynein molecules recruited to individual RNA molecules, further suggesting that it is a key factor underlying mRNA transport [196]. The recruitment of multiple copies of similar (but not identical) motors can also promote mRNA localization, as shown in *Xenopus* oocytes where an association of both kinesin I and kinesin II is required for the targeting of *Vg1* mRNA to the vegetal pole [208].

A Combination of Motors Modulates RNP Transport
The transport of many mRNAs appears to be regulated by a combination of different cytoskeletal motors that either mediate distinct steps in the localization pathway, or cooperatively interact to fine-tune directional transport events. Two consecutive steps of *Drosophila oskar* mRNA transport, for example, appear to be mediated by different motors: while the transport of *oskar* to early oocytes is likely dynein-dependent [51, 212], its transport to the posterior pole during mid-oogenesis is kinesin-dependent [106, 205, 213]. Furthermore, this later step has been suggested to be coordinately regulated by the activity of both MT-based kinesin I and actin-based myosin V [214].

As demonstrated by live-imaging experiments performed in higher eukaryotes, localizing mRNAs frequently undergo bidirectional transport characterized by long directed runs interspersed with short reversals, suggesting the recruitment and the activity of opposite polarity motors [158, 189, 215, 216]. Consistent with this hypothesis, the transport of *β-actin* to the dendrites of rat neurons has been shown to require both KIF5A and dynein activity

[217]. Furthermore, the RNA-binding protein FMRP, which is responsible for the transport and translation control of various neuronal transcripts, associates both with the kinesin I subunit KLC and with dynein in mammalian neurons [158]. Although the need for bidirectional motion may not be obvious at first, it may help RNPs to navigate through a crowded cytoplasm and ensure a constant reassessment and fine-tuning of directional transport [218]. How the activity of minus- and plus-end motors is orchestrated remains to be elucidated, however.

4.3
Anchoring of mRNAs at the Final Destination

Once at the destination, mRNA molecules must be retained to specific cellular structures to prevent them from diffusing away. Although studying this final anchoring step independently of mRNA transport events can be challenging, live-imaging studies performed in different model organisms have revealed that a variety of mechanisms underlies the maintenance of mRNAs at their destination.

First, a retention of mRNAs can be achieved either via continuous rounds of active short-range transport events, or via static anchoring. For example, the maintenance of *bicoid* transcripts at the anterior pole of late-stage *Drosophila* oocytes relies on an active translocation of *bicoid* particles toward the anterior cortex [219]. In contrast, *run* and *grk* transcripts become statically anchored after transport in the *Drosophila* embryo and oocyte, respectively [220, 221].

Second, a range of cellular structures appears to be implicated in mRNA anchoring, although their precise nature remains to be determined in most cell types. Maternal mRNAs, for example, frequently associate with the oocyte and/or the egg cortex. As revealed by high-resolution microscopy, anchoring to the cortex is mediated in ascidian eggs by association with a subdomain of the cortical endoplasmic reticulum (cER) [222], while it is mediated at the anterodorsal corner of *Drosophila* oocytes by electron dense "sponge-like" bodies [221].

Third, depending on the cell types and the mRNA, anchoring at the destination requires either intact actin filaments, or MTs. In *Drosophila* and *Xenopus* oocytes, indeed, the anchoring of maternal mRNAs at the posterior and vegetal pole, respectively, is disrupted upon destabilization of the actin network [90, 201, 223–226]. In the *Drosophila* oocyte and embryo, in contrast, anchoring of the *grk* and *run* transcripts relies on MTs, and more specifically on a motor activity-independent function of dynein [220, 221]. Moreover, it has been suggested that the tumor-suppressor adenomatous polyposis coli (APC) may be required for MT-dependent anchoring of transcripts in protrusions of mammalian migrating cells [14].

Finally, translation of the mRNA at the destination may – at least in some cases – be required for correct anchoring. In yeast, indeed, the introduction of a stop codon next to the AUG codon, or a deletion of the AUG codon, impairs anchoring of *ASH1* mRNA at the bud [107, 227]. Whether the key factor is Ash1p product itself, or the recruitment of translating ribosomes to the mRNA, remains to be determined. In the *Drosophila* oocyte, however, the long form of Oskar protein has been shown to maintain *oskar* mRNA at the posterior [228].

5 Coupling mRNA Localization with Translational Control

The translation of localized mRNAs at their destination implies the presence of a functional protein synthesis machinery at these cellular sites. Consistent with this, ribosomes have been detected in various compartments by biochemical or ultrastructural studies [12]. Selective synthesis at the destination also implies that the translation of localizing mRNA molecules is repressed. Indeed, localizing mRNAs do not usually cosediment with fractions corresponding to translating ribosomes [97, 99, 229]. Furthermore, transport RNPs contain various translation repressors, some of them being released from the mRNA at the destination (see Sect. 5.2).

As described in the following paragraphs, the emerging view is that localizing mRNAs are repressed via a combination of complementary mechanisms, and are translated either directly upon arrival at their cellular destination, or in response to specific external signals [230, 231].

5.1 Translational Repression of Localizing mRNAs

Translational repressors recruited to transport RNPs prevent protein synthesis, most frequently by blocking the rate-limiting initiation step [230]. Different examples of translation repressors associating with the 3′UTR of localizing mRNAs, and preventing the recruitment of the initiation factor eIF4G by interfering with its binding to cap-associated eIF4E have been described [232]. In *Drosophila* oocytes, the eIF4E-binding protein (eIF4E-BP) Cup is recruited to *oskar* mRNA through direct interaction with the RNA-binding protein Bruno, and represses *oskar* translation by binding eIF4E to the exclusion of eIF4G [233]. A similar repression mechanism has been described for dendritically localized mRNAs in mammalian neurons. *CaMKIIα* mRNA, indeed, is selectively bound by the RNA-binding protein FMRP that further recruits the eIF4E-BP CYFIP1, thereby preventing translation initiation [234].

miRNAs are major translational repressors, and have recently been shown to regulate the translation of mRNAs accumulating in the dendrites of neuronal cells [235–237]. *mir-125a*, for example, is present in dendrites and represses the translation of the postsynaptic scaffolding protein encoding *PSD-95* mRNA [235]. Interestingly, an association of the RNA-binding protein FMRP with *PSD-95* mRNA is required for the binding of *mir-125a*-containing RNA-induced silencing complexes (RISCs) to *PSD-95* mRNA, suggesting that FMRP may help recruiting the miRNA machinery on specific mRNA targets.

The control of poly(A) tail length is also a mechanism allowing precise translational control, as long poly(A) tails promote translation initiation, while short poly(A) tails rather repress it [238]. Indeed, the translation of localized mRNAs at the destination appears to correlate with increased poly(A) tail length [87, 239, 240]. Conversely, the CCR4-NOT deadenylase complex has been shown to be recruited to nonlocalized *nanos* mRNA in the *Drosophila* embryo, thereby inducing a selective shortening of poly(A) tail length [87].

It is becoming increasingly clear that multiple mechanisms cooperate to ensure the robust and flexible translational repression of localizing mRNAs. In yeast, for example, different translation repressors with different functions are recruited to *ASH1* mRNA. While Khd1 represses translation initiation by directly interacting with the C-terminal domain of eIF4G, Puf6 rather blocks 60S ribosomal subunit joining, most likely through competitive binding to eIF5B [241, 242]. In *Drosophila* oocytes, the translation repressor Bruno represses *oskar* in two ways: (i) by recruiting the eIF4E-BP protein Cup; and (ii) by promoting the multimerization of *oskar* molecules [99, 233]. The RNA-binding protein Smaug also has a dual function in this system: it is both blocking translation initiation through recruitment of Cup on *nanos* mRNA, and keeping *nanos* poly(A)tail short by directly binding to a subunit of the CCR4–NOT complex [87, 243].

5.2
Translational Derepression upon Arrival

The translational repression of localizing mRNAs must be relieved once mRNAs have reached their destination. But, how is the transition from translation repression to translation activation achieved? The results of several studies have suggested that post-translational modifications of translation repressors at the destination promote their dissociation from target mRNAs, thereby triggering local protein synthesis [241, 242, 244–246]. In yeast, for example, casein kinase-II (CK2) is present at the bud cortex, and phosphorylates the translation repressor Puf6 both *in vitro* and *in vivo* [241]. The phosphorylation of Puf6 by CK2 decreases Puf6 affinity for *ASH1* mRNA, and reduces its translation repression activity. Similarly, it has been proposed that the type I casein kinase Yck1 phosphorylates the translation repressor Khd1 at the yeast bud cortex, thus triggering a dissociation of Khd1 from the *ASH1* complex and a local synthesis of Ash1p [242].

In these examples, the spatially restricted distribution of active kinases, and/or their spatially restricted interaction with their substrates, ensures that translation is activated specifically at the destination. As described in the next section, a temporal control of translation derepressor activity can also provide another level of regulation.

5.3
Translational Derepression in Response to Specific Signals

mRNA localization represents an efficient means to restrict protein synthesis in space, and also to control gene expression in time. In some cell types, indeed, localized mRNAs are stored in a dormant state at their cellular destination, their translation being activated only in response to specific external stimuli. β-*actin* mRNA, for example, is transported to the growing axons of cultured vertebrate neurons, and is translated in growth cones in response to attractive cues such as Netrin-1 or BDNF [65, 66]. Several lines of evidence support a model in which the synthesis of β-actin in growth cones is triggered by a cue-induced phosphorylation of the β-*actin* mRNA-associated protein ZBP1:

- First, ZBP1 represses β-*actin* translation *in vitro* and *in vivo* [244].
- Second, ZBP1 is phosphorylated by the Src kinase *in vitro* and *in vivo* [244].
- Third, a nonphosphorylatable form of ZBP1 shows a significantly higher affinity for β-*actin* mRNA, and suppresses the BDNF-induced increase in β-actin protein in growth cones [244, 247].

- Finally, stimulation with BDNF activates Src kinases and induces ZBP1 phosphorylation *in vivo* [66, 247].

Translation activation in response to neuronal activity has also been reported for dendritically localized mRNAs, and the molecular mechanisms underlying this process are starting to emerge [68, 69, 248]. For example, neuronal stimulation has been suggested to release the translation repression of dendritically localized mRNAs by triggering dissociation of the inhibitory eIF4E-BP CYFIP1 from eIF4E at synapses [234]. In addition, activity-induced dephosphorylation of the translation repressor FMRP has been proposed to act as a switch between translation repression and translation activation. As demonstrated by Bassell and colleagues, FMRP negatively regulates *PSD-95* translation, partly by enhancing the recruitment of the microRNA *mir125-a* to *PSD-95* mRNA [235]. Furthermore, the stimulation of metabotropic Glutamate receptors (mGluRs), which triggers the translation of dendritically localized mRNAs such as *PSD-95*, also induces FMRP de-phosphorylation [249]. Finally, the expression of a phosphomimetic form of FMRP promotes recruitment of the *mir125-a*–RISC inhibitory complex on *PSD-95* mRNA, and blocks *PSD-95* translation in response to mGluR stimulation [235]. A change in the phosphorylation status of CPEB, another translation repressor associated with dendritically localized mRNAs, is also observed in response to neuronal stimulation. Glutamate stimulation, indeed, induces the phosphorylation of CPEB by the kinase Aurora, a process associated with poly(A) tail elongation and translation activation of *CaMKIIα* mRNA [250].

Thus, the activity of translational repressors can be modulated locally, in response to extrinsic signals, providing a means to couple mRNA localization with inducible activation of translation, and thus to spatially and temporally regulate gene expression.

6
Integrated Examples of Localizing mRNAs

In the following paragraphs, three integrated examples of mRNA localization processes will be presented in detail. First, the localization of *ASH1* mRNA to the yeast bud tip will be described, as the *trans*-acting factors regulating this process have been comprehensively identified, and the mechanisms underlying the different steps of localization (from cotranscriptional assembly of complexes to anchoring and translational activation at the final destination) have been well characterized. This example also illustrates the importance of asymmetric mRNA targeting in cell fate specification.

Second, the localization of maternal transcripts within *Drosophila* oocytes will be presented, as the powerful combination of genetic analyses and live-imaging experiments in this system has led to the discovery of important concepts in the field. In addition, it provides an example where different mRNAs are targeted to distinct intracellular locations via specific mechanisms, and illustrates the requirement of mRNA targeting for patterning early embryos.

Finally, the localization of mRNAs within mammalian dendrites will be presented, as the translation of dendritically localized mRNAs is controlled by external stimuli (e.g., synaptic activity), thus allowing a precise spatiotemporal control of gene expression.

The aim of the following paragraphs is not to present in great detail the mechanisms underlying the selected examples, as most of these have been described previously. Rather, the aim is to present the different actors and steps in a linear and integrated fashion.

6.1
mRNA Localization in the Budding Yeast: Focus on ASH1 mRNA

During mitosis, the yeast *Saccharomyces cerevisiae* divides asymmetrically, producing a larger mother cell and a smaller daughter cell. More than 30 transcripts localizing to the bud of dividing yeast cells have been identified so far [21, 251, 252]. Of these, *ASH1* mRNA was the first to be discovered [43, 44], and is still the most extensively studied. An efficient transport of *ASH1* mRNA to the daughter cell, and subsequent local translation of Ash1p, are functionally essential, as they allow for the generation of progenies with distinct (and thus compatible) mating types [150, 253].

The transport of *ASH1* mRNA to the bud tip is controlled by four *cis*-localization elements found in *ASH1* coding sequence and 3′UTR [105, 107], and three core *trans*-acting factors: the RNA-binding proteins She2p; the type V myosin motor Myo4p; and She3p [21, 150]. Whilst She3p has long been considered as an adapter between *ASH1*-bound She2p and Myo4p [151, 152], its capacity to bind RNA and to form a complex of high selectivity together with She2p has only recently been demonstrated [153]. Additional RNA-binding proteins are involved in the localization of *ASH1* mRNA *in vivo*, including Puf6p and Khd1p – two proteins that repress the translation of unlocalized mRNAs [227, 241, 242, 254].

Based on the unique combination of tools available in yeast, the sequence of events leading to the asymmetric sorting of Ash1p to the daughter cell has largely been elucidated (Fig. 1) [21]. In the nucleus, She2p is recruited to nascent *ASH1* transcripts by the transcription elongation factor Spt4/5 [181]. Although it remains to be fully demonstrated, the results of different studies have suggested that the *ASH1*:She2p complex may then transit through the nucleolus. First, She2p and *ASH1* mRNA accumulate in the nucleus upon inhibition of nuclear export [179]. Second, the nucleolar RNA binding protein Loc1p has been shown to associate with *ASH1* mRNA, and is required for its efficient localization [255]. At least two other factors – the translation repressors Puf6p and Khd1p – are recruited to *ASH1* mRNA before nuclear export [227, 254]. Once in the cytoplasm, the She3p:Myo4p complexes selectively associate with She2p-bound *ASH1* mRNAs [153], triggering transport along actin filaments [44, 211]. Upon arrival at the bud tip, Khd1p and Puf6p become phosphorylated by the membrane-associated kinases Yck1p and CK2, respectively [241, 242]. Phosphorylated Khd1p and Puf6p are then released from *ASH1* mRNA, thereby allowing a translational activation of *ASH1* mRNA [241, 242]. How *ASH1* mRNA is maintained at the bud tip is still unclear, although *ASH1* anchoring (but not transport) has been shown to rely on the Bud6p/Aip3p and Bni1p/She5p proteins [256], and on *ASH1* translation [107, 227].

6.2
mRNA Localization in *Drosophila* Oocyte: Focus on *oskar* mRNA

In *Drosophila*, the establishment of embryonic anteroposterior and dorsoventral

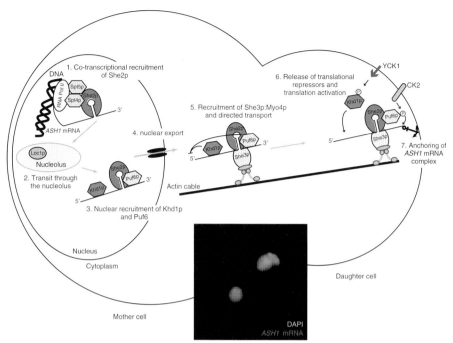

Fig. 1 Localization of *ASH1* mRNA in the budding yeast *Saccharomyces cerevisiae*. Schematic representation of the main steps underlying the assembly of transport-competent *ASH1* RNPs, the transport of *ASH1* mRNA to the bud tip, and the translational control of *ASH1*. The inserted image shows the accumulation of *ASH1* mRNA at the bud tip. Upon division, *ASH1* mRNA will be inherited exclusively by the daughter cell. Red: *ASH1* mRNA; blue: DAPI. Image courtesy of R.P. Jansen, Institute of Biochemistry, Tubingen, Germany.

axes relies on the asymmetric distribution of four maternal mRNAs within the oocyte [26]. *bicoid* mRNA, which encodes a transcription factor required for the formation of head structures, localizes at the oocyte anterior pole [3, 31]. *oskar* and *nanos* mRNAs, required for the formation of abdominal structures and the specification of germ cells, localize at the posterior pole [32–34]. *gurken*, a transforming growth factor-beta (TGF-β) encoding mRNA, accumulates at the anterodorsal corner [36]. These mRNAs, transcribed in nurse cells, are transferred to the oocyte where they localize via different mechanisms: while the anterior localization of *bicoid* and *gurken* mRNAs relies on the MT-minus end-directed motor dynein [190, 212, 257, 258], the posterior localization of *oskar* mRNA relies on the activity of kinesin I [205, 213]. The posterior accumulation of *nanos* mRNA in the late oocyte is largely independent of MTs, and is achieved by diffusion coupled with local entrapment [90].

The localization of *oskar* mRNA to the posterior pole of *Drosophila* oocytes is a highly regulated and multistep process (Fig. 2). The formation of *oskar* transport-competent RNPs starts in the nucleus with the splicing of *oskar* and the concomitant deposition of core components of the EJC (eIF4AIII, Y14, and Mago), all required for the cytoplasmic

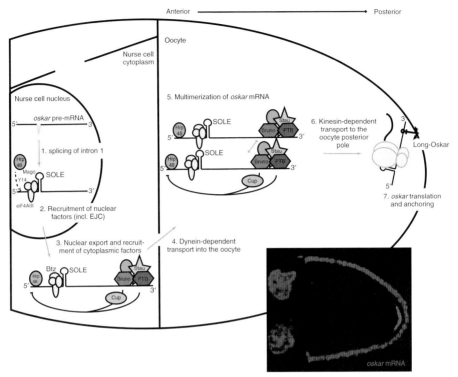

Fig. 2 Localization of *oskar* mRNA to the posterior pole of *Drosophila melanogaster* oocyte. Schematic representation of the main steps underlying the assembly of transport-competent *oskar* RNPs, the transport of *oskar* mRNA to the posterior pole, and the anchoring of *oskar* mRNA at the destination. The inserted image shows the accumulation of *oskar* mRNA at the oocyte posterior pole. Red: *oskar* mRNA, blue: DAPI. Image from S. Lopez de Quinto, F. Besse and A. Ephrussi, EMBL, Heidelberg, Germany.

localization of *oskar* [185–188]. The splicing of *oskar* intron 1 also triggers the assembly of a functional localization element (the spliced *oskar* localization element or SOLE), which is composed of intron 1 flanking sequences and is essential for the motility of *oskar* particles [106]. Various *trans*-acting factors identified by biochemical purifications [259–262], genetic screens [170, 172, 187] or visual screens [161, 187, 263, 264] are recruited to *oskar* transcripts, many of them via selective binding to 3'UTR sequences. Among these factors, Hrp48 [259, 262], Hrp40/Squid [261], and PTB [161] shuttle between the nucleus and the cytoplasm, and may associate with *oskar* in the nucleus. RNA-binding proteins recruited to *oskar* mRNA are involved in the translation repression of localizing *oskar*, such as PTB [161, 259], Bruno [258], Me31B [264], and Hrp48 [261], or in the posterior transport of *oskar* mRNA, such as Staufen [172], Barentsz [187], or Hrp48 [170]. In the cytoplasm, *oskar* mRNA appears to be packaged into large particles containing multiple mRNA copies, as *oskar* transcripts are able to dimerize *in vivo* and *in vitro* [193], and nonlocalizing *oskar* molecules can hitchhike on wild-type molecules for their transport to the oocyte posterior pole [188]. As demonstrated by live-imaging analyses, the posterior transport of *oskar* particles is blocked in *kinesin* mutant contexts [106, 205], and may be achieved via a random walk on a weakly polarized cytoskeleton [204, 205]. Once at the posterior pole, *oskar* mRNA becomes translated. Although the translation activation of *oskar* mRNA likely involves release of the RNA-binding protein Bruno and its binding partner Cup [233], how this process is molecularly regulated remains unclear. The translation of Oskar and production of the long isoform Long-Oskar is in turn required, together with an intact cortical actin cytoskeleton, for static anchoring of *oskar* mRNA at the posterior pole, and local accumulation of both Long-Oskar and Short-Oskar isoforms [223, 225, 226, 228]. The posterior accumulation of short-Oskar then nucleates formation of the germ plasm, and recruits the components required for both germline formation and embryonic posterior patterning [32, 228, 265, 266].

6.3
mRNA Localization in Mammalian Dendrites: Focus on CaMKII mRNA

As shown in both vertebrates and invertebrates, dendritic protein synthesis is essential for various forms of long-term synaptic plasticity [69]. While dendritically localizing mRNAs have long been identified [7, 139, 267, 268], the enormous potential for local dendritic translation has emerged only recently with the discovery that thousands of transcripts – a large proportion of them encoding known synaptic proteins – are localized in synaptic regions [15].

CaMKIIα mRNA targeting to dendrites and translation regulation have been particularly studied in mammalian neurons, both at the functional and at the molecular level (Fig. 3). As suggested by independent studies, the dendritic localization of *CaMKIIα* mRNA appears to be mediated by different 3′UTR-located *cis*-acting elements that may act synergistically to ensure efficient localization [269–271]. Removal of these elements prevents the dendritic accumulation of *CaMKIIα* mRNA *in vivo*, and impairs long-term potentiation and long-term memory [78]. CPEB and FMRP – two RNA-binding proteins with dual function in mRNA transport and translation repression – have been shown to associate with *CaMKIIα* mRNA and to regulate its expression. Both proteins, indeed, assemble into actively moving particles and promote the targeting of *CaMKIIα* mRNA to dendrites [158, 168]. Furthermore, consistent with previous studies revealing that the transport of *CaMKIIα* mRNA to dendrites is kinesin-dependent [95], Bassell and colleagues have shown that FMRP mediates the association of *CaMKIIa* mRNA with kinesin, thereby promoting the activity-induced transport of *CaMKIIα* mRNA [158]. CPEB and FMRP also repress *CaMKIIα* translation during its transport, thus ensuring that CaMKIIα protein is produced in dendrites, in response to specific stimuli. Indeed, *CaMKIIα* mRNA is found in a complex with the FMRP-binding CYFIP1 protein, which prevents the recruitment of eIF4G by binding to eIF4E [234]. As suggested by Napoli *et al.*, the translation activation of *CaMKIIα* mRNA would be triggered by a dissociation of CYFIP1 from eIF4E upon stimulation [234]. CPEB may regulate *CaMKIIα* translation independently, by regulating poly(A) tail length. CPEB has been shown in different

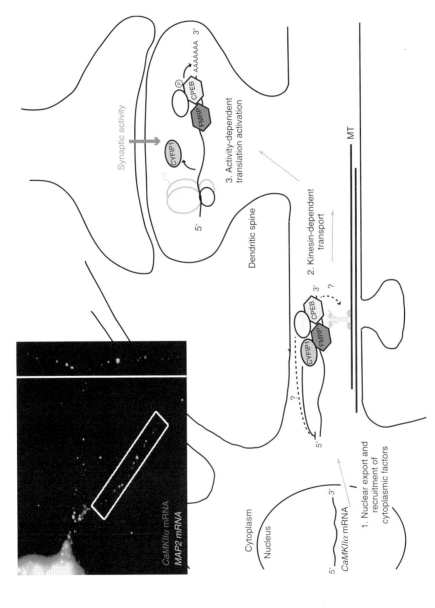

Fig. 3 Localization of *CaMKIIα* mRNA to the dendrites of mammalian neurons. Schematic representation of the main steps underlying the transport of *CaMKIIα* mRNA to dendrites, and its translation activation in response to synaptic activity. The inserted image shows the localization of *CaMKIIα* (red) and *MAP2* (green) mRNAs in mammalian cultured neurons. Reproduced with permission from Mikl et al. [198].

Tab. 3 Main methods used to detect mRNAs.

Tools	Principle	Pros	Cons	Primary Reference(s)
Antisense RNA probes	Hybridize on complementary sequences in target mRNA	Detection of endogenous transcripts Highly specific and sensitive (down to the single molecule resolution)	Usable on fixed tissues only	[282–284]
Molecular beacons	Hairpin-shaped oligonucleotide probes that become fluorescent upon binding to complementary sequences in target mRNAs	Allow dynamic visualization of endogenous mRNAs	Non-membrane-permeable (must be injected) May impair target mRNA behavior upon binding Non-specific signals can be observed	[285, 286]
Fluorescent sense RNAs	*In-vitro* synthesized RNAs emitting a fluorescent signal	Allow dynamic visualization of injected RNAs Strong and specific signals	Must be injected RNAs have not matured *in vivo* and may thus not recapitulate the behavior of endogenous mRNAs	[93]
RNA tags/RNA binding proteins	Tags inserted into RNAs of interest are recognized by RNA binding proteins fused to fluorescent proteins	Allow dynamic visualization of tagged mRNAs	RNA tags have to be inserted into a reporter constructs (not endogenous RNAs) Insertion of the tag may impair mRNA behavior	[207, 211, 287]
Aptamers/dyes	Aptamers inserted into RNAs of interest are recognized by fluorogenic dyes that become fluorescent upon binding to the aptamer	Allow dynamic visualization of tagged mRNAs Low photo-bleaching and small tag	Sensitivity and specificity of the signal remains to be tested Aptamers must be inserted into a reporter constructs (not endogenous RNAs)	[288]

systems to repress the translation of its target mRNAs when nonphosphorylated and, conversely, to promote translation by inducing poly(A) tail elongation when phosphorylated [232, 272]. Interestingly, glutamate stimulation of synaptic fractions induces both *CaMKIIα* poly(A) elongation and CPEB phosphorylation [250], suggesting that CPEB might promote *CaMKIIα* translation in response to stimulation.

7
Conclusions and Perspectives

The past few years have witnessed an explosion in the number of asymmetrically distributed mRNAs, cellular structures used as mRNA destinations, and functions associated with localized mRNAs. As a result, mRNA localization is now recognized as a prevalent mechanism that, when coupled to local protein synthesis, allows for the accumulation of proteins at specific cellular sites. Studies performed on model organisms have considerably improved the present knowledge regarding the molecular players involved in this process, and have more recently revealed that the repertoire of localized mRNAs varies in response to developmental programs [18, 52, 273], and chemical [76, 158, 274, 275] or mechanical [276] signals. How the activity of different mRNA transport machineries is orchestrated and regulated in response to these signals is still largely unknown, but this question can now be tackled using a variety of tools that has been developed to follow the behavior of localizing mRNAs in real time (Table 3). Another exciting current challenge is to unravel the combinatorial codes underlying mRNA behavior, and to test the prediction that transcripts with similar fates share combinations of *cis*-regulatory elements. As revealed by the results of recent studies [117, 277], predicting the fate of a given mRNA based on its primary sequence appears to be a reachable objective. Finally, it will be interesting to investigate whether the different populations of noncoding RNAs that have recently been discovered also exhibit specific subcellular distributions. Consistent with this view, microRNAs accumulating in specific compartments [278, 279], as well as long noncoding RNAs with asymmetric distribution patterns [280, 281], have been identified and their local functions now remain to be elucidated.

Acknowledgments

The author is very grateful to I. Gaspar and M. Simonelig for helpful discussions, and to C. Medioni, F. Degrave, I. Gaspar, and H. Lipshitz for critical reading of the manuscript (or parts of the manuscript). Thanks are also given to the authors mentioned in the figure legends for generously sharing images of their work. F.B. acknowledges funding from the ATIP/CNRS program, the HFSP organization, the FRM and ARC foundations, and the ANR agency.

References

1 Jeffery, W.R., Tomlinson, C.R., Brodeur, R.D. (1983) Localization of actin messenger RNA during early ascidian development. *Dev. Biol.*, **99** (2), 408–417.
2 Rebagliati, M.R., Weeks, D.L., Harvey, R.P., Melton, D.A. (1985) Identification and cloning of localized maternal RNAs from *Xenopus* eggs. *Cell*, **42** (3), 769–777.
3 Berleth, T., Burri, M., Thoma, G., Bopp, D. *et al.* (1988) The role of localization of bicoid RNA in organizing the anterior pattern of the *Drosophila* embryo. *EMBO J.*, **7** (6), 1749–1756.
4 Frigerio, G., Burri, M., Bopp, D., Baumgartner, S. *et al.* (1986) Structure

of the segmentation gene paired and the *Drosophila* PRD gene set as part of a gene network. *Cell*, **47** (5), 735–746.

5 Lawrence, J.B., Singer, R.H. (1986) Intracellular localization of messenger RNAs for cytoskeletal proteins. *Cell*, **45** (3), 407–415.

6 Trapp, B.D., Moench, T., Pulley, M., Barbosa, E. et al. (1987) Spatial segregation of mRNA encoding myelin-specific proteins. *Proc. Natl Acad. Sci. USA*, **84** (21), 7773–7777.

7 Garner, C.C., Tucker, R.P., Matus, A. (1988) Selective localization of messenger RNA for cytoskeletal protein MAP2 in dendrites. *Nature*, **336** (6200), 674–677.

8 Bashirullah, A., Cooperstock, R.L., Lipshitz, H.D. (1998) RNA localization in development. *Annu. Rev. Biochem.*, **67**, 335–394.

9 Holt, C.E., Bullock, S.L. (2009) Subcellular mRNA localization in animal cells and why it matters. *Science*, **326** (5957), 1212–1216.

10 Martin, K.C., Ephrussi, A. (2009) mRNA localization: gene expression in the spatial dimension. *Cell*, **136** (4), 719–730.

11 Medioni, C., Mowry, K., Besse, F. (2012) Principles and roles of mRNA localization in animal development. *Development*, **139** (18), 3263–3276.

12 Lecuyer, E., Yoshida, H., Krause, H.M. (2009) Global implications of mRNA localization pathways in cellular organization. *Curr. Opin. Cell Biol.*, **21** (3), 409–415.

13 Lecuyer, E., Yoshida, H., Parthasarathy, N., Alm, C. et al. (2007) Global analysis of mRNA localization reveals a prominent role in organizing cellular architecture and function. *Cell*, **131** (1), 174–187.

14 Mili, S., Moissoglu, K., Macara, I.G. (2008) Genome-wide screen reveals APC-associated RNAs enriched in cell protrusions. *Nature*, **453** (7191), 115–119.

15 Cajigas, I.J., Tushev, G., Will, T.J., tom Dieck, S. et al. (2012) The local transcriptome in the synaptic neuropil revealed by deep sequencing and high-resolution imaging. *Neuron*, **74** (3), 453–466.

16 Moccia, R., Chen, D., Lyles, V., Kapuya, E. et al. (2003) An unbiased cDNA library prepared from isolated *Aplysia* sensory neuron processes is enriched for cytoskeletal and translational mRNAs. *J. Neurosci.*, **23** (28), 9409–9417.

17 Andreassi, C., Zimmermann, C., Mitter, R., Fusco, S. et al. (2010) An NGF-responsive element targets myo-inositol monophosphatase-1 mRNA to sympathetic neuron axons. *Nat. Neurosci.*, **13** (3), 291–301.

18 Gumy, L.F., Yeo, G.S., Tung, Y.C., Zivraj, K.H. et al. (2011) Transcriptome analysis of embryonic and adult sensory axons reveals changes in mRNA repertoire localization. *RNA*, **17** (1), 85–98.

19 Zivraj, K.H., Tung, Y.C., Piper, M., Gumy, L. et al. (2010) Subcellular profiling reveals distinct and developmentally regulated repertoire of growth cone mRNAs. *J. Neurosci.*, **30** (46), 15464–15478.

20 Crofts, A.J., Washida, H., Okita, T.W., Satoh, M. et al. (2005) The role of mRNA and protein sorting in seed storage protein synthesis, transport, and deposition. *Biochem. Cell Biol.*, **83** (6), 728–737.

21 Heym, R.G., Niessing, D. (2011) Principles of mRNA transport in yeast. *Cell. Mol. Life Sci.*, **69** (11), 1843–1853.

22 Zarnack, K., Feldbrugge, M. (2010) Microtubule-dependent mRNA transport in fungi. *Eukaryot. Cell*, **9** (7), 982–990.

23 Keiler, K.C. (2011) RNA localization in bacteria. *Curr. Opin. Microbiol.*, **14** (2), 155–159.

24 Kumano, G. (2011) Polarizing animal cells via mRNA localization in oogenesis and early development. *Dev. Growth Differ.*, **54** (1), 1–18.

25 Abrams, E.W., Mullins, M.C. (2009) Early zebrafish development: it's in the maternal genes. *Curr. Opin. Genet. Dev.*, **19** (4), 396–403.

26 Becalska, A.N., Gavis, E.R. (2009) Lighting up mRNA localization in *Drosophila* oogenesis. *Development*, **136** (15), 2493–2503.

27 Sardet, C., Dru, P., Prodon, F. (2005) Maternal determinants and mRNAs in the cortex of ascidian oocytes, zygotes and embryos. *Biol. Cell*, **97** (1), 35–49.

28 King, M.L., Messitt, T.J., Mowry, K.L. (2005) Putting RNAs in the right place at the right time: RNA localization in the frog oocyte. *Biol. Cell*, **97** (1), 19–33.

29 Momose, T., Derelle, R., Houliston, E. (2008) A maternally localised Wnt ligand required for axial patterning in the cnidarian *Clytia hemisphaerica*. *Development*, **135** (12), 2105–2113.

30 Momose, T., Houliston, E. (2007) Two oppositely localised frizzled RNAs as axis

determinants in a cnidarian embryo. *PLoS Biol.*, **5** (4), e70.

31 Driever, W., Nusslein-Volhard, C. (1988) The bicoid protein determines position in the *Drosophila* embryo in a concentration-dependent manner. *Cell*, **54** (1), 95–104.

32 Ephrussi, A., Dickinson, L.K., Lehmann, R. (1991) Oskar organizes the germ plasm and directs localization of the posterior determinant nanos. *Cell*, **66** (1), 37–50.

33 Kim-Ha, J., Smith, J.L., Macdonald, P.M. (1991) oskar mRNA is localized to the posterior pole of the *Drosophila* oocyte. *Cell*, **66** (1), 23–35.

34 Gavis, E.R., Lehmann, R. (1992) Localization of nanos RNA controls embryonic polarity. *Cell*, **71** (2), 301–313.

35 Wang, C., Dickinson, L.K., Lehmann, R. (1994) Genetics of nanos localization in *Drosophila*. *Dev. Dyn.*, **199** (2), 103–115.

36 Neuman-Silberberg, F.S., Schupbach, T. (1993) The *Drosophila* dorsoventral patterning gene gurken produces a dorsally localized RNA and encodes a TGF alpha-like protein. *Cell*, **75** (1), 165–174.

37 Birsoy, B., Kofron, M., Schaible, K., Wylie, C. et al. (2006) Vg 1 is an essential signaling molecule in *Xenopus* development. *Development*, **133** (1), 15–20.

38 Melton, D.A. (1987) Translocation of a localized maternal mRNA to the vegetal pole of *Xenopus* oocytes. *Nature*, **328** (6125), 80–82.

39 Zhang, J., Houston, D.W., King, M.L., Payne, C. et al. (1998) The role of maternal VegT in establishing the primary germ layers in *Xenopus* embryos. *Cell*, **94** (4), 515–524.

40 Zhang, J., King, M.L. (1996) *Xenopus* VegT RNA is localized to the vegetal cortex during oogenesis and encodes a novel T-box transcription factor involved in mesodermal patterning. *Development*, **122** (12), 4119–4129.

41 Houston, D.W., King, M.L. (2000) A critical role for Xdazl, a germ plasm-localized RNA, in the differentiation of primordial germ cells in *Xenopus*. *Development*, **127** (3), 447–456.

42 Lai, F., Zhou, Y., Luo, X., Fox, J. et al. (2011) Nanos1 functions as a translational repressor in the *Xenopus* germline. *Mech. Dev.*, **128** (1-2), 153–163.

43 Long, R.M., Singer, R.H., Meng, X., Gonzalez, I. et al. (1997) Mating type switching in yeast controlled by asymmetric localization of ASH1 mRNA. *Science*, **277** (5324), 383–387.

44 Takizawa, P.A., Sil, A., Swedlow, J.R., Herskowitz, I. et al. (1997) Actin-dependent localization of an RNA encoding a cell-fate determinant in yeast. *Nature*, **389** (6646), 90–93.

45 Jedrusik, A., Parfitt, D.E., Guo, G., Skamagki, M. et al. (2008) Role of Cdx2 and cell polarity in cell allocation and specification of trophectoderm and inner cell mass in the mouse embryo. *Genes Dev.*, **22** (19), 2692–2706.

46 Lambert, J.D., Nagy, L.M. (2002) Asymmetric inheritance of centrosomally localized mRNAs during embryonic cleavages. *Nature*, **420** (6916), 682–686.

47 Takatori, N., Kumano, G., Saiga, H., Nishida, H. (2010) Segregation of germ layer fates by nuclear migration-dependent localization of Not mRNA. *Dev. Cell*, **19** (4), 589–598.

48 Broadus, J., Fuerstenberg, S., Doe, C.Q. (1998) Staufen-dependent localization of prospero mRNA contributes to neuroblast daughter-cell fate. *Nature*, **391** (6669), 792–795.

49 Hughes, J.R., Bullock, S.L., Ish-Horowicz, D. (2004) Inscuteable mRNA localization is dynein-dependent and regulates apicobasal polarity and spindle length in *Drosophila* neuroblasts. *Curr. Biol.*, **14** (21), 1950–1956.

50 Li, P., Yang, X., Wasser, M., Cai, Y. et al. (1997) Inscuteable and Staufen mediate asymmetric localization and segregation of prospero RNA during *Drosophila* neuroblast cell divisions. *Cell*, **90** (3), 437–447.

51 Bullock, S.L., Ish-Horowicz, D. (2001) Conserved signals and machinery for RNA transport in *Drosophila* oogenesis and embryogenesis. *Nature*, **414** (6864), 611–616.

52 Horne-Badovinac, S., Bilder, D. (2008) Dynein regulates epithelial polarity and the apical localization of stardust A mRNA. *PLoS Genet.*, **4** (1), e8.

53 Li, Z., Wang, L., Hays, T.S., Cai, Y. (2008) Dynein-mediated apical localization of crumbs transcripts is required for Crumbs activity in epithelial polarity. *J. Cell Biol.*, **180** (1), 31–38.

54 Nagaoka, K., Udagawa, T., Richter, J.D. (2012) CPEB-mediated ZO-1 mRNA localization is required for epithelial tight-junction assembly and cell polarity. *Nat. Commun.*, **3**, 675.

55 Schotman, H., Karhinen, L., Rabouille, C. (2008) dGRASP-mediated noncanonical integrin secretion is required for *Drosophila* epithelial remodeling. *Dev. Cell*, **14** (2), 171–182.

56 Simmonds, A.J., dos Santos, G., Livne-Bar, I., Krause, H.M. (2001) Apical localization of wingless transcripts is required for wingless signaling. *Cell*, **105** (2), 197–207.

57 Van de Bor, V., Zimniak, G., Cerezo, D., Schaub, S. et al. (2011) Asymmetric localisation of cytokine mRNA is essential for JAK/STAT activation during cell invasiveness. *Development*, **138** (7), 1383–1393.

58 Mingle, L.A., Okuhama, N.N., Shi, J., Singer, R.H. et al. (2005) Localization of all seven messenger RNAs for the actin-polymerization nucleator Arp2/3 complex in the protrusions of fibroblasts. *J. Cell Sci.*, **118** (Pt 11), 2425–2433.

59 Condeelis, J., Singer, R.H. (2005) How and why does beta-actin mRNA target? *Biol. Cell*, **97** (1), 97–110.

60 Liao, G., Simone, B., Liu, G. (2011) Mis-localization of Arp2 mRNA impairs persistence of directional cell migration. *Exp. Cell. Res.*, **317** (6), 812–822.

61 Gu, W., Katz, Z., Wu, B., Park, H.Y. et al. (2012) Regulation of local expression of cell adhesion and motility-related mRNAs in breast cancer cells by IMP1/ZBP1. *J. Cell Sci.*, **125** (Pt 1), 81–91.

62 Taylor, A.M., Berchtold, N.C., Perreau, V.M., Tu, C.H. et al. (2009) Axonal mRNA in uninjured and regenerating cortical mammalian axons. *J. Neurosci.*, **29** (15), 4697–4707.

63 Campbell, D.S., Holt, C.E. (2001) Chemotropic responses of retinal growth cones mediated by rapid local protein synthesis and degradation. *Neuron*, **32** (6), 1013–1026.

64 Lin, A.C., Holt, C.E. (2007) Local translation and directional steering in axons. *EMBO J.*, **26** (16), 3729–3736.

65 Leung, K.M., van Horck, F.P., Lin, A.C., Allison, R. et al. (2006) Asymmetrical beta-actin mRNA translation in growth cones mediates attractive turning to netrin-1. *Nat. Neurosci.*, **9** (10), 1247–1256.

66 Yao, J., Sasaki, Y., Wen, Z., Bassell, G.J. et al. (2006) An essential role for beta-actin mRNA localization and translation in Ca2+-dependent growth cone guidance. *Nat. Neurosci.*, **9** (10), 1265–1273.

67 Wu, K.Y., Hengst, U., Cox, L.J., Macosko, E.Z. et al. (2005) Local translation of RhoA regulates growth cone collapse. *Nature*, **436** (7053), 1020–1024.

68 Doyle, M., Kiebler, M.A. (2011) Mechanisms of dendritic mRNA transport and its role in synaptic tagging. *EMBO J.*, **30** (17), 3540–3552.

69 Sutton, M.A., Schuman, E.M. (2006) Dendritic protein synthesis, synaptic plasticity, and memory. *Cell*, **127** (1), 49–58.

70 Wang, D.O., Martin, K.C., Zukin, R.S. (2010) Spatially restricting gene expression by local translation at synapses. *Trends Neurosci.*, **33** (4), 173–182.

71 Aakalu, G., Smith, W.B., Nguyen, N., Jiang, C. et al. (2001) Dynamic visualization of local protein synthesis in hippocampal neurons. *Neuron*, **30** (2), 489–502.

72 Grooms, S.Y., Noh, K.M., Regis, R., Bassell, G.J. et al. (2006) Activity bidirectionally regulates AMPA receptor mRNA abundance in dendrites of hippocampal neurons. *J. Neurosci.*, **26** (32), 8339–8351.

73 Ju, W., Morishita, W., Tsui, J., Gaietta, G. et al. (2004) Activity-dependent regulation of dendritic synthesis and trafficking of AMPA receptors. *Nat. Neurosci.*, **7** (3), 244–253.

74 Muddashetty, R.S., Kelic, S., Gross, C., Xu, M. et al. (2007) Dysregulated metabotropic glutamate receptor-dependent translation of AMPA receptor and postsynaptic density-95 mRNAs at synapses in a mouse model of fragile X syndrome. *J. Neurosci.*, **27** (20), 5338–5348.

75 Ouyang, Y., Rosenstein, A., Kreiman, G., Schuman, E.M. et al. (1999) Tetanic stimulation leads to increased accumulation of Ca(2+)/calmodulin-dependent protein kinase II via dendritic protein synthesis in hippocampal neurons. *J. Neurosci.*, **19** (18), 7823–7833.

76 Steward, O., Wallace, C.S., Lyford, G.L., Worley, P.F. (1998) Synaptic activation causes the mRNA for the IEG Arc to localize

selectively near activated postsynaptic sites on dendrites. *Neuron*, **21** (4), 741–751.

77 Wang, D.O., Kim, S.M., Zhao, Y., Hwang, H. et al. (2009) Synapse- and stimulus-specific local translation during long-term neuronal plasticity. *Science*, **324** (5934), 1536–1540.

78 Miller, S., Yasuda, M., Coats, J.K., Jones, Y. et al. (2002) Disruption of dendritic translation of CaMKIIalpha impairs stabilization of synaptic plasticity and memory consolidation. *Neuron*, **36** (3), 507–519.

79 An, J.J., Gharami, K., Liao, G.Y., Woo, N.H. et al. (2008) Distinct role of long 3′ UTR BDNF mRNA in spine morphology and synaptic plasticity in hippocampal neurons. *Cell*, **134** (1), 175–187.

80 Lyons, D.A., Naylor, S.G., Scholze, A., Talbot, W.S. (2009) Kif1b is essential for mRNA localization in oligodendrocytes and development of myelinated axons. *Nat. Genet.*, **41** (7), 854–858.

81 Karakozova, M., Kozak, M., Wong, C.C., Bailey, A.O. et al. (2006) Arginylation of beta-actin regulates actin cytoskeleton and cell motility. *Science*, **313** (5784), 192–196.

82 Wang, J., Boja, E.S., Tan, W., Tekle, E. et al. (2001) Reversible glutathionylation regulates actin polymerization in A431 cells. *J. Biol. Chem.*, **276** (51), 47763–47766.

83 Baj, G., Leone, E., Chao, M.V., Tongiorgi, E. (2011) Spatial segregation of BDNF transcripts enables BDNF to differentially shape distinct dendritic compartments. *Proc. Natl Acad. Sci. USA*, **108** (40), 16813–16818.

84 Ding, D., Parkhurst, S.M., Halsell, S.R., Lipshitz, H.D. (1993) Dynamic Hsp83 RNA localization during *Drosophila* oogenesis and embryogenesis. *Mol. Cell. Biol.*, **13** (6), 3773–3781.

85 Semotok, J.L., Cooperstock, R.L., Pinder, B.D., Vari, H.K. et al. (2005) Smaug recruits the CCR4/POP2/NOT deadenylase complex to trigger maternal transcript localization in the early *Drosophila* embryo. *Curr. Biol.*, **15** (4), 284–294.

86 Semotok, J.L., Luo, H., Cooperstock, R.L., Karaiskakis, A. et al. (2008) *Drosophila* maternal Hsp83 mRNA destabilization is directed by multiple SMAUG recognition elements in the open reading frame. *Mol. Cell. Biol.*, **28** (22), 6757–6772.

87 Zaessinger, S., Busseau, I., Simonelig, M. (2006) Oskar allows nanos mRNA translation in *Drosophila* embryos by preventing its deadenylation by Smaug/CCR4. *Development*, **133** (22), 4573–4583.

88 Tadros, W., Goldman, A.L., Babak, T., Menzies, F. et al. (2007) SMAUG is a major regulator of maternal mRNA destabilization in *Drosophila* and its translation is activated by the PAN GU kinase. *Dev. Cell*, **12** (1), 143–155.

89 Chang, P., Torres, J., Lewis, R.A., Mowry, K.L. et al. (2004) Localization of RNAs to the mitochondrial cloud in *Xenopus* oocytes through entrapment and association with endoplasmic reticulum. *Mol. Biol. Cell*, **15** (10), 4669–4681.

90 Forrest, K.M., Gavis, E.R. (2003) Live imaging of endogenous RNA reveals a diffusion and entrapment mechanism for nanos mRNA localization in *Drosophila*. *Curr. Biol.*, **13** (14), 1159–1168.

91 Bullock, S.L. (2011) Messengers, motors and mysteries: sorting of eukaryotic mRNAs by cytoskeletal transport. *Biochem. Soc. Trans.*, **39** (5), 1161–1165.

92 Gagnon, J.A., Mowry, K.L. (2011) Molecular motors: directing traffic during RNA localization. *Crit. Rev. Biochem. Mol. Biol.*, **46** (3), 229–239.

93 Ainger, K., Avossa, D., Morgan, F., Hill, S.J. et al. (1993) Transport and localization of exogenous myelin basic protein mRNA microinjected into oligodendrocytes. *J. Cell Biol.*, **123** (2), 431–441.

94 Elvira, G., Wasiak, S., Blandford, V., Tong, X.K. et al. (2006) Characterization of an RNA granule from developing brain. *Mol. Cell. Proteomics*, **5** (4), 635–651.

95 Kanai, Y., Dohmae, N., Hirokawa, N. (2004) Kinesin transports RNA: isolation and characterization of an RNA-transporting granule. *Neuron*, **43** (4), 513–525.

96 Kiebler, M.A., Bassell, G.J. (2006) Neuronal RNA granules: movers and makers. *Neuron*, **51** (6), 685–690.

97 Krichevsky, A.M., Kosik, K.S. (2001) Neuronal RNA granules: a link between RNA localization and stimulation-dependent translation. *Neuron*, **32** (4), 683–696.

98 Arn, E.A., Cha, B.J., Theurkauf, W.E., Macdonald, P.M. (2003) Recognition of a bicoid mRNA localization signal by a protein complex containing Swallow, Nod,

and RNA binding proteins. *Dev. Cell*, **4**, 41–51.

99 Chekulaeva, M., Hentze, M.W., Ephrussi, A. (2006) Bruno acts as a dual repressor of oskar translation, promoting mRNA oligomerization and formation of silencing particles. *Cell*, **124** (3), 521–533.

100 Wilhelm, J.E., Mansfield, J., Hom-Booher, N., Wang, S. et al. (2000) Isolation of a ribonucleoprotein complex involved in mRNA localization in *Drosophila* oocytes. *J. Cell Biol.*, **148** (3), 427–440.

101 Chabanon, H., Mickleburgh, I., Hesketh, J. (2004) Zipcodes and postage stamps: mRNA localisation signals and their trans-acting binding proteins. *Brief. Funct. Genomics Proteomics*, **3** (3), 240–256.

102 Jambhekar, A., Derisi, J.L. (2007) Cis-acting determinants of asymmetric, cytoplasmic RNA transport. *RNA*, **13** (5), 625–642.

103 Bi, J., Tsai, N.P., Lin, Y.P., Loh, H.H. et al. (2006) Axonal mRNA transport and localized translational regulation of kappa-opioid receptor in primary neurons of dorsal root ganglia. *Proc. Natl Acad. Sci. USA*, **103** (52), 19919–19924.

104 Meer, E.J., Wang, D.O., Kim, S., Barr, I. et al. (2012) Identification of a cis-acting element that localizes mRNA to synapses. *Proc. Natl Acad. Sci. USA*, **109** (12), 4639–4644.

105 Chartrand, P., Meng, X.H., Singer, R.H., Long, R.M. (1999) Structural elements required for the localization of ASH1 mRNA and of a green fluorescent protein reporter particle in vivo. *Curr. Biol.*, **9** (6), 333–336.

106 Ghosh, S., Marchand, V., Gaspar, I., Ephrussi, A. (2012) Control of RNP motility and localization by a splicing-dependent structure in oskar mRNA. *Nat. Struct. Mol. Biol.*, **19** (4), 441–449.

107 Gonzalez, I., Buonomo, S.B., Nasmyth, K., von Ahsen, U. (1999) ASH1 mRNA localization in yeast involves multiple secondary structural elements and Ash1 protein translation. *Curr. Biol.*, **9** (6), 337–340.

108 Serano, J., Rubin, G.M. (2003) The *Drosophila* synaptotagmin-like protein bitesize is required for growth and has mRNA localization sequences within its open reading frame. *Proc. Natl Acad. Sci. USA*, **100** (23), 13368–13373.

109 Van De Bor, V., Hartswood, E., Jones, C., Finnegan, D. et al. (2005) gurken and the I factor retrotransposon RNAs share common localization signals and machinery. *Dev. Cell*, **9** (1), 51–62.

110 Buckley, P.T., Lee, M.T., Sul, J.Y., Miyashiro, K.Y. et al. (2011) Cytoplasmic intron sequence-retaining transcripts can be dendritically targeted via ID element retrotransposons. *Neuron*, **69** (5), 877–884.

111 Andreassi, C., Riccio, A. (2009) To localize or not to localize: mRNA fate is in 3'UTR ends. *Trends Cell Biol.*, **19** (9), 465–474.

112 Kislauskis, E.H., Zhu, X., Singer, R.H. (1994) Sequences responsible for intracellular localization of beta-actin messenger RNA also affect cell phenotype. *J. Cell Biol.*, **127** (2), 441–451.

113 Ainger, K., Avossa, D., Diana, A.S., Barry, C. et al. (1997) Transport and localization elements in myelin basic protein mRNA. *J. Cell Biol.*, **138** (5), 1077–1087.

114 Munro, T.P., Magee, R.J., Kidd, G.J., Carson, J.H. et al. (1999) Mutational analysis of a heterogeneous nuclear ribonucleoprotein A2 response element for RNA trafficking. *J. Biol. Chem.*, **274** (48), 34389–34395.

115 Serano, T.L., Cohen, R.S. (1995) A small predicted stem-loop structure mediates oocyte localization of *Drosophila* K10 mRNA. *Development*, **121** (11), 3809–3818.

116 Chao, J.A., Patskovsky, Y., Patel, V., Levy, M. et al. (2010) ZBP1 recognition of beta-actin zipcode induces RNA looping. *Genes Dev.*, **24** (2), 148–158.

117 Patel, V.L., Mitra, S., Harris, R., Buxbaum, A.R. et al. (2012) Spatial arrangement of an RNA zipcode identifies mRNAs under post-transcriptional control. *Genes Dev.*, **26** (1), 43–53.

118 Bullock, S.L., Zicha, D., Ish-Horowicz, D. (2003) The *Drosophila* hairy RNA localization signal modulates the kinetics of cytoplasmic mRNA transport. *EMBO J.*, **22** (10), 2484–2494.

119 Cohen, R.S., Zhang, S., Dollar, G.L. (2005) The positional, structural, and sequence requirements of the *Drosophila* TLS RNA localization element. *RNA*, **11** (7), 1017–1029.

120 Ferrandon, D., Koch, I., Westhof, E., Nusslein-Volhard, C. (1997) RNA-RNA interaction is required for the formation of specific bicoid mRNA 3′ UTR-STAUFEN ribonucleoprotein particles. *EMBO J.*, **16** (7), 1751–1758.

121 Olivier, C., Poirier, G., Gendron, P., Boisgontier, A. et al. (2005) Identification of a conserved RNA motif essential for She2p recognition and mRNA localization to the yeast bud. *Mol. Cell. Biol.*, **25** (11), 4752–4766.

122 Bullock, S.L., Ringel, I., Ish-Horowicz, D., Lukavsky, P.J. (2010) A′-form RNA helices are required for cytoplasmic mRNA transport in *Drosophila*. *Nat. Struct. Mol. Biol.*, **17** (6), 703–709.

123 Tiedge, H. (2006) K-turn motifs in spatial RNA coding. *RNA Biol.*, **3** (4), 133–139.

124 Macdonald, P.M., Struhl, G. (1988) cis-acting sequences responsible for anterior localization of bicoid mRNA in *Drosophila* embryos. *Nature*, **336** (6199), 595–598.

125 Macdonald, P.M., Kerr, K., Smith, J.L., Leask, A. (1993) RNA regulatory element BLE1 directs the early steps of bicoid mRNA localization. *Development*, **118** (4), 1233–1243.

126 Chartrand, P., Meng, X.H., Huttelmaier, S., Donato, D. et al. (2002) Asymmetric sorting of ash1p in yeast results from inhibition of translation by localization elements in the mRNA. *Mol. Cell*, **10** (6), 1319–1330.

127 Gautreau, D., Cote, C.A., Mowry, K.L. (1997) Two copies of a subelement from the Vg1 RNA localization sequence are sufficient to direct vegetal localization in *Xenopus* oocytes. *Development*, **124** (24), 5013–5020.

128 Gavis, E.R., Curtis, D., Lehmann, R. (1996) Identification of cis-acting sequences that control nanos RNA localization. *Dev. Biol.*, **176** (1), 36–50.

129 Mowry, K.L., Melton, D.A. (1992) Vegetal messenger RNA localization directed by a 340-nt RNA sequence element in *Xenopus* oocytes. *Science*, **255** (5047), 991–994.

130 Deshler, J.O., Highett, M.I., Schnapp, B.J. (1997) Localization of Xenopus Vg1 mRNA by Vera protein and the endoplasmic reticulum. *Science*, **276** (5315), 1128–1131.

131 Bubunenko, M., Kress, T.L., Vempati, U.D., Mowry, K.L. et al. (2002) A consensus RNA signal that directs germ layer determinants to the vegetal cortex of *Xenopus* oocytes. *Dev. Biol.*, **248** (1), 82–92.

132 Lewis, R.A., Kress, T.L., Cote, C.A., Gautreau, D. et al. (2004) Conserved and clustered RNA recognition sequences are a critical feature of signals directing RNA localization in *Xenopus* oocytes. *Mech. Dev.*, **121** (1), 101–109.

133 Hilgers, V., Perry, M.W., Hendrix, D., Stark, A. et al. (2011) Neural-specific elongation of 3′ UTRs during *Drosophila* development. *Proc. Natl Acad. Sci. USA*, **108** (38), 15864–15869.

134 Ji, Z., Lee, J.Y., Pan, Z., Jiang, B. et al. (2009) Progressive lengthening of 3′ untranslated regions of mRNAs by alternative polyadenylation during mouse embryonic development. *Proc. Natl Acad. Sci. USA*, **106** (17), 7028–7033.

135 Shepard, P.J., Choi, E.A., Lu, J., Flanagan, L.A. et al. (2011) Complex and dynamic landscape of RNA polyadenylation revealed by PAS-Seq. *RNA*, **17** (4), 761–772.

136 Smibert, P., Miura, P., Westholm, J.O., Shenker, S. et al. (2012) Global patterns of tissue-specific alternative polyadenylation in *Drosophila*. *Cell Rep.*, **1** (3), 277–289.

137 Bassell, G.J., Zhang, H., Byrd, A.L., Femino, A.M. et al. (1998) Sorting of beta-actin mRNA and protein to neurites and growth cones in culture. *J. Neurosci.*, **18** (1), 251–265.

138 Zhang, H.L., Eom, T., Oleynikov, Y., Shenoy, S.M. et al. (2001) Neurotrophin-induced transport of a beta-actin mRNP complex increases beta-actin levels and stimulates growth cone motility. *Neuron*, **31** (2), 261–275.

139 Tiruchinapalli, D.M., Oleynikov, Y., Kelic, S., Shenoy, S.M. et al. (2003) Activity-dependent trafficking and dynamic localization of zipcode binding protein 1 and beta-actin mRNA in dendrites and spines of hippocampal neurons. *J. Neurosci.*, **23** (8), 3251–3261.

140 Shan, J., Munro, T.P., Barbarese, E., Carson, J.H. et al. (2003) A molecular mechanism for mRNA trafficking in neuronal dendrites. *J. Neurosci.*, **23** (26), 8859–8866.

141 Hamilton, R.S., Davis, I. (2011) Identifying and searching for conserved RNA localisation signals. *Methods Mol. Biol.*, **714**, 447–466.

142 Hamilton, R.S., Davis, I. (2007) RNA localization signals: deciphering the message with bioinformatics. *Semin. Cell Dev. Biol.*, **18** (2), 178–185.

143 Rabani, M., Kertesz, M., Segal, E. (2008) Computational prediction of RNA structural motifs involved in posttranscriptional regulatory processes. *Proc. Natl Acad. Sci. USA*, **105** (39), 14885–14890.

144 Jambhekar, A., McDermott, K., Sorber, K., Shepard, K.A. et al. (2005) Unbiased selection of localization elements reveals cis-acting determinants of mRNA bud localization in *Saccharomyces cerevisiae*. *Proc. Natl Acad. Sci. USA*, **102** (50), 18005–18010.

145 Ross, A.F., Oleynikov, Y., Kislauskis, E.H., Taneja, K.L. et al. (1997) Characterization of a beta-actin mRNA zipcode-binding protein. *Mol. Cell. Biol.*, **17** (4), 2158–2165.

146 Oleynikov, Y., Singer, R.H. (2003) Real-time visualization of ZBP1 association with beta-actin mRNA during transcription and localization. *Curr. Biol.*, **13** (3), 199–207.

147 Hoek, K.S., Kidd, G.J., Carson, J.H., Smith, R. (1998) hnRNP A2 selectively binds the cytoplasmic transport sequence of myelin basic protein mRNA. *Biochemistry*, **37** (19), 7021–7029.

148 Cote, C.A., Gautreau, D., Denegre, J.M., Kress, T.L. et al. (1999) A Xenopus protein related to hnRNP I has a role in cytoplasmic RNA localization. *Mol. Cell*, **4** (3), 431–437.

149 Deshler, J.O., Highett, M.I., Abramson, T., Schnapp, B.J. (1998) A highly conserved RNA-binding protein for cytoplasmic mRNA localization in vertebrates. *Curr. Biol.*, **8** (9), 489–496.

150 Jansen, R.P., Dowzer, C., Michaelis, C., Galova, M. et al. (1996) Mother cell-specific HO expression in budding yeast depends on the unconventional myosin myo4p and other cytoplasmic proteins. *Cell*, **84** (5), 687–697.

151 Bohl, F., Kruse, C., Frank, A., Ferring, D. et al. (2000) She2p, a novel RNA-binding protein tethers ASH1 mRNA to the Myo4p myosin motor via She3p. *EMBO J.*, **19** (20), 5514–5524.

152 Long, R.M., Gu, W., Lorimer, E., Singer, R.H. et al. (2000) She2p is a novel RNA-binding protein that recruits the Myo4p-She3p complex to ASH1 mRNA. *EMBO J.*, **19** (23), 6592–6601.

153 Muller, M., Heym, R.G., Mayer, A., Kramer, K. et al. (2011) A cytoplasmic complex mediates specific mRNA recognition and localization in yeast. *PLoS Biol.*, **9** (4), e1000611.

154 Dienstbier, M., Boehl, F., Li, X., Bullock, S.L. (2009) Egalitarian is a selective RNA-binding protein linking mRNA localization signals to the dynein motor. *Genes Dev.*, **23** (13), 1546–1558.

155 Hoogenraad, C.C., Wulf, P., Schiefermeier, N., Stepanova, T. et al. (2003) Bicaudal D induces selective dynein-mediated microtubule minus end-directed transport. *EMBO J.*, **22** (22), 6004–6015.

156 Navarro, C., Puthalakath, H., Adams, J.M., Strasser, A. et al. (2004) Egalitarian binds dynein light chain to establish oocyte polarity and maintain oocyte fate. *Nat. Cell Biol.*, **6** (5), 427–435.

157 Yoon, Y.J., Mowry, K.L. (2004) Xenopus Staufen is a component of a ribonucleoprotein complex containing Vg1 RNA and kinesin. *Development*, **131** (13), 3035–3045.

158 Dictenberg, J.B., Swanger, S.A., Antar, L.N., Singer, R.H. et al. (2008) A direct role for FMRP in activity-dependent dendritic mRNA transport links filopodial-spine morphogenesis to fragile X syndrome. *Dev. Cell*, **14** (6), 926–939.

159 Bianco, A., Dienstbier, M., Salter, H.K., Gatto, G. et al. (2010) Bicaudal-D regulates fragile X mental retardation protein levels, motility, and function during neuronal morphogenesis. *Curr. Biol.*, **20** (16), 1487–1492.

160 Elisha, Z., Havin, L., Ringel, I., Yisraeli, J.K. (1995) Vg1 RNA binding protein mediates the association of Vg1 RNA with microtubules in *Xenopus* oocytes. *EMBO J.*, **14** (20), 5109–5114.

161 Besse, F., Lopez de Quinto, S., Marchand, V., Trucco, A. et al. (2009) *Drosophila* PTB promotes formation of high-order RNP particles and represses oskar translation. *Genes Dev.*, **23** (2), 195–207.

162 Ma, S., Liu, G., Sun, Y., Xie, J. (2007) Relocalization of the polypyrimidine tract-binding protein during PKA-induced

neurite growth. *Biochim. Biophys. Acta*, **1773** (6), 912–923.

163 Oberstrass, F.C., Auweter, S.D., Erat, M., Hargous, Y. et al. (2005) Structure of PTB bound to RNA: specific binding and implications for splicing regulation. *Science*, **309** (5743), 2054–2057.

164 Dreyfuss, G., Kim, V.N., Kataoka, N. (2002) Messenger-RNA-binding proteins and the messages they carry. *Nat. Rev. Mol. Cell Biol.*, **3** (3), 195–205.

165 Glisovic, T., Bachorik, J.L., Yong, J., Dreyfuss, G. (2008) RNA-binding proteins and post-transcriptional gene regulation. *FEBS Lett.*, **582** (14), 1977–1986.

166 Roegiers, F., Jan, Y.N. (2000) Staufen: a common component of mRNA transport in oocytes and neurons? *Trends Cell Biol.*, **10** (6), 220–224.

167 Yisraeli, J.K. (2005) VICKZ proteins: a multi-talented family of regulatory RNA-binding proteins. *Biol. Cell*, **97** (1), 87–96.

168 Huang, Y.S., Carson, J.H., Barbarese, E., Richter, J.D. (2003) Facilitation of dendritic mRNA transport by CPEB. *Genes Dev.*, **17** (5), 638–653.

169 Norvell, A., Kelley, R.L., Wehr, K., Schupbach, T. (1999) Specific isoforms of squid, a *Drosophila* hnRNP, perform distinct roles in Gurken localization during oogenesis. *Genes Dev.*, **13** (7), 864–876.

170 Huynh, J.R., Munro, T.P., Smith-Litiere, K., Lepesant, J.A. et al. (2004) The *Drosophila* hnRNPA/B homolog, Hrp48, is specifically required for a distinct step in osk mRNA localization. *Dev. Cell*, **6** (5), 625–635.

171 Goodrich, J.S., Clouse, K.N., Schupbach, T. (2004) Hrb27C, Sqd and Otu cooperatively regulate gurken RNA localization and mediate nurse cell chromosome dispersion in *Drosophila* oogenesis. *Development*, **131** (9), 1949–1958.

172 St Johnston, D., Beuchle, D., Nusslein-Volhard, C. (1991) Staufen, a gene required to localize maternal RNAs in the *Drosophila* egg. *Cell*, **66** (1), 51–63.

173 Vessey, J.P., Macchi, P., Stein, J.M., Mikl, M. et al. (2008) A loss of function allele for murine Staufen1 leads to impairment of dendritic Staufen1-RNP delivery and dendritic spine morphogenesis. *Proc. Natl Acad. Sci. USA*, **105** (42), 16374–16379.

174 Kwon, S., Abramson, T., Munro, T.P., John, C.M. et al. (2002) UUCAC- and vera-dependent localization of VegT RNA in *Xenopus* oocytes. *Curr. Biol.*, **12** (7), 558–564.

175 Welshhans, K., Bassell, G.J. (2011) Netrin-1-induced local beta-actin synthesis and growth cone guidance requires zipcode binding protein 1. *J. Neurosci.*, **31** (27), 9800–9813.

176 Perycz, M., Urbanska, A.S., Krawczyk, P.S., Parobczak, K. et al. (2011) Zipcode binding protein 1 regulates the development of dendritic arbors in hippocampal neurons. *J. Neurosci.*, **31** (14), 5271–5285.

177 Giorgi, C., Moore, M.J. (2007) The nuclear nurture and cytoplasmic nature of localized mRNPs. *Semin. Cell Dev. Biol.*, **18** (2), 186–193.

178 Marchand, V., Gaspar, I., Ephrussi, A. (2012) An intracellular transmission control protocol: assembly and transport of ribonucleoprotein complexes. *Curr. Opin. Cell Biol.*, **24** (2), 202–210.

179 Du, T.G., Jellbauer, S., Muller, M., Schmid, M. et al. (2008) Nuclear transit of the RNA-binding protein She2 is required for translational control of localized ASH1 mRNA. *EMBO Rep.*, **9** (8), 781–787.

180 Shen, Z., Paquin, N., Forget, A., Chartrand, P. (2009) Nuclear shuttling of She2p couples ASH1 mRNA localization to its translational repression by recruiting Loc1p and Puf6p. *Mol. Biol. Cell*, **20** (8), 2265–2275.

181 Shen, Z., St-Denis, A., Chartrand, P. (2010) Cotranscriptional recruitment of She2p by RNA pol II elongation factor Spt4-Spt5/DSIF promotes mRNA localization to the yeast bud. *Genes Dev.*, **24** (17), 1914–1926.

182 Kress, T.L., Yoon, Y.J., Mowry, K.L. (2004) Nuclear RNP complex assembly initiates cytoplasmic RNA localization. *J. Cell Biol.*, **165** (2), 203–211.

183 Pan, F., Huttelmaier, S., Singer, R.H., Gu, W. (2007) ZBP2 facilitates binding of ZBP1 to beta-actin mRNA during transcription. *Mol. Cell. Biol.*, **27** (23), 8340–8351.

184 Powrie, E.A., Zenklusen, D., Singer, R.H. (2011) A nucleoporin, Nup60p, affects the nuclear and cytoplasmic localization of ASH1 mRNA in *S. cerevisiae*. *RNA*, **17** (1), 134–144.

185 Hachet, O., Ephrussi, A. (2001) Drosophila Y14 shuttles to the posterior of the oocyte and is required for oskar mRNA transport. *Curr. Biol.*, **11** (21), 1666–1674.

186 Palacios, I.M., Gatfield, D., St Johnston, D., Izaurralde, E. (2004) An eIF4AIII-containing complex required for mRNA localization and nonsense-mediated mRNA decay. *Nature*, **427** (6976), 753–757.

187 van Eeden, F.J., Palacios, I.M., Petronczki, M., Weston, M.J. et al. (2001) Barentsz is essential for the posterior localization of oskar mRNA and colocalizes with it to the posterior pole. *J. Cell Biol.*, **154** (3), 511–523.

188 Hachet, O., Ephrussi, A. (2004) Splicing of oskar RNA in the nucleus is coupled to its cytoplasmic localization. *Nature*, **428** (6986), 959–963.

189 Bullock, S.L., Nicol, A., Gross, S.P., Zicha, D. (2006) Guidance of bidirectional motor complexes by mRNA cargoes through control of dynein number and activity. *Curr. Biol.*, **16** (14), 1447–1452.

190 MacDougall, N., Clark, A., MacDougall, E., Davis, I. (2003) Drosophila gurken (TGFalpha) mRNA localizes as particles that move within the oocyte in two dynein-dependent steps. *Dev. Cell*, **4** (3), 307–319.

191 Rand, K., Yisraeli, J. (2001) RNA localization in *Xenopus* oocytes. *Results Probl. Cell Differ.*, **34**, 157–173.

192 Wilkie, G.S., Davis, I. (2001) Drosophila wingless and pair-rule transcripts localize apically by dynein-mediated transport of RNA particles. *Cell*, **105** (2), 209–219.

193 Jambor, H., Brunel, C., Ephrussi, A. (2011) Dimerization of oskar 3′ UTRs promotes hitchhiking for RNA localization in the *Drosophila* oocyte. *RNA*, **17** (12), 2049–2057.

194 Lange, S., Katayama, Y., Schmid, M., Burkacky, O. et al. (2008) Simultaneous transport of different localized mRNA species revealed by live-cell imaging. *Traffic*, **9** (8), 1256–1267.

195 Reveal, B., Yan, N., Snee, M.J., Pai, C.I. et al. (2010) BREs mediate both repression and activation of oskar mRNA translation and act in trans. *Dev. Cell*, **18** (3), 496–502.

196 Amrute-Nayak, M., Bullock, S.L. (2012) Single-molecule assays reveal that RNA localization signals regulate dynein-dynactin copy number on individual transcript cargoes. *Nat. Cell Biol.*, **14** (4), 416–423.

197 Batish, M., van den Bogaard, P., Kramer, F.R., Tyagi, S. (2012) Neuronal mRNAs travel singly into dendrites. *Proc. Natl Acad. Sci. USA*, **109** (12), 4645–4650.

198 Mikl, M., Vendra, G., Kiebler, M.A. (2011) Independent localization of MAP2, CaMKI-Ialpha and beta-actin RNAs in low copy numbers. *EMBO Rep.*, **12** (10), 1077–1084.

199 Latham, V.M., Yu, E.H., Tullio, A.N., Adelstein, R.S. et al. (2001) A Rho-dependent signaling pathway operating through myosin localizes beta-actin mRNA in fibroblasts. *Curr. Biol.*, **11** (13), 1010–1016.

200 Clark, I., Giniger, E., Ruohola-Baker, H., Jan, L.Y. et al. (1994) Transient posterior localization of a kinesin fusion protein reflects anteroposterior polarity of the *Drosophila* oocyte. *Curr. Biol.*, **4** (4), 289–300.

201 Yisraeli, J.K., Sokol, S., Melton, D.A. (1990) A two-step model for the localization of maternal mRNA in *Xenopus* oocytes: involvement of microtubules and microfilaments in the translocation and anchoring of Vg1 mRNA. *Development*, **108** (2), 289–298.

202 Zhang, H.L., Singer, R.H., Bassell, G.J. (1999) Neurotrophin regulation of beta-actin mRNA and protein localization within growth cones. *J. Cell Biol.*, **147** (1), 59–70.

203 Schejter, E.D., Wieschaus, E. (1993) Functional elements of the cytoskeleton in the early *Drosophila* embryo. *Annu. Rev. Cell Biol.*, **9**, 67–99.

204 Parton, R.M., Hamilton, R.S., Ball, G., Yang, L. et al. (2011) A PAR-1-dependent orientation gradient of dynamic microtubules directs posterior cargo transport in the *Drosophila* oocyte. *J. Cell Biol.*, **194** (1), 121–135.

205 Zimyanin, V.L., Belaya, K., Pecreaux, J., Gilchrist, M.J. et al. (2008) In vivo imaging of oskar mRNA transport reveals the mechanism of posterior localization. *Cell*, **134** (5), 843–853.

206 Bullock, S.L. (2007) Translocation of mRNAs by molecular motors: think complex? *Semin. Cell Dev. Biol.*, **18** (2), 194–201.

207 Bertrand, E., Chartrand, P., Schaefer, M., Shenoy, S.M. et al. (1998) Localization of

ASH1 mRNA particles in living yeast. *Mol. Cell*, **2** (4), 437–445.

208 Messitt, T.J., Gagnon, J.A., Kreiling, J.A., Pratt, C.A. et al. (2008) Multiple kinesin motors coordinate cytoplasmic RNA transport on a subpopulation of microtubules in *Xenopus* oocytes. *Dev. Cell*, **15** (3), 426–436.

209 Fusco, D., Accornero, N., Lavoie, B., Shenoy, S.M. et al. (2003) Single mRNA molecules demonstrate probabilistic movement in living mammalian cells. *Curr. Biol.*, **13** (2), 161–167.

210 Dunn, B.D., Sakamoto, T., Hong, M.S., Sellers, J.R. et al. (2007) Myo4p is a monomeric myosin with motility uniquely adapted to transport mRNA. *J. Cell Biol.*, **178** (7), 1193–1206.

211 Chung, S., Takizawa, P.A. (2010) Multiple Myo4 motors enhance ASH1 mRNA transport in *Saccharomyces cerevisiae*. *J. Cell Biol.*, **189** (4), 755–767.

212 Clark, A., Meignin, C., Davis, I. (2007) A Dynein-dependent shortcut rapidly delivers axis determination transcripts into the *Drosophila* oocyte. *Development*, **134** (10), 1955–1965.

213 Brendza, R.P., Serbus, L.R., Duffy, J.B., Saxton, W.M. (2000) A function for kinesin I in the posterior transport of oskar mRNA and Staufen protein. *Science*, **289** (5487), 2120–2122.

214 Krauss, J., Lopez de Quinto, S., Nusslein-Volhard, C., Ephrussi, A. (2009) Myosin-V regulates oskar mRNA localization in the *Drosophila* oocyte. *Curr. Biol.*, **19** (12), 1058–1063.

215 Knowles, R.B., Sabry, J.H., Martone, M.E., Deerinck, T.J. et al. (1996) Translocation of RNA granules in living neurons. *J. Neurosci.*, **16** (24), 7812–7820.

216 Kohrmann, M., Luo, M., Kaether, C., DesGroseillers, L. et al. (1999) Microtubule-dependent recruitment of Staufen-green fluorescent protein into large RNA-containing granules and subsequent dendritic transport in living hippocampal neurons. *Mol. Biol. Cell*, **10** (9), 2945–2953.

217 Ma, B., Savas, J.N., Yu, M.S., Culver, B.P. et al. (2011) Huntingtin mediates dendritic transport of beta-actin mRNA in rat neurons. *Sci. Rep.*, **1**, 140.

218 Jolly, A.L., Gelfand, V.I. (2011) Bidirectional intracellular transport: utility and mechanism. *Biochem. Soc. Trans.*, **39** (5), 1126–1130.

219 Weil, T.T., Forrest, K.M., Gavis, E.R. (2006) Localization of bicoid mRNA in late oocytes is maintained by continual active transport. *Dev. Cell*, **11** (2), 251–262.

220 Delanoue, R., Davis, I. (2005) Dynein anchors its mRNA cargo after apical transport in the *Drosophila* blastoderm embryo. *Cell*, **122** (1), 97–106.

221 Delanoue, R., Herpers, B., Soetaert, J., Davis, I. et al. (2007) *Drosophila* Squid/hnRNP helps Dynein switch from a gurken mRNA transport motor to an ultrastructural static anchor in sponge bodies. *Dev. Cell*, **13** (4), 523–538.

222 Paix, A., Yamada, L., Dru, P., Lecordier, H. et al. (2009) Cortical anchorages and cell type segregations of maternal postplasmic/PEM RNAs in ascidians. *Dev. Biol.*, **336** (1), 96–111.

223 Babu, K., Cai, Y., Bahri, S., Yang, X. et al. (2004) Roles of Bifocal, Homer, and F-actin in anchoring Oskar to the posterior cortex of *Drosophila* oocytes. *Genes Dev.*, **18** (2), 138–143.

224 Cha, B.J., Serbus, L.R., Koppetsch, B.S., Theurkauf, W.E. (2002) Kinesin I-dependent cortical exclusion restricts pole plasm to the oocyte posterior. *Nat. Cell Biol.*, **4** (8), 592–598.

225 Jankovics, F., Sinka, R., Lukacsovich, T., Erdelyi, M. (2002) MOESIN crosslinks actin and cell membrane in *Drosophila* oocytes and is required for OSKAR anchoring. *Curr. Biol.*, **12** (23), 2060–2065.

226 Polesello, C., Delon, I., Valenti, P., Ferrer, P. et al. (2002) Dmoesin controls actin-based cell shape and polarity during *Drosophila melanogaster* oogenesis. *Nat. Cell Biol.*, **4** (10), 782–789.

227 Irie, K., Tadauchi, T., Takizawa, P.A., Vale, R.D. et al. (2002) The Khd1 protein, which has three KH RNA-binding motifs, is required for proper localization of ASH1 mRNA in yeast. *EMBO J.*, **21** (5), 1158–1167.

228 Vanzo, N.F., Ephrussi, A. (2002) Oskar anchoring restricts pole plasm formation to the posterior of the *Drosophila* oocyte. *Development*, **129** (15), 3705–3714.

229 Schratt, G.M., Nigh, E.A., Chen, W.G., Hu, L. et al. (2004) BDNF regulates the translation of a select

group of mRNAs by a mammalian target of rapamycin-phosphatidylinositol 3-kinase-dependent pathway during neuronal development. *J. Neurosci.*, **24** (33), 7366–7377.

230 Besse, F., Ephrussi, A. (2008) Translational control of localized mRNAs: restricting protein synthesis in space and time. *Nat. Rev. Mol. Cell Biol.*, **9** (12), 971–980.

231 Vardy, L., Orr-Weaver, T.L. (2007) Regulating translation of maternal messages: multiple repression mechanisms. *Trends Cell Biol.*, **17** (11), 547–554.

232 Richter, J.D., Sonenberg, N. (2005) Regulation of cap-dependent translation by eIF4E inhibitory proteins. *Nature*, **433** (7025), 477–480.

233 Nakamura, A., Sato, K., Hanyu-Nakamura, K. (2004) Drosophila cup is an eIF4E binding protein that associates with Bruno and regulates oskar mRNA translation in oogenesis. *Dev. Cell*, **6** (1), 69–78.

234 Napoli, I., Mercaldo, V., Boyl, P.P., Eleuteri, B. et al. (2008) The fragile X syndrome protein represses activity-dependent translation through CYFIP1, a new 4E-BP. *Cell*, **134** (6), 1042–1054.

235 Muddashetty, R.S., Nalavadi, V.C., Gross, C., Yao, X. et al. (2011) Reversible inhibition of PSD-95 mRNA translation by miR-125a, FMRP phosphorylation, and mGluR signaling. *Mol. Cell*, **42** (5), 673–688.

236 Siegel, G., Obernosterer, G., Fiore, R., Oehmen, M. et al. (2009) A functional screen implicates microRNA-138-dependent regulation of the depalmitoylation enzyme APT1 in dendritic spine morphogenesis. *Nat. Cell Biol.*, **11** (6), 705–716.

237 Siegel, G., Saba, R., Schratt, G. (2011) microRNAs in neurons: manifold regulatory roles at the synapse. *Curr. Opin. Genet. Dev.*, **21** (4), 491–497.

238 Piccioni, F., Zappavigna, V., Verrotti, A.C. (2005) Translational regulation during oogenesis and early development: the cap-poly(A) tail relationship. *C.R. Biol.*, **328** (10-11), 863–881.

239 Castagnetti, S., Ephrussi, A. (2003) Orb and a long poly(A) tail are required for efficient oskar translation at the posterior pole of the Drosophila oocyte. *Development*, **130** (5), 835–843.

240 Wu, L., Wells, D., Tay, J., Mendis, D. et al. (1998) CPEB-mediated cytoplasmic polyadenylation and the regulation of experience-dependent translation of alpha-CaMKII mRNA at synapses. *Neuron*, **21** (5), 1129–1139.

241 Deng, Y., Singer, R.H., Gu, W. (2008) Translation of ASH1 mRNA is repressed by Puf6p-Fun12p/eIF5B interaction and released by CK2 phosphorylation. *Genes Dev.*, **22** (8), 1037–1050.

242 Paquin, N., Menade, M., Poirier, G., Donato, D. et al. (2007) Local activation of yeast ASH1 mRNA translation through phosphorylation of Khd1p by the casein kinase Yck1p. *Mol. Cell*, **26** (6), 795–809.

243 Nelson, M.R., Leidal, A.M., Smibert, C.A. (2004) Drosophila Cup is an eIF4E-binding protein that functions in Smaug-mediated translational repression. *EMBO J.*, **23** (1), 150–159.

244 Huttelmaier, S., Zenklusen, D., Lederer, M., Dictenberg, J. et al. (2005) Spatial regulation of beta-actin translation by Src-dependent phosphorylation of ZBP1. *Nature*, **438** (7067), 512–515.

245 Tsai, N.P., Bi, J., Wei, L.N. (2007) The adaptor Grb7 links netrin-1 signaling to regulation of mRNA translation. *EMBO J.*, **26** (6), 1522–1531.

246 White, R., Gonsior, C., Kramer-Albers, E.M., Stohr, N. et al. (2008) Activation of oligodendroglial Fyn kinase enhances translation of mRNAs transported in hnRNP A2-dependent RNA granules. *J. Cell Biol.*, **181** (4), 579–586.

247 Sasaki, Y., Welshhans, K., Wen, Z., Yao, J. et al. (2010) Phosphorylation of zipcode binding protein 1 is required for brain-derived neurotrophic factor signaling of local beta-actin synthesis and growth cone turning. *J. Neurosci.*, **30** (28), 9349–9358.

248 Bramham, C.R., Wells, D.G. (2007) Dendritic mRNA: transport, translation and function. *Nat. Rev. Neurosci.*, **8** (10), 776–789.

249 Narayanan, U., Nalavadi, V., Nakamoto, M., Thomas, G. et al. (2008) S6K1 phosphorylates and regulates fragile X mental retardation protein (FMRP) with the neuronal protein synthesis-dependent mammalian

target of rapamycin (mTOR) signaling cascade. *J. Biol. Chem.*, **283** (27), 18478–18482.
250 Huang, Y.S., Jung, M.Y., Sarkissian, M., Richter, J.D. (2002) N-methyl-D-aspartate receptor signaling results in Aurora kinase-catalyzed CPEB phosphorylation and alpha CaMKII mRNA polyadenylation at synapses. *EMBO J.*, **21** (9), 2139–2148.
251 Aronov, S., Gelin-Licht, R., Zipor, G., Haim, L. et al. (2007) mRNAs encoding polarity and exocytosis factors are co-transported with the cortical endoplasmic reticulum to the incipient bud in *Saccharomyces cerevisiae*. *Mol. Cell. Biol.*, **27** (9), 3441–3455.
252 Shepard, K.A., Gerber, A.P., Jambhekar, A., Takizawa, P.A. et al. (2003) Widespread cytoplasmic mRNA transport in yeast: identification of 22 bud-localized transcripts using DNA microarray analysis. *Proc. Natl Acad. Sci. USA*, **100** (20), 11429–11434.
253 Sil, A., Herskowitz, I. (1996) Identification of asymmetrically localized determinant, Ash1p, required for lineage-specific transcription of the yeast HO gene. *Cell*, **84** (5), 711–722.
254 Gu, W., Deng, Y., Zenklusen, D., Singer, R.H. (2004) A new yeast PUF family protein, Puf6p, represses ASH1 mRNA translation and is required for its localization. *Genes Dev.*, **18** (12), 1452–1465.
255 Long, R.M., Gu, W., Meng, X., Gonsalvez, G. et al. (2001) An exclusively nuclear RNA-binding protein affects asymmetric localization of ASH1 mRNA and Ash1p in yeast. *J. Cell Biol.*, **153** (2), 307–318.
256 Beach, D.L., Salmon, E.D., Bloom, K. (1999) Localization and anchoring of mRNA in budding yeast. *Curr. Biol.*, **9** (11), 569–578.
257 Duncan, J.E., Warrior, R. (2002) The cytoplasmic dynein and kinesin motors have interdependent roles in patterning the *Drosophila* oocyte. *Curr. Biol.*, **12** (23), 1982–1991.
258 Januschke, J., Gervais, L., Dass, S., Kaltschmidt, J.A. et al. (2002) Polar transport in the *Drosophila* oocyte requires Dynein and Kinesin I cooperation. *Curr. Biol.*, **12** (23), 1971–1981.
259 Kim-Ha, J., Kerr, K., Macdonald, P.M. (1995) Translational regulation of oskar mRNA by bruno, an ovarian RNA-binding protein, is essential. *Cell*, **81** (3), 403–412.
260 Mansfield, J.H., Wilhelm, J.E., Hazelrigg, T. (2002) Ypsilon Schachtel, a *Drosophila* Y-box protein, acts antagonistically to Orb in the oskar mRNA localization and translation pathway. *Development*, **129** (1), 197–209.
261 Yano, T., Lopez de Quinto, S., Matsui, Y., Shevchenko, A. et al. (2004) Hrp48, a *Drosophila* hnRNPA/B homolog, binds and regulates translation of oskar mRNA. *Dev. Cell*, **6** (5), 637–648.
262 Wilhelm, J.E., Mansfield, J., Hom-Booher, N., Wang, S. et al. (2000) Isolation of a ribonucleoprotein complex involved in mRNA localization in *Drosophila* oocytes. *J. Cell Biol.*, **148** (3), 427–440.
263 Munro, T.P., Kwon, S., Schnapp, B.J., St Johnston, D. (2006) A repeated IMP-binding motif controls oskar mRNA translation and anchoring independently of *Drosophila melanogaster* IMP. *J. Cell Biol.*, **172** (4), 577–588.
264 Nakamura, A., Amikura, R., Hanyu, K., Kobayashi, S. (2001) Me31B silences translation of oocyte-localizing RNAs through the formation of cytoplasmic RNP complex during *Drosophila* oogenesis. *Development*, **128** (17), 3233–3242.
265 Ephrussi, A., Lehmann, R. (1992) Induction of germ cell formation by oskar. *Nature*, **358** (6385), 387–392.
266 Markussen, F.H., Michon, A.M., Breitwieser, W., Ephrussi, A. (1995) Translational control of oskar generates short OSK, the isoform that induces pole plasma assembly. *Development*, **121** (11), 3723–3732.
267 Burgin, K.E., Waxham, M.N., Rickling, S., Westgate, S.A. et al. (1990) In situ hybridization histochemistry of Ca2+/calmodulin-dependent protein kinase in developing rat brain. *J. Neurosci.*, **10** (6), 1788–1798.
268 Link, W., Konietzko, U., Kauselmann, G., Krug, M.S. et al. (1995) Somatodendritic expression of an immediate early gene is regulated by synaptic activity. *Proc. Natl Acad. Sci. USA*, **92** (12), 5734–5738.
269 Blichenberg, A., Rehbein, M., Muller, R., Garner, C.C. et al. (2001) Identification of a cis-acting dendritic targeting element

in the mRNA encoding the alpha subunit of Ca2+/calmodulin-dependent protein kinase II. *Eur. J. Neurosci.*, **13** (10), 1881–1888.
270 Mori, Y., Imaizumi, K., Katayama, T., Yoneda, T. *et al.* (2000) Two cis-acting elements in the 3′ untranslated region of alpha-CaMKII regulate its dendritic targeting. *Nat. Neurosci.*, **3** (11), 1079–1084.
271 Subramanian, M., Rage, F., Tabet, R., Flatter, E. *et al.* (2011) G-quadruplex RNA structure as a signal for neurite mRNA targeting. *EMBO Rep.*, **12** (7), 697–704.
272 Mendez, R., Richter, J.D. (2001) Translational control by CPEB: a means to the end. *Nat. Rev. Mol. Cell Biol.*, **2** (7), 521–529.
273 Crino, P.B., Eberwine, J. (1996) Molecular characterization of the dendritic growth cone: regulated mRNA transport and local protein synthesis. *Neuron*, **17** (6), 1173–1187.
274 Tongiorgi, E., Righi, M., Cattaneo, A. (1997) Activity-dependent dendritic targeting of BDNF and TrkB mRNAs in hippocampal neurons. *J. Neurosci.*, **17** (24), 9492–9505.
275 Willis, D.E., van Niekerk, E.A., Sasaki, Y., Mesngon, M. *et al.* (2007) Extracellular stimuli specifically regulate localized levels of individual neuronal mRNAs. *J. Cell Biol.*, **178** (6), 965–980.
276 Chicurel, M.E., Singer, R.H., Meyer, C.J., Ingber, D.E. (1998) Integrin binding and mechanical tension induce movement of mRNA and ribosomes to focal adhesions. *Nature*, **392** (6677), 730–733.
277 Pique, M., Lopez, J.M., Foissac, S., Guigo, R. *et al.* (2008) A combinatorial code for CPE-mediated translational control. *Cell*, **132** (3), 434–448.
278 Kye, M.J., Liu, T., Levy, S.F., Xu, N.L. *et al.* (2007) Somatodendritic microRNAs identified by laser capture and multiplex RT-PCR. *RNA*, **13** (8), 1224–1234.
279 Natera-Naranjo, O., Aschrafi, A., Gioio, A.E., Kaplan, B.B. (2010) Identification and quantitative analyses of microRNAs located in the distal axons of sympathetic neurons. *RNA*, **16** (8), 1516–1529.
280 Kloc, M., Wilk, K., Vargas, D., Shirato, Y. *et al.* (2005) Potential structural role of non-coding and coding RNAs in the organization of the cytoskeleton at the vegetal cortex of *Xenopus* oocytes. *Development*, **132** (15), 3445–3457.
281 Tiedge, H., Fremeau, R.T. Jr, Weinstock, P.H., Arancio, O. *et al.* (1991) Dendritic location of neural BC1 RNA. *Proc. Natl Acad. Sci. USA*, **88** (6), 2093–2097.
282 Gall, J.G. (1968) Differential synthesis of the genes for ribosomal RNA during amphibian oogenesis. *Proc. Natl Acad. Sci. USA*, **60** (2), 553–560.
283 Raap, A.K., van de Corput, M.P., Vervenne, R.A., van Gijlswijk, R.P. *et al.* (1995) Ultra-sensitive FISH using peroxidase-mediated deposition of biotin- or fluorochrome tyramides. *Hum. Mol. Genet.*, **4** (4), 529–534.
284 Raj, A., van den Bogaard, P., Rifkin, S.A., van Oudenaarden, A. *et al.* (2008) Imaging individual mRNA molecules using multiple singly labeled probes. *Nat. Methods*, **5** (10), 877–879.
285 Bratu, D.P., Cha, B.J., Mhlanga, M.M., Kramer, F.R. *et al.* (2003) Visualizing the distribution and transport of mRNAs in living cells. *Proc. Natl Acad. Sci. USA*, **100** (23), 13308–13313.
286 Tyagi, S., Kramer, F.R. (1996) Molecular beacons: probes that fluoresce upon hybridization. *Nat. Biotechnol.*, **14** (3), 303–308.
287 Daigle, N., Ellenberg, J. (2007) LambdaN-GFP: an RNA reporter system for live-cell imaging. *Nat. Methods*, **4** (8), 633–636.
288 Paige, J.S., Wu, K.Y., Jaffrey, S.R. (2011) RNA mimics of green fluorescent protein. *Science*, **333** (6042), 642–646.

16
tRNA Subcellular Dynamics

Tohru Yoshihisa
University of Hyogo, Graduate School of Life Science, Kamigori-cho, Ako-gun, Hyogo 678–1297, Japan

1	Introduction	515
2	**Arrangement of Processing Machinery in a Eukaryotic Cell**	**515**
2.1	Transcription	515
2.2	Trimming of 5′- and 3′-Extension Sequences	517
2.2.1	5′-Trimming and RNase P	517
2.2.2	3′-Trimming with Two Modes	517
2.3	CCA Addition	518
2.4	tRNA Splicing: Conserved in the Mechanism but Diverse in Localization	519
2.4.1	Enzymes Required for tRNA Splicing	519
2.4.2	Variable Intracellular Distribution of Splicing Enzymes among Eukaryotes	519
2.5	Modifications: Separation and Collaboration of Enzymes in the Nucleus and the Cytoplasm	521
2.6	Quality Control of tRNAs	522
2.6.1	tRNA Quality Control by TRAMP and Nuclear Exosome	522
2.6.2	Rapid tRNA Degradation	524
3	**Nuclear–Cytoplasmic Transport of tRNAs**	**524**
3.1	Intranuclear Transport of tRNAs	525
3.2	Nuclear Export of tRNAs	525
3.2.1	Characterization of tRNA Export	525
3.2.2	Structural Basis of the Export Mechanism	526
3.3	Nuclear Import of tRNAs	528
3.4	Regulation of tRNA Export/Import Balance	529

RNA Regulation: Advances in Molecular Biology and Medicine, First Edition. Edited by Robert A. Meyers.
© 2014 Wiley-VCH Verlag GmbH & Co. KGaA. Published 2014 by Wiley-VCH Verlag GmbH & Co. KGaA.

4	tRNA Import into the Mitochondria 530
4.1	tRNA Import in S. cerevisiae: Not Essential but a Few tRNAs are Imported 531
4.2	tRNA Import in Plants: More tRNAs Are Imported 532
4.3	tRNA Import in Protozoa: Every tRNA is Imported 532
4.4	Mammalian Import of tRNAs and Other Small RNAs 534
4.5	Mitochondrial RNA Import and Its Possible Therapeutic Application 535

References 535

Keywords

Nuclear export of tRNA
The process by which tRNAs are delivered from the nucleus to the cytosol. Both, newly transcribed tRNAs approaching maturation and mature tRNAs imported from the cytosol are subjected to this delivery.

Nuclear import of tRNA
The process by which tRNAs are delivered to the nucleus from the cytosol. The level and functionality of cytosolic tRNAs are thought to be maintained by a combination of nuclear import and export.

Mitochondrial import of tRNA
The process by which cytosolic tRNAs are delivered to the matrix of the mitochondria, where proteins encoded on organellar genomes are translated. Many eukaryotes do not have complete sets of tRNA genes on the organellar genomes, so that some or all of mitochondrial tRNAs are supplied from the cytosol.

Quality control of tRNA
Systems employed to eliminate abnormal or unstable tRNAs during or after the maturation of tRNAs. Alternatively, a system that rejects the supply of such tRNAs to the cytosol where translation proceeds.

Transfer RNAs (tRNAs) are classical noncoding RNAs required for protein translation. tRNAs are unique among classes of RNAs in that they are heavily processed despite their small size. Recent analyses have revealed interesting aspects in the subcellular dynamics of tRNAs, from their birth to death. During the maturation of tRNAs, a tight collaboration between the nucleus and cytoplasm is required. When tRNAs become mature, they are shuttled between the cytosol and nucleus, and eukaryotic cells utilize this shuttling to control the quality of the tRNAs and to regulate their translation. Furthermore, translation in the mitochondria relies heavily on tRNAs imported from the cytosol by various modes of transport. In this chapter, tRNA subcellular dynamics is discussed on the basis of their biogenesis, function, and quality control.

1
Introduction

A typical transfer RNA (tRNA) consists of 76 nt, and its cloverleaf-like secondary structure folds into an L-shaped tertiary structure (Fig. 1). tRNAs are essential adaptors between a nucleotide sequence and an amino acid sequence in translation. Each tRNA carries a three-nucleotide anticodon (positions 34–36) that matches a cognate codon on an mRNA and a corresponding amino acid residue on its 3′ terminus. tRNAs not only present amino acid residues to the catalytic center of peptidyl transfer in the ribosome as substrates, but also provide a critical hydroxyl group as a part of the catalytic center. tRNAs are unique, when compared to other classes of RNAs, in that they are extensively modified to account for the above-described activities. During the biogenesis of tRNAs, these modifications must be applied specifically at an appropriate timing and subcellular location. All tRNAs must travel from their gene loci in the nucleus (their birthplace) to the cytosol (the site of their function), but the cytosol is not the final destination. Some tRNAs need to be further delivered to the mitochondria or chloroplasts for organellar translation, while the remainder of them shuttle between the cytosol and the nucleus throughout their lives. Such intracellular movement is regulated by environmental conditions. The quality control of tRNAs is also important for their correct functions. As the elimination of aberrant tRNAs relies on nucleases and other factors that are strictly localized in certain intracellular compartments, the aberrant tRNAs must be correctly delivered to the compartments for degradation. The above-described facts emphasize the importance of the subcellular dynamics of tRNAs during their lives. In this chapter, many aspects of tRNA subcellular dynamics are discussed on the basis of their biogenesis, function, and quality control.

2
Arrangement of Processing Machinery in a Eukaryotic Cell

In order to understand the subcellular dynamics of tRNAs, it is important first to appreciate the cellular arrangement of the machineries responsible for their maturation and function. How eukaryotes arrange such machineries in their cells is discussed in the following section, with attention focused mainly on tRNA biogenesis. The processing events that newly transcribed pre-tRNAs (transfer RNA precursors) experience during their maturation are summarized in Figure 2.

2.1
Transcription

Typically, tRNAs account for some 15% of the total RNAs such that, in an actively dividing yeast cell, approximately three million molecules must be transcribed per generation [1]. Therefore, many eukaryotes have multiple genes

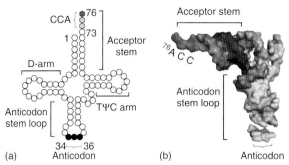

Fig. 1 Structure of tRNA. (a) The secondary structure of a typical tRNA, shown schematically. Important nucleotide positions are numbered starting from the 5′ terminus of position 1 and ending of the 3′ terminus of position 76. Circles represent nucleotides defined by the tRNA gene, and three hexagons represent the CCA end of the tRNA (C: light gray, A: dark gray). The three black circles represent the anticodon; (b) A surface drawing of the three-dimensional structure of yeast tRNA$^{Phe}_{GAA}$ (PDB: 1TRA).

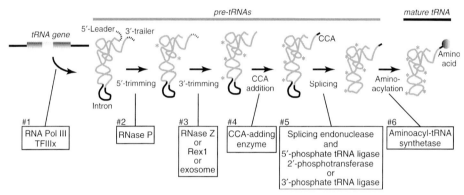

Fig. 2 Maturation processes of a newly transcribed tRNA (illustrated schematically). Enzymes or proteins responsible for each process (#1 to #6) are indicated in the boxes. Modified parts of the pre-tRNAs are highlighted in black. Asterisks appearing on the tRNA molecules represent various nucleotide modifications introduced at various timings.

encoding an isoacceptor tRNA (a tRNA with a certain anticodon) and also possess a special enzyme, RNA polymerase III (Pol III), and other factors for the transcription of tRNAs and other highly expressed small RNAs [2] (Fig. 2, #1). These tRNA genes are usually dispersed on eukaryotic nuclear genomes. tRNA genes form clusters on chromosomes in some organisms, such as *Leishmania* [3]. However, tRNA genes dispersed on the chromosomes are not localized randomly in the nucleus. For example, in the budding yeast *Saccharomyces cerevisiae*, a total of 274 tRNA genes, scattered on all 16 chromosomes, are localized in the nucleolus [4]. This spatial clustering depends on the presence of *condensing*s; these are large protein complexes that are required for the correct assembly of chromosomes and their segregation during nuclear division [5, 6]. In another yeast, *Schizosaccharomyces pombe*, general transcription factors for Pol III are found as five to six foci on the nuclear periphery, and these are colocalized with centromeres [7]. In both cases, such clustering may facilitate the efficient recycling of Pol III and other factors for transcription, and may also assist in handing primary transcripts to the subsequent maturation processes.

2.2 Trimming of 5'- and 3'-Extension Sequences

2.2.1   5'-Trimming and RNase P

A newly transcribed pre-tRNA contains 5'-leader and 3'-trailer sequences. The 5'-leader sequence is cleaved by an endoribonuclease called *RNase P*, an RNA–protein complex in which the catalytic center usually resides in the RNA subunit [8] (Fig. 2, #2). In *S. cerevisiae*, the RNA subunit of RNase P is localized in the nucleolus, where the tRNA genes are clustered and pre-tRNAs are accumulated [9]; therefore, 5'-trimming is thought to proceed just after transcription in the nucleolus [9] (Fig. 3). Although the localization of mammalian RNase P is complicated [10], fluorescence *in-situ* hybridization (FISH) analysis of the RNA subunit, RNA H1, showed that RNA H1 exists largely in the nucleoplasm, while part of the pool is located in the nucleolus [11]. On the other hand, nine out of 10 proteinaceous subunits are accumulated in the nucleolus [12, 13]. Thus, in mammalian cells, some of the RNase P activity should be present in the nucleolus and may contribute to a streamlined processing of newly transcribed pre-tRNAs, as is the case for *S. cerevisiae*.

2.2.2   3'-Trimming with Two Modes

Removal of the 3'-trailer sequence is carried out via two modes, an endonucleolytic mode and an exonucleolytic mode [14]. The endonucleolytic 3'-trimming is driven by RNase Z, encoded by an essential gene *TRZ1* in *S. cerevisiae*, and in humans by *ELAC2*, the mutation of which enhances prostate cancer susceptibility [15] (Fig. 2, #3). Systematic green fluorescent protein (GFP) fusion analysis in *S. cerevisiae* revealed that RNase Z is found in the nucleus and mitochondria, and a similar localization was reported in mammalian cells [16, 17] (Fig. 3). The dual localization of tRNA processing enzymes is not surprising, as the mitochondria also need to process tRNAs transcribed from their own genome. Usually, mRNAs of such dual-localizing enzymes have two translational initiation sites, one of which extends the open reading frame to encode an N-terminal extra-sequence for mitochondrial targeting [18]. The exonucleolytic 3'-trimming is catalyzed by two exonucleases, Rex1 and the nuclear exosome (Fig. 2, #3), both of which are localized in the nucleoplasm [19]. Under normal conditions, the endonucleolytic cleavage is predominant for the 3'-trimming.

Removal of the 5'-leader sequence usually precedes that of the 3'-trailer [20]. This order of trimming is mainly determined by RNA chaperones, La and the Lsm complex, which are widely found in eukaryotes [21–23]. Especially, La recognizes a 3'-oligo(U) tract of the trailer sequence, and protects the 3' terminus from exonucleolytic enzymes [24]. The 3'-terminal oligo(U) is a common feature of Pol III primary transcripts including pre-tRNAs [2]. Although La in *S. cerevisiae* and *S. pombe* (Lhp1 and Sla1, respectively) is not essential for viability, its deletion leads to a significant alteration in tRNA processing characteristics: the 3'-trimming precedes the 5'-trimming, and the exonucleolytic 3'-trimming becomes dominant over the endonucleolytic 3'-trimming [22].

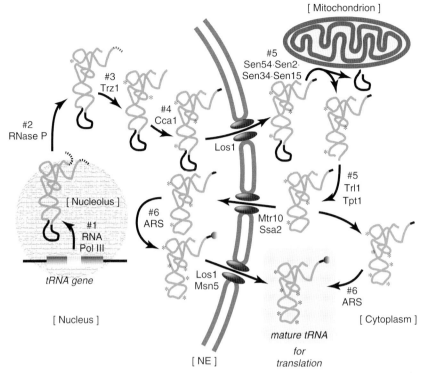

Fig. 3 Subcellular localization of tRNA processing enzymes in S. cerevisiae. Maturation processes of tRNAs and subcellular localization of enzymes responsible for each processing step in S. cerevisiae are summarized schematically. Names of subcellular compartments are shown in brackets, and names of enzymes or carrier proteins required for the processing steps represented by arrows are shown in boldface. Numbers on the top of the enzyme/protein names correspond to those in Fig. 2. NE, nuclear envelope.

Yeast Lhp1 is localized in the nucleoplasm, but is accumulated in the nucleolus where RNase P and tRNA genes are concentrated [25]. Most likely, La not only protects the 3′-trailer from exonucleolytic cleavage by steric hindrance but also retains the pre-tRNAs near the nucleolus, where RNase P resides to control the trimming order in the yeasts. In mammals, phosphorylated La – which is active in RNA binding – is found in the nucleoplasm, while the nonphosphorylated or inactive form is in the nucleolus [26]. As the nucleolus is involved in the processing of ribosomal RNA (rRNA) and small nucleolar RNA (snoRNAs), La accumulated in the mammalian nucleolus may contribute to the processing of these noncoding RNAs. Although the mammalian intranuclear arrangement of La and tRNA trimming enzymes differs from that in the yeasts, the early processing of pre-tRNAs is most likely tuned according to such intranuclear arrangements.

2.3
CCA Addition

When both the 5′-leader and 3′-trailer sequences of pre-tRNAs are processed, a

CCA sequence that is essential for aminoacylation by aminoacyl-tRNA synthetase (ARS) and amino acid residue-positioning in the ribosome is placed at the 3′ terminus of the pre-tRNAs [27] (Fig. 2, #4). The CCA-adding enzyme is required not only for the maturation of newly synthesized tRNAs but also for the repair of mature tRNAs [28]. Therefore, the enzyme localizes both in the nucleus and in the cytoplasm in *S. cerevisiae*, and this is also true in vertebrates [28, 29]. The enzyme is also localized in the mitochondrion, as is RNase Z.

2.4
tRNA Splicing: Conserved in the Mechanism but Diverse in Localization

2.4.1 Enzymes Required for tRNA Splicing

Some tRNA genes have introns. Most of the introns in the eukaryotic tRNA genes interrupt the anticodon loop 1 nt 3′ to the anticodon [30]. The splicing of such introns is a protein-catalyzing process, unlike mRNA splicing where RNA subunits in the spliceosomes have pivotal roles [30] (Fig. 2, #5). The *in vitro* analyses with extracts from various organisms revealed that two or three enzymes are required to achieve the splicing, namely splicing endonuclease (Sen/TSen), tRNA ligase, and 2′-phosphotransferase [31–33]. Splicing endonuclease, which is responsible for the cleavage of splice sites on pre-tRNAs, consists of four different subunits (Sen54, Sen2, Sen34, and Sen15 in *S. cerevisiae*), and is conserved among eukaryotes [34–36]. The catalytic subunits Sen2 and Sen34, which are responsible for cleavage of the 5′ and 3′ splice sites, respectively, show considerable sequence homology to archaeal splicing endonucleases [34]. On the other hand, there are two different modes of ligation;

5′-phosphate ligation and 3′-phosphate ligation [37]. 5′-Phosphate ligation, which requires the 5′-phosphate tRNA ligase system and 2′-phosphotransferase, is found in many eukaryotes from fungi to vertebrates and has been studied extensively in yeasts and higher plants [38–41]. Alternatively, 3′-phosphate ligation is catalyzed by one enzyme, HSPC117/RtcB, and has been identified in a wide variety of animals, and even in prokaryotes [42].

2.4.2 Variable Intracellular Distribution of Splicing Enzymes among Eukaryotes

An intriguing fact is that an intracellular localization of the splicing enzymes is not conserved among eukaryotes. In *S. cerevisiae*, the splicing endonuclease is associated with the cytosolic surface of the mitochondria, and such mitochondrial localization is required for its function [43] (Fig. 3). Furthermore, a small cytoplasmic pool of intron-containing pre-tRNAs proved to be a vital intermediate for splicing [44]. The second and third enzymes required for tRNA splicing, tRNA ligase (Trl1) and 2′-phosphotransferase (Tpt1), are also localized in the yeast cytoplasm [45, 16]. Although the three enzymatic steps of splicing are tightly coupled *in vivo*, no stable interaction among the three enzymes in *S. cerevisiae* has been detected.

On the other hand, tRNA splicing in vertebrates is carried out in the nucleus. The vertebrate tRNA splicing activity was demonstrated by assays with nuclear extracts of HeLa cells *in vitro*, and by microinjection experiments with *Xenopus* oocytes *in vivo* [33, 37, 20]. Mammalian homologs of yeast Sen subunits are localized in the nucleus [39], indicating that tRNA splicing is initiated in the nucleus. Mammalian cells can catalyze both modes of ligation. Although a full description of the 5′-phosphate ligase has still not been reported, part of the ligation reaction – 5′-phosphorylation of the 3′-exon – is catalyzed by hClp1 [46]. hClp1 was originally identified as a subunit of the cleavage and polyadenylation factor I to process the 3′ terminus of mRNAs [47]. Interestingly, hClp1 was next identified as a component of the mammalian tRNA splicing endonuclease complex [35], and then revealed to participate in tRNA ligation [46]. Therefore, hClp1 can phosphorylate the tRNA 3′-exon just after its release from the pre-tRNA in the nucleus to initiate ligation. The remainder of the ligation steps seem to be catalyzed by unidentified enzymes existing as separate polypeptides. Indeed, a chordate, *Branchiostoma floridae*, has a ligase system whose burden is shared by two independent polypeptides [48]. Nevertheless, tRNA splicing in mammalian cells relies primarily on the 3′-phosphate ligase, HSPC117 [42]. Although, to date, the intracellular localization of HSPC117 has not been directly tested, the HSPC117-associated proteins DDX1 and CGI-99 are known to be nuclear proteins, supporting the proposal that HSPC117 functions in the mammalian nucleus [49, 50].

Plants have adopted tRNA splicing enzymes similar to those of yeasts. For example, *Arabidopsis* and rice have 5′-phosphate ligases that show weak but apparent homology to yeast tRNA ligase throughout the sequences [39]. However, two catalytic subunits splicing endonuclease, tRNA ligase and 2′-phosphotransferase are localized to the nucleus when expressed as GFP fusions, which suggests that tRNA splicing proceeds in the nucleus [51]. It should be noted that in the same localization analysis a part of the splicing enzymes was identified in chloroplasts, where no tRNA gene on the organellar genome has a eukaryote-type intron. Indeed, these proteins have targeting signals for chloroplasts in their N termini. Most likely, plant cells produce both chloroplastic and nuclear forms of the enzymes by alternative translational initiation, as described previously for RNase Z.

The nuclear splicing of pre-tRNAs found in vertebrates and plants seems to be more straightforward than cytoplasmic splicing in *S. cerevisiae*. However, in most eukaryotes, tRNA ligase has another pivotal role in the processing of a unique mRNA in the cytosol. Eukaryotes are equipped with a signal transduction system termed unfolded protein response (UPR) to cope with protein folding stress in the endoplasmic reticulum (ER) [52]. The key event for the UPR is an unconventional cytoplasmic splicing of *HAC1* mRNA in yeasts and *XBP1* mRNA in animals, both of which encode master transcription factors regulating UPR-related genes [53–55]. In *S. cerevisiae*, tRNA ligase catalyzes the ligation of *HAC1* exons in this cytoplasmic splicing [56]. Therefore, it is reasonable to have tRNA ligase in the cytoplasm for the other eukaryotes, even adopting the nuclear splicing of pre-tRNAs. However, the ligase for *XBP1* exons has not yet been identified in any other organism.

Although the true reason why tRNA splicing occurs close to the mitochondria in the yeast is not fully understood, it has been indicated in one report that tRNA splicing may be carried out in the nucleus, and that unidentified extra functions of tRNA splicing endonuclease in the cytoplasm are required for full growth of the yeast [57]. Similar extra-functions of tRNA splicing endonuclease in mammals have also been speculated. Although all cells in a multicellular organism require tRNA splicing, certain mutations on human TSen subunits lead to neurodegenerative diseases, namely pontocerebellar hypoplasia types 2 and 4 [58]. To date, however, no cellular processes have been shown to be specifically impaired in tissues affected by this disease.

2.5
Modifications: Separation and Collaboration of Enzymes in the Nucleus and the Cytoplasm

Currently, more than 90 types of modifications on tRNAs have been identified among various organisms [59]. These modifications are provided at various timings during tRNA maturation, and the modification enzymes reside at various subcellular locations. Even a burden of the same type of modifications at different targets is shared by several enzymes at different locations. For example, a total of 100 characterized dihydrouridine (D) formations at positions 16 and 17, 20, 20a, and 20b (insertion in between positions, 20 and 21), and 47 in S. cerevisiae, are introduced by Dus1, Dus2, Dus3, and Dus4, respectively [60]. A systematic localization of the Dus proteins has indicated that Dus1 and Dus2 are restricted to the nucleus, while Dus4 is located both in the nucleus and the cytoplasm [16]. This localization of Dus proteins may reflect the sites of their actions, and suggests a flexibility of spatial assignment for modifications in eukaryotes.

Another interesting example is pseudouridine (Ψ) formation on tRNA$^{Ile}_{UAU}$ in S. cerevisiae. The anticodon of tRNA$^{Ile}_{UAU}$, which is encoded by intron-containing genes, is post-transcriptionally modified into $\Psi_{34}A\Psi_{36}$ from UAU by pseudouridylated synthase Pus1 in the yeast [61]. The Ψ modification is thought to enhance adenine selectivity [62]. Yeast Pus1 also modifies uridines at positions 27 and 67 of tRNA$^{Ile}_{UAU}$ in addition to the above positions [63]. Interestingly, Pus1 only modifies U_{34} and U_{36} on the intron-containing pre-tRNA$^{Ile}_{UAU}$ but not on the spliced pre-tRNA, while Pus1 can modify U_{27} and U_{67} on both the intron-containing and spliced pre-tRNAs [61, 63]. As Pus1 localizes in the nucleus [64], any intron-dependent modification of these tRNAs must be carried out before their nuclear export. Thus, the subcellular dynamics of pre-tRNAs can restrict the available time-window for certain modification enzymes.

Some complex modifications require multiple chemical reactions and active movement among intracellular compartments. Wybutosine (yW) is a guanosine-derived nucleotide with a tricyclic base attached with a large side chain [65] (Fig. 4). tRNA$^{Phe}_{GAA}$ of S. cerevisiae receives this modification at position 37, and this modification prevents -1 frameshifting [66]. The formation of yW is catalyzed by sequential actions of Trm5, and Tyw1–Tyw4 in the yeast [65]. Trm5 is a nuclear protein, and the other Tyw enzymes are cytoplasmic (Fig. 4); however, Trm5 only accepts spliced pre-tRNA$^{Phe}_{GAA}$ as a tRNA substrate, while tRNA splicing occurs outside the nucleus in S. cerevisiae (see Sect. 2.4.2). Indeed, it was shown that the spliced pre-tRNA$^{Phe}_{GAA}$ is imported into the nucleus to initiate yW formation and then re-exported to undergo the remainder of the processes ([67]; see also Sect. 3.3). Although this type of complex detour has not been demonstrated in other eukaryotes, such close collaboration between the nucleus and the cytoplasm for pre-tRNAs processings may be common among other organisms. The tRNAs that received all processings are

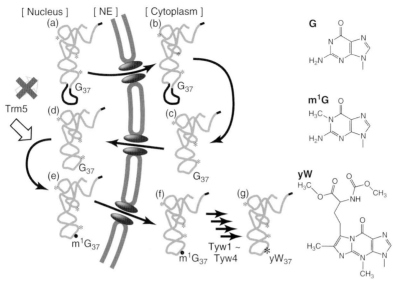

Fig. 4 The formation of wybutosine on tRNA$^{Phe}_{GAA}$ in *S. cerevisiae*. Since the initial step catalyzed by nuclear enzyme Trm5, which only recognizes spliced (**b**) but not unspliced (**a**) pre-tRNA$^{Phe}_{GAA}$ in the nucleus, so that spliced and immature tRNA$^{Phe}_{GAA}$ in the cytosol (**c**) must be imported to the nucleus (**d**). The m^{1}G$_{37}$-bearing tRNA$^{Phe}_{GAA}$ (**e**) is then re-exported to the cytoplasm (**f**) and receives the rest of the modification steps catalyzed by Tyw1 ~ Twy4 locating in the cytoplasm to complete yW$_{37}$ formation (**g**). The base structures of G, m^{1}G, and yW are shown at the right.

now aminoacylated by cognate ARSs and serve as adaptors with amino acid residues for translation (Fig. 2, #6).

2.6 Quality Control of tRNAs

Several layers of quality control exist to eliminate abnormal or unstable tRNAs during their lifetime (Fig. 5). At least three stages have been identified where the quality of tRNAs is inspected:

- The first inspection is applied at the early stage of tRNA maturation in the nucleus, and is governed by a collaboration of the TRAMP (Trf4/Air2/Mtr4p polyadenylation) complex and the nuclear exosome.

- The second inspection is applied just before the nuclear export by ARS and nuclear export factors.
- The final inspection, termed the rapid tRNA degradation (RTD), can be applied to eliminate fully matured tRNAs but requires components in both the cytosol and the nucleus.

The nuclear export-coupled quality control is discussed in Sections 3.1 and 3.2, while the other two controls are described in detail in the following subsections.

2.6.1 tRNA Quality Control by TRAMP and Nuclear Exosome

The elimination of abnormal tRNAs during their biogenesis in eukaryotes was first identified in *S. cerevisiae*. 1 Methyladenosine formation at position 58 is one

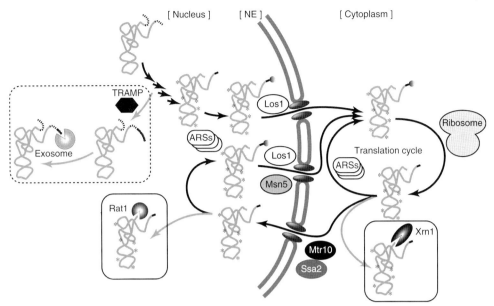

Fig. 5 Summary of the nuclear–cytoplasmic dynamics of tRNA and quality control in *S. cerevisiae*. Biosynthetic and recycling flows of tRNAs are shown by black arrows, and degradative flows by gray arrows. The TRAMP-exosome system and the RTD system are encircled with dashed and solid boxes, respectively. Export and import carriers are represented by ovals. The first round of nuclear export of tRNAs is catalyzed by Los1, while recycling export is carried out both by Los1 and by Msn5.

of the few essential modifications for tRNAs; indeed, mutations on *TRM6/GCD10* or *TRM61/GCD14* encoding the m^1A_{58} methyltransferase lead to a severe instability of $tRNA_i^{Met}$, and disruption of these genes are lethal for the yeast [68]. Hypomodified $tRNA_i^{Met}$ is degraded in the nucleus with the TRAMP complex and the nuclear exosome [69].

TRAMP is a nuclear surveillance complex that polyadenylates various abnormal transcripts including tRNAs [69–71]. Among its four subunits, namely Air1, Air2, Mtr4, and Trf4, Trf4 has polyadenylation activity, while Mtr4 is an ATP-dependent $3' \rightarrow 5'$ RNA helicase [70]. Short poly(A) tails, 16–30 nt in length, are attached by TRAMP and act as a degradation tag in contrast to usual poly(A) of mRNAs. TRAMP has an ability to distinguish abnormal or unstable tRNAs from their normal counterparts [71].

The nuclear exosome is a complex of exoribonucleases and endoribonucleases responsible for the degradation and processing of various RNAs in the nucleus [72, 73]. Two types of exosome have been identified, namely nuclear and cytoplasmic; both types share a core unit with 10 subunits and also have specific subunits for each type [72]. Among the core subunits, Rrp44/Dis3 has both exonucleolytic and endonucleolytic activities [74]. The nuclear exosome is associated with nucleus-specific factor Rrp6, while the cytoplasmic exosome functions with cytoplasm-specific factors, Ski2, Ski3, Ski7, and Ski8 [75]. The nuclear exosome – but not the cytoplasmic exosome – degrades unstable tRNAs polyadenylated by TRAMP [69]. Whilst TRAMP is distributed throughout the yeast nucleus, TRAMP is partly concentrated in the nucleolus, especially when ribosomal biosynthesis pathway is compromised [16, 76] (Fig. 5). Nucleolar enrichment of the nuclear exosome is consistent with the fact that hypomodified pre-tRNA$_i^{Met}$ with 5′ and 3′ extensions is a target of the TRAMP–exosome system [77]. As such pre-tRNAs are concentrated in the nucleolus, tRNAs seem to be under the inspection of the quality control system from the first stages of their lives.

2.6.2 Rapid tRNA Degradation

Another quality control system is employed to inspect the quality of mature tRNAs in the cytoplasm. Single deletions of *TRM4*, encoding tRNA methyltransferase for m^7G_{46} formation, or *TRM8*, which encodes tRNA methyltransferase for m^5C formation in *S. cerevisiae*, show almost no effect on tRNA stability and cell growth *in vivo*. However, the double deletion of *TRM4* and *TRM8* leads to a temperature-sensitive growth and to an instability of tRNA$^{Val}_{AAC}$ upon an upshift of the culture temperature [78]. This degradation of tRNA$^{Val}_{AAC}$ is not related to the TRAMP and nuclear exosome, and pre-exiting mature tRNAs are degraded immediately after the temperature shift; thus, the novel degradation was named RTD. The latter system relies on two 5′ → 3′ exonucleases, nuclear Rat1 and cytoplasmic Xrn1 [79] (Fig. 5). The RTD system monitors the structural integrity of the acceptor and T-stems of mature tRNAs, and most likely also their thermodynamic stability [80]. Whilst *in vitro* analyses have suggested that Xrn1 in the cytoplasm is responsible for this substrate recognition, the substrate recognition of RTD in the nucleus remains an open question.

Recently, it was reported that CCA-adding enzymes can add a CCACCA sequence instead of a CCA sequence to the 3′ terminus of tRNAs or tRNA-like RNAs destabilized by hypomodification or by mutations of tRNAs, both in mammals and in *S. cerevisiae* [81]. Such tRNAs are degraded by the RTD pathway in the yeast, and yeast Rrp44 – one of the exosome core subunits (in addition to Xrn1) – contributes to this degradation. As discussed in Section 2.3, the CCA-adding enzyme resides in both the nucleus and the cytoplasm. The cytoplasmic pool of the CCA-adding enzyme may be responsible for adding a CCACCA sequence to tag unstable tRNAs. Such tagged tRNAs are sequestered from the translation system to be degraded in the cytosol or to be imported back to the nucleus (see Sect. 3.3).

3 Nuclear–Cytoplasmic Transport of tRNAs

The delivery of newly synthesized tRNAs to the cytoplasm across the nuclear envelope (NE) is essential for eukaryotes, and this is achieved through the nuclear pore complex (NPC). An effective nuclear organization and hand-over mechanism of tRNAs from processing factors to export factors also support the efficient delivery. Previously, tRNAs had been thought to cross the NE only once in their lifetime to become aminoacylated and to serve for translation in the cytosol. However, this dogma was totally altered when tRNAs were shown to be aminoacylated in the nucleus to prove their functionality before export whilst, in *S. cerevisiae*, intron-containing pre-tRNAs were shown to be exported to the cytoplasm where the splicing factors reside. Furthermore, yeast tRNA$^{Phe}_{GAA}$ must return to the nucleus after its cytoplasmic splicing in order to acquire yW modification. Even after full maturation, the tRNAs are shuttled between the nucleus and the cytoplasm throughout their lifetime (Fig. 5). Thus, eukaryotic cells are equipped with both nuclear export and import machineries for tRNAs, and with regulatory systems to control the nuclear–cytoplasmic balance of tRNAs.

3.1
Intranuclear Transport of tRNAs

As noted in Section 2, tRNAs generated close to the nucleolus must be delivered to the nuclear periphery in order to gain access to the NPC; moreover, the tRNAs must also pass qualification for nuclear export. To date, several nuclear factors supporting this transfer and qualification have been identified. In *S. cerevisiae*, one such factor is Utp8, which was originally isolated as a protein to support 18S rRNA maturation in the nucleolus [82, 83]. Utp8 binds directly to tRNAs, to the tRNA-export carriers Los1/Exportin-t and Msn5/Exportin-5, to Gsp1/Ran, and to certain ARSs (see below) [82, 84, 85]. While Utp8 is stably localized in the nucleolus, its depletion causes a nuclear retention of tRNAs, and consequently Upt8 is thought to transfer tRNAs from processing factors to the nuclear export carriers by employing a channeling mechanism. Until now, the intranuclear transport of tRNAs has not been examined in detail in other eukaryotes.

Another more common group of factors that function in the nucleus for tRNA export are the ARSs. Although their primary role is aminoacylation of tRNAs during the translation cycle of tRNAs in the cytosol, they are also able to select correctly processed tRNAs just before export from the NPC [86–88]. Indeed, most of the mature tRNAs in the nucleus are aminoacylated in various organisms, and aminoacylated tRNAs are more efficiently exported to the cytoplasm in *Xenopus* oocytes [87, 88]. Furthermore, there is a small pool of ARSs in the nucleus, while many eukaryotic ARSs harbor nuclear localization signals (NLSs). In the case of Tyr-tRNA synthetase of *S. cerevisiae*, its NLS is not essential for its aminoacylation activity but is required for an efficient nuclear export of cognate tRNA, tRNA^{Tyr}, but not of other tRNAs [88]. Consequently, each ARS hands "aminoacylatable" cognate tRNAs to the export factors at the NPC.

3.2
Nuclear Export of tRNAs

3.2.1 Characterization of tRNA Export
Genetic analyses conducted in *S. cerevisiae* and biochemical analyses in mammals have identified factors required for the tRNA export from the nucleus, and the export has been revealed to consist of several parallel pathways with certain different characteristics. Los1/Exportin-t and Msn5/Exportin-5 are export carriers that bind and enable tRNAs to cross the NPC [89, 90].

Los1 was originally identified through the analysis of tRNA splicing mutants [91], after which its mammalian homolog, Exportin-t, followed by Los1 itself, was shown to be the major tRNA export carrier [92–94]. The *Arabidopsis thaliana* homolog, Paused, which was identified through genetic screening of developmental abnormalities, was also confirmed as contributing to tRNA export [95–97].

The second carrier, Msn5/Exportin-5, was first identified as a tRNA-binding exportin in mammalian cells [98, 99], and its ability to export tRNAs was confirmed in S. cerevisiae [100]. Msn5 was known as a bidirectional nuclear transporter for proteins in the yeast [101], while the primary function of mammalian Exportin-5 is the nuclear export of pre-miRNAs [102, 103]. An Exportin-5 homolog in A. thaliana, Hasty, was also shown to serve as an export carrier for pre-miRNAs, but not for tRNAs [97, 104].

These two export carriers are not alone, since the *los1 msn5* double deletion that has a striking impact on tRNA export still lead to a minor defect in yeast growth [100]. These situations vary among organisms; for example, in A. thaliana the *paused hasty* double mutant shows stronger growth defects than the single deletions, while the double mutant is still viable, as is the case for the yeast [96]. *Drosophila melanogaster* does not possess Exportin-t but rather uses Exportin-5 as a primary tRNA exporter, while *Caenorhabditis elegans* has only Exportin-t but not Exportin-5 [105, 106]. However, such organisms with only one of the two exportins may possess another tRNA exporter to compensate for a lack of the second exporter.

From the view of tRNA shuttling between the nucleus and cytoplasm, these parallel pathways are classified into two categories according to their export tasks; one for newly synthesized pre- and mature tRNAs; and the other for the re-export of imported tRNAs. In S. cerevisiae, Los1 acts as the primary export carrier for the initial nuclear export of intron-containing and intronless pre-tRNAs, and also for the re-export of mature tRNAs [43, 107]. On the other hand, Msn5 specifically bears a responsibility for the re-export of imported mature tRNAs [85, 107]. Related to this functional differentiation, Los1/Exportin-t and Msn5/Exportin-5 bind different spectra of tRNA species [99].

3.2.2 Structural Basis of the Export Mechanism

Many proteins and RNAs employ importins and exportins (also known as *karyopherins*) as carriers for their nuclear–cytoplasmic transport, and Los1/Exportin-t and Msn5/Exportin-5 belong to this protein family. Importin/exportin-dependent transport across the NPC is driven by a GTPase cycle of a small GTP-binding protein, Ran [89, 90]. Importins bind their cargo molecules in the cytoplasm to facilitate passage through the NPC. When the importin–cargo complexes reach the nucleoplasm, the GTP-bound form of Ran (RanGTP) disassembles the complexes to release the cargos. In contrast, exportins form trimeric complexes with their cargos and RanGTP in the nucleus, and allow transport of the complexes through the NPC. The exportins then release the cargos and Ran when the complexes encounter Ran GTPase-activating protein (RanGAP) and Ran BP1 on the cytoplasmic side of the NPC, and RanGTP is converted into the GDP-bound form, RanGDP [89]. Both, importins and exportins consist of approximately 20 tandem HEAT repeats to form large coil-like or snail-like structures [90]. Although Los1/Exportin-t and Msn5/Exportin-5 can each bind a variety of tRNAs, the patterns of tRNAs associated with Exportin-t or Exportin-5 *in vivo* are different, as noted above [99]. Such differences in tRNA specificity seems to derive from variations in the tRNA-binding modes. Based on the recently resolved crystal structures of Los1 from *S. pombe* and Exportin-5 from human, these two exportins hold

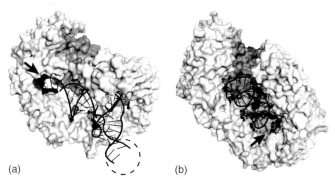

Fig. 6 Structures of tRNA export carriers. (a) The crystal structure of *S. pombe* Los1 complexed with RanGTP and tRNA (PDB: 3ICQ). SpLos1 and RanGTP are shown as surface models in light and dark gray, respectively. tRNA is shown as a cartoon in black. A protein region interacting with the 3′-terminal region of tRNA is marked by an arrow. A part of anticodon stem–loop cannot be solved in this structure (marked by a dashed circle); (b) The crystal structure of human Exportin-5 holding RanGTP and pre-miRNA (PDB: 3A6P). Expressions of the molecules are the same as in panel (a). In both structures, the N-terminal and C-terminal portions of these exportins locate upper left and middle right, respectively.

the tRNAs in opposite directions [108, 109]. Los1–RanGTP interacts with a nonaminoacylated tRNA molecule with contacts around the acceptor arm, the TΨC loop and the D loop in the crystal structure [108] (Fig. 6a). The acceptor stem of the tRNA is held by the N-terminal HEAT repeats 1–9 of Los1 to accommodate the 4-nt single-stranded 3′ terminus with enough space for an amino acid residue. On the other hand, the Exportin-5·RanGTP·pre-miRNA complex holds a double-stranded body of the pre-miRNA with its baseball mitt-like overall structure, and captures the 2-nt single-stranded 3′ terminus with a tunnel formed by C-terminal HEAT repeats 12–15 [109] (Fig. 6b). A tRNA has been proposed to be held by Exportin-5 in the same orientation as the pre-miRNA. The Los1-tRNA structure explains its substrate specificity. First, it was shown previously in biochemical experiments that Los1/Exportin-t binds primarily end-matured tRNAs [93], and that the structure of Los1 has no space for the 5′-leader or 3′-trailer sequences of the pre-tRNAs. Second, Los1 can function as an export factor for intron-containing pre-tRNAs in *S. cerevisiae*. Even mammalian Exportin-t was shown to bind intron-containing pre-tRNAs *in vitro* [93]. Indeed, Los1 from *S. pombe* does not touch the anticodon stem–loop, where an intron may be inserted, in the crystal structure [108] (Fig. 6a). Finally, Los1 can accommodate aminoacylated tRNAs, although the crystal structure was solved with a deacylated tRNA [108]. Thus, Los1 itself can take a pivotal role in the quality control of tRNAs supplied to the cytoplasm. In contrast, other biochemical experiments have revealed that Msn5/Exportin-5 also exports aminoacylated tRNAs, and that Exportin-5 (and probably also Msn5) exports eEF1A as a piggyback cargo of the tRNA export [85, 98, 99]. Based on the crystal structure of Exportin-5, a possible structural arrangement for eEF1A bound to the acceptor arm of the tRNA cannot be elucidated [109].

3.3
Nuclear Import of tRNAs

Nuclear import, or retrograde transport, of tRNAs was first identified in *S. cerevisiae* and further confirmed in mammalian cells [100, 110, 111]. Notably, a small pool of mature tRNAs was identified in the nucleus that had been considered as an intermediate state of mature tRNAs before export to the cytoplasm. However, it was found that splicing and other modifications are not finalized in the nucleus (as described in Sect. 2). Rather, this pool was shown to have been imported from the cytoplasm, based on the following points:

- If a yeast cell expressing heterogeneous tRNAs is fused to a nonexpressing cell, the tRNAs expressed from one nucleus is also found in the other nucleus [100, 110].
- Nuclear accumulation of tRNAs in export mutants is canceled by depleting ATP from the cells, and can be re-established by replenishing the intracellular ATP, even in the absence of *de novo* transcription of tRNAs [100].
- Nuclear accumulation of tRNAs is observed in wild-type cells experiencing nutrient stress, such as the deprivation of amino acids [110].

Currently, only limited information is available relating to the machinery of tRNA import. As noted above, the import is energy-dependent, and there appear to be two different pathways for tRNA import, namely a Ran-dependent pathway and a Ran-independent pathway. The nuclear accumulation of tRNAs under nutrient stress conditions requires the Ran GTPase cycle [110] whereas, under nonstressed conditions, tRNAs are imported into the nucleus even if the Ran GTPase cycle is compromised [100]. tRNA accumulation in the nucleus under nutrient stress conditions is impaired in an *mtr10* mutant, which encodes another importin that is known to import RNA-binding proteins such as Npl3 [110, 112]. As $msn5\Delta$ $mtr10\Delta$ double mutant cells do not accumulate tRNAs in the nucleus, even under nonstressed conditions, it is suggested that the Mtr10-dependent import operates even under nonstressed conditions. Thus, a part of the nuclear import is driven by the Ran- and Mtr10-dependent mechanisms, while the direct recognition of tRNAs by yeast Mtr10 has not been demonstrated. The present authors' group recently identified Ssa2 [113], one of the major cytoplasmic Hsp70 in *S. cerevisiae*, as a novel tRNA-binding protein. A mutant of this showed a tRNA accumulation defect under nutrient stress conditions, indicating that Ssa2 is a tRNA-binding protein implicated in nuclear import (A. Takano, T. Kajita, T. Endo, and T. Yoshihisa, unpublished results). These two factors are most likely essential constituents of the independent import pathways, and tRNAs are imported by more than two parallel pathways with different characteristics, as in the case of tRNA export.

In human cells, tRNA import into the nucleus is closely related to the nuclear import of retroviruses. During human immunodeficiency virus (HIV) proliferation in human cells, the HIV reverse transcription complex (RTC) must first be imported into the nucleus; this process is driven by a tRNA-mediated mechanism, but the Ran GTPase cycle is not implicated here [114]. Interestingly, Transportin-3 (TNPO-3), a human homolog of Mtr10, is required for HIV maturation, and this binds tRNA directly [115]. However, TNPO-3 has been shown rather to accelerate the export of tRNA, but not its import, *in vitro* [115]. Clearly, further studies are required to determine full details of the tRNA import machinery in this case.

Whilst the question of the physiological roles of tRNA import remain, several have been either proven or are expected:

- Quality control of the cytoplasmic tRNAs.
- The suppression of translation under conditions that are unfavorable for growth.
- Regulation of the modification order by separating the chemical steps with the NE.
- Nuclear translation.

Although, at present, the third of these roles is known only for yW formation in *S. cerevisiae* (as discussed in Sect. 2.5), there may be other similar cases. The expected role of nuclear translation is a long-standing issue of debate [116, 117], but recent reports have provided solid evidence for nuclear translation in mammals [118]. Nuclear translation does not necessarily provide the amounts of proteins required for nuclear functions. The ribosome, which is one of the most complex machineries in living cells, should be subject to quality control, and a fully active pool of mature tRNAs in the nucleus may be used for trial translation [119].

The first of the above-listed roles, namely the implication in tRNA quality control, is to be expected based on the fact that various tRNAs species are imported into the nucleus (as noted above; see Fig. 5). It is well known that the nucleus incorporates a variety of tRNA processing enzymes and systems to exclude abnormal tRNAs. In addition, gatekeeping by ARSs assures the export of fully functional tRNAs in the case of re-export, as mentioned above. On the other hand, the tRNA import system can bring a variety of tRNA species into the nucleus. Until now, aminoacylated and deacylated mature tRNAs, CCA-less tRNAs, and spliced but not fully modified tRNAs, such as yeast tRNA$^{Phe}_{GAA}$ (see Sect. 2.5), have been shown to be imported into the nucleus [67, 100, 110]. This combination of import with a broad specificity and export with strict qualification mechanisms can function as a "molecular sieve" to maintain a functional pool of tRNAs within the cytosol. Furthermore, the RTD system partially relies on nuclear exonuclease Rat1 [79] (see Sect. 2.6.2). Indeed, as the RTD system degrades pre-existing tRNAs in the cytosol, a portion of the destabilized tRNAs may be imported to the nucleus for degradation mediated by Rat1. The specificity of tRNA import, as described above, is also compatible with the transport of RTD substrates, including CCACCA-added tRNAs.

Today, accumulated lines of evidence support the suppression of translation under conditions that are unfavorable for growth with regards to tRNA transport for translational regulation. This issue is closely related to regulation of the nuclear–cytoplasmic balance of tRNA distribution, and is discussed in the following subsection.

3.4
Regulation of tRNA Export/Import Balance

Alterations in tRNA distribution according to environmental conditions have attracted much attention to the relationship with translation. It is plausible that the cytoplasmic level of total tRNAs is actively regulated to achieve a correct translational level, and that the nuclear–cytoplasmic transport of tRNAs is the target of regulation.

Both, tRNA export and import are thought of as constitutive processes. Thus, in order to accumulate tRNAs in the nucleus under nutrient stress conditions, a cell should either suppress tRNA export, enhance tRNA import, or perform a combination of both. S. cerevisiae responds to the starvation of various nutrients, such as amino acids, glucose, and phosphate [110, 120]. The response against nutrient deprivation is rapid (as quickly as 15 min in S. cerevisiae after the medium shift), such that the regulation appears to be applied directly to the transport machineries. When amino acids are depleted, the TOR (target of rapamycin) pathway – which serves as a central regulator to sense nutrient conditions and to control various cellular processes [121] – sends certain signals to the transport machineries, whereas the Gcn2 pathway – which senses a shortage of amino acids to upregulate amino acid biosynthetic pathways [122] – provides very little contribution to the nuclear accumulation of tRNAs [120]. On glucose depletion, Snf1 kinase and protein kinase A (PKA) [123] are required for the nuclear accumulation of tRNAs, while the activation of Snf1 and PKA results in positive and negative effects, respectively, on nuclear accumulation [120]. Furthermore, tRNA dynamics is coordinated with mRNA dynamics in the cytoplasm; when glucose is depleted from the medium, the mRNAs on polysomes are rapidly relocated to the P-body, a cytoplasmic storage compartment for untranslated mRNAs [124]. Some components of the P-body, such as Dhh1 and Pat1, are involved in P-body formation [125]. Upon glucose depletion, *pat1* and *dhh1* mutant cells cannot accumulate tRNAs in the nucleus, which indicates that a crosstalk exists between the sequestration of mRNAs and that of tRNAs from ribosomes to suppress global translation [126]. Indeed, such environmental changes, which are sensed by several signaling pathways, affect the localization of Los1 and Msn5 in S. cerevisiae [127]. Whether the exportins or other auxiliary factors are the targets of the upstream kinases remains an open question, however. It should be noted that, at present, contradictory reports exist regarding the generality of tRNA redistribution on the basis of amino acid depletion. Several groups have reported the nuclear accumulation of tRNAs under such conditions in mammalian cells, which suggests that tRNA redistribution is general in eukaryotes [111, 128]. However, one group has reported that that this type of tRNA redistribution is restricted to certain types of fungi [129]. While further studies will clearly be required to reconcile this discrepancy, at least some eukaryotes utilize the nuclear–cytoplasmic dynamics of tRNAs to maintain homeostasis in the cytoplasm.

4
tRNA Import into the Mitochondria

Mitochondria possess small – but their own – genomes and can translate some proteins within the organelle. Mitochondrial translation in protozoa, such as *Trypanosoma brucei* and *Leishmania*, relies completely on the import of all tRNAs for 20 amino acids from the cytosol [130, 131], whereas human and *S. cerevisiae* retain a complete set of tRNA genes on their mitochondrial genomes [132, 133]. Intermediates also exist between the above two forms [134–136]. Consequently, eukaryotes with mitochondria that lack at least one tRNA gene for a certain amino acid must include an import system that translocates such tRNAs to the mitochondrial matrix across the outer and inner mitochondrial membranes. However, mitochondria in human and yeasts also have the ability to import certain tRNAs from the cytosol [132, 133]. The total number and variation of isoacceptor tRNAs that are imported, or which must be imported, from the cytosol vary even among closely related species [135, 136]. In fact, it is highly probable that requirements for the mitochondrial import of certain tRNAs have emerged independently on many occasions during evolution, and import specificity to satisfy such requirements has coevolved rapidly.

In contrast to the tRNAs for mitochondrial translation, most mitochondrial proteins are encoded by the nuclear genomes in every eukaryote. Thus, such nuclear-encoded mitochondrial proteins are synthesized in the cytosol and imported into the mitochondria. Mitochondrial import machinery for proteins has been studied extensively in *S. cerevisiae* [137, 138]. The mitochondrial matrix proteins utilize two translocators for their import: (i) the translocator of the outer mitochondrial membrane (TOM); and (ii) the translocator of the inner mitochondrial membrane (TIM). The precursors of mitochondrial proteins usually have a cleavable N-terminal extension called a "presequence" that has mitochondrial target information, and the presequence is first recognized by receptors in the TOM complex. The import of precursors is driven by both the membrane potential across the inner membrane and the hydrolysis of ATP by mitochondrial Hsp70 (mtHsp70) in the matrix. As the import of proteins and tRNA is similar in that biopolymers cross double membranes in an energy-dependent and substrate-specific manner, tRNA import has been investigated in view of its relationship with the protein import machinery.

4.1
tRNA Import in *S. cerevisiae*: Not Essential but a Few tRNAs are Imported

S. cerevisiae imports three tRNAs – $tRNA^{Lys}_{CUU}$, $tRNA^{Gln}_{CUG}$, and $tRNA^{Gln}_{UUG}$ – into the mitochondria. The tRNA import has a strict substrate specificity, and an alternative isoacceptor for Lys, $tRNA^{Lys}_{UUU}$, cannot be imported into the mitochondria [139]. The import properties differ between $tRNA^{Lys}$ and $tRNA^{Gln}$; however, by using *in vitro* assays with isolated mitochondria it was shown that the import of $tRNA^{Lys}_{CUU}$ requires not only a preprotein receptor Tom20 (though not another receptor Tom70) but also Tim44, a TIM complex component [139]. The import requires both a membrane potential and ATP hydrolysis (Fig. 7a), and these features resemble those of protein import. On the other hand, some features of $tRNA^{Lys}_{CUU}$ import are not seen in protein import. $tRNA^{Lys}_{CUU}$ requires two cytosolic factors for its import: (i) a precursor of mitochondrial Lys-tRNA synthetase (pre-Msk1); and (ii) an isoform of enolase, Eno2, a glycolytic enzyme [140, 141]. In particular, pre-Msk1 can discriminate $tRNA^{Lys}_{CUU}$ from $tRNA^{Lys}_{UUU}$ [142], and it has been proposed that pre-Msk1 and $tRNA^{Lys}_{CUU}$ are cotranslocated across the mitochondrial membranes [140]. It is difficult to imagine that a structured protein complexed with its substrate, such as pre-Msk1 holding $tRNA^{Lys}_{CUU}$, can traverse the narrow pores of the TOM/TIM complexes, which usually can accept only loosely folded proteins [137, 138, 143]. Most likely, pre-Msk1 and $tRNA^{Lys}_{CUU}$ pass simultaneously through different pores that are closely located to each other.

On the other hand, the import of tRNA$^{Gln}_{CUG}$ and tRNA$^{Gln}_{UUG}$ is somehow different from that of tRNA^{Lys}, and remains a matter of debate. It has been reported by one group that both tRNA$^{Gln}_{CUG}$ and tRNA$^{Gln}_{UUG}$ are present and function within the yeast mitochondria, despite the fact that the mitochondrial genome contains its own tRNA$^{Gln}_{UUG}$ [144]. Although both tRNA$^{Gln}_{CUG}$ and tRNA$^{Gln}_{UUG}$ are imported into isolated mitochondria, the import is independent of soluble factors and of a membrane potential (Fig. 7a). Another group argued that neither the nuclear-encoded tRNA$^{Gln}_{CUG}$ nor tRNA$^{Gln}_{UUG}$ was detected in the mitochondrial fraction [145]. It was also mentioned that, for the glutaminylation of mitochondrial tRNA$^{Gln}_{UUG}$, the tRNA is first charged with glutamic acid by a nondiscriminating Glu-tRNA synthetase, Gus1, after which an amidotransferase consisting of Pet112, Gtf1, and Her2 modifies the Glu-tRNA$^{Gln}_{UUG}$ (tRNA$^{Gln}_{UUG}$ charged with Glu) into Gln-tRNA$^{Gln}_{UUG}$, in a manner similar to the bacterial GatBCD system [145, 146]. Although further studies are required to confirm these proposals, yeast mitochondria may use both imported and organellar-encoded tRNA^{Gln}, though the latter is the major form.

In summary, *S. cerevisiae* has two types of import system: (i) tRNA$^{Lys}_{CUU}$ import, which is dependent on a membrane potential and soluble factors; and (ii) tRNA^{Gln} import, which is independent of the membrane potential and soluble factors. These are most likely parallel pathways for each tRNA, which found its own way to mitochondria independently during evolution.

4.2
tRNA Import in Plants: More tRNAs Are Imported

Comparisons between nuclear and mitochondrial genomes and direct RNA analyses of isolated mitochondria have predicted and revealed that various plants import subsets of nuclear-encoded tRNAs into their mitochondria [135, 147, 148]. *In vivo* analyses with transgenic tobacco BY2 systems showed that the determinants for plant tRNA import into mitochondria are scattered around the whole molecules of tRNAs [149, 150]. tRNA import into isolated mitochondria was first reproduced *in vitro* with potato [151]. Plant mitochondria import tRNAs in an ATP-dependent manner, such as yeast tRNA$^{Lys}_{CUU}$, but no soluble factor is required (Fig. 7b). The import depends on the TOM complex components Tom20 and Tom40, while import requires the presence of another outer membrane protein VDAC (voltage-dependent anion channel), a pore-forming protein that conducts nonspecific solute transport across the outer membrane [152]. As tRNA import and protein import are not inhibited by excess amounts of opposite substrates, respectively, it has been postulated that the TOM complex is required only for the early steps of tRNA import and that tRNAs most likely cross the outer membrane through VDAC.

4.3
tRNA Import in Protozoa: Every tRNA is Imported

Protozoa, such as *Trypanosoma* and *Leishmania*, are of interest from two aspects in the mitochondrial import of tRNAs. First, as mentioned above, all of the tRNAs for mitochondrial translation

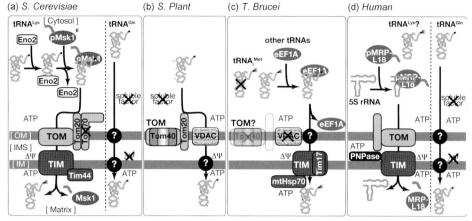

Fig. 7 Various modes of tRNA import into mitochondria. (a) Mitochondrial import systems of tRNAs in *S. cerevisiae*. There are two independent pathways for tRNA^{Lys} and tRNA^{Gln}. Import of tRNA^{Lys} requires cytosolic soluble factors, the TOM on the outer membrane (OM) and the TIM on the inner membrane (IM), and is driven by ATP and $\Delta\Psi$. Import of tRNA^{Lys} is supposed to be coupled with import of pre-Msk1 (pMsk1). In contrast, the import of tRNA^{Gln} does not require any soluble factor nor $\Delta\Psi$, but utilizes ATP as an energy source. IMS, intermembrane space; (b) A tRNA import system in plants. Plants use the TOM complex and VDAC for import across the OM, while no soluble factor is required. The translocator of the IM is not known. Import is driven by ATP and $\Delta\Psi$; (c) An import system in *T. brucei*. This organism does not have the usual TOM complex, and its VDAC is dispensable for tRNA import. eEF1A selects tRNAs for mitochondrial import in the cytosol. The import is ATP- and $\Delta\Psi$-dependent. For translocation across the IM, the TIM complex and mtHsp70 are essential; (d) Human import systems for tRNAs and 5S rRNA. There are two pathways to import small RNAs in humans. One pathway utilizes the TOM and perhaps TIM complexes, and requires ATP and $\Delta\Psi$ to import 5S rRNA, tRNA^{Lys}, etc. For import of 5S rRNA, the MPR-L18 precursor (pMPR-L18) escorts the RNA, as in the case of yeast tRNA^{Lys}. In addition, RNAs other than tRNAs require PNPase in the IMS. In contrast, the import of tRNA^{Gln} requires only ATP, as does tRNA^{Gln} import in yeast.

should be imported from the cytosol [130, 131]. Second, the mitochondrial protein import machinery on the outer mitochondrial membrane is different from the typical TOM complexes that are found in fungi and mammals [153].

Although protozoal mitochondria import most types of cytosolic tRNAs, the initiator tRNA^{Met} (tRNA$_i^{Met}$) and tRNA for selenocysteine are excluded [154]. This is consistent with the fact that bacteria-type translation in the mitochondria does not require eukaryotic tRNA$_i^{Met}$ and that there is no mitochondrial protein with selenocysteine. Comparisons between elongator tRNA^{Met} (tRNA$_e^{Met}$) and tRNA$_i^{Met}$ have revealed that the unmodified nucleotide pair 51:63 is the main determinant of tRNAs to be imported in *T. brucei* [155]. This base-pair also acts as an antideterminant to prevent eEF1A from recognizing tRNA$_i^{Met}$ [156]. Indeed, eEF1A knockdown in *T. brucei* reduces the recovery of newly synthesized tRNAs in mitochondria [156].

As in the case of tRNA import in yeast, *in vitro* analyses with isolated mitochondria from *T. brucei* have revealed that the import is dependent on some proteins on the mitochondrial surface, externally added ATP, and the membrane potential [154, 157] (Fig. 7c). In *Leishmania*, ATP both inside and outside of the mitochondrion is required, as with *S. cerevisiae*. The dissipation of a membrane potential across the mitochondrial inner membrane inhibits the complete transfer of tRNAs to the matrix, but a considerable portion of tRNAs remains with the inner membrane. This suggests that tRNAs first traverse the outer membrane in an ATP-dependent but membrane potential-independent manner, and are then further translocated into the matrix in a membrane potential-dependent manner [158, 159].

Protozoa possess no apparent homolog of Tom40, a central subunit of the TOM complex to form the protein-conducting channel [153]. In addition, VDAC required for tRNA import in plant is dispensable for tRNA import [160] (Fig. 7c). The only candidate for a channel of macromolecules including tRNAs at present is ATOM, which is a homolog of Omp85 in Gram-negative bacteria [153]. Indeed, ATOM forms a TOM complex-like protein assembly and plays pivotal roles in protein import into the mitochondria [153]. To date, no direct evidence is available to support ATOM function in tRNA import, and the mechanism of tRNA passage through the outer membrane remains an open question. On the other hand, mitochondrial protein translocator components on the inner membrane, namely Tim17 and mtHsp70, are involved in the import of tRNAs in *T. brucei*, which suggests that the protozoal TIM complex is involved in the translocation of tRNAs across the inner membrane and that mtHsp70 is a factor utilizing ATP in the mitochondrial matrix for tRNA import [161] (Fig. 7c).

In summary, considerable evidence exists to indicate a contribution of the protein import machinery to tRNA import in protozoal mitochondria.

4.4
Mammalian Import of tRNAs and Other Small RNAs

Although, as in the case of *S. cerevisiae*, the human mitochondrial genome harbors a full set of tRNAs for organellar translation, human mitochondria can import specific tRNAs, tRNA^{Lys} and tRNA^{Gln}, *in vitro* [133, 162]. Isolated human and rat mitochondria also contain nuclear-encoded tRNA$^{Gln}_{CUG1}$ and tRNA$^{Gln}_{UUG1}$, but not tRNA$^{Gln}_{CUG2}$ or tRNA$^{Gln}_{UUG2}$ [133]. Interestingly, mammalian mitochondria contain another Pol III transcript, 5S rRNA, which is an essential component of the ribosome, and RNA subunits of RNase P and RNase mitochondrial ribosomal protein (MRP) [163, 164]. Mitochondrial RNase P and RNase MRP are each involved in the processing of mitochondrial transcripts, while the function of mitochondrial 5S rRNA is not known precisely. In the *in vitro* analysis, the import of tRNA^{Lys} and 5S rRNA has similar characteristics; the import is ATP-dependent and is accelerated by soluble factors; in particular, 5S rRNA requires a MRP-L18 precursor in the cytosol [162, 165]. Some proteins are also required on the mitochondrial surface, and tRNA import is blocked when the protein-conducting channel of the TOM complex is fully occupied by a poor import substrate [162]. One the other hand, the import of tRNA^{Gln} *in vitro* proceeds without a membrane potential across the inner membrane if sufficient ATP is supplied [133]. It is possible that at least two parallel pathways exist in mammalian mitochondria, as in *S.*

cerevisiae, with different energetic requirements (Fig. 7d). Recently, polynucleotide phosphorylase (PNPase), an ATPase localized in the intermembrane space of the mitochondria, was shown to have a role in the import of 5S rRNA and RNA subunits of RNase P and RNase MRP [166]. However, the implication of PNPase in tRNA import has not been established extensively, and to date no reports have been made of the involvement of specific TOM or TIM components in tRNA import to mammalian mitochondria.

4.5 Mitochondrial RNA Import and Its Possible Therapeutic Application

Human diseases known to be caused by mutations in the mitochondrial genome include myoclonic epilepsy with ragged-red fibers (MERRFs) and mitochondrial myopathy, encephalopathy, lactic acidosis, stroke-like episodes (MELAS). As about a half of these diseases are related to mitochondrial tRNA genes [167], the artificial introduction of functional tRNAs from outside the mitochondrion offers a possible strategy for therapy. Indeed, several reports have been made that this new strategy has been proved promising both *in vitro* and in cell culture [168, 169]. In the extreme case, improvements in the import of substrates have enabled the import of mRNAs into the mitochondrion [170]. Indeed, it is likely that further improvements in this system will lead not only to therapeutic success, and the fundamental understanding of RNA import into the mitochondrion in various biological systems will hopefully accelerate the innovation of these therapeutic strategies.

References

1 Warner, J.R. (1999) The economics of ribosome biosynthesis in yeast. *Trends Biochem. Sci.*, **24**, 437–440.
2 Geiduschek, E.P., Kassavetis, G.A. (2001) The RNA polymerase III transcription apparatus. *J. Mol. Biol.*, **310**, 1–26.
3 Shi, X., Chen, T.D.-D., Suyama, Y. (1994) A nuclear tRNA gene cluster in the protozoan *Leishmania tarentolae* and differential distribution of nuclear-encoded tRNAs between the cytosol and mitochondria. *Mol. Biochem. Parasitol.*, **65**, 23–37.
4 Thompson, M., Haeusler, R.A., Good, P.D., Engelke, D.R. (2003) Nucleolar clustering of dispersed tRNA genes. *Science*, **302**, 1399–1401.
5 D'Ambrosio, C., Schmidt, C.K., Katou, Y., Kelly, G. et al. (2008) Identification of *cis*-acting sites for condensin loading onto budding yeast chromosomes. *Genes Dev.*, **22**, 2215–2227.
6 Haeusler, R.A., Pratt-Hyatt, M., Good, P.D., Gipson, T.A. et al. (2008) Clustering of yeast tRNA genes is mediated by specific association of condensin with tRNA gene transcription complexes. *Genes Dev.*, **22**, 2204–2214.
7 Iwasaki, O., Tanaka, A., Tanizawa, H., Grewal, S.I.S. et al. (2010) Centromeric localization of dispersed Pol III genes in fission yeast. *Mol. Biol. Cell*, **21**, 254–265.
8 Jarrous, N., Gopalan, V. (2010) Archaeal/eukaryal RNase P: subunits, functions and RNA diversification. *Nucleic Acids Res.*, **38**, 7885–7894.
9 Bertrand, E., Houser-Scott, F., Kendall, A., Singer, R.H. et al. (1998) Nucleolar localization of early tRNA processing. *Genes Dev.*, **12**, 2463–2468.
10 Jarrous, N. (2002) Human ribonuclease P: subunits, function, and intranuclear localization. *RNA*, **8**, 1–7.
11 Jacobson, M.R., Cao, L.-G., Taneja, K., Singer, R.H. et al. (1997) Nuclear domains of the RNA subunit of RNase P. *J. Cell Sci.*, **110**, 829–837.
12 Lygerou, Z., Pluk, H., van Venrooij, W.J., Séraphin, B. (1996) hPop1: an autoantigenic protein subunit shared by the human RNase P and RNase MRP ribonucleoproteins. *EMBO J.*, **15**, 5936–5948.

13 Jarrous, N., Wolenski, J.S., Wesolowski, D., Lee, C. et al. (1999) Localization in the nucleolus and coiled bodies of protein subunits of the ribonucleoprotein ribonuclease P. *J. Cell Biol.*, **146**, 559–571.

14 Maraia, R.J., Lamichhane, T.N. (2011) 3' Processing of eukaryotic precursor tRNAs. *Wiley Interdiscip. Rev. RNA*, **2**, 362–375.

15 Takaku, H., Minagawa, A., Takagi, M., Nashimoto, M. (2003) A candidate prostate cancer susceptibility gene encodes tRNA 3' processing endoribonuclease. *Nucleic Acids Res.*, **31**, 2272–2278.

16 Yeast Resource Center Public Data Repository. Available at: http://www.yeastrc.org/pdr/pages/front.jsp (accessed 30 May 2012).

17 Rossmanith, W. (2011) Localization of human RNase Z isoforms: dual nuclear/mitochondrial targeting of the *ELAC2* gene product by alternative translation initiation. *PLoS ONE*, **6**, e19152.

18 Yogev, O., Pines, O. (2011) Dual targeting of mitochondrial proteins: mechanism, regulation and function. *Biochim. Biophys. Acta*, **1808**, 1012–1020.

19 Copela, L.A., Fernandez, C.F., Sherrer, R.L., Wolin, S.L. (2008) Competition between the Rex1p exonuclease and the La protein affects both Trf4p-mediated RNA quality control and pre-tRNA maturation. *RNA*, **14**, 1214–1227.

20 Melton, D.A., De Robertis, E.M., Cortese, R. (1980) Order and intracellular location of the events involved in the maturation of a spliced tRNA. *Nature*, **284**, 143–148.

21 Stefano, J.E. (1984) Purified lupus antigen La recognizes an oligouridylate stretch common to the 3' termini of RNA polymerase III transcripts. *Cell*, **36**, 145–154.

22 Yoo, C.J., Wolin, S.L. (1997) The yeast La protein is required for the 3' endonucleolytic cleavage that matures tRNA precursors. *Cell*, **89**, 393–402.

23 Kufel, J., Allmang, C., Verdone, L., Beggs, J.D. et al. (2002) Lsm proteins are required for normal processing of pre-tRNAs and their efficient association with La-homologous protein Lhp1p. *Mol. Cell. Biol.*, **22**, 5248–5526.

24 Teplova, M., Yuan, Y.R., Phan, A.T., Malinina, L. et al. (2006) Structural basis for recognition and sequestration of $UUU_{OH}3'$ termini of nascent RNA polymerase III transcripts by La, a rheumatic disease autoantigen. *Mol. Cell*, **21**, 75–85.

25 Rosenblum, J.S., Pemberton, L.F., Blobel, G. (1997) A nuclear import pathway for a protein involved in tRNA maturation. *J. Cell Biol.*, **139**, 1655–1661.

26 Intine, R.V., Tenenbaum, S.A., Sakulich, A.L., Keene, J.D. et al. (2003) Differential phosphorylation and subcellular localization of La RNPs associated with precursor tRNAs and translation-related mRNAs. *Mol. Cell*, **12**, 1301–1307.

27 Betat, H., Rammelt, C., Mörl, M. (2010) tRNA nucleotidyltransferases: ancient catalysts with an unusual mechanism of polymerization. *Cell. Mol. Life Sci.*, **67**, 1447–1463.

28 Wolfe, C.L., Hopper, A.K., Martin, N.C. (1996) Mechanisms leading to and the consequences of altering the normal distribution of ATP(CTP):tRNA nucleotidyltransferase in yeast. *J. Biol. Chem.*, **271**, 4679–4686.

29 Solari, A., Deutscher, M.P. (1982) Subcellular localization of the tRNA processing enzyme, tRNA nucleotidyltransferase, in *Xenopus laevis* oocytes and in somatic cells. *Nucleic Acids Res.*, **10**, 4397–4407.

30 Heinemann, I.U., Söll, D., Randau, L. (2010) Transfer RNA processing in archaea: unusual pathways and enzymes. *FEBS Lett.*, **584**, 303–309.

31 Ogden, R.C., Beckmann, J.S., Abelson, J., Kang, H.S. et al. (1977) In vitro transcription and processing of a yeast tRNA gene containing an intervening sequence. *Cell*, **17**, 399–406.

32 Stange, N., Beier, H. (1987) A cell-free plant extract for accurate pre-tRNA processing, splicing and modification. *EMBO J.*, **6**, 2811–2818.

33 Perkins, K.K., Furneaux, H., Hurwitz, J. (1985) Isolation and characterization of an RNA ligase from HeLa cells. *Proc. Natl Acad. Sci. USA*, **82**, 684–688.

34 Trotta, C.R., Miao, F., Arn, E.A., Stevens, S.W. et al. (1997) The yeast tRNA splicing endonuclease: a tetrameric enzyme with two active site subunits homologous to the archaeal tRNA endonucleases. *Cell*, **89**, 849–858.

35 Paushkin, S.V., Patel, M., Furia, B.S., Peltz, S.W. et al. (2004) Identification of

a human endonuclease complex reveals a link between tRNA splicing and pre-mRNA 3' end formation. *Cell*, **117**, 311–321.

36 Akama, K., Junker, V., Beier, H. (2000) Identification of two catalytic subunits of tRNA splicing endonuclease from *Arabidopsis thaliana*. *Gene*, **257**, 177–185.

37 Popow, J., Schleiffer, A., Martinez, J. (2012) Diversity and roles of (t)RNA ligase. *Cell. Mol. Life Sci.*, **69**, 2657–2670.

38 Phizicky, E.M., Schwartz, R.C., Abelson, J. (1986) *Saccharomyces cerevisiae* tRNA ligase: purification of the protein and isolation of the structural gene. *J. Biol. Chem.*, **261**, 2978–2986.

39 Englert, M., Beier, H. (2005) Plant tRNA ligases are multifunctional enzymes that have diverged in sequence and substrate specificity from RNA ligases of other phylogenetic origins. *Nucleic Acids Res.*, **33**, 388–399.

40 Culver, G.M., McCraith, S.M., Consaul, S.A., Stanford, D.R. *et al.* (1997) A 2'-phosphotransferase implicated in tRNA splicing is essential in *Saccharomyces cerevisiae*. *J. Biol. Chem.*, **272**, 13203–13210.

41 Harding, H.P., Lackey, J.G., Hsu, H.-C., Zhang, Y. *et al.* (2008) An intact unfolded protein response in Trpt1 knockout mice reveals phylogenic divergence in pathways for RNA ligation. *RNA*, **14**, 225–232.

42 Popow, J., Englert, M., Weitzer, S., Schleiffer, A. *et al.* (2011) HSPC117 is the essential subunit of a human tRNA splicing ligase complex. *Science*, **331**, 760–764.

43 Yoshihisa, T., Yunoki-Esaki, K., Ohshima, C., Tanaka, N. *et al.* (2003) Possibility of cytoplasmic pre-tRNA splicing: the yeast tRNA splicing endonuclease mainly localizes on the mitochondria. *Mol. Biol. Cell*, **14**, 3266–3279.

44 Yoshihisa, T., Ohshima, C., Yunoki-Esaki, K., Endo, T. (2007) Cytoplasmic splicing of tRNA in *Saccharomyces cerevisiae*. *Genes Cells*, **12**, 285–297.

45 Mori, T., Ogasawara, C., Inada, T., Englert, M. *et al.* (2010) Dual functions of yeast tRNA ligase in the unfolded protein response: unconventional cytoplasmic splicing of *HAC1* pre-mRNA is not sufficient to release translational attenuation. *Mol. Biol. Cell*, **21**, 3722–3734.

46 Weitzer, S., Martinez, J. (2007) The human RNA kinase hClp1 is active on 3' transfer RNA exons and short interfering RNAs. *Nature*, **447**, 222–226.

47 Minvielle-Sebastia, L., Preker, P.J., Wiederkehr, T., Strahm, Y. *et al.* (1997) The major yeast poly(A)-binding protein is associated with cleavage factor IA and functions in premessenger RNA 3'-end formation *Proc. Natl Acad. Sci. USA*, **94**, 7897–7902.

48 Englert, M., Sheppard, K., Gundllapalli, S., Beier, H. *et al.* (2010) *Branchiostoma floridae* has separate healing and sealing enzymes for 5'-phosphate RNA ligation. *Proc. Natl Acad. Sci. USA*, **107**, 16834–16839.

49 Xu, L., Khadijah, S., Fang, S., Wang, L. *et al.* (2010) The cellular RNA helicase DDX1 interacts with coronavirus nonstructural protein 14 and enhances viral replication. *J. Virol.*, **84**, 8571–8583.

50 Pérez-González, A., Rodriguez, A., Huarte, M., Salanueva, I.J. *et al.* (2006) hCLE/CGI-99, a human protein that interacts with the influenza virus polymerase, is a mRNA transcription modulator. *J. Mol. Biol.*, **362**, 887–900.

51 Englert, M., Latz, A., Becker, D., Gimple, O. *et al.* (2007) Plant pre-tRNA splicing enzymes are targeted to multiple cellular compartments. *Biochimie*, **89**, 1351–1365.

52 Walter, P., Ron, D. (2011) The unfolded protein response: from stress pathway to homeostatic regulation. *Science*, **334**, 1081–1086.

53 Cox, J.S., Walter, P. (1996) A novel mechanism for regulating activity of a transcription factor that controls the unfolded protein response. *Cell*, **87**, 391–404.

54 Yoshida, H., Matsui, T., Yamamoto, A., Okada, T. *et al.* (2001) XBP1 mRNA is induced by ATF6 and spliced by IRE1 in response to ER stress to produce a highly active transcription factor. *Cell*, **107**, 881–891.

55 Kawahara, T., Yanagi, H., Yura, T., Mori, K. (1997) Endoplasmic reticulum stress-induced mRNA splicing permits synthesis of transcription factor Hac1p/Ern4p that activates the unfolded protein response. *Mol. Biol. Cell*, **8**, 1845–1862.

56 Sidrauski, C., Cox, J.S., Walter, P. (1996) tRNA Ligase is required for regulated

mRNA splicing in the unfolded protein response. *Cell*, **87**, 405–413.

57 Dhungel, N., Hopper, A.K. (2012) Beyond tRNA cleavage: novel essential function for yeast tRNA splicing endonuclease unrelated to tRNA processing. *Genes Dev.*, **26**, 503–514.

58 Budde, B.S., Namavar, Y., Barth, P.G., Poll-The, B.T. *et al.* (2008) tRNA splicing endonuclease mutations cause pontocerebellar hypoplasia. *Nat. Genet.*, **40**, 1113–1118.

59 Phizicky, E.M., Alfonzo, J.D. (2010) Do all modifications benefit all tRNAs? *FEBS Lett.*, **584**, 265–271.

60 Xing, F., Hiley, S.L., Hughes, T.R., Phizicky, E.M. (2004) The specificities of four yeast dihydrouridine synthases for cytoplasmic tRNAs. *J. Biol. Chem.*, **279**, 17850–17860.

61 Szweykowska-Kulinska, Z., Senger, B., Keith, G., Fasiolo, F. *et al.* (1994) Intron-dependent formation of pseudouridines in the anticodon of *Saccharomyces cerevisiae* minor tRNA$_{Ile}$. *EMBO J.*, **13**, 4636–4644.

62 Agris, P.F., Vendeix, F.A.P., Graham, W.D. (2007) tRNA's wobble decoding of the genome: 40 years of modification. *J. Mol. Biol.*, **366**, 1–13.

63 Motorin, Y., Keith, G., Simon, C., Foiret, D. *et al.* (1998) The yeast tRNA: pseudouridine synthase Pus1p displays a multisite substrate specificity. *RNA*, **4**, 856–869.

64 Simos, G., Tekotte, H., Grosjean, H., Segref, A. *et al.* (1996) Nuclear pore proteins are involved in the biogenesis of functional tRNA. *EMBO J.*, **15**, 2270–2284.

65 Noma, A., Kirino, Y., Ikeuchi, Y., Suzuki, T. (2006) Biosynthesis of wybutosine, a hyper-modified nucleoside in eukaryotic phenylalanine tRNA. *EMBO J.*, **25**, 2142–2154.

66 Waas, W.F., Druzina, Z., Hanan, M., Schimmel, P. (2007) Role of a tRNA base modification and its precursors in frameshifting in eukaryotes. *J. Biol. Chem.*, **282**, 26026–26034.

67 Ohira, T., Suzuki, T. (2011) Retrograde nuclear import of tRNA precursors is required for modified base biogenesis in yeast. *Proc. Natl Acad. Sci. USA*, **108**, 10502–10507.

68 Anderson, J., Phan, L., Cuesta, R., Carlson, B.A. *et al.* (1998) The essential Gcd10p–Gcd14p nuclear complex is required for 1-methyladenosine modification and maturation of initiator methionyl-tRNA. *Genes Dev.*, **12**, 3650–3662.

69 Kadaba, S., Krueger, A., Trice, T., Krecic, A.M. *et al.* (2004) Nuclear surveillance and degradation of hypomodified initiator tRNA^{Met} in *S. cerevisiae*. *Genes Dev.*, **18**, 1227–1240.

70 LaCava, J., Houseley, J., Saveanu, C., Petfalski, E. *et al.* (2005) RNA degradation by the exosome is promoted by a nuclear polyadenylation complex. *Cell*, **121**, 713–724.

71 Vanácová, S., Wolf, J., Martin, G., Blank, D. *et al.* (2005) A new yeast poly(A) polymerase complex involved in RNA quality control. *PLoS Biol.*, **3**, e189.

72 Schmid, M., Jensen, T.H. (2008) The exosome: a multipurpose RNA-decay machine. *Trends Biochem. Sci.*, **33**, 501–510.

73 Mitchell, P., Petfalski, E., Shevchenko, A., Mann, M. *et al.* (1997) The exosome: a conserved eukaryotic RNA processing complex containing multiple 3'–5' exoribonucleases. *Cell*, **91**, 457–466.

74 Schaeffer, D., Tsanova, B., Barbas, A., Reis, F.P. *et al.* (2009) The exosome contains domains with specific endoribonuclease, exoribonuclease and cytoplasmic mRNA decay activities. *Nat. Struct. Mol. Biol.*, **16**, 56–62.

75 van Hoof, A., Lennertz, P., Parker, R. (2000) Yeast exosome mutants accumulate 3'-extended polyadenylated forms of U4 small nuclear RNA and small nucleolar RNAs. *Mol. Cell. Biol.*, **20**, 441–452.

76 Dez, C., Houseley, J., Tollervey, D. (2006) Surveillance of nuclear-restricted pre-ribosomes within a subnucleolar region of *Saccharomyces cerevisiae*. *EMBO J.*, **25**, 1534–1546.

77 Kadaba, S., Wang, X., Anderson, J.T. (2006) Nuclear RNA surveillance in *Saccharomyces cerevisiae*: Trf4p-dependent polyadenylation of nascent hypomethylated tRNA and an aberrant form of 5S rRNA. *RNA*, **12**, 508–521.

78 Alexandrov, A., Chernyakov, I., Gu, W., Hiley, S.L. *et al.* (2006) Rapid tRNA decay can result from lack of nonessential modifications. *Mol. Cell*, **21**, 87–96.

79 Chernyakov, I., Whipple, J.M., Kotelawala, L., Grayhack, E.J. et al. (2008) Degradation of several hypomodified mature tRNA species in *Saccharomyces cerevisiae* is mediated by Met22 and the 5'-3' exonucleases Rat1 and Xrn1. *Genes Dev.*, **22**, 1369–1380.

80 Whipple, J.M., Lane, E.A., Chernyakov, I., D'Silva, S. et al. (2011) The yeast rapid tRNA decay pathway primarily monitors the structural integrity of the acceptor and T-stems of mature tRNA. *Genes Dev.*, **25**, 1173–1184.

81 Wilusz, J.E., Whipple, J.M., Phizicky, E.M., Sharp, P.A. (2011) tRNAs marked with CCACCA are targeted for degradation. *Science*, **334**, 817–821.

82 Steiner-Mosonyi, M., Leslie, D.M., Dehghani, H., Aitchison, J.D. et al. (2003) Utp8p is an essential intranuclear component of the nuclear tRNA export machinery of *Saccharomyces cerevisiae*. *J. Biol. Chem.*, **278**, 32236–32245.

83 Grandi, P., Rybin, V., Bassler, J., Petfalski, E. et al. (2002) 90S pre-ribosomes include the 35S pre-rRNA, the U3 snoRNP, and 40S subunit processing factors but predominantly lack 60S synthesis factors. *Mol. Cell*, **10**, 105–115.

84 Strub, B.R., Eswara, M.B.K., Pierce, J.B., Mangroo, D. (2007) Utp8p is a nucleolar tRNA-binding protein that forms a complex with components of the nuclear tRNA export machinery in *Saccharomyces cerevisiae*. *Mol. Biol. Cell*, **18**, 3845–3859.

85 Eswara, M.B.K., McGuire, A.T., Pierce, J.B., Mangroo, D. (2009) Utp9p facilitates Msn5p-mediated nuclear reexport of retrograded tRNAs in *Saccharomyces cerevisiae*. *Mol. Biol. Cell*, **20**, 5007–5025.

86 Ibba, M., Söll, D. (2000) Aminoacyl-tRNA synthesis. *Annu. Rev. Biochem.*, **69**, 617–650.

87 Lund, E., Dahlberg, J.E. (1998) Proofreading and aminoacylation of tRNAs before export from the nucleus. *Science*, **282**, 2082–2085.

88 Azad, A.K., Stanford, D.R., Sarkar, S., Hopper, A.K. (2001) Role of nuclear pools of aminoacyl-tRNA synthetases in tRNA nuclear export. *Mol. Biol. Cell*, **12**, 1381–1392.

89 Güttler, T., Görlich, D. (2011) Ran-dependent nuclear export mediators: a structural perspective. *EMBO J.*, **30**, 3457–3474.

90 Cook, A., Bono, F., Jinek, M., Conti, E. (2007) Structural biology of nucleocytoplasmic transport. *Annu. Rev. Biochem.*, **76**, 647–671.

91 Hopper, A.K., Schultz, L.D., Shapiro, R.A. (1980) Processing of intervening sequences: a new yeast mutant which fails to excise intervening sequences from precursor tRNAs. *Cell*, **19**, 741–751.

92 Kutay, U., Lipowsky, G., Izaurralde, E., Bischoff, F.R. et al. (1998) Identification of a tRNA-specific nuclear export receptor. *Mol. Cell*, **1**, 359–369.

93 Arts, G.-J., Kuersten, S., Romby, P., Ehresmann, B. et al. (1998) The role of exportin-t in selective nuclear export of mature tRNAs. *EMBO J.*, **17**, 7430–7441.

94 Hellmuth, K., Lau, D.M., Bischoff, F.R., Künzler, M. et al. (1998) Yeast Los1p has properties of an exportin-like nucleocytoplasmic transport factor for tRNA. *Mol. Cell. Biol.*, **18**, 6374–6386.

95 Li, J., Chen, X. (2003) *PAUSED*, a putative exportin-t, acts pleiotropically in *Arabidopsis* development but is dispensable for viability. *Plant Physiol.*, **132**, 1913–1924.

96 Hunter, C.A., Aukerman, M.J., Sun, H., Fokina, M. et al. (2003) *PAUSED* encodes the *Arabidopsis* exportin-t ortholog. *Plant Physiol.*, **132**, 2135–2143.

97 Park, M.Y., Wu, G., Gonzalez-Sulser, A., Vaucheret, H. et al. (2005) Nuclear processing and export of microRNAs in *Arabidopsis*. *Proc. Natl Acad. Sci. USA*, **102**, 3691–3696.

98 Bohnsack, M.T., Regener, K., Schwappach, B., Saffrich, R. et al. (2002) Exp5 exports eEF1A via tRNA from nuclei and synergizes with other transport pathways to confine translation to the cytoplasm. *EMBO J.*, **21**, 6205–6215.

99 Calado, A., Treichel, N., Müller, E.-C., Otto, A. et al. (2002) Exportin-5-mediated nuclear export of eukaryotic elongation factor 1A and tRNA. *EMBO J.*, **21**, 6216–6224.

100 Takano, A., Endo, T., Yoshihisa, T. (2005) tRNA actively shuttles between the nucleus and cytosol in yeast. *Science*, **309**, 140–142.

101 Yoshida, K., Blobel, G. (2001) The karyopherin Kap142p/Msn5p mediates nuclear import and nuclear export of different cargo proteins. *J. Cell Biol.*, **152**, 729–739.

102 Yi, R., Qin, Y., Macara, I.G., Cullen, B.R. (2003) Exportin-5 mediates the nuclear

export of pre-microRNAs and short hairpin RNAs. *Genes Dev.*, **17**, 3011–3016.

103 Lund, E., Güttinger, S., Calado, A., Dahlberg, J.E. et al. (2004) Nuclear export of microRNA precursors. *Science*, **303**, 95–98.

104 Bollman, K.M., Aukerman, M.J., Park, M.Y., Hunter, C. et al. (2003) HASTY, the *Arabidopsis* ortholog of exportin 5/MSN5, regulates phase change and morphogenesis. *Development*, **130**, 1493–1504.

105 Shibata, S., Sasaki, M., Miki, T., Shimamoto, A. et al. (2006) Exportin-5 orthologues are functionally divergent among species. *Nucleic Acids Res.*, **34**, 4711–4721.

106 Büssing, I., Yang, J.-S., Lai, E.C., Grosshans, H. (2010) The nuclear export receptor XPO-1 supports primary miRNA processing in *C. elegans* and *Drosophila*. *EMBO J.*, **29**, 1830–1839.

107 Murthi, A., Shaheen, H.H., Huang, H.-Y., Preston, M.A. et al. (2010) Regulation of tRNA bidirectional nuclear-cytoplasmic trafficking in *Saccharomyces cerevisiae*. *Mol. Biol. Cell*, **21**, 639–649.

108 Cook, A.G., Fukuhara, N., Jinek, M., Conti, E. (2009) Structures of the tRNA export factor in the nuclear and cytosolic states. *Nature*, **461**, 60–65.

109 Okada, C., Yamashita, E., Lee, S.J., Shibata, S. et al. (2009) A high-resolution structure of the pre-microRNA nuclear export machinery. *Science*, **326**, 1275–1279.

110 Shaheen, H.H., Hopper, A.K. (2005) Retrograde movement of tRNAs from the cytoplasm to the nucleus in *Saccharomyces cerevisiae*. *Proc. Natl Acad. Sci. USA*, **102**, 11290–11295.

111 Shaheen, H.H., Horetsky, R.L., Kimball, S.R., Murthi, A. et al. (2007) Retrograde nuclear accumulation of cytoplasmic tRNA in rat hepatoma cells in response to amino acid deprivation. *Proc. Natl Acad. Sci. USA*, **104**, 8845–8850.

112 Pemberton, L.F., Rosenblum, J.S., Blobel, G. (1997) A distinct and parallel pathway for the nuclear import of an mRNA-binding protein. *J. Cell Biol.*, **139**, 1645–1653.

113 Werner-Washburne, M., Stone, D.E., Craig, E.A. (1987) Complex interactions among members of an essential subfamily of *hsp70* genes in *Saccharomyces cerevisiae*. *Mol. Cell. Biol.*, **7**, 2568–2577.

114 Zaitseva, L., Myers, R., Fassati, A. (2006) tRNAs promote nuclear import of HIV-1 intracellular reverse transcription complexes. *PLoS Biol.*, **4**, e332.

115 Zhou, L., Sokolskaja, E., Jolly, C., James, W. et al. (2011) Transportin 3 promotes a nuclear maturation step required for efficient HIV-1 integration. *PLoS Pathog.*, **7**, e1002194.

116 Iborra, F.J., Jackson, D.A., Cook, P.R. (2001) Coupled transcription and translation within nuclei of mammalian cells. *Science*, **293**, 1139–1142.

117 Dahlberg, J.E., Lund, E., Goodwin, E.B. (2003) Nuclear translation: what is the evidence?. *RNA*, **9**, 1–8.

118 David, A., Dolan, B.P., Hickman, H.D., Knowlton, J.J. et al. (2012) Nuclear translation visualized by ribosome-bound nascent chain puromycylation. *J. Cell Biol.*, **197**, 45–57.

119 Pederson, T., Politz, J.C. (2000) The nucleolus and the four ribonucleoproteins of translation. *J. Cell Biol.*, **148**, 1091–1095.

120 Whitney, M.L., Hurto, R.L., Shaheen, H.H., Hopper, A.K. (2007) Rapid and reversible nuclear accumulation of cytoplasmic tRNA in response to nutrient availability. *Mol. Biol. Cell*, **18**, 2678–2686.

121 Kim, J., Guan, K.-L. (2011) Amino acid signaling in TOR activation. *Annu. Rev. Biochem.*, **80**, 1001–1032.

122 Hinnebusch, A.G. (1993) Gene-specific translational control of the yeast GCN4 gene by phosphorylation of eukaryotic initiation factor 2. *Mol. Microbiol.*, **10**, 215–223.

123 Santangelo, G.M. (2006) Glucose signaling in *Saccharomyces cerevisiae*. *Microbiol. Mol. Biol. Rev.*, **70**, 253–282.

124 Parker, R., Sheth, U. (2007) P bodies and the control of mRNA translation and degradation. *Mol. Cell*, **25**, 635–646.

125 Teixeira, D., Parker, R. (2007) Analysis of P-body assembly in *Saccharomyces cerevisiae*. *Mol. Biol. Cell*, **18**, 2274–2287.

126 Hurto, R.L., Hopper, A.K. (2011) P-body components, Dhh1 and Pat1, are involved in tRNA nuclear-cytoplasmic dynamics. *RNA*, **17**, 912–924.

127 Quan, X., Yu, J., Bussey, H., Stochaj, U. (2007) The localization of nuclear exporters of the importin-β family is regulated by

128. Snf1 kinase, nutrient supply and stress. *Biochim. Biophys. Acta*, **1773**, 1052–1061.
128. Miyagawa, R., Mizuno, R., Watanabe, K., Ijiri, K. (2012) Formation of tRNA granules in the nucleus of heat-induced human cells. *Biochem. Biophys. Res. Commun.*, **418**, 149–155.
129. Chafe, S.C., Pierce, J.B., Eswara, M.B.K., McGuire, A.T. et al. (2011) Nutrient stress does not cause retrograde transport of cytoplasmic tRNA to the nucleus in evolutionarily diverse organisms. *Mol. Biol. Cell*, **22**, 1091–1103.
130. Hancock, K., Hajduk, S.L. (1990) The mitochondrial tRNAs of *Trypanosoma brucei* are nuclear encoded. *J. Biol. Chem.*, **265**, 19208–19215.
131. Simpson, A.M., Suyama, Y., Dewes, H., Campbell, D.A. et al. (1989) Kinetoplastid mitochondria contain functional tRNAs which are encoded in nuclear DNA and also contain small minicircle and maxicircle transcripts of unknown function. *Nucleic Acids Res.*, **17**, 5427–5445.
132. Rubio, M.A.T., Rinehart, J.J., Krett, B., Duvezin-Caubet, S. et al. (2008) Mammalian mitochondria have the innate ability to import tRNAs by a mechanism distinct from protein import. *Proc. Natl Acad. Sci. USA*, **105**, 9186–9191.
133. Tarassov, I.A., Entelis, N.S. (1992) Mitochondrially-imported cytoplasmic tRNA^{Lys}(CUU) of *Saccharomyces cerevisiae*: in vivo and in vitro targetting systems. *Nucleic Acids Res.*, **20**, 1277–1281.
134. Schneider, A. (2011) Mitochondrial tRNA import and its consequences for mitochondrial translation. *Annu. Rev. Biochem.*, **80**, 1033–1053.
135. Marienfeld, J., Unseld, M., Brennicke, A. (1999) The mitochondrial genome of *Arabidopsis* is composed of both native and immigrant information. *Trends Plant Sci.*, **4**, 495–502.
136. Kumar, R., Maréchal-Drouard, L., Akama, K., Small, I. (1996) Striking differences in mitochondrial tRNA import between different plant species. *Mol. Gen. Genet.*, **252**, 404–411.
137. Chacinska, A., Koehler, C.M., Milenkovic, D., Lithgow, T. et al. (2009) Importing mitochondrial proteins: machineries and mechanisms. *Cell*, **138**, 628–644.
138. Neupert, W., Herrmann, J.M. (2007) Translocation of proteins into mitochondria. *Annu. Rev. Biochem.*, **76**, 723–749.
139. Tarassov, I., Entelis, N., Martin, R.P. (1995) An intact protein translocating machinery is required for mitochondrial import of a yeast cytoplasmic tRNA. *J. Mol. Biol.*, **245**, 315–323.
140. Tarassov, I., Entelis, N., Martin, R.P. (1995) Mitochondrial import of a cytoplasmic lysine-tRNA in yeast is mediated by cooperation of cytoplasmic and mitochondrial lysyl-tRNA synthetases. *EMBO J.*, **14**, 3461–3471.
141. Entelis, N., Brandina, I., Kamenski, P., Krasheninnikov, I.A. et al. (2006) A glycolytic enzyme, enolase, is recruited as a cofactor of tRNA targeting toward mitochondria in *Saccharomyces cerevisiae*. *Genes Dev.*, **20**, 1609–1620.
142. Entelis, N.S., Kieffer, S., Kolesnikova, O.A., Martin, R.P. et al. (1998) Structural requirements of tRNA^{Lys} for its import into yeast mitochondria. *Proc. Natl Acad. Sci. USA*, **95**, 2838–2843.
143. Eilers, M., Schatz, G. (1986) Binding of a specific ligand inhibits import of a purified precursor protein into mitochondria. *Nature*, **322**, 228–232.
144. Rinehart, J., Krett, B., Rubio, M.A.T., Alfonzo, J.D. et al. (2005) *Saccharomyces cerevisiae* imports the cytosolic pathway for Gln-tRNA synthesis into the mitochondrion. *Genes Dev.*, **19**, 583–592.
145. Frechin, M., Senger, B., Brayé, M., Kern, D. et al. (2009) Yeast mitochondrial Gln-tRNA^{Gln} is generated by a GatFAB-mediated transamidation pathway involving Arc1p-controlled subcellular sorting of cytosolic GluRS. *Genes Dev.*, **23**, 1119–1130.
146. Curnow, A.W., Hong, K.-W., Yuan, R., Kim, S.-I. et al. (1997) Glu-tRNA^{Gln} amidotransferase: a novel heterotrimeric enzyme required for correct decoding of glutamine codons during translation. *Proc. Natl Acad. Sci. USA*, **94**, 11819–11826.
147. Ramamonjisoa, D., Kauffmann, S., Choisne, N., Maréchal-Drouard, L. et al. (1998) Structure and expression of several bean (*Phaseolus vulgaris*) nuclear transfer RNA genes: relevance to the process of

tRNA import into plant mitochondria. *Plant Mol. Biol.*, **36**, 613–625.

148 Glover, K.E., Spencer, D.F., Gray, M.W. (2001) Identification and structural characterization of nucleus-encoded transfer RNAs imported into wheat mitochondria. *J. Biol. Chem.*, **276**, 639–648.

149 Delage, L., Duchêne, A.-M., Zaepfel, M., Maréchal-Drouard, L. (2003) The anticodon and the D-domain sequences are essential determinants for plant cytosolic tRNA^{Val} import into mitochondria. *Plant J.*, **34**, 623–633.

150 Laforest, M.-J., Delage, L., Maréchal-Drouard, L. (2005) The T-domain of cytosolic tRNA^{Val}, an essential determinant for mitochondrial import. *FEBS Lett.*, **579**, 1072–1078.

151 Delage, L., Dietrich, A., Cosset, A., Maréchal-Drouard, L. (2003) *In vitro* import of a nuclearly encoded tRNA into mitochondria of *Solanum tuberosum*. *Mol. Cell. Biol.*, **23**, 4000–4012.

152 Salinas, T., Duchêne, A.-M., Delage, L., Nilsson, S. *et al.* (2006) The voltage-dependent anion channel, a major component of the tRNA import machinery in plant mitochondria. *Proc. Natl Acad. Sci. USA*, **103**, 18362–18367.

153 Pusnik, M., Schmidt, O., Perry, A.J., Oeljeklaus, S. *et al.* (2011) Mitochondrial preprotein translocase of trypanosomatids has a bacterial origin. *Curr. Biol.*, **21**, 1738–1743.

154 Tan, T.H.P., Pach, R., Crausaz, A., Ivens, A. *et al.* (2002) tRNAs in *Trypanosoma brucei*: Genomic organization, expression and mitochondrial import. *Mol. Cell. Biol.*, **22**, 3707–3716.

155 Crausaz Esseiva, A., Maréchal-Drouard, L., Cosset, A., Schneider, A. (2004) The T-stem determines the cytosolic or mitochondrial localization of trypanosomal tRNAs^{Met}. *Mol. Biol. Cell*, **15**, 2750–2757.

156 Bouzaidi-Tiali, N., Aeby, E., Charrière, F., Pusnik, M. *et al.* (2007) Elongation factor 1a mediates the specificity of mitochondrial tRNA import in *T. brucei*. *EMBO J.*, **26**, 4302–4312.

157 Yermovsky-Kammerer, A.E., Hajduk, S.L. (1999) *In vitro* import of a nuclearly encoded tRNA into the mitochondrion of *Trypanosoma brucei*. *Mol. Cell. Biol.*, **19**, 6253–6259.

158 Mukherjee, S., Bhattacharyya, S.N., Adhya, S. (1999) Stepwise transfer of tRNA through the double membrane of *Leishmania* mitochondria. *J. Biol. Chem.*, **274**, 31249–31255.

159 Rubio, M.A., Liu, X., Yuzawa, H., Alfonzo, J.D. *et al.* (2000) Selective importation of RNA into isolated mitochondria from *Leishmania tarentolae*. *RNA*, **6**, 988–1003.

160 Pusnik, M., Charrière, F., Mäser, P., Waller, R.F. *et al.* (2009) The single mitochondrial porin of *Trypanosoma brucei* is the main metabolite transporter in the outer mitochondrial membrane. *Mol. Biol. Evol.*, **26**, 671–680.

161 Tschopp, F., Charrière, F., Schneider, A. (2011) *In vivo* study in *Trypanosoma brucei* links mitochondrial transfer RNA import to mitochondrial protein import. *EMBO Rep.*, **12**, 825–832.

162 Entelis, N.S., Kolesnikova, O.A., Dogan, S., Martin, R.P. *et al.* (2001) 5 S rRNA and tRNA import into human mitochondria. Comparison of *in vitro* requirements. *J. Biol. Chem.*, **276**, 45642–45653.

163 Puranam, R.S., Attardi, G. (2001) The RNase P associated with HeLa cell mitochondria contains an essential RNA component identical in sequence to that of the nuclear RNase P. *Mol. Cell. Biol.*, **21**, 548–561.

164 Kiss, T., Marshallsay, C., Filipowicz, W. (1992) 7-2/MRP RNAs in plant and mammalian cells: association with higher order structures in the nucleolus. *EMBO J.*, **11**, 3737–3746.

165 Smirnov, A., Entelis, N., Martin, R.P., Tarassov, I. (2011) Biological significance of 5S rRNA import into human mitochondria: role of ribosomal protein MRP-L18. *Genes Dev.*, **25**, 1289–1305.

166 Wang, G., Chen, H.-W., Oktay, Y., Zhang, J. *et al.* (2010) PNPASE regulates RNA import into mitochondria. *Cell*, **142**, 456–467.

167 Howell, N. (1999) Human mitochondrial diseases: answering questions and questioning answers. *Int. Rev. Cytol.*, **186**, 49–116.

168 Kolesnikova, O.A., Entelis, N.S., Jacquin-Becker, C., Goltzene, F. *et al.* (2004) Nuclear DNA-encoded tRNAs targeted into mitochondria can rescue a mitochondrial DNA mutation associated with the MERRF syndrome in cultured

human cells. *Hum. Mol. Genet.*, **13**, 2519–2534.

169 Sieber, F., Duchêne, A.-M., Maréchal-Drouard, L. (2011) Mitochondrial RNA import: from diversity of natural mechanisms to potential applications. *Int. Rev. Cell Mol. Biol.*, **287**, 145–190.

170 Wang, G., Shimada, E., Zhang, J., Hong, J.S. *et al.* (2012) Correcting human mitochondrial mutations with targeted RNA import. *Proc. Natl Acad. Sci. USA*, **109**, 4840–4845.